COSMOS

The frontispiece to Johannes Kepler's Rudolphine Tables (1627), symbolizing the history of astronomy. Present in this temple of Urania are Hipparchus, Copernicus, Tycho Brahe, and Ptolemy, and their various instruments. The map (lower center) is of Tycho's island of Hven. Note the Keplerian telescope (upper left) and a Keplerian diagram on which the imperial eagle disgorges gold pieces. Kepler calculates by candlelight (lower left panel), while printers are working at the lower right. Tycho's heirs objected to an earlier sketch, which omitted the ermine robes and royal medallion that Tycho is wearing here.

COSMOS

AN ILLUSTRATED HISTORY OF ASTRONOMY AND COSMOLOGY

✳ ✳ ✳ ✳ ✳ John North ✳ ✳ ✳ ✳ ✳

16510/

The University of Chicago Press   *Chicago & London*

JOHN NORTH is professor emeritus of the history of philosophy
and the exact sciences at the University of Groningen.

The University of Chicago Press, Chicago 60637
The University of Chicago Press, Ltd., London

An earlier version was published as *The Norton
History of Astronomy and Cosmology*.

17 16 15 14 13 12 11 10 09 08   1 2 3 4 5

ISBN-13: 978 0 226 59440 8 (cloth)
ISBN-13: 978 0 226 59441 5 (paper)
ISBN-10: 0 226 59440 8 (cloth)
ISBN-10: 0 226 59441 6 (paper)

North, John David.
Cosmos : an illustrated history of astronomy and cosmology / John North.
p. cm.
Includes bibliographical references and index.
ISBN-13: 978-0-226-59440-8 (cloth : alk. paper)
ISBN-13: 978-0-226-59441-5 (pbk. : alk. paper)
ISBN-10: 0-226-59440-8 (cloth : alk. paper)
ISBN-10: 0-226-59441-6 (pbk. : alk. paper)
1. Astronomy—History. 2. Cosmology—History. I. Title.
QB15.N67 2008
520.9—dc22
2007035579

FOR MARION

Contents

Illustrations

Plates

Credits

Preface to the
First Edition (1993)

It is no exaggeration to say that astronomy has existed as an exact science for more than five millennia. Throughout all of that time it has touched on matters of deep human concern. Writing its history presents us with innumerable problems. We begin with a period known to us largely by inference; we continue into times from which much of the evidence is known to have been lost; and we end with the last decades of a century that has provided astronomers with unprecedented attention and economic resources. From a typical century in the Hellenistic period, a golden age of astronomy, we might be left with a mere handful of texts. By contrast, there are now more than 20,000 astronomical articles being published every year, and over a five-year period the number of astronomers under whose names they appear is of the order of 40,000.

If this history begins as a sketch, therefore, of necessity it ends as a silhouette, defining its material as much by what it leaves out as by what it includes. It gathers pace in such a way that the space devoted to a dozen recent books of the highest importance might be a small fraction of that given at the outset to what would now seem a quite trivial piece of doctrine. This is not by accident. I have tried to pace the book in part by reference to the intellectual challenge facing successive generations of astronomers, and in part by the rational and social repercussions of their work, so that by the time I reach the twentieth century a certain sleight of hand is required. The design of a rocket gyroscope that would have been a technological miracle in antiquity will pass entirely unnoticed, as will, alas, every one of the hundred people who made it possible. And similarly at a theoretical level. There is no royal road to an understanding of Einstein, Eddington, and Hawking, for instance, short of reading their scientific writings.

This book is a history, and cannot even begin to serve as a substitute for astronomical treatises, although I should not feel cheated if it were to increase the five-year phalanx to forty thousand and one. It is a history, and histories are usually written, if not for historians, at least for readers in a historical frame of mind. Of course no professional historian ever admits to being pleased by universal history, but then, the sort of history that can only satisfy professional historians is hardly worth writing.

The book owes its existence to the intuition of my wife, Marion North. She persuaded me that it was needed, and hers was the crucial discovery of a way of creating the time needed to write it, that is, of prizing open crevices in the rocks of faculty administration.

Preface to the
Present Edition

The text of the present edition includes most of the first, but with some rearrangement, as well as the addition of many substantial passages. The most conspicuous change is that in the number of illustrations, which is increased by more than two hundred. Most of the new textual material touches on recent developments in astronomy, but a few minor adjustments have been made in the light of recent historical research. In keeping with the spirit of the series in which the book first appeared, it follows a broadly chronological order and covers the whole time-span of its subject. Touching as it does on the astronomy of many different cultures over that very long period of history, it faces us with a dilemma. The very definitions of astronomy and cosmology have changed with time and have varied from one culture to another. In a comprehensive survey of the present kind, it would have been extremely confusing had I constantly changed cultural perspectives. My viewpoint is a product of Western science, which on occasion I have no doubt tacitly used as a standard of comparison—although never intentionally as a means of reinterpreting non-Western ideas or of judging their human value.

There are cultures that I have ignored, not because they are without value or historical interest, but because they do not fit into the scheme of the book. In some cases, the omission is only apparent. There is no chapter on Jewish astronomy, which assumed great importance at various stages of history, usually within the traditions of the people with whom the Jews were in immediate contact. (The Jews of medieval Spain are an important case in point, and that is the context in which they will be found.) On the other hand, I have said nothing about the numerous cultures of Africa—the obvious exception being Egypt—or about those of aboriginal Australia, the Maoris, the Polynesians, or the numerous peoples of the 13,000 islands and 300 ethnic groups of the Indo-Malay archipelago. By and large, the native peoples of North America share their fate, and there are others. In all of these cases it would have been possible to trace signs of familiarity with common celestial phenomena, especially concerning the Sun, Moon, and stars. Those phenomena are likely to have been identified by name, or by using analogies with familiar objects, and it is a barren society that has no myth-making surrounding the heavens. In Africa, history is greatly complicated by the absence of early written records, and by interactions with Islamic and European societies. I suspect that it would be possible to link African and European prehistory through primitive astronomical practices, but to do so would require a very different kind of book. Such material would not fit well into the overall plan of the book, lacking as it usually does any clear sign of the theoretical element that turns myth into science. It is true that traditional Polynesian knowledge of the sky had a theoretical component of sorts, going beyond mythology: their navigators, who were of crucial importance for the inhabitants of thousands of scattered islands within the great expanse of the Pacific, used the stars as an aid. How they proceeded, however, is still not well understood, and a book of this kind is not a place for a lengthy discussion of rival historical interpretations. Brevity requires dogmatism, and more even-handed authors will find much to debate.

This book does not presuppose specialist astronomical knowledge, and it is hoped that it will be more or less self-sufficient—not for everyone, but for those who are likely to read it. A few unfamiliar physical principles may be found in places where I have dealt with modern developments, and the material naturally becomes more difficult in some of the later chapters. The relevant passages will be easily recognized—and may either be sidestepped or supplemented, for instance, by works in the long bibliographical survey at the end. The text is untrammeled with footnotes, but again, the bibliography is meant to point the way to more detailed sources. In a subject where historical writing has snowballed in recent years, it would have been impossible to compile a truly comprehensive bibliography; and mine is in any case strongly biased towards reasonably accessible sources in English.

I have benefited in various ways from the experience of having the first edition translated into Polish, German, and Spanish, and I should like to thank Edward N. Haas, James Morrison, Gaston Fischer, and Rainer Sengerling for their kindness in supplying me with corrections to earlier printings. I am especially grateful to Noel Swerdlow for prompting me to prepare this new edition, and to him, Owen Gingerich, Julio Samsó, and an anonymous referee, for their astute comment and generous criticism. Finally, having in mind my previous remarks about the exponential growth in the numbers of people who have been engaged in astronomy and cosmology in recent decades, I can only beg the indulgence of those whose research I have not found space to mention. Mine is no more than a series of selected topics—representative, I hope, of its overall theme, but essentially incomplete.

John North
Oxford, 2006

On Numbers and Units

Many very large numbers, and a few very small, are here expressed only in words, with the word *billion* always denoting a *thousand million*. The metric system has usually been preferred for units of mass, length, and time, except in a few cases where doing so might have been confusing. (Thus many telescopes, including the famous 200-inch at Palomar, are widely known by the diameters of their mirrors in inches.) It is useful to remember that a foot is about 30 centimeters and that 5 miles are equal to about 8 kilometers. Familiarity takes precedence over consistency, and no rigorous attempt has been made, for example, to introduce so-called SI units. Weights are usually given in grams, kilograms, or tonnes, according to context. (A metric tonne of 1,000 kilograms is 2,204.6 pounds or 0.984 imperial tons.) Temperatures are usually quoted in degrees Kelvin (°K), which can be roughly defined as degrees Celsius (°C, Centigrade) from an absolute zero of about −273°C.

Most readers will be familiar with the division of degrees (angular measure) into sixtieths. We owe the sexagesimal system to the Babylonians, as explained in chapter 3, and need a convenient modern notation for it. Sixtieths of a degree are minutes of arc, sixtieths of these are seconds of arc, and so forth. The common notation for an angle of, say, 23 degrees 5 minutes and 14 seconds would have been 23°5′14″. Such a notation may be extended in an obvious way to thirds (sixtieths of seconds), fourths (sixtieths of thirds) and so on, but then it becomes very cumbersome. It is now conventional in historical works to replace it (to take this same example) with $23;05,14°$. The same notation can be used with our familiar—but rather inconsistent—units of time, as in the mixed example $5^{d}\,21;56,07,16,34^{h}$, where here "d" denotes days and "h," hours.

The following astronomical constants might prove useful:

1 astronomical unit (or AU, the mean distance of the Sun from the Earth's center) = 149,674,000 kilometers = 93,003,000 miles.

1 parsec (a distance corresponding to a parallax of 1″) = 206,265 AU.

1 light-year (the distance light travels in a year) = 9,460 billion kilometers = 5,878 billion miles.

Introduction

When calculation called for personal rather than electronic powers and numbers had a greater capacity to mystify, there was a widespread feeling that a science that quoted its results to ten places of decimals certainly merited the description "exact." It was thought that astronomy must in some sense be superior to sciences that counted petals, that mixed *A* and *B* to get *C*, or that predicted the death of a patient "within a week or two." Judged in this limited way, for well over two thousand years, astronomy has had no equal among the empirical sciences. In a far more important sense, however, astronomy has been an exact science for a much longer period of time, for it has been set out in a highly logical and systematic way, with its patterns of argument modeled on—and occasionally helping to create—those of mathematics. So highly regarded were they, in the past, that astronomy and her sister geometry became accepted as models, prototypes for the empirical sciences generally, helping to provide them with form and structure.

If astronomers are to take a special pride in the ancient origins of their subject's exactness, then it should be in this second sense of the word. Astronomy was not altogether fitted for such a responsible role. After all, it differs from most other sciences in at least one important respect: it studies objects that cannot for the most part be manipulated for purposes of experiment. The astronomer observes, analyzes what is seen, and devises principles to explain what has been seen and what will be seen tomorrow. Even in these days of interplanetary rockets, the subject remains chiefly analytical rather than experimental. This quality no doubt explains in part why astronomy became the first highly formalized science.

How and where this came about we cannot say. The answer will depend to some extent on how generously we define our terms. It has been claimed that sequences of moon-shaped marks cut into bone artifacts, found from cultures as widely separated in time as 36,000 BC and 10,000 BC, represent the days of the month. The length of the month, from new moon to new moon, is approximately twenty-nine and a half days, but any primitive tally would naturally have introduced extra days when the Moon was invisible. Since in some cases counts might have been made from the new crescent to the last visible crescent, and in others up to the next new crescent, we should not be too disdainful of the fact that groupings found on these pieces of bone, ranging from 27 to 31, have been claimed as evidence of lunar counting. There is much variation in the numbers of marks in groupings that have been said to distinguish between the four quarters of the month. Such evidence is intrinsically difficult to handle, even statistically. The thesis is not implausible; the marks on these bones do often seem to have been gouged out to resemble a lunar crescent; and more than that we cannot say with confidence.

It is tempting to see in the counting of lunar days a first step towards a mathematics of the heavens. Keeping a tally of days of the month would have been useful to anyone who valued the Moon's light by night, but there are also obvious connections with human fertility. We must be wary of introducing our own preconceptions into prehistory. It is often said that a primitive calendar requires the counting of days, and that the movements of the Sun were first intensively studied in order to establish the sort of calendar the earliest agricultural peoples needed. The seasons, however, obviously mattered and were known to the hunter long before the introduction of farming. One may be fairly certain that solar calendars—in the broadest sense of devices for keeping a check on the seasons—did not originally have anything to do with the counting of days. They were based rather on the changing pattern of rising and setting of the Sun over the horizon throughout the year.

The epoch at which the change from a hunting to a farming culture took place varied considerably according to geographical region. There were settled agricultural communities in southwest Asia before 8000 BC, and communities may have been just as early in southeast Asia. Farming spread into the Mediterranean well over a thousand years before it reached Britain, at the outer edge of Europe, which it did around 4400 BC. No doubt some astronomical ideas were developed locally in all regions at one time or another. Some were undoubtedly transmitted from one center to another, and not always with a corresponding migration of peoples. It is difficult to decide on the relative importance of these tendencies, but it does seem that in the fourth millennium BC some remarkable changes were taking place in the sheer intensity with which the heavens were being studied in northern Europe. The evidence for these changes comes from archaeological remains. Perhaps time will prove them to have been anticipated or paralleled elsewhere, but those from northern Europe are remarkable enough in themselves to be taken as a starting point for our history.

There is every reason for introducing this early European activity into our account. It was truly scientific, in the sense that it reduced what was observed to a series of rules. This is not to say that its motivation resembled our own. I have no doubt that the main reasons for rationalizing the appearances of the heavens were religious or mystical in character, and that they remained so, well into historical times. Most peoples have in the course of their development treated the Sun, Moon, and stars as alive, and even as human in character. They have told stories of the heavenly bodies with a strong reliance on an analogy between them and human life. In such analogies we might claim to detect the beginnings of science, but let us not exaggerate this element. Astronomy has always been strongly allied with religion. Both were often concerned with the same objects: the Sun, Moon, and stars were divinities in many societies. To have done full justice to this alliance with religion, a book of a very different sort would have been needed, but to have omitted it entirely, in favor of the "exact" parts of the science, would have been to mistake the forest for the trees.

Throughout history, omens and signs have been carefully sought as a means of foretelling, or even forestalling, those superhuman powers that seem to govern the Universe. It is not surprising that the heavenly bodies were among the objects to which most attention was paid. They were assigned divine natures in part because they were of such obvious importance in themselves—the Sun in particular. As material for omens, they have the virtue of behaving with a certain degree of regularity. Here it is surely the magical conception of nature that comes first, to be followed at a later stage by a realization that it is easier to be systematic and precise about the stars than about the livers of sheep, the weather, and flocks of birds. Astrology flourished in the Near East, for instance, once this principle was recognized. This may not entirely explain the very first systematic astronomical theories, but it would be foolish to pretend that astronomy is somehow too noble a subject to have depended on anything of the sort.

The science of the stars was not at first pursued for its own sake, but so potentially strong is the human feeling for system and order that in time astronomy did acquire a measure of self-sufficiency. Astronomers managed eventually to tear up most, if not all, of its ancient roots. It was studied increasingly as an independent system of explanations, a system that did not have to be justified by its usefulness—whether in astrology, or navigation, or elsewhere, or even by its relevance to God's creation of the world. Admittedly, in comparison with the science itself, its nonastronomical context was often far more important to common humanity. I hope that I have not ignored it unduly in favor of the more formal side of the science, but if I have done so, this is because astrology and cosmic religion are commonplace, viewed as symptoms of the human condition. The long record of achievement in astronomy, on the other hand, has very few intellectual parallels in the whole of human history.

Prehistoric Astronomy

PALEOLITHIC SYMBOLS

The patterns of stars in the night sky are likely to have been far more familiar to prehistoric peoples than they are us, and it would be no cause for surprise to find them depicted in some way at an early date. Some of the most astonishing archaeological discoveries of recent centuries have been those of highly sophisticated cave paintings, especially in southwest France and Spain. Paleolithic cave art was first discovered in 1856, in a cave at Niaux, but there was much skepticism about its origins until indisputably genuine examples were found elsewhere. The Lascaux cave, discovered by children in 1940, is one of the most important of all Paleolithic sites. It has an astonishing collection of more than two thousand images, mostly of animals (horses, auroch bulls, and deer), in paintings dated by radio-carbon methods to around 15,000 years ago. One of the commonest of themes is that of bulls, and one such painting has aroused much interest in recent years, on account of painted spots adjacent to a bull's head and shoulders (fig. 1).

The spots have been interpreted in many different ways, and the question of their meaning is likely never to be convincingly settled. Some have insisted that they are records of success in hunting, but against this it has been pointed out that most of the bones remaining in the cave, the remnants of the hunters' diet, were of deer. More convincing are the various astronomical interpretations, mostly put forward at the end of the twentieth century. The spots above the bull's shoulder have been seen, for example, as the Pleiades, or even as the brightest stars in the constellation of Taurus; and those on the bull's face have been seen as the Hyades, also in Taurus. Both Pleiades and Hyades are star clusters, easily made out by those with good eyesight. They attract interest simply because they are clusters, and not single stars, and an enormous amount of attention has been paid to them, in literature and art, throughout recorded history. Many extravagant claims have been made on the strength of such astronomical interpretations of the Lascaux bull. We find the painting described as the first planetarium rather than the first zoological treatise, and even as a sign of the extreme

1. This instance of Upper Paleolithic cave art is one of many painted images of bulls in the Lascaux cave in France. The painted dots have been interpreted as stars. The cave contains more than two thousand images, mostly of animals. Radiocarbon methods date the collection at ca. 15,000 BC.

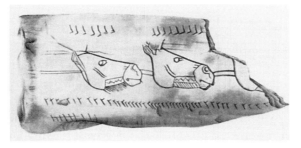

2. Upper Paleolithic engraved bone (Magdalenian period, ca. 9000 BC–15,000 BC), from Le Mas-d'Azil, France. Reindeer, wild horses (pictured here), and bison were often engraved on such bones, but rows of marks—suggestive of counting—often appear alone on bones large and small. They have been found throughout Europe, as far afield as Russia and even in parts of Africa, from the same period. Such bones have been interpreted as hunting tallies, but Alexander Marshack argued that they were counts of days grouped in months. A Paleolithic naked Venus figure from Laussel, in the same general area of France, holds aloft a horn with similarly striated markings. That she was no huntress strengthens the feeling that there was a link between such markings and the menstrual cycle.

sophistication of Paleolithic peoples. There is certainly great skill evident in the animal paintings of the period, most of them done in semidarkness; and since hunting was often done by night, it seems natural enough that stars and animals should be recorded together in paint. The association of the Pleiades with the constellation group that we know as Taurus, the Bull—an association deriving from the "V" shape of bright stars that suggests a bull's horns—offers one intriguing interpretation of this particular image. It fits well with the more general supposition that the night sky was intimately known. It is difficult to go beyond that point, on such slender evidence.

No less familiar than the stars, of course, were the Sun and Moon. Among surviving artifacts from the Paleolithic period are bones scratched in ways suggesting that some sort of counting was taking place, and on the basis of the grouping of incisions into sets of thirty, or thereabouts, claims have been made that at least some of the counting was of the days of the Moon (fig. 2). It is conceivable that the female menstrual period was somehow connected with these tallies, but the light of the Moon is also a matter of great concern to hunters. Anthropologists have established that the links between hunting, sexual rituals, and the Moon, found today in sub-Saharan Africa, go back perhaps forty thousand years or more. Even granting the basic lunar interpretation of the incised bones, however, it is hard to see how any more precise conjecture about their use as tallies could ever be proved. Did they have a practical or a ritual function? It is a sign of our own mentality that we draw a distinction between the two.

EUROPEAN NEOLITHIC, BRONZE AGE, AND IRON AGE ASTRONOMY

Painted star patterns are hardly astronomy, although they might point to a tradition of storytelling about the heavens. Counting off the days of the Moon has a more intellectual look about it, but whether any such simple activities as these should be described as "astronomical" is a moot point. It seems reasonable to restrict that word to activities that are supported by some sort of rational system—if possible, one that allowed of prediction. The recording of risings and settings over the horizon of the Sun, Moon, and stars, through material markers, and recognition of the fact that those markers are of use day after day (for the stars), or at yearly intervals (for the Sun), is an activity that we can loosely describe as astronomical in this stronger sense. Such activities certainly began in prehistoric times, and continued well into our own era. As seen from a particular place, the stars rise and set over at the same points of the horizon every day of the year. (Of course, those risings and settings may not be visible to us at certain seasons—that is, when the Sun is in such a position relative to the stars that they occur during the hours of daylight.) For the time being we may ignore the slow drift of star positions, which—using horizon markers—is scarcely detectable in less than a century or so. As for recognizing the fixity of the places where stars seem to rise and set, on the other hand, this is something that is far more likely to have happened in a settled community than by hunter-gatherers.

The places on the horizon where the Sun seems to rise and set, as seen from a particular point, change from day to day. In the Northern Hemisphere, the Sun rises in the east and sets in the west at the spring and autumn equinoxes. In spring, its points of rising and setting move progressively further to the north as the year advances, reaching a maximum at the summer solstice. The points then move southward, through the autumn equinox until a maximum southern azimuth (horizon direction) is reached at the winter solstice. The points then begin to move northward again until the spring equinox is reached and the year is completed (see fig. 3). The very language with which we describe these events carries with it certain information—such as that day and night are of equal length at certain times of the year, the *equinoxes*. That information was perhaps not always known to early peoples, but the broad truth of the cyclical movements of the points of rising and setting, and their correlation with the cycle of growth in nature,

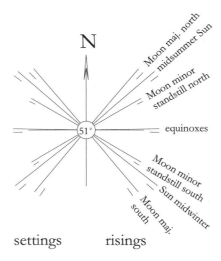

3. The extremes of rising and setting of the Sun and Moon, for a geographical latitude of 51° (southern Britain), and an epoch of 2000 BC. The Sun's extremes are reasonably constant over long periods of time, but the Moon's fluctuate between upper and lower limits, as marked. The directions depend on the definition of rising and setting. Here it is assumed that the upper limb of the Sun or Moon is on the horizon, and that the first or last glint of the object was what mattered. The directions also depend on the altitude of the local horizon in the directions in question. A horizon of zero altitude is assumed here. Even then, the equinoctial solar directions are not perfectly east-west.

was certainly known from very early times. This we can see from the many prehistoric archaeological remains that are indisputably directed toward the points of midsummer and midwinter risings and settings of the Sun. What significance was attached to the Sun's behavior is a far more difficult question than the simple fact of its embodiment in monuments. It seems likely that in some cultures the winter solstice was connected with the idea of death or rebirth (see also p. 239 below). At all events, the winter and summer extremes were evidently what counted most; the equinoxes were seemingly of lesser importance to early peoples.

In the earliest periods during which structures were oriented on the heavens, directions seem to have been toward the risings and settings of a handful of bright stars rather than toward the Sun and Moon. From, say, half a millennium before 4000 BC until very roughly 3000 BC, people in northern Europe were buried in long barrows—elongated and wedge-shaped communal graves, mostly mounds of earth flanked by ditches. These barrows took on different shapes in different regions. There are many different local styles in evidence in Britain, Ireland, northern France, the Low Countries, Germany, Denmark, and Poland, and yet they all have much in common and share one property of particular significance: they all have carefully planned geometrical forms, usually incorporating at least one right angle. (For an external view of one of very many different types of long barrow, see fig. 4.) They were never constituted of randomly piled

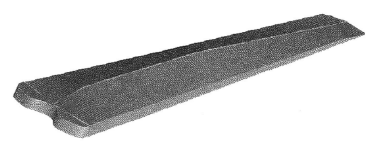

4. The original form of the long barrow at Hazleton North, Gloucestershire, England, one of many different types of English long barrow. In this case, the internal structure was created by stone walls, defining 19 cells, 2 of which contained mortuary houses. When needed, they were entered through doors in the middle of the sides. More often, mortuary houses were at the head of the barrow. There is good reason to think that such barrows served as artificial horizons, across which the risings or settings of an honored bright star would have been observed in some sort of ritual. These barrows were flanked by ditches, and it is likely that ritual observations were made by people standing in them.

earth. They were used to bury the dead and so must surely have been associated with fitting rituals, but what were those?

There is excellent reason to associate such monuments with the risings and settings of stars. Just as their shapes were carefully planned—in all three dimensions—so too were their connections with the sky. It is often extremely difficult to decide when and where a star crosses the distant natural horizon, especially if that horizon is covered by forest. Neolithic people surmounted this difficulty in numerous ways, all showing extraordinary intelligence and dedication. The long barrows were often so fitted into the landscape that, by looking along their slope, the observer's line of sight was elevated very slightly above the natural horizon beyond, and toward a particularly bright star. The smooth back of the barrow thus provided an artificial horizon. Long barrows could also serve as artificial horizons when viewed transversely; and there is good reason to think that the viewing of risings and settings was often done by people standing in the ditches on each side of the barrow. Altitudes and directions were planned with one of a handful of bright stars in mind.

Neolithic people built other ditched monuments mainly of earth, often on a very grand scale, and these too followed similar principles of alignment. One example is the Greater Cursus at Stonehenge, about three kilometers long, but now barely visible from the ground. There are many other long banked and ditched enclosures.

Arguments for the ritual sightings of stars in the Neolithic period are necessarily lengthy and cannot be rehearsed here, but there is one piece of evidence that is both simple and convincing. Figure 5 shows the positioning of long barrows over an extensive area in the Stonehenge district—long barrows that mostly predated the Stonehenge monument itself. It is an astonishing fact that these are almost all in alignments of three or four barrows, and that the directions of the alignments were toward the risings or settings of a few very bright stars. (The directions are no longer what they

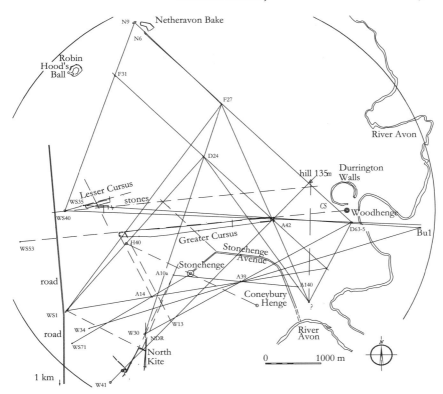

5. The positions of long barrows on Salisbury Plain, in the Stonehenge region, England, over an area roughly seven kilometers across. The standard notation for the barrows (N9, N6, and so on) may be ignored (it is based on parish names). The figure reveals a remarkable property of the barrows: they are almost all in alignments of three or four, and the directions of the alignments are toward the risings or settings of a few very bright stars. The same is true of barrows in neighboring regions, although the stars favored are often different, as though they have a totemic value. Notice the conspicuous parallels, especially in the northern area. These could not have been arranged by land surveying, over such vast distances, with woods and other obstacles between barrows. Aligning two triples of barrows on bright stars, the parallels appear automatically. Note how no fewer than five alignments pass through a barrow (A42) at the end of the Stonehenge Cursus. It is now virtually destroyed, but its former importance cannot be doubted, once the star alignments are appreciated.

would be today, but the differences can be easily calculated.) The conspicuous parallel alignments, especially in the northern area, could not have been arranged by means of land surveying over such vast distances, nor would there have been any point in making them parallels for the sake of being parallels. They are a simple consequence of the aligning on bright stars of distinct triples of barrows. The parallels appeared quite automatically.

When we reach the third millennium BC, there is a clear sign that more attention is being paid to the Sun and Moon. In northern Europe, a new vogue for circular monuments developed, embodying the Sun's solstitial positions. Stonehenge is the best known example of these circular monuments, but there were very many others, made of timber, of comparable complexity, although not as grand in engineering terms. The monument on

6. A fair approximation to the overall appearance of the most conspicuous elements of the Stonehenge monument, as it was around the end of the third millennium BC. (After E. H. Stone, 1924.)

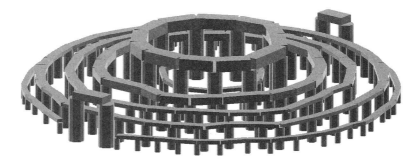

7. A schematized version of the likely appearance of the Durrington Walls timber henge (southern circle), based on the sizes and depths of the excavated post-holes. Approaching from the nearby river Avon (see fig. 5), through the gap in the bank and ditch, one would have faced the entrance to the circle (shown here on the left). Surveying the skyline, one would have found that the horizon, set by the enormous surrounding bank half a kilometer across, was constructed so that its altitude was largely constant. Timber monuments of much the same type were numerous, and they provide the Neolithic context for stone circles such as Stonehenge.

the Stonehenge site evolved significantly for well over a thousand years. Its now-familiar structure represents the form it took a century or so before and after 2000 BC. Figure 6 gives a fair idea of the monument's appearance when at its most spectacular, and plate 1, an impression of its present sad state. During the millennium or so during which its various component rings were added or modified, there was much comparable activity—although on a less ambitious scale—at numerous other centers in Britain and northern Europe. When circles were of stones, they were often a more permanent version of circles of timber posts that had gone before, and that had carried cross-members (lintels) in much the same way. For a stylized reconstruction of such a timber monument, see figure 7. They were not primarily tombs, although they are occasionally found to contain isolated burials. In the middle of the third millennium, however, we find the technique of timber-ring construction being transferred to grave architecture, some of the so-called round barrows being then surrounded by linteled post-rings with orientations toward the Sun at the solstices. In this form, the custom lasted for well over a thousand years, and in some areas of Europe for twice as long.

Why these particular phases in prehistoric astronomical activity are so important to us is that they are witness to the early marriage of astronomy with geometry. The key to an understanding of the circular monuments, for example, is their intricate three-dimensional structural form. In plan, they may show concentric circles of posts, or rings that were accurately drafted in oval outline. Their elevations, however, are just as important as their plans, and only by considering the paths of lines of sight in three dimensions can they be properly understood. There are very many different varieties of circular monument, but always the line of sight to the Sun's extreme positions (summer or winter solstice) was marked by trapping the Sun's image in a "window" created by at least two uprights, one far and one near, and at least two lintels, one far and one near. It was not just any part of the Sun that was observed, but (usually) its upper limb, so that the first glint of the Sun was seen at sunrise and the last glint at sunset, midsummer or midwinter.

These linteled structures can thus be considered as artificial horizons, like the long barrows and other earthen structures before them. Viewing was done from carefully prepared places, often in a circular ditch surrounding the monument, occasionally over the bank that surrounded it. The usual interpretation of Stonehenge would make its center the place from which the midsummer Sun was observed over the Heel Stone. This is almost certainly mistaken. The viewing position was at the Heel Stone itself, outside the sacred space, and the chief celebration was that of the setting of the midwinter Sun, seen through the narrow central corridor (see fig. 8). Stonehenge is a skeleton through which light can pass from numerous directions, as in the timber monuments before it, but all of these were carefully planned so as to present a solid appearance against the sky when viewed from suitable positions— and the Heel Stone is just such a position. Sight of the last glint of winter sunlight through the center of the black edifice must have been deeply moving. Even this sight, however, was the culmination of a carefully orchestrated ritual, in which the Heel Stone was approached by a rising avenue, such that sight of the setting midwinter Sun could be held for a time in the window through the center of the monument. That there was "calculation" in some sense is undeniable, although that is as loaded a word as "astronomical" and certainly does not require more than could be planned by simple diagrams or conveyed by word of mouth. A standard unit of length was often used in the design, not only for the plan but for the heights of the posts or stones. The size of this unit was derived from measurements on relatively late stone monuments by Alexander Thom, who called it the *megalithic yard* (MY), but the use of the unit predates the period of the great megalithic (stone) monuments. There is reason to think that angles of solar altitude were fixed by ratios set in terms of this unit. At Woodhenge, near

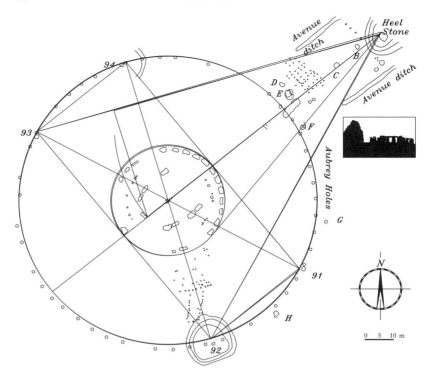

8. The geometry of the plan of Stonehenge, with surviving stones mapped. A key to understanding the monument is the Heel Stone, from which a clear sight line through the center of the monument allowed the rays of the setting midwinter Sun to pass, filling the center of the otherwise silhouetted monument with light. Post circles and stone circles both have skeletal structures, with light passing through them, but there are usually several viewpoints from which they look like solid buildings. The Heel Stone is at such a point. The inset figure shows the present view of the silhouetted monument from the Heel Stone, but gaps have now opened up as a result of the loss of stones. The peripheral circles mark the positions of the so-called Aubrey holes (named after their discoverer, John Aubrey), connected with the first phase of building (ca. 3000 BC). Four modest stones around the periphery, the "station stones," are at the corners of the large rectangle drawn here, and are related to the Moon's extremes of rising and setting.

Stonehenge, there stood a monument comprising six ovals of timber posts, now long vanished. Its form can be reconstructed to a large extent from the remaining post-holes, and it seems that a favored angle appropriate to the rising of the solstitial Sun in winter was set there by a rise of one in sixteen. There is evidence for several other simple ratios with a similar use.

At some monuments there is clear evidence that the Moon was also observed. The Moon has a somewhat similar pattern of behavior to the Sun's, except that its extremes of rising and setting follow several cycles, the most important of course being the monthly cycle. Briefly, one may say that each month it reaches to certain extremes that may themselves be treated as objects, with their own movements along the horizon. They in turn reach to certain extremes. These absolute extremes seem to have attracted prehistoric interest. In all strictness they are not correctly described as "absolute" extremes, since they also fluctuate and can be treated as objects at a still higher level, with their own extremes (again, see fig. 3). It has been held that even this third level of fluctuation was known to the people of the Bronze Age, say, from the end of the third millennium onward. In Brittany, Britain, and Ireland, there are more than seven hundred recorded avenues or rows of stones, some aligned on the stars, some on the Sun, but others on the Moon. Comparable traces of rows of wooden posts have been found in the Low Countries, where stone is scarce. The impressive multiple rows at Carnac, in Brittany, have much in common with others in Scotland, more than a thousand kilometers away; and one example on Dartmoor, in the west of England, is even longer than the Carnac rows, although not so dramatic. Alexander Thom surveyed the Carnac region in great detail and showed how the sightlines to various tumuli and menhirs marked the extreme positions of the Moon, several of those lines even crossing inlets and bays in the sea to the south of the Morbihan district. At the center of that vast system stood the huge stone known as Le Grand Menhir Brisé—now shattered, as its name implies.

There is no doubt that the Moon's extremes of rising and setting were also recorded to some degree of accuracy in more compact monumental form, presumably for ritual and religious reasons, rather than for the creation of a calendar in any complex sense. The religious dimension is all-important. The grave of a three-year-old girl was found at Woodhenge, her skull cleaved so as to suggest a ritual sacrifice. It turns out that there are three key rays set by the monument, rays that cross at the center of the child's grave. A line to midsummer sunrise is at right angles to it; a line to midwinter sunrise is very nearly at right angles to the first ray; and a line to the Moon's extreme northern setting is in exactly the reverse direction (in plan) of the midwinter ray. Those with a partial awareness of the underlying astronomy may suspect a mistake here, for are the lunar and solar lines of rising and setting not in markedly different directions, at the latitude of Woodhenge? The answer is that if suitable choices are made of the *altitudes* set by the artificial horizons of the monument, the directions (azimuths) of the rays can be brought into exact agreement. The builders of Woodhenge did this, and they related what they had done to the grave of the sacrificial victim.

There is much more to be said about the intellectual achievements of these northern peoples of the Neolithic and Bronze ages. We can say far more of their purely astronomical and geometrical achievements than of the way they fitted their view of the heavens into their social and religious life. Among the stars that seem to have attracted most attention were Deneb, which set in a fairly constant direction to the north, Rigel, the foot of the constellation we know as Orion, and Aldebaran, which has long been regarded as the eye of Taurus. It is probable that the sacrifice of bulls was related to the risings and settings of Aldebaran. Whether the entire constellation of Taurus was viewed as a bull it is impossible to say. Perhaps an early constellation figure was the precursor of a figure that still survives, cut into the chalk downs at Uffington in southern England. The "Uffington White Horse" is a creature of indeterminate species, but that the star Aldebaran rose over it, as seen from a curved gallery on a nearby track, is determinate enough. Another chalk figure, this time cut into the south downs at Wilmington in East Sussex, is human in form and seems to correspond to our constellation of Orion, which from a certain point to the north of the figure would have seemed to walk along the curved natural horizon, with the daily turning of the heavens (plate 3).

These phenomena raise a question that must be introduced sooner or later. The places over which the stars rise change very slowly, but in most cases that fact must have become appreciable over a century or two. The cause is the so-called precession of the equinoxes, discovered—although described differently—by Hipparchus in the second century BC. At many Neolithic sites, religious activity lasted for well over a millennium, and it cannot be doubted that the drift in question would have been appreciated, sooner or later. This is of course not to say that the precession of the equinoxes was discovered, much less that it was quantified, in Neolithic times. There was a "Pan-Babylonian" movement in the early years of the twentieth century, and some of its followers wished to ascribe the discovery to a much earlier culture than Hipparchus's. Indeed, the idea of a pre-Hipparchan discovery goes back to Enlightenment writers on the history of astronomy, from the end of the eighteenth century, men such as Jean Sylvain Bailly and Charles François Dupuis, who wanted to show that religious forms, and Christianity in particular, had their roots in astronomy, or rather in astronomical myth-making. There is no evidence, however, to support the idea that anyone before Hipparchus both recognized the slow drift in star positions and also described it with reference to a precise astronomical coordinate system. As we see later, Hipparchus discovered that the stars increase in ecliptic longitude, keeping their ecliptic latitude constant.

Many attempts have been made to reconstruct the belief systems of the early northern peoples responsible for these astonishing astronomical monuments, working from surviving artifacts, of course, but often also from much later written accounts—for instance from Roman accounts of the Druids and from early Scandinavian literature. Dubious as many of these attempts are, the Scandinavian material is often interesting in its own right. There are numerous indications of cults of the Sun and Moon, not all of them stemming from the orientation and planning of large monuments. One of the most interesting finds was that made in 1902 at Trundholm on Zealand (Sjælland, Denmark), of a Bronze Age horse-drawn disk, dating perhaps from roughly 1400 BC (fig. 9). There can be little doubt that this had solar significance. The Sun is shown being pulled by a horse in several crude Swedish rock carvings of much the same date. An equally rich discovery, this time from Germany, was of a disk of bronze 32 centimeters in diameter, studded with gold shapes that related to the heavens in some way (fig. 10). Found near Nebra at the end of the twentieth century and now known as the Nebra disk, it came more specifically from Mittelberg—a modest hill in the Ziegelroda Forest, between Halle and Erfurt. It seems to have been discovered within a pit inside what had once been a Bronze Age palisade

9. One side of the Trundholm disk (shown at the right mounted on a model horse-drawn chariot), which dates from perhaps the fourteenth century BC. The other side is very similar but still has much of its gold overlay, into which a pattern was stamped. Overall length about 60 centimeters. It is technologically remarkable: the wheels and shafts are of solid bronze, but the horse, and what is presumed to be a Sun disk, were made around clay cores. The model was turned up in 1902 by a farmer as he was plowing former bogland in Trundholm, Denmark. (Now in the National Museum, Copenhagen.)

10. The disk from Nebra, Germany, datable to ca. 1600 BC. Of bronze, with gold additions (the lighter areas), it is approximately 32 centimeters in diameter. A missing gold strip on the left mirrored that on the right reasonably closely. The two arcs they define are very close to the arcs of the horizon covered in the course of a year by the risings and settings of the Sun, for the place where the disk was found. It plainly related to beliefs of some kind about the heavens; that the Sun and stars are represented seems certain. The crescent might indicate the Moon or an eclipsed Sun.

and complex of defensive ditches. Its context has been hard to establish, since it was found by grave robbers, and was only recovered after a sting operation, helped by an archaeologist to whom it had been offered for sale.

The shapes of the gold pieces on the Nebra disk suggest three lines of inquiry. The dots seem likely to represent stars, 32 in all, perhaps including the Pleiades cluster of 7. The Sun, and possibly the crescent Moon, seem to be represented, although the gold crescent could equally represent an eclipsed solar disk. Some have even suggested that both crescent and full circle—despite being of gold—represent the Moon. And third, the arcs of gold on the edge (one is lost) seem likely to have represented, with fair accuracy, the arcs of the horizon within which the Sun rose and set in the course of the year. (Compare figure 3 with the border decoration in figure 10. It fits well with the ranges of rising and setting at the latitude of Mittelberg, which is not far from that of Stonehenge, although we must not fail to take horizon altitudes into account.) There are also holes around the edge in need of an explanation—or of none, if they were merely for decoration. The Nebra disk has had numerous extravagant claims made for it, many purporting to prove that it somehow outshone ancient Egypt, and others that it anticipated the calendrical techniques of the Babylonians. The lower circular arc has been interpreted as a Sun boat, partly by analogy with northern rock carvings, but also with Egypt in mind. Dating from around 1600 BC, it is by no means as old as many a northern solar monument at which risings and settings of the Sun and Moon were systematically observed, but it does seem to confirm what has long been known from them, namely, that northern Bronze Age peoples were deeply concerned with celestial matters.

EARLY ASTRONOMICAL ALIGNMENTS IN THE MEDITERRANEAN

The same is almost certainly true of the peoples of the more southerly parts of Europe and the Mediterranean, although there traditions were substantially different, and surviving monuments are often much harder to interpret. When the involvement of stars is suspected, it is usually difficult to decide on the method by which they were observed. It may well be the case, for instance, that observers noted the rising or setting of stars—the horizon directions of which do not change on a daily basis—with their backs to a flat wall or megalith. The lack of precision in any measurements we may make to test this hypothesis is such that we can rarely be confident, except on a statistical basis. A good example of this problem is to be seen in Sardinia, with its 452 *nuraghes*, towers and complexes of towers that were constructed in the second millennium BC out of large, mostly untrimmed stones, and that still dominate the landscape. Individually, almost nothing could have been deduced with confidence, but statistically it has become clear that entrances and corridors of many nuraghes have solar and perhaps lunar orientations and that attention was very probably being paid to the stars α and β Centauri and to the Southern Cross. It has been suggested, also on the basis of statistics, that those who built the *taula* sanctuaries of Menorca oriented them on those same stars, at a slightly later period. In Menorca, the T-shaped megaliths, achieved by the placement of one enormous stone on an even larger upright, are carefully worked, but they do not lend themselves individually to very precise measurement.

The situation is much simpler when a monument is made of worked

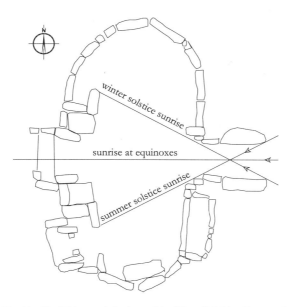

11. Plan of the Mnajdra Neolithic temple in the parish of Qrendi, Malta. The temple, dating from ca. 2850 BC, was first excavated in 1836. Rays of the rising Sun pass into the chamber throughout the year, but at the quarters (and perhaps also at intermediate points in the year) its rays fall at significant places on a screen of stones. At the solstices, only a narrow beam enters, just reaching to the edge of one or other of the two large side stones. At the equinoxes, both doorways are filled with light. (Based on a plan by Chris Micallef.)

stones that set clear directions toward the Sun at one of the solstices in such a way as to illuminate the interior of a building through some deli berately designed aperture. There is a fine example of this at Newgrange, in Ireland. A massive tumulus there, roughly 75 meters across and 9 meters high, structured internally and surrounded by stone walling, has an entrance on the southeastern side. That entrance was blocked at an early date, but even after it was blocked, light was allowed to enter through an aperture above it. At winter solstice sunrise, and a few days on either side of it, the rays of the Sun passed through this aperture and down a passage about 19 meters long, illuminating a vast central chamber in which the bones of the dead reposed. This was clearly not something that the living were meant to see, except possibly in its earliest history, around 3000 BC.

Such illuminated passages were very probably common elsewhere. On the Mediterranean island of Malta, for instance, there are impressive megalithic monuments from much the same period, although in entirely different architectural styles, sporting grand doorways leading down corridors into oval chambers, some of them arranged on elaborate clover-leaf plans. The alignments of the entrance passages have given rise to a variety of claims for their astronomical use. Some seem to be directed to the Sun's solsticial risings and settings, and some of these are more sophisticated that the simple unidirectional corridor of Newgrange. For an example, see the plan of the Mnajdra temple in figure 11. It is unfortunate

12. The unique style of megalithic building in Gozo and Malta is well illustrated at Ħaġar Qim, although its highly complex, many-lobed plan cannot be appreciated from the outside. The only trace of a burial was that of a child, perhaps not connected with the sanctuary function of the monuments on the site, of which the present temple is the third, dating from ca. 2900 BC. Some of the seven former entrances are possibly connected with lunar alignments.

13. Limestone fragment from the small Neolithic temple of Tal-Qādī, near Burmarrad, Malta (29 × 24 centimeters, 5 centimeters thick). Since the surface is eroded, the stars and lunar crescent are here overdrawn for the sake of clarity. (Not perfectly to scale.)

that all have lost the greater part of their uppermost stones, but it seems quite possible that in the grandest of all of the Maltese Neolithic temple complexes, one with a six-lobed plan at Hajar Kim (Ħaġar Qim, fig. 12), there are alignments to the extremes of the Moon's southern (and possibly northern) risings and settings. Why this is especially interesting is that a fragment of a limestone slab from Tal-Qādī—which like the other Maltese temples mentioned here is dated to within a century or so of 2900 BC—is incised with what are surely astronomical symbols (fig. 13). The fragment is divided into five segments, four of them displaying stars and a few short strokes, and one segment including nothing but a crescent, presumably of the Moon. One might have imagined that that it was part of a semicircular primitive sundial, had it not been marked with images from the night; and it would not be difficult to find more profound scientific explanations for it. It seems just as likely, however, that it had a symbolic function, like so many other artifacts uncovered in the astonishing megalithic monuments of Malta and the neighboring island of Gozo.

Prehistory is littered with symbols, many of them undoubtedly solar or lunar, but most of them open to many different interpretations, and the modern cult of Freud does not make matters any easier. Even drawing inspiration from ancient literature is fraught with difficulties. Greek evidence about northern culture, for instance, is often third or fourth hand. The invading Romans interpreted what they found, but in terms of their own experience. Medieval Scandinavian sources, which in many ways seem close to prehistory, are often contaminated with an alien Christian prejudice. And all of these literary sources, of course, provide very late testimony indeed, when we are considering the second, third, and fourth millennia before the Christian era.

SOLSTICIAL ALIGNMENTS IN THE AMERICAS

Europe does not have a monopoly of Neolithic monuments, and evidence from elsewhere is likely to accumulate with time. In 2006, archaeologists found an ancient stone structure at Calcoene, 390 kilometers from Macapa, the capital of the district of Amapá, in northern Brazil. Based on finds of pottery, it is provisionally described as being about two millennia old. Situated on a hilltop, what is assumed to have been a place of worship comprises 127 large granite stones, evenly spaced in three circles, the largest stone being about three meters tall. The monument has been compared to Stonehenge, although it is considerably younger and has an utterly different plan. One thing it shares with Stonehenge, however, and with so many other European monuments of prehistory, is the fact that it is directed toward the winter solstice. One of the stones has evidently been pierced with a hole to admit the Sun at that season.

This discovery of a hitherto unsuspected type of activity of such an early date near the Atlantic coast of South America came hot on the heels of an announcement that an even earlier astronomically aligned monument had been found in the Peruvian Andes. This is a rock-built pyramid, ten meters high, originally covered with plaster and painted in red and white. It has been dated to about 2200 BC—three millennia before the emergence of the Incas, and at least eight centuries earlier than any known building of similar form in the Americas. The pyramid was built around a platform from which a person looking through a certain small window would have seen a carved head, 2.4 meters tall, on a ridge 60 meters away, aligning on a notch in the hills beyond. This sightline headed toward the rising Sun at summer solstice. The site, on a parched and rock-covered hillside called Buena Vista, is about 40 kilometers inland in the Rio Chillon Valley, just north of Lima. A second building on the same site houses a massive clay sculpture with frowning face, flanked by two animals (fig. 14). The frowning face is said to be facing the place of the setting midwinter Sun.

14. The clay frowning face sculpture in the southern temple at Buena Vista in the Peruvian Andes. Said to have faced towards the setting Sun at the winter solstice. Nearby organic remains are dated at ca. 2200 BC.

It would be very surprising if time does not turn up yet more traces of similar prehistoric concerns in the vast unexplored regions of South America. North America has revealed nothing of comparable grandeur. Studies of astronomy in the northern continent have tended to concentrate on the prehistory of the Southwest and the Pueblos in historic times, the latter being known largely through ethnographical reports—as when Alexander M. Stephen, at the end of the nineteenth century, gave a detailed account of the horizon calendar used by the Hopi Sun Chief. There are clues to be found in rock art, which is especially rich in California, although the interpretations offered for it are inevitably open to dispute. There are monuments of a modest and ephemeral sort. Of these, the Bighorn Medicine Wheel, found above the treeline on Wyoming's Medicine Mountain, is perhaps the best known, if only by virtue of a headline "Stonehenge U.S.A." in *Time* magazine in 1984. The "wheel" amounts to a large ring of small stones with 27 or 28 "spokes," and it resembles several others, especially over the Canadian border. Of many properties claimed for it, a summer solstice alignment is the most plausible. Even its date is uncertain, some having claimed that it is as late as the seventeenth century, although several of its fellows are certainly older. The peoples of North America used and revered the sky in their lives as farmers, hunters, and gatherers. Like others across the globe, they took symbols from the sky and worked them into their religious rituals, but their lifestyles did not naturally prompt them to give astronomy expression in the form of grandiose stone monuments. At least, if the idea ever crossed their minds, it kindled no obvious enthusiasm.

✳ 2 ✳

Ancient Egypt

While in northern Europe we are forced to rely mainly on archaeological remains for our knowledge of prehistoric culture in general and astronomy in particular, in Egypt there are written historical records taking us back three millennia before the Christian era, that is, to the period of perhaps the first important phase of activity on the Stonehenge site. Our familiar image of Egypt belongs to the third and second millennium BC. It is an image of the Pharaohs, the pyramids, and the sphinx, the treasures of Tutankhamun and the Egyptian gods—Osiris, Isis, Ptah, Horus, Anubis, and the rest. There was a strong practical and spiritual concern with the heavens from even before this time, although there are no technical astronomical writings from this period to be compared in quality with those produced in Mesopotamia (Babylonia and Assyria) in the later part of it. Egypt had to wait for the Persian conquests of the first millennium BC for the stimulus that cosmological ideas from the Near East could provide.

More intensive activity belongs to the period after Alexander the Great won the Egyptian throne from Persia in 332 BC This last period falls under the dynasty of the Ptolemies, who were Macedonian Greek in origin, and who gained control of the country in the wars that divided Alexander's empire after his death in 323. When this happened, there was a blending of imported and native cosmology—for example, of the kind found in a *Cosmology* associated with the rulers Seti I (1318 BC–1304 BC) and Ramses IV (1166 BC–1160 BC). The Egyptians had long been adept at measuring time and designing calendars based on simple astronomical techniques. And text or no text, they too aligned their buildings on the heavens, although their custom of creating alignments toward the river Nile frequently complicates a modern analysis. In the upper Egyptian Nile Valley, most of the axes of temple buildings were at right angles to the river, while some were parallel to it. It is now well established that when the river does not dominate the surroundings of ancient temples—and it obviously does not, in the case of inland oases—they tend to be astronomically aligned. Some were directed to the cardinal points—predominantly to the meridian—and some, especially in the area of ancient Thebes, to the rising of the Sun at the winter solstice. Even alignments that are difficult to explain seem to cluster around favored directions, which might have been those of stellar risings or settings, or of solar risings and settings at important dates in the calendar. There is similar uncertainty about the significance of the date—currently around 22 February—of what must be counted as the most spectacular of all surviving Egyptian contrivances for illuminating an interior by the rays of the rising Sun. This is to be seen at the innermost sanctuary of the main temple of Ramses II at Abū Simnel. On and around the chosen date, the sculpted figures of Amun-Ra and Ra-Horakhty, solar divinities, are illuminated by the Sun's rays, while the figure of Ptah, god of the underworld, remains in darkness.

ORIENTATION AND THE PYRAMIDS

Some early Egyptian sources speak of a cult relating the Sun god Ra and an earlier creator god Atum. At first the cult was centered on a temple to the north of the old Egyptian capital of Memphis. The place was later to be known appropriately by the Greeks as Heliopolis, "City of the Sun," but the Egyptians knew it as On. (In the book of *Genesis*, Potipherah is said to be a priest of On.) By historical times, the priests of Heliopolis had laid down a cosmogony that held Ra-Atum to have generated himself out of Nun, the primordial ocean. His offspring were the gods of air and moisture, and only after them, and as their offspring, were Geb, the earth god, and Nut, the sky goddess, created (fig. 15). The nine deities of Heliopolis (the Great Ennead) were made up with Osiris, Seth, Isis, and Nephthys, the offspring of Geb and Nut.

The cult of Ra-Atum was well established by the time of the first great pyramids, that is, about 2800 BC. For all their outward simplicity, there seems to have been a relationship of sorts between the pyramids and the Sun and stars. The architect of the first great stone pyramid, the step pyramid built for King Djoser (who ruled from 2667 BC to 2648 BC), was Imhotep, who passed into Egyptian history as an astronomer, as well as a magician and physician. He was later deified, not without reason. When he built his enormous step pyramid, it was without equal anywhere in the world—and its complex subterranean structure distinguishes it even from its successors. Some of these, however, particularly the Great Pyramid of Khufu in the Giza group, were much more accurately Sun-orientated, in the simple sense of being directed toward the four points of the compass—something they had in common with many Egyptian temples. Some of the most interesting plans are those of the Osiris temples at Abydos, in which a series of reconstructions can be traced from the first to the twenty-sixth dynasty. At first, the entrance faced south; in the next four structures it faced north; and in the last three, east.

About eighty royal pyramids have been found in Egypt, most of them small and in a ruined state. The three best known are those enormous members of the Giza group, on the west bank of the northern Nile, belonging to the

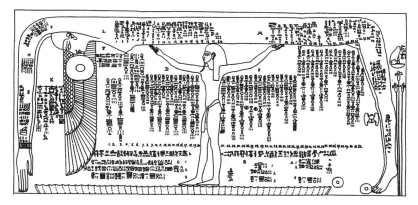

15. The figure of the goddess Nut, with accompanying texts, from the ceiling of the Sarcophagus Chamber in the tomb of Rameses IV (Valley of the Kings, Western Thebes). The vaulted figure of Nut represents the northern sky, supported by the air god Shu. With her head to the west, Nut gives birth to the Sun and stars in the east. They move along her body in the course of the day. (After H. Brugsch, 1883–1891.)

fourth dynasty (ca. 2575 BC to ca. 2465 BC). Built for the kings Khufu, Khafra, and Menkaura, they are often known by the equivalent Greek names—Cheops, Chephren, and Mykerinos. The founder of the fourth dynasty was Snefru (in Greek, Soris). He was the father of Khufu, Egypt's most famous pyramid builder, but had many comparable architectural achievements of his own. Snefru was responsible for completing the pyramid at Meidum for his own father Huni, after which he built his own step pyramid there. An earlier example of his is at Seila. Of modest size, and now in ruins, the Seila step pyramid might well be the oldest Egyptian pyramid to have been aligned on the cardinal points. The Meidum pyramid is the earliest with smooth sides: while it had started life under Huni as a step pyramid, Snefru later converted it into a true pyramid, by infilling the steps and finally adding smooth casing stones.

While these pyramids are more astonishing as works of engineering than as examples of complex orientation—the massive Khufu Great Pyramid is about 146 meters high—the sheer precision of their simple alignments is remarkable. Where reasonably accurate estimates of the original stonework are possible, average errors along some edges seem to have been of the order of only a few minutes of arc. The leveling of the bed of rock on which the Great Pyramid sits is such that the difference in average levels between the north-west corner and the south-east is of the order of two or three centimeters. The Great Pyramid is estimated to contain nearly six million metric tons of stone. Explaining the economic basis of such a colossal enterprise is considerably harder than the astronomical problems it poses.

The pyramids are in several cases pierced by upward-slanting shafts that are usually described as ventilation shafts, but that some writers regard as having been directed toward selected stars at their upper culminations (that is, when they cross the extended meridian, beyond the pole). Khufu's pyramid has an entrance passage on the northern side, sloping steeply down toward the center (in this case an underground chamber) at an angle of 26;31,23°, and apart from a slight variation of a fraction of a degree in the angle, it shares this property with six of the nine surviving pyramids at Giza, and with the only two well-preserved pyramids at Abu Sir. At the Khafra (Chephren) pyramid, the angle is 25;55°. Was there any important star culminating at these altitudes at the time? (Of course the geographical latitudes, and hence the altitudes, all differ very slightly.) If so, whatever star it was, it had to be a star circling the north pole and at *lower* culmination (that is somewhat south of the pole), since the altitude of the pole above the horizon is nearly 30°. A star of moderate brightness in the neighborhood of the true pole was then α Draconis (Thuban), and this, at lower culmination, seems to have been the star to which the passages were directed. Why was a star observed at its *lower* culmination? Perhaps because the constellation we know as the Great Bear, then commonly portrayed in Egypt as the foreleg of a bull, was in a reasonably upright position to the side of it. The Egyptians identified several constellations around the pole: the most significant seems to have been that foreleg of the bull (not, of course, connected with our Taurus), but other constellations in that region of the sky included a hippopotamus, a crocodile, and a mooring post.

SOLAR AND LUNAR RITUAL

The ruins of the temple of Amun-Ra at Karnak, across the Nile from Thebes, have a corridor running northwest and southeast through the middle of the buildings on the site for more than four hundred meters. The central courtyard and chambers of this date from the time of the Middle Kingdom (2052 BC–1756 BC), but the most impressive parts were due to Thutmose III (1490 BC–1436 BC), and additions were regularly made right into the Christian era. There has been much controversy about the significance of the precise direction of this axis, but here too it is seems likely that use was made of an artificial horizon, marginally higher than the line of the distant hills, rather as had been northern practice so long before. For well over a millennium, the last glint of the setting Sun would have been visible from the Holy of Holies, without the need to move far from a position on the axis of the temple. Yet again, religion was a proven stimulus to simple astronomy.

The Egyptian pantheon grew steadily over the centuries, with the incorporation of different local traditions into the official religion. As a result, the behavior ascribed to the gods became filled with inconsistencies. There was always a strict hierarchy of importance, however, and after the fourteenth century BC the dynastic god Amun, "the hidden one," was generally regarded as supreme. (The god's association with the Sun gave rise to the name Amun-Ra.) When the heretical king Amenhotep IV decided that the nation should be converted to the worship of something more visible than Amun, he settled on the Globe of the Sun, Aton, as deserving of elevation to the status of the one true god, and built a new capitol, Amarna, filled with Sun-inspired art (see fig. 16 for a specimen of this). This eighteenth-dynasty

16. Akhenaton (eighteenth dynasty) making offerings to the Sun. This small limestone relief is only about 32 centimeters high. (Now in the Cairo Museum.)

king, who ruled from 1353 BC to 1336 BC, assumed the name Akhenaton, "one useful to Aton." (He is often known by the Greek name, Amenophis.) When he built his own tomb at his new city of Akhetaton—now Tel el Amarna— all corridors and rooms leading up to the burial chamber were along a single axis, directed southeast and northwest. This "pure" solar convention, reminiscent of those that had been adopted in the chambered tombs of northern Europe two millennia earlier, was in violation of a tradition of the burials of his forebears and successors. In them, either the burial chamber was entered by a corridor at right angles to the main corridor of approach, or the main corridor was staggered, so as to break the directness of the approach.

The extreme form of Akhenaton's solar religion was short-lived, but solar rituals lived on in many ways, for instance in the installation of rulers. We are very fortunate in the Egyptian remains from this historical period, for we have a number of actual illustrations of solar ritual, especially from the walls of the royal tombs at Tel el Amarna. Some idea of Egyptian solar symbolism may be had from the tomb of Tutankhamun, a king now chiefly remembered because his tomb escaped pillage. The son-in-law of Akhenaton, he reigned from 1333 to 1323. The walls of the entrance passage to his tomb, and four of its chambers, are all strictly north-to-south and east-to-west. Entry is from east to west. Each of the four chambers had a known ritual purpose, as depicted on its walls, and no doubt these have something in common with their more ancient counterparts. One (keyed to the south) was a chamber of eternal kingship, another (east) a chamber of rebirth, a third (west) a chamber of "departure to sepulchral destinies," and the fourth (north) that of the reconstitution of the body. The dead Osiris would, it was thought, after his arduous

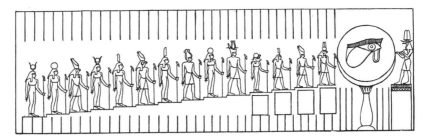

17. In the temples at Edfu and Dendera, the fourteen gods of the waxing Moon climb a staircase with fourteen steps. The stairs end at the crescent and the symbolic form of the Moon, the left eye of the sky god Horus. The right eye of Horus, used to represent the Sun, was a much more common image. (After E. A. W. Budge, 1904.)

quest for rebirth, reappear in the form of the rising Sun, Ra. Those who were responsible for ordering solar rituals must have had a sound intuitive awareness of the Sun's behavior, but this is a relatively simple astronomical affair by comparison with the behavior of the Moon, which seems never to have been studied with the same care as in northern Europe at roughly the same time.

There were lunar myths in plenty, of course. The sixth day of the lunar month was associated with a feast for Osiris, and importance was also ascribed to the day of the full moon. In some Pyramid Texts, the deceased is identified with the Moon. Coffin Texts from Deir el-Bersheh give a place in the after-world to the lunar god Thoth, next to Osiris and Ra. As in so many other cultures, the Moon was long associated with fertility. The horns of its crescent—symbolic of cattle—and the length of the female menstrual cycle no doubt reinforced this association. There was a myth which told of a battle between Horus and Seth, in which the eye of Horus was stolen and damaged, but eventually returned as the Moon. The theft was by Onuris, Thoth, or Osiris. The Thoth version of the story has him aided by fourteen gods, he and they representing the fifteen days leading up to the full moon (fig. 17); and that myth was continued with an account of fourteen gods connected with the waning moon. There were other variants of the myth. At Edfu it was said that Horus assumed control of the sky at full moon, a time known as "the uniting of the two bulls." The same occasion was described in the New Kingdom temple of Osiris at Abydos. A ritual in later temples was celebrated with the offering of two mirrors, symbolizing Sun and Moon. At Thebes, the moment was said to symbolize the rejuvenation of the Sun god Amun-Ra. Rich though such mythology was, judged as astronomy it will certainly not bear comparison with what was being done in other cultures at the same period of history.

THE CALENDAR

Some time after nomadic tribes of North Africa first settled as farmers in the Nile valley, they realized that there was a correlation between the pattern of the river's behavior and that of the star Sirius (called Sothis in Egyptian), the brightest in the sky. The rising of the Nile, important because its flood waters irrigated the valley, was seen to coincide with the first sighting of Sirius on the eastern horizon shortly before sunrise, after a long period of invisibility. This event, now known as its *heliacal rising*, took place in mid-July, and not at an especially notable point in the solar year. The Sun's connection with the seasons tends to be less obvious in low latitudes than in high. The three seasons for the Egyptians were more often related to the behavior of the river, and the names of the months that began them were "Flood," "Emergence," and "Low Water" or "Harvest." The other month names came from lunar festivals. The Sun was at first important for them, it seems, mainly as an indicator of the yearly cycle.

To reconcile the three interlocking systems—stellar, lunar, and solar—was one of the main tasks of Egyptian astronomy, and it remained at the center of astronomy as a problem of religious importance until modern times. The festival of Sirius/Sothis more or less followed the solar year of about 365 ¼ days, but the length of twelve months, each of 29 or 30 days, averaged at only about 354 days. From at least the middle of the third millennium, therefore, the Egyptians devised one of the earliest calendar rules known to history: an extra ("intercalary") month named *Thoth,* after the lunar deity, was added to the year only if Sirius/Sothis rose heliacally in month twelve.

As Egyptian society became steadily better organized, the calendar was developed further. The length of the year was set at 365 days, and the "months" were standardized at 30 days, each of them divided into three "weeks" of ten days. The system, which dates from perhaps the twenty-ninth or thirtieth century BC, has many advantages. The week, whether of ten days or seven, is a conventional matter with no astronomical significance, but there is an obvious ambiguity in the word "month" used here. Those for whom the visible Moon is a matter of religious importance will not find here a source of accurate information about it. The main problems that this calendar solves are problems of bookkeeping, and in this respect astronomers have often been grateful for its ease of historical use. The Egyptian year of 365 days in particular is one that they have found attractive, for it simplifies the conversion of long periods of time to days. Even Copernicus, followed Hellenistic-Egyptian practice, and used the Egyptian year for his astronomical tables.

Not everyone shared the bookkeeping mind of the astronomers, of course, and the new calendar inevitably ran into difficulties as the slight error in the length of the year accumulated, and the seasons drifted through it. This seems to have been considered as less serious than the lack of precision in the lunar calendar. A new lunar year was devised to run in harness with the civil year. New rules of intercalation were drawn up around 2500 BC, and for over two millennia Egypt had three calendars in use, side by side.

STARS AND THE HOURS OF DAY AND NIGHT

The Egyptians' wish to divide the night into smaller parts combined in a curious way with their civil calendar to give us our division of the day into 24 hours. Any society that carries out ritual acts by night is likely to devise ways of judging the passing of the night. The Egyptians wrote copiously on the passage of the Sun god Ra on his night-boat through the Other World, between sunset and sunrise, and the stages of the journey were marked by the movements of the stars. To find how they were marked we must turn back for a moment to the calendar.

A key Egyptian calendar notion was that of heliacal rising—the first morning visibility after a period during which the star was above the horizon only in daylight. Any bright star could be chosen to mark a point in the calendar, and we have already seen that Sirius/Sothis was the most important marker of all. Each day subsequent to the heliacal rising of a star, it rises a little more in advance of the Sun than before, until another suitable marker-star rises heliacally, and counting starts from that date. Who was to select those stars, and on what principle? The solution was reasonably straightforward. We have seen that the civil calendar divided the year into thirty-six "weeks" of ten days each. Thirty-six stars or constellations were therefore sought, to mark by their heliacal risings the beginnings of those thirty-six weeks. (It seems that they were chosen to be as much like Sirius/Sothis as possible, each being invisible for about seventy days in the year.)

Forget, now, the reasons for which the stars were chosen, that is, reasons having to do with the division of the year, *calendar* reasons. We simply have thirty-six stars, or groups of stars, recognized as being of great importance, and during any night their risings will occur at moderately regular intervals. Turning as they do, once in roughly a single solar day, we might imagine that eighteen of the thirty-six stars would rise successively during an average night, but in fact the problem is complicated in several respects. During most of twilight, most stars are invisible: night was reckoned as the period of total darkness. The calendar stars chosen were to the south of the celestial equator, that is, they crossed the horizon south of east. (They were in fact in a belt roughly parallel to, and to the south of, the Sun's path through the stars, the ecliptic.) Not all nights are in any case of equal length: at Egyptian latitudes a midwinter's night is nearly half as long again as a midsummer's night. In short, without going further into the theoretical reasons for it, during much of the year the night was not divided into exactly twelve parts by these stars. Nevertheless, *it was finally regarded as being so divided.* Evidence for this fact, this way of dividing the night, comes from diagrams on the inside of coffin lids from the Eleventh Dynasty (twenty-second century BC).

Daylight was later divided into twelve hours, the number of divisions being taken by analogy with the number of hours in the night. In this way, the world was given the twenty-four hours of the day. Even the endeavors of revolutionary France did not manage to replace this system with one that was considered more rational—that is, one based ultimately on the number of our fingers.

It is often said that the coffin lids are concise versions of representations on the ceilings of known tombs of rulers of the Middle Kingdom and after. They are both far too intricate to be described, let alone compared, in a brief space, but the ceilings are too important to be ignored, for they provide

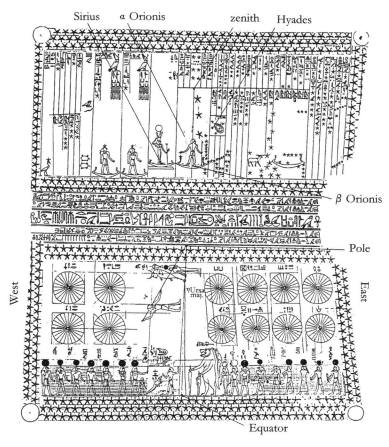

Sirius   α Orionis   zenith   Hyades

β Orionis

Pole

West

East

Equator

18. Astronomical ceiling from the secret tomb of Senmut, near the temple of Hatshepsut at Dar el-Bahri, Western Thebes. Senmut was Steward of the Estate of Amun-Ra under Queen Hatshepsut and directed the building of the temple. Discovered by H. E. Wenlock, 1925–1927. The tomb dates from ca. 1473 BC. The upper half of the figure is southern, the lower, northern. The narrow columns represent the decans, each named together with one or more stars; and there are also gods and planets. The arrangement is not that of the "diagonal clocks" on the coffins. Marshall Clagett considered this extremely complex depiction of the sky to be the oldest representation of the ancient Egyptian celestial diagram.

numerous examples of simple celestial mapping. The earliest of them is in an unfinished tomb of Senmut, vizier of Queen Hatshepsut (fig. 18). This is in the Valley of the Kings, in Western Thebes, and dates from around 1473 BC. It has perhaps the oldest known images that can be reasonably described as celestial diagrams with identifiable elements. A few of its constellation images are shown in figure 19, where they are compared with equivalent images of almost two centuries later. The latter are drawn from the underground cenotaphs of Seti I; and there are others resembling them in the cenotaphs of Ramses IV, and later rulers.

Egyptian traditions were long lasting. Almost a thousand years separate the Seti ceiling from a notable papyrus manuscript, Carlsberg 1, that nevertheless amounts to a commentary on it. This funerary text also contains

19. A detail of the previous ceiling (upper figure), compared with a corresponding section of a ceiling in Hall K, in the tomb of Seti I (who ruled from 1306 BC to 1290 BC), also in Western Thebes (lower figure). Both depict northern constellations, the older in a more abstract way.

instructions for making a shadow-clock, with four divisions on its base (fig. 20). Another such device, from the time of Thutmose III (1490 BC–1436 BC), has five divisions. How aware early peoples were of the pattern of variation in the lengths of day and night it is impossible to say, but this awareness must have been heightened with the invention of the water clock. The earliest example dates from the time of Amenhotep III (1397 BC–1360 BC), but relates to calendar reckoning during the reign of Amenhotep I (1545 BC–1525 BC), and it is from an inscription in a tomb from that earlier period that we know of attempts to express the ratio of the lengths of longest and shortest nights. The ratio given (14 to 12) is not particularly accurate, but it is the principle that counts; and we are in the unusual position of knowing the name of the man responsible, Amenemhet.

As early as the fifteenth century BC, it had been recognized in Egypt that the risings of the decanal stars provided a poor way of regulating civil and religious life, and another set of stars was chosen. These were observed, not at the horizon, but as they crossed the meridian. Again this was no exact matter: these meridian transits were observed with reference to the head, ears, and shoulders of a sitting man. Three royal Ramesside tombs (ca. 1300 BC–1100 BC) were decorated each with twenty-four tables (two to a month) that made it possible to judge the hour in this way (fig. 21). It has been conjectured

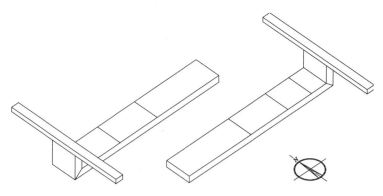

20. An Egyptian shadow clock, in its morning (right) and afternoon (left) positions. The shadow of the crossbar marks the time on the horizontal scale. There were many other types of Egyptian shadow clocks, some much more complex than this.

21. Copy of one panel of a typical Ramesside star clock from the hall of a tomb in the Valley of the Kings (twentieth dynasty). Each panel covers the night hours for the first or sixteenth day of the month, and the fourteen days following.

that these star clocks were drawn up with the help of water clocks, since several water clocks survive with astronomical material engraved on them.

EGYPT, GREECE, AND ROME

Despite the great cultural wealth and length of time over which the heavens were scrutinized by the Egyptians, not to mention the respect in which they held many celestial objects, except in the case of the calendar it does not seem to have occurred to them to seek for any deeply systematic explanation of what they observed. For all that they were in possession of a script, they seem to have produced no systematic records of planetary movements, eclipses, or other phenomena of a plainly irregular sort. The Egyptians abstracted legends more easily than mathematics from the stars. Decorated monuments—of which more than eighty are known that could be somehow classified as astronomical—represent the cosmic deities of mythology, including the solar and lunar deities, the planets, the winds, the constellations, earth, air, sky, the cardinal points, and so forth. They show a close familiarity with the constellation patterns. (Theirs were not identical to ours, of course.) The great reputation for astronomy that the Egyptians

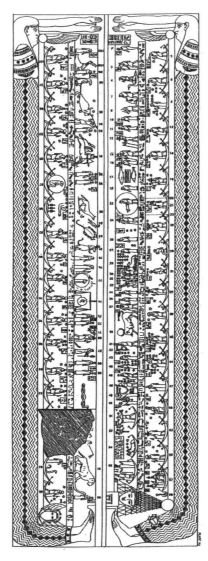

22. Rectangular zodiac from the temple of
Khnum, near Esna, dating from ca. 200 BC.
(This is no longer extant; drawing after
J. Bentley, 1825.)

have enjoyed for most of the last two thousand years is based, however, on a confusion going back to the period when Rome ruled Egypt.

To the Romans, "the Egyptians" were simply those who lived in Egypt at that time, and the description included those of Greek culture. Almost invariably, when Egyptian astronomy or astrology was mentioned by Latin authors, they had in mind *Hellenistic* Egypt, the Egypt ruled by the successors of Alexander the Great. Zodiacs in Egyptian temples and tombs were Hellenistic, although based on Mesopotamian models that had been introduced after the Alexandrian conquests succeeded in Hellenising Egypt in the fourth century BC. The first Egyptian zodiac known to us is rectangular in form, and was on a ceiling from the temple of Khnum, near Esna, dating from around 200 BC (fig. 22). A drawing was made of it

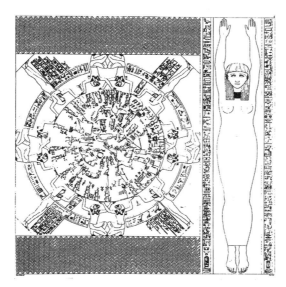

23. An outline drawing of the sculpted zodiac which is now in the Louvre, but which came from the ceiling of the small eastern chapel of Osiris on the roof of the Hathor temple at Dendera. From the fourth volume of C. L. F. Panckoucke's published account of the researches made during Napoleon's Egyptian campaign (*Description de l'Egypte . . . l'expédition de l'armée Française*, ca. 1820–1830). The drawing contains some small errors. The temple at Dendera today has a plaster replica of the original.

during the Napoleonic expedition to Egypt, but along with the temple, it was destroyed in 1843, the stone blocks being used to build a canal. The best known zodiac—indeed, the oldest complete zodiac in a now familiar form—is to be seen in the Louvre, in Paris. It is on two blocks of sandstone, removed from the temple of Hathor, at Dendera—with the help of saws and gunpowder—during the Napoleonic expedition. Once part of the ceiling of a chapel on the roof of the temple, it dates from around 30 BC (figs. 23 and 24).

The origins of these zodiacs are usually judged from the constellation images they contained, such as those of our Capricorn and Sagittarius, which have been found on much more ancient Babylonian boundary stones. While both rectangular and circular forms of the zodiac are to be found in Egypt, they all ultimately have their pictorial origins in Babylonian—rather than Egyptian or Greek—astronomy. Having said this, we should not overlook a less obvious Egyptian element in their design. In addition to the signs of the zodiac and the planets, they include the old decans, and a few Egyptian stars and constellations. It is possible that the decans were no longer in use to mark the hours of the night but had simply become viewed as standard divisions of the signs—most signs having three decans, but some four. Taken over into Greek and Roman zodiacs, the decans became ten-degree segments of the signs, and as such they passed into Greek (Hellenistic), Roman, Byzantine, Indian, Islamic, and Western astrology (in Western astrology, they are often known as "faces").

24. The inner star map (about 1.5 meters in diameter) of the Dendera ceiling shown in the previous figure. The offset superimposed circle (the ecliptic) threads through the signs of the zodiac, most of which are easily recognized. To the right of center are Pisces, that is, the Fishes, with their tails tied together by a V-shaped cord. Below them are Aries and Taurus—the Ram and the Bull. Next Gemini, Cancer, and Leo—the Twins, the Crab, and the Lion. (Cancer is not represented by a Scarab beetle, as is often claimed.) Virgo, the Virgin with corn ear, follows next; then Libra and Scorpio, the Balance and the Scorpion. (The lion near Libra is not the zodiacal lion, but represents our constellation Centaurus.) Sagittarius, Capricorn, and Aquarius follow, namely the Archer, the Goat-Fish, and the Waterman. The ecliptic circle and other construction lines reveal a debt to the Greek tradition of planispheres (for which see chapter 4). Elsewhere on the map are several characteristically Egyptian constellations, such as the Bull's Foreleg (part of Ursa Major) and the Hippopotamus (Draco). There is still room for controversy about other elements in this sculpted scheme. There are certainly planets in it, placed in constellations where Greek astrology gave them high significance. The four female supporting figures have been claimed as goddesses of the cardinal points of the compass; and alternatively, as four images of the sky goddess Nut.

If we are to narrow our focus and consider only mathematical, theoretical progress, then we may take calendar schemes as the strongest native tradition, so that it comes as no surprise to find that in perhaps the fourth century BC the Egyptians recognized a 25-year lunar cycle. This equated 25 Egyptian years (9,125 days exactly) with 309 months ("synodic" months, new moon to new moon). This is an excellent approximation, which slips by a day in only about five centuries. It provides a parameter worthy to be placed side by side with the more numerous parameters derived by the mathematical astronomers of Mesopotamia. How original was it? From the sixth century BC, a time of Persian domination in Egypt, the

Egyptians had been adapting the Babylonian lunar months to their own civil calendar, and this fact perhaps goes some way toward explaining a result that is conspicuous by its rarity.

During the later Roman Empire, Egyptian preeminence in astronomy was secured by the reputation of one man, Claudius Ptolemy, the greatest astronomer of antiquity. Ptolemy, of whom we shall say more in chapter 4, was an Alexandrian—and Roman citizen—of the second century AD. Before long, the political history of Egypt would undergo yet another great transformation, with the founding of Constantinople in the year 330, as an imperial capital and an Eastern counterpart to Rome itself. Thereafter, Alexandria gradually lost its position as the chief intellectual center of the Greek-speaking world. After the death of Theodosius in 395, Rome finally split into Eastern and Western empires, and fifteen years later the city of Rome was sacked by the Visigoths. A century before this happened, however, Egypt had become largely Christianized, with the result that the newer "Egyptian" astronomy was passed on in two different directions. That it was absorbed into the intellectual culture of Constantinople is less surprising than its journey southward, when Coptic literature, using a form of Greek script for native Egyptian dialects, carried a superficial version of astronomical knowledge down to Ethiopia. In both cases, the astronomy was at heart Hellenistic, and the debts it owed to ancient Egypt were fewer by far than those it owed to ancient Mesopotamia.

* 3 *

Mesopotamia

The sort of astronomy developed in Mesopotamia, at least during the long period from which we have written evidence, differs so radically from that of the Egyptians before it, and of the Greeks and Hindus after, that it seems natural enough to separate them. To do so is to risk encouraging the mistaken belief that these cultures were themselves simple and monolithic. Matters of general cosmological concern enter into the mythology of all peoples, even though local variations are legion. Language forms a natural obstacle to the diffusion of ideas, and the Near East knew many languages, but such was their pride, and sense of identity, that even neighboring cities sharing a common language would often worship different planetary deities. Under these circumstances it is impossible in a brief space to do more than rough historical justice to the region, but one point, at least, should be kept constantly in mind. All of those sciences in the modern world that make use of mathematical methods are indebted to the astronomers of Babylon, especially those from the last five or six centuries before the Christian era, which saw the culmination of the Mesopotamian achievement. This being so, we must look beneath the diversity for what was peculiar to Mesopotamian astronomical practice as a whole.

In the sixth century BC, Babylon enjoyed a brief period of independence after having long been dominated by the Assyrians. With the overthrow of the Assyrian Empire, the Babylonians set to work to revive their own lost glories—that is, from a thousand years before, when the great Hammurabi had been ruler (perhaps 1792 BC–1750 BC). The power of the Persians was also growing rapidly, however, and in 539 BC, Cyrus the Great succeeded in defeating Nabonidus, the last Babylonian king. The repercussions of these events are familiar to many through the Biblical account of the release of the Israelites who had been brought to Babylon after the destruction of Jerusalem by Nebuchadrezzar II.

At length, the Persian Empire was defeated by Alexander the Great, and links with Greek culture became greatly strengthened. Alexander entered

Babylon in 331 BC. After his death in 323 and the division of his empire, Babylon came to fall, after various reverses, under the rule of Alexander's most politically able successor, Seleucus, the founder of a new dynasty. Seleucus ensured a steady stream of Greek immigration into this new Greek and Macedonian Empire in Asia. When we speak generally of "Greek learning" after this time, we must bear in mind not only the city-states of the Balkan Peninsula and surrounding islands, but also this vast cultural area in the Middle East—not to mention Alexandria and a Hellenized Egypt.

With these events in mind, it is useful to distinguish between four historical periods: first, that of the Hammurabi dynasty, and what went before it; second, the Assyrian period (1000–612 BC); third, the period of independence (612 BC–539 BC), followed by subservience to Persia (539 BC–331 BC); and last, the Seleucid period (331 BC–247 BC).

BABYLONIAN ASTRONOMY UNDER THE HAMMURABI DYNASTY

Over the last century and a half, archaeologists have excavated several Mesopotamian temples with their massive associated towers, known as *ziggurats*. At Uruk, for instance, one complex was found to go back to the fourth millennium, while at the old Sumerian city of Eridu some of the buildings might be as old as 5000 BC. The oldest written records are from the temples, and it seems likely that political power was largely centered on them. They owned most arable land and played a central role in the control of irrigation. Although it is not altogether clear how the religious power base was related to the power of the secular rulers, there is no doubt that an extraordinarily complex bureaucracy grew up within the priesthood. At some time before 3000 BC, the Sumerians invented the cuneiform script, where marks were impressed by a wedge-shaped stylus in soft clay, which was then allowed to dry out in the sun. The Babylonians—a Semitic people—used this technique and adapted it to their own language, using a mixture of phonetic spelling and Sumerian ideograms.

They also took over the Sumerian number system, a fact of great scientific importance, since this used a place-value notation, just as our own system does (unlike that of roman numerals). The difference is that where we work to a scale of ten, they took a scale of sixty. This sexagesimal system goes back to the third millennium. Numbers up to 60 were built up in a simple way, reminiscent of our roman numerals, based on the repetition of a wedge-mark for 10 and a vertical stroke for unity. (See fig. 25, for example, for the number 23.) Beyond 60, numbers were separated with spaces. Representing these spaces by commas, for instance, 2,9,14 would be taken to signify:

$$2 \times 60^2 + 9 \times 60^1 + 14$$

Potential complications in this system arose when the same arrangement was used for fractions. It is as though we, in our decimal notation, were to

25. A cuneiform representation of the number 23.

use a string of numbers like 3546 to mean, indifferently, 3546, or 345.6, or 35.46, and so on. We need some sort of punctuation to separate the fractional part from the rest, and the convention among modern writers is to use the semicolon. Thus, the string of numbers 2,7,17;52,13 would now be used to represent the following mixed sexagesimal number:

$$2 \times 60^2 + 7 \times 60^1 + 17 \times 1 + 52 \times \frac{1}{60} + 13 \times \frac{1}{60^2}$$

We may perhaps feel uneasy at grafting our decimal notation on such a powerful system as theirs, but the mixture ought to give rise to no great confusion, for we have inherited the Babylonian legacy whenever we write a time in hours, minutes, and seconds, or an angle in degrees, minutes, seconds, and so on. As for describing their system as "powerful," it is so simply because 60 has so many prime factors. (Had there been an evolutionary advantage in finger reckoning, we might have found ourselves endowed with 12 or even 30 fingers.)

The Sumerians, and the Babylonians after them, became expert calculators with this system of numeration, and to help with calculation they devised one of the most useful of all scientific inventions, that of tables of numbers. They had multiplication tables, tables for reciprocals, for squares, and even for square roots. The Babylonians were adept at carrying out procedures that we should not hesitate to describe as algebraic. They could solve linear and quadratic equations, and even special cases of equations of higher degrees. They used geometrical proofs of algebraic formulas, much as do we when we follow in the Greek tradition. These techniques flourished under the first Babylonian dynasty, which began with Hammurabi, "the Lawgiver," and lasted for three centuries. (The dynasty ended no later than 1531, and some would argue as early as 1651 BC.) All the evidence points to relatively many members of the educated priesthood being highly expert in arithmetic, and this fact was of great importance for the astronomical style they gave to the world.

The earliest documented period of Mesopotamian history speaks for a great variety of local cults, spreading with changes in political power, even from the older Sumerian population to their Semitic successors. Many of the gods had nothing to do with the heavens, but the city of Ur favored a local deity called Sin, a Moon-god; the cities of Larsa and Sippar both followed Shamash, a Sun-god (fig. 26); and several accepted Ishtar, later identified with the planet Venus, but perhaps once a personification of fertility. Babylon itself was loyal to Marduk, a god who was given supremacy as "Creator," when Hammurabi BC combined the gods of the city-states into a single pantheon. The Babylonian creation myth *Enūma eliš* ("When on high") was recited annually on the fourth day of the New Year festival held in Babylon

26. Drawing of the upper part of the "Sun god tablet" from Sippar, now in the British Museum. In a long cuneiform text below this image (not shown), Nabu-apla-iddina records his restoration of the ancient image of the Sun god and of his temple (ca. 870 BC). The king is being led towards an altar on which the Sun-disk of Shamash rests.

during the first twelve days of the month of Nisan. It describes how the god Marduk rose to kingship among the gods by defeating the forces of chaos, personified by the sea, Tiamat. He was then supposed to have organized the universe, building the city of Babylon as the gods' earthly home.

Not all gods, by any means, can be identified with stars: the three highest—Anu, Enlil, and Ea—corresponded to heaven, earth, and water. There is, however, a good case to be made for claiming most of the old Babylonian gods as being in some way *cosmic*. A deep concern with cosmic affairs is apparent in the oldest literature to have come down to us. The Babylonian Gilgamesh epic has in it hints of a ritual of observing the Sun, Moon, and planets over the tops of distant peaks.

After the trauma of his companion's death, Gilgamesh arrived on his journey at a mountain, Mâshu, "whose peaks reach to the banks of heaven, and whose breast reaches down to the underworld." He there prayed to the Moon (Sin) to preserve him. There was on the mountain a gate through which the Sun passed on its daily journey, guarded by two scorpion-people, man and wife, who allowed him to pass. Gilgamesh traveled eleven hours in utter darkness, and a twelfth in what clearly represented twilight, before coming out on the other side in a beautiful garden with jeweled shrubs. He discovered a plant that can bestow immortality, but it was snatched away by a serpent that ate it, shed it skin, and renewed its life—presumably an allegory of the Sun's motion. The Sun, like the serpent, was supposed to have the power of renewing its life, but Gilgamesh was refused the option. The gloomy message of the epic as a whole is that man must die.

Old Babylonian seal cylinders—engraved cylinders about the size of a pen cap which are rolled over soft clay to reproduce a picture—support the idea that the elements of this sort of story were well known. They often show the Sun-god stepping through a mountain pass between two gateposts, sometimes brandishing the key to the gate. There are lions on the gates, a common solar attribute, and the solar rays issuing from the god's arms. There are no mountains in Babylonia visible from the valley of the Tigris, and any mountain Sun-god worshipped there might have come from Elam in the east (and the region around the city of Susa) or from the ranges that run northwestward from there through present-day

27. Even in the fifth millennium BC, traders along the Tigris and Euphrates used stamped clay to seal doors and containers. Around 3500 BC, elaborately carved cylinder seals were used, rolled over the clay to mark it. This drawing of the result of rolling out a seal of the Akkadian dynasty (third millennium), now in the British Museum, shows a common image, related to the Gilgamesh epic. Here the Sun god Shamash—identifiable by the rays from his shoulder and the saw he carries—rises between the twin peaks of Mount Mashu.

Turkey. In the earliest period, the Sun-god is depicted holding a serrated weapon, and if not lifting himself by his hands, then he stands with one foot on a mountain (fig. 27). As represented later in the Middle Empire—say 2800 BC, but still earlier than the literary sources—a stool takes the place of the mountain, in keeping with the features of the newly inhabited terrain. This shows the importance of local circumstance. Just as the mythology is partly determined by it, so too must the religious techniques for observing the heavens have been.

Under Hammurabi, not only the pantheon but also the calendar was unified, and Babylonian names were imposed on the months. Since the length of the month is near to, but not exactly, 29.5 days, and since the number of months in a solar year is not a whole number (it is about 12.4), rules of intercalation are needed in any calendar that is to take both into account. Such rules were found, decreeing whether a month was to be counted as 29 or 30 days, or whether a year should have 12 or 13 months. Although they were far from perfect and not uniformly applied, they remained in use until 528 BC. Throughout history we find examples of an intimate relationship between calendars and cultures, each dynasty—or in some cases, each religion or religious sect—wanting its own calendar. It is therefore interesting to find a letter from Hammurabi, informing the recipient of his decision to intercalate a month. The letter noted that this would affect the collection of taxes. The ruse seems to have no modern parallel.

Not always for the best of reasons, the Babylonians have been very commonly associated with astrology. The subject usually understood by this word made its appearance at a relatively late date and owes far more to Greek influence, but it is true that much Babylonian divination assumed a cosmic character at a very early date. By no means were all portents astral. The constellations, the Sun, the Moon, and Venus, for example, were linked with gods, but there were earthly deities in plenty. There is a very large surviving series of omens, about seven thousand in all, containing originally many thousands of phenomena of very many different kinds that were all open to interpretation. The series is known from its opening words as "Enūma Anu Enlil . . ." (When the gods Anu and Enlil . . .). The omens survive in tablets from the time of the Cassite rule over the Babylonians (approximately 1500 BC–1250 BC), but many of them were very probably taken from sources as early as the Akkadian dynasty (about 2300 BC). Even as late as a few centuries BC, there was a class of people known as "the scribes of Enūma Anu Enlil." Many of the omens concerned the astrological significance of the position and appearance of the planet Venus. The sixty-third tablet of the series of omens is in fact one of the most important of all early astronomical documents, for it deals with methods for calculating the appearance and disappearance of Venus and of interpreting astrologically those phenomena.

The Venus tablets (various copies survive) are associated with the reign of the King Ammizaduga, whose rule began 146 years after the beginning of Hammurabi's. In chapter 2 we introduced the idea of heliacal risings and settings of fixed stars. In the Ammizaduga tablet, a virtually complete list of heliacal risings and settings of the planet Venus is given for a period of 21 years, each event being offered an astrological interpretation concerning the fortunes of climate and war, famine and disease, kings, and nations.

PLANETARY MOTIONS: A DIGRESSION

To discuss these remarkable tables we must first make a digression, to explain from a modern point of view the motions that are involved. Although not true to an ancient view of these matters, to remember the pattern of the motions it is helpful to think of them in terms of a Sun-centered planetary system. The general sense of rotation round the Sun of all the planets visible to the naked eye is the same: looking down on the solar system from the northern half of the sky, the rotations are anti-clockwise. (These are the long-term motions and have nothing to do with the apparent daily motions, risings and settings, that are due to the spin of the Earth on its axis. Figure 28 illustrates the general points being made here.) A planet that is observed from a place in the Northern Hemisphere of the Earth will of course rise and set daily, but seen against the background of fixed stars over a long period of time it will *generally* move slowly leftwards, against the sense of the daily motion. In this case it will rise slightly later each day. This is true even of the Sun, for—as a moment's thought will show—just as the Earth goes counterclockwise round the Sun, so, with reference to the Earth, the Sun goes counterclockwise round the Earth. This type of motion against the stellar background is called "direct." For future reference, it is the sense in which coordinates of the "longitude" type are always taken to increase, when the positions of the planets and stars are recorded in relation to the ecliptic or equator.

Planets move in general in relation to the stars with direct motion, but there are times for any planet when it moves in the opposite sense, with what is called "retrograde" motion. Venus is nearer to the Sun than are we, and so may lie between us and the Sun, or beyond it. When Venus is beyond, it moves with direct motion, for it shares the Sun's direct motion and

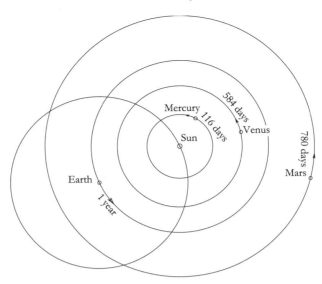

28. From a purely geometrical standpoint, the solar system may be considered as heliocentric or geocentric—although not with the geocentric arrangement assumed in antiquity. The Earth and planets all revolve in the same sense around the Sun. (For the sake of simplicity, only four orbits are shown here, all assumed to be circular.) Without compromising this arrangement, we may relate the system to a fixed Earth. We then assume that the Sun revolves around the Earth (in the same counterclockwise sense, on this diagram), the other planets being carried round with it, that is, with their orbits still centered on the Sun.

has its own in addition. When it is nearer, however, it can have a retrograde motion of its own (a movement from left to right, for our northern observer) that, when viewed from the Earth, appears faster than the Sun's direct motion (right to left). There are certain points at which, from the Earth, it will seem to have no motion at all. They are marked as *MS* and *ES* on figure 29 (where the Sun-Earth line is regarded as fixed). At the former stationary point, the planet will be seen ahead of sunrise, in the morning, but in the evening it will have moved below the horizon before the Sun, and so will not be visible then. When it is at its other stationary point, *ES*, it will be seen only in the evening, after sunset, while in the morning it will rise after the Sun and be hidden in the Sun's rays.

When Venus is in the neighborhood of superior conjunction, namely *SC* in the figure, or at inferior conjunction, *IC*, it will not be seen, for here too it will be lost in the Sun's rays. It has to be of the order of 10-degree distance from the Sun to be seen at all. (This angle depends on many factors, and we need not go into more detail here.) The points *MF* and *ML* in the figure are those at which Venus is seen for the *first* time and the *last* time in its cycle, as a *morning star;* and *EF* and *EL* are its points of *first* and *last* appearance as an *evening star.*

Figure 29 is drawn approximately to scale. Venus takes 584 days in its complete circuit of the Sun. How long it is lost to view depends on various factors, such as the geographical latitude and the time of year. As a very

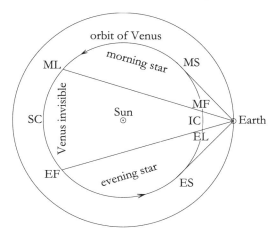

29. Risings and settings of Venus, an inferior planet (that is, within the Earth's orbit). The orbits are roughly to scale. The other inferior planet, Mercury, shows similar behavior.

rough guide, a dozen weeks of invisibility at superior conjunction, and a fortnight at inferior conjunction, are not unusual.

Of the planets known before the discovery of Uranus in the eighteenth century, the only other planet between us and the Sun, Mercury, has a pattern of behavior similar to that of Venus. The superior planets (Mars, Jupiter, and Saturn were the only planets known before Uranus), that have orbits outside our own, behave differently. They can be seen overhead at midnight, for example, which is clearly out of the question for Mercury and Venus; but they too can have retrograde motions, and they too have periods of invisibility, that is, when their angular separation from the Sun at (superior) conjunction is small.

The phenomena of first and last sightings of the planets were of great interest to the Babylonians. Presumably attention was drawn to them only after a tradition had been established of observing the heliacal risings and settings of the fixed stars. First and last sightings were to be of great importance for the foundation of theories capable of predicting planetary positions, for they provided reference points with respect to which planetary positions could be specified at particular times.

RELIGION AND ASTROLOGY

The Venus tablets of Ammizaduga give the years, months, and day-numbers on which the planet reached the positions labeled *EL*, *MF*, *ML*, and *EF* over a period of about 21 years. It might have been evident then, as it certainly was later, that the sequence of phenomena began to repeat itself almost perfectly every eight years (more precisely, every 99 Babylonian months less 4 days). It did so after 5 complete cycles of the 4 phenomena. We ourselves can see this equivalence as a consequence of the fact that the synodic period of Venus (the time taken to orbit the Sun once, with reference to the Earth-Sun line) is 583.92 days: 5 such periods amount to 2,919.6 days, while 8 years of 365.25 days equal 2,922.0 days. Astronomers made much use of this near-equivalence throughout later history, recognizing that it is relatively easy to draw up a tolerably good ephemeris for the planet Venus after the first eight years of observations, since the planet's coordinates then repeat themselves on almost the same (solar) calendar dates. There are similar cyclical relationships for all the planets, but none so simple as this.

In one version of the tablets it is clear that we are no longer dealing with a set of unmodified observations, for there is a recognizable pattern in the data. Periods of invisibility, for example, are always put at either three months or seven days, and it can be seen that the dates are grouped in sets involving equal increments from set to set. (Thus it goes from one *MF* at month 1, day 2, to a later *EF* at month 2, day 3, to a later *MF* at month 3, day 4, and so on.) This is a clear proof that the periodicity of the phenomena was recognized, and so is a very important milestone in the history of scientific astronomy. It is unfortunate that the tablet cannot be dated with certainty. It predates the destruction of Assurbanipal's library by the Medes in 612 BC, but might be as much as eight or nine centuries earlier.

The precise sequence of dates in the basic Ammizaduga tablets allowed B. L. van der Waerden to date them, and he consequently rejected two out of the three Babylonian chronologies previously favored by historians. In his chronology, Hammurabi's dynasty is set at 1830 BC–1531 BC, his own reign at 1728 BC–1685 BC, and Ammizaduga's at 1582 BC–1562 BC. The whole subject remains highly controversial. Part of the problem is that the Venus tablets were copied repeatedly over many centuries, for religious and astrological reasons, and even their association with Ammizaduga is not universally accepted. The planet Venus was identified with the goddess Ishtar, as mentioned earlier, and the calendar that charted her appearances was based on the movements of the Sun and Moon, that were in turn related to the gods Sin and Shamash. Astrological and religious motives cannot be separated at this early date. The deities that were worshipped were, through their celestial behavior, thought to be capable of determining what happened in matters of love, war, and other forms of human behavior. ("When Venus is high in the sky, copulation will be pleasurable . . .," and so on.) The third aspect of the case, the foretelling of events by the planets, was one that followed naturally as soon as the prediction of their positions became scientifically possible. Of course one had to believe in their efficacy as gods. With the passage of time, this religious belief dwindled or was lost entirely, leaving behind, however, a belief in the possibility of prediction on the basis of planetary positions alone. Astrology without religion might have become more scientific, but in a sense it had become less rational by making human actions depend on celestial objects rather than gods with human attributes. In classical antiquity, and then in the Middle Ages and after, astrologers worked hard to try to restore some of the subject's lost rationality, by inventing theories of celestial influence.

BABYLONIAN ASTRONOMY IN THE ASSYRIAN PERIOD

Babylon fell to the Hittites in 1530 BC but was soon incorporated in the Cassite Empire. This period came to an end around 1160 BC, but during that time astronomical traditions were gradually strengthened. This was the period during which the *Enūma Anu Enlil* omen list was written—the series of tablets that we mentioned earlier as preserving the Ammizaduga Venus. Babylon already had a reputation for learning: even before the advent of Assyrian supremacy, the Assyrians were using the Babylonian dialect in their inscriptions. Star lists were developed that connected the passage of the months with the heliacal risings of stars. They are often misleadingly called "astrolabes," but are better known by their Assyrian name, "three stars each." They contained typically 36 stars in all, 12 "stars of Anu" that were near the celestial equator, 12 "stars of Ea" that were to the south of it, and 12 "stars of Enlil" to the north. (One peculiarity is that planets' names also occasionally occur. This is odd, since their heliacal risings are not at fixed dates.) It now seems probable that the "paths" of the three gods are not actually bands of the sky, but sections of the eastern horizon within which the stars rose. Known examples of the "three stars each" come from Babylon, Assur, Nineveh, and Uruk. Some of these particular clay tablets were rectangular in form, but others were circular, resembling a dartboard with twelve divisions.

Certain numerals on two of them (one rectangular and one circular) have been interpreted as showing that the Babylonians divided daylight into twelve equal parts, and night likewise. (Compare what we said on p. 28 above, of the Egyptian division of night and day into twelve.) This division, if strictly interpreted, makes the lengths of hours vary with the season of the year. It became customary throughout the Near East, and eventually throughout Europe. Writing in the fifth century BC, the Greek historian Herodotus tells us that the Greeks had taken their twelve-part day from the Babylonians. It became the method of time-reckoning of the populace as a whole, despite its probable astronomical ancestry. By contrast, there is the method of "equal hours," for instance, as measured precisely by the daily rotation of the heavens or by a water clock. Even when using a water clock, however, there was a demand for unequal, seasonal hours, so that we find clay tablets with tables giving the weight of water discharged during a watch of the night at five-day intervals throughout the year. (There were three watches in the night.) An ivory prism from Nineveh, dated from the eighth century BC or later and now in the British Museum, shows that conversion between the two systems of time-reckoning, equal and seasonal hours, was then conceived of as an astronomical problem. Babylonian astronomers used a system of 12 double-hours (*bēru*), each divided into 30 parts (USH), one of which was thus 4 minutes of our time.

Out of the "three stars each" tradition grew another, a series of clay tablets written on both sides and containing the text of an astronomical-astrological compendium known as MUL.APIN (plate 2). What is known of this series has been pieced together from texts spanning several centuries, the oldest dating from the seventh century BC. The title simply means "the Plough" and is taken from the opening words of the list of stars and constellations with which the text begins. The basic set seems to have been just one pair of tablets. From the scores of surviving fragments, almost all of the content of the first and most of the second is known. They contain what seems to have been a substantial selection of Babylonian astronomical knowledge, and they end with celestial omens, all from the beginning of the first millennium BC. Modern scholars have viewed the collection in very different ways, some treating it as a series of astronomical commentaries for many different purposes, others as a set of ideal schemes for purposes of divination, and for nothing else. Whatever their application, they reveal a high degree of astronomical sophistication, for at some points they seem to have been designed to allow for adjustment when what was observed in the heavens did not fit with what the schemes predicted.

There are improvements to the older lists of the "stars of Ea, Anu, and Enlil"; there are lists of stars that rise while others set and of periods of visibility of certain stars. The tablets allow a picture to be built up of the constellation names and patterns adopted by the Babylonians, which in some cases were taken over from the Sumerians. They are not quite ours, although ours ultimately owe much to them, through Greek intermediaries. APIN, for instance, while referring to a plough, is our constellation of Triangulum, but inclusive of the star γ Andromedae. Even if their meaning has shifted somewhat, the Sumerian names for the constellations of the Bull, the Lion, and the Scorpion have all passed down to our own day. Producing a map of the fixed stars is a fairly natural thing to do, and an early example from the library of Assurbanipal at Nineveh, despite its fragmentary condition, depicts nine or ten recognizable groups (fig. 30).

In real life, the stars do not rise and set over the desert in a clear-cut way, as they tend to do in Hollywood's rendering of Araby. Just as water vapor, over and above other atmospheric absorption, affects visibility appreciably in cooler latitudes, so the horizon in the Near East is often obscured by atmospheric turbulence and dust. Assyrian records show that at the first sighting of the new moon, it was often high in the sky. Since the observation of risings was so often a hit-and-miss affair, and might in any case have been obscured by buildings, the MUL.APIN tablets gave lists of secondary stars (the *ziqpu* stars) that culminated (crossed the meridian) at the same time as the more fundamental stars were rising. By "fundamental star," we mean the star by whose horizon position time and the calendar were reckoned. This list of *ziqpu* stars is scientifically important, for it represents a step toward a more reliable measure of time. There are reports of lunar eclipses, for instance, no later than the seventh century BC, that measure the time with reference to the culmination of *ziqpu* stars, supplemented by water clocks for the slight difference. (As mentioned earlier, the time unit used was the USH of 4 minutes, on our reckoning.)

The MUL.APIN series does not make use of the signs of the zodiac as a way of dividing up the path of the Sun through the stars into twelfths, but it does use a system that is somewhat similar, for it lists the constellations in the path of the *Moon*. There seem to be eighteen constellations named,

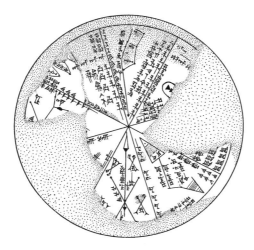

30. Drawing based on the remains of a star map from the library assembled at Nineveh by Assurbanipal, the last of the great kings of Assyria (who reigned from 668 BC to 627 BC. This clay disk was inscribed with cuneiform names and schematic drawings of stars and constellations (the stars being joined by lines, in a manner still current). Counterclockwise from the bottom: Sirius (Arrow), Pegasus and Andromeda (Field and Plough), Aries, the Pleiades, Gemini, Hydra with Corvus and Virgo, and Libra. Notice the eight divisions of the sky. (British Museum tablet K 8538, drawn by Leonard King [1912], corrected by Johannes Koch [1989].)

the gods in the path of the Moon, rather than the later twelve of the zodiac. Of course, the Moon's path through the sky is more or less coincident with the Sun's—the paths are inclined at five degrees or so—so that the longer list contains the shorter. The text makes it plain that the Sun, Moon, and five planets were regarded as moving on the same path. Despite the existence of an eighteen-part division of the sky, the solar year was at this date certainly divided into twelve months.

Another example of the way the MUL.APIN series shows a growing mathematical involvement in astronomy is its list of the times when the shadow of a vertical rod (gnomon) one cubit high is 1, 2, 3, 4, 5, 6, 8, 9, or 10 cubits long on the ground, at various seasons. Again, this is not strictly a report on observations, but rather a rationalized list, embodying certain rough rules of proportionality between time from sunrise and shadow length. In short, although the list is primitive, it is genuinely scientific, and we find the same systematic reduction of observation to rational order elsewhere in the tablets. They give rules, for instance, for calculating the times of rising and setting of the Moon in relation to its phase. (Thus at new moon, setting is just before sunset; thereafter setting is delayed by one fifteenth part of the night each successive day; and after the fifteenth night there are parallel rules provided for its rising.) The same rules were still in use in Roman times. As well as being based on arithmetical procedures, this material is highly practical. One simple scheme for inserting extra days into the calendar is based on the date in the first month of the year on which the Moon passes the Pleiades. In an "ideal year"—so goes

the rule—this happens on day 1. If it turns out to be day 3, then intercalation must be performed.

By comparison, the planets are accorded much slighter treatment, and
even the lunar theory contained in this work is elementary by comparison
with what was available at the time of copying of many of the surviving
tablets. They had a long history, but probably stemmed from somewhere
near the latitude of Nineveh at the beginning of the first millennium.
Whatever their precise date, they are witness to a deep concern with providing a theoretical framework for astronomy.

BABYLONIAN ASTRONOMY DURING INDEPENDENCE
AND PERSIAN RULE

Throughout the period that saw the fall of the Assyrian Empire, the
revival of Babylon's fortunes under Chaldean rule, and then the Persian
conquests, intellectual life and religious pursuits seem to have continued
much as before. Texts continued to be written in cuneiform script in the
Sumerian and Akkadian languages, even after both had given way to the
Aramaic language and alphabet in much of civil life. Omens gave way to
a new style of divination based on the horoscope, that is, on the arrangement of the heavens at whatever was deemed the significant moment—
the commencement of a journey, a battle, the birth of an individual, or
whatever. There was systematic observation of planetary phenomena, as
well as of the Moon and eclipses, not sporadically, but without serious
interruption until late Seleucid times. The eclipse records are surprisingly
detailed, numerically and otherwise, in some cases even providing information about wind direction. The records that were deemed valuable survive in texts from Seleucid archives. The two main categories are of what
Abraham Sachs called the "astronomical diaries" and (much less common)
of collections of data for particular types of astronomical phenomena, as
observed over several years.

The diaries (those known date from 652 BC onward) record many different sorts of "meaningful" events—planetary positions in relation to
the fixed stars, the weather, solar halos, earthquakes, epidemics, water
levels, even market prices. Angular positions were specified in "fingers"
and "cubits" (each of 24 fingers), both being common measures of length.
Where it is legitimate to equate such cubits with degrees in interpreting
observational data, they seem to be the equivalent of about 2.2 degrees, but
in computation they were often taken as equivalent to 2 degrees.

The collections of observations were to prove of great importance in
the long term. In fact, the Alexandrian astronomer Ptolemy chose the
year of Nabonassar's accession (747 BC) as the starting point of his calendar, expressly because the old observations were preserved from that
time onward. Some of the eclipse data in collections available today go
back to 731 BC. Lunar and solar eclipses, both arranged in 18-year periods,
observations of Jupiter (12-year periods), Venus (8-year periods, for reasons

explained earlier), and others for Mercury and Saturn, are here included. As time progressed, more detail was added—for instance in recording conjunctions with, and distances from, fixed stars. The recording of planetary periodicities was also a new occurrence, for although clearly known much earlier, not even in the MUL.APIN texts were they explicitly recorded. They were used increasingly as an aid to prediction, short periods for rough use, and longer periods where more accuracy was called for.

The 18-year period for the Sun and Moon is of great importance in the later history of calendar-reckoning. The Babylonians discovered a period over which the cycle of eclipses begins to repeat itself, more or less, not only in character but in the time of day. They discovered that this happens after 18 years, or more exactly, 6585 ⅓ days (18.03 years), which is a whole number (223) of *synodic months* (new moon to new moon). They were fortunate in this discovery, for the time of day of an eclipse depends heavily on the motion of the Moon "in anomaly," to use the modern technical term. The *anomalistic month* is reckoned from perigee to perigee (*perigee* is the nearest point of the Moon's orbit to the Earth), and it turns out that 223 synodic months are a fair approximation to 239 anomalistic months.

This is a suitable point to introduce the *draconic month* (or *nodical month*), which is measured with reference to the Moon's nodes, the points in the sky where the Moon's path crosses the Sun's. Bearing in mind the reasons for the two kinds of eclipse, it should be obvious that the Sun and Moon must both be near to a node simultaneously for an eclipse to take place. (The Moon can wander five degrees or so away from the Sun's path, and the apparent angular sizes of the two bodies are only about half a degree. The angular size of the Earth's full shadow (*umbra*) at the Moon's distance, into which the Moon passes during a lunar eclipse, is about 84´.) Now it so happens that 223 synodic months are approximately 242 draconic months, and the fact that again we have a whole number is what makes the period of 18 years work so well in eclipse reckoning.

The period of 223 months is still misleadingly called a *saros* by some modern writers, using a Greek word that had formerly been used to denote a much longer Babylonian period of time (3,600 years). The modern usage has a long and convoluted history, but goes back only to AD 1000 or so and the encyclopedia of Suidas. The ancient Greeks themselves, to get rid of the ⅓ day, sometimes took a period of 669 months, the so-called *exeligmos*. This too was earlier known to the Babylonians.

It is sometimes suggested that the 223-month period was primarily used to bring solar and lunar calendars into correspondence, but its value to astronomers was clearly much more than that. In the long term, it helped in the search for fundamental astronomical periods, those of the Sun and Moon, on which so much precise planetary theory depends. The earliest use of the 223-month period, however, was in identifying the dates on which an eclipse was possible, without giving a guarantee that the eclipse would occur. It is likely that it was first used for lunar eclipses, and only later for eclipses of the Sun. The Babylonians discovered that eclipse possibilities mostly occur every 6 months, but that occasionally they occur after 5 months. Suppose that in a single 223-month period there are x intervals of 6 months and y of five. Clearly $6x + 5y = 223$, and this allows of only one solution, with x equal to 33 and y equal to 5. A tabular scheme for the dates of possible lunar eclipses incorporating these numbers was drawn up by the Babylonians at least as early as 575 BC, but the earliest known eclipse records begin in 747 BC, and they already include predicted eclipses. From a later period, there are even eclipse predictions with the time of day or night at which the eclipse was expected to begin.

Those who wish to check the accuracy of the relationships mentioned above, in connection with the 223-month period, may use these modern figures for the lengths of the three sorts of month named: synodical, 29.5306 days; anomalistic, 27.5546 days; and draconic, 27.2122 days. For future reference, a fourth type of month is the *sidereal*, defined by reference to the Moon's circuit of the stars as seen from the Earth, its mean value being 27.3217 days. (All days are here mean solar days.) Looking at the 223-month period through modern eyes, however, and accepting the fact that no such periodic relationships will ever be exact, we obscure the fact that a scheme that repeated after whole numbers of days was what was wanted, and that this could not be had. Dealing with average values also obscures the fact that particular months vary slightly in length, with the result that the lengths of actual 223-month periods are in excess of a whole number of days by a variable amount (between about 6 and 9 hours). There are two Babylonian schemes known for modeling the length of the fluctuating 223-month period, representing a remarkable achievement on the part of those responsible.

By comparison with these discoveries, finding the periodicities required by a lunisolar calendar (one combining months and years) must seem a relatively trivial affair, but it was one that was nevertheless judged to be a matter of great importance in all Near Eastern religions, in Ancient Greece, and later, in the Islamic and Christian worlds. For a time in the sixth century BC, the Babylonians used an 8-year period (99 synodic months), and later, a 27-year period (334 months), but the most widely used period equated 19 years to 235 months. The excellence of this relationship can easily be verified approximately using the length of the synodic month as quoted above, but in doing so, we might very easily miss an important point. What, on this basis, should we take as the length of the year?

If we take 365.25 days, we shall find the discrepancy to be about 0.06 day. As we know, 365.25 is itself only an approximation, but again we must distinguish between two ways of defining our unit of time, in this case the solar year. Just as we have a sidereal month, so we have a *sidereal year,* measured by the referring the Sun to the fixed stars. The more common astronomical definition, at least since Greek times, measures the year (the *tropical year*) differently, namely by the Sun's return to one of the equinoxes or one of the solstices. Today, reference is now usually made to the Sun's return to spring equinox, or rather to that highly abstract point which will later be used as the origin of astronomical coordinates, the *vernal point* (or "Head of Aries"). The ecliptic and celestial equator meet there, and the Sun passes through the point in its passage from south to north at the vernal equinox.

Why should the two definitions of a year give different results? That they did so was discovered effectively only by the Greek astronomer Hipparchus, in the second century BC. To take a longer view, it can be said to be a consequence of the fact that the Earth's axis does not hold a constant orientation in space. As a result, the vernal point—a reference point which the Earth's axis helps to fix—is moving with respect to the stars. The length of the year considered as the interval between the Sun's returns to the vernal point—namely, the tropical year—is therefore different from the sidereal year. The former is in fact (today) 365.2422 mean solar days, while the sidereal year is 365.2564 mean solar days.

The difference is slight, but the accuracy of Babylonian astronomy was high, and it is possible to see from several of the cyclical relationships quoted by the Babylonians that they were derived using the *sidereal year.* Since early Greek astronomers generally used the *tropical year,* we have here one of several fundamental differences in the approaches adopted by these two important groups of astronomers.

Just as it is with the Sun and Moon, so too with the Sun and the planets: there are simple periodic relations that were found at an early date by the Babylonians. We have already discussed Venus, which makes 5 circuits of the Sun and 8 returns to the same place in the stars in 8 years. Abbreviating this relationship as [5c, 8r, 8y], we can say that some at least of the following relationships were used long before they were recorded in Seleucid times: Mercury, [145c, 46r, 46y]; Mars, [37c, 42r, 79y] and [22c, 25r, 47y]; Jupiter, [76c, 7r, 83y] and [65c, 6r, 71y]; and Saturn, [57c, 2r, 59y]. How these periods were used is explained in a type of text that Sachs called "Goal-Year texts."

The Babylonians knew that these relationships were not exact, and used additional rules for correcting them. Venus, for example, was said to complete 2 ½ degrees less than 8 revolutions with respect to the stars, in 5 circuits of the Sun. This sort of consideration led to the statement of longer periodicities. In the case of Venus, since the deficit is $1/144$ part of a circle, in 720 circuits of the Sun ($720 = 144 \times 5$), Venus will complete 1,152 revolutions, and this in 1,151 solar years ($8 \times 144 - 1$).

The long-period relations often quoted for the other planets were found in a similar way, and a certain delight in the calculation of these long periods of time led to this becoming almost a subject in its own right, finding favor in India to the east and—in more restrained forms—in Greece to the west. According to one account (by the first-century Roman writer Seneca), the Babylonian Berossos, who was a priest of Bel and who founded an astronomical school on the Greek island of Kos in the third century

BC, taught that when the planets all conjoin in the last degree of Cancer, there would be a world conflagration, and then a flood. Here is the idea of periodicity. Given the periodicities described, the recurrence of planetary influences easily becomes a part of religious and astrological dogma and fits well with the idea that history generally, and even human existence, is a recurrent phenomenon. All one has to do, to find the interval of time after which history repeats itself, is to find the least common multiple of the long planetary periods. This was perhaps too much to ask. In some accounts of the Bel story, simple round numbers of years are quoted—for instance, 2,160,000 years (600 × 3,600). The much later Hindu period of time known as the *Mahāyuga* is just double this amount, showing a clear connection with Babylonian time-reckoning. When the Greeks set out their concept of a Great Year, that, too, tended to have large multiples of 360 in it.

The figure of 360, which we recognize as the number of degrees in the circle, is almost as good as a Babylonian signature. By the early fifth century BC, the Babylonians had the makings of a coordinate system, for they had by then begun their division of the zodiac into twelve "signs" of equal length, naming them after the constellations or important star-groups—Aries, Taurus (or the Pleiades), Gemini, Cancer, and so forth. As in later periods, however, this left the risk that constellation and sign would be confused, sharing as they did the same name. The potential confusion became more serious as time went by, and as the precessional drift of the equinoxes moved the stars out of their old signs completely. The Babylonian coordinate system, fixed in relation to individual stars rather than the vernal point, was for this reason not ideal, although it was easy to understand—since the stars can be seen and the vernal point cannot.

That the systems differed, but that the differences were often not appreciated is nowhere better illustrated than in the common Babylonian statement that the vernal point is at 8° Aries. This remark found its way into some second-rank medieval astronomy, and when the context was lost it made no sense whatsoever. (It was later provided with a new meaning in the context of the theory of access and recess, a subject to which we shall return in chapter 8.)

The first Greek text showing the astronomical use of degrees is by Hypsicles, from the mid-second century BC. (Earlier writers specified angle in terms of fractions of a circle or quadrant.) Babylonian degrees were adopted by Hipparchus, the most influential astronomer before Ptolemy. Strabo, however, writing a century later, says that Eratosthenes divided the circle into sixty parts.

BABYLONIAN ASTRONOMY IN THE SELEUCID PERIOD

The establishment of a system of celestial coordinates—in this case the division of the zodiac into twelve signs of thirty degrees each—was of the greatest importance for the advance of mathematical astronomy. Accurate planetary periods can be found without it, from observations made over long periods of time, but for an analysis of the finer points of planetary motions such a system is essential. The motives for making that analysis must have been in part intellectual, but they also had much to do with religion and astrological prediction.

The old Mesopotamian stellar religions had encouraged only a crude astrology of simple omens. Various religions from the Near East, such as Orphism and Mithraism, supported a slightly more developed zodiacal astrology, and some of these were brought into the Latin and Greek worlds with the spread of the Persian Empire. Of these oriental religions, one in particular, Zoroastrianism, deserves mention here. This religious doctrine, ascribed to the prophet Zoroaster (or Zarathushtra), gradually became the dominant religion in Iran and is still practiced by isolated communities there and in India. Its doctrines involve a moral dualism of good and evil spirits, and it readily became intermingled with old Babylonian myth, for instance, the myth describing the contest between Marduk and Tiamat. Its relevance to astronomy, however, is less obvious. In its later versions it helped to spread the doctrine that the natural place for the human soul is in the heavens—or more specifically, in its Western manifestations, the spheres of the planets. Some have argued for a link between Zoroastrian doctrines and the rise of birth horoscopes, especially in Greece. There was an idea that the soul, coming from the heavens where it partakes of the rotation of the stars, when united with a human body continues to be governed by the stars. (That notion is to be found, for instance, in the *Phaedrus* of the Athenian philosopher Plato.) This philosophical motive might not fully explain the phenomenal rise of astrology in the later Hellenistic world, but whether or not it does so, the rise was real enough, and put astronomical prediction at a premium. The Greeks had heard of Zoroaster by the fifth century BC, a century before Plato, but it is interesting to learn that it was one of the greatest of Greek astronomers, a contemporary of Plato, who was largely responsible for introducing Zoroaster's philosophical ideas to the Greeks. The scholar in question was Eudoxus of Cnidos, whom we shall meet again in the next chapter.

Whatever the philosophical impact of Zoroastrianism, it was astronomically naive. There were some routines that would have benefited from astronomical knowledge, such as predicting the harvest from the sign in which the Moon stands in the morning of the first visibility of Sirius, but there is no reason to think that the Persians made any great advance in actually *predicting* such things. And even that astrological doctrine was probably borrowed. The mathematical astronomy necessary for the practice of horoscopic astrology owed almost everything to the Babylonians. The oldest known cuneiform horoscope is datable to 410 BC and is from a Babylonian temple. In Plato's century, the Greeks were already giving due credit to the "Magi" or the Chaldeans, and throughout the world of classical antiquity these epithets stuck as synonyms for "astrologer." This fact should not, as it has so often done in the past, obscure the brilliant mathematical contribution made by the Babylonians, by which their astrology was underpinned.

Over three hundred cuneiform tablets survive in this category, many of them damaged, and some even in fragments divided between museums in different countries. They are usually now distinguished as "procedure texts" (in which the methods of calculation are explained) and "ephemerides" (in which the results of the calculations are listed over a period of time, rather as they are in the modern *Nautical Almanac*). Ephemerides (the Greek word *ephemeris* simply means "daily") are more than three times as numerous as procedure texts. All come from Babylon (excavated 1870–1890) and Uruk (excavated 1910–1914), so that even now we may be very ignorant of the total achievement of these people.

It is possible that concentrating on the problem of the Moon led to comparable solutions for the planets. How many days are there in a month? A Babylonian month began with the first visibility, after sunset, of the thin lunar crescent. Days were likewise counted from the evening. On this definition, a month has a whole number of days, and experience taught that the number was either 29 or 30. But how do we decide in advance? Today we have a general picture of the situation, a model to which we can apply standard geometrical procedures to derive an answer. Even now, it is far from easy to do so. The Babylonians, having been provided with no such model by their forebears, were obliged to work in reverse. Let us first try to appreciate the intrinsic difficulties they faced by showing how our own analysis would proceed.

We begin with the crude approximation that a month is just 30 days in length. The Sun covers about 1° of the zodiac per day, so that the Sun, in moving from conjunction to conjunction with the Moon, will have moved 30°. They are in conjunction, so the more rapidly moving Moon will have moved 390°, or 13° per day over the 30-day period. (A more accurate average figure is 13.176°, but the Moon's velocity varies appreciably.) Now, to predict when the crescent Moon will first be seen, several factors have to be taken into account:

(1) The brightness of the Sun means that they have to be separated by a certain minimum distance, or it will not be possible to see the Moon's crescent.

(2) The relative speed of the Moon and Sun decides how long it takes the Moon to cover this distance. The relative speed is on average roughly 12° per day, but this "daily elongation" can vary by two or three degrees either way.

(3) The critical distance is affected by the brightness of the background sky, which is in turn affected by the angle between the horizon and the line joining the setting Sun and the crescent Moon (see fig. 31). This can in turn be resolved into various factors: (a) the season of the year, which is another way of talking about the Sun's place on its own path, that is, the "ecliptic" through the middle of the zodiac; (b) the Moon's divergence from that path, its "ecliptic latitude," which can exceed 5°; and

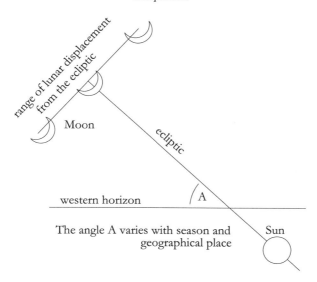

range of lunar displacement from the ecliptic

Moon

ecliptic

western horizon

A

The angle A varies with season and geographical place

Sun

31. First visibility of the lunar crescent. The Sun is below the western horizon, its line of descent not shown. (It is not the line of the ecliptic, but is decided by the rotation of the heavens around the pole.) The Moon falls within a range of a little over 5° on each side of the ecliptic.

(c) the geographical place of the observer (latitude) that fixes the angles at which the stars generally cut the horizon in rising and setting.

This summarizes the procedure we should probably follow. The astonishing thing is that the Babylonians somehow managed to extract the various factors by analyzing their observations of the beginnings of their months. They did so using wholly arithmetical methods, that is, without resorting to geometrical models, and if we represent their results here in graphical form, that is only because it is much more economical to do so. (To get the full flavor of the originals, one should look at such a work as Otto Neugebauer's *Astronomical Cuneiform Texts*, where the material is transcribed and analyzed.)

Two chief systems for the representation of varying solar, lunar, and planetary movement are recognized. The first, known as "System A," assumes that a velocity (say that of the Sun) is held constant at a particular value over a certain substantial range of the zodiac, then changes to another value, which it holds for an appreciable time before changing again, and so on. Rules were needed for the changeover. If we think of velocity as plotted against time, this gives us a graph that looks like a castle battlement, generally an irregular one, often said to describe a "step function." "System B" is on the face of things more sophisticated. It assumes that successive rows in the tabulation of positions, or whatever, are different, but that they differ by a constant positive or negative amount, except where this would take us beyond a predetermined maximum or minimum. When that is the case, the direction of change (increase or decrease) is reversed. If we produce a graph from the table of values, the result has an irregular sawtooth appearance, a zigzag function. In practice, System A was found to be

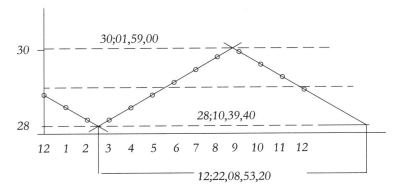

32. The Babylonian zigzag function, represented in a modern graphical way.

more flexible, for it could be easily used with many sizes of step, making it more accurate than a simple System B arrangement.

The reversals of direction found in zigzag functions follow strict rules that are most easily explained by reference to figure 32, with an example drawn from an ephemeris for the year 179 of the Seleucid era (133 BC–132 BC). The horizontal scale is of monthly units, in some sense, these being named in the first column of the tablet. They turn out to be those points in the month at which mean conjunctions of the Sun and Moon occur. The vertical scale corresponds to the second column on the tablet, which contains sexagesimal numbers of the order of 28 or 29. Another column on the tablet, not reproduced on the graph, can be shown to give the longitudes of Sun and Moon at the time of their conjunction. Only after the numbers in the second column were analyzed by modern scholars could their significance be understood. Since it emerges that the second column contains the differences between successive entries in the third column, that second column evidently contains, as we should say, solar velocity (the change of longitude per month). Using our graphical way of describing purely arithmetical quantities, we can say that the zigzag created out of straight lines was their approximation to something for which we should now require, at the very least, a sine curve. All told, theirs was an extraordinary achievement.

The period of the zigzag function in months can be easily found—it is marked on figure 32 as 12;22,08,53,20. This is the value accepted for the year, measured in synodic months. It was clearly not obtained by direct observation, but from one of those cyclical relationships we referred to earlier, in this case seemingly equating 810 years with 10,019 months. Of course, observations must ultimately have been involved, but numerical convenience too must have entered into the equation. Unfortunately, only the end results of the enterprise are known.

Other equivalences are also found, for instance, that of 225 years and 2,783 months, yielding 12;22,08 months to the year. This number is found in tablets drawn up using both System A and System B. One of the more surprising discoveries of those who have labored with these cuneiform tablets is that, although System A is older, the two systems were neverthe-less in use during the entire period from which they have been preserved, say 250 BC–50 BC, both in Babylon and Uruk.

Relatively few lunar ephemerides cover more than one year. Most have columns for lunar and solar velocity and position. Some list the length of daylight or night, corresponding to the Sun's position in an earlier column. For us, this is a matter for calculation using the methods of spherical geome-try, but for the Babylonians, only arithmetical methods were available. There were in some cases columns for the Moon's latitude, and in others columns for the magnitudes of eclipses. A procedure—a formula—for finding eclipse magnitude was applied every month, whether or not an eclipse was due. This might be regarded as contrary to the spirit of an empirical science, but it cer-tainly speaks for a high level of abstraction, and a clear grasp of the notion of a mathematical function. Among the procedures followed were some for cor-recting results on account of a solar velocity that had been accepted as con-stant at an earlier stage in the calculation, but that was known to be variable. On System B, this correction was bound to be more difficult than on System A, which goes some way to explaining the survival of the latter system.

As explained on p. 49, it was realized that lunar and solar eclipses are impossible if the potentially eclipsed object has too great a latitude at the time of new or full moon. The problem of solar eclipses is much more dif-ficult than that of lunar eclipses, and all that could be said in their case was that an eclipse was impossible. To say in advance that it would be observed, they would have needed much more information about the distances and sizes of the Earth, Sun, and Moon. There is no firm evidence that patterns in the recurrence of solar eclipses—another route to their prediction—were known, although some have argued that they were.

The tablets mentioned so far deal with long periods of solar and lunar movement, but there are others using similar methods for *daily* changes, and from them, for example, an equivalence of 251 synodic months and 269 anomalistic months may be derived. Here, 29;31,50,08,20 days was taken to be the length of the synodic month, and 27;33,20 days the anomalistic. The high accuracy of these figures is not likely to be obvious, but the val-ues quoted today are identical to them to within one and four parts in six million, respectively. (There have been certain much smaller changes in those periods in the meantime, so the comparison is not absolutely strict, but it can hardly fail to impress.) A more interesting historical comparison is between the length of the Babylonian synodical month quoted here and the length used in the high Middle Ages in Europe, in the so-called Tole-dan Tables. The two parameters are to all intents and purposes identical, separated though they are by well over a thousand years.

When the Babylonians of the Seleucid period turned their attention to the planets, their arithmetical procedures—to use our graphical analogy once more—came one step nearer to the ideal sinusoidal curve. There are also tablets that may be said to give the Moon's latitude, where the lines of

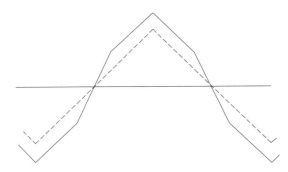

33. A graphical representation of a Babylonian solution to what may be described as the problem of lunar latitude. (We are dealing here only with broad principles. In all strictness, this was probably intended only as an auxiliary function for use in eclipse calculation, but the underlying idea can be expressed in terms of latitude.)

the simple zigzag had already been modified, bent nearer to the sinusoidal ideal, as it were, as shown in figure 33. Before explaining how the Babylonians proceeded, it will be as well to have an approximate picture in our minds of the way the planets actually move and are seen to move by an observer on the Earth. The following nonhistorical aside is therefore offered as background material for this and the following chapters, which deal with other classical theories of planetary motion.

TWO APPROACHES TO PLANETARY MOTIONS: A NONHISTORICAL DIGRESSION

Of the planets recorded before the eighteenth century, all in orbits around the Sun, Mercury is the nearest to the Sun, followed by Venus, and then the Earth. Mars, Jupiter and Saturn have orbits outside of the Earth's (fig. 34). The "inferior" and "superior" planets (or "inner" and "outer") have somewhat different patterns of behavior as seen from the Earth. Each orbit is an ellipse, with the Sun at a focal point, as Johannes Kepler was able to show; but to a first approximation the orbits may be taken as circular, with the Sun at the center of all. We imagine a diagram of the system with a pin through the point representing the Sun. If we remove the pin and put it through the point representing the Earth, the relative positions of the planets will be unaltered, and yet it should be clear that now we may regard the planets as moving on circles centered on the moving Sun.

Consider a simple case, that of the inferior planet Mercury, which is now to be thought of as a satellite of the Sun. The Sun's circle may be called by the traditional name of *deferent circle*, literally a "carrying circle," and the satellite's orbit, the carried circle, will be an *epicycle*. The other inferior planet, Venus, will travel on a larger epicycle. With the superior planets, the roles of epicycle and deferent will be reversed, but we need not go into more detail for the time being.

The description of individual planetary motions in terms of epicycles is of great importance in the history of astronomy, although it has to be

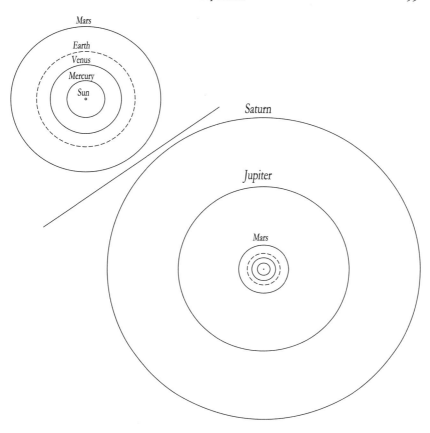

34. Planetary orbits, assumed to be circular. In each drawing the sizes of the orbits are approximately to scale.

emphasized that each planet was treated separately, its epicyclic movement being appreciated, but that it was a long and painful trek to an appreciation of the fact that the Sun is present in the epicycle-deferent system for each planet. Only when this was fully realized by Copernicus was it possible to bind the planets into a single system, putting a pin, so to speak, through the Sun's position on each separately drawn system.

This type of explanation is capable of great refinement; for instance, by giving the epicycles a slight tilt to the plane of the Sun's orbit, it is possible to account for the fact that the planets are not precisely in that plane, but have movement in latitude. "Ecliptic longitude" is measured around the ecliptic from a zero-point where the ecliptic, the Sun's path, meets the celestial equator. This zero-point is the vernal point, which we encountered earlier. "Ecliptic latitude" is then the coordinate measured toward the poles of the ecliptic, northern or southern, starting from the ecliptic.

The epicyclic description was typical of later Greek astronomy, but not of Babylonian. Once it was appreciated, the astronomer's task was much simplified. The strategy was then to assume a geometrical model and to draw consequences from it, for instance, about the patterns of rising and setting of the

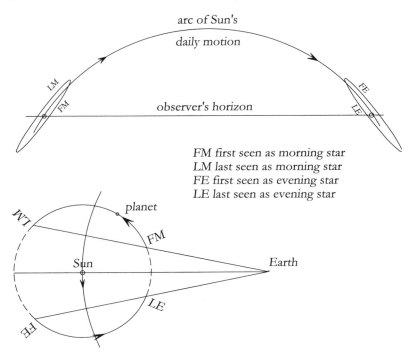

arc of Sun's

daily motion

observer's horizon

FM first seen as morning star
LM last seen as morning star
FE first seen as evening star
LE last seen as evening star

planet

Sun

Earth

35. Morning and evening risings and settings of Mercury and Venus. It is not difficult to draw a comparable diagram for the superior planets, with orbits outside the Earth's.

planets. Having done so, those theoretical consequences could be compared with observation. If the model seemed to be worth retaining, then those and further observations would allow the astronomer to determine, or improve, the numerical properties of the model—such parameters as the relative sizes of the circles, the angular speeds in the circles, and when the planets passed a suitably chosen starting point. The Babylonians, arriving on the scene first, worked in more or less the reverse order: what was to the Greeks—as it is to us—a derived consequence was to them a starting point, a datum. Consider, for example, their concern with risings and settings over the horizon. It is a very natural thing to pay attention to first appearances, and concern with them was shared by most early cultures. Such observations yield very un-promising information for an astronomer, however, and the miracle is that powerful theories emerged from them when they did.

We have already mentioned the first and last appearance of the fixed star Sirius, which spends part of the year lost in the Sun's rays, and the fact that the planets too may be lost to view for a time, for similar reasons. The inferior planets, Mercury and Venus, which never wander so far from the Sun as to be in opposition to it, have patterns of behavior touched on earlier (see fig. 35 and our earlier fig. 29). Once again we may consider the matter from a modern perspective. When an inferior planet is on that part of the orbit represented by a broken line, its angular separation from the Sun is so small that it is lost in the glare of the Sun. At the point *FM* it becomes visible, in this case for the

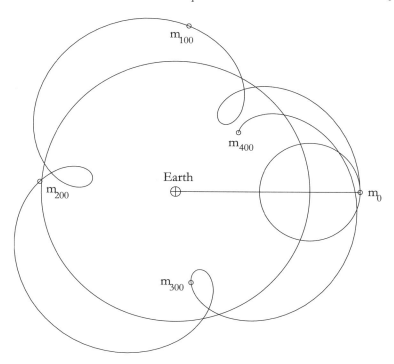

36. The spiral path of Mercury with respect to the Earth. The Sun moves round the Earth as shown, and the orbit of Mercury around the Sun is carried with it. The spiral drawn here is covered in 300 days.

first time as a morning star. As seen from the Earth, the planet will be moving round the sky in its daily path as though carried by the Sun. The fact that it will be seen in the morning, shortly before the Sun rises, should be obvious from the upper part of the figure, where the orbit is roughly drawn nearly edge-on to the observer. *LM* is the point at which it will be last seen as a morning star, and *FE* and *LE* are the points of first and last appearance as an evening star.

When dealing with the inferior planets, which have to be near the same horizon as the Sun, our fourfold division (*FM, LM, FE, LE*) is unambiguous. Heliacal rising is then, as we have seen, the first visible appearance on the eastern horizon before sunrise (*FM*), and heliacal setting is the last visible setting just after sunset (*LE*). Mars, Jupiter, and Saturn, however, can be seen rising just after the Sun has set, and setting just before the Sun has risen, so that to discuss them we might need an extra qualification, such as "first morning setting." The words "acronychal" (not to be spelled "achronycal") and "cosmical" are often used of risings and settings in ways that are ambiguous when they overlook this fact. It is better not to define them at all, but to remember—when reading others who do—that the first adjective concerns evening sightings (first or last) and the second concerns morning events.

We may consider the Earth to be fixed and the Sun to go around the Earth. We know that the planets go round the Sun, so that—if we accept this convention—a spiraling motion of the planets will result, as shown in somewhat different ways for Mercury and Venus in figures 36 and 37. The

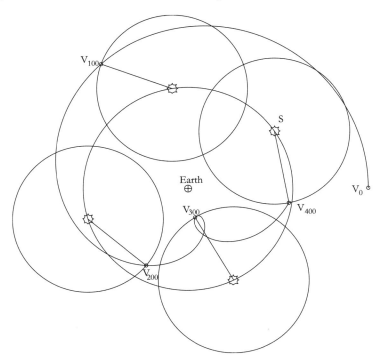

37. The spiral path of Venus (V) with respect to the Earth. The diagram is of the same general character as figure 36, but now the Sun (S) and the corresponding place of the planet's orbit are shown at intervals of 100 days, up to 400. As before, we start for convenience from a situation where the Sun is on the line joining the Earth and Venus. (That is not a common situation. The rough symmetry with 100-day intervals is a matter of chance, not principle.)

planets never appear very far from the plane containing the path of the Sun through the stars (the ecliptic). As already explained, it is along that apparent path that we measure (celestial) longitude. For an alternative picture of a planet's motion, dispensing with our geometrical schemes, we may simply tabulate its celestial longitude against time, or better still, plot it on a graph. This is done for Mercury in figure 38, where the horizontal axis spans approximately one year and the vertical axis gives the longitude of Mercury, measured from the vernal point. The Sun's longitude is also plotted: it is the diagonal running through the middle of Mercury's graph, the line around which Mercury seems to oscillate. Mercury goes around the Sun about four times in a year. (Its sidereal period is 0.24 tropical years.) Marked on the graph are the points of first and last visibility (with the notation of the last paragraph) and the so-called *planetary stations,* where the planet seems to stand still in relation to the background of fixed stars, that is, when its motion changes from direct to retrograde (S_1), or conversely (S_2).

The case of the superior planets may be illustrated by Mars (figs. 39 and 40). The rectangular graph here covers a period of over six years, and since the vertical axis represents longitude, a coordinate that is essentially cyclical, the lines representing both the Sun (roughly straight) and Mars

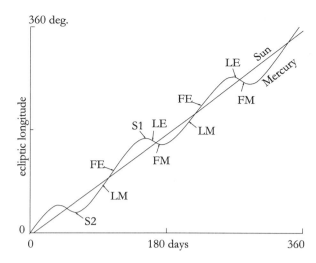

38. Graph of the changing longitudes of Mercury and the Sun over a period of approximately one year. The notation used (*FM*, *LM*, etc.) is explained earlier in chapter 3.

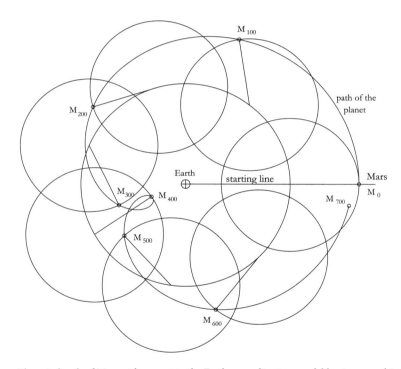

39. The spiral path of Mars with respect to the Earth, according to a model having something in common with that for the inferior planets. The positions of the planet are drawn for 100-day intervals. It will be seen that the planet comes back to its starting point in approximately two years. The diagram is included here in anticipation of the longer discussion of epicyclic models in chapter 4. Its relation to the modern view of the solar system (even in its simplified earlier form) will not be easily appreciated at this stage, since there are no obvious points corresponding to the place of the Sun, around which Mars revolves. The link with the Sun is present, however, through the radii of the epicycles which indicate the places of Mars in the epicycles. Those radii will be seen to change direction steadily. They do so in such a way as to complete a full circle in a year, and are analogs of the line in space joining the Earth to the Sun.

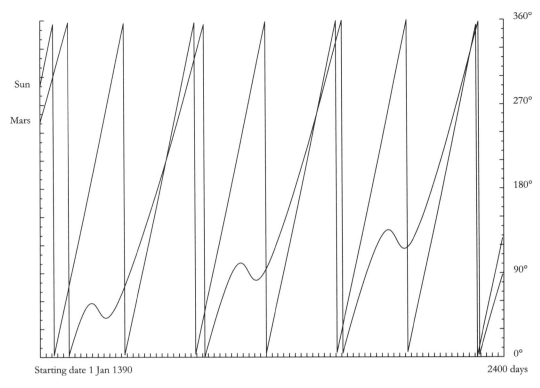

Sun

Mars

360°

270°

180°

90°

0°

Starting date 1 Jan 1390   2400 days

40. The changing longitude of Mars (the kinked lines) and the Sun, over a period of 2,400 days (compare the earlier graph for Mercury and the Sun). Such graphs may be considered simply as records of observed planetary positions in longitude. The astronomer's task is to explain them.

have to be repeatedly broken. Some general principles will be obvious, even so. Proximity to the Sun, making the planet difficult to see, occurs in the middle of long-term, moderately regular, direct motion, while during the period of retrograde motion the planet is approximately in opposition to the Sun (180 degrees of longitude distant from it). The lines for the Sun are of course spaced at yearly intervals, and from them it is evident that the kinks in the graph for Mars are a little under two years apart. The sidereal period of Mars is in fact 1.88 tropical years. This periodicity will also be roughly appreciated from figure 39, in which the spiral representing the path of Mars over a period of 700 days almost brings the planet back to its starting point.

So much by way of our nonhistorical aside.

BABYLONIAN PLANETARY THEORY

We have already seen how a concern with horizon phenomena was important to the principles underlying the Venus tablets of Ammizaduga. It is interesting to see how, when the Babylonians analyzed the behavior of first and last appearances, morning and evening, they treated them as separate phenomena. It was as though each had its own existence as an object residing in the ecliptic. Consider, for instance, the points on the Mercury graph (fig. 38, above) marked as *FM*. Taking a large number of such points in isolation from the rest of the graph, they fall on a moderately straight line parallel to the Sun's graph, but by no means a perfect one. The Babylonians used the same sort of arithmetical methods as they had used for the Sun and Moon to account for deviations from (as we are representing it) the straight-line graph. The problem, expressed graphically, is one of breaking the line into segments, that is, finding the breaking points and gradients of suitable component parts. Those gradients (angular speeds) were seemingly expressed by the Babylonians in terms of convenient whole-number relationships, such as "Mercury undergoes 1,513 phenomena in 480 years," or, to quote a secondary result that we can understand more immediately, "Mercury rises 2,673 times in 848 years."

Finding such relationships was not easily done, and the fact that the Babylonians were tied to a lunar calendar added an additional complication, for of course the planetary periods have virtually nothing to do with the Moon's motion. In the end, this did not matter greatly, for although they did not express their dates in days, they used a unit of one thirtieth of a *mean* synodic month. This is now usually called a *tithi,* a word that comes from later Hindu astronomy, where the same unit was in use. The *tithi* is roughly a day in length. The Moon's motion varies in intricate ways, but since the *tithi* is by definition an average value, it can in principle be used in a perfectly workable astronomical system. This is not to say that it is an astronomically well-chosen unit, but it had clear advantages for those whose religion bound them to the old lunisolar calendar.

Having produced a set of rules for the behavior of phenomena of type *FM,* analogous rules were found for the behavior of *LM, FE* and *LE.* The general principle was that of finding the amounts (in longitudes and *tithis* of time) to be added to the base values (for *FM*), in subsequent phenomena. The resulting graph by which we can represent these arithmetical procedures is a plausible approximation to the sinusoidal form of those we have given above.

One highly sophisticated element in the exercise was the introduction of certain "phenomena" into the scheme of calculation that could not in fact have been observed. The situation can perhaps be best explained by analogy with the case of full moon. Much of the time, at the precise moment at which this occurs, the Moon will be below the horizon, but the calculation proceeds regardless of whether it is to be seen or not. The other "phenomena" we have been discussing (*FM, LM, FE,* and *LE*) similarly began life as observable events, but were eventually turned into computational ideals—ideal points on our ideal graph, so to speak.

Broadly similar procedures were followed for all the planets, although here the theories were less ambitious. The general aim was to determine the longitudes and the dates of the principal planetary phenomena. For the superior planets, these were the first appearance after a period of invisibility, and the disappearance; the opposition of the planet to the Sun; and the two stations, when the planet pauses before changing direction. Much Jupiter material survives. A Mars tablet from Uruk is remarkable for the plausible representation of the highly variable velocity of that planet. For the inner planets, appearances and disappearances are differentiated as between those in the western sky and those in the eastern. For Venus, the stations are also heeded. (Those of Mercury were rarely visible.) All of the planets then known are represented by tablets from Babylon and Uruk. It is not possible to say which of these two places was the birthplace of the methods, or when they were first devised, but System A is likely to have been developed within fifty years of 400 BC, before or after, and System B, slightly later. In both places, several different methods of arithmetical representation were in use, and remained in use until at least the first century of the Christian era. To say that the general principles were much the same in all cases is rather like saying that all aircraft fly in much the same way. In a work of this kind it is not possible to explain the two systems in more detail. What should at least be mentioned, however, is that there are a few surviving tablets which show that "points on the graph" intermediate between the key phenomena (*FM*, *FE*, first station, and so on) were supplemented by intermediate points which were decided, not by a straightforward interpolation (not by straight-line segments, to use our graphical language), but by schemes based on second-order and even third-order differences. There are sciences that even today have not advanced to that level of sophistication.

The Greek and Roman Worlds

ASTRONOMY IN HOMER AND HESIOD

Babylonian astronomical documents contain evidence of two complementary processes, one in which theories capable of representing and predicting observation were created, the other involving the use of those theories to predict phenomena. The second process is of course what is usually encountered in surviving tablets, and the first usually has to be reconstructed from it. That second type of activity required a set of skills that could be exercised by people well drilled in routine procedures of which they needed little understanding. They occasionally added their names to the tablets they produced, and even the names of their forefathers, together with a date, and the name of the ruler. This all suggests a level of professionalism that might hint at some sort of formal education encompassing the reasoning that lay behind the quantitative fundamental theories.

In the case of ancient Greek cultures and civilizations, matters were not very different, once the art of processing large sets of observational data had been learned from Eastern sources, but this happened at a relatively late date. The most significant influences took effect only in the second century BC, and the man deserving most of the credit for this change was Hipparchus. By his time, however, the Greeks had developed a geometrical method of their own that was to assume an extraordinary importance in subsequent history. They modeled the heavens on a sphere, with stars, planets, and circles on it, and they learned to explain the simple daily and annual movements in terms of the rotation of the celestial sphere.

A very common modern view of Greek astronomy is one very heavily influenced by Ptolemy, its greatest practitioner, in whose shade all later astronomers worked. It is important to recognize, however, that by Ptolemy's time, the second century AD, Babylonian arithmetical methods had been very efficiently grafted on to Greek geometrical astronomy. This fact tends to obscure just how cavalier the first of the great Greek astronomers were about observational data. As we shall see, this is true of even the greatest of them, Eudoxus, of the fourth century BC.

The Greeks developed a notable picture of the universe as a whole, and explained its workings on a rational mathematical and philosophical basis that by the time of Eudoxus was tearing itself free from the creation myths and legends of earlier ages. The Greeks shared in some of the traditions of prehistoric astral religion that we have described already, and there was some cultural interchange with the great neighboring cultures—for instance, with Egyptians and Persians—but very little is known of the state of astronomical knowledge in early Greece, even in Minoan or Mycenean times. Names of the months occur in the famous Linear B tablets, and worship of the Sun and Moon in some form seems likely, judging from their art. Some four or five centuries after the Mycenean period came the age we know from the poetry of Homer (perhaps from the mid-eighth century BC) and Hesiod (around 700 BC).

Homer's *Iliad* and *Odyssey* contain only fragments of any relevance to our subject, but they are of great interest. In the first, Achilles' shield is likened to the Earth, which is surrounded by an ocean-river, the source of all water and of the gods. In the *Odyssey*, the starry heaven is said to be of bronze or iron and supported on pillars. Several star groups are named—the Pleiades and Hyades (groups in Taurus), Orion, Boötes, and the Wagon—which is unusual in that it does not rise out of the ocean (it is too close to the pole to rise and set at all). There are references to the Evening Star and the Morning Star, presumably Venus, possibly not then recognized as a single planet. The "turnings of the Sun" seems to refer to the solstices. The phases of the Moon are often alluded to; the winds and seasons are personified; and Athena is once likened to a shooting star. All told, although this was heroic poetry meant for the courts of princes, the astronomy is very homespun, and that which is found in Hesiod's *Works and Days* is only marginally less so. His is a manual in verse, relating the farmer's year to the changing seasons, as judged by the Sun and stars—heliacal risings and so forth. There are here no echoes whatsoever of Babylonian expertise.

COSMOLOGICAL IDEAS IN THE SIXTH CENTURY BC

It was Aristotle, the greatest philosopher of antiquity, who in the fourth century BC established a tradition of collecting together the opinions of previous thinkers and subjecting them to criticism, as vigorously as if they were still alive. Some of his material goes back to the late sixth century BC, but like others who collected early teachings he was largely dependent on unreliable intermediaries. This is particularly true of the four earliest philosophical thinkers of note—Thales, Anaximander, Anaximenes, and Pythagoras, all of the sixth century BC. Thales was considered by Aristotle to be the founder of Ionian natural philosophy, knowledge of the physical world. Stories were told of his intense practicality—for example, of his having used his astronomical knowledge to predict a surplus in the olive crop. He obtained a monopoly of the presses, and so made a fortune. On the other hand, he was also presented as a visionary: in the reported words of a Thracian servant girl, he was so engrossed in a study of the heavens that he fell down a well, unable to see what was at his feet. (Aristotle retailed the first story, and Plato, the second.) Thales is supposed to have predicted an eclipse of the Sun that took place during a battle between the Lydians and the Persians, which is now usually set at 28 May 585 BC. The truth of this story has been much debated, but it can almost certainly be disregarded, except as a symptom of myth-making at the time of Aristotle.

It was Aristotle who set his pupils the task of writing a compendious history of human knowledge. Eudemus of Rhodes was assigned astronomy and mathematics, and from him we hear that Thales, after a visit to Egypt, brought these studies to Greece. Others have argued that Thales was a borrower from Babylon. If true, no signs of his borrowing remain. It has been claimed that he introduced his fellow countrymen to the method of geometrical proof, no less; but the evidence is extremely tenuous and disregards the strong possibility that European traditions of semiformal geometrical reasoning are much more ancient.

Anaximander and Anaximenes held cosmological views that are almost as similar as their names. The second man was possibly the pupil of the first, around the time of the fall of Sardis (546 BC). Like Thales, both came from Miletus, the southernmost of the great Ionian cities of Asia Minor (like Sardis, at the western extreme of modern Turkey), a fact that serves to remind us of the great spread of the civilization that we describe as Greek. Among the greatest of ancient Greek astronomers and mathematicians, Eudoxus was from Cnidos, Apollonius was of Perge, Aristarchus of Samos, and Hipparchus of Nicea and Rhodes—locations that are all in, or off the coast of, Asia Minor. Euclid and Ptolemy taught in Alexandria, albeit more than four centuries apart; and Archimedes lived and worked at Syracuse, in Sicily.

Anaximander is said to have made a map of the inhabited world and to have invented a cosmology that could explain the physical state of the Earth and its inhabitants. The infinite universe was said to be the source of an infinity of worlds, of which ours was but one, a world that separated off and gathered its parts together by their rotatory motion. (This analogy with vortex motion has perhaps more to do with the observation of cooking vessels than of slings. Somewhat similar theories were being advanced in Newton's century.) Masses of fire and air were supposedly sent outward, to become the stars. The Earth was some sort of floating circular disc, and the Sun and Moon were ring-shaped bodies surrounded by air. The Sun acted on water to produce animate beings, and men and women were descended from fish.

Bizarre though these ideas now seem, we catch glimpses in them of a type of scientific reasoning that is by no means trivial. When Anaximenes elaborates Anaximander's ideas and argues that air is the primeval infinite substance from which bodies are produced by condensation and rarefaction, he produces logical arguments based on everyday experience. (Admittedly they were not always well chosen. He considers breathing through pursed and open lips, breathing into cold air, and so forth.) As Anaximander had done, he introduces rotatory motion as the key to understanding how the heavenly bodies could have been formed out of air and water. Such attempts at a physics of creation are characteristic of much Greek thought, and the fact that in these early centuries they are combined with a rather weak grasp of the circulation of the heavenly bodies and how they proceeded when they passed below the horizon, the rim of the world, does not take away their importance for the later history of cosmological thought. Questions have to come before answers.

The name of Pythagoras needs no introduction, although the famous geometric theorem for which he is remembered certainly had little to do with him, at least in the Euclidean form in which it is—or was until recently—taught. He lived in the late sixth and early fifth centuries BC. Despite his large religious following, nothing he wrote has come down to us except by report, but it seems that he took the cosmic ideas of Anaximander and Anaximenes one stage further, saying that the universe was produced by Heaven inhaling (note the metaphor) the Infinite so as to form groups of numbers. Why numbers? He is said to have maintained that all things are numbers. His greatest claim to fame is his discovery of the arithmetical basis of musical intervals, a discovery which gave rise to mystical forms of numerology that even now have adherents. Pythagoras seems to have been convinced that everything—from opinions, opportunities, and injustices, to the most distant stars—is rooted in arithmetic and has a corresponding place in the structure of the universe as a whole. Whether or not this mystical belief can be defended, there has been scarcely any period in history since his time when it has not had important repercussions on scientific thought.

Aristotle tells us of a cosmological system proposed by the Pythagoreans involving a central fire around which the celestial bodies, including the Earth, move in circles. The system is often ascribed to the Pythagorean Philolaus of Croton. It has often been confused with that of Copernicus, but the central fire was not the Sun, which was also supposed to circulate around it, in an orbit beyond the Earth's. In an orbit inside the Earth's, another object, a "counter-earth," was postulated, in order to account for lunar eclipses. The entire scheme was a product of thought rather than of observation. It was drafted, however, in a characteristically Greek style, rational and physical, that would yield important dividends when eventually it was grafted on to sound observational data.

GREEK CALENDAR CYCLES

The zodiac, having originated in Mesopotamia early in the first millennium BC, seems not to have been known to the Greeks before the fifth century BC. We know that the Greek version of it was in use by the second half of that century, since it was used then in the parapegmata—star calendars that used zodiacal signs for the division of the year. (A parapegma is essentially a public notice board, but the word can mean almost any information displayed on it.) Meton and Euctemon, who flourished around 430 BC, were Athenian astronomers who were often cited in the parapegmata. Since both Greeks and Babylonians were subject to Persian rule before the Greco-Persian wars of the fifth century BC, the use of the Babylonian zodiac in Greece—with the addition of the signs of Aries and Libra—is not surprising. Babylonian influences seem to have been at work in other ways too. The solstices had been under observation from time immemorial, but now attention was being given to the careful recording of seasonal events, with a view to improving either the civil calendar or the calendrical scheme into which astronomical observations were fitted. In the century or so before the golden age of Eudoxus, Plato, and Aristotle, there was much Greek concern with improving the civil calendar, but that this was not viewed as belonging only to the province of astronomy is clear from the somewhat random way in which magistrates took corrective action whenever the solar and lunar cycles slipped out of alignment. Or perhaps it was simply that magistrates missed the point.

Whether or not it is correct to speak of a "school" of astronomy founded by Meton and Euctemon, both seem to have collaborated in proposing a regular calendar cycle of 19 years, the so-called Metonic cycle. Indeed, Meton is said to have set up an instrument for observing the solstices on the hill of the Pnyx in Athens, and according to Ptolemy—six centuries later, admittedly—the two men made observations in Athens, the Cyclades, Macedonia, and Thrace. We have already seen that the Babylonians knew the properties of the 19-year period, that brings the solar and lunar cycles back into agreement—235 months being in fact very close to 19 years. It is now generally accepted that the 19-year cycle was known in Mesopotamia before 500 BC. An observation of the summer solstice by Meton and Euctemon, albeit one that had its reliability called into question by Hipparchus and Ptolemy long afterwards, had been made on 27 June 432 BC. Babylonians and Greeks alike made the cycle the basis of rules for intercalation, using the insertion of extra days (like our leap-year day) to correct the calendar's slip: according to Ptolemy, 235 months were equated by Meton with 6,940 days. In his *Introduction to Astronomy* from the first century BC, Geminus of Rhodes makes it clear that the Hellenic goal was to divide solar years into months in such a way as to make traditional festivals fall on the right day of the right month. He tried several traditional schemes and pointed out their failings. Seemingly interpolated in his book is a passage with a Babylonian value for the length of the synodic month. (If this was a later interpolation, then Ptolemy, two centuries after Geminus, was the first Greek-speaking astronomer known to have been in possession of the Babylonian figure. Through Ptolemy, it entered into the Arabic, Latin, and Hebrew traditions.) The Geminus text gives the value of the synodic month as 29 + ½ + ⅓₃ days. In his version of the calendar, out of the 235 months, 110 were "hollow" months of 29 days, and 125 were "full" months of 30 days. Presumably, Meton had earlier drawn up some such scheme of his own for intercalation, but if so, there is no firm evidence that it was ever used for the Athenian civil calendar. The history of the calendar is a morass of alternatives, however, and it would be wise to keep an open mind here.

Whether or not the Babylonians knew of the 19-year cycle before the Athenians, they did make use of other intercalation rules, based on the risings of Sirius, from much the same time as the first references to the 19-year cycle, and it seems probable that they preempted the Greeks on both scores. Calendar cycles nevertheless became something of a Greek astronomical specialty. A century after Meton and Euctemon, Callippus improved their cycle further by taking four periods (76 years) and removing one day (making 27,759 days). The Callippic cycle was then used later by Hipparchus and Ptolemy in a modified form. Hipparchus's refinement (equating 304 years to 111,035 days and 3,760 synodic months) does not seem to have been much used in practice. The simpler cycles of 19 and 76 years were for most purposes enough, and the former eventually became enshrined in the Easter computation (*computus*) of the Christian Church, where it remains in use to this day.

THE GREEKS AND THE CELESTIAL SPHERE

Greek astronomy of the fifth century BC, like that of the Near East, was intertwined with a study of meteorological phenomena generally—with clouds, winds, thunder and lightning, shooting stars, rainbows, and so on. This component remained, with an astrological underpinning, until modern times, but far more important in the long term were the seeds of the geometrical methods that early Greek procedures contained. The discovery that the Earth is a sphere was traditionally assigned to Parmenides of Elea in southern Italy, who was born around 515 BC. Parmenides was also credited with the discovery that the Moon is illuminated by the Sun. A generation later, Empedocles and Anaxagoras seem to have given a correct qualitative account of the reason for solar eclipses, namely the obscuration of the Sun's face by the intervening Moon. Astronomy was spread very thinly through the period leading to the first great age of mathematical advance, the fourth century BC, which began with the remarkable planetary scheme of Eudoxus and ended with the first extant treatises on spherical astronomy, those by Autolycus and Euclid. Small but important developments were taking place before then, however. We are told that a catalog of stars was drawn up by Democritus in the fifth century BC, and others followed his lead, although most are known only by reputation. Most were probably no more than descriptive star lists. Not until Hipparchus do we have clear evidence of a consistent Greek scheme of coordinates on the sphere, by which star catalogs could be given real astronomical value.

An important step on this road, a change from the listing of stars by reference to the zodiacal constellations to a system of numerical ecliptic longitudes, had been taken by the Babylonians around 500 BC. Reckoning as we now do, from a zero-point where the equator and ecliptic meet, did not come for another six centuries. Ptolemy introduced it, for his definition of the (tropical) year. The Babylonians had reckoned from the zero-points of each zodiacal sign, measuring in each from 0° to 30°. That system long remained—we might even say remains—in astrological use. The Babylonian signs were shifted 8° or 10° from where a Ptolemaic astronomer places them (on systems B and A, respectively), and we have already noted that traces of this discrepancy are to be found in medieval Western sources, where the idea was repeated by scholars who had no idea what it was all about. Unpolished though this system might seem, the Greeks of the fifth century BC had nothing comparable.

THE HOMOCENTRIC SYSTEM OF EUDOXUS

The discovery of the sphericity of the Earth, and of the advantages of describing the heavens as spherical, captured the imagination of the Greeks of the time of Plato and Aristotle, in the fourth century BC, and of one man in particular: Eudoxus of Cnidos (ca. 400 BC–347 BC) produced a very remarkable planetary theory based entirely on spherical motions. In terms of its predictive power, this theory cannot bear comparison with the Babylonian arithmetical schemes, but it was in many ways more important. First, it showed posterity the great power of geometrical methods; and second, by an accident of history—its adoption by Aristotle—it was for two thousand years instrumental in shaping philosophical views on the general form of the universe.

Eudoxus was born in Cnidos, an ancient Spartan city on a peninsula at the southwest corner of Asia Minor. In his youth he studied music, arithmetic, and medicine, and he was taught geometry by the notable mathematician Archytas of Tarentum. On a first visit to Athens he studied with Plato, thirty years his senior. He later visited Egypt, perhaps on a diplomatic mission, and is said to have composed an eight-year calendar cycle (the *octaëteris*) while studying with the priests at Heliopolis. Returning to Asia Minor, he founded a school at Cyzicus, rivaling Plato's Academy in Athens—which he again visited at least once more. (Cyzicus was a Greek city that had been abandoned to the Persians in 387 BC.) He professed the principle that pleasure is the highest good, and it is likely that Plato had him in mind when writing on this subject in his work *Philebos*. Whether or not this is so, Eudoxus's influence in arithmetic, geometry, and astronomy was considerable. He was largely responsible for some of the finest sections—Books V, VI, and XII—of Euclid's *Elements of Geometry,* one of the most influential compositions in the history of the world of learning. The merits of Eudoxus's rigorous definitions of numbers were not fully appreciated until recent times, when it was realized that they bear a strong resemblance to those of Dedekind and Weierstrass from the nineteenth century. His planetary theory, however, attracted much attention from the beginning, and it aroused sporadic scientific interest as late as the sixteenth century.

In the penultimate section of the previous chapter, we discussed the motions of the planets from a simple modern perspective. It has often been said—on the basis of a much later authority, Simplicius, through Sosigenes—that it was Plato who set to posterity the problem of explaining how the observed movements of the planets may be explained in terms of "uniform and orderly" motions in the heavens. (Geminus, incidentally, tells us that the Pythagoreans were first with this demand, made on the grounds that it would be indecent to suppose that the planets move in any other way.) While the views of such an influential philosopher as Plato are bound to be of much interest, his influence on mathematics and astronomy is easily exaggerated. His contributions to both subjects were not direct; rather, they stemmed from his concern that both be a part of the education of the ruling class, the Guardians, and the ordinary citizens. As a propagandist, his influence is still a force to be reckoned with: he saw these studies as a means of training the soul to look beyond the transitory things of this world to a true reality that only thought can grasp. As an astronomer, we can see Plato as little more than a straw in the wind, but his overall importance should not be underestimated. Having eloquently insisted that the universe operates according to mathematical laws that can only be understood by a properly trained intelligence, he helped to create an entire educational climate favorable to the subject.

The discovery of the spherical form of the Earth, extended to that of the heavens, had clearly taken hold of Athenian thinking by Plato's time. In the tenth book of one of his finest works, *The Republic*, Plato introduces a myth, told with much use of poetic imagery by his teacher Socrates. It is the story of Er, a man killed in battle whose soul visits the land of the dead, only to return after his miraculous revival. Socrates tells how Er's soul went first to a certain magic place, described in some detail, and how eventually he saw what amounted to the mechanism of the entire planetary system, with nested whorls—"bowls," according to one of several possible interpretations of the text, and "hoops" following another—turning around a spindle of steel, and each carrying a planet. The spindle rested in turn on the knees of Necessity, in whose charge the daily rotation, as well as the planetary motions, were thus left. The whorls were turned with their various characteristic speeds by the Fates (daughters of Necessity), and on each a Siren sang a single note, so that together they made a harmonious sound.

The counter-earth of the Pythagoreans is nowhere to be seen in this account. There is no indication that the zodiac is inclined at an angle to the equator, but it would be foolish to look too deeply into the myth of Er for such niceties. What seems to be suggested very strongly by this myth is that physical models of the universe were actually being constructed at this time, and not merely described. To describe such a universe as Er's without a real model, one would surely have spoken of *complete* spherical shells; a real model, however, would have had to be open-topped to allow people to see into its workings. In a later work of Plato's, the *Timaeus*, he describes how the Demiurge creates the universe out of the four basic elements, writing in terms that show even more clearly that he has a real model in mind. What he describes there, however, is no longer a system of concentric whorls, but a simple armillary sphere—a model of the celestial sphere of the astronomers, made with hoops which would not have hidden the interior.

There is in another of Plato's writings, the *Laws*, an Athenian Stranger who says that he was no longer a young man when he realized that each of the planets moves on a single path, and that it is wrong to call them "wanderers." Plato had called them erratic in his *Republic*, so this is perhaps an autobiographical statement, and it is tempting to suppose that it was Eudoxus who persuaded him to change his mind. The occasional retrogradations of the planets make them "wanderers," but their conformity to geometrical order, if that can only be discovered, will show that they are not truly erratic.

No writings by Eudoxus survive, but his system can be pieced together from the writings of two others in particular: Aristotle—a late contemporary—and Simplicius. Simplicius was a Platonist who wrote influential commentaries on Aristotle's work, but he was no mathematician. Since he was born around AD 500, and died after AD 533, his testimony—nine centuries after the event—would be dubious were it not for two or three invaluable remarks. He describes the shape of the planetary path that results from the Eudoxan construction as a *hippopede*, a horse fetter, a figure eight; and he speaks of an attack made on Eudoxus for having thereby ascribed breadth to each planet's path. Taken in conjunction with the general form of the theory, on which he and Aristotle are more or less in agreement, this tells us a great deal, as we shall see.

Eudoxus's system is built up from concentric spheres, spheres centered on the Earth. They are inside one another, but his is a mathematician's universe in which their differences of size are ignored. The idea that spheres are needed was always as obvious as it now seems; but given that spherical models, real or imaginary, were then under discussion, it would have been clear that to describe the Sun one needed at least two spheres, one for the rapid daily rotation, the other for the Sun's annual motion in a contrary direction. The second sphere needs to be pivoted around the poles of the ecliptic circle, of course. The Moon could have been roughly described along the same lines. (In both cases the object is imagined to be situated midway between the poles of the sphere on which it resides.) In fact, for both the Sun and Moon, Eudoxus added a third sphere. For the Moon this could well have been meant to take care of the fact that the Moon's orbit is inclined at about five degrees to the ecliptic, which it intersects in points (nodes) that move slowly backward around the zodiac. (They circle the sky in about 18.6 years, as explained in the previous chapter.) A rudimentary knowledge of eclipses

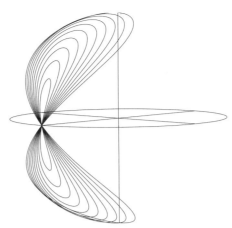

41. An assortment of hippopedes. Each of Eudoxus's planetary models requires only one, but we can see how, by choosing from this assortment, he would have been able to account for a wide range of movements in latitude and longitude.

could have supplied this insight. If that is the reason for the Moon's third sphere, then Aristotle and Simplicius seem to have the order of the Moon's second and third spheres wrong, but otherwise their accounts make reasonable sense. What is puzzling is the fact that Eudoxus seems to have added a third sphere for the *Sun's* motion too, in the belief that, at the winter and summer solstices, the Sun did not always rise at the same point of the horizon. Simplicius says that those who preceded Eudoxus had thought this. The idea is found repeated by several later writers.

It was in his explanations of the direct and retrograde motions for the planets that Eudoxus's pivoted spheres came into their own. He proceeded to show how a point could describe a figure eight, which in turn could be carried around the sky with the planet's long-term motion, more or less in the zodiac. To create that figure, the hippopede, he simply took a pair of spheres, one turning in one direction, and the other turning with the *same speed* in the *opposite direction* around an axis that was carried by the first sphere—but that did not coincide with its own axis. Ten specimens of the mathematical curve in question, corresponding to different inclinations of the two axes, are shown for convenience on the same figure (fig. 41). One is to imagine the planet moving round the figure eight in time. It is easy enough to see how, pivoted around another axis at right angles to the length of the hippopede, it can be carried round the zodiac (or a path near the zodiac) producing occasional retrograde motions. To that third motion one will add the daily rotation of the sky, the "rotation of the fixed stars."

Omitting that daily rotation, the path followed might have the general character of the line drawn in figure 42, a figure accurately drawn but with arbitrary parameters for the speeds and inclinations. We may put aside for the time being the question of accurately representing the planetary motions that are actually observed.

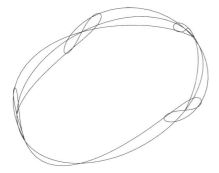

42. The general character of a Eudoxan planetary path, qualitatively acceptable, but in reality impossible.

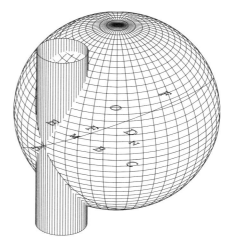

43. The hippopede as the curve of intersection of a sphere and a cylinder tangential to it. The lettering is that of figure 44.

In approximating in this way, at least qualitatively, to the motions of the planets, their seemingly erratic motions had been reduced to law. Plato was no doubt pleased to discover that fact. But what was Eudoxus's real aim? There is good reason to suppose that the delight taken by the Greeks in the explanation he had provided had less to do with its precise predictive power than with its geometrical virtues. To appreciate the real character of Eudoxus's achievement, it is necessary to sketch, however briefly, the geometrical reconstruction offered by the talented Italian astronomer Giovanni Virginio Schiaparelli, in the 1870s. Using simple theorems of Greek geometry available in Eudoxus's time, he showed that the hippopede is a curve of intersection of a cylinder tangential to the sphere on which the curve lies (fig. 43).

This beautiful geometrical result, at which the descriptions by Aristotle and Simplicius hint only very darkly, is of a sort not altogether unknown from the period. Archytas, Eudoxus's teacher, in solving the problem of the duplication of the cube, considered the intersections of *three* surfaces of revolution, a torus (anchor-ring), cone, and cylinder. Those who feel that Eudoxus should not be outdone, but who are reluctant to talk in this context of transcendental curves of the fourth order, might like to add to the

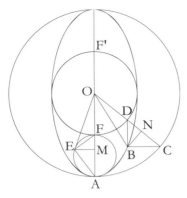

44. An aid to understanding the geometry of the hippopede. The diagram fits into the central plane of the figure 43.

sphere and cylinder another simple surface on which the hippopede lies. This is a surface with a parabola as its uniform section. (Imagine a sheet of paper, curved so that two opposite edges are identical parabolas, with the line of the hippopede lying entirely on that sheet of paper.) We have no good reason to suppose that Eudoxus knew of this property of his hippopede, but the same goes in all strictness for the intersecting cylinder. It seems highly probable that he introduced at least that, although when medieval and Renaissance astronomers investigated similar models, they showed no knowledge of it either.

This Eudoxan model is so important in the history of geometrical astronomy that a bare outline of a proof is called for, if only to show the sophistication of an astronomical doctrine now more than 23 centuries old. We distinguish between the carrying sphere and the carried sphere. In figure 44 we are looking down along the axis of the former, and of the cylinder (on which are points *F*, *E*, and *A*) lying parallel to it. (It is instructive to ask why it is not parallel to the other axis; or, for example, symmetrically placed between them.) *A* is the starting point of the planet, and the arc *AB* is its motion around the equator of the carried sphere, in a certain time. Looking down on this, it seems to be an ellipse, and angle *AOB* on the figure is smaller than the angle in three dimensions. This is in fact equal to angle *AOC* on the figure, where *C* is a point starting out from *A* at the same time around the other circle. Clearly *B* and *C* will be on the same level (*CB* perpendicular to *OA*). Consider now the planet's compound movement at the time in question, as seen against the plane of the diagram (that is, in orthogonal projection on it). It moves up to *B* with the carried motion, and then swings round with the carrier sphere's motion, bringing *OB* to *OE*; and here the angle *BOE* is equal to *AOC*. It has to be proved that *E* lies on a circle, the section of the cylinder. If angle *CBD* is a right angle, with *D* on *OC*, then it is enough to show that *CD* is of constant length; for then all points of type *D* (including *F*) will lie on a circle centered at *O*. The angle *FEA* will be a right angle, so that *E* will lie on a circle with diameter *FA*, the section of the cylinder.

To prove the constant length of *CD* by making use of the properties of an ellipse is more immediate, but considering the relevant part of the

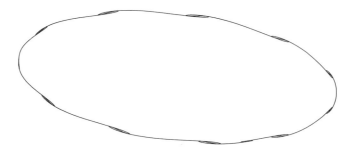

45. An accurately drawn Eudoxan model for Jupiter. This is a perspective view of the three-dimensional path.

diagram in three dimensions, the proof is not difficult on the basis of proportions of sides in similar triangles. This is easier than the initial act of visualization; and considerably easier than proving the theorem of the paraboloidal sheet. I should add that the focus of this parabola is a quarter of the distance from *A* to *F*.

We have here the makings of a powerful geometrical model for the planetary motions, but alas, as it stands it suffers from severe limitations. These are sometimes misrepresented. It is not true to say that all the loops in the planetary retrogradations are identical, as we see from figure 42, nor is it true that the planet's motion in latitude is necessarily great. The retrogradations of Saturn and Jupiter may be fairly plausibly represented without introducing undue latitude (fig. 45 is for Jupiter). Unfortunately, without adding more spheres, the model has only two main parameters that can be varied, the relative speeds within and of the hippopede, and the size of that hippopede—which depends on the inclination of the pivoted sphere. These parameters are simply not enough to accommodate the real motions of Mars, Venus, or Mercury. If the speeds are roughly right, then the length of the retrograde arc will be wildly wrong; and the converse is also true.

From a modern standpoint, the relative *speeds* in and of the hippopede are related to the planet's and our own angular speeds in orbit round the Sun, while the *size* of the hippopede in relation to the sphere is related to the relative sizes of the planet's orbit round the Sun and our own. First, without entering into any details, it should be obvious that the facts might require the hippopede to move as a whole so rapidly in relation to the movement of the planet on it that this can never enter a retrograde phase at all. This happens in the cases mentioned. And second, if we fix the length of the retrograde arc in the model on the strength of observation, we are obliged to accept whatever breadth the hippopede turns out to have as a consequence. It is not only that in the case of Mars and Venus these happen to be excessive, but that planetary movement in latitude has little to do with orbital sizes. It is chiefly a consequence of the fact that the planetary orbits, and our own, are all in slightly different planes.

ARISTOTELIAN COSMOLOGY

There are many unanswered and unanswerable questions about the Eudoxan schemes, not least about Eudoxus's motivation in drafting them. Since he taught in a Greek colony in Asia Minor—Cyzicus is on the southern shore of the sea of Marmara, across the water and southwest of modern day Istanbul—it is not unlikely that he was aware of astronomy's astrological and religious affiliations. The intellectual attitudes of the Greeks at this time, however, did not coincide with those of their Asian neighbors. Star worship may not have been entirely alien to the Greeks, but it took a secondary place in their religion, as did even worship of the Sun and Moon, divinities though they were—in the persons of Helios and Selene. The great poet and playwright Aristophanes, who died at about the time of Eudoxus's birth, characterized the difference between the religion of the Greeks and that of foreigners by noting that while the latter sacrificed to the Sun and Moon, the Greeks made offerings to personal gods, such as Hermes. Hellenic religion had long been moving away from the old and simplistic celestial religions, even though some centuries later the trend was checked with the advent of Eastern astrology.

In Eudoxus's time, the philosophers were happy to make a place for the heavenly bodies in their pantheon. For Pythagoras, these had been regarded as divine, and Plato shows that he was shocked by Anaxagoras's atheistic claim that the Sun is a burning mass, and the Moon a sort of Earth. For Plato, the stars were visible gods in which the supreme and eternal Being had put life. His was no longer a celestial religion for the populace, but a religion for intellectual idealists, and in the hands of his numerous successors, many of them Christian, the Platonic view of the heavens proved highly influential. Even his rival Aristotle defended the divinity of the stars, representing them as eternal substances in unchanging motion. Divinities they may have been, but all this is a far cry from the doctrines of those Chaldeans who claimed to predict from heavenly signs the life and death of nations and individuals, as well as the weather and what depended on it.

We do not know with any certainty Eudoxus's attitudes to these things, but there can be little doubt that in his astronomy he was largely driven by the intellectual pleasures of the geometer—a commodity of which many social historians seem to be entirely unaware. Although the testimony of the Roman statesman and scholar Cicero in his work *On Divination* is relatively late—he died in 43 BC—it is that Eudoxus maintained that "no credence should be given to the Chaldeans, who predict and mark out the life of every man according to the day of his birth." By Cicero's time, the Roman world was deeply conscious of these practices, and some have therefore considered this reference to be anachronistic, but there is no reason why it should have been so. In fact, it might have come from a source that was referring to a non-astronomical way of predicting human life, for there were Babylonian techniques for doing so that were known in Egypt long before Eudoxus, which rested only on the calendar. If this is so, however, then it takes away the force of the argument that Eudoxus was deliberately spurning astrology in its better-known forms.

Whatever his motives, we cannot consider his achievement to have been complete when measured by later astronomical ambitions or by those of the Babylonians. The fact that we can fit the behavior of Jupiter and Saturn to his model does not mean that Eudoxus did so at all accurately. That we ourselves find it easy to vary the schemes, for instance, by changing the speeds of carried and carrying spheres, does not mean that others did so in ancient times. For the most part, doing so produces unpalatable consequences. Among the geometrically curious results, we may mention the following: in the basic two-sphere system, by making the angular speed of the carrier sphere double that of the carried sphere, the resulting curve is simply a circle with tilt opposed to that of the equator of the carried sphere. This sort of possibility warns us against speculating on the nature of the next step in the development of the general theory, at the hands of Callippus of Cyzicus, around 330 BC.

Callippus was, as it happens, the pupil of Polemarchus, who had been a pupil of Eudoxus; and he followed Polemarchus to Athens, where he stayed with Aristotle, "correcting and completing, with Aristotle's help, Eudoxus's discoveries." So speaks Simplicius, who tells us that Callippus increased the number of spheres by two each for the Sun and Moon, and by one each for the planets, excepting Jupiter and Saturn. This suggests that there was a measure of satisfaction with those two planets, the very pair that we ourselves can best fit into the Eudoxan model. Under these circumstances it might be uncharitable to suggest that the Greeks of this period were concerned only with a *qualitative* model of retrogradation. Simplicius goes so far as to say that Eudemus had listed the phenomena that had induced Callippus to expand the system.

Did the Greeks see any merit in the economy of one scheme as against another? It is customary to note that the total number of spheres in Eudoxus was 26, and in Callippus 33, but totals quoted in this way should not mislead us into thinking that either man was aiming as a unified scheme, a single system for all of the planets. As far as can be seen, the two were content to advocate a separate scheme for each planet or luminary. Whatever progress Callippus made, however, we do know that Aristotle enlarged on his ideas and turned what had presumably been a set of abstract geometrical theories into a unified mechanical system. And as such, the theory held an important place in natural philosophy for two thousand years.

Aristotle was by far the most influential ancient philosopher of the sciences. He was born in Stagira in 384 BC into a privileged family: his father had served as a personal physician to the grandfather of Alexander the Great, and Alexander was, in turn, Aristotle's own pupil. Aristotle studied under Plato in Athens until the latter's death in 348 BC, and after moving to Mysia, Lesbos, and Macedonia, he returned to Athens, where he founded his own school of philosophy, the Lyceum. His very extensive writings are highly systematic and coherent and cover a large part of human knowledge. Since they were written over a long period, there are naturally a few minor inconsistencies. The most important single source for Aristotle's cosmology, his *De caelo* (*On the Heavens*—but the Latin title is usually used) was an early treatise and does not contain all of what in his work was most influential. It does not, for instance, have the theory of the unmoved mover, for which we must consult his *Physics*. This unmoved mover, or *Primum mobile*, at the outermost part of the universe, was taken to be the source of all movement of the spheres within it.

Aristotle writes in a semihistorical vein, reviewing the main arguments of his predecessors. The longest chapter in *De caelo* concerns the celestial sphere—a construction by this time generally accepted by the Greeks—and the Earth of similar shape at its center. He mentions the theories of the Pythagoreans and of an unnamed school according to which the Earth *rotates* at the center of the universe. He dismisses the idea, as well as that of an orbital motion of the Earth. Both of these we now of course accept. He seems to have been persuaded by Eudoxus's theory that, if accepted, they would imply that the stars are subject to "deviations and turnings," and that these we do not, in fact, experience. Eudoxus had unwittingly scored a hit for the fixed-Earth doctrine. Had he been alive, he might have pointed out that if the stars are at great distances, the argument fails.

Aristotle offers various arguments for the spherical nature of Earth and the universe. The natural movement of earthly matter is from all places downward, to a center, around which a sphere of matter will inevitably build up. There is also the observed fact that the line dividing light from dark regions of the Moon's surface during a lunar eclipse is always convex—not a perfect argument by itself, of course. He refers to mathematicians who try to measure the Earth's circumference—Archytas or Eudoxus?—and who put it at 400,000 stadia, approximately 74,000 kilometers. This is much too high, but is the oldest estimate known to us. (The equatorial circumference of the Earth is now known to be about 43,613 kilometers.)

The sphere, says Aristotle, is the most perfect solid figure, in the sense that when rotated about any diameter it continues to occupy the same space. He conceives the universe to be built layer on layer over a spherical Earth. Only circular motion is capable of endless repetition without a reversal of direction, and rotatory motion is prior to linear because what is eternal, or at least could have always existed, is prior, or potentially prior, to what is not. Circular motions are for Aristotle a distinguishing characteristic of perfection. In this way, the heavens acquired a special place in most later discussions of perfection. On the Earth, natural motion was up (for smoke and so forth) or down (for earthy material), whereas in the heavens, natural motion was circular, allowing of no essential variation—which would have been a mark of imperfection, incapacity. The heavens are simple and unmixed bodies, made not of the four elements that we know close at hand—namely earth, air, fire, and water—but of a fifth element, the ether. Its purity varies, being least where it borders on the air, which reaches to the sphere of the Moon. (It is the idea of this *fifth,* incorruptible element, or *essence,* that gives us our word "quintessential.")

Aristotle had, then, a celestial realm that was in sharp contrast with the sublunary world of change and decay. It was unique, ungenerated, and eternal—qualities that would provide future Christian Aristotelian apologists with problems. He was here opposing the beliefs of the Greek atomists, Democritus and Leucippus, who had argued for void space—which Aristotle rejected on philosophical grounds—and for a multitude of worlds. He was opposing Heraclitus, who had said that the world was periodically destroyed and reborn, as well as Plato, who had held that the world was created by the Demiurge.

It is surprising to find that Aristotle even opposed the notion of celestial harmony, such as we found in the myth of Er. There is, he says, the absurdity of the claim that we hear no sound because it has been in our ears since our birth. And what of the general principle that the greater the object the greater the sound? Thunder would be as nothing by comparison if the vast heavens emitted sounds. But Aristotle did not manage to stem the idea of celestial harmony entirely, and his insistence on the relative perfection of the etherial regions helped to keep in play a Platonic belief in the divinity of the heavenly bodies.

For the technicalities of Aristotle's planetary system, we must turn to his *Metaphysics.* There he seems to be accepting the theory of Callippus, but he tells us that "if all the spheres put together" are to account for what we see, then for each of the planetary bodies—the carrying and carried spheres, to use my previous words—there must be other "unrolling" spheres to counteract the effects of the spheres above them that do not belong to the planet in question. For Jupiter, for instance, its own spheres are enough to explain its motion, apart from the star sphere. This being so, since all the spheres for Saturn are outside its own, they must be neutralized by giving to Jupiter counteracting spheres, with the poles appropriate to Saturn's but equal and opposed angular velocities. When we come to Mars, we shall have to neutralize Jupiter's spheres but not Saturn's, which have already been taken care of; and so for the rest. The spheres in Callippus are as follows, with the required numbers of counteracting spheres in parentheses: Saturn, 4 (3); Jupiter, 4 (3); Mars, 5 (4); Venus, 5 (4); Mercury, 5 (4); the Sun, 5 (4); the Moon, 5 (none). The total is thus 55 spheres, and Aristotle actually quotes this number. He adds a puzzling remark that has never been convincingly explained, to the effect that omitting the extra movements for the Sun and Moon makes the total 47. I suspect that at an earlier stage he had given 4 counteracting spheres to the Moon, to secure the fixity of the Earth.

Aristotle's is thus a mechanistic view of a universe of spherical shells with various functions, some carrying planets. Motions were no longer being postulated as though they were mere items in a geometry book, nor were they justified in terms of Platonic intelligences, but rather, in terms of a physics of motion, a physics of cause and effect. The first sphere of all, the first heaven, shows perpetual circular movement, which it transmits to all lower spheres; but what moves the first heaven? What moves it must be unmoved and eternal. There is much theological interpretation of this prime mover, whose activity is the highest form of joy, with pure contemplation with itself as object, a natural condition for something divine. One might then imagine that as far as ultimate causes are concerned, Aristotle has all that he needs. Certain later commentators speak as though the first mover of the outermost sphere is enough for the system. Aristotle nevertheless speaks as though each planetary motion of Eudoxan type has its own prime mover, so that there will be 55 (or 47) of them altogether. It does seem that in the end he accepts them as gods. Those who in late antiquity and the Middle Ages found the idea repugnant usually spoke instead of "intelligences" or of angels.

In Simplicius, we read that a system of concentric spheres continued to be taught, and that it was accepted by Autolycus of Pitane, who lived around 300 BC. Autolycus wrote works of "spherical astronomy," namely, the geometry of the (celestial) sphere, and these had a certain currency in Arabic, Hebrew, and Latin, well into the Middle Ages; but they contain no theory of the Eudoxan type. Autolycus defended the theory, however, against a certain Aristotherus, who is known to history as the teacher of the astronomer-poet Aratus. The theory was recognized as deficient in failing to account for the changes in brightness of the planets. Simplicius suggests that this was something of which Aristotle was aware.

HERACLEIDES AND ARISTARCHUS

Heracleides, a near-contemporary of Aristotle in Athens, was a man whose fame in the history of astronomy probably far outstrips his achievements. A colorful figure whose greatly admired literary works have failed to survive, he is said to have died suddenly while being presented with a golden crown in the theater. The justice in this lay in his having obtained it by subterfuge: he had—so one story goes—persuaded envoys from the Delphic oracle to say that the gods had promised to lift a plague on his city, Heraclea, if he were to be crowned while alive and given a hero's cult after death.

While he failed in that ambition, he seems to have achieved more success with his audience of historians. He is supposed to have maintained that the orbits of Mercury and Venus have the Sun at their center, whilst the Sun is in orbit round the Earth. That this would have been a step in the direction of Copernicanism is what gives it its special interest. In any event, it is reasonably certain that he did believe in the rotation of the Earth on its axis, a doctrine to which Aristotle alluded, and he is the earliest astronomer known to have held to it. Copernicus actually mentions his name in this connection. Perhaps a silver crown is called for.

The idea that the Sun is the center of the orbits of Venus and Mercury would have been very difficult to work into astronomy during its Eudoxan period, that is, during the lifetime of Heraclides. In an epicyclic theory, the question occurs more naturally. It is mentioned in this context by Theon of Smyrna, but he lived in the early second century AD. A commentary on a passage from Plato, by an even later writer, Chalcidius, mentioning Heraclides, has been thought to report the doctrine, but when it is said there that Venus is "sometimes above and sometimes below the Sun," it is clear from some numerical data that the meaning is simply "ahead of the Sun in the zodiac" and "behind the Sun in the zodiac."

The first astronomer to have clearly put forward a true Sun-centered theory was Aristarchus of Samos. He was born around 310 BC on the island of Samos, off western Asia Minor, near to Miletus, and he lived until perhaps as late as 230 BC. From the same center of Ionian culture came another astronomer and mathematician, Conon of Samos, in the following century, and he, in turn, was a friend of Archimedes—who lived from ca. 287 BC to 212 BC. It is from Archimedes that we learn of Aristarchus's heliocentric theory, for the only surviving work by Aristarchus himself is his treatise *On the Sizes and Distances of the Sun and Moon*, and this naturally enough takes the Earth at the center from which distances are measured.

According to Archimedes, near the beginning of his book *The Sand Reckoner*, Aristarchus's hypotheses are that the fixed stars and the Sun are stationary, that the Earth is carried in a circular orbit around the Sun, which lies in the middle of its orbit, and that the sphere of fixed stars, having the same center as the Sun, is so great in extent that the circle on which the Earth is supposedly carried is in the same ratio to the distance of the fixed stars as that of the center of the sphere to its surface.

Archimedes criticized Aristarchus for the last meaningless statement, which involves the ratio of a point to a surface, and supposed that he meant rather that the ratio of the diameters of Earth and Sun was equal to the ratio of the sphere in which the Earth revolves to the sphere of fixed stars. Some modern interpreters accept this reading, while others take the ratio of point to surface to mean no more than "an extremely great ratio," so great that we should not expect to observe stellar parallaxes, changes in apparent star positions, as the Earth goes round the Sun.

Whatever his intentions, there is no doubt that Aristarchus believed in the motions that we now generally associate with the name of Copernicus—who certainly knew of his predecessor (see p. 306 below, on that point). The strange thing is that the only astronomer known to have supported these ideas in antiquity was Seleucus of Seleuceia. Seleucus is said to have tried to prove the hypothesis. He lived in the mid-second century BC, flourishing about eighty years after the death of Aristarchus in 230 BC. Seleuceia is on the Tigris, but the fact that Seleucus was later described by Strabo as a Chaldean probably implies more than merely a Mesopotamian origin: it suggests that he practiced astronomy in the style of the Babylonians. He was certainly no lightweight, for Strabo says of him that he discovered periodical variations in the tides of the Red Sea, which he realized were related to the Moon's position in the zodiac.

If Aristarchus did indeed believe that the Sun was at the precise center of the Earth's orbit, then it is very unlikely that he could account for the variation in the Earth's motion throughout the year, that is, for the inequality of the seasons. As we shall see, this inequality was known and explained, on a geocentric hypothesis, by Hipparchus in the following century. Aristarchus's failure in such technical respects as this are unlikely to have been the prime reason for the failure of his heliocentric theory to become popular. Far more important would have been the influence of Aristotle's geocentric cosmology, with its appealing doctrine of the natural movement of bodies toward or away from the center of the world, identified by Aristotle as the center of the Earth. There was also a religious dimension to the question, and according to Plutarch, the Stoic philosopher Cleanthes thought that Aristarchus should be put on a charge of impiety for maintaining that the Earth moved. Cleanthes is notorious for the fervor with which he introduced religion into philosophy, but his attitude to heliocentrism is a strange one, bearing in mind his belief that the universe is a living being, with God its soul and the Sun its heart.

THE INTERPLAY OF GREEK GEOMETRY AND ASTRONOMY

To mention Aristarchus only in connection with his heliocentric theory would be to overlook an important aspect of early Greek astronomy, that is, its practical side. The first-century Roman architect Vitruvius is one of our most important sources of information on dialing. He was convinced that a good architect should know philosophy, music, medicine, history, and all of the sciences with a bearing on building. Under these he included astronomy and time reckoning, which he investigated at some length in the ninth of his *Ten Books on Architecture*. He included much on the geometry of dialing, and ended with a list of different types of dial and their supposed inventors. He there tells us that Aristarchus invented the *scaphe*, a hemispherical bowl in which stood a gnomon (pointer), the shadow of which marked the time on a network of hour lines. He also ascribed to Aristarchus a dial in the form of a flat disc. It is impossible to say whether or not these ascriptions are correct, and we cannot even be sure what Vitruvius had in mind, but we can say that hemispherical dials became extremely popular, judging by the numerous surviving examples, especially those in which much of the hemisphere is cut away, as redundant. The gnomon's shadow can only reach a limited portion of the hemisphere, and a cut-away surface has the advantage of not filling with rainwater in the way that a hemisphere does.

An even more popular sundial was the so-called conical dial (fig. 46), which at a glance is easily confused with the cut-away hemispherical version. The call to design all such dials provided an important stimulus to astronomy and geometry. The problems posed became especially acute when the surface on which the shadow was allowed to fall was a plane surface, whether horizontal or vertical—and here, if we are to believe Vitruvius, Aristarchus

46. The remains of a "conical" sundial, from what had been the public square in the Doric town of Kamiros on the island of Rhodes. The type was common in the ancient Greek world, and nearly a hundred examples are extant. Vitruvius ascribed its invention to Dionysodorus of Caunus (250 BC–190 BC), whom we know to have been a brilliant geometer. The concavity on which the hour lines are engraved is a section of a cone (vertex *V*, in the reconstruction, its base *C* being circular). The gnomon (*G*) is horizontal, and the hour has to be read from its shadow according to the season. The tracks along which the hours are to be read are only marked for the seasonal extremes, *s* for the summer solstice, *w* for the winter solstice, and *e* for the equinoxes. The section of the cone on the upper stone surface is elliptical. It is likely that designing such dials was a stimulus to the mathematical study of conic sections.

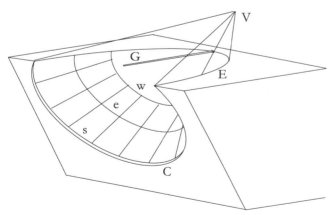

made some sort of advance. A typical example of an early type of plane dial is shown in figure 47. The Athenian populace, two centuries after the death of Aristarchus, became familiar with some fine examples of plane dials on the Tower of the Winds in their Agora marketplace. Dating from around 50 BC, the tower is still standing, and many of the hour lines of those dials are still visible. Undoubtedly, the most spectacular plane dial in antiquity was that

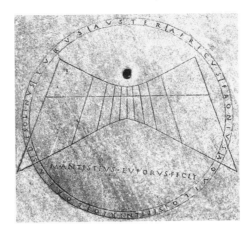

47. Horizontal plane Roman sundial from a temple to Jupiter in Aquileia, perhaps dating from the first century BC. Seen as by a person looking south. The hole held the gnomon. The names of the eight winds are around the edge in a circle about 66 centimeters across. Inscribed with the maker's name, Marcus Antistius Euporus. While easier to make than conical dials, plane dials are now far less common, since flat slabs of stone were often reused.

set up in Rome to celebrate the victories of Augustus in Egypt in 30 BC. Its gnomon, now in the Piazza de Montecitorio, was an obelisk from Heliopolis, 22 meters high, surmounted by a bronze globe. What remains of the vast inscription, inlaid in bronze letters on a marble pavement, is now submerged under later layers of building, but parts of it have been excavated, and it is significant that the letters found were from the Greek alphabet.

Not all sundials strike us immediately as exercises in the geometry of the celestial sphere, but all must ultimately have that character—even a dial in the form of a ham, with pig's tail as gnomon (fig. 48). Studying the art of geometrical projection needed for all of these dial types must have paid intellectual dividends, not only in gnomonics but in geometry more generally, and perhaps in the design of the first plane astrolabe. This we shall mention shortly in connection with Hipparchus, and then again in connection with the Tower of the Winds, the main purpose of which was to house a water clock with a dial of astrolabe type. The early history of techniques for representing the celestial sphere on a plane surface is unclear, but here again Vitruvius gives us a little help. What he called his *analemma* was a geometrical construction with this purpose, not meant to be shown on sundials but used as an intermediary in their construction. (He did not explain exactly how it was to be used, although it is not difficult to supply the missing explanation.) The origins of this particular analemma are unknown, but if we were to guess at its inventor, Aristarchus and Hipparchus would be among the leading candidates. The analemma was later used in a variety of ways, to solve certain mathematical and astronomical problems. Hero (or Heron) of Alexandria, for example, in the first century AD, seems to have made use of it, directly or indirectly, when establishing a method of finding the shortest distance between two cities, based on observations of

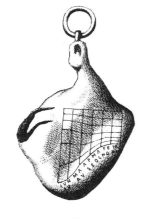

48. Portable Roman sundial found at Pompeii, and therefore made no later than the eruption of Vesuvius in AD 79. It was of silvered bronze in the form of a ham, the tail forming the gnomon. Hanging freely, the dial would have been turned until the tip of the gnomon reached the vertical line for the appropriate date. Each vertical was labeled with an abbreviated month name, and so corresponded to a particular ecliptic position of the Sun. Each of the lines crossing the verticals marked an hour of the day twice over, for morning and afternoon. The curved hour lines may have been plotted with the help of a table, rather than from actual observations over the year. They were certainly not derived by some fundamental geometric-astronomical argument, as in the case of hemispherical, conical, plane, and cylindrical dials. Vitruvius remarked that "many have left instructions for making hanging dials for travelers." The lower figure shows yet another, much more sophisticated type of Roman portable dial with moving parts, and in principle capable of accurate time telling. (From perhaps the third century AD; found in Bratislava and now in the Museum of the History of Science, Oxford.)

a lunar eclipse made from the two places. Our chief witness to the graphical techniques of the analemma is Ptolemy, a century later, but its roots certainly antedate Hero. At a much later date, in the Muslim world, it was used to support some methods for finding the direction of Mecca.

The case of sundial design underscores the importance of the mutual interaction of theory and practice in Greek science, and this was not something confined to timekeeping. While we must not exaggerate the accuracy of Aristarchus's astronomical observations, his work *On the Sizes and Distances of the Sun and Moon* gives an indication of a different kind of interplay between mathematical and observational methods in Greek astronomy, which has often been misrepresented. It is natural enough that we should judge the early situation in the light of modern, or even Ptolemaic, astronomical ambitions, but the ambitions of Aristarchus were almost certainly different and were more appropriate to pure geometry than to observational astronomy. The point may be well illustrated by the famous series of deductions relating the sizes and distances of the Sun, Moon, and Earth, as set down in his book. The way in which Aristarchus set out his basic assumptions explicitly was something he learned from his training in geometry. His work went beyond existing geometrical tradition by introducing certain techniques that heralded the trigonometry of later centuries. He tackled questions that did not admit to precise numerical answers, and the fact that he could only give answers within a quoted range of uncertainty, and that he was ostensibly dealing with the real world, has left some readers with the impression that his were empirical approximations. In some of his extremely involved geometrical proofs he has even been taken to task for not ignoring certain lines that were of insignificant length. What amounts to a pedantic attention to small quantities in the case of an empirical astronomer, however, is not so in the case of a geometer. Aristarchus's treatise belongs to a geometrical tradition which looks to the real world for its examples. Eudoxus belonged to the same world, albeit half a century earlier. Aristarchus has been scorned for quoting the angular diameters of the Sun and Moon at 2°, four times too large. (While neither he, nor the Greeks in general at this period, used Babylonian degrees for angular measure, it is convenient to use them here.) We see, however, that this objection is an irrelevance if he was playing the same sort of game as the modern applied mathematician who poses questions about the motion of balls on circular billiard tables. Historically speaking, there was a gradual slide from the sort of problem set by Aristarchus to the sort of problems astronomers—in our modern sense of the word—would ask. "How far away is the Moon, the real Moon?"

Aristarchus set the stage for such questions without providing satisfactory empirical answers. Among his basic assumptions were that the Moon gets its light from the Sun; and that when the Moon is seen by us to be exactly half-illuminated, the observer's eye is on the great circle dividing light from dark regions. (See fig. 49, in which T is the Earth, S is the Sun, and M is the Moon.) He later assumed that the Sun and Moon have the same angular diameters—something he considered evident from their exact overlap at the time of an eclipse of the Sun by the Moon. (This followed from his claim that there is no surrounding ring of the edge of the eclipsed Sun, and that total obscuration lasts for no significant time. This is not always true, of course.) His treatise is geometrically complicated when it comes to the question of relative distances and sizes of the luminaries, but it contains one proposition which lends itself to a simple discussion. This is now often called "the dichotomy," from the Greek word for a cutting into two parts. From the stated assumption about the half-illuminated Moon, we ourselves should not have any hesitation in writing down the ratio of TS to TM as the reciprocal of cosine of the angle MTS. He stated baldly that the angle in question is $^{29}/_{30}$ of a quadrant (87°). Accepting this figure, our trigonometrical tables give us immediately 19.11 as the required ratio of solar to lunar distance. Aristarchus lacked the concept of a cosine, let alone tables of values of the function, but by a rather tedious geometrical argument he was able to derive a value of "more than 18 and less than 20." To obtain that result he made use in his proof of a theorem equivalent to what we ourselves should write down as:

$$\tan A \div \tan B > A \div B > \sin A \div \sin B$$

Whatever the ratio of solar and lunar distances, since he considered the angular sizes of Sun and Moon to be equal, the ratio of their true diameters must—as he saw—be more or less the same. (One should feel uneasy at that "more or less," since he aimed at exactness, and was even inclined to take

49. The method of Aristarchus for the relative distances of the Sun and Moon from the Earth.

into account the fact that when we look at a sphere we do not see exactly half of its surface.) Other theorems as to the relative volumes of the luminaries followed easily and are not remarkable. What is far more important is that we do not misrepresent the nature of his achievement, which was to have found an upper and a lower limit between which a certain trigonometrical ratio (as we perceive it) must lie. Historically speaking, the astronomical implications of his argument were of secondary importance, and he would probably have agreed. It is virtually impossible to decide when the Moon's disk is divided equally as between light and dark, even to the nearest day. It was quite impossible, given the instruments then available, to measure the vital angle—which should have been approximately 89.8°.

The dichotomy is certainly not the only proposition of historical interest in Aristarchus's treatise *On the Sizes and Distances*, for he went on to develop there a geometrical account of what may in principle be discovered about the *absolute* sizes of the Sun and Moon, taking the diameter of the Earth as unit. While he started from what may in principle be found from an observation of an eclipse of the Moon, his own "data" are clearly not derived from careful observation. It seem highly probable that he chose them with a view to illustrating his method. This was done often by Greek geometers, even by Ptolemy, but not all who took up Aristarchus's method left it as a pedagogical device. It prepared the way for a more incisive analysis by Hipparchus in the following century, and for Hipparchus it showed the way to an empirical result.

Here again, when deriving absolute distances, Aristarchus used a kind of prototrigonometry, leading him, as before, to upper and lower limits within which his answers had to lie. Viewed as geometry, his was an *exact* procedure: when he said, for instance, that the Sun's diameter is between $19/3$ and $43/6$ Earth diameters, his statements of the upper and lower limits were in a sense exact. Only when his data are considered to be observational data can we describe his method as approximative, and this, once again, is almost certainly to misunderstand him. We know from Archimedes that Aristarchus had said that the angular diameter of the Moon was half a degree. Why take a figure four times as large as this? It was the method that counted, and his method hints strongly at the fact that he had tried his hand at deriving what *we* should describe as the trigonometrical ratios of selected small angles. Whether or not this is so, his method in this important example cannot be summarized briefly. By substituting trigonometrical ratios for his discursive geometrical procedures, his general strategy of proof is easily grasped, as explained under figure 50. To appreciate his geometrical achievement, the reader could do worse than try to solve his problem by

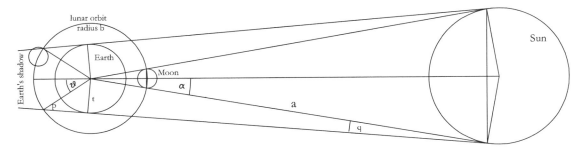

50. Aristarchus's diagram for the absolute distances (or sizes) of the Sun and Moon. (To accommodate the diagram, its proportions are grossly exaggerated.) He assumed the relative distances, found earlier, and postulated that we know the angular size of the Earth's shadow at the lunar distance (discoverable in principle from the time between the Moon's entering and leaving the shadow). His lengthy discussion presented a geometrical method rather than true empirical parameters. The trigonometrical ideas in our brief account encapsulate his techniques only remotely. Consider the angles here marked α and ρ, and the two angles between them, altogether making two right angles. That same unmarked pair makes up two right angles with the angles p and q, so that $(α + θ)$ is equal to $(p + q)$. Angle α is easily measured (although Aristarchus gave the nonsensical value of 1° for it in his treatise), and θ he set at 2°, so the sum is 3°. He had already presented an argument for the *ratio* of the solar and lunar distances ("greater than 18 and less than 20"). If we make the usual approximation that small angles are proportional to their sines (or tangents), this amounts to saying that p is between $18q$ and $20q$, so that 3° lies between $19q$ and $21q$. Taking the average (to shorten the discussion), q is 3°/20 and p is 57°/20. In Earth radii, therefore, the solar distance (a) is the reciprocal of sin 0.15° (about 382.0) and the lunar distance (b) the reciprocal of sin 2.85° (about 20.1). Aristarchus did not actually give the final results. The sizes of the Sun and Moon (in terms of the Earth) follow using simple geometrical procedures.

elementary methods, before even consulting our figure. Given the angular size of the Earth's shadow at the lunar distance, the (equal) angular sizes of Sun and Moon, and the ratio of their distances from the Earth, the problem is to find their absolute distances, in units of Earth radii.

APOLLONIUS AND THE SHIFT TO EPICYCLIC ASTRONOMY

Apollonius of Perga (now commonly spelled "Perge," an ancient Greek city in southern Asia Minor), lived in the second half of the third century BC and into the following century. He visited Alexandria. It is doubtful whether he studied long with the pupils of Euclid there, as Pappus claimed six centuries later, but he was certainly one of the greatest of Greek mathematicians in antiquity, to be compared perhaps only with Archimedes. He did for the geometry of conic sections (parabola, hyperbola, line-pair, circle, and ellipse) what Euclid had done for elementary geometry. He set out his own work, and much of that done by his predecessors, in a strikingly logical way. He also showed how to generate the curves using methods strongly reminiscent of those used in modern algebraic geometry. Those methods were to prove enormously important to astronomy, in the century of Kepler, Newton, and Halley—who studied Apollonius's text closely.

Apollonius's interest in astronomy is known from various oblique references. One writer tells us that he was known as Epsilon, since that Greek

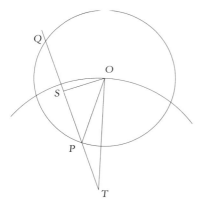

51. Illustrating a theorem of Apollonius relating to epicyclic motions.

letter (ε) is shaped like the Moon, which he studied most intensively. Another source quotes the distance of the Moon from the Earth according to him as 5 million stadia (about 0.96 million kilometers), which is about two and a half times too great. Another writer, the astrologer Vettius Valens, who flourished around AD 160, claims to have used tables of the Sun and Moon drawn up by Apollonius; but this might be another man of the same name. The most intriguing statement relating to his astronomical interests, however, concerns a theorem of his in the theory of planetary motion. According to Ptolemy, writing a couple of decades earlier, Apollonius found a relationship between the velocity of a planet moving in an epicycle, the velocity of the center of that epicycle round the deferent circle, and two distances on the figure representing the situation when the planet appears stationary, passing between direct and retrograde motion. (For our terminology, see the penultimate section of the previous chapter, where these ideas were introduced in advance of their historical place.)

This situation is shown in figure 51, where O is the center of the epicycle and P is the planet. The latter appears stationary to an observer on the Earth, here taken to be at a point T. The motion of P at right angles to the line of sight TQ must have two equal and opposite components: one is due to the fact that it is carried with the velocity of O, and the other is the result of its velocity around O and along the tangent to the epicycle at P. Resolving these velocities using elementary modern methods leads easily to the following theorem: the ratio of the angular velocity in the deferent to that in the epicycle relative to OT is equal to the ratio of PS to PT. (Here, PS is half the chord QP.)

The same result is obtainable by using the methods of limits with classical geometry, as it was derived in Ptolemy's *Almagest*, more than three centuries later. No matter what method Apollonius of Perga used, it is clear that he was well capable of handling motions in two dimensions. This is a matter of some importance, since it seems that he is a key figure in the early development of the idea of epicyclic motion. Ptolemy says that when he proved the relationship explained above, he did so both for an epicyclic arrangement (as here) and for an equivalent arrangement where the planet moves on what is now called a movable eccentric circle.

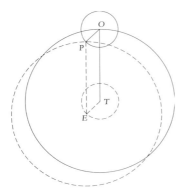

52. The equivalence of certain types of eccentric and epicyclic motions.

The equivalence of the two schemes can be readily seen in figure 52, where the solid lines show the epicyclic arrangement, and the broken lines, the alternative. Ignore the broken circles for the time being. To reach *P* from *T*, one may clearly go from *T* to *O* and thence to *P*, or one may go from *T* to *E*, where *TE* is equal and parallel to *OP*, and thence to *P*, where *EP* is equal and parallel to *TO*. The equality of the lengths mentioned here means that *E* and *P* lie on the broken circles, as shown. *E* is usually referred to as an *eccentric* ("off center") *point,* and the larger broken circle as an *eccentric* circle. It should be noted that it is a movable circle.

From a strictly geometrical point of view, where the difference between large and small circles is of no consequence, the two constructions are equivalent, and the only compelling reason for attaching different words to them is historical.

The eccentric circles we shall meet with in later constructions are simply fixed circles centered on points away from the Earth. This is a suitable place at which to note that they too can be equivalent to epicyclic motions, albeit of a special sort. Suppose that the large continuous circle in figure 52 is fixed, with *T*, the Earth, at its center. If, as *O* goes round the large circle, *OP* remains always parallel to the fixed line *TE*, then *P* will lie on a fixed eccentric circle (the large broken circle in the figure), with center at *E*.

HIPPARCHUS AND THE MOVEMENT OF THE SPHERE OF STARS

Apart from Vettius's tantalizingly ambiguous remark about certain tables drawn up by Apollonius, we know nothing of any attempt he might have made to relate his epicyclic theories of planetary position to observation. His theoretical work, however, led others to do just that, and all the evidence suggests that an infusion of Babylonian methods was essential. The first Greek astronomer who is known to have systematically applied arithmetical methods to geometrical astronomical theory was Hipparchus, who flourished between 150 BC and 125 BC. Born in Nicea in northwest Asia Minor (modern-day Iznik, Turkey), Hipparchus seems to have worked mainly on the island of Rhodes. The importance of his contribution to astronomy was very great, and his numerical methods are known, often in detail, from the *Almagest* of Ptolemy—who quotes similar work by no other astronomer.

We can better appreciate how important Babylonian influence was to the development of astronomy if we bear in mind that, if we were to list all that we know from the Greek world before Ptolemy, we should find scarcely more than twenty reports of precise observations antedating Hipparchus. The earliest is of the Athenian summer solstice observation in 432 BC, mentioned earlier; and the others are all Alexandrian, beginning with a series of lunar occultations of stars by Timocharis. This is not to say that there were no other observations, in a loose sense, for there are many events that it would be hard to avoid observing. Often, reports are no more than writers marveling at dubious predictions without even specifying the date or time of the occurrence. Among the most often trumpeted example is the observation of an eclipse supposedly predicted by Thales, a man whose reputation was already great in the fourth century BC, when Aristotle called him the first natural philosopher and cosmologist. In fact, the historian Herodotus tells us that Thales foretold the year of a preternatural darkness, which actually came to pass, coinciding with the ending of a battle on the Cappadocian Plain between Medes and Lydians. Since a somewhat questionable interpretation of Herodotus I.74 by G. B. Airy in an 1853 article, this has commonly been taken to have been a prediction of the eclipse of 28 May 585 BC. We cannot even be sure, however, that Herodotus was referring to an eclipse, let alone that Thales would have been able to predict it. The same doubt extends to the supposed prediction of a solar eclipse by a friend of Plato, Helicon of Cyzicus, who was reputedly rewarded for his pains by the king of Syracuse. This eclipse is set by some scholars at 12 May 361 BC, and by others at 29 February 357 BC. In the *Bibliotheca historica*, Diodorus Siculus reports an occasion when "day turned to night" during a military encounter between Agathocles and the Carthaginians, an occasion identified as the solar eclipse of 15 August 310 BC. Not one of these is a clear example of an observation made to refine or support an astronomical theory. More promising is a vague report that Archimedes, who died in 212 BC, observed the solstices; but we have no detailed information concerning this event, and even the oft-repeated story of his having measured of the diameter of the Sun at half a degree ($\frac{1}{720}$ circle) is not substantiated.

These few examples present a picture very different from the Near Eastern scene. There were numerous points of contact between Greek and Eastern cultures, and we have already come across some of an astronomical sort, relating to the calendar and the zodiac. We have seen Cicero's claim that he had seen a written statement by Eudoxus, giving an adverse opinion on the astrological predictions of the Chaldeans. Measuring angle in degrees, and sexagesimal arithmetic, first appear in Greek in the *Anaphorics* of Hypsicles, not long before Hipparchus, but Hipparchus clearly had access to Babylonian data and theory of a much more intricate sort than any to be found in earlier Greek sources. It was F. X. Kugler, at the very end of the nineteenth century, who first realized that Hipparchus had taken fundamental period-relations for his lunar theory (so many months equal so many years) from Babylonian lunar theory, on what we call System B. Since then, many other lesser examples of indebtedness have been found, and it seems likely either that an abstract of the Babylonian archives made for local use was translated into Greek, or that a Greek astronomer, perhaps bilingual, had access to the archive and made such an abstract. Babylonian methods continued in use in traditional form until after Ptolemy's time, even in Roman Egypt, and Hipparchus himself might have learned them at source.

Essential to any program that linked geometrical models to observational data was something equivalent to what we would now call trigonometry. Hipparchus played an important part in the foundation of this subject. He wrote a work on chords (a chord is the line joining two points of a circle) and drew up a simple table of chords. If we take the radius as our unit, in our terminology it is, of course, equal to twice the sine of half the angle it subtends at the center, so that a table of chords will have much the same sort of use as a table of sines. Hipparchus, following Babylonian practice, took his circumference to be divided into 360 degrees, each of 60 minutes, and he took his standard radius to be divided into the same number of units and subdivisions. Ptolemy later set the radius at 60 units, a standard that began to fall into disuse only in the sixteenth century. Indian astronomy long continued to use the Hipparchan norm, however, and the Indians also followed Hipparchus in calculating chords for successively halved angles, starting with simple chords such as those for 90°, and 60°. This explains why angles of 22½°, 15°, and 7½° are often mentioned in later astronomical texts as fundamental.

As we know from the work of Eudoxus, Greek three-dimensional geometry was highly developed, and it is very probable that Hipparchus broke down problems on the surface of a sphere—for instance, problems concerning the risings and settings of the Sun and stars—into problems involving circles and triangles in a plane. (We have already alluded to this method in connection with the drafting of the hour lines on a plane sundial, using a construction known as an analemma.) Hipparchus seems often to have solved similar problems arithmetically, no doubt enlarging on Babylonian techniques. Another geometrical method is one that requires the three-dimensional celestial sphere, with its appropriate great circles, to be projected on a plane in much the same way as the Earth's surface is projected on terrestrial maps. There is no doubt that Hipparchus exploited this technique successfully, using projections of different sorts.

One of these, the projection we know as "stereographic," was especially important for its influence on the design of astronomical instruments, even instruments that are in use today. It makes possible the mapping of stars on a plane surface. It also makes possible the mapping of lines representing our local horizon, the meridian line, and various other coordinate lines that are fixed in relation to the local observer. The instrument known as the plane

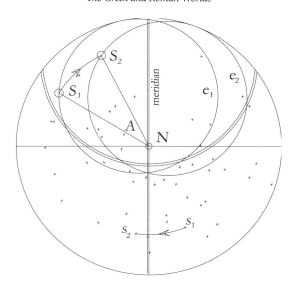

53. This diagram illustrates the broad principle of the plane astrolabe. S_1 and S_2 are two positions of the Sun, as it turns around the poles (N is shown here at the center). The corresponding positions of the ecliptic, the path of the Sun over the year, are also shown (e_1 and e_2), as are those of the stars. The movement of just one star is marked. A more extensive disk of stars could have been shown, including stars nearer to the South Pole, but it was usual to construct a star map just big enough to include the complete ecliptic, since for northern observers this was likely to include most visible stars. The movement shown, covering the angle A, would take about a couple of hours. Not moving with the Sun and stars are the local horizon and the meridian, represented here by double lines. The Sun at S_1 obviously rose above the horizon about half an hour previously. The placing of the circular arc representing the horizon depends on the geographical latitude for which the astrolabe is constructed. For a better idea of the instrument as a whole, see figures 65–68.

(or planispheric) astrolabe superimposes the first of these maps on the second, traditionally putting the first on a pierced metal plate so that the second is visible below it. A pivot at the center (corresponding to the pole) makes it possible to rotate the star maps, and so simulate the daily rotation of the heavens with respect to the local horizon and meridian. A rudimentary illustration of the pattern of the most important lines on this important instrument is given in figure 53. Here a part of the daily rotation of the Sun is represented. The Sun has a position in the ecliptic on the star map. We may suppose its position fixed for the day, although this is not of course strictly true. As the star map rotates about the central pin of the instrument, representing the pole, the angle on the instrument between the two positions of the Sun (the angle A in the figure) will be the same as the angle of daily rotation of the heavens in general and the Sun in particular. The angle would have been measurable on a scale at the rim of the instrument, a scale graduated either in degrees or in hours. (There would of course have been twenty-four hours to the complete circle, but the hour-divisions were in practice often given letters of the alphabet, rather than numerals.)

The evolution of the astrolabe continued for two millennia, and it would require a treatise of its own to explain its many uses. Some brief

54. The *Farnese Atlas*. (See also figure 55.)

additional remarks on its history are included in the last section of the present chapter. It seems probable that we owe its invention to Hipparchus—our source here is the Byzantine astronomer Synesius, although he was writing more than five centuries later. Certainly Ptolemy knew the theory of stereographic projection, and if Synesius was right, then we can speculate as to how Hipparchus managed to complete the mammoth task of computing so many simultaneous risings and settings of stars, as set down in his various writings—including his sole surviving work, *A Commentary on the Phenomena of Aratus and Eudoxus.*

Aratus, following Eudoxus, had written a poem, *Phenomena*, on which the mathematician Attalus of Rhodes wrote a commentary shortly before Hipparchus followed suit. This was no new tradition: a Babylonian text from around 700 BC, which lists twenty sets of stars that culminate simultaneously, shows that such matters had long been a human concern. The difference between Hipparchus's work and that of his predecessors was that he listed the points (degrees) of the ecliptic that culminated at the same time as the stars. (We can call this quantity the *mediation* of the star.) What might at first seem a pointless exercise was in fact quite expressly intended to allow astronomers to tell the time by night when making observations. It paid far greater dividends to Hipparchus himself, and he probably made much use of an instrument of the astrolabe type to help him perform the necessary calculations. We do at least know that he had a three-dimensional globe with the constellations indicated on it. It has been suggested that the Farnese globe (figs. 54 and 55), seemingly deriving from a Greek original of the second century BC, may have been based on the star positions established by Hipparchus. A remote connection with Hipparchus's catalog is possible, perhaps through intermediary

55. Renderings of engravings of the *Farnese Atlas* celestial globe, from a work of 1750 by Gianbattista Passeri. The astronomer Francesco Bianchini had made a careful study of the globe in the 1690s. In Greek mythology, Atlas was condemned by Zeus to support the sky, or "to hold up the pillars of the Universe." The marble sculpture is so named because it was acquired by Cardinal Alessandro Farnese in the early sixteenth century and then shown in the Farnese Palace in Rome (It is currently in the Museo Nazionale Archeologico in Naples). Art historians believe the statue to be a Roman copy, made in the second century AD, of a Greek original dating perhaps from the second century BC. The remarkable globe (about 65 centimeters in diameter) is the earliest known representation of its type. We should not necessarily reject the idea of high accuracy in a copy, but neither can we be absolutely certain that it is a copy. Marked on it are the zodiac, equator, and the Arctic and Antarctic circles. The last pair tell us in principle which star regions were always, and which were never, visible from the place of manufacture, and so depend on latitude. Unfortunately, second-hand information on visibility cannot be ruled out. The date should in principle be deducible from the equinoxial points (intersections of the ecliptic and equator) relative to the stars, points which change with time as a result of precession. The intended star positions can unfortunately be only roughly estimated, from the constellation images. The north polar region is damaged, so that the constellations Ursa Major and Ursa Minor are missing, but forty-two others are intact. As is usual on globes, the constellations are shown, quite logically, not as seen from the Earth but as from "the outside." It is probable that the original globe predated the catalog of Hipparchus, and came from a latitude within a degree or so of those of Samos, Athens, the toe of Italy, and northern Sicily.

catalogs that are now lost. There are many significant differences, however, between the star positions on the Farnese globe and those specified by Hipparchus in his *Commentary on Aratus*, and a direct link seems very unlikely.

This work of Hipparchus marks the beginning of a system of rigorously applied star coordinates. Hipparchus did not have our "pure" systems—either that of ecliptic latitude and longitude, or that of declination and right ascension. These gradually evolved from his system, that is, from a system of declination and mediation—one that, as it happens, was soon to pass into Indian astronomy. Hipparchus built up his own catalog of stars, not necessarily with coordinates for each, however, but in some cases perhaps just giving statements about stars that were in line, and estimates of distances. Aristyllus and Timocharis had listed a few declinations, in the third century BC. According to Pliny the Elder in his *Natural History*, Hipparchus noticed a "new star"—whatever that may have been. Realizing that it was moving, he asked himself whether others did likewise, and so—if we follow this route to the discovery—he found that indeed *all* stars have small motions parallel to the ecliptic. Their ecliptic longitudes increase.

Until the age of Copernicus, this was regarded as a "movement of the eighth sphere," the sphere supposedly carrying the stars. As we should now say, taking a Copernican perspective, it is the *reference system* that is moving. There is a slow conical motion of the Earth's axis that makes it seem that the equinoxes are moving round the ecliptic from east to west. This "precession of the equinoxes" we know to be a little over 50″ per year, or 1° in 72 years. Hipparchus put the figure as *at least* one degree in a century, a very remarkable discovery. But was it found purely from star positions?

The movement of the equinoxes obviously affects the relationship between the length of the year, measured as the return of the Sun to a particular star, and as measured by its return to one of the equinoctial (or solstitial) points. The latter, the tropical year, is shorter than the former, the sidereal year, as we saw in the last chapter. Hipparchus realized as much, and although he certainly tried to find the slow movement by considering star positions as quoted by Timocharis, his most accurate findings fairly certainly came from a comparison of the sidereal with the tropical year. His data for the latter cover equinox observations between 162 and 128 BC, and lunar eclipse observations, which are useful because they give an accurate Moon-Earth-Sun line. He settled on a moderately accurate figure for the tropical year of 365¼ days minus ¹⁄₃₀₀ day. The last fraction should have been about ¹⁄₁₂₈, but Ptolemy continued to accept the earlier fraction. We do not know Hipparchus's figure for the sidereal year, and can only make an estimate, based on the upper limit he gave to the precessional motion. (The estimate turns out to be 365¼ days plus ¹⁄₁₄₄ day, if we reckon precession as one degree per century.)

It is instructive to see how ignorant we are of the sequence, and hence the motivation, of so much of this astronomical work. Was it the year or the star positions, or time by night, that set Hipparchus off on the track of what we call precession? There is reason to think that Timocharis was investigating the length of the lunar *month* when he gave his star positions. His lunar observations include no arc measurements: they are simply observations of occultations of stars, with time in seasonal hours.

There has been much nonsense talked in our own time by the Pan-Babylonians about a Near Eastern "discovery of precession." As pointed out in our first chapter, in a certain sense, a "knowledge of precession" was in the possession of any prehistoric observer who found that the risings and settings of stars were not quite as marked out by his ancestors. In a sense, the movement was also known to those Babylonian astronomers who first realized that there is a difference between the tropical and sidereal mean longitudes of the Sun. But to make such statements is not to say that any of those early observers could rationalize the discrepancy, as did Hipparchus. It is highly significant here that Hipparchus reached an appreciation of the universality of the slow drift of the stars only after a period during which he thought it to be restricted to stars in the zodiacal belt.

HIPPARCHUS AND THE SUN, MOON, AND PLANETS

Hipparchus made good use of the two geometrical devices used earlier by Apollonius, the eccentric and the epicycle. The former is, in principle, enough to account for the Sun's motion quite accurately, and from data for the lengths of the four seasons Hipparchus derived parameters to fit the observational data available to him. He decided that the eccentricity was ¼₄ of the eccentric circle's radius, and that the direction of apogee, its furthest point from Earth, was Gemini 5½°. The latter result is commendable, but the former figure is substantially too great. (In round numbers, the eccentricity is about ¼₀.) What was notable here, however, was not Hipparchus's accuracy, but the very fact of his having fitted Babylonian-style observational data to Greek models. He made similar attempts on the Moon's motion, but there he was confronted by far greater problems, even though he was able to draw on Babylonian sources for extremely accurate values of the principal component motions of the Moon, the four different sorts of month (the synodic, the sidereal, the draconic, and the anomalistic).

Hipparchus was particularly concerned to find eclipse periods, presumably for their own sake but also because they help to give accurate positions—and hence motions—for the Sun and Moon. He was fortunate in being able to compare his own eclipse data with those of the Babylonians; and three centuries later, Ptolemy was able to do the same. No Greek astronomer before Hipparchus is known to have borrowed such materials, but again, more important still is the use to which he put them. He devised a simple, epicyclic, lunar model, notable for the way its motions were made to match the Moon's observed motions. He made the motion of the epicycle around the Earth follow the Moon's known average motion in ecliptic longitude, while the motion of the Moon in the epicycle he made keep time with the Moon's observed "motion in anomaly." (The anomalistic month is the period after which the Moon returns to the same velocity, and is to all intents and purposes its time from perigee to perigee.) He found a geometrical procedure that allowed him to derive the relative sizes of the circles and motions around them, based on observations of the times of three lunar eclipses. He applied his method with two different trios of eclipses, once using the epicyclic model as explained, and once using the equivalent eccentric model. (See the earlier section on Apollonius for this equivalence, at p. 93 above.)

His calculations were flawed, but the method was an excellent one that displayed great originality. Ptolemy, nearly three centuries later, developed it further. Hipparchus's model can account quite well for the Moon's returns to opposition and conjunction. From Ptolemy's words it seems that Hipparchus himself realized that for intervening positions it is less acceptable, but he does not appear to have improved on it.

Hipparchus did not restrict himself to a model for predicting only the Moon's longitude. Again using his own and Babylonian data, he established the maximum latitude of the Moon from the ecliptic at 5°. He had a

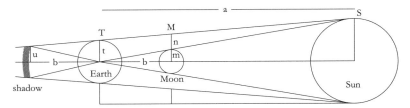

56. Hipparchus followed the lead given by Aristarchus (fig. 50) when trying to determine the distances of the Sun (*a*) and Moon (*b*), relative to the radius of the Earth (*t*). He had reasonable values for their apparent angular sizes (assumed to be identical). He took the angular size of the Earth's shadow at the lunar distance to be 2 ½ times as great as the Moon's. This was still not enough: he estimated the solar parallax to be 7 minutes of arc, equivalent to a distance of 490 Earth radii. (For the concept of parallax, see p. 103) His explanation is lost, though it has been reconstructed from an account by Ptolemy. Like that of Aristarchus, it seems to have been geometrical and discursive, but Hipparchus was apparently prepared to make approximations that were numerically acceptable with small angles. They are made here without comment. The distance labeled *n* in the figure is unimportant, and will later be eliminated. Since the Earth is midway between Moon and shadow, *t* is the average of *u* and (*n* + *m*). The lines marked *n* and *t* are the bases of similar triangles, and (*n* ÷ *t*) is equal to (*SM*/*ST*). The latter ratio is equal to (*a* − *b*) ÷ *a*, again since the lines named are corresponding sides of similar triangles. (To make this obvious, but to avoid confusion, the triangles are drawn in, unlabeled, on the *lower* side of the figure.) We now have all that we need, two equations from which to eliminate *n*:

$$2t = m + n + u \text{ and } n \div t = (a - b) \div a.$$

Hipparchus believed that the angle subtended by the Moon's radius at its average distance (that is, by length *m*) was ¹⁄₁₃₀₀ part of its orbit (that is, of 2π*b*). From all this, and his values for *u* and *a* (given above), the lunar distance (*b*) is found to be about 67.2 earth radii (he found 67 ⅓). What is much more commendable is that he considered the effects of error in his working. What if the Sun were more distant? If *infinitely* distant, the reciprocal of 490 in the last calculation will be replaced by zero, making the minimum distance of the Moon marginally greater than 59 Earth radii. The approximate modern value is 60.27 Earth radii. This result must be counted as one of the finer achievements of ancient astronomy, despite the weak observational data on which it rested.

clear grasp of the three-dimensional arrangement of Sun, Moon, and Earth during eclipses, and he developed geometrical procedures for calculating the actual distances of the Sun and Moon from the Earth that would best account for the observations available to him. His results were very imperfect, and it is much to his credit that he stated them in terms of upper and lower limits. Thus the average lunar distance was set at between 59 and 67⅓ Earth radii. No previous astronomer had come so close to the correct value—a little over 60 Earth radii. For the solar distance he gave a figure that was less than a fiftieth of the true value, but at least he knew how helpless he was: he could not measure the Sun's parallax, but had to guess at a figure. He took seven minutes of arc, although the figure is in fact close to nine seconds. (For the geometrical configuration on which he based his calculations, see fig. 56.)

"Parallax" in general refers to the angle through which an object seems to be displaced when viewed from two different positions (see fig. 57). For future reference, we note that as the Earth moves in its orbit round the Sun, the displacement of a nearby star will so change that the star will gradually describe a tiny ellipse in the sky relative to the background of distant stars (see the left-hand

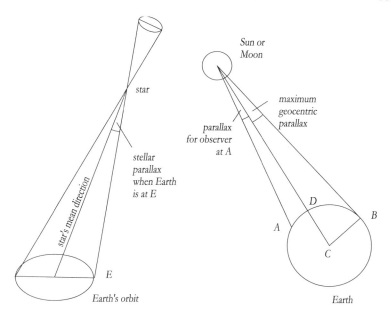

57. "Parallax" in general refers to the angle through which an object seems to be displaced when viewed from two different positions. For future reference, note that as the Earth moves in its orbit round the Sun, the displacement of a nearby star will so change that the star will gradually describe a tiny ellipse in the sky relative to the background of distant stars (see the left-hand figure). The complete ellipse will be traversed in a year. Such *stellar parallax* (or "annual parallax") must be distinguished from the *geocentric parallax* that is so important in correcting predicted solar and lunar positions. Those, when calculated from planetary models, are often referred to the center of the Earth. Our observations of the Sun and Moon, however, are from a point more than 6,350 kilometers from that center. If we observe either object when it is directly overhead, the parallax is zero, since we (at D in the right-hand figure), the center of the Earth, and the object, are then in line. The parallactic angle clearly increases to a maximum when the object is near our horizon and we are at A in the figure. When we speak loosely of solar or lunar parallax, we refer to these *maximum* values. They are obviously directly related to the distances of the bodies in question, and to the Earth's radius, and in quoting accurate figures since the eighteenth century it has been usual to take the *equatorial* radius. This usage gave rise to the full-fledged notion of "mean equatorial horizontal parallax."

figure). The complete ellipse will be described in a year. Astronomers were not able to detect it before the nineteenth century. Such stellar parallax (or "annual parallax") must be distinguished from the geocentric parallax that is so important in correcting predicted solar and lunar positions. These, when calculated from planetary models, are referred to the center of the Earth. Our observations of the Sun and Moon, however, are from a point more than 6,350 kilometers from that center. If we observe either object when it is directly overhead, clearly the parallax is zero, since we (at D in the right hand figure), the center of the Earth, and the object, are all then in line. The parallactic angle clearly increases to a maximum when the object is near our horizon and we are at A in the figure. When we speak loosely of solar or lunar parallaxes, we refer to these maximum values. They are obviously directly related to the distances of the bodies in question, and to the Earth's radius, and in quoting accurate figures since the eighteenth century it has been usual to take the equatorial

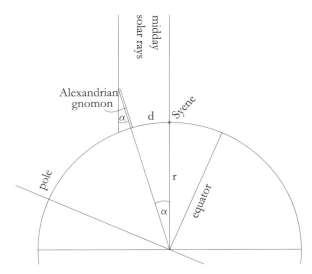

58. Working from reports that the Sun at Syene was directly overhead at noon at the summer solstice, so that gnomons cast no shadow there, and the Sun's reflection was visible at the bottom of deep wells, Eratosthenes reputedly measured the Sun's angle (α) at Alexandria, supposedly at a distance (*d*) of 5000 stadia from Syene. Finding that α had a value of ¹⁄₅₀ circle, he concluded that the Earth's circumference was 250,000 stadia. The Greek stade was always equivalent to 600 feet, but the foot was only loosely standardized. Estimating his result at 48,000 kilometers, it would have been about a fifth-part too high. He apparently lacked confidence in it, and made other estimates. The reports of sailors, who insisted that the true distance of Syene was one-fifth less than he had first assumed, led him to a revised figure of 180,000 stadia.

radius. This therefore gave rise to the full-fledged notion of "mean equatorial horizontal parallax," a complex phrase describing a simple quantity.

According to Ptolemy, Hipparchus did not establish particular models for planetary motion but did criticize those of his predecessors. (Ptolemy's was a presumption based on an absence of writings by Hipparchus on planetary theory. The idea was that Hipparchus, lover of truth, would not accept a defective model.) That Hipparchus compiled digests of Babylonian planetary data, however, perhaps mixing them with some of his own, allowed Ptolemy to exploit them to the fullest. Hipparchus's critical acumen was put to use in another connection. In the middle of the third century BC, Eratosthenes had given a description of the inhabited world, and a lost work *On the Measurement of the Earth* is assumed to have been related to it. An astronomical treatise by Cleomedes, written six centuries later, provides an account of rather dubious historical value, describing how Eratosthenes estimated the Earth's circumference. The simplest method described supposedly led to a value of 250,000 stadia (fig. 58). Hipparchus was a severe critic of many points in his work. None of Eratosthenes' writings survives, however, and some have questioned whether he ever found either the circumference of the

Earth, or—as is often stated—the obliquity of the ecliptic, on the basis of measurements.

Hipparchus was responsible for changing the direction of Greek astronomy, away from qualitative geometrical description and toward a fully empirical science. He never composed a systematic treatise covering the whole of the science, and his many short works were probably lost because they were too difficult for ordinary readers. His reputation in the ancient world was nevertheless considerable. Ptolemy made much use of his writings—although a modern tradition that Ptolemy was little more than a plagiarist of Hipparchus is hardly worth refuting. His influence is fairly certainly to be found in Indian astronomy, as intimated earlier, and from India it traveled back westward, as we shall see, to mingle with other writings produced in a later (Ptolemaic) tradition. Hipparchus was represented twice over, therefore, in this curious amalgam.

THE ALEXANDRIANS

At much the same time as Babylonian influences were making themselves felt in the work of Hipparchus, Mesopotamian astrology was beginning to thrive in Egypt. Egypt was by then at least superficially Hellenized, having been conquered by Alexander the Great (356–323 BC) and ruled by his associates and their descendents after his death. While Alexander had been tutored by Aristotle, it is the course of his conquests, rather than any learning he might have had, that best explains the intellectual movements that interest us here. Alexander has a good claim to be described as the greatest military commander from the ancient world. Succeeding to the throne at the age of 20 years, he secured Macedonia, Greece, and his northern frontiers before crossing the Hellespont in 334 BC, ostensibly to free the Greek cities of Asia Minor. Having defeated the Persian armies, he delayed a thrust eastward into Mesopotamia until he had occupied Phoenicia, Palestine, and Egypt. He then moved eastward, defeated the Persians, under Darius, on their own territory, and then drove on into what is now Turkestan. From there he went on to India, extending the eastern bounds of his empire to the lower Indus. After he died—of a fever, at the age of only 33—his generals divided up, and fought over, the territories he had conquered.

We have already spoken of the subsequent rule of the Seleucids, in Babylonia. (The Seleucids were members of the dynasty founded by Seleucus Nicator, one Alexander's generals. They reigned over a wide Syrian region from 312 BC to 65 BC.) The city of Alexandria had been founded by Alexander himself, perhaps as a future capital. His friend and general, Ptolemy Soter, became satrap of Egypt, and finally declared himself king in 304 BC. "Ptolemy" was the name of all the Macedonian kings of Egypt. Under their rule, the old seat of government at Memphis was moved to Alexandria, which grew in importance to become one of the most influential cities in the history of the ancient world. We touched on some of its history in chapter 2.

Alexandria was important as a center not only of commerce but of learning, and it held its preeminent position in the region throughout the period of Roman rule. Under Soter, two great institutions had been founded near his palace—the Museum and the Library. The Museum, which soon became the most famous of the many places that took that name from their connection with the Muses, housed a group of salaried scholars under the presidency of a priest. Lectures and symposia were held there, and in these the Ptolemies often took part, even down to the time of the famous Cleopatra, the last of their number. A great fire ravished its Library in 47 BC, during Julius Caesar's siege of the city, but the collection was built up again under Roman rule. The later misfortunes of the Museum fall mostly after the period of its intellectual importance, namely, the second century of the Christian era. It suffered many reversals in the third century, but there were scholars of distinction there until the very end of the fourth, when Theon, father of the famous scholar Hypatia, was the last member. Both father and daughter were schooled in astronomy and the sciences, and both wrote commentaries on the astronomer Ptolemy's work.

Throughout these centuries, the city had served to channel ideas from its eastern neighbors into a Mediterranean mold. The Arab conquests would eventually make use of Alexandria's eastward-looking intellectual orientation, so that at length it became a largely Islamic center. Even the ruling Ptolemies had become Egyptianized, and much of the old Egyptian religion reappeared, but with a Greek vocabulary. The native language survived under the veneer of a Greek ruling class, however, especially outside the towns, and this language eventually sprang back into life as Coptic.

Remarkably little is known of the development of Greek astronomy between the time of Hipparchus and that of Ptolemy; and since Ptolemy usually treats Hipparchus as though he were his only significant astronomical predecessor, we can only suppose that little theoretical progress was made during that long period. Astronomy was certainly not dead in the interval. Without listing the numerous small pieces of evidence that it was very much alive, we should not omit mention of an invaluable item of archaeological evidence, in the form of a geared astronomical mechanism made of wood and bronze, remains of which were found in 1900. It was part of the treasure aboard a wrecked ship, found by sponge fishermen on the sea bed off Antikythera, an island lying between Crete and the Peloponnese. The wreck contained much more treasure of archaeological interest, especially bronze and marble statues, but the mechanism was of a different order of importance, for it is an entirely unique survival and of an astonishing degree of complexity. It has been much studied, especially since the 1960s, and while the various interpretations of its corroded mass of gears and inscriptions do not all agree, some of its properties are beyond all doubt.

The Antikythera device is now thought to date from around the end of the second century BC, though the vessel is dated from between 80 BC and 60 BC. Hand-driven, it was in a wooden case of approximately

315×190×100 millimeters in size, with front and back doors containing astronomical and mechanical inscriptions, fragments of which it has been possible to transcribe. There were at least 30 known gear-wheels in the mechanism, the largest nearly as wide as the box, and the smallest less than a centimeter across. They had teeth in triangular form, which was typical of most geared devices before the Renaissance, the numbers of teeth ranging from 15 to 223. The device calculated and displayed lunar and solar information of a calendrical sort, and its gears allowed for the use of the eclipse cycle of 223 lunar months as well as the cycle that equates 19 years with 235 months. (We met these earlier, under the misleading names of *saros* and Metonic cycle.) It also appears to incorporate the Callippic cycle, which is one of four Metonic cycles less a single day—another cycle that we saw earlier, improving the agreement between whole numbers of years and months. The 235-month sequence was displayed on a spiral scale, with an ingenious arrangement whereby a pin in a spiral groove caused a pointer to move over the correct part of the spiral scale. Not only did the mechanism show the shifting positions of Sun and Moon in the ecliptic, which were duly labeled, and the lunar phases, and lunar as well as solar eclipse possibilities, but it is thought to have been geared to display some planetary positions.

Almost inevitably, attempts have been made to associate the parameters embodied in the mechanism with Hipparchus, and one of the most remarkable properties of the device is its use of an epicyclic arrangement of wheels. Inscribed on its casing, the instrument had what would have been known as a *parapegma*—a word we encountered in connection with the calendars posted in Athens. There are many kinds of parapegmata. Some are known from documents written on papyrus, and some from inscriptions in public places, such as on water clocks. A parapegma may present the public with a list of weather changes, correlated with the first and last appearances of constellations and bright stars throughout the solar year. Some of the deciphered words seem to indicate wind directions, which were also occasionally linked with eclipse records.

Somewhat similar devices are known from much later periods in Byzantium and Islam, so that a tenuous tradition may be suspected, but no truly comparable mechanical device is known from before the fourteenth-century St. Albans clock (p. 262 below). Such devices are the products, rather than the cause, of theoretical astronomy, which seems not to have been making very notable progress at the time the Antikythera device was built or in the following two centuries. We must not, of course, conclude from our ignorance of significant astronomical activity that there was no such thing. There was one person in particular whom we should certainly not overlook from a generation before Ptolemy, namely, Menelaus of Alexandria, who flourished around 100 BC. Very little is known of his life, beyond the fact that he visited Rome and made some astronomical observations, but he is now chiefly remembered as a mathematician, and as such

59. A common misunderstanding: Ptolemy as a king. Detail from a woodcut by Gregorius Reisch, *Margarita philosophica* (1503). The astronomer is being instructed in the use of a quadrant by a personified Astronomy.

he proved a theorem of inestimable value to all who wished to perform serious calculations in spherical astronomy. Those who know the theorem of Menelaus only for a plane triangle intersected by a transversal may not realize that it is a special case of a much more powerful analogous theorem, where great circles on a sphere replace the straight lines. Whereas in the theorem for the plane figure we have simple lengths of lines, for the theorem in the spherical case we have chords of arcs. Ptolemy made good use of the theorem of Menelaus for the sphere. Indeed, he had a similar appetite for all that was best in the work of his predecessors. He can hardly be blamed for the unfortunate fact that the aura surrounding his name has so often served to hide their contribution to it.

Astronomer, mathematician, astrologer, and geographer, Claudius Ptolemy was born around AD 100 and died about seventy years later. His name, Ptolemaeus, shows that he was an Egyptian descended from Greek, or at least Hellenized, ancestors, while his first name, Claudius, shows that he held Roman citizenship. His astronomical works are dedicated to an otherwise unknown "Syrus," and his immediate teachers probably included one Theon, from whom he acknowledges having received records of planetary observations. Beyond these simple facts we know virtually nothing about him of a personal sort. (This Theon is not to be confused with Hypatia's father. Theon, Ptolemy, even Cleopatra, were common Egyptian names. Overlooking this fact, medieval Arabic and Latin writers often mistakenly represented the astronomer Ptolemy as a king, and frequently portrayed him wearing a crown, as in fig. 59.)

Ptolemy's extensive writings suggest that he was engaged in assembling an encyclopedia of applied mathematics. Of his books on mechanics, only the titles are known. Much of his *Optics* and his *Planetary Hypotheses* can be pieced together from Greek or Arabic versions. Some minor works on geometrical projection (his *Analemma* and *Planisphere*), as well as the monumental *Geography*, survive in Greek, as does his great treatise on astronomy, the *Almagest*.

The title of this, his finest work, is itself an interesting indicator of cultural movements. It began in Greek as "Mathematical Compilation," and then became "The Great (or Greatest) Compilation." When the Arabs

translated it in the ninth century, only the word "Greatest" was kept in its title, but this in an approximation to the Greek word *megiste,* so that it now
became *al-majisti.* From there to the Latin *Almagesti* or *Almagestum,* in the
twelfth century, and thence to our *Almagest,* were small steps.

This work in thirteen books begins with a statement of reasons for holding
to a largely Aristotelian philosophy—but one that shows the influence of the
Stoics, too. We may all attain moral insight in the ordinary course of our
affairs, Ptolemy remarks, but to attain a knowledge of the universe we must
study theoretical astronomy. He follows Aristotle's lead in placing physics
on a lower plane, as it deals with the changing and corruptible lower world.
Astronomy, by contrast, helps theology, for it draws our attention to the
First Cause of celestial motions, the divine Prime Mover. From such relatively brief philosophical beginnings, Ptolemy turns to some rather general cosmological arguments of a qualitative sort, concerning the heavenly
sphere and the various motions observed in it. Again he follows Aristotle,
more or less, in his physical arguments for the spherical shape, central position, and fixity of the Earth. Ptolemy also considers the Earth's insignificant
size in relation to the heavens. He does not refer to the discussions of the
Earth's size by Eratosthenes or Posidonius—the latter an extremely influential Greek astrologer and Stoic philosopher, born around 135 BC.

Ptolemy's failure to mention these earlier authorities is interesting, because Cleomedes, a near contemporary of Ptolemy's, does inform us of Eratosthenes' measurements, and Cleomedes in the same work writes of the
refraction of rays of light passing down through the Earth's atmosphere. It
seems likely that Ptolemy was simply unaware of this writer. It is thought
that Cleomedes might have been the discoverer of atmospheric refraction,
a phenomenon of great importance to astronomy. In his *Almagest,* Ptolemy
considers refraction only as an influence on the sizes of heavenly bodies
when seen near the horizon. In the *Optics,* he considers atmospheric refraction in more theoretical detail—but that was a later work.

A mathematical introduction now follows, with Menelaus's theorem, and
a table of chords to three sexagesimal places, as well as other items that we
should now classify as "trigonometry." His table, drawn up for half-degree intervals, is based on a value of the chord of 1°, which he evaluates by a clever approximation procedure. He is soon at work in Books I and II of the *Almagest,*
applying his mathematical techniques to astronomical problems, and one item
that has repercussions throughout the books that follow is his calculation of
the obliquity of the ecliptic, the angle between it and the celestial equator.

From the extremes of the Sun's declination he found the value of this
fundamental parameter to be between 23;50° and 23;52,30°. He finally settled on 23;51,20°, which falls in this range but is a relatively poor figure.
(A better would have been 23;40,42°.) Some have said that here he was
taking this value because Hipparchus—even Eratosthenes—had accepted
it. It would be surprising if Eratosthenes had known anything better than
24°. Ptolemy's instruments were imperfect, and he probably suspected as

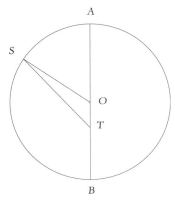

60. A model for the motion of the Sun, using a simple eccentric circle on which the Sun is taken to move at constant speed (that is, constant angular velocity around the center of the circle).

much. One inevitably wonders whether he allowed his admiration for Hipparchus to sway his judgment—or even his instruments. Evidence for a debt to Hipparchus on this point, however, is simply unavailable.

In Book III of the *Almagest*, Ptolemy accepts Hipparchus's solar theory. He compared his own observation of the dates of the equinoxes with those made by Hipparchus; and he compared a solstice observation with that made by Meton and Euctemon in 432 BC, that is, nearly six centuries earlier. Here he made a calendar error of one day, but even that was enough to throw his figure for the tropical year out of joint, and to persuade him yet again to accept Hipparchus's figure of 365¼ days minus ⅟300 day. This was over 6 minutes of time too large—in fact 6;26 minutes—but the theory accounted for most solar phenomena reasonably well, and he evidently had little incentive to change it. From the ninth century onward, Muslim astronomers made several estimates closer to the truth, and medieval Europe added others, all hovering around the correct figure. It was unfortunate that Ptolemy's figure for the solar motion was inevitably bound up with his parameters for lunar and planetary motion, and for the precessional motion of the stars. Such an interlocking of parameters has always been a serious problem for astronomers, who have rarely had the privilege of starting with a clean slate, but have been forced to rely heavily on data from past history.

PTOLEMY AND THE SUN'S MOTION

Ptolemy added tables to allow for the rapid calculation of two angles that are needed to settle the Sun's position. The techniques he used were later extended by him to the more complicated motions of the planets, and will serve to give an idea of these theories of heavenly motion generally. Two parameters are needed initially for the Sun, and we shall add a third shortly. Taking the simple eccentric model (but recalling its equivalence with an epicyclic model), these parameters are (1) the mean motion of the Sun on the deferent circle, that is, around its center; and (2) the eccentricity *OT* as a fraction of *OS*, in fig. 60). The angle we ultimately wish to know is the angle *ATS*. Here *O* is the deferent center and *T* is the observer on the Earth, the latter being supposedly of insignificant size. The angle *ATS* is the mean motion (angle *AOS*) minus the angle *OST*. Now it should be obvious that the angle *AOS* (the mean motion) is easy to tabulate against the time, say in days, or hours, or both, for it increases at a constant rate. Using trigonometry, the angle *OST* can be easily enough expressed as a function of the mean motion and the eccentricity. (Ptolemy called the angle the *prosthaphairesis*, "an angle to be added or subtracted"; we should call it an *equation* or an *anomaly*. The idea is that it corrects a mean position and turns it into a true one, where "true" here means what we see.) Ptolemy therefore drew up a table allowing it to be found quickly from the mean motion, this having been found from the first table.

We should perhaps emphasize here that when past astronomers talked about "mean motion" they meant an *angle*, for instance, the angle moved per day, or per hour. They could also be referring to an angle accumulated over a long period of time, or to a final position, after such a movement. Of course we should still describe an angle covered in a given unit of time as a motion, but they did not view matters in the way we do, with our notion of instantaneous velocity.

One parameter remains, if Ptolemy is to enable us to fix the Sun's position. We need to know the date at which it passed some base point, such as the apogee or perigee; or alternatively, we might give its position at a particular date. Ptolemy chose as his standard epoch day one of year one of the Babylonian king Nabonassar, which was in fact 26 February 747 BC. There was much to be said for taking such an early date: it meant that he did not have to deal with years counted backward, negative years.

Had he been in possession of more accurate data, Ptolemy could have added another parameter, for the line of symmetry, *AB*, does in fact move. (This apse line, or line of apsides, joins apogee to perigee.) He was convinced that the seasons were of the same lengths in his own day as they had been in Hipparchus's time, and so concluded that the line of apsides was fixed.

One subtlety he did not miss was what we now call the *equation of time*. The Sun's daily motion across the sky has been used throughout most of history as a basis for the measurement of short time-intervals. This motion, however, is doubly irregular. There is the annual variation in the Sun's speed along the ecliptic, as explained in terms of the eccentric model; but the motion around the poles (the motion measured with reference to the equator) is variable for a different reason: the Sun is moving in a plane (the ecliptic plane) that is inclined to the equator at more than 23°. Ptolemy explained how to compensate for both of these factors. To this day, the best sundials carry a table to allow for the equation of time, and this correction term is in direct descent from Ptolemy's.

PTOLEMY'S THEORY OF THE MOON'S MOTION

Book IV of the *Almagest* contains a careful discussion of the lunar theory of Hipparchus, accepting a concentric deferent, with new parameters obtained from observation. In Book V, when he came to compare it with his own

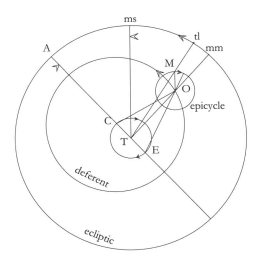

61. Ptolemy's lunar model. *T* is the Earth, *C* the (moving) center of the deferent circle, *M* is the Moon, and *E* an "equant" point around which the center of the epicycle (*O*) moves at constant angular speed. Note the unusual direction of the revolution of the Moon on its epicycle. The epicycles of the planets all have "direct" motion, that is, turn in the opposite sense. The mean ecliptic position of the Moon is *mm*, while its final true longitude is *tl*.

observations, Ptolemy found that it fitted well only when the Sun, Earth, and Moon, were in line (at conjunctions and oppositions, or syzygies, as they are collectively called). This is not surprising, in view of the fact that eclipses had always been the most important factor in settling the details of the simple model. At right angles to these points (at "quadratures") the error was several lunar diameters—not at all a satisfactory situation. Ptolemy had here found a new variation in the lunar motion, now known as *evection*, and its discovery was a great achievement, but his way of accounting for it was no less remarkable.

The details of his arguments are not something for a short account, but his final model may be briefly explained. As Hipparchus had done, Ptolemy supposed the Moon to move with a retrograde motion on an epicycle, but now he supposed the deferent's center *C* in fig. 61) to be eccentric to the Earth, and also to move around a small circle centered on the Earth *T*). He had to choose, now, velocities that would effectively pull the epicycle nearer to the Earth when it was at quadrature with the Sun. This he did by making the line to the mean Sun (*ms*) the bisector of the angle between *TO* and *TC*. Another refinement was that he reckoned the constantly increasing angle on the epicycle not from the line *TO* but from the line *EO*. This amounts to adding yet another (a third) inequality. It is a mark of Ptolemy's genius that he could add what amounted to new parameters to the model in such ways. Those who dwell on the Greek obsession with circular motions should note the ways in which Ptolemy found it possible to rise above the restrictions they imposed.

This model produced reasonably good results for the Moon's longitude, and certainly better than any before it. The ecliptic has been added to our figure, to show how the key longitudes change. There *mm* is the mean Moon, *A* is the moving apogee of the deferent, and *tl* is the final true longitude of the Moon. As the model stands, however, it has one clear blemish: there is an enormous variation in the distance of the Moon *M*) from the Earth, implying that its apparent diameter should vary by a factor of almost two during a single revolution. One did not need to be an astronomer to know that this was untrue, and that variations in the size of the Moon's disk are relatively insignificant. Ptolemy kept silent on this point. He had explained longitude well enough, and by placing the deferent and epicycle in a plane at 5° to the ecliptic plane he could explain the Moon's changes in latitude, as well.

It has often been said that he did not regard his model as describing the motion of real bodies in space, that it was no more than a device for calculating coordinates, and that he therefore would not have cared about the predicted variation in the Moon's disk. From his work *Planetary Hypotheses*, however, we know that he cared deeply about creating a planetary system in which all the epicyclic apparatus for all the heavenly bodies was contained without superfluous empty space. If he noticed his model's prediction of the Moon's variation—and he could hardly have failed to do so—it must have been a great disappointment to him.

Almagest Book v ends with a discussion of the distances of the Sun and Moon, and includes the earliest extant theoretical discussion of parallax, that is, of the correction it is necessary to apply to the Moon's apparent position to obtain its position relative to the Earth's center. (For the meaning of parallax and the findings of Hipparchus on this theme, refer back to p. 103 and figures 56 and 57. The Earth's radius is a significant fraction of the Moon's distance. Ptolemy's distance for the Sun in Earth-diameters was too small by a factor of about twenty.) From there he could pass on to give a geometrical account of eclipses, starting with the theoretically known motions of the Sun and Moon, and then deriving the circumstances of eclipses, rather than simply hoping to spot patterns in their recurrences. Ptolemy was fortunate in being able to make use of Babylonian observations of eclipses going back to the beginning of the reign of Nabonassar in 747 BC. He was not able to chart the geographical limits within which solar eclipses are visible, a difficult problem that was not really mastered until studied by Cassini in the mid-seventeenth century. Ptolemy's mathematical abilities were certainly equal to the task, but he had no access to a widely spread astronomical community, which might have provided him with a motive for developing this subject further.

PTOLEMY AND THE FIXED STARS

Before dealing with the planets, Ptolemy turned to the longitudes, latitudes, and magnitudes of the fixed stars. These he placed in six classes of *megethos*, a word better translated as "size" rather than "brightness" (the technical term "magnitude" seems to have changed from the former meaning to the latter by imperceptible steps only in the eighteenth century). His stars of magnitude 6 are the faintest visible, and today we still follow his classification, broadly speaking, although not his assumption about stellar sizes. His catalog of 1,022 stars in 48 constellations, and a handful of nebulae, provided the framework for almost all others of importance in the Islamic and Western worlds until the seventeenth century. It was based to a large extent on materials by Hipparchus that are no longer extant, and of course took into account his theory of precession, "the motion of the eighth sphere." Where Hipparchus had merely fixed one degree per century to be a lower limit, Ptolemy took this as an exact figure. He did not, as is often said, merely add precession to coordinates to update a similar catalog by Hipparchus, for his predecessor left his data in a very different form, with descriptions, alignments of stars, co-risings, and so on. Again, Ptolemy's was a remarkable feat, even though his stellar longitudes are on the low side.

The reason for this last, and very slight, blemish was the high degree of interrelatedness of what were superficially different parts of Ptolemy's book. He judged stellar longitudes in many cases with reference to the *Moon*, but an error in the *Sun's* motion—which we have just seen enters the lunar model—upset his measurements by small amounts. Most of those who required star positions in later centuries were content to add precession to his longitudes, and so bring his catalog up to date. The best astronomers incorporated their own observational measurements, but Ptolemy's thoroughness was for long unequaled.

PTOLEMY AND THE PLANETS

Books IX, X, and XI, of the *Almagest* account for the longitudes of the planets—inferior (Mercury and Venus) as well as superior (Mars, Jupiter, and Saturn). Two different arrangements of the epicycle in relation to the deferent are needed, as we saw in chapter 3, and since Mercury gives rise to difficulties of its own, further refinements are needed for that planet. Again, we shall give only the end results of Ptolemy's labors. Here he had much less reliable material from his predecessors than for the Sun and Moon. He had the concept of the epicycle, of course, and—through Hipparchus—some Babylonian period-relations of the type "in 59 years Saturn returns twice to the same longitude and 57 times to the same anomaly (the same stationary point in its retrogradation)." From such period relations he could construct tables of mean motions, although he needed to trim these later in the light of the models he developed from them.

It is perhaps worth adding here that Ptolemy gave two different accounts of how he found his very precise planetary mean motions. In addition to the one here explained, elsewhere he claimed that they were obtained directly, from observations widely separated in time. They could in principle have been found in this way, but it can be shown that in practice they were not. As for the adjustment of parameters obtained from his period relations, this was in some cases done on the basis of the observations he

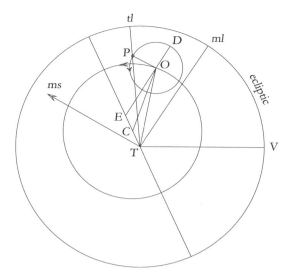

62. Ptolemy's model for
a superior planet.

records, but in the case of Mercury and Saturn, his quoted observations do not account for the mean motions he claimed to have derived from them.

We have already seen that the Sun enters into the epicyclic theories. (For our modern perspective on this matter, see p. 41 above.) Broadly speaking, for the inferior planets the mean Sun is the center of the epicycle, while for the superior planets the epicycle radius carrying the planet *OP* in fig. 62) is always parallel to the line from the Earth to the mean Sun (*ms*). It will be noted that in this figure, where *C* is the center of the deferent circle, an extra point *E* has been added to the line joining *T* to *C*, at the same distance but to the other side of *C*. This, the so-called *equant* point, was Ptolemy's device for introducing yet another anomaly. It had always previously been assumed that the epicycle moved uniformly around the center of the deferent. (It is conceivable that Apollonius thought otherwise, but this is a moot point.) Having tried to derive the size of the epicycle, Ptolemy found that it seemed to vary in a way that did not fit well with the simple assumption of an eccentric deferent circle. He adjusted its angular speed, therefore, by the device of making it constant not around *C* but around *E*. (In figure 62, the line *EO* is parallel to the line from *T* to *ml*, the mean longitude.)

This introduction of the notion of an equant was all the more commendable because it meant breaking with the traditional dogma that all must be explained in terms of uniform circular motions. Ptolemy introduced an equant circle (not shown on fig. 62) on which a point moved round at constant speed, as a prolongation of the line *EO*. That should have saved him from criticism, but it did not, and fourteen centuries later we find that even Copernicus found the equant distasteful. Taste is of course something that can be taught.

When it came to Venus and Mercury, the roles of epicycle and deferent are interchanged, for reasons we have already seen. Venus has a large

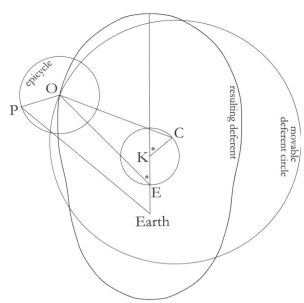

63. In Ptolemy's rather complicated model for Mercury, the center (*C*) of the deferent circle moves in such a way that the angles at *E* and *K* marked with black spots are equal. The center of the epicycle, following this rule, traces out an oval, and had he wished he could have posited this as a unique and static deferent curve. Some medieval instrument designers looked at the matter in this way, but reverence for circles as the elements of an acceptable theory generally worked against the idea.

epicycle, but otherwise a relatively simple motion. The model for Mercury, however, shows Ptolemy at his most ingenious. It embodies all the ideas we have encountered thus far. The equant center, for instance, is *E* in figure 63, and there is an epicycle moving around a deferent circle, but now the center *C* of the deferent circle is made to move. We have come across a similar device with the model for the Moon's longitude, but here the small circle on which *C* moves is centered not at *T* but at a point *K*, beyond *E*, so that *KE* equals *TE*. The position of *C* for a particular time is fixed by making the two angles marked with small circles equal. In other words, they move round at a constant rate in opposite senses. Ptolemy arrived at this complex model on the basis of faulty observations that led him to suppose that Mercury had two perigees, neither of them opposite the apogee, but at points about 120° from where perigee would normally have been expected. Whatever the merits of his observations, he here effectively provided planetary astronomy with its first oval. For every position of *C*, there is a single corresponding position of *O*, and the path followed by *O* is essentially the resulting deferent curve on which the epicycle moves. Its shape is shown by the heavy line in the figure (which is not to scale). Some thirteenth-century astronomers called it a "pinecone": it is an oval pulled in at the waist, and for small eccentricities is not far removed from an ellipse.

Ptolemy was concerned to account for the changing planetary motions that were in fact observed, but he also wanted to make it easy to calculate them for any moment of time, past, present, or future. For this, he devised a series of rules that could be applied in a routine way, even by people lacking in great expertise. Running through all the models here described, we have a situation where for every "mean motion"—that is, an angle increasing at a steady rate—there is another angle, slightly different from it, that has to be used when we come to combine the component angles in the final true longitude. We first met the small differences, the so-called equations, in the case of the solar model. To simplify practical computation, Ptolemy drew up tables of mean motions, supplementing them with other special tables of equations. Some were simple functions of the mean motions, but others were more complicated, calling on intermediate terms in the calculation. In the end, however, the astronomer was asked only to add or subtract angles, to get the final result. Even so, the calculation of a full set of planetary positions for any one particular time would have occupied an hour or two of a competent astronomer's time, and much more if planetary latitudes were needed.

In *Almagest* Book XIII, latitudes were introduced into Ptolemy's scheme, rather as they had been for the Moon, so introducing a third dimension to what otherwise was a two-dimensional treatment. He made the plane of the planet's deferent inclined to the plane of the ecliptic. In the case of the superior planets the inclination was fixed, but for the inferior planets he made it oscillate, following rules that must have cost him much trouble in formulating. He then required the epicycles to lie in still different planes, and yet again, for the inferior planets, he devised rules for making their inclinations oscillate, now with reference to the plane of the deferent.

We can easily appreciate why the latitude problem was intrinsically so difficult for Ptolemy and all who adopted an *Earth*-centered system. It was that the physical planes of the planetary orbits pass not through the Earth but through the *Sun* (because the gravitational forces acting on the planets are toward the Sun). It would have been some compensation for this invisible theoretical obstacle had he made the planes of the epicycles parallel to the ecliptic plane. To his very great credit, it is clear that he had made precisely this simplification by the time he came to compose a later work, the *Handy Tables*, where for Mercury and Venus fixed the inclinations of the epicycles and improved the accuracy of the values he gave them. Unfortunately he also fixed the inclinations of the epicycles for the superior planets, leading to certain unavoidable errors, although this mistake he later corrected in his *Planetary Hypotheses*. (Muslim, and later, Western, astronomers followed the rules he had given in *Almagest*, here as in most other respects, so that his changes of heart had little historical impact.) In the *Handy Tables* we have only the procedures we are to follow to apply the models, and no justification for the models themselves, so we cannot say how he made his discovery. Here, as in so many other ways, however, we have evidence for Ptolemy's high genius in the selection and analysis of astronomical observations for theoretical purposes. Astronomy has many other sides to it, but in this supremely important respect, Ptolemy simply had no equal until Johannes Kepler came to analyze the observations of Tycho Brahe.

It is impossible in a short space to show how the parameters of a particular model can be derived from actual observations, but some very brief general remarks are in order. First, it is important to realize how vital such procedures are, for any well-founded empirical science, and yet how rare they are in surviving documents from an earlier period. Ptolemy makes much use of observational data, throughout his *Almagest*, but he did not start with a blank sheet, so to speak, for many of his data were inherited from his predecessors, and it is not always easy to decide on those which he himself supplied. There are some instances where what he presents as his own observational data were not strictly that, but were adopted to fit with a preconceived end result. Matters are also complicated by virtue of the fact that he occasionally had more data than were strictly needed. Setting aside all such considerations, we can say that the methods he taught were of enduring value. Having built up a picture of the sort of model he would require for a particular planet, he had to separate out the angular motions—in the deferent and epicycle, for example—and to establish the relative scale of the circles. Using the assumption that motions are on circles and are uniform, so that angles from the center are proportional to times, finding the parameters of the model involves at the very least solving the following geometrical problem: given three points on a circle, find a point—inside or outside the circle—from which lines drawn to the three points will form given angles (these, in the astronomical case, will be observed angles). Apollonius is thought to have solved this general geometrical problem, and this not merely in an empirical fashion. Hipparchus certainly applied it to the cases of the Sun and Moon. Later astronomers realized the advantages of making observations at specially chosen times, to give a simpler solution. For instance, if one observes the Sun at the equinoxes and solstices, the angles are multiples of a right angle. In the case of the Moon, following Hipparchus's lead, but correcting some errors of calculation, Ptolemy made use of trios of lunar eclipses to pin down the lunar parameters for the case when the Earth lies on the line joining the Sun and Moon. Ptolemy made a masterful selection of many other similarly special cases, and it was a minor—but understandable—tragedy that later astronomers all too often paid scant attention to his methods, accepting his numerical answers without question.

PTOLEMY'S INFLUENCE: ASTROLOGY

So large does the reputation of Ptolemy loom in our view of late antiquity that it is easy to forget that astronomy continued to be practiced on a much lower intellectual plane. Of this, by the nature of things, we know relatively little. There is, for example, a papyrus (located in the collections at the University Library, Heidelberg [*P. Heid. Inv.* 4144] and University of Michigan [*P. Mich.* 141]) dating from the century after Ptolemy, that gives evidence for the use of a crude scheme for finding the position of Mars. This scheme evidently made use of an epicyclic model, but one that was blended with the so-called zones of what we now call the Babylonian System A. There are other indications that Babylonian schemes were known in Roman Egypt. Hellenistic astrology was flourishing, and methods that were easier to apply than Ptolemy's, however inaccurate, were called for. In something approximating to the Greco-Babylonian techniques, astrologers seem to have found what they needed, and there are fragments of texts relating to the Sun and planets and—especially numerous—to the Moon. One innovation in these crude schemes is the treatment accorded there to lunar latitude.

This strange blend of arithmetical and geometrical methods makes one thing clear: it is a mistake to imagine that between the time of Hipparchus and Ptolemy there must have been a steady advance in theoretical astronomy. The methods employed even by Hipparchus were a patchwork of elements, geometrical and arithmetical techniques—and under that last heading we should distinguish the "zigzag" techniques and the methods of cycles of time after which phenomena repeat. One can say with some confidence that Ptolemy was *uniquely* responsible for establishing a tradition of building up astronomy from a coherent set of first principles. With the help of his predecessors' ideas, he was able to conjecture as to how the heavenly bodies moved in space. Having found the parameters of the models by fitting them to observation, he could then predict the phenomena that would be seen, as the consequences of his geometrical assumptions. In short, where others had found patterns of repetition, Ptolemy gave reasons for those patterns—and here he undoubtedly benefited from his familiarity with the methodological writings of Aristotle, not to mention the Greek geometrical tradition. With Ptolemy, astronomy had come of age.

The lesser astronomers practiced their trade, none the less, and those of a more academic disposition began to write commentaries on the *Almagest* and *Handy Tables* of Ptolemy—beginning with Pappus and Theon of Alexandria in the fourth century. There were other commentaries written at about the same time, for example one by Theon's daughter Hypatia, but virtually nothing is known of their contents. The *Almagest* was first translated into Arabic around 800, but improved versions followed quickly. It reached western Europe in two Latin translations, one done from the Greek around 1160, the other—far better known—done from the Arabic by Gerard of Cremona in 1175.

There were two classes of scholar in particular whose needs it failed to meet, the astrologers and the natural philosophers—or cosmologists, as we might now call them. For astrology, Ptolemy wrote what again became a standard text, the *Tetrabiblos* ("a work in four parts"). His *Planetary*

Hypotheses went far toward providing a more sophisticated version of Aristotelian cosmology. It was based on the assumption that there are no empty spaces in the universe, but that neither can there be overlapping of matter with matter, so that the outermost point reached by a planet in its epicycle must be equal to the minimum distance reached by the planet next above it. This assumption allowed Ptolemy to turn his separate planetary models into a universal system. A moment's thought will show that, since the relative sizes of the circles in any planet's geometrical model is laid down by Ptolemaic astronomy, and since the scale of the circles of one planet now fixes the scale of the circles of the planet next above it, the entire scale of the universe is fixed (up to Saturn), in terms of the lowest sphere, the innermost limit of the Moon's possible motion. Since Ptolemy could quote a distance for the Moon, it became possible to write down distances for all the planets. The answers are plausibly large—they are in millions of miles—but of course they do not correspond to reality. The scheme was seized upon by Islamic writers, and through an indifferent but much-copied summary of the *Almagest* written by al-Farghānī (flourished 850), it became a standard part of the curriculum in European universities in the Middle Ages. It helped to inspire some of the details in Dante's *Divine Comedy*, for example.

The *Tetrabiblos* likewise entered the European consciousness through Islam, but in this case it acquired much extra astrological baggage on voyage. Although its subject matter is certainly not to modern scientific taste, it is nevertheless a masterly book, and in many respects a scientific one. As we have seen, astrology had—among others—Babylonian roots, and we can even trace specific points of astrological contact between the Hellenistic world and Babylon. Most famous is the migration of Berosos, a priest of Bel, from Babylon to Ionia, where he founded an astrological school on the island of Kos around 280 BC. Greek scholars often declared themselves disciples of the Chaldeans and boasted of having received instruction in their schools. Even when they took over what they thought were Egyptian ideas, they were often taking over Babylonian material secondhand. It was possibly at Alexandria around 150 BC that treatises were written purporting to be from the pen of the king—mythical, as we know—Nekauba (or Nechepso, as he is also known) and his priest Petosiris. These books acquired great authority in the Roman world, as did other writings attributed to the god Thoth, the "thrice great Hermes" of the Greeks, Hermes Trismegistus. These works are to *Tetrabiblos* as a crystal ball is to a professional economist: neither is wholly reliable, both may be wrongly motivated, and there is a world of difference between their techniques, but both fill a niche in society.

Where Babylonian and Assyrian divination had mostly concerned public welfare and the life of the ruler, the Greeks applied the art in large measure to the life of the individual. The activity was unintentionally encouraged by the teachings of Plato and Aristotle on the divinity of the stars, and in late antiquity many astrologers regarded themselves as interpreting the movements of the gods. With the rise of Christianity, this attitude was of course repressed, although it flourished as a literary device throughout Roman antiquity, and has been a characteristic of Christian Europe almost until the present day. Ptolemy's *Tetrabiblos* was thus a handbook for people of many different persuasions.

It opens with a defense of astrology, and is ostensibly written around the idea that the influences of the heavenly bodies are entirely physical. In the end, however, it amounts to a codification of unjustified superstition, largely inherited from Ptolemy's predecessors. Book II deals with cosmic influences on geography and the weather, the latter a popular and spiritually safe subject in later centuries. Books III and IV deal with influences on human life, as deduced from the state of the heavens, but oddly lacking in any of the mathematics of casting the houses that so obsessed astrologers in later centuries. (Something more on this theme will be found in chapter 10.)

In late Roman antiquity, the so-called Chaldaei and mathematici—words we can interpret simply as "astrologers"—were very numerous, judging by the frequent criticism leveled at them by Roman magistrates and satirists. Astrology, in its various guises, had reached Rome chiefly through the agency of Hellenized Greeks. One influential scholar in particular deserves mention here, if only to show how mobile scholarship could be. Posidonius (ca. 135 BC to ca. 51 BC) was born in Apamea, in present-day Syria. He was educated in Athens but finally settled in the free city of Rhodes. On one occasion he was sent on an embassy to Rome, but later seems to have been led by his curiosity to tour Spain, Africa, Italy, southern Gaul (France), Liguria, and Sicily. A school he later set up in Rhodes became a center for Stoic philosophy, a magnet for Roman intellectuals, including the powerful political figures of Cicero and Pompey. Five works he wrote on astrology became, in this way, transmitted to Rome. Astrology arrived there, however, in another and shadier form, through its association with Eastern mystery religions, notably Mithraism and the cult of Isis. Both were helped on their way by their popularity in the Roman armies. The cult of Mithras brought with it worship of the Sun, and several Roman rulers liked to present themselves as personifications of the Sun. This atmosphere, with which of course early Christianity had to contend, was at times highly favorable to professional astrologers, but there were many obvious reasons for the unpopularity of astrologers. A number of expulsions from Rome and Italy are known from before the first century, and there were edicts in force against them in the fourth century, when the Christian emperors added their own religious scruples to old political objections. In AD 357, Constantius II made divination a capital offense, and the ban was repeated in 373 and 409.

The ancient tradition of astrological divination had a marked influence on the practice of medicine. Latin literary style was also much influenced from an astrological quarter, for instance by the first-century poem known as the *Astronomica* of the Stoic philosopher Marcus Manilius. The Stoics were a philosophical sect with a long history, beginning around 300 BC, and one of its chief doctrines was that the philosopher's aim should be to live in harmony with Nature through the use of reason. As time went on, the sect became increasingly concerned with ethical questions, and it is not surprising that the Babylonian idea of a stellar necessity ruling the world found a sympathetic audience among Stoic philosophers generally. Manilius put about the idea that human life is absolutely determined by the stars, but he did so in the course of a work that was no doubt valued more for its astrological technicalities than for the philosophical ideas underpinning it. The philosophers helped to give respectability to ideas that were broadly astrological, however. Around AD 265, Plotinus—the founder of Neoplatonism—proposed a related doctrine that magic, prayers, and astrology, are all possible because each part of the universe affects the rest through a kind of mutual sympathy. Such ideas gave much comfort to later generations of scholars anxious to play with fire.

A literary work that helped to counter such influences was the *City of God* by St. Augustine (354–430), in which he warned that astrologers could enslave the free human will, by claiming to predict a person's life from the stars. If predictions come true, he said, this is due to chance or to demons. He had in fact once been a believer in both astrology and sacrifice to demons, and his testimony was well argued and compelling to many medieval churchmen. However, they continued to believe, as did he, in God's foreknowledge and in celestial influence, and so were presented with a dilemma. How could man be free, if all was preordained, either by God—who can only know what is to come if indeed it is to come—or by the influence of utterly predictable planetary motions? The usual way out was to say that the stars force us in a certain direction, but do not compel us to act against our free will. They "incline but do not compel." Prayer would help people to resist. Other Church fathers touched on these questions. Origen, for instance, tried desperately to purge astrology of fatalism.

These facts have an obvious relevance to the practice of astronomy. Regardless of any real astrological association, this was regarded with deep suspicion by those classes of people—that is to say almost everybody— who did not understand its potential independence of astrology. There are some famous astrological names from the Roman world: Vettius Valens from the second century, Palchus, Eutocius, and Rhetorius from the fifth, and no doubt many materials have disappeared without trace from the intervening period, but after them we come to a period in which astrological practice was firmly suppressed in the West until something of a revival came in the eighth century. And from late antiquity onward, such Western astrology as we find being practiced tends to be thoroughly derivative.

BYZANTINE ASTRONOMY

Astronomy continued to be practiced in Byzantium, the Eastern Roman empire. (This takes its name from the refounding of the old city of Byzantium as "New Rome," by Emperor Constantine in 330. From him came the city's name "Constantinopolis," which we know as "Constantinople," and is today named Istanbul) Indeed, Ptolemaic methods were adopted there and continued in use, even alongside older arithmetical methods of Babylonian type, up to and beyond the fall of Constantinople to the Turks in 1453. The new Byzantium produced a number of scholars whose work filtered out into Islam and the Christian West. In its early history, the most famous of them was perhaps Proclus (ca. 410–485), although he owed most for his education to Alexandria and Athens. Proclus was the last of the influential Neoplatonist philosophers of the ancient world, and had a sound understanding of Euclidean geometry and Ptolemaic astronomy. He was familiar with Ptolemy's instrumentation, something we do not normally associate with Neoplatonism, and he was familiar with Ptolemy's quasi-Aristotelian cosmology of nested spheres, a theory with a strong philosophical element. Despite his knowledge, Proclus was nevertheless critical of the arbitrary character—as he mistakenly judged it—of Ptolemy's hypotheses, and he preached against the way in which some astronomers spoke of their models as picturing real entities. As for astrology, he was an adept, not above making a paraphrase of *Tetrabiblos*, where there was surely much more room for scepticism.

Constantinople had seen the foundation of a sizable school in AD 425, but not until two centuries later do we hear much of serious study in astronomy or astrology. It was in the early seventh century that the Emperor Heraclius brought in an astronomer from Alexandria, a certain Stephanus, and the emperor himself apparently composed a guide to Ptolemy's *Handy Tables*. This imperial interest does not appear to have paid high dividends. There was at best a fitful study of astronomy in Constantinople and a few outlying places in the empire over the next two centuries; and from the ninth century onward there was considerable activity in the copying and study of old Alexandrian materials, and more modern works composed in the Islamic world, for instance, in Baghdad and Damascus. Before the catastrophic sack of Constantinople in 1204 by the Venetians at the time of the fourth crusade, astronomical texts seem to have been much copied, but there is surprisingly little evidence that they were well comprehended. Astrology was more obviously successful, finding its way into Byzantine literature, albeit at a low technical level. That there were competent astronomers who have now sunk without trace, however, is suggested by occasional fragments of evidence, one example of which is a fine astrolabe, dated July 1062. Lettered in Greek, it is signed by one Sergius, a "Persian." A century later, we find a treatise on the same instrument, announcing its "Saracen" origins. That Constantinople had a minor role in the much earlier history of this type of instrument had no doubt been long forgotten.

THE ASTROLABE

Synesius of Cyrene (d. ca. 412–415) was from a Greek colony in what is now Libya, and had gone from there to Alexandria, where he was a pupil of Hypatia. Synesius is now better known for the part he played in the spread of Christianity than for his astronomy, but he presents us with some useful evidence on the history of instrumentation. Having married a Christian woman, he was with great reluctance persuaded to accept the bishopric of Ptolemais in 410, and then baptism—in that unusual sequence. He spent three years on an embassy to Constantinople, and while his fame rests more on his oratory and poetry, he appears to have found time to make improvements to some kind of astrolabe. To a friend in Constantinople he presented a silver astrolabe, described only as "instrument," together with a letter referring to a full description of it (this portion is no longer extant). While the idea behind the instrument was centuries old when he was writing, from a rather convoluted remark he makes in his letter it seems that he knew of evidence that the idea went back to Hipparchus.

Whatever the truth of the matter, there is a cluster of historical material which requires no great degree of speculation, centering on the widely disseminated works of the architect Vitruvius. In his *Ten Books of Architecture*, written near the beginning of the Christian era, he described a water clock which was capable of showing the seasonal hours of day or night, and which had a sort of astrolabe as its dial. It is significant that when Synesius referred to "sixteen stars on the instrument by Hipparchus" he added that "they sufficed for the night clock." The Vitruvian mechanism as a whole is called an "anaphoric clock," from the Greek word for "bearing back" or "repeating," the repetition in this case being one of risings. Fragments of later Roman anaphoric clocks have been found in France, and there are other reasons for supposing that the instrument was not altogether rare in the ancient world, although early literature is surprisingly scarce. The most precious evidence for it, however, is the Tower of the Winds in the old marketplace in Athens, to which Vitruvius actually refers (fig. 64). It was constructed according to designs by the Macedonian astronomer Andronicus of Cyrrhus, around 50 BC. On each of the eight sides of the building, which was topped by a wind-vane, there is a relief sculpture of a god of the appropriate wind—Boreas for the north wind, and so on. Beneath them, there is a series of plane sundials, as mentioned earlier. To one side of the building, however, there is a water tank, the significance of which was only appreciated when, in the 1960s, a study of water channels inside the tower showed convincingly that there had been an anaphoric clock there, of the type described by Vitruvius.

While there are several different instruments to which the name "astrolabe" has at various times been attached, and innumerable stylistic variations among most of them, there is one fundamental type of plane astrolabe that far outnumbers all others. To distinguish it from Ptolemy's

64. The Tower of the Winds, erected in the old Agora in Athens early in the first century BC by the Macedonian astronomer Andronikos of Kyrrhos. The building's eight sides face the points of the compass and are decorated with a frieze of figures in relief, representing the winds. Below them, on the sides facing the Sun, are the remains of plane sundials. The tower was surmounted by a weathervane in the form of a bronze Triton. Inside the building there was a water clock, which was built with a dial of astrolabe type, according to the techniques devised earlier by Ctesibios and Philo. Water was piped from a spring on the side of the Acropolis to the reservoir for the clock, which is all that survives of it (illustrated below).

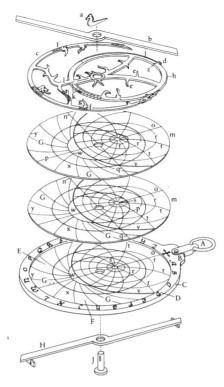

65. An exploded view of a typical medieval astrolabe. The items shown are all held together by the slotted pin (shown at the bottom), through which the wedge (horse) is pushed. The uppermost perforated disk, the rete, is essentially a rotatable star map, conventionally (not necessarily) bounded by the tropic of Capricorn. Its inner (eccentric) ring is for the ecliptic. Each plate over which it may rotate carries a horizon line and other coordinate and hour lines. Only one plate is needed for a given geographical latitude, but an astrolabe might have several plates made for it, each catering for a different latitude, or for two latitudes when engraved on both sides. The plates and rete sit in the hollow of the mater, itself usually engraved in the same way as the plates, to save metal. The rule at the top is not essential for all uses of the astrolabe, but it helps with many. The sighting rule (alidade) with pierced vanes at each end allows angular measurements to be made when the instrument is used for observation. (It had many uses for calculation, independently of observation.) The ring and shackle on the rim of the mater allow the instrument to hang vertically from the observer's thumb. The altitudes of celestial objects could then be measured by means of the alidade and a scale on the back (not here visible). There were other scales on the back, typically a calendar scale for correlating the day of the year with the ecliptic position of the Sun.

large spherical astrolabe, in an early treatise—now lost—by Theon of Alexandria, this portable instrument was simply called "the little astrolabe." As mentioned earlier, it comprises one or more circular disks overlaid with a pierced circular disk of the same material, usually of brass. The pierced disk is known as the "rete"—the Latin word *rete* simply denotes a net. Essentially a star map, it turns around a pin at the common center representing the pole (see figs. 53 and 65). Schematically, we may represent the rete by figure 66, on which the crosses mark stars and the graduated circle represents the ecliptic. The rete, a fretwork of metal or other material,

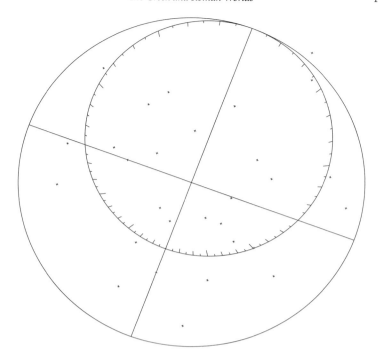

66. The chief elements on the rete of an astrolabe. The graduated circle is the ecliptic, usually the most conspicuous ring. Note that its degree divisions are not uniform. The crosses are the places of bright or familiar fixed stars, usually about twenty in all. In brass, they are usually represented by the tips of pointers.

is pierced, so as to allow the various reference circles engraved on the disk beneath it to be seen. (Had transparent materials been available, it would not have been necessary to pierce it.) These circles, including the meridian, the horizon, lines at altitude 5°, 10°, and so on above the horizon, and others, are shown schematically in the next three figures. It is with reference to those lines that the positions of the heavenly bodies are judged.

In size, portable versions of the instrument are typically between ten and twenty centimeters across, although smaller and larger examples survive in plenty. For an impression of a richly decorated Persian astrolabe of a relatively late period, included here only to give a general idea of the layout of the instrument, see figure 68. The portable astrolabe is meant to be used for observation as well as for computation. For the former purpose, a ring and shackle are provided, so that the instrument can hang vertically from the thumb of one hand while observations of objects in the heavens are made with the help of the sighting vanes carried on the centrally pivoted rule. This, the *alidade*, is on the back of the instrument. It is distinguished from a second pivoted rule, without sights, that is often to be found incorporated on the front, to assist in taking readings from the rete. It should be appreciated that it is not necessary that the pierced plate represent the star sphere and the plate represent the horizon, meridian, and so forth. While this became the almost universal preference, the roles of the rete and plate

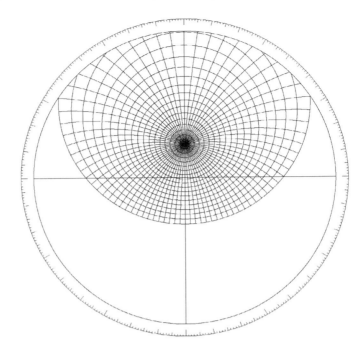

67. The principal lines on an ordinary astrolabe plate, in this case drawn for the latitude of
Oxford (51.8°). The lines of constant altitude and azimuth (known as "almucantars" and
"azimuths," respectively) are bounded below by the horizon circle. Each set is here drawn
at 5-degree intervals, but 10-degree intervals are quite common, and intervals of one degree
are not unknown on the finest instruments. Compare with figure 53. The outer scale of degrees
is usually inscribed on the mater, and so would have been outside the plate proper. (It is often
replaced by a scale of hours, twenty-four to the circle.) For the hour lines that are conventionally
drawn on the plate, but not shown here, see figure 71.

could be perfectly well reversed, as they usually were on anaphoric clocks,
and on the dials of many medieval mechanical clocks.

So much for the overall appearance of the instrument, but what of
its rationale? The plane astrolabe makes use of certain properties of
stereographic projection likely to have been appreciated by Hipparchus
himself. One may get a rough idea of how it functions, however, without
understanding the intricacies of its geometry. We begin by imagining ourselves at the center of the hemisphere of the sky, which—if we live north
of the equator—seems to turn around the north pole with a daily motion.
Turning with the stars we may imagine the ecliptic, and in fact anything
else associated with the star sphere, and that moves with the daily rotation
around the poles. Separate from this, we may imagine that we are at the center of a hemispherical fixed grid, with circular arcs representing our meridian and our horizon, and above the latter a parallel circle, say five degrees in
altitude above it, and then another at altitude ten degrees, and so on up to the
zenith. Astronomers frequently built a three-dimensional model of this double system, or parts of it. The circles of the two systems were usually made out

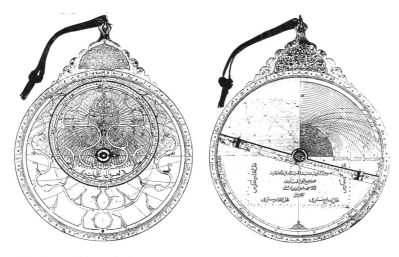

68. A Persian astrolabe made for the Safavid Shāh Abbas II, showing (right) the front, with the movable rete and the fixed plate beneath it, and (left) the back, with alidade (observing rule) and various trigonometrical scales. It is signed by Muḥammad Muqim al-Yazdi, and dated the equivalent of AD 1647–AD 1648. For its inscriptions, see p. 213. (Museum of the History of Science, Oxford. Inv. 45747.)

of simple rings or bands—hence the name "armillary sphere," from the Latin for "ring," *armilla*. Such a piece of apparatus was even used for observation, since it lends itself to easy visualization of its correspondence with reality. Its meridian ring is first set in the local meridian, its horizon ring is set horizontal, and its pole is made to point to the pole of the local sky. A correctly placed point of the model—such as a star marker—being then set in the direction of the object in the heavens which it truly represents, the entire system, with all other markers for celestial objects—stars, circles, and so forth—will be correctly set for that same moment of time. Since time is reckoned by relating the Sun to the meridian, if we can identify the Sun's place on the model—bearing in mind that it changes from day to day—we shall have an entry to the general problem of timekeeping. An armillary may be used to obtain the time in other ways, and in expert hands it has many other astronomical uses, but it is really little more that a three-dimensional diagram of the heavens that is being put into correspondence with the heavens. Such an object is often wrongly described as Ptolemaic or pre-Copernican, but there is nothing illogical about the idea of a celestial sphere centered on the observer, as long as it is regarded as a representation of the angles between directions to celestial objects, rather than of distances in any other sense.

Models of the celestial sphere could be used for purposes of instruction but also, as hinted above, for observation, as was the case with Ptolemy's armillary instrument. (See fig. 69, and for an interpretation of Ptolemy's armillary sphere by the Renaissance scholar Regiomontanus, see fig. 70.) Armillaries were occasionally made with great sophistication, by adding extra movements and rings. Ptolemy, in his *Almagest*, even explains how

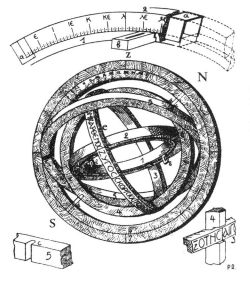

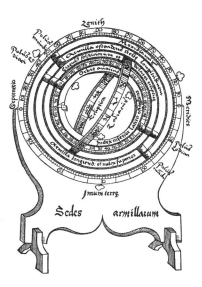

69. The armillary sphere described by Ptolemy in his *Almagest*. The labeled ring *e* is the ecliptic ring. The ring inside it, and at right angles to it, carries the adjustable sights (see also the upper figure). *N* and *S* are the north and south celestial poles (note the pivots on which the inner rings all turn with the daily motion), and *z* is the zenith. (After a drawing by the engineer Pierre Rome for a work by his brother Adolphe Rome, 1927.)

70. An interpretation of Ptolemy's "armillary astrolabe" by Regiomontanus (1436–1476). From his posthumous collected works (1544).

to make a celestial sphere that exhibits precession, and a few designers of armillaries later followed his lead. The armillary sphere was much used by teachers of astronomy throughout the Middle Ages and after, and as a result was often used as a symbol for astronomy in general. Familiar as it must have been to many students of the subject, it was not easy to make, and accurate examples must have been rare.

The plane astrolabe provided one golden road out of this difficulty, by turning the three-dimensional instrument into a two-dimensional equivalent. While our imagined armillary sphere may be mapped on to a plane in many different ways, the simplest arrangements will be those that project the north (or south) pole into the center of the map, so that the moving part of the map may pivot round it. This is precisely what is done in the case of the plane astrolabe. The plane on which the circles are projected is *any* plane parallel to the equator. The projected image of the ecliptic, which is inclined to the equator and does not share its pole, will not be centered on the pivot representing the pole of the equator, but will still be a circle—and not an ellipse, for example. The same is true for the horizon and parallels of altitude—called almucantars, from an Arabic word. One way of imagining the projection would be to suppose an intensely bright point-source of light to be at one of the poles of the armillary, casting shadows of the two sets of rings, fixed and moving, on any screen parallel to the celestial equator. Another way would be to follow an Arab writer, who asked us to imagine that a camel had trodden on an armillary sphere, squashing it flat between its poles, but so as to allow the two parts still to move relatively to one another.

The astrolabe has numerous uses—a typical text will explain over forty—but many of them require us to position the moving rete correctly in relation to the fixed plate by reference to the observed altitude of a star. In this case, the rete is rotated until the star's marker falls against the correct altitude line (almucantar) on the plate below. The Sun is effectively a star moving along the ecliptic on the rete. To know its position, one needs its longitude. This may be found from a calendar, listing the solar longitude for each day of the year. A simpler way is to consult a so-called calendar scale, which is to be found on the back of most astrolabes, correlating longitude with day—but not so accurately. Since the Sun's position in relation to the meridian is a measure of the time of day, the astrolabe was always useful as a time-measuring device, by day or night. By night the rete is positioned with the help of a star, the Sun's place on the rete being known as explained. By day, the rete can be positioned from an observation of the Sun, yet again known in longitude as before. For an idea of how the hour (equal or unequal) can be judged, see the brief explanation with figure 71.

The oldest surviving treatise with a systematic account of the theory of stereographic projection was Ptolemy's *Planisphere*. The Greek original of this is lost, but the work survives in Arabic translation, and a revision of that was made by an Islamic Spanish astronomer in the tenth century, finally reaching western Europe in a Latin version in 1143. If Synesius was right in claiming to have been the first after Ptolemy to have written on the geometrical theory of astrolabe projection, then a more comprehensive work that has been ascribed to Theon, father of Hypatia, "revered teacher" of Synesius, can only have been written a few years later than his. Only the table of contents of the latter survives, but it fits closely with a later work by Philoponos (d. ca. 555) and even more closely with a Syriac treatise on the same theme by the bishop Severus Sebokht (d. 665). (It has to be said that some recent scholars have insisted that the Theon treatise is a phantom.) Ptolemy's was most certainly not the sort of treatise that could ever have made the astrolabe popular, but in the later Islamic and Christian worlds, many

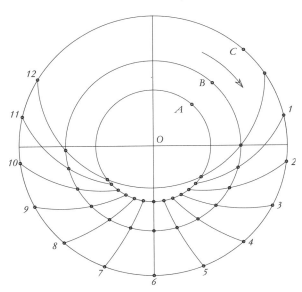

71. One of the principal uses of an astrolabe was for time telling. On any particular date, the Sun has a certain place in the ecliptic, that is, on the rete. As the rete is turned with the daily motion, that point sweeps out 15 degrees every (equal) hour. Since it crosses the meridian at noon, if the rete has been correctly set for the present moment (no matter how that is achieved), the angle separating the Sun from the meridian is easily found, and so, too, the present time. Those who wanted the time in "unequal" (or "seasonal") hours needed another technique. The horizon line on the plate is crossed by the Sun at sunset (on the right) and at again at sunrise (on the left). The angle traversed in the interval, divided into twelve equal parts, is what is meant by an unequal hour by night. The lines on an astrolabe plate cover the whole year. The Sun is nearest the pole at midsummer (and describes the circle through *A*, tropic of Cancer) and is furthest at midwinter (circle through *C*, tropic of Capricorn). Astrolabes are usually bounded by the Capricorn circle. There is no reason why they should not go further, although they cannot cover the whole sky in a finite space.

scores of alternative texts were eventually produced. Medieval European writings, although numerous, fall into only three main families, all radiating from Muslim Spain. In no center was there any other scientific instrument of comparable importance, artistically or symbolically. Admittedly, as an instrument for precise observation the astrolabe was of no great value, while for computation it was usually too small to give more than an approximate answer to complex problems. As a teaching device, however, and for clarifying problems in positional astronomy, it has had few equals.

The identity of the oldest surviving portable astrolabe is a matter of dispute, but from eastern Islam we have an early copy of a ninth-century example, and a dated original from about AD 928. There is a fine Persian example dated at the year 374 of the Hijra (AD 984) that bears witness to a long craft tradition, confirming what we know from eighth-century literary references from Baghdad and Damascus. Within a few centuries we find examples of texts and instruments stemming from every important center of civilization between India and the Atlantic. Persian craftsmanship remained of consummate quality, and was widely rivaled in Europe only in the sixteenth and seventeenth centuries. Apart from much superficial variation in style and artistry, however, the principles on which most astrolabes were based remained the same for more than a millennium.

One consequence of the most common astrolabe design—that described here—was that the horizon plate was of use for only a single geographical latitude. For that reason, most astrolabes had a whole collection of different plates stored in the body of the astrolabe, the mater (or mother). The user who traveled far afield or who wished to perform a calculation for a remote latitude, simply chose the most appropriate plate. This was not a very satisfactory state of affairs, and new types of universal instruments were developed, the most innovative being those from Muslim Spain. Such astrolabes, designed for use at any location, often comprised only a single plate. Despite their strong economic advantage, such "universal astrolabes" were too difficult for most people to understand, so that far fewer were made.

At first, most astrolabes were made by astronomers for their own use, but as they gradually became more familiar in educated circles, many were made for patrons and men of substance, richly ornamented and exquisitely engraved. The astrolabe was beyond the reach of most ordinary people, although a poor scholar could make a passable instrument out of parchment or wood, good enough for computation if not for the observation of altitudes. In the later Middle Ages, the general appearance of an astrolabe became known even outside educated circles, since it was to be seen as the front of many an astronomical mechanical clock. Indeed, true to its Vitruvian ancestry, it was the prototype clock face. Whether or not the mechanical clock originated in England in the late thirteenth century, as seems likely, its invention had much to do with a desire to represent the moving heavens in a material form, and for this purpose the astrolabe, with its 24-hour dial, was ideal. The hour pointer was then the line to the point on the rete representing the Sun: it was moved round the rete mechanically. In due course, all but that hour hand disappeared from common clocks; and when its original rationale had been more or less forgotten, the 24-hour dial was reduced to one covering only 12 hours.

In the East and West, well into the seventeenth century, the astrolabe remained both a working tool of the astronomer and a powerful symbol. Like its cousin the armillary sphere, it was often used to symbolize both the cosmos and astronomy itself. To those with the advantage of an education in the seven liberal arts, it was a tangible reminder of the key to the exact sciences, the Greek genius for combining astronomy with geometry.

China and Japan

Ancient China developed a form of writing with pictographs at a very early stage in its history. Written Chinese has kept its ideographic form for well over three millennia, and in fact, the first great dictionary was produced as long ago as AD 121. Although linguistic changes have been continuous over time and space, our historical knowledge of very ancient Chinese affairs is far more reliable than that of most other ancient cultures. The richness of the historical record, however, does not make it any easier to mark the boundaries between early Chinese religion, astrology, and astronomy. Just as in Mesopotamia, several methods of forecasting were developed on the basis of interpretations of many different sorts of signs. Among these were celestial signs, and indeed the oldest record of a new star, a nova, is on an oracle bone dating from about 1300 BC (fig. 72). Some of the techniques adopted for divination were remarkably similar to those used in the West. One example is that of interpreting cracks in the heated shoulder-blades of animals (scapulomancy), the main difference being that in China, with luck, the cracks would take on the appearance of a standard pictograph. At one level, astrology was no more than a form of divination. As in the case of people in lands to the west, those who practiced this form were less concerned with the fate of isolated individuals than with those of prince and state, harvest, war, and the commonwealth. There are strong parallels, for instance, between a text from the library of Assurbanipal (seventh century) and the Chinese *Shih Chi* (*Historical Record*) of Ssuma Chhien (completed in 90 BC), both of them interpreting the movements of planets through the stars in terms of the destiny of the king and his enemies.

One striking difference between celestial observation in these cultures was that, whereas in countries to the west of China attention was at first generally focused on the *horizon*, in China great importance was attached to the constellations around the northern celestial pole and those around the celestial equator. (There are of course many exceptions on both sides. Did not the sentries at the siege of Troy change guard by referring to the

72. The earliest known record of a nova. The two central columns of an inscription on an oracle bone of around 1300 BC read "On the seventh day of the month, a *chi-issu* day, a great new star appeared in company with Antares."

direction of the tail in the Great Bear?) Having concentrated their attention on the stars around the northern celestial pole, stars that never rise and set, the Chinese learned how to position celestial objects relative to the Sun, in particular, in opposition to the Sun when the Sun was not above the horizon. They also mentally positioned stars by imaginary lines through circumpolar stars to stars that did indeed rise and set. In the Northern Hemisphere we all know the method of locating the Pole Star by the same technique of following a line from the stars at the end of the Dipper. The Chinese pinpointed stars by pairs of such lines.

Perhaps as early as 1500 BC, and certainly not later than the sixth century BC, the Chinese divided the heavens into 28 lunar mansions (*hsiu*), each regarded as a section of the equator, or as the stars within its limits, in which the Moon happened to be of interest at the time. Here, historical influence is difficult to determine. There is an oracle bone that dates from before 1281 BC and that mentions stars by name, some but not all of which have been identified. Another oracle bone refers to a solar eclipse identifiable as that of 1281 BC It is possible that the Chinese in the first millennium BC took from the Mesopotamian culture some of the broad principles of divination using the stars and Moon but that they interpreted it in accordance with their own patterning of constellations, based on an equatorial rather than an ecliptic system.

The Chinese continued to absorb Western astrological ideas, but rarely took them over in quite their original forms. In the fourteenth century we find horoscopes in Chinese sources—for instance in the *Thu Shu Chi Chhêng*—that are remarkably similar in all but outward shape to European horoscopes. The reason in this case is simply that both had one or more common source—namely, Islamic, and occasionally still earlier, Greek—astrology. The Chinese in such cases tended to increase greatly the number of alternative interpretations. They did so under the influence of other ancient systems of divination already in place, for instance, some based on biological theories and others on the calendar. Even here, when we find what appears to be Western influence—such as in the matter of lucky and unlucky days—there is often the possibility of a similar earlier tradition on which the later was grafted.

In ancient China, astronomy was intimately connected with government and civil administration. Perhaps the most famous illustration of this concerns an incident in the eighth century BC, reported in the *Shu Ching* (*Historical Classic*). It concerns a commission sent by a legendary emperor Yao to six astronomers, of whom two were named—the brothers Hsi and Ho. They were instructed to move to various places from which they were to observe the rising and setting of the Sun to determine the solstices and make other observations of importance to the drafting of a calendar. In a later chapter of the *Shu Ching* there is an account of an expedition led by the prince Yin to punish yet other astronomers for failing to foresee or prevent an eclipse. These legends, considered for three millennia to be the official account of the origin of Chinese astronomy, have helped to create the image of a venerable branch of learning, but they are now known to stem from a still earlier mythological tradition in which Hsi-Ho is the name of either the mother or the chariot driver of the Sun. The brothers Hsi and Ho have disappeared from history.

Early Chinese interest in cosmological matters was not markedly scientific, in the Western sense of the word. It did not develop any great deductive system of a character such as we meet in Aristotle or Ptolemy, for example. The great scholar we know as Confucius (551 BC–478 BC) did nothing to help this situation—if in fact it needed help. Primarily a political reformer who wished to ensure that the human world mirrored the harmony of the natural world, he wrote a chapter on their relation, but it was soon lost, and a number of stories told of him give him a reputation for having no great interest in the heavens as such. It is worth remembering here a definition of Confucianism as "the worship of the universe through the worship of its parts," a program very different from that of the great system-builders in the West.

The all-pervading Chinese view of nature as animistic, as inhabited by spirits or souls, gave to their astronomy a character not unknown in the West, but at a scholarly level made it markedly less well structured. At a concrete level, we come across such Chinese doctrines as that there is a cock in the Sun and a hare in the Moon—the hare sitting under a tree, pounding medicines in a mortar, and so forth. At a more abstract level there is the notorious all-encompassing doctrine of the *yin* and the *yang*, a form of cosmology that is to Aristotelian thinking as *yin* is to *yang*. Like so much in China, its origins are interwoven with government. Before the early Han dynasty (207 BC to AD 9) there had been numerous ruthless attempts, in the interests of efficient state regulation, to unify all the rival schools of Chinese philosophy that had developed over previous centuries. There had been earlier laws, for example, forbidding even reference to Confucius, Lao Tzu, or

Mo Tzu. Under the Han, a different approach was adopted, attempts being made to reconcile the rival schools and amalgamate them into a single system. The Han philosophers focused attention on the so-called Five Classics, particularly the *I Ching* (*Book of Changes*), from which they purported to derive the single great principle underlying the universe. The new doctrine was added as an appendix to the *I Ching*, and marks the beginning of what is known as the *yin-yang* school of Chinese thought.

The *yin-yang* school held that the single principle guiding the universe, the Tao, is divided into two opposing principles, *yin* and *yang*, to which all the opposites we perceive can be reduced. These principles make use of five material agents, or *wu hsing*, all phenomena being supposedly understood with the help of these concepts—the movements of the stars, the nature of foods and medicines, the workings of the body, the qualities of music, the moral qualities of individuals, and even the workings of history. The doctrine—which has some interesting parallels with certain brands of nineteenth- and twentieth-century philosophy, for which Hegel carries much of the blame—was taken to imply that all things can be related to one another in some way. From this, it was supposed to follow that the stars may be used to decide on courses of future action, including government action. Astrologers were thus in demand. Techniques of magical divination were developed, which continue to assume great importance in modern times—in both the East and West. Without entering into the simplistic workings of the doctrine of the *yin* and *yang*—the female and the male, Moon and Sun, night and day, dark and light, completion and creation, submission and dominance, and so forth—we should at least note that it produced a mentality highly receptive to ideas from India and further to the west. Indian notions of a cyclical world fit well, for instance, with the idea that the production of *yin* from *yang* and *yang* from *yin* occurs cyclically and constantly, making for an eternal cycle of reversal.

The Chinese rarely spoke in terms of a god beyond the world, a supreme maker. Heaven was their highest god, and the emperor was Heaven's son and head of the state religion. The most important sacrifice to Heaven took place on the night of the winter solstice, when the *yang* was held to begin to increase again after reaching its lowest ebb. Of course there were many other and related imperial rites. It is clear that the record of them set down in the book *Chou Li*, from the second century BC, describes a situation that had long been evolving. The tasks of the imperial astronomer and the imperial astrologer are there distinguished. Planetary observation belonged to the second office—observations such as those of Jupiter's twelve-year cycle that corresponded, it was said, to the cycle of good and evil in the world. As in other cultures, the astrologer observed the weather—here the colors of five types of clouds, the state of the twelve winds, and so on. For more than two millennia, these (hereditary) officials headed government departments with a large staff. They were the keepers of time, even of the hours, and the Chinese developed a whole series of water clocks, some of them extremely intricate hydromechanical affairs that drove astronomical and other models by water power. It is said that Chang Hêng, about AD 132, was the first to devise a means of turning an armillary sphere by a water wheel so that it kept pace with the heavens. The general idea was further developed, although sporadically.

By the time of the Northern Sung dynasty (960–1126) there were two separate observatories in the capital, one imperial and the other belonging to the Hanlin Academy. Each was equipped with water clocks and instruments for observation. The two groups were meant to make independent observations and then compare the results. When Phêng Chhêng became Astronomer Royal in 1070, he discovered that the two sets of astronomers had for years been simply copying each other's reports, even taking the positions of heavenly bodies from old tables. His successor Shen Kua discovered that the examiners in the state examinations were still no more competent than their predecessors had been.

Since calendars were important symbols of dynastic power and were often totally revised with changes of dynasty, old calendars could be politically sensitive material, and perhaps this explains why relatively few old astronomical documents survive, in comparison with mathematical materials, for example. Secrecy was enjoined, which explains some of the difficulties encountered by Jesuit missionaries visiting China at the end of the sixteenth century. The political character of calendars meant that copies of them were numerous: no fewer than 102 were produced in the period between 370 BC and AD 1851, and many of them have star tables and planetary ephemerides, in addition to solar and lunar material, making them an excellent historical indicator of the progress of astronomical theory. As we shall see, some of the numerous calendars were of course only slightly modified versions of their predecessors. What mattered was the new title page, so to speak.

From the fourth century BC onward, there was a steady increase in the number of works devoted to the visible universe of stars and their grouping. Although there were no great cosmological systems, in the Western style, there were simple cosmological pictures that went beyond the contrasting of *yin* and *yang*. For example, in the *Kai Thien* cosmology, which is of uncertain age but might predate the fourth century BC, the heavens were pictured as a hemispherical bowl placed over a hemispherical Earth— perhaps somehow meant to be trimmed square. This arrangement might conceivably have come from Babylonian sources. In places, the idea was taught well into the sixth century AD, and figures were sometimes quoted for the dimensions of the bowl, although they seem rather arbitrary.

The oldest surviving Chinese description of the heavens as completely spherical is by Chang Hêng in the late first century AD. Its circumference was divided into 365¼ units, each unit the distance traversed by the Sun in a day. This corresponds to a length for the year that was known at least as early as the thirteenth century BC. There was a school of thought, the

Hsüan School, older than Chang, which teaches that the heavens were end-lessly extended, and this view found favor with neo-Confucian philoso-phy well after the twelfth century AD. It fit well with Buddhist ideas and did not conflict head-on with the spherical picture in the same way as that had done with the *Kai Thien* cosmology. Yet another theory was that when the Sun is on the meridian it is five times more distant than when rising and setting, suggesting that the sky was conceived of as a highly elliptical dome.

There are other "philosophical" systems of cosmological thought, but it is easy to find ourselves searching for them merely to draw parallels with what was happening at the same time in Greece. Unlike Platonic and Aristotelian thought, Chinese thought was not overtly philosophical, but rather, it was historical. Joseph Needham, a well-known authority on the history of science in China, has suggested that the reason for this is that Chinese religion had no lawgiver in human guise, so that the Chinese did not naturally think in terms of laws of nature. An important instance of writing in a historical style is Ssuma Chhien's *Shih Chi*. One chapter of the book provides a survey of current astronomical doctrine, including much on natural meteorological phenomena. The work is by no means unique. There are twenty-eight generally recognized dynastic histories of China, and very many of them have astronomical chapters, often including as-trological portents with a possible bearing on the future of the state. They typically contained instructions for calculating planetary positions using only moderately accurate (synodic) periods, and instructions on simple lu-nar eclipse calculation. As we shall see shortly, astronomical styles changed materially, as more use was made of the ecliptic, in the first century AD.

CHINESE MATHEMATICAL ASTRONOMY

Earlier generations of Western historians paid relatively little attention to Chinese traditions of mathematical astronomy, partly from a lack of in-terest, but partly because they seemed primitive, differing as they did so greatly from the more coherent western tradition, with its Near Eastern and Greek ancestry. At the court of the Eastern Han dynasty (AD 25–AD 220), however, there was an expert on mathematical astronomy who was revered after his death as a man who had been without equal "since the time of Liu Xin." The newer genius was Liu Hong (ca. AD 135–AD 210); his distin-guished predecessor was Liu Xin (ca. 50 BC to AD 23), who had created a system of mathematical astronomy and cosmology known as the *San tong li*; and Xu Yue, the writer who was paying compliments to Liu Hong, was himself a mathematician and astronomer of merit.

Liu Hong is the earliest Chinese astronomer to have left us with a complete theory of the Sun and Moon: it is to be found in a seemingly complete text, known as *Qian Xiang li*. Here the astronomer took into account the Moon's motion in latitude, as well as a certain "inequality" by which its motion in longitude was to be adjusted. In the vocabulary we use here, there is a need to consider our perspective carefully, if we wish to draw comparisons between China and the West. Following Ptolemy, and until relatively modern times, solar, lunar, and planetary motions were referred primarily to the ecliptic—west of China, that is. An inequality (or anomaly, or equation) was then an adjustment to a uniform motion along the ecliptic. The Chinese, however, began by referring almost all celestial phenomena to the equator, so they considered it natural to apply any adjustment to equatorial motion. Liu Hong was unusual—and helped to set a fashion—in referring the motions of the Sun and Moon to the ecliptic, the Sun supposedly moving at constant speed, but the Moon not.

Liu Hong's treatise provided a series of computational rules aimed at making the required predictions, but it contained no justification for his rules, nor did it explain how his supporting data were acquired. He began by providing a starting point, an epoch at which all elements in the system take on simple initial values. He next provided a set of constants, governing changes over time, his basic unit of time being the day as defined by the Sun and beginning at midnight. (There were, of course, well-established calendrical systems for months and years, to which his parameters were easily adapted.) In two cases—lunar speed and lunar latitude—he supplied tables, giving values at daily intervals. His motions were based on certain assumptions about their cyclical nature, starting from his "Ultimate Origin," which is the equivalent of 21 January 7172 BC. Finally, he presented his bald rules for computation—for lunar positions in longitude and latitude, lunar phases, and eclipses. When we apply them, we find that his rules for latitude give a commendably sinusoidal curve, although one that implies too great a value for the inclination of the Moon's orbit from the ecliptic (6°).

The Han dynasty sanctioned various different astronomical systems during its four centuries of existence. In Liu Hong's time, a system was in place that had been inaugurated in AD 85. To assist with a calendar reform that had been overseen by Chia Khuei around that year, debates had been held among astronomers, and new instruments had been constructed— for example, a device to measure the ecliptic's obliquity. The new system remained in place until some time near the end of the Eastern Han dynasty, when Liu Hong wrote his *Qiang Xiang li*. When the Han dynasty disintegrated into the so-called Three Kingdoms, the southern state of Wu (AD 220–AD 80) adopted his system and kept it until Wu was eventually taken over by a temporarily reunified Chinese state under the Jin.

Early in the fourth century AD, the astronomer Yü Hsi (who was active between 307 and 338) discovered the changing longitudes of the stars, our "precession of the equinoxes," seemingly independently of western knowledge of it, as first found by Hipparchus. By and large, however, the character of Chinese astronomy continued with the compilation of data, with little attention being paid to the improvement of mathematical theories. In other words, Liu Hong seems to have been quite untypical of his profession. For the rest, while many simple regularities were appreciated, when they broke down in a particular instance there was an inclination simply to regard the situation as "irregular," and to label it as such. (One's thoughts naturally turn here to the role of miracles in religion, as a way of explaining something that is seemingly irrational.) Many Chinese writers betrayed a belief that, while broad analogies are to be found in the world, reality is essentially too subtle to be encoded in general principles. Others found explanations in terms of astrology, or the workings of the Emperor's virtue.

These attitudes had at least one very fortunate consequence. In Western astronomy, phenomena—such as comets, novae, and oddities in the appearance of the Sun—that did not readily lend themselves to treatment in terms of laws, were taken far less seriously than those that were. The history-conscious Chinese, on the other hand, kept detailed and plentiful records of all such phenomena—interspersed in some cases with records of events that could not possibly have happened. Duly filtered, these records later proved—as they continue to prove—an important source of astronomical information. Sunspots, which in European terms were a telescopic discovery of the seventeenth century, were already recorded in China in the time of Liu Hsiang, in 28 BC, and perhaps long before that. Between then and AD 1638, there are well over a hundred references to sunspots in official histories, and many more in local records. The Chinese knew the art of looking at the Sun through smoky crystal or jade, and haze and dust storms could also be turned to advantage. It is interesting to discover that when Thomas Harriot first turned his telescope toward sunspots in seventeenth-century England, he too took advantages of natural haze, to cut down the Sun's intensity.

ECLIPSE OBSERVATION AND PREDICTION

Eclipses, lunar and more especially solar, were of importance in the regulation of the Chinese calendar and were also seen as portents of events on Earth. It is not hard to see why those that were not predicted were deemed to be more important than those that were, but failure does not seem to have dampened the spirits of astronomers, who tended to predict more eclipses than could have possibly occurred. Predicted eclipses that failed to materialize were again explained away as having been prevented by the emperor's powers. Two treatises belonging to the dynastic histories, the *Five-Phases Treatise* and the *Astrological Treatise*, record many hundreds of solar eclipse reports, mostly brief, and in perhaps a quarter of all cases completely unreal. It has been suggested that political motives might be responsible for some of the latter, and even for the omission of some that were almost certainly experienced in the Chinese capital itself. By the third century AD, predictive techniques were yielding fair results, and by the tenth century they seem to have been improved as far as their general character allowed. Lunar eclipse observations came relatively late, say around AD 400, which is surprising in view of the fact that several earlier texts are known that gave methods for predicting them.

The earliest methods of predicting eclipses were based on analyses of past observational records, done in the hope that cycles of eclipses could be discovered. We have already seen that the Babylonians identified a useful cycle for predicting lunar eclipses, that which is now usually called the saros, of 223 synodic months. Chinese astronomers chose a cycle of 135 synodic months. By the third century AD, on the other hand, a different technique had been found. It is in the *Ching-ch'u-li* calendar of the astronomer Yang Wei. It was adopted by the Wei dynasty in AD 237 and remained in use by the Liu-Sung dynasty until 444. Here it was assumed that an eclipse, solar or lunar, would occur when the Sun and Moon were both within 15° of the node (or nodes) of the Moon's orbit. (For a solar eclipse the objects are near one node; for a lunar eclipse they are near opposite nodes.) The calendar includes predictions of such eclipse properties as magnitude, duration, and angle of approach.

In time, as inequalities in the motions of the Sun and Moon, as well as lunar parallax, were taken into account, predictive accuracy improved. Details in the basic rules were changed on several occasions during the Sung dynasty (960–1279), but at the beginning of the following Yuan dynasty a calendar was produced—with its eclipse rules—that remained in use from 1279 until the end of the Ming dynasty in 1644, with only a minor change in the interval. This was a calendar based on the system of Kuo Shou-Ching, which can be found in his *Shou-shih-li* (*Calendar Providing the Seasons*). The resulting calendar includes predictions of times of conjunction and duration and all the main phases of eclipses lunar and solar. As a rough guide to this, we may say that errors in the times predicted lessened fairly steadily between 400 and 1300, starting at around three hours and ending at around twenty minutes. Not until the arrival of the Jesuits, with their more accurate western methods and parameters, was there any further advance in accuracy. The Babylonians of the last seven centuries BC had predicted lunar eclipses with an accuracy of just over an hour, and solar eclipses to just under two hours.

STAR MAPS

There is a long tradition of drafting star maps in China. This is not surprising, bearing in mind that omens in the heavens were thought to have relevance to the region of the stars in which they were sighted, and these regions were associated with regions of the Earth—whether foreign territories, provinces, cities, or even divisions of the Imperial Palace. One of the most famous divisions of the sky is flanked by two chains of stars, seen as walls of the Imperial Palace that enclose the Purple Forbidden Enclosure. Solar halos, sunspots, colored clouds, the aurora borealis, or indeed almost any other astronomical or meteorological event, might be heeded—especially before battles, with their territorial implications. There

73. Box lid with an image of the Great Bear encircled by the positions of the twenty-eight lunar mansions. Found in 1987 in the tomb of a nobleman of the Warring States period (ca. 433 BC). The Bear is close to the North Pole, and in the course of the year the tail of the Bear revolves round the pole (as judged by a given time of day), so indicating the seasons. Each season corresponds to a "heavenly palace" (*gong*) of seven lunar mansions.

are several instances of grave goods with representations of parts of the sky, and they too presumably had regional connotations. (For an example of an item with a representation of the all important Great Bear, see fig. 73.)

There were star catalogs at least as early as the fourth century BC, when Shih Shen, Kan Tê and Wu Hsien drew up theirs. The work of those three continued in use for a thousand years, although only seven centuries had passed when Chhen Cho amalgamated them in a single star map, to be followed by Chhien Lo-Chih, with a planispheric star map in the fifth century. On Chhien's map, colors were assigned to the stars, indicating the astronomical tradition responsible. (The precise details of earlier maps are unknown at first hand, unless we accept a recent claim that will be discussed further at the end of this section.) The three lists, when combined, included 1,464 stars in all, grouped into 284 "chairs." Clearly, "constellations" would be a misleading translation. These numbers may be compared with those in Ptolemy's *Almagest*, with its 1,022 stars in only 48 constellations. The ancient catalogs listed chairs by name, the numbers of stars within them, the positions of the lesser stars in a group with respect to the chief star, and the coordinates of the latter. Coordinates were of an equatorial type, but not quite that used in Western countries. The catalogers used a sort of right ascension in Chinese degrees (but from the first point in the *hsiu*, the lunar mansion, in which the star lies, rather than the vernal point) and polar distance (rather than declination). This type of activity had interesting repercussions on the style of instruments being built. Traditionally, Chinese astronomy used armillary spheres rather than solid celestial globes. Of course, the former could be and was used for observation, but in all probability, most armillaries were used only for demonstration. It was apparently only in the time of Chhien Lo-Chih, after the accurate mapping of stars was well established, that complete celestial globes were built. In AD 435, in fact, he built both and armillary and a globe, the globe based on Chhen Cho's map. Some globes were made of wood and some of metal, but they are known only by report until we come to the magnificent examples made by the Jesuits in China, which survive, and which will be mentioned at a later stage.

For well over a millennium, the coordinate system in use was equatorial, but after the introduction of ecliptic coordinates—perhaps ultimately from the Greek world through India—one notable discovery, or pseudodiscovery, was made in consequence. Around the year 725, the monk I-Hsing, using instruments built by the engineer Liang Ling-Tsan, found star coordinates differing from those in the old lists. Precession could of course account for most of the discrepancies, but not for ten and more cases of movements that also changed the ecliptic latitudes. We now know that the stars have their own "proper motions" in addition to precession, a fact that was to be discovered eventually by Halley in the eighteenth century. Such proper motions across the celestial sphere are chiefly the result of their movements relative to the Sun. An enthusiast may say that I-Hsing had "discovered the proper motions" of individual stars, since the shifts he recorded cannot be explained away by the precessional drift in longitude, but it seems likely that his instrumental errors accounted for many—not necessarily all—of the observed effects. He offered no systematic statement or argument, and nothing came of his potential discovery.

Many stories are told of I-Hsing that reflect on his reputation for magic of an intellectual brand. The occasion of one story, found in a collection dating from AD 855, was the imprisonment of a friend on a charge of murder. In the Temple of the Armillary Sphere, with his hundreds of assistants, I-Hsing ordered seven pigs to be caught and put into a pot. Subsequently the Emperor complained that the Head of the Astronomical Bureau had found that the constellation of the Great Bear was missing. I-Hsing said that he could recall only one remotely similar occurrence, when Mars had once been lost. The new mishap he interpreted as a dire warning, perhaps of frost or drought, but this could, he said, be averted if the Emperor followed Buddhist preaching and issued a general amnesty. The amnesty did indeed restore the stars to the heavens; and when the pot was opened, the pigs had disappeared. Putting aside the undoubted talents of the astronomer, the fact that the story could be told with a fair degree of seriousness goes far to distinguishing between Chinese and Western styles of astronomizing at this period.

Texts are one thing, artifacts another, but actual star maps have a foot in both worlds. The oldest known manuscript star map of the whole sky is that which was found in 1907 by Sir Marc Aurel Stein. Now in the British Museum, it had been found buried with a rich collection of sundry unrelated materials, in a desert cave at Dunhuang (or Tunhuang), on the ancient Silk Road entering China from the west. The chart, written on paper, encompasses some 257 "chairs" of stars, numbering 1,585

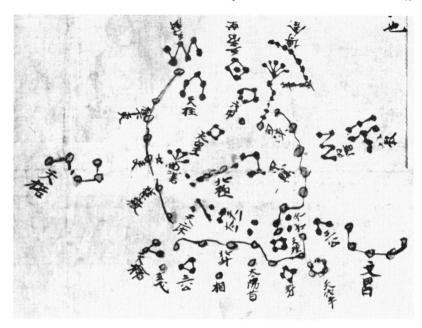

74. Part of a paper manuscript star map found in the cave at Dunhuang. This perhaps dates from the seventh or eighth century of our era, but has much earlier antecedents. The stars on the map are in three colors—white, black, and yellow—corresponding to the three ancient schools of positional astronomy. To the left of the section are the stars of the Purple Palace and Great Bear (compare the previous figure), the key to Chinese cosmology. The Chinese consistently gave sacred status to the North Pole, as the pivot of the sky, and were much concerned to identify the true Pole Star, which we know changes its position with time as a result of precession.

in all, again outstripping the catalog in Ptolemy's *Almagest*. Many of the stars are difficult to see with the naked eye. (For an idea of its appearance, see the fig. 74.) The date assigned to this document by modern scholars has marched steadily backward: Joseph Needham suggested AD 940, others pushed the date back more than two centuries, and then Jean-Marc Bonnet-Bidaud and Françoise Praderie argued for a date as early as the start of the Tang period (AD 618). They went so far as to insist on some stylistic resemblances to the *Yüeh Ling* (*Monthly Ordinances of the Chou*), astronomical texts that probably date from no later than 300 BC.

CHINA BETWEEN THE TENTH AND SIXTEENTH CENTURIES

Chinese astronomy had a strong cohesion: bound together as it was by language and script, it was widely regarded as the work of a single national group. It was admittedly enriched from time to time by contact with astronomers from countries to the west, and it was passed on, with an admixture of Korean astronomy, to Japan between the sixth and eighth centuries. Very early contacts with Babylon have been suspected, since by the mid-second millennium BC the Chinese were using a count of days by sixties. Persian

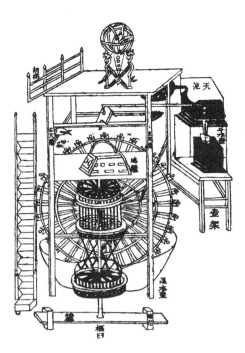

75. The astronomical water clock sponsored by the statesman and scientist Su Sung, completed in AD 1095. The water wheel and escapement are shown (left), with steps leading to the roof, where there was an armillary sphere. The housing (above) was seven or eight meters high, and the armillary added about three meters. From the third chapter of the *Hsin I Hsiang Fa Yao* (1092).

astronomers certainly visited China in the eighth and ninth centuries of the present era and carried with them the Babylonian and Greek methods of computation. By this time, great political changes were taking place that would eventually weaken the old pyramid of power, with the emperor at its peak and aristocratic land-owning families coming immediately below. The elitist civil service, despite its supposedly rigorous system of examination, was really accessible only to the upper echelons of society, and the system did not encourage high enterprise. For three centuries the Tang dynasty gradually declined until it was reduced to a series of rival kingdoms. Something approaching a new "universal state" eventually came with the Northern Sung (960–1126). Taxation began to weaken the old families; the center of economic power moved from the north—the old center of civilization of the Han people—to the lower reaches of the Yangtze valley; and society became more open to innovation.

This situation is reflected in the fortunes of astronomy. The second Sung emperor is known to have had a large astronomical library, and one work that has survived from it is of great interest, for it seems to parallel the Greek and Arab tradition of water-driven representations of the cosmos. This "astronomical clock" was described in the *Hsin I Hsiang Fa Yao* (*New Description of an Armillary Clock, 1092*) by Su Sung (1020–1101). He was first privy councilor during a hectic political period, but managed to guide a vast imperial plan to amalgamate medical writings and print the ancient medical classics. The clock (fig. 75), with an escapement that used a technique of tipping buckets on a chain drive, was built between 1088 and 1095 under his direction, at K'ai-fêng in Honan province, then

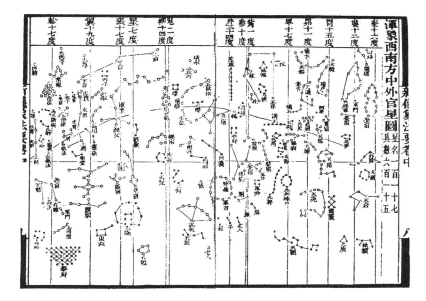

76. Star maps for half of the celestial globe from the *Hsin I Hsiang Fa Yao*. The equatorial coordinates of the stars are plotted as rectangular coordinates, somewhat as in a Mercator projection. The central horizontal is the celestial equator, and half of the ecliptic is in an arc above it. The vertical lines mark the divisions into fourteen *hsiu*, half of the twenty-eight lunar mansions.

capital of the empire. The description of the mechanism also includes a star map based on a new survey of the heavens, in fact the oldest extant printed star map. Since so much of Chinese learning in general was spread by books printed with wooden blocks, it is not surprising to find evidence for star maps in this medium. For a late eleventh-century example, see figure 76.

A scholar of still greater repute who served the same emperor and became the emperor's confidant, only to die in disgrace, was Shen Kua (1031–1095). He was a good mathematician who applied mathematics to a number of physical problems—to music and harmonics, for instance—and who made a number of attempts to do the same for astronomy. What is especially interesting is that whereas earlier Chinese astronomers, like the Babylonians before them, had used primarily arithmetical methods to account for the retrogradations of the planets, Shen Kua proposed a geometrical model. By Greek standards, his might now be thought a primitive theory, but it serves to remind us that a solution in terms of circular motions alone was not an automatic choice for all people. Shen Kua thought that each planet followed a circle until it came to a "willow leaf" part of its path, that might be inside or outside the main orbit, and that then it performed a detour, before returning to the main path (fig. 77).

One of Shen's early services to the court was to commission improved instruments—including a gnomon for solstice measurements and an armillary sphere. For the alignment of the polar axis of the latter, a faint star near the north celestial pole was used. Since it was not quite at the pole, it

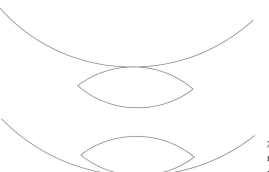

77. Shen Kua's willow-leaf
model, descriptive of planetary
retrogradation.

described a small nightly arc of a circle. Shen made a tube of such a size
that, looking through it toward the faint star, and fixing it in position,
one saw a circle of open sky of just the size described by the star's path.
The star in question was one of a succession of stars used as a "pole star."
The drift in star positions due to precession was of course what led people
to change their choice from time to time. It is a remarkable fact that very
many stars have Chinese names that indicate their use as pole stars, taking
us back perhaps even as far as 3000 BC. In Shen's day the star of choice was
our 4339 Camelopardi. In the fifth century this had been only a little over
a degree from the true pole, but Shen found the distance in his own day to
be in excess of three Chinese degrees (each of 360°/365.25, as mentioned
above), not a very acceptable state of affairs.

Another of Shen's reforms was to discard the ring on the armillary
sphere representing the Moon's path, for it could not be made to move
so as to reflect the backward motion of the Moon's nodes. Such changes
as these were not great reforms, but they helped to simplify a system that
had been hidebound by tradition for centuries. The same would have
been true of his suggested reform of the calendar, had it been adopted.
He proposed a purely solar calendar in place of the lunisolar calendar that
had been in use from ancient times. He was right in supposing that it
would give offense, and radical calendar reform was not to come until the
mid-nineteenth century. Even that, instituted after the Taiping rebellion,
was short-lived. China finally adopted the Western (Gregorian) calendar
for most public purposes only in 1912.

The period of Mongol domination, known in China as the Yuan
dynasty, lasted from 1260 to 1368 and saw a revival of standards in astronomy as a consequence of Persian and Arab influence. The greatest
Chinese astronomer of the period was Kuo Shou-Ching (1231–1316), a
great mathematician and a designer of sophisticated water clocks into
the bargain. We have already met him in connection with the theory of
eclipses and his important *Shou-shih-li*, the basis of a calendar adopted
by the new dynasty. A magnificent armillary made under his direction
for the latitude of Phin-Yang in Shansi around AD 1276 still survives, not
in original form, but as a replica dating from 1437. Still also surviving is

the Tower of Chou Kung, a Ming renovation of a building set up by Kuo Shou-Ching around the year 1276, for use with a 12-meter gnomon for measuring the length of the shadow cast by the Sun, and so determining the times of the solstices. This magnificent instrument had a stone scale, over 36 meters long, flanked by parallel troughs carrying water for purposes of leveling. As ever, the collection of data remained a matter of the highest priority, and the urge to produce new and improved theories remained weak.

During what in Europe are treated as the late Middle Ages, when astronomy there was rapidly gathering momentum, affairs in China seem to have been showing signs of decline. That great specialty of Chinese workmanship, the grand armillary sphere cast in bronze, with its supporting lions, dragons, and other symbolism, reached a high point with the Northern Sung in the eleventh century. The fall of the capital to the Chin Tartars put an end to this period of practical expertise. Instruments of various sorts were produced for common use. There were water clocks of various ingenious sorts, and even incense clocks, on which time was judged by the burning of incense pressed into a spiral groove, or maze of grooves, but the involvement of astronomy in telling the hour was left to rather primitive sundials. The Ming dynasty (1368–1644), which produced so much fine art, would have been unmemorable from a more serious astronomical point of view had it not been for a remarkable historical episode—the coming to China of the Jesuits. This will be the subject of a later section.

KOREA AND JAPAN

In the ancient mythology of Japan, the Sun goddess Amaterasu has a central role. The Moon god, her brother Tsuki-Yomi, is relatively unimportant, except in some stories of the birth of the Japanese archipelago, which make this out to have been born of the union of Sun and Moon. The stars seem to have had an even less significant place. There are early traces of festivals of certain stars, but that idea seems to have come from China: such records begin to multiply at the time of the introduction of Buddhism from China in the sixth century AD. That cultural invasion immediately gave rise in Japan to a discussion of the relationship of the old Japanese gods and those of the Buddhist pantheon. The most striking parallel was between the Japanese Sun goddess and the Sun myth used to explain the Buddha's personality. The notion of Buddha Vairochana (The Illuminator) was the result, and this notion continued to influence worship in Japan until the nineteenth century, when it was forbidden.

Korea was in many respects a staging ground for Chinese astronomy on its way to Japan. An incident that well illustrates the drift of ideas across Asia, ending with a sea journey to Japan, begins with no fewer than three schools of Indian experts serving in the Tang national observatory in the eighth century. In the seventh and eighth centuries two Indian astronomers, Qutan Luo and Gautama Siddhartha, actually rose to the post of director of the Astronomical Bureau in China. In 729 a new calendar was put into effect, the *Dayan li*, designed by I-Hsing. Three years later there was a dispute at court about its accuracy, led by an Indian astronomer Qutan Zhuan, who felt that he should have been consulted. He accused its author of plagiarizing Indian work, and of adding mistakes of his own. A competition was therefore held between three different calendars, one Indian (the *Navagrāha*, known in Chinese as *Jiuzhi li*), one the old Chinese version, and one the new. In this competition the *Dayan li* proved to be much the best. The Indian *Jiuzhi li* thus came to exert little influence on official Chinese practice, but despite this fact it was taken to Korea, and there was adopted for a long period. It also influenced Korean mathematics to some extent, since it contained trigonometrical tables and explanations of their use—a subject in which Indian astronomers were well versed.

In addition to astronomy, many other technical and scientific professions were instituted in Japan after the immigration of peoples from Korea and China in the sixth, seventh, and eighth centuries. Until the first influx of European science (1543), Japanese astronomy was based almost wholly on that of the Chinese and Koreans. In 607, during the Sui dynasty in China (581–618), the Japanese emperor sent an embassy to China, and they returned with reports of the role of astronomy in court culture. Over the next century, Korean masters were frequently invited to teach their art in Japan, and as the Chinese view of nature began to take hold there, new institutional patterns were created on the Chinese model. Following the style of the Astronomical Bureau in China, a Japanese "Board of *Yin-Yang*" was given tasks in astrology, *yin-yang* divination, and calendar-making—especially for ordering court ceremonies. In terms of analytical skill, we should now regard the last class of tasks as the most demanding, and yet in the scale of human wisdom in China, astrology and alchemy were always rated higher. Even during the Tokugawa period (1600–1867), when a Confucian set of values persuaded the Japanese shogunate (the military government) to place a higher value on the mathematical aspects of astronomy, the hereditary family of *yin-yang* diviners continued to be ranked above them. This did not augur well for astronomical theory.

Before the military class gained power, the Japanese government had been monopolized by a hereditary court aristocracy that had eventually managed to destroy the bureaucratic pattern that had been borrowed from China, and that had been their rival for power. Hereditary systems do little for intellectual advancement, and responsibility for astrology and calendar-making was largely controlled by two families, the Abe and the Kamo families. This did nothing to encourage scientific standards, and even the sober mathematical science of the calendar, a simple enough subject for trained astronomers, became increasingly arcane. There was one compensating advantage: this saved the calendar from the sort of repeated revision that had been known for centuries in China. In Japan, a much greater interest was shown in auspicious and inauspicious days than in the calendar's astronomical quality. The Japanese did not make or use many instruments comparable with those of the Chinese astronomers, and at some periods there were even heavy legal penalties for the private use of timekeeping instruments. As official standards declined, however, an element of competition crept into calendar-making. Unauthorized agricultural calendars were produced in large numbers, and the Buddhist *sukuyo dō* school challenged and competed with court calculators in predicting eclipses. We might see in this fact a glimmer of hope for the future of the subject, but little came of the competition. Astronomical knowledge remained moribund, being almost entirely dependent on antiquated Chinese sources until both cultures were stirred into a new kind of activity by the extraordinary impact of a European tradition in the sixteenth century.

THE JESUIT MISSIONS TO JAPAN

The Jesuits, members of the Society of Jesus, were an order founded by Ignatius Loyola. Converted during convalescence following a wound received in battle, he gathered together a number of companions and created a religious order that received the approval of Pope Paul III in 1540. In style, the order took much of its stern discipline from military models, but many of its members soon began to regard themselves also as an intellectual elite. Almost at once they began to carry the Christian message to all quarters of the known world, first notably with St. Francis Xavier's journeys to India and Japan (1541–1552). The next great Catholic mission was based in Macao, an island on the southeast coast of China that had been colonized by the Portuguese. Again, the mission was run by the Jesuit order, and after a slow start their apostolic ambitions were greatly helped by their astronomical skills. In both Japan and China, the long-term effect of the work of the Jesuits was great, even though between 1600 and 1640 every missionary in Japan was put to death or deported, while a similar fate met missionaries in China in 1665. After 1638, the only foreigners who were allowed to remain in Japan were the Chinese and the Dutch, and they were obliged to remain in Nagasaki for purposes of trade. Their Japanese official interpreters continued to read Western works, however, and through them western learning infiltrated Japanese culture indirectly and slowly.

Francis Xavier landed in Japan in 1549, and despite initial difficulties of language, his message was eventually found acceptable to many Japanese. He found them eager to learn about cosmic phenomena, planetary motions, and eclipse calculations, for example, especially when they recognized the superiority of Western methods over those that had come from China. In many instances, astronomy became a means of converting the elite classes to the Christian faith. Once the elite had been converted, the lower orders of society followed suit wholesale in a time-honored pattern.

As early as 1552, Francis Xavier was teaching the sphericity of the Earth and other Aristotelian ideas in Japan. An excellent insight into Japanese attitudes to all this can be had from a point-by-point commentary on a Western work published by the Confucian physician Mukai Gensho around 1650 (his book, *Kenkon Bensetsu*, is a collection of commentaries on a cosmological work by Cristóvão Ferreira). He contrasted the views of "those who write vertically and eat with chopsticks" and "those who write horizontally and eat with their bare hands." He thought Westerners ingenious in matters of techniques dealing with appearances and utility, but poor on metaphysics, especially in the understanding of heaven and hell. He considered Indian ideas to have only spiritual meaning, and to be fantastic and incomprehensible. As for Chinese and Japanese traditions, he remained a loyal neo-Confucian in his admiration for them. Whether or not it was against his better judgment, he and other native commentators nevertheless showed that they had learned much from those western purveyors of appearance and utility—and astronomy.

The first official astronomer to the Japanese shogun was a competent writer on mathematical astronomy, Shibukawa Harumi, who was responsible for the first important native calendar reform in Japan. He used chiefly a Chinese calendar, the *Shou-shih-li* of 1282, but referred to two others, one of them originating with Chinese Jesuits—the *Shih-hsien* of 1644. He used no new observations, but at least was competent enough to adapt the calendar to a Japanese longitude. After much controversy, his *Jōkyō* calendar was accepted in 1684, and at last Japan had something it could call its own. Although traditional in many respects, it incorporated some ingenious new mathematical techniques, and had it been presented at a session of the Royal Society in London, some aspects of it would not have been deemed uninteresting—its interpolation techniques, for example.

A man who carried much greater responsibility for turning Japanese astronomy toward European models was Asada Gōryū (1734–1799; this was his later pen name—his original name was Ayube Yasuaki). A member of a family of Confucian civil administrators under the Kizuki fief government, he had access to Chinese and Jesuit-Chinese works, and earned something of a reputation when his calculation of the solar eclipse of 1763 was much closer than the official predictions to the truth. Employed as physician to a feudal lord, who refused him leave to pursue astronomy as he wished, he fled to Osaka and supported himself by practicing medicine there among the wealthy merchant class. He also taught astronomy, and with the help of new instruments—many of them, including telescopes, made by Asada himself— his school began to collect new data of an accuracy that was entirely unprecedented in Japanese science. When he published a theory of planetary motion, based largely on that of the long-outdated system of Tycho Brahe, it was with sound new parameters that he and his pupils had evaluated themselves.

Many of these pupils were of the samurai class, and Asada was offered preferment by other lords, and by the shogunate itself, but his shame at his earlier desertion led him to refuse the offers. In later life he assisted the then well-established movement of translating Dutch scientific works into Japanese. He helped produce a synthesis of astronomy that was curious because it was such a chronological jumble, with elements taken from Newton, Kepler, Copernicus, and Ptolemy, without any regard to the sequence of their discovery. He was at heart an algebraist and never fully appreciated the advantages of geometrical models. He did not master the Newtonian theory of gravitation. Analyzing existing European data, however, he developed a number of useful techniques, and managed to devise quite a passable formula for the length of the tropical year.

The last great traditional astronomer in Japan, who shared Asada's ambitions to turn the direction of his subject, was Shibukawa Kagesuke (1787–1856). By his time, Jesuit influence hardly mattered any more, but foreign relations were not without their somber side to a man whose brother had been executed for helping a German traveler to smuggle forbidden materials out of the country. Shibukawa, working in the Astronomical Bureau, was privileged to read all the foreign works he could obtain. His struggle to rectify the inferior data that were circulating in his time, much of it forged by astronomers who had no feeling for the need to represent reality, makes tragic reading. In the end, he carried the day on technical matters. He saw that his compatriots would soon have to make themselves familiar with the doctrines of Copernicus and Newton, and there too he prepared the ground, although with less sympathy for the end result. "Let us melt down the mathematical principles of the West and recast them in the mold of our own tradition," he wrote—using a Chinese adage.

THE JESUIT MISSIONS TO CHINA

Matteo Ricci was born in 1552 in Macerata, Italy, the son of a pharmacist. Ricci joined the Jesuit order and studied astronomy among other subjects at the Collegio Romano in Rome. He was much influenced by Christoph Clavius who taught there, who was a friend of Galileo Galilei, and who was one of the most respected of European astronomers in his day—although he did not accept the Copernican idea of a Sun-centered universe. Ricci left Rome with his Jesuit mission in 1577 and sailed from Lisbon to Goa and thence to Macao, which he reached in 1582. In 1583 he arrived in China, and joined up with another Jesuit astronomer, Michael Ruggerius. Ricci settled at Ch'ao-ching in Kwantung province, but traveled much. Having established several missions throughout the empire, in 1601 he finally settled in Beijing (then Peking) under the protection of the emperor Wan-li. There he remained until his death in 1610.

Having won their way into official Chinese circles by their expertise in calendar computation, Ricci and his companions published a well-chosen selection of European materials that was carefully chosen to put rival Chinese material in the shade. In astronomy, that was not a difficult thing to do. Ricci's writings in Chinese were largely on theology and ethics, but he also translated or abbreviated Clavius's writings on the astrolabe, the calendar, spherical trigonometry, and mathematical subjects, including the first six books of Euclid. He was assisted in all this by a pupil Hsu Kuang-ch'i. They published an enormous map of the world (179 by 69 centimeters) in various editions, giving the Chinese their first awareness of the distribution of lands and seas across most of the globe. Among other useful astronomical techniques, the Jesuits taught those of the new European algebra, and later they introduced logarithms and the logarithmic slide rule for help in calculating.

Needless to say, the impact of all this activity was considerable and changed drastically the direction of a Chinese science that during the Ming and early Qing (Chi'ing) dynasties had become virtually static. In their letters home, however, the Jesuits presented a rosier picture of Chinese science. They gave Europeans a sense of the great value of the astronomical culture they had discovered. What impressed them so much? Among other things, they found in China a rich mine of potentially valuable astronomical data from the distant past, and as such it attracted—as it continues to attract—attention that has more to do with astronomy than with history. In short, there were exchanges in both directions. The first telescope to reach China was brought by Johann Schreck (Father Terrentius) in 1618 and was given to the emperor in 1634. Schreck was a talented friend of Galileo, who corresponded with him and with Kepler. It was during Schreck's time that the Jesuits began to make a monumental compilation of the scientific knowledge of the day, traditional and western, a project of both scientific and historical value that was augmented well into the eighteenth century.

The fortunes of the Jesuits in China changed dramatically in the wake of political change. While they had found some sympathy in the Ming palace after 1600, they appear to have lost their powers of persuasion, for two generations went by without their having managed to convert more than two or three high officials. They began to fail in their attempts to have their Western ideas published. The most sudden change came about in 1644, when Manchu invaders from beyond the Great Wall swept aside the old order and founded the Qing dynasty. The Jesuits now seized the opportunity of serving their new masters with two skills in particular: cannon casting and calendar reform. A new Chinese dynasty needed a new calendar to underscore its importance, or so it was believed. No sooner had the invading army marched into Beijing than the new regime appointed a European as head of its Directorate of Astronomy. The new man, J. Adam Schall von Bell, was prepared to provide traditional astrological services for the new court, but he nevertheless insisted that his native staff—Chinese and Muslim—practice only Western methods.

Adam Schall's autocratic rule lasted until 1665, in which year his many local enemies plotted very effectively against him, on astronomical and religious grounds. Some members of the Jesuit mission were put to death while others were exiled, and all Christian churches were closed for a time. Schall died in 1666, but within three years he was succeeded by another Jesuit, a Flemish priest with a more subtle tongue, Ferdinand Verbiest. Verbiest reached China in 1669. The great strength of the Jesuit order was its learning. Verbiest had been singled out for missionary service by his predecessor in China, Schall, who had taught for eight years in the Jesuit College in Ghent, and who appreciated Verbiest's skill in mathematics and astronomy. These Western scholars had not only skill but a phenomenal amount of energy. Schall, for example, when he died in 1666, had been working with twelve Chinese assistants on a vast translation project since 1631. With Terrentius (Johann Schreck) and Jacobus Rho, Schall had translated substantial sections of a hundred and fifty Western books; and Verbiest continued to write for the Chinese in the same impassioned spirit.

With the Jesuit order in disarray, Verbiest managed to worm his way into the Emperor Kangxi's favor. He began by proving to the young ruler that his native astronomers were incompetent. In this way he persuaded the emperor to return the Directorate to Jesuit control. Kangxi (K'ang Hsi), who reigned from 1661 to 1722 and who was the second emperor of the long-lived Qing dynasty, became so entranced by Verbiest's astronomy that he studied the subject himself in some depth, and encouraged his courtiers to do the same. The Imperial Observatory, founded in 1279, had been largely neglected, and Verbiest began to plan for its renewal a task he had more or less completed by 1673. His diplomatic skills were exceptional, and his published astronomical correspondence is an impressive collection in Latin, Portuguese, Spanish, Russian, French, and Dutch. The most important of his surviving petitions to the emperor runs to more than a thousand pages of Chinese text.

The calendar is of much interest in this context, in view of its extraordinary political dimension in China. There were rival factions at the Chinese court, which were divided on the subject, one group even defending Muslim calendar techniques. Cosmological questions seem to have been of less concern to the Chinese than to their European contemporaries. It is interesting to observe how, during the second half of the seventeenth century, Jesuit treatises as a whole—here following the example set by Athanasius Kircher—typically described a number of different planetary systems, as if to leave the final choice to the reader. There were the geocentric models of "Plato" and Ptolemy; the geo-heliocentric models ascribed to "the Egyptians" or Martianus Capella, as well as to Tycho Brahe (a general favorite with the Jesuits), and Giambattista Riccioli; and finally there were the dangerous heliocentric models of Aristarchus and Copernicus. Jesuit planetary astronomy in China was at root that of Tycho Brahe, Earth-centered but sharing many of the advantages of the earlier Copernican system. (Their relative merits will be touched on again in chapter 12.) The astronomical instruments Verbiest introduced into his writings were also based on Tycho's—sound, but certainly not at the cutting edge of astronomical research by the 1670s. The Jesuits carried Tycho not only to Asia but to Spain and to South America, where in some quarters his system reigned supreme for two centuries.

Verbiest wrote a highly influential Latin work, *Astronomia Europaea* (*European Astronomy*), written before 1680 but not published until 1687,

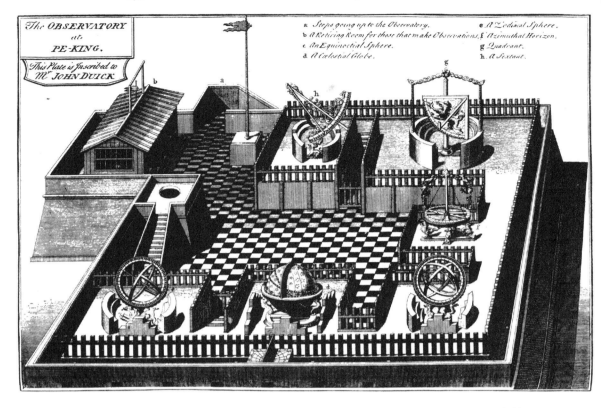

78. The Imperial Observatory, on the eastern wall of Beijing, with the instruments provided by Ferdinand Verbiest. Redrawn in 1794 from Verbiest's *Astronomia Europaea* (1687).

when it was printed in Dillingen, Germany. Despite its title, it reproduced a series of observations made in China in 1668 and 1669, which had already been published in Chinese. The book included an illustration of the newly refitted Imperial Observatory in Beijing, with the many fine instruments that had been made for it under Verbiest's direction, instruments again owing much to Tycho for their design (fig. 78). One of the book's obvious aims was to demonstrate the superiority of European learning and technology over its Chinese equivalents, and to recruit young and technically qualified priests to the missionary cause. Instrument technology, calendar theory, optics, gnomonics, pneumatics, music, horology, and meteorology, for instance, rub shoulders with fundamental astronomy. Running through it there is an amusing undercurrent of boastfulness about the fact that they, the Europeans in China, were casting fine cannons, building clocks, automatic organs, and telescopes, sketching in perspective, and so forth, all in the best European tradition. Like most Jesuit writings, this book whetted the appetites of a number of Western scholars to learn more about a system of astronomy that was both exotic and respectably ancient. It had some merits that the Jesuits did not perceive, however, bound as they were by the Catholic Church's rejection of a Sun-centered

79. Bronze armillary sphere constructed in Beijing under the direction of the Jesuits in 1744.

and potentially infinite world. It is ironic that other scholars in Europe were propagating the idea of infinite space at the very time when the Jesuits were turning the Chinese away from the Hsüan Yeh doctrine of heavenly bodies floating in endless space. In fact so different was it, in overall character, that much of Chinese cosmology escaped their notice, and remained little appreciated by historically minded Europeans until long after the exodus of the Jesuits from China.

The Jesuit mission failed in its attempt to turn China into a Christian country, but its continued to guide Chinese astronomical practice, without ever really coming out of the mindset of the sixteenth century. An armillary sphere built for the Imperial Observatory in 1744, for instance, is a fine example of bronze casting, but as a precision instrument it would not have astonished European astronomers a century before (fig. 79). Jesuit intellectual control of Chinese astronomy lasted until as late as 1790—in other words, it survived well beyond 1773, when the European suppression of the Jesuit order was beginning. (After being expelled from various Catholic countries in Europe, the order was reinstated by papal authority in 1814.) We have already seen how hostile Japan was to Western scholars at an earlier date. Western books, however, reached not only remote parts of China, but also Japan. To a large extent, they reached Japan through translations done in China, especially after a relaxation of the strict ban in the time of the eighth shogun, Yoshimune, in 1720. Here is one of the many hidden consequences of the long period of Jesuit missions to the Far East.

TIBETAN ASTRONOMY

It is natural to treat of Tibetan astronomy in connection with China, in view of Tibet's long but often bittersweet relations with that country, but Tibet has been influenced from other directions too. The region is not only bordered by three Chinese provinces, but also by Burma, India, Bhutan, Nepal, and Kashmir. The ancient nomadic and pastoral tribes of the region moved around over great distances, mingling moderately easily, forming inter-tribal alliances, and inevitably sharing many of their traditions. As a result, Chinese, Burmese, Turkic, Dardic, and Indian cultural strains were combined in a loosely knit whole, which might have been even more varied had the world's highest mountains not formed so many natural obstacles to travel.

From the seventh century to the ninth, the peoples of the region were ruled by powerful kings, but gradually the kingdom fell apart. For four or five centuries afterwards, loosely organized trading was the most important binding force. This was not a propitious environment for science or scholarship, but a Tibetan form of Buddhism had begun to develop as early as the seventh century. Buddhism was at first imported from China, but at the end of the eighth century a deliberate decision was taken to adopt the Indian Buddhist tradition. Monks began to travel to and from India, Indian teachers were occasionally recruited, and—especially in the eleventh and twelfth centuries—an enormous program of translating hundreds of items of Indian literature into the Tibetan language was carried through.

The greatest political achievement of the Tibetans was a result of the creation of an important religious school by the fifteenth-century Buddhist reformer Tsong-kha-pa. In 1578, monks from this school converted the Mongol Altan Khān, who duly rewarded them with political power, effectively allowing their Dalai Lama to rule Tibet as a theocracy. Succeeding dalai lamas were considered to be reincarnations of the Boddhisattva Avalokitesvara, the second great hero of Buddhism, who renounced Nirvana to save all sentient beings. Ruling from the capital city of Lhasa, the dalai lamas held their powerful position until the Chinese takeover of 1959, although there had been many internal rivalries during the previous three centuries.

There are four main branches of traditional Tibetan astronomy, two of which derive from India, and two from China. From India there are translations of more general Buddhist works, in which reference is made to Indian astronomy and astrology of the Vedanga period, the post-Vedic period before the bulk of Greek influence arrived on the scene. The Greek ideas written into the *Siddhantas*, books of "established tenets" (which will be considered in more detail in chapter 7), were introduced piecemeal, possibly beginning in the eleventh century. This type of astronomy, called *grub-rtsis* in Tibetan, is not set out in a single place in a coherent fashion, but like the Indian texts, it is of great interest since it uses a few parameters from pre-Ptolemaic Greek astronomy.

There are more sophisticated Hindu methods to be found, set out in great detail in works on "star calculation," *skar-rtsis*. More will be said about these Indian methods in due course, especially about their methods for predicting planetary positions, but broadly speaking it can be said that they follow the Greek technique of beginning with the calculation of mean motions, and then applying corrections (equations) to them. Four types of day are used: one is from sunrise to sunrise; a second is a thirtieth part of a synodic month; a third is the time it takes for the Moon to move ahead of the Sun by 12°, the two bodies being assumed corrected for the equation of center; and the fourth is one 360th part of a tropical year. All of these come from Indian sources, as do the systems of signs of the zodiac and lunar mansions used in Tibet. There is much Indian astronomy in a classic fourteenth-century Tibetan treatise written by the scholar Bu-ston Rin-chen-grub (1290–1364), a man famed for an encyclopedia he compiled. The material he assembled was copied into several later treatises, even as late as the eighteenth century, but not changed significantly for the better. The planetary calculations were done using rather strange mathematical techniques for handling proper fractions: one feels almost as though in the presence of someone who has reinvented the wheel, but whose wheels have turned out square. The astronomy is not trivial, however. It has recently been shown that the astronomical constants used in Bu-ston's work are close to those of the Ārdharātrika school of Hindu astronomy. A calendar based on this type of astronomy is still used by the Tibetan people.

According to tradition it was in the seventh century that Tibet received from China a system of Chinese astrology known as the *nag-rtsis*, or "black calculation." This has continued in use in modern times, and there are popular modern Western works which purport to be based on it, since the very idea of wisdom from the roof of the world still apparently has a Shangri-la cachet. With the passage of time, some of the underlying principles have been obscured, but the general character of the system is true to its source. There are the five elements of Chinese natural philosophy: wood, fire, earth, metal, and water. There are the twelve animals, one of which gives its name to a year in the cycle: rat, ox, tiger, rabbit, and so forth; and to a month, by another rule; and to twelve parts of a day by a third. There is material from the famous *I Ching*; and an interpretation is included of the Chinese lunar mansions—28 in this system. We came across some of these ideas earlier in this chapter.

Material of these kinds, Indian and Chinese, often reached Tibet by roundabout ways. The *rgya-rtsis* ("Chinese calculation"), for instance, explains the theory of the Shixian calendar, the last lunisolar Chinese calendar. What was to become a favored Tibetan calendrical system based on the latter did not arrive directly, but was based on a Mongolian translation dated AD 1711 of an original work of 1669 (during the Qing dynasty). Even material that arrived directly is often found to have roots stretching far back into other cultures. As an instance of the complexity of transmission, consider the doctrine of Sun, Moon, and *six* planets set out in the *Chinese* astrological text mentioned earlier. What is the sixth? Its name is Rāhu, a name taken directly from *Indian* sources. In Indian myth it is familiar as the name of an ill-fated demon. Denounced to the gods by the Sun and Moon for drinking the amrita drink, Vishnu severed Rāhu's head; but the amrita drink gave him immortality, and so an endless opportunity for revenge on the luminaries.

Astronomically, this makes good sense, for Rāhu personified the ascending node of the Moon, the point where the lunar orbit on the celestial sphere cuts the ecliptic circle. There is a corresponding term *Ketu*, for the descending node, diametrically opposite the first—in myth, it is the body of the headless Rāhu. The two points on the celestial sphere are crucial in Greek eclipse theory, for as we have seen on more than one occasion, it is in their neighborhood that eclipses occur. At some early stage, however, they acquired a curious association with the idea of a monster responsible for the damage inflicted in eclipses. It is even possible that the dragon of the Biblical Apocalypse, symbol of Satan, is connected with the idea. It is the *Tinnin* of the Arabs. It is the *Jawzahr* of the Persians, which some Latin writers transcribed as *Geuzahar*. The concept also found its way into Latin proper, with a trace of mythology attached: *Caput Draconis* and *Cauda Draconis*, the Head and Tail of the Dragon, have been terms used for the lunar nodes (if only by old fogeys) within living memory. Astronomically, the nodes have retrograde motion, covering the ecliptic in about 18.6 years. It was natural enough to treat them as a generally invisible planet, and the Indians did exactly that, as did the Chinese, and as did the Tibetans in their version of that Chinese astrological work. The sky is common property, so that astronomy travels well. It seems that the same can be said of dragons.

Pre-Columbian America

A region bounded by Arizona and New Mexico to the north, and Honduras and El Salvador to the south, Central America saw a number of advanced city-cultures rise and fall in the two millennia ending around the time of Columbus's discovery of America. The four most notable are the Olmec, Zapotec, Aztec, and Maya, of which at least the Maya developed the ability to analyze astronomical events using mathematical techniques. Somewhere along the way, all of them seem to have shared a theory of a layered universe, each layer containing only one sort of celestial body. First above the Earth was the Moon's layer, and then followed one for the clouds, another for the stars, then those for the Sun, Venus, the comets, and so on, until the thirteenth, where the creator-god resided. If nothing else, this scheme gives an answer to those who suppose the Greek system of the spherical universe to be self-evident or to mark a necessary stage in human development. The cosmological beliefs of most South American peoples were utterly different from those in the Near Eastern and Mediterranean cradles of Western astronomy, but the mathematical analysis of planetary motions were another matter. We can only view with astonishment certain extraordinary similarities between the Mayan analysis of the behavior of Venus and their Greek and Babylonian counterparts.

The Maya survive even now in parts of Guatemala and the Yucatán peninsula. Columbus first met some inhabitants of Yucatán paddling a large canoe on the high seas, and he visited their homeland briefly. Others followed his example, and seizures of gold and reports of stone-built cities soon led to exploration and conquest by Cortés and others. One of the great tragedies of the entire episode is that—according to Diego de Landa, first archbishop of Yucatán—"a great number" of the books of the Maya of Yucatán were burned on account of the devilish superstitions they supposedly contained. Before expostulating, however, it is just as well to remember that De Landa had been appalled by witnessing the sacrifice of children, some even inside his own churches, by supposedly converted Maya.

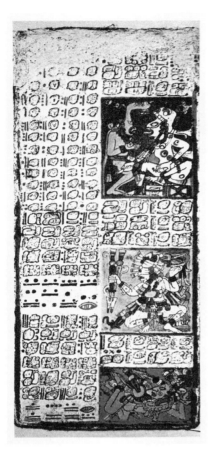

80. Part of the *Dresden Codex*, a Mayan
astronomical text that deals with the
revolutions of Venus. Probably a thirteenth-
century copy of a much older original. The
middle scene represents the planet's heliacal
rising. Spears (also in the lower scene)
represent the planet's death-dealing shafts
of light.

De Landa left a substantially correct account of the Mayan calendar,
and Mayan dates can be stated today with reasonably high accuracy. Apart
from books that were destroyed, others have disappeared through neglect.
It seems that only five Mayan manuscript books or fragments now survive,
one of them a congealed block of pages, but those that are readable include
lunar and solar calendars and a Venus calendar of great interest. (Each of
the Mayan books is made of a single sheet of bark-cloth paper up to 6.7
meters long and 20 to 22 centimeters high, pleated into folds that make
pages about half as wide as they are high.) The best of them is the *Dresden
Codex*—named after its present location—which also has many drawings
of the gods of the Maya (fig. 80). It is also one of the oldest Mayan books,
dating from the thirteenth, or possibly fourteenth, century, and was probably sent to Emperor Charles v by Cortés in 1519, that is, soon after his exploration and conquest of Yucatán.

THE MAYAN CALENDARS

The surviving Mayan books have glyphs and pictures on both sides that
give evidence of extraordinarily complex calendrical conventions and of
the sophisticated use of astronomy in religious ritual and divination. There

are almanacs of various sorts, including farmers' almanacs, and multiplication tables to help in their use. In some, the luck to be expected on every day in cycles of 260 and 364 days is marked. Another cycle of fortune, in evidence from the first century AD, was the *katun*, a period of 7,200 days. These cycles will be explained shortly. They did not use a cycle incorporating leap years; rather, they simply let the cycles run their course through the seasons. Their time-reckoning took no account of fractions of days, and in this respect we are reminded of the Egyptian year of exactly 365 days, which had so much to recommend it. (The system of Julian day numbers, used by astronomers almost everywhere, continues in a similar tradition, although decimal fraction of those are of course often used for continuously varying quantities.)

Whereas most cultures in the subcontinent had at best primitive pictographic books, the Maya had an elaborate system of hieroglyphs, that is, pictures denoting syllables of speech. There was an elite class of scribes, who were expected not only to be literate but also to have a sound knowledge of important celestial cycles, which were considered essential to the ordering of affairs of state. In addition to ordinary glyphs and hieroglyphs, they had an arithmetical notation using dots and bars, in a system using bases 20 and 5. All told, they had 3 different calendrical systems, with some overlapping content. These are known by scholars as the *long count*, the *haab*, and the *tzol kin*. The long count is a counting of days from the beginning of a cycle with a period of 2,880,000 days (this being very close to 7,885 years). It was assumed that the universe is destroyed and reborn at the end of every cycle. When counting days, they favored a multiplier of 20, except that in the second place they preferred 18, to approximate to a year. In other words, they took units of 20 days, 360, 7,200, and 144,000 (half of a cycle), and when the philosophical mood took hold of them, they even went beyond a full cycle into the realm of past universes. They consistently counted on a scale of 20 thereafter, until reaching a span of 23,040 million days (more than 63 million years). Theirs was virtually a place-value notation; thus, the number they might have written as 1.8.13.17.3 (if they had been using our decimal notation for values) is the equivalent of the following:

$$3 + 7 \times 20 + 13 \times 360 + 8 \times 7{,}200 + 1 \times 144{,}000$$

This amounts to 206,623. The cycle in which we now live, according to the calculation accepted by most modern scholars, began on 6 September 3114 BC (Julian). Why that date was used is not known.

In addition to the long count, the Maya used a civil calendar, the *haab*, using a year comprising 18 periods of 20 days, with 5 more "unlucky" days at the end. The names and pictographs for the 20-day periods are known. There are some oddities in the method of counting, one of which was shared, strangely enough, with that used for an important Indian calendar. Explained with reference to our own calendar, it is as though we were to label 31 Janu-ary with the name 0 February. For correlation of the *haab* calendar with the long count there was a standard Mayan procedure.

The third Mayan calendar, the *tzol kin*, was a religious calendar with two cycles, one of 13 days and the other of 20 names. The names and the numbers were cycled simultaneously, in what to our eyes seems a bizarre system, although we seem not to think it strange that we cycle our week-day names with dates within months. The Maya frequently used the *haab* and *tzol kin* dates together, for example, resulting in a cycle with factors 365 and 260, that is, 18,980 days (51.96 years). It is clear that these people had developed a familiarity with their calendars that has had few parallels in other parts of the world.

MAYAN AND AZTECAN WORSHIP OF VENUS

The 260-day period—which is still in use by some communities—was pos-sibly related to the movement of Venus, or possibly to the human period of gestation. The Maya had accurate tables giving the synodic revolution of the planet Venus, again accompanied by glyphs showing the fate of mankind according to the day of the planet's heliacal rising. They had studied Venus to such effect that they knew the importance of a period of 2,920 days—eight years of 365 days, after which Venus begins to repeat its movements in rela-tion to the Sun. (Heliacal risings, and so forth, thereafter repeat themselves in a Venus calendar, as we have already seen in connection with Babylonian astronomy.) When the real planet Venus did not conform with the Mayan tables of predictions, the astronomers seem to have made corrections to the latter, rather as we do with our leap years.

Astronomy was in the service of the gods, and gods are frequently men-tioned, including—in rough order of frequency—gods of rain, the Moon, death, creation, maize, and the Sun. Their almanacs of given sets of days variously arranged (4×65, 5×52, 10×26, and so on) were for different purposes, such as net making, fire drilling, maize planting, marriage and childbear-ing. Some of the computations deal in millions. There is material on New Year ceremonials and on the weather. It seems to be by mere chance that the surviving books deal with the subjects that most concern us here, for other materials from the Mexican region cover a much wider range of subjects.

Many of the astronomical practices and beliefs of the Maya are a ques-tion of inference from archaeological remains. Those of the Aztecs of cen-tral Mexico are better known from their literature, in particular from a work known as the *Codex Mendoza* that was written at the time of the con-quest. The Aztecs were relatively recent invaders of the valley of Mexico when the Spanish arrived. Their loathsome habit of human sacrifice is ex-plained in terms of their wish to placate the Sun god Tonatiuh and keep the heavens moving—and the victims were only prisoners of war, after all. The Aztecs offered incense to certain stars at appropriate times of the night. In this their king, the famous Montezuma II, still participated after the Spanish conquest. He was supposedly born on the same day of the calendar ("Nine Wind") as the god Quetzalcoatl—the Morning Star, our Venus, but seen as a male deity. There was always some form of Sun worship in all cultures, and examples are too numerous to mention. A previous Aztec king, in the middle of the fifteenth century, had set up a large porphyry pillar halfway up the great temple staircase in the town of Tlatelolco (now part of Mexico City) covered with solar symbols, which some would have us believe were used in eclipse calculation. Certainly the Aztecs are known to have sacrificed hunchbacks at solar eclipses.

The worship of Venus in Central America was no less universal and important than that of the Sun, and gave rise to laudable astronomical techniques of prediction. The planet was regularly and closely observed. There used to be a relevant picture—now destroyed by the tourists' habit of throwing bottles at it—near a sacred underworld lake at the Mayan city of Chichén Itzá, a lake into which sacrificial victims were thrown. This showed a square Sun rising over the horizon, and it carried a date equivalent to AD 15 December 1145. Modern calculations show that a rare transit of Venus across the face of the Sun was indeed to be seen on that day. The *Dresden Codex* contains other impressive tokens of astronomical skills that must have required centuries to acquire. There is a table for predicting solar and lunar eclipses, and ephemerides for Venus and Mars. Other documents show their ability to place the Moon in relation to the stars.

The early Spanish historian Juan de Torquemada, who was writing a century after the conquest, was able to discuss with native people astronomical ceremonies that were still taking place. It is said that most divination was then done by other means than by observing the heavens, although Atahuallpa's general said that the coming of the Spaniards had been foreseen astrologically. The point here is not that it was true, but that it was not out of character with the practices of the time. Among a number of astronomical omens is one quoted from Atahuallpa himself, the last native ruler of Peru, who attributed the death of a man to a comet in the sword of our constellation Perseus.

Torquemada reported having seen rods placed in holes on the roof of the palace of the Aztec king of Texcoco, the rods having on them balls of cotton or silk to help in measuring celestial motions. In some pictographs from Mexican sources, it seems that men are using crossed rods, over which they are taking sightings. But to what end? Torquemada's informant said that it was to aid the king, who with his astrologers viewed the heavens and the stars. While one of the chief concerns of these peoples must have been the regulation of the agricultural calendar, no doubt there were also rituals that had taken on a life of their own, their origins having been forgotten, and yet other rituals that were derived by analogy. Venus does not provide an immediate guide to the seasons, although by relating its motion to the Sun's it can be made to do so. The historian Fr. Bernardino de Sahagún tells us how the Aztecs sacrificed prisoners to Venus when the planet made its first appearance in the east, splashing blood toward the star. There is no doubt that Venus was of extraordinary importance to the peoples of Central America, and the surviving Mayan books give us an insight into what was achieved on this score.

Probably more often than not, what we are inclined to identify as "astronomy" was an art of an informal and descriptive kind, even when entering into the rituals of which most must have been aware. There is no doubt that at this level it was a part of the general consciousness. A good example is in the *Popol Vuh* (*Book of the People*), a Mayan story of what happened to the hero, who was the Morning Star, in the underworld. Captured and decapitated, his skull spat at the daughter of the Lord of Death, who conceived a child that became once again the Morning Star. The *Popol Vuh* is one of several sources from which we know the cosmic nature of a ball game that was played right across the continent. The ball, of rubber, represented the Sun, and victory went to the team that first put the ball through a stone ring, about 6 meters above the ground. Several courts still survive. Board games in use often had a cosmic meaning. The Mexican game of *patolli*, for instance, moves a stone representing a heavenly body through four divisions of the board that represent divisions of the sky.

The stone rings and walls down the side of the courts used for ball games were in some cases clearly aligned on astronomical events, but there are numerous other stone monuments throughout southern, central and north America, with pillars, entrances, and windows, all showing clear alignments on sunrise and sunset at the solstices. Some have hieroglyphic marks that have left ample scope for the imagination of their interpreters. One stone found at Chapultepec as long ago as 1775 had under it three crossed arrows pointed accurately to sunrise at equinox and solstices. Alignments on the cardinal points north, south, east and west are commonplace, especially on the pyramids of Mexico and Central America. Alignments on the solstitial risings and settings of the Sun are also common. One example is the Aztec Templo Mayor, in the center of Mexico City, the remains of which were excavated thoroughly only after workers on a subway broke through the wall of a richly endowed offertory, by chance, in the 1980s. The Caracol of Chichén Itzá (plate 7), on the other hand, a Mayan building of the ninth or tenth centuries that has been studied since the 1930s with a view to finding potential astronomical alignments, has been said to reveal alignments to the extremes of Venus's horizon position. This seems unlikely, but similar claims have been made for the governor's Palace at Uxmal, famed also for its depiction of the Mayan zodiacal constellations. These claims have been even more strongly disputed than those for the Caracol, and yet the existence of some sort of cosmic element in the architecture is not improbable.

There is little written evidence about the astronomical aspects of religion in Peru, but there are monuments there with alignments on the Sun's position at the solstices. In the southern region there are linear constructions on the hills behind the site of Nasca, with its straight rows of whitish stones that might possibly have been astronomically aligned. Across the lines there are very extensive outlines of birds that cannot be seen from the ground, but that resemble birds on Nasca textiles dating from a little before or after the beginning of the Christian era.

In Europe, the alignments of the monuments of prehistory are the only testimony to their astronomical character. In Central America, there has been ample living testimony. An early Spanish writer, Fr. Toribio Motolinía, telling of the Aztecs, reports a festival that took place in the Aztec capital at the equinox. The chief religious building, a double pyramid now called the Templo Mayor, was "slightly out of true," so Montezuma wished to pull it down and rebuild it correctly. The pyramid is surmounted by twin temples, between which sunrise at the equinoxes would have been observed from an observation tower on the Temple of Quetzalcoatl to the west of it. Many urban centers have an appearance of equally careful planning. The most famous is the temple complex at Teotihuacán, which arose in a relatively short space of time ending around 50 BC. This, the largest and most influential of all Central American cities of the pre-Columbian period, includes a vast Pyramid of the Sun and a lesser Pyramid of the Moon. The city was certainly laid out with great accuracy on an axis a little over 15° east of north (and the same south of east, of course, looking in the reverse direction). The first direction was in all probability connected with a sighting of the rising of the Pleiades. Two pecked petroglyphs (diagrams in stone, 2.4 kilometers apart along an east-west line) that were found there had very probably a calendrical purpose—a claim which would make sense of some notes by a sixteenth century Spanish priest, Bernardino de Sahagún, on the Aztec use of the Pleiades for checking on a certain 52-year calendar cycle. The dating of the foundation might be the second century AD, which fits the arrangement well, and there are other buildings in southern and central America that seem to follow a similar convention.

As for the stars, throughout most of the American continent we find attention being given to the Pleiades. As so often in the Babylonian and Greek astronomy, these were linked with Venus. They were also linked with harvest and rain, from Peru to the Eskimo, so that in Aztecan astronomy they had a name denoting a marketplace, elsewhere maize, doves, or a granary. Seven-fold items were obviously favored for our "seven sisters." To the Algonquin, the Pleiades were the seven heated stones from a ritual bath. Legends concerning them are legion: in many they are boy or girl dancers. To the Maya, as well as to the Micmac, they were the rattle of a rattlesnake, and this last idea makes understandable the association with the Way of the Dead at Teotihuacán, for from some of the mounds rattlesnake figures have been excavated.

The Great Bear too has its large collection of stories, but they, and the mystery of the many elements shared with legends from the Eurasian land mass, are problems more of anthropology than history.

THE INCAS

The culmination of a Central and South American skill in binding together tribes into kingdoms and empires was reached with the Incan culture of Peru. Gold had been known from the second century BC and became a symbol of gods and kings and of the Sun, whom the Incas conceived to be their ancestor. Spanish greed for gold hastened the downfall of this remarkable civilization, although it had begun from within, when Atahuallpa rose in revolt. Powerful though the Inca empire was, its rule over a vast area of the Andes had lasted for little more than a century when the Spaniards arrived. Its center was the Andean city of Cuzco, which was laid out in a carefully contrived plan, with roads leading out to the limits of the empire from each of the four corners of a central square. The resulting divisions of their territory were assigned a social ranking, as were their inhabitants. Religion, society, and cosmography, were integrated in this way, and astronomy provided a binding force, much as it did in neighboring cultures.

The city had a wheel-like structure, at the hub of which was the Coricancha, a center for ancestor worship, today covered by the church of Santo Domingo. Along the spokes of the wheel, stretching across the landscape, were strings of *huacas*, 328 in all, considered to be openings into the body of Mother Earth, places where the people communed with their deities. There were monuments—for example, arrays of pillars—indicating specific astronomical events on the horizon, solar, lunar, stellar, and planetary events—yet again Venusian. The rising of the Pleiades star cluster was given great importance, and was used in the regulation of the Incan year. (Whereas in most parts of the world most people can make out only six or seven stars in the Pleiades, the clear, high-altitude atmosphere allowed sharp-sighted Cuzco residents to make out twelve or thirteen.) Among the many celestial alignments related to the strings of *huacas* are those on the stars α and β Centauri—"The Eyes of the Llama." Theirs was a more pragmatic cosmology than most others in the Americas, being closely linked with agricultural practice. At a social level, the sky helped to provide the code by which people lived, structuring the cityscape, the landscape outside it, and the hierarchy of kinship. Through agriculture, it was also a tool for the control of the rich Andean economy.

NORTH AMERICA

In North America, there was no civilization comparable to those of Central and South America, although many of the simple astronomical rituals of observing rising and settings were shared. There was no native written language, and occasional records discovered by archaeologists engraved on rock are too rudimentary to be interpreted with certainty. The lunar crescent was a favorite symbol, sometimes together with a star, especially in the southern part of North America. (According to one ambitious interpretation, one such has been said to symbolize the supernova of AD 1054, in Taurus, supposedly first seen when the Moon was nearby.) A nineteenth-century Canadian Ojibwa woman tells of how her father kept a tally of days of the Moon by notching a stick that would last him a year. This may not be high astronomy, but it does bring notched Paleolithic bones to mind—and indeed, Alexander Marshack used this example in support of a lunar reading of the bones mentioned at the beginning of our first chapter.

As a link between the end of that chapter and the affairs of pre-Columbian northern peoples, we should note their custom of creating mounds of earth, some of them no more than a couple of meters across, others many hectares in extent. These are chiefly the work of three different cultures. The Adena culture, from about a millennium BC, made some mounds in the forms of animals. An example is the Great Serpent Mound in Adams County, Ohio. The Hopewell culture that followed preferred geometrical shapes. The Mississippian culture, from about AD 1000, built large mounds with platforms, often pyramidal in shape, on which buildings of some sort stood. These were still in use when sixteenth-century Europeans first explored the Mississippi River valley. There is some reason to think that alignments, not only on the solstices but on the Moon's extremes of rising and setting, are indicated by these monuments.

The Hopi of northeast Arizona, the westernmost group of Pueblo Indians, were solstice watchers, marking the horizon artificially if they could not locate suitable natural indicators. They called the solstices "houses," and some of their horizon markers were small shrines, erected by their priests. Some of the shrines have apertures to allow sunlight to illuminate the interior at appropriate seasons. Prayer sticks were deposited in these shrines by the Sun Priest who was in charge of the calendar, to welcome the Sun and help him on his way. A central observing point had to be established, from which the seated priest could follow the Sun's horizon movements, and anticipate the solstices. He was expected to be able to give four days warning, and was criticized if he failed. Having so prepared the ordinary people to congregate, a nine-day ceremony was held to mark the solstice.

Stars too were perhaps involved in ritual sightings. The Bighorn Medicine Wheel in Wyoming, mentioned in chapter 1, and the Moose Mountain Medicine Wheel in Saskatchewan, Canada, both have similar arrangements of cairns round them. They seem to align on summer solstice sunrise and on three bright stars visible at dawn in summer. The uncertainties here are very much greater than in the vaguely similar prehistoric monuments of Europe and the grand monuments of Central America, for the North American structures were generally both cruder and flimsier than either. There is also the problem that in all the literature reporting conversations with native Americans—who were still using some of these as centers of ritual activity—there is very little to suggest that their astronomical associations were known at the time.

Whatever the truth of the matter, only by an act of intellectual charity can we at present say that the pre-Columbian inhabitants of North America ever developed an astronomy with more than a token theoretical element. In Central America, on the other hand, we find peoples who had discovered, independently of their contemporaries on the other side of the world, that—to quote Galileo—"the book of nature is written in mathematics." Their use of mathematics makes their astronomy doubly intriguing, but we should not exaggerate its intellectual depth. Theirs was a mathematics of integral numbers, not using fractions or anything but the simplest of geometrical constructions. And as for their horizon-based view of the world above, it was closely tied to the needs of the observer, and contrasts strongly with the Greek view of a cosmos complete in itself, and independent of mere human beings.

Indian and Persian Astronomy

VEDIC ASTRONOMY

Like all early peoples, those of the Indian subcontinent allied their accounts of divine and supernatural powers with accounts of what they observed in the heavens. The Vedic religion, the source of the modern Hindu religion, is of great historical interest, because it is among the earliest of all religions to have been recorded in literary form—in this case, in Sanskrit—and it shows us this interaction between the cosmic and the divine at work.

The oldest of the Vedic writings, the *Rigveda*, is thought to have been composed over a period of five or six centuries, ending no later than the eleventh century BC, but often edited later. It contains more than one account of the creation of the world, the main version being that the world was made by the gods, as a building of wood, with heaven and earth somehow supported by posts. Later it was suggested that the world was created from the body of a primeval giant. This last idea gave rise to the principle, found in later Vedic literature, that the world is inhabited by a world-soul. Various other cosmogonies followed, with the creation of the ocean sometimes being given precedence, and place being made for the creation of Sun and Moon. There is a certain circularity in it all, however, since Heaven and Earth are generally regarded as the parents of the gods in general; and water was sometimes introduced into the parentage.

There are numerous myths of astral gods, for instance, of the Sun, husband of Dawn, drawn in a chariot by seven horses; and there are simple computational rules for timing Vedic rituals, as early as the twelfth century BC. Vedic literature gives no clear indication that mathematical techniques for describing the motions of heavenly bodies were discussed in India before the fifth century BC. There is much earlier evidence for contact with Mesopotamia, however, in the neo-Assyrian period, for instance in the matter of omens. Some statements in the Vedic texts can be traced back to statements in MUL.APIN. Contact from this direction eventually proved decisive in forming the character of Indian astronomy.

The Vedic texts make much use of periods of time of various lengths, the so-called *yugas* of two, three, four, five, and six years; periods of twelve months of thirty days; periods of half-months with fourteen or fifteen days. No clear evidence of any sophisticated calendar schemes can be found here, but one highly developed aspect of lunar observation during the last few centuries before the Christian era is the scheme of *naksatras*. These are 27 stars (sometimes 28 stars, sometimes groups of stars) which mark the passage of the Moon through the sky in a month. Each is associated with a different deity. The system—or rather two systems—had a long and involved history, and in the Middle Ages went as far afield as Europe, where it entered into astrology and geomancy. We earlier touched upon versions of the doctrine in China.

INFLUENCES ON INDIA FROM MESOPOTAMIA AND GREECE

Mesopotamian astronomy reached India in the late fifth century BC, with the conquest of northwest India by the Achaemenids. (This was the dynasty that held power in Persia from 558 to 330.) Contact is evidenced by the writer Lagadha's use of Mesopotamian, Greek, Egyptian and Iranian calendar techniques, that is, the use of such "period-relations" as when 5 years are equated with 1,860 *tithis* (the Sanskrit word for a Mesopotamian unit that we discussed on p. 65 above); or, to take another example, when 25 years are equated with 310 synodic months.(The Egyptians equated 25 years with 309 months, using a different year.) Lagadha also took over a Babylonian doctrine concerning the length of daylight, using not only their arithmetical technique of the "zigzag function" but also the water clock by which time at night could be measured.

Another Mesopotamian instrument taken over by the Indians was the gnomon, the vertical pillar that could indicate the time of day by its shadow. They usually divided the gnomon into twelve units, and this convention was curiously persistent, surviving even in Western astronomy through the use of tables that had their origin in this part of the world. Tables correlating time with shadow-length are of course dependent on geographical latitude, a fact not always appreciated either in India or elsewhere.

Although there was no doubt some small flow of astronomy into India in the centuries that followed, the next significant stage was in the Seleucid period, when Babylonian methods, now modified by the Greeks, found their way eastward. This movement was helped greatly by the enhanced level of trade between western India and the Roman Empire. In AD 149/150, a long Greek astrological treatise was translated into Sanskrit prose, and part of this was given over to mathematical astronomy. In 269/270, this was in turn versified by Sphujidhvaja under the title *Yavanajātaka*. The length of the solar (tropical) year in it turns out to be exactly the same as that accepted by Hipparchus and Ptolemy.

Sphujidhvaja, whose motivation was plainly astrological, uses "linear zigzag" techniques for solar and lunar positions, and a Babylonian System A method for the rising times of the signs of the zodiac, known also from Greek texts. The twelve-part gnomon is now introduced into the rules, but here, and for the procedures given for finding planetary positions, Sphujidhvaja gives what are clearly Greek versions of Babylonian methods.

There are other, parallel, instances of the same tendency, where a Greek intermediary text may be inferred from the style of the procedure being advocated. Periodic times for the motions of the planets are repeatedly found, identical to others found in much earlier cuneiform texts. That the accompanying procedures are often garbled shows conclusively that the parameters were not found independently.

With the "Roman" *Siddhanta* (*Romakasiddhānta*) of the third or fourth century, the doctrine of precession is in evidence, and again the same length is given for the tropical year as that used by Hipparchus. This same work has material on solar eclipse calculation that uses Greek geometrical models, although now there are many signs of slight adaptations of them and their parameters. The same is true of another text of Greek origin, more or less contemporaneous, the *Paulisasiddhanta*—although this uses a length for the solar year that is said by a later Arab writer (al-Battānī) to be due to "the Egyptians and the Babylonians." The text is known through a mediocre summary included in a work entitled *Pañcassiddhāntikā*, into which additional matter has been introduced by its famous sixth-century author, the astrologer Varahamihira.

These works are invaluable for the insight they give us into Greek astronomy during the period before Ptolemy, a period from which few original texts survive. The Indian writers make valiant attempts at problems in spherical trigonometry, using Greek methods, inclusive of methods involving projection on to a plane surface (compare the astrolabe). They introduce the sine function in place of the Greek chord function, and in a sense we may regard their gnomon-shadow tables as tables of the tangent function. They reveal a strong interest in problems relating to solar and lunar eclipse prediction, and in some cases here there are traces of Babylonian omen literature. Their working is often blemished in ways that are to be expected, bearing in mind the crossing of so many cultural barriers, and yet it is clear that Indian astronomy, for all its faults, was by this time far from being a purely passive science. As an example of the level of competence reached by Indian astronomers, there is a statement in the *Pañcassiddhāntikā* concerning the time differences between Alexandria (called Yavanapura) and two Indian towns, Ujjayini and Vārānasī. To establish what amounts to a difference in terrestrial longitudes—the Sun of course covers 360° of terrestrial longitude in 24 hours—a comparison was made of the local times at which an identifiable lunar eclipse was observed. This implies some sort of intellectual contact. However they were achieved, the results obtained were remarkably good: if we put their longitude differences into degrees, we find that in two examples, one of 44° and the other of 54°, the correct modern values are 45;50° and 53;07° respectively. This was not the work of dilettanti.

It is worth mentioning that the town with which others were compared in longitude in these examples was Ujjain, (also spelled Ujjayn or Uzzayn). It acquired a certain celebrity, and even appeared in Ptolemy's *Geography* as Ozēnē. In medieval Latin works, as a result of many misreadings, it was usually rendered Arin. Ujjain had from early times been regarded as a sacred site for Hindus, but its importance to astronomy stems from the fact that it came to be treated as the Indian prime meridian—a fact mentioned in many a work later transmitted to Western culture in medieval times, where it was sometimes described as "the middle of the world." It was considered to be on the equator and at a distance of 90° from both the eastern and western limits of the inhabited world. (Its use as a prime meridian reminds us of the fact that many Christians tried to give that status to Jerusalem—with minimal success.) After a very stormy history, in the eighteenth century Ujjain was governed on behalf of the emperor Muḥammad Shāh, who around 1730 built one of his five observatories there, which went some way toward reviving its old reputation.

Astronomy was actively and persistently pursued in India over the centuries. The Indian Subcontinent was a natural assembly point for numerous different cultural influences, as the star map in plate 9 well illustrates. In 1825, Lieutenant Colonel John Warren published a long and remarkable study of calendars and astronomy in southern India. He told of a calendar maker in Pondicherry who showed him how to compute an eclipse using shells on the ground as his counters and using tables that he had memorized with the help of artificial words and syllables. His Tamil informant, who knew nothing of Hindu astronomical theories, was in this way able to compute a lunar eclipse for 1825 with an accuracy of +4 minutes for its onset, −23 minutes for its middle, and −52 minutes for its end. The tradition in which this man was working went back to the *Pañcassiddhāntikā*, and beyond that to Seleucid Babylonian astronomy—in other words, more than two millennia in all.

It is easy to forget how much learning was done by rote in the past: a versified table of the sines of the planetary equations (by Haridatta) is only an extreme example of a common phenomenon, a human equivalent of electronic means of recall. In Europe, the much simpler task of learning the ecclesiastical calendar was achieved in similar ways by all members of church and university.

Another small point perhaps worth putting on record is the occasional Indian use of 57;18 parts as the length of the radius of a standard circle, chosen because it ensures a circumference of 360 parts. The analogy with our measurement of angles in radians should be obvious. (When, as is usual, we take the radius as unity, the angle subtended at the center by an arc of unit length, the radian, is equal to 57;18°.) Whereas Ptolemy used a standard radius of 60 parts, some Indian astronomers preferred 150 parts. These last two conventions both passed into the Western tradition, eventually being overridden by the decimal system in the Renaissance and early-modern periods.

COSMOLOGICAL INFLUENCES: THE DOUBLE EPICYCLE MODEL

Although influenced by texts going back to Vedic times and by Iranian sources too, the *Puranas* are writings with cosmological sections dating from the early centuries of the Christian era. The Earth at this point was represented as a flat circular disc, with a mountain, Meru, at its center. The mountain was surrounded by alternating rings of sea and land, so that there were seven continents and seven seas. Wheels were conceived to carry the celestial bodies, these turning around the star at the north pole by Brahma, using cords made of wind. This cosmology was taken over by Jainism, a monastic religion which, like Buddhism, denies the authority of the Veda, but from the fifth and sixth centuries onward it was undermined by the influx of a new form of Greek cosmology, with pre-Ptolemaic roots. In short, Aristotelianism reached India.

The sources of these influences from Greece are never likely to be pinned down, but they coincided with the conquests in northern and western India by the Gupta dynasty. Another factor might be persecution of the Nestorian Christians by Emperor Zeno in the fifth century, for this led to an eastward migration, including that of scholars with Greek and Syriac texts. Many settled in Gondeshāpūr. At any event, there was an adoption of simple planetary theory, with its period relations. The Indians translated these into their own system of time reckoning with *yugas* of considerable size. (They are based on the Babylonian figure, which dates from 4,320,000 years ago, called now the Mahāyuga, and certain multiples and submultiples of it, each with its own name. Thus a *Kalpa* is a thousand *Mahāyugas*, or ten thousand *Kaliyugas*, each of 432,000 years.)

This was all distantly related to the Greek notion of a Great Year, and the two traditions were recombined during the Middle Ages. The Greek idea had roots stretching back to a time before Plato, but his notion of a Great Year, after which the planets were taken to return to their starting places, became a standard of reference, and the rather different idea that human history was subject to periodical recurrences, seemed to many to lend it support. (Yet another of Plato's ideas, that planets move at the same speed along their paths—so that their distances from us vary inversely as their angular speeds—is to be found in Indian writings at least as early as the first century, when it was introduced into the earliest known version of the *Paitāmahasiddhānta*.) The Great Year—the length of which Plato never actually specified—caught the imagination of the Neoplatonists, although Augustine was critical of the idea of a periodic return on religious grounds; he thought that it would not allow such sacred historical events as the Redemption their unique importance. Much of the material had never-theless crept into Western thought by the ninth century, and eventually it was bolstered by the writings of the astrologer Abū Maʿshar, in which some of the Indian periodicities were to be found. This subject has a rich history to which no brief account can do justice, but it is worth mentioning that the general idea of a cyclical world history gained credence from the common belief that it was based on wisdom predating the biblical Flood.

Not only is the Indian doctrine reminiscent of Pythagorean and Stoic teaching, but by their numerical values, the *yugas* show yet again that India was being influenced by cultures to the west. These several vast periods of time are almost all divisible by the third power of 60, that is, 216,000. Commonly found are 2, 4, 6, and 8 times this figure, as well as these multiplied by powers of ten. Since the Indian number system was from its beginnings purely decimal, the Babylonian-Greek influence is plainly revealed even here. The *yuga* system might have been developed as early as the third century BC.

The Indians quoted the planetary periods in such a form as "5,775,330 sidereal months are equal to one Kaliyuga," and then provided a date at which the current Kaliyuga began. (It was a certain day in the Western year 3102 BC.) This is enough to give *mean* positions at later dates.

Indian astronomy did not fight shy of the problem of planetary inequalities, but the Indians made matters difficult for themselves by taking very seriously the Aristotelian cosmology they inherited, with its insistence on a purely concentric set of spheres. This, however, they blended with Greek epicyclic astronomy in an extremely interesting way. The basic astronomy was seemingly pre-Ptolemaic. (There appears to be no sign of Ptolemy's equant, for example, although it has to be admitted that the equant principle might somehow have been smuggled in, and concealed in the new geometrical schemes.) The Sun and Moon are each provided with a single epicycle on their deferent circles, which, like all the other deferents, are centered on the Earth. As for the planets, each has two epicycles, with a single center, this center moving round with the mean planetary velocity. (The epicycles were called the *manda* and *shighra* epicycles.) Each epicycle is taken to have a point traveling round it, in one case at a speed similar to that in the epicycle of the corresponding Greek model. The other case is more complicated. Now it moves in such a way that the radius joining the point to the epicycle center lies in the direction of the head of Aries, the zero of longitude. These points on the two epicycles are simply auxiliary points in the model, however, and neither is identifiable with the moving planet as such. Where, then, is the planet supposed to be?

Of course it is safe to say that the planet was simply meant to be at the longitude determined by the procedural rules laid down in the text, and leave the matter there. In other words, we can simply do what an Indian calculator did, apply the rules found in the texts, and produce a figure for the planetary longitude. Alternatively, however, we can begin from the

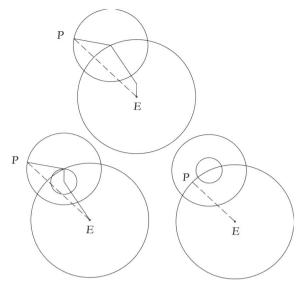

81. Stages in the evolution of Indian planetary models. The unbroken lines are of constant length.

Greek epicyclic model, and try to work out how the Indian procedures could have been derived from it, how the one model could have been turned into the other.

The first, "unthinking" approach, is not without its interest. The complex of computational rules was applied to yield a first approximation to the longitude, and this was repeatedly re-evaluated, until repetition produced no significant change in the result. This "iterative" procedure was one of great sophistication. Similar procedures are often found in later mathematics and astronomy, but this was a very early example, and a powerful one. It is already found in the *Paitāmahasiddhānta*, in a version dating from the fifth century. (The work has much other elaborate mathematical material of Greek type concerning trigonometry and plane projection.) In this procedure, the first approximation was made by halving each of two independent corrections (see the following paragraph), and for this reason the Arab followers of the Indians used a name meaning "the method of halving the equation."

To start from the Greek model and justify these complicated procedures would take us too far afield, but a very cursory account of the situation may be given. Taking first the Greek scheme with a simple eccentric epicycle (fig. 81, uppermost), this may be transformed into a scheme with two epicycles, not concentric with one another, the second epicycle being carrier for the main epicycle (lower left of the figure). (The transformation makes use of the equivalence enunciated by Apollonius, touched on in chapter 4, where figure 52 could be consulted.) Each epicycle has a radius that subtends at the center a certain angle (not drawn in the present figure) that serves as a correction ("anomaly" or "equation"), and that has to be added to, or subtracted from, the mean longitude of the planet. The two epicycles in question will not be concentric, but as far as the task of calculating those correction-angles is concerned, they can be made concentric (lower right of the figure) as long as further rules are introduced for correcting the corrections. Of course it is not possible in a short space to explain how this was done.

This is a far cry from a physical model of the planetary circles. Even though the Indians placed the Earth at the centers of the planetary deferents for purely physical ("philosophical") reasons and perhaps made the epicycles concentric for the same reason, yet in the last analysis they had followed a bizarre path. We find that a Greek model had been accepted, then geometrically transformed, and then distorted yet again before its "correctness" was restored by purely computational devices.

The *Paitāmahasiddhānta* is a text based on the *Brāhmapakṣa*, revelations supposedly from Brahma. There were many other texts in the same tradition, but one influential work in particular followed it closely. It was written—in southern Rājasthān, in Sanskrit still—by Brahmagupta in the year AD 628, and was of considerable historical importance. Known originally as the *Brāhmasphuṭasiddhānta*, it was introduced to the Arabs in Baghdad at some date between AD 771 and 773 by a member of an embassy from Sind. The Arabs gave it—or a work intimately based on it—the name *Zīj al-Sind-hind*. Under this name, it had a new and influential career, in the Eastern and Western Islamic, Byzantine, and Western Christian worlds.

Brahmagupta included various shortcuts in the computation of planetary positions. They included new and intricate rules for planetary longitudes and eclipses. They made use of, among other things, a pulsation in the sizes of some of the epicycles. (He was not the first to make use of the idea, as we shall see.) Of course, if the observer is not at the center of the deferent, the epicycle will—in the mind's eye—seem to fluctuate in size as it moves round the deferent. This way of adding a correction to the planetary longitude is lost when the observer is placed at the deferent center, but whether Brahmagupta had it in mind to compensate for what amounts to the loss of a parameter in the Indian models is a moot point.

Indian astronomers continued to write in this tradition into the seventeenth century, and even beyond. In the eleventh century, extensive changes were made to the planetary parameters, and material also flowed back to India from Islamic (and ultimately Ptolemaic) astronomy, but the techniques used were much as before.

ĀRYABHAṬA

The earliest full descriptions of Indian planetary models are known from sixth-century copies of writings by Āryabhaṭa I, who is said to have been born in AD 476, probably near what is today Patna, in Bihar. His astronomy was held to have been based on a revelation from Brahma. Āryabhaṭa too used the *Paitāmahasiddhānta* as his main source. He wrote various astronomical works, one of which had many imitators in the Arabic-speaking world, from the beginning of the ninth century onward. (The Arabs called it the *Zīj al-Arjabhar*.) It is said that in his home territory in southern India his methods are still in use, a millennium and a half after they were composed. Āryabhaṭa tampered with his inheritance in ways later unpopular with Brahmagupta. He reduced the length of the *yugas*, after which the planets were supposed to return to their original positions, bringing the duration down to 4,320,000 years—one-thousandth of its original length. This was too short a period to allow the planets to start the *yuga* and end it at the head of Aries, and he therefore juggled with various parameters in the planetary theory, fixing the planetary apogees on astrological and numerological principles. His system was finally taken to be one in which the returns to a state of conjunction happened after only 1,080,000 years, a quarter of the full period (known as a *Cataryuga*). His numerological cavortings are of dubious scientific merit, but they were always performed with half an eye on the final parameters of the planetary motions and usually produced plausible final positions. Āryabhaṭa introduced into his account a theory of pulsating planetary epicycles, as Brahmagupta was later to do. That he knew the Greek models is apparent from the fact that he discussed in general terms the relation of the eccentric epicycle to the double epicycle.

At this point it is necessary to draw attention to a controversy which has divided historians of Indian astronomy in recent decades, concerning the precise nature of Indian dependence on Greek astronomy. Earlier in this chapter, I followed the majority view in supposing that Indian algorithms were based on the Greek eccenter-plus-epicycle models. Others have tried to demonstrate that those algorithms fit better a certain equant model. (Matters are somewhat complicated by the slight differences between Ptolemy's parameters and those of the Indians, but the differences can be allowed for.) If this second group of historians is right, then which came first: Ptolemy's *Almagest* or some earlier Greek equant model that served as a common ancestor? As an argument in favor of the second option, numerous sophisticated elements in Ptolemaic astronomy are missing in Indian astronomy. What can be deduced about the Indian use of the equant suggests that it, too, was less refined than that of *Almagest*. The choice would then seem to be between a poor line of transmission of the latter, and a Greek pre-Ptolemaic invention of the equant principle. One can only hope that when future generations of historians decide between the alternatives they will not start from the tacit premise that Ptolemy did little that was truly original.

The elder Bhaskara (his dates of birth and death are unknown, but he was writing in the year AD 629), a notable mathematician, was one of the many exponents of Āryabhaṭa's astronomy, and his writings were in turn the subject of much commentary. He wrote much on questions of lunar and planetary visibility, and he introduced a way of accounting for the planets' movements in latitude by supposing that both of the epicycles were tilted. A direct follower in the same tradition was Latadeva (ca. 500), who helped revise, and contributed to, the *Sūryasiddhānta*. Such revisional activity to this influential work went on for over a thousand years, and to do justice to it one would have to consider not only the numerous contributions themselves, but the local rivalries—between places and even families—to which they bear witness. Astronomers frequently claimed to have corrected the work of their predecessors from observation, having often done no more than put it into different arithmetical dress. For example, the astronomers Keshava (writing in 1496) and Ganesha (writing in 1520) divided time into 11 periods of 4,016 days, with some supposed advantages, and the system swept from Gujarāt across northern India, but to no obvious astronomical advantage.

THE CHARACTER OF INDIAN ASTRONOMY

Indian computational activity had few parallels anywhere in the world between, say, the third century and the ninth, and even after this time it was extraordinary in its sheer extent. It was marked not by theoretical originality on a large scale, but rather by a delight in modifying computational techniques and supplying revised data. In many cases, it seems, changes were made simply for the sake of change. There are many isolated pieces of astronomical doctrine of interest, but it is rare to find any astronomy that is fundamentally new. There are intriguing rules for finding the evection of the Moon—how the Moon's mean place differs form the actual place at first and last quarter—in the writings of Muñjala in the tenth century and Srīpati in the eleventh, but their sources are obscure. In this example, both rules point back somehow to Ptolemy. Indian work in mathematics, and especially in trigonometry, was generally far more original, with results that often found their way to Islam and the West. Indian astronomers developed several advanced techniques for successive approximation. There was no Western mathematician alive in 1400 capable of handling power series for trigonometric functions ($\sin x$, $\cos x$, and $\arctan x$) in a way comparable with Madhava's. The highest achievements of this scholar from Kerala in southern India seem not to have belonged to astronomy, although he is known to have written on the subject (all that is known of his work has to be deduced from reports by pupils or later followers). Between 1393 and 1432, however, a pupil of his, Paramesvara, made some fundamental astronomical observations with the help of an astrolabe, his aim being to improve solar and lunar parameters.

Such activity was rare. Indian religious tradition was a powerful controlling force, not only of content, but also of form and of the ways of learning by rote. As a result, a typical work of eighteenth-century astronomy can be easily mistaken for one of the previous millennium. We are reminded of the situation in China, markedly different from that in the West. When new material worked its way into India from Arab and Persian sources, especially after the tenth century, it often arrived with an admixture of ancient materials, the outdated character of which seems to have been no great

82. Monumental stone instruments in the eighteenth-century observatory built by Jai Singh II at Benares. His much grander observatory in Delhi is better known, although the instruments are of the same type. The large triangular structure on the left, the hypotenuse of which is directed to the pole, is part of a sundial, its structure perhaps owing a little to the way in which Fakhrī sextants were built. (This uncommon term will be explained on p. 210.) The drum-shaped building on the right provides an artificial horizon for observations from inside it—reminiscent of Neolithic practice, five thousand years earlier. (The photograph was taken in the 1860s, possibly by Samuel Bourne.)

cause for concern. When Ptolemy's *Almagest* was translated into Sanskrit in 1732, it was not as a historical exercise. There was something of a renaissance of astronomy at Jaipur (earlier called Amber) in eastern Rājasthān around the same time, and several vast monumental instruments were built there by the ruler Jai Singh II, but they were of a style that Europe had long since put aside (fig. 82). Such a strange mixture of elements appearing within a couple of decades—monumental dials, a Ptolemy translation, and the beginnings of much more advanced European astronomy—is only surprising if we forget the vast size of India, and its great social complexity.

The fact that Indian astronomers' motives were largely religious and astrological helps to explain Indian conservatism, especially in the earlier centuries. Efforts were directed, for instance, toward the preparation of calendars for settling the times of religious observances. Works from Arabic and Persian slowly brought home the message that theory had to survive the test of observation, but still there was no strong drive to link astronomy with other systems of knowledge. The link with physics was rarely attempted, even long after European contacts with India. Europeans, of course, did not in the first instance take their scientific knowledge and books with them with the intention of making Indian converts to Western science, but by slow degrees that happened. Reuben Burrow is a good example of a British mathematics teacher and surveyor in India in the late eighteenth century who helped to create a new two-way traffic in astronomical ideas. He was also a writer on Indian astronomical history, although he was more famed for his love of the bottle than for scholarship. (On the title page of his copy of Newton's *Principia*, Burrow has scurrilous things to say about Queen Anne.) Lancelot Wilkinson was an even more important ambassador for Western science, who ensured that much modern European work was translated into Sanskrit. There are many others, some from France, and they will be discussed later. Activity of this new sort finally bore fruit, and much work of distinction of a new and original kind was done—for instance in astrophysics, at the observatories of Kodaikanal and Madras. It was there in 1909 that John Evershed discovered the radial motions of material, parallel to the Sun's surface, in sunspots. In the twentieth century, in Subrahmanyan Chandrasekhar, India produced one of the world's leading authorities on the physics of stars. There have been many others, but to try to produce a "list of Indian astronomers" would prove only that the single word "astronomy" covered, and continues to cover, a multitude of subjects, different in content and fundamentally different in purpose.

THE INTERPLAY OF EARLY EASTERN CALENDARS

It is tempting, in view of the remarkable Islamic contribution to astronomy, to forget the astronomical traditions of those peoples over whom the Arabs gained political domination. Nevertheless, in the early years of Islam, Arab astronomy was relatively primitive, and most of the civilized peoples who were conquered by the Arabs between the seventh and ninth centuries—Indians, Persians, Syrians, Copts, and Greeks—had reached higher levels of sophistication. This is true, for instance, if we are to judge them by their mathematical planetary theory, their claims to evaluate the ways in which the stars influence human events, or their calendar systems with astronomical content. Jewish astronomy, too, was generally superior, although it was often seen primarily as a tool for settling the religious calendar.

There were, broadly speaking, three main phases in the history of the Jewish calendar: the biblical, the Talmudic, and the post-Talmudic. In all of them the driving force was the need to settle the dates of religious festivals, and all of the chief examples are lunisolar. At first, key points in the calendar year were settled through observation of the Sun and Moon. Gradually, more and more theory was introduced, until, in the final stage, calculation alone held sway. The fact that the weather often made the new moon invisible was a spur to the development of sound theoretical rules about the length of the months. The chief external influence was the Babylonian tradition of lunisolar years, and the Talmud makes it clear that the Jews decided on their names for the months during the Babylonian exile. There remained a powerful Jewish community in Babylon, however, and its members were not without influence.

From before the destruction of the Herodian Temple of Jerusalem in AD 70, there were several reasonably clear calendrical rules, the complex details of which we need not mention here. After the destruction of the Temple, the Sanhedrin—the highest court of justice and supreme council at Jerusalem—was moved to Jabneh (modern-day Yibnah, south of Tel Aviv), and decisions as to the new moon were left to this body, which met on the twenty-ninth day of each month to assess the judgments of suitable witnesses. The patriarch assembled scholars to assist in the decision, which was then announced by letter to the main Jewish communities. There was much dissension between the Babylonian and the Palestinian communities, and there were even reports of incidents in which the Samaritans were said to have lit fires to simulate the new moon, and so lead the Jews into error.

In time there was a clear drift away from the Sanhedrin's use of empirical evidence in the direction of sound lunar and solar astronomical theory. At the end of the second century, for example, the director of an academy at Nehardea in Babylon, Samuel Yarhinaah (nicknamed "The Astronomer," ca. 177–257), composed a calendar of the feasts so as to be independent of Judea. Others followed in the same direction, and by the early fourth century, expertise was such that the appearance of the new moon was settled entirely by calculation, despite conservative protests from those who preferred the old ways.

The final stage in settling procedures came with the reign of the Roman emperor Constantius (337–361), a notorious persecutor of the Jews. Splits were already developing between Jewish and Christian practice. At the Council of Nice in the year 325, when—amidst much dissension—rules were being developed for the timing of Easter, there was wide agreement on one point, namely, that the Christian Easter should not coincide with the Jewish Passover. The actions of Constantius were of a different order of seriousness: he forbade all religious exercises, including the computation of the calendar. It was then that, in an act of defiance, the Patriarch Hillel II (330–365) decided to establish firm and publicly visible rules for calendar computation, rules that were meant to hold for all Jews. By poetic justice they were passed back to the Babylonian community. Most scholars place the date of the established calendar at AD 359, although some suppose that it was a later development, even as late as the year 500.

A thorough discussion of calendars would require a book in itself, but the finished form of the Jewish calendar may be used here to illustrate briefly some of the complexities facing those whose religious writings led them to adopt a lunisolar measure of time. In this case, the calendar months were strictly set by the Moon, while Passover was a feast that had to be held in the spring, that is, with reference to the solar year as an empirical matter. (The feast was held on the evening of the fourteenth day of the (first) month Nisan, and commemorated the "passing over" of the houses of the Israelites whose doorposts were marked with the blood of a lamb, when, according to Exodus 12:29, the Egyptians were smitten with the death of their firstborn.)

The problem of synchronizing in some way the two measures of time, solar and lunar, has been faced in many cultures—in the Chinese and Indian, for instance. The solutions arrived at have differed most radically in the rules for intercalation. The closest that most of those who use a purely solar calendar get to the practice of intercalation is with the familiar rule for the number of days in February, a rule that adjusts, more or less, for the fact that the true solar year is a quarter of a day in excess of a whole number of days. Those who want to bring the Moon into the picture are faced with a much more complicated problem. To take the case of the Jewish calendar mentioned here, there were rules for the number of days (29 or 30) in each month, and there were rules for when we should assign thirteen months to a year, rather than the standard twelve in "common years." A Jewish common year thus turns out to have between 353 and 355 days, while a leap year has between 383 and 385 days. There had to be a rule for leap years, and in fact they were taken to be years 3, 6, 8, 11, 14, 17, and 19 of the 19-year "Metonic" cycle. We need not go deeper into the strange world of calendar juggling, to appreciate that more is at stake than the truths of astronomy. No less a person than Maimonides—the great twelfth-century Jewish scholar, whose ambition was to reconcile Jewish law with rational philosophy—insisted that the Hebrew calendar was first revealed by God to Moses. In sixteenth-century Europe, when the passion for studying these things was at its height, the Hillel calendar was highly esteemed, not primarily on astronomical grounds, but as the solution to a problem of reconciling religious belief with celestial realities. Jewish calendar traditions relating to Passover were of course responsible for influencing the Christian tradition of Easter, despite consciously created differences in the relevant rules, and this explains in part why so much interest was taken in the former.

Not all calendars from the Near East were linked to the Moon. The Christian Copts of Egypt, for instance, kept a calendar based on the ancient Egyptian solar calendar. (Their months were all of 30 days, followed by a "small month" of five days, or of six in every fourth year.) If they are people of the Sun, then Muslims are people of the Moon, for the Islamic calendar is straightforwardly lunar—not lunisolar. Since it is independent of the solar cycle, its months are unrelated to the seasons, and gradually drift through them. There are slightly variant conventions about the beginning of the count of years: most Muslims now accept sunset on 15 July 622, the first day of Muharram in the year of the Hijra (the flight of the Prophet to Medina), although the precise date of the Hijra is not known. There are also differences between the civil calendar, with its standard lengths for the months, and the more commonly accepted religious calendar, which in principle accepts months based on observation of the new moon. (Some present-day astronomers tell of being telephoned with requests as to when the lunar crescent may first be seen, the religious authorities then basing their proclamations on the often dubious answers provided. There are Muslims in various parts of the world who receive the timing of visibility by fax or e-mail from Mecca.) Despite the many minor variants—which make the dating of past historical events uncertain to a day or two—the pattern of the year is generally agreed: there are twelve months, alternating between 30 and 29 days each. A rule is then added to modify the length of the twelfth month, making it 30 days, rather than 29, eleven times in every 30 years. The length of a year is thus normally 354 days, and less often, 355. These years with an extra day were often regarded as a foible of the calculators, and it was ignored by most Arabs, when they based their calendar on observation.

For a rough rule of thumb, correlating years in the Western calendar with those in the Islamic, it is enough to note that the latter are about 3 percent shorter than the former, and begin from the year AD 622, the year of the Hijra. As for the Muslim week, it is like that of the Jews, with seven days (numbered rather than named, in the Muslim world), and ending with the Sabbath as the seventh.

There is a cultural component in the Islamic choice of a lunar calendar, one that is all too easily overlooked, as we plough through the often clumsy numerical schemes. Disputes over its origins tend to concentrate on a tradition that it was introduced by Caliph 'Umar in 638. Many ascribe it to Muḥammad himself, but it is clear that its roots go much further back into the traditions of pre-Islamic Arabia. There were already, in that region, lunar and lunisolar calendars in use, and the Prophet's role was simply to forbid the use of the latter. The Koran has it that since God's law decreed that there are twelve months, intercalating a thirteenth (as in the lunisolar calendar) would only encourage unbelief. The Prophet was also conscious of the threat posed by worship of the Sun, then common in Arabia.

There were, however, other deciding factors of a cultural sort: quite simply, a lunar calendar is a perfectly plausible choice for the nomadic peoples of Arabia, to whom seasonal changes are not critical, whereas there are obvious advantages in a solar or lunisolar calendar for agriculturalists, for whom the seasons mean everything. Those who used lunar calendars certainly had ways of keeping an eye on the seasons. They used, and in some agricultural calendars in the Arabian peninsula still use, a system better known from Chinese and Indian astronomy, in which the ecliptic is divided up into twenty-eight lunar mansions. The mansions were of course marked out by selected stars, and it is thought that the Arabs took a local weather-predicting system of *anwā*', based on the star groups which rose just ahead of the Sun at a given time of the year, and combined it with the Indian system of mansions. The English term comes from the Latin *mansio* (dwelling), chosen by early translators to represent the Arabic *manazil al-qamar*. The latter, used for a camel train's resting place rather than a dwelling, corresponds better to the idea of the Moon's resting place on successive nights of the sidereal month.

There are South Arabian inscriptions revealing different local calendars, and while their details are obscure, some of them were certainly lunisolar, and from outside the peninsula. The fact that some calendrical knowledge had been imported from further east is evident from another consideration. In the early texts, the years were not numbered within a given era—such as the Christian or Muslim eras—but were given the names of appointed officials, eponyms. The Assyrians used the same system, the name for the functionary being *limu*, and so did the Sabeans, and later the Athenians, the official there being called the *archon eponymos*. The system slowly died, for it was difficult to match it to the numerical systems that astronomically-minded scholars preferred. Several different eras were introduced in later pre-Islamic Arabia, the best known being the Himyarite era, commencing in 110 (or possibly 115) BC.

Calendars are pedestrian things in themselves, but they provide a framework for human life, and there are many practical reasons for taking them seriously. After the dramatic expansion of Islam during its first centuries, correspondence with officials in the newly conquered lands had to be dated, but local calendars were wildly different, in style and in the epoch from which years were counted. The Sasanids, the ruling dynasty of Persia who had been in power since AD 226, used 16 June AD 632, the date of the accession of the last Sasanid monarch, Yazdagird III. Syria, where the caliphate was based, but which had been part of the Byzantine empire, used a calendar of the Roman Julian type, but with an epoch of 1 October 312 BC. The Coptic epoch was 3 August AD 284. A unifying calendar, no matter what its form, was clearly needed. An added advantage was that it helped to wipe out the memory of those religions with which old calendars were associated. There remained a problem of converting dates between different systems, especially the Julian calendar of the Roman empire. In this they developed much expertise; but their religion, codified and unchangeable, continued to hold them to a system best suited to desert nomads.

PERSIAN INFLUENCES ON ISLAMIC ASTRONOMY

What distinguishes subsequent Islamic astronomy from that of neighboring cultures is not to be sought in the rules of the calendar. The flourishing of Islamic scholarship generally was spectacular, and began quite simply with a desire to amass the superior knowledge of the outside world. There were numerous astronomical tributaries to that process. Iran had perhaps less to offer than India at this time, but it was a vital intermediary, and Persian culture was ancient and rich. There are many surprising clues as to influence from this direction. There are two Persian calendars, for example, that were started from the year 503 BC, the nineteenth year of the reign of Darius I. It emerges that one of them had the tropical year as its base, the other the sidereal year. An older Persian calendar had been lunisolar, and did not follow Babylonian rules for intercalation. Of the new calendars, however, the religious one used a year of length so close to the Babylonian sidereal year (System B) that borrowing seems certain. Incidentally, the existence of the two different years in the new calendars shows that the ground had already been prepared for Hipparchus's discovery of precession, long before that explicit discovery had taken place, and a millennium or so before the rise of Islam.

Looking beyond the calendar, and further afield, we find several pre-Islamic writings in Pahlavī which later became influential in Islamic astronomy, or rather more often in astrology. (Pahlavī can be roughly described as Old Persian, but written in a Semitic script which hides the ways in which the words were pronounced.) It was soon after Ardashir I founded the Sasanian Empire in AD 226 that translators were set to work on a program of turning Greek and Indian astrological writings into Pahlavī. There were translations into that language of the astrologers Teucer of Babylon, who probably flourished in the first century BC, and Vettius Valens, an influential Greek astrologer from Antioch, of the second century AD. There were translations of Dorotheus of Sidon, an astrologer-poet of much the same period as Vettius Valens, and of the mythical Hermes, as well as of an Indian who went under the name of "Farmasp." An Iranian fashion developed for Hellenistic writing in general, but it was modified by the addition of ideas taken over from Indian religious authors. Sasanian astrologers were not entirely without ideas of their own. They influenced later Muslim and Christian thinking materially by the application to large-scale world history of the more limited Greek astrological doctrines of personal horoscopes and the prediction of what might happen in the coming year. The Sasanians had bigger ideas, and their source is not hard to find: they simply fell under the spell of the Zoroastrian religion, which divided history, past and future, into twelve millennia. To this day, millenarianism is a force to be reckoned with in some parts of the world—among Christians, Jews, and Muslims.

These Persian books were especially important as the bearers of astrology to the Arabs, and astronomy was necessary to them by way of support. In one of the classic Arab texts, by Abū Ma'shar, the writer even goes so far as to present astrology as "the teaching of the Persians." There were other translations of Indian books and "the Roman *megiste*." All of these were available at least as early as AD 250. Gradually the Persians developed their own versions of astronomy, but they were strange hybrids, suspended as their culture was between Greece and India. Around AD 550, what was to become the most influential Persian work of all was revised, to appear as the *Zīj-i Shatro-ayār*. Known after its translation into Arabic (around 790) as the *Zīj al-Shāh*, the work was heavily dependent on Indian sources. It was still in use in Spain in the eleventh century, chiefly for its table of star positions. The work is lost, but much is known about it from others, especially from the great eleventh-century Muslim scholar al-Bīrūnī.

This great astronomer, mathematician, geographer, and historian— whom we shall meet again in the next chapter—was born in the region south of the Aral Sea and died in Ghazni (which is now in Afghanistan). He traveled widely, and was very well placed to write a full-length study of Indian customs and learning, but he also has interesting things to say about the Persians. Of the "restorer of the Persian empire," Ardashir, he writes that the king restored the classes or castes of the population, and that first in order of importance he made knights and princes, second he placed monks, fire-priests, and lawyers, and that he put "physicians, astronomers, and other men of science" in the third class (E. C. Sachau, *Alberuni's India*, London, 1910). (Husbandmen and artisans followed. As a Muslim, al-Bīrūnī was hostile to a distinction of caste.) Of the *Zīj al-Shāh*, he tells us that its zero meridian was taken to be at Babylon, and not, as in Indian tables, at Ujjain. Babylon was probably chosen because the capital of the Sassanids was at Ktesiphon nearby.

When Baghdad was founded by al-Manṣūr on a carefully chosen day in 762—the day was 30 July in the Julian calendar—a horoscope was cast for the event, the propitious moment having been chosen by two astrologers. One was the converted Jew Māshā'allāh, born in Basra, the other a Persian whose family did much translation into Arabic, Naubakht. The Pahlavī work from which the *Zīj al-Shāh* was translated was evidently used for the calculations. Māshā'allāh, who was familiar with Pahlavī, went on to use it when he cast horoscopes for several of his writings in the last decades of the eighth century. We have already seen how, at some time between 771 and 773, a member of an embassy from Sind to al-Manṣūr's court brought with him an important Sanskrit astronomical work. It was said to be the Caliph himself who insisted that it be translated, and this was done by the bearer, working together with al-Fazārī. The result eventually spawned a whole series of similar works in Arabic, most of them at first highly derivative.

Translation of other astronomical and astrological writings into Arabic began to be pursued feverishly at this time, without much attention to the quality of the originals. Side by side with the written authorities, the oral traditions of conquered peoples also found their way into astrological writings. Astrology in particular, where inconsistencies are less obvious than in mathematical astronomy, ran wild. Where Christianity had earlier taken hold, astrological activity had been frowned upon and even suppressed. This was so in Syria, for instance, although there it had to contend with the rampant pagan astrology practiced in Ḥarrān. Among the Jews in the region, the subject was discouraged by the orthodox; yet, astrology was not only being practiced by many of them in the early years of Islam, but they also provided Islam with some of its principal astrological writers. Among them were Sahl ibn Bishr, Sanad ibn'Alī, Rabban al-Ṭabarī, and one of the most influential astrologers of all time, Māshā'allāh, The Arabs had begun a political conquest, but intellectual domination was largely of them, by others.

Eastern Islam

The Islamic faith stems from the teachings of Muḥammad. Persecuted in his native town of Mecca, his fortunes changed dramatically after he fled to Medina with two hundred followers in 622. Through preaching and warfare, the new faith spread rapidly throughout the Arabian peninsula. Within a decade of his death in 632, Armenia, Mesopotamia, and much of Persia to the east, and Egypt to the west, were in the hands of his followers. Within a century of his death, the conquest of the north African coast was complete, much of Spain was taken, and attacks were being made on the Mediterranean coast of Europe. A Muslim advance that penetrated far into what is now France was stemmed by the victory of Charles Martel near Poitiers (in 732). To the east, by this time, all of Persia had been taken, as had much of Kashmir and the Punjab, and conquests beyond those limits continued. The faith of Islam spread to northern India, with the establishment of the Mogul empire. Not since the days of Rome had the world seen anything comparable.

At first, Islam might have seemed irresistible, but gradually, in the East and West, ground was lost. There had been a fundamental split in Islam thirty years after Muḥammad's death, between the Sunni and Shī'ī, followers of the third and fourth of his successors (caliphs). The Syrian-based caliphate, the Umayyads, ruled an empire from the Atlantic to China. Overthrown by the 'Abbāsids in 750, the seat of government was then moved to Baghdad. Gradually, these Arabs were forced to share power with dynasties elsewhere—for instance in Persia and Turkey (the Seljuks). The first Western places to gain independence from Baghdad were in Spain, where an Umayyad prince, fleeing from the east, set up an emirate in 756. The Umayyad caliphate of Cordoba lasted until 1031. The Muslims were finally expelled from Spain in 1492, but their contribution to European astronomical learning, especially in earlier centuries, had been very great indeed, as we shall see in chapter 9.

Between the tenth and thirteenth centuries a succession of invasions by heathen Mongols took away much of the territory that had been conquered by the Arabs, creating a new empire that had its capital first in Mongolia, later in Beijing. The Mongols went as far in their conquests as Anatolia, and were eventually converted to Islam, founding various Mongol-Turkish states along the way. From the opposite direction came the attacks on Islam made by the various Crusades of the Christian Church, which were concentrated on a region small in area but not in spiritual importance to the Muslims. Viewed from a modern perspective, however, the Latin kingdoms of Jerusalem were no more than a minor check on the influence of Islam over the entire area, which was completed by the conquest of Constantinople in 1453, bringing to an end the old Byzantine empire.

From its beginnings, Islam was a remarkable movement in spiritual imperialism, and during its first five or six centuries, the arts and sciences of the ancient world were fostered and then developed under its protection to an extraordinary degree. The peoples who adopted Islam—or who had it thrust upon them—spoke a great variety of languages. Greek was of course the language of most of that inherited learning that concerns us, but many Greek texts had been translated, notably into Syriac, in the fifth and sixth centuries. The Arabs, like most other peoples, had their own native astronomical lore, but it was as yet relatively primitive, and without any profound mathematical content. There was much lore concerning cosmic and heliacal risings and settings of stars, and at some stage this was woven into the Indian doctrine of lunar mansions—those divisions of the zodiac which were used as a measure of the passage of the Moon. The focus of this kind of astronomy was agricultural and meteorological. At first it was mainly transmitted orally, but was eventually codified in written almanacs, and for centuries these continued to be produced alongside more sophisticated astronomical treatises.

For a long time, there was a tendency for theologians—who are to be viewed as scholars of sacred law rather than as a priesthood—to be content with the old folk astronomy, but with the rapid influx of new learning in the first two centuries of Islam, pressure was brought to bear on even them. Astronomical techniques were so plainly capable of serving the needs of religion that they were hard to resist: they could help to determine more precisely the hours of prayer, the direction in which to pray—the so-called *qibla*, the direction of Mecca—and the periods of fasting. Calculating the times of prayer became almost an end in itself, feeding off the cleverness of Muslim astronomers, in the way astrology had always done. Respect for the new sciences was encouraged in the newly established administrative centers, and schools were founded where the teaching of them was very efficiently coordinated. We have already seen a few of the ways in which astronomy was brought to the heartland of Islam from Alexandria and India, from Syria and Ḥarrān, from Persia and the fringes of Byzantium. Of course astronomical activity was only one aspect of a much wider intellectual movement, but astronomy had certain advantages, over and above its religious value. It was allied with astrology and so could satisfy a common human desire for certainty about the future. It was regarded as useful in medicine, and since every ruler had his physicians, astrology was to be found being practiced at the top of the social pyramid. At Baghdad, a hospital was built with great pomp, the prototype of many others. The market for astronomy was growing apace. It is worth noting, incidentally, that this hospital was modeled on one at Jūndīshāpūr, in the present-day province of Khuzistan, in southwest Iran. This was a well-respected academic center, for around the year 560 the Persian leader Khosru I had established an institution there on the lines of the Alexandrian Museum, where instruction was mostly conducted in Syriac.

Not only was technical astronomy cultivated in Baghdad, and the mathematics needed to support it, but also Greek cosmology, as a special aspect of Greek philosophy. A Syriac neo-Platonic philosophy was developed that spoke much of celestial influence, and that took over Ḥarrānian theological ideas. Jewish and Christian learning was imbibed just as avidly. Translations of ancient texts were at first often done second- and third-hand, since so much had been taken from the Greeks at an early date by Persians, Indians, Jews, and Syrians. Notable patrons of the sciences included the ʿAbbāsid caliphs at Baghdad, in particular AbūJaʿfar al-Manṣūr, Hārūn al-Rashid, and ʿAbdallāh al-Maʾmūn. Although the description "Islamic" is often applied to the learning that resulted, this obscures its diverse origins. Within a century or two, however, research centers came into being throughout the Islamic world, where genuinely new and remarkable advances were made in astronomy, and indeed in many other sciences.

Al-Maʾmūn was the seventh ʿAbbāsid caliph, well aware of the world outside the palace, for he was the son of a Persian concubine and spoke her language as well as Arabic. It was during his reign (813–833) that a government-supported library at Baghdad, already established for the translation of scientific works into Arabic, reached the pinnacle of its importance. He is said to have acquired Greek manuscripts from Byzantium and Cyprus. His translators worked in teams, comparing different manuscript versions and checking them where possible against earlier Syriac translations. This sort of work continued for about two centuries, after which time it became redundant and died away.

Some of the translations have already been mentioned—those of Ptolemy's *Almagest*, of the *Sindhind*, and the *Zīj al-Shāh*, for example. Thābit ibn Qurra was a distinguished mathematician and astronomer who also acted as translator (I shall discuss him more later). What mattered here was not the simple act of making a few important works available to scholars who would otherwise have had no access to them but the creation of a huge linguistic empire, comparable with those of Greek, Latin, Sanskrit, Chinese, and later English, capable of binding together intellectually large numbers of people of different outlook and belief, so that it became scarcely necessary for them to learn more than a single language to practice their science. "Arabic science" was far more than the science of the Arabs, but with time it became increasingly easy to forget that fact.

THE ZĪJ

We have mentioned several works with the name zīj in the title without explaining the word, although we encountered the broad principle in connection with Ptolemy. His *Almagest* contained a full complement of tables that allowed the practicing astronomer—whose ultimate concern was no doubt in a majority of cases astrological—to perform more or less all of the ordinary calculations needed in his day-to-day work. Ptolemy's tables were interleaved with their theoretical underpinnings, and so were not very convenient for regular computation, where their justification was irrelevant. In the *Handy Tables*, Ptolemy therefore issued a new version, somewhat changed, with an introduction explaining their use. (Apart from his introduction, which survives in the original Greek, and in a copy of some muddled Latin fragments assembled from a Greek original in the sixth century, this work is now only available to us in a revised form due to Theon of Alexandria, who lived about two centuries after Ptolemy.)

The first astronomical works of a mathematical character written in Arabic—although known only through citations in later writings—seem to have been of this character. They came from the area of what are today Afghanistan and the Sind region of Pakistan. The usual word for a single table in this style was zīj. At an early date, the word was applied to a complete set of tables, and this soon became its standard meaning. The word had entered Arabic from Persian, and was handed on to Latin and its vernaculars in such forms as *azig*, and *açig*. Another word with this meaning is the Latin *canon*, which comes from the Greek, often through the Arabic intermediary *qānūn*. All of these words have at least two meanings, first that of a *thread* in a fabric—observe the analogy with the parallel rules marking out the columns of a written table—and second that of a *model*, something to guide one's actions. In the Latin languages, therefore, *canones* was the name given to introductory instructions as to how the tables were to be used.

Some of the tables in a zīj were purely arithmetical or trigonometrical aids. Most zījes had tables for the sine and shadow (tangent or cotangent) functions, listing values at degree, half-degree, or quarter-degree intervals, and typically quoting results to three sexagesimal places. (Ptolemy's table of chords does the same, implying an ideal accuracy of better than five parts in a million.) As time progressed, Muslim astronomers improved on the detail in such tables. Al-Samarqandi, for instance, in the tenth century, had tangent tables drawn up to three sexagesimal places still, but listing values at intervals of a minute of arc. Before the century was out, Ibn Yūnus prepared sine tables to five sexagesimals, computed for intervals of a minute of arc: he did not achieve the accuracy he was tacitly claiming, but he was setting a target at which others could aim. In this particular case, it

83. Despite the great variety of types of astronomical tables within collections (zījes), most share the same general layout—inherited from Greek and Sumerian astronomy. This page is from a zīj compiled by the Andalusian astronomer Ibn al-Raqqām around AD 1300. The mainly sexagesimal system of numeration found in typical zījes uses letters of the Arabic alphabet. Integers could be on the scale of sixty or of ten. It is likely that many astronomers used finger reckoning when calculating. Numbers would then be stated in words, calculations being mostly done mentally (but also with the help of multiplication tables). Intermediate results would have been stored on the hand in ways familiar to non-astronomers, using conventional bending of the fingers. The table shown is for the solar equation. For adepts who wonder why the latter is here called *juz'ī* ("partial"): this is because a more complex table follows, assuming not a fixed but a variable eccentricity.

was a target that was more or less reached in the Samarqand zīj of Ulugh Beg in the early fifteenth century.

Some of the tables in a zīj were required for calendar computation— often involving the conversion of dates from one calendar to another. Others had to do with the time of day, and the related problems of risings and settings of Sun, Moon, and planets. There were tables to evaluate the daily or even hourly changes in position of these bodies; more specifically, they covered mean motions, planetary equations (correction terms based on the geometrical models then accepted), the stationary points in the planets' paths as they move forward and backward in the zodiac, and planetary latitudes. (For a specimen opening in an Arabic zīj containing the solar equation and mean motion of the Moon, see figure 83.) It should be emphasized that while the planetary models used in these tables could be more or less strictly Ptolemaic, deriving from the tradition of *Almagest* and the *Handy Tables*, many of them were based on Indian models, in the *Sindhind* tradition. In the last analysis, most astronomers wanted zījes to bring them as quickly as possible to the end result—a set of solar, lunar, or planetary positions for a particular occasion or problem. Ephemerides, in which complete sets of such positions were tabulated at (typically) daily intervals, were much more welcome than ordinary zījes, but they were relatively rare. Computing them was extremely tedious work, although it could be simplified with the use of certain auxiliary tables, and zījes occasionally included those as well. The great eleventh-century scholar al-Bīrūnī, when writing his lengthy *Instruction in the Elements of the Art of Astrology*, included instructions on the method of calculating ephemerides. It is hard to decide how many were produced, in relation to the zījes on which they were based, for the simple reason that the latter had a lasting value, whereas ephemerides became more or less redundant after the years for which they were written had passed.

Other tables were added to zījes, for lunar parallax—a subsidiary calculation—and for calculating the circumstances of eclipses of the Sun and Moon. Because in most Eastern cultures the date of first visibility of the crescent Moon was of prime religious importance—for instance in settling the beginning and end of Ramadan, the month of fasting—tables were also included to help settle the date. (Legal scholars were usually happy to rely on direct sightings, but astronomers produced a rich variety of techniques for solving this surprisingly difficult problem.) Star coordinates were almost always listed. In simple cases, they could be there to assist with the creation of devices for time-keeping, for instance the quadrant or the astrolabe. In such cases, the lists were short, but often they were in the form of a revised version of the catalog of 1,022 stars found in the *Almagest* of Ptolemy. The Arabic star names used in such lists were to be an overwhelming influence on European nomenclature, and very many remain in use to this day, albeit with bizarre spellings. Tables were usually included to allow for precessional change in star positions. This might be according to the simple theory of Hipparchus and Ptolemy—which added one degree to the longitudes every century—or it might use an improved parameter with the same simple theory. An increase of 1° in 66 ⅔ years is a value often found for a simple constant rate of precession, although some astronomers came even closer to the correct value. (In the ninth century this was in reality 1° in about 71.9 years.) In some zījes, use was made of the more complex theory of what was known as "access and recess." This theory, later known as the theory of "trepidation," made the increase in the ecliptic longitudes of the stars a variable quantity.

Zījes often included collections of tables for drawing up horoscopes, and there were other tables to help in applying such esoteric astrological doctrines as those of the "projection of rays," of "aspects," and of "the revolution of the year." Other tables were included for deriving the length and quality of a person's life, on astrological principles. Many of the tables were appropriate to only a single geographical latitude and longitude. Geographical tables were therefore often found with the rest, being chiefly lists of cities with their terrestrial coordinates. These were often needed when a calculation—astrological or astronomical—related to a place other than that for which the primary table had been set up in the first instance.

Canons, instructions for the use of zījes, were almost invariably added to the tables when they were first compiled, although of course the canons have often been lost over the centuries. They are rarely very long, but they were certainly important, to the extent that they provided those who used them with a smattering of the basic principles of astronomy, something they could not always easily obtain elsewhere. Zījes often therefore became detached from tables and circulated as texts in their own right.

From the middle of the eighth century to the end of the fifteenth, well over two hundred recognizably distinct zījes were produced, and perhaps more than twenty of them incorporated new parameters that were recapitulated on the basis of original observations. If nothing else, this figure should give an idea of the high importance accorded to astronomy in the Islamic world. The basic theory in most instances was that of the *Almagest*, although in some cases, as already indicated, Indian or Iranian theories were used. A notable surviving example of this eastern influence is the zīj of al-Khwārizmī (ca. 840), of which much more will be said later. Baghdad was long the chief new center of activity, the first true successor to Alexandria in matters of science. From the mid-tenth century onward, Iran took over the lead in zīj production in the east. The Jews later played an important part, especially in Muslim Spain. Wherever astronomy was practiced, however, zījes were composed, working tools that the average astronomer no doubt valued far more than the underlying theory.

ABŪ MAʿSHAR

Abū Maʿshar (787–886) was born in or near Balkh in Khurasan, and although he eventually entered the service of the ʿAbbāsids in Baghdad, the conquerors of his people, he was intellectually inclined toward Iran and the Shiʾa sect. In his forty-seventh year, he was persuaded by a renowned Neoplatonist philosopher, al-Kindī, that he must study mathematics in order to understand philosophy, and in this way his interests turned to astrology. The reputation he acquired in this subject was extraordinary and has lasted in some circles to the present day. In a sense, it is a cause for regret that this man stood at the crossroads of so many different traditions—the Greek, the Indian, the Iranian, and the Syrian, as well as their various composite forms—and that he understood so many languages. When making his synthesis, the breadth of his reading was unfortunately not matched by any regard for consistency—the besetting sin of most practicing astrologers, if not all. His philosophical awareness was unusual, even so, and his use of Aristotelian and Platonic writers in his attempts to justify astrology became an influential document in later debate, in the Islamic and Christian worlds. His debased Aristotelianism entered Europe, strangely enough, before much of Aristotle's own work was available there.

Abū Ma'shar's earlier training in religious exegesis had made him an expert in calendar work and chronology, and it is not surprising to find him later arguing for the idea that scientific knowledge has been handed down, imperfectly, from a divine source which we can come to know from revelation. He wrote his *Zīj al-hazārāt* to recover the lost knowledge of the true astronomy, and in this zīj he made use of Indian planetary parameters and mean motions (using the *yugas*), but with a Ptolemaic model. If ever proof was needed of the progressive debasement of knowledge, here it is; and yet, the zīj was supposedly based on a manuscript buried at Iṣfahān before the biblical Flood.

Abū Ma'shar's historical approach to science is clearly visible in a doctrine that struck a chord in many a later writer both in the East and West. The idea was that human institutions—for example, religious sects and secular powers like his masters in the caliphate—rise and fall according to a timetable set by certain types of conjunction of the planets Saturn, Jupiter, and Mars. This was a doctrine of hope for those who looked forward to an Iranian revival, and of apprehension in later centuries for those who awaited the end of the world or the coming of the Antichrist.

The more discerning astronomers of Islam were severe critics of Abū Ma'shar. One of the greatest of them was al-Bīrūnī, and it is a symptom of the real motivation of later "astronomers" that Biruni's work is relatively rare, and was indeed unknown in medieval Europe, while copies of Abū Ma'shar's writings are legion.

AL-KHWĀRIZMĪ

Another highly influential astronomer who worked under the patronage of the caliphs of Baghdad was al-Khwārizmī (d. before 850), renowned for his zīj, a mixture of largely inconsistent Hindu, Persian, and Hellenistic elements. The fundamental era of the zīj—the base date from which calculations were done—was the Yazdajird (or Yazdagird) era, and the calendar was Persian. In due course it was revised by the Muslim astronomer al-Majrītī from Córdoba, Spain, who not unnaturally changed the era to the Islamic Hijra, and recalculated the tables to his own meridian. The sheer length of this intellectual chain is remarkable: parts of it have been mentioned already, but its sinuous path across half the globe is worthy of note. Pre-Ptolemaic Hellenistic astronomy first passed to India, and eventually from there into a Pahlavi treatise, in Sasanian Iran. There it was revised several times, ending with the version of the Sasanian King Yazdigird III. From that last Pahlavi treatise it was translated into Arabic as the *Zīj al-Shāh*. Through intermediaries (perhaps al-Fazārī, whom we encountered earlier as translator of an Indian work) it reached al-Khwārizmī, who based his own tables on it. These tables, remarkable more for their influence than their quality, in due course passed along the length of the Mediterranean to al-Andalus, Muslim Spain. Even then, the long journey of what had become somewhat superannuated knowledge was far from complete. By the early twelfth century, a version of the tables had reached England, with the help of Adelard of Bath. Whether or not Adelard was working in Spain is a question for debate, but he was certainly using at best only al-Majrīṭī's revision, and perhaps even a revision of that. While al-Khwārizmī zīj is the earliest Arabic astronomical treatise of any size to survive, today it is known only through the Latin translation by Adelard.

Even without piecing together all the stages of that long journey, the Hindu connection is easily proved today from the parameters underlying the tables. Some evidently come from the Persian source of the *Zīj al-Shāh*. Many of the procedural rules laid down in the explanatory canons, for example the method of "halving the equation," are also Indian in character. There are reasons for suspecting that al-Khwārizmī might have made use of the equant in his working, which has led some to suspect that the equant was a pre-Ptolemaic invention; but the possibility remains that it might have arrived from a Ptolemaic source independently of the Indian material. This is a historical problem to which I alluded earlier (see p. 179).

Al-Khwārizmī's zīj did not go uncriticized. From the very first, its shortcomings were noted by al-Farghānī, a young contemporary—a fact noted by al-Bīrūnī. If this adverse publicity was echoed in Spain, the echo was very faint indeed, whereas al-Khwārizmī's zīj had an enthusiastic reception. From Spain it was launched into a highly successful European career. One of the strangest proofs of al-Khwārizmī's capacity for survival is the fact that his zīj was still in use in Samaria in the eighteenth century, and in Cairo, as late as the nineteenth—a fact known from documents preserved in the Jewish Geniza there.

Al-Khwārizmī wrote an influential work on algebra, and indeed, the word "algorithm" comes from his name, but he also wrote what seems to be the oldest extant treatise on the astrolabe in the Arabic-Islamic tradition. It is at present known from only a single manuscript.

AL-BATTĀNĪ

During the ninth century, the *Almagest* and the *Handy Tables* became available in Arabic translation, and the general quality of astronomical work improved greatly as the superiority of Ptolemy's system became recognized. The two centuries following the death of al-Khwārizmī saw five great Islamic astronomers: al-Battānī, al-Ṣūfī, Abū'l Wafā, Ibn Yunis, and al-Bīrūnī. Far from being the products of a single center of activity, they worked in places as far afield as al-Raqqa, Baghdad, Cairo, and Afghanistan. The Islamic world was beginning to dissolve into separate states, in a movement that we sketched at the beginning of this chapter.

Al-Raqqa is on the left bank of the Euphrates, in the northern part of modern Syria. In other words, although he was a Muslim, al-Battānī (ca. 858–929) came from the region around the city of Ḥarrān, where an astral religion was still practiced, and even tolerated by its Muslim rulers. Thābit ibn Qurra, a generation before him, had adhered to it.

Here al-Battānī composed his zīj—a name that does not do justice to such a solid text—basing it essentially on Ptolemy's superior methods. Despite its fame, it is now known in Arabic from only a single manuscript. There were translations of the text into Latin in the middle of the twelfth century, and later into Spanish and Hebrew, here based on the meridian of Jerusalem.

Between Ptolemy and the end of the eighth century, very few astronomers had a clear conception of their science as one that required observation as a test of theory. In the preface to his zīj, al-Battānī made it clear that he at least had understood this precept, implicit in the *Almagest*, and he set something of a fashion for observation. Admittedly he was not the first to recognize the urgency of the matter. Thus during the reign of of al-Ma'mūn (d. 833), a group of astronomers had set up a new zīj based on new observations made at Baghdad and Damascus, and had given it the name of *Mumtahan* (tested) *Zīj*. (It became known in the West as the *Tabulae probatae*, the *Tested Tables*.) Also under the ʿAbbāsids, Ḥabash al-Hasib (d. 862) had made use of extensive observations of planetary positions, solar and lunar eclipses, observed from the same two places and his native Sāmarrāʾ. We shall come across more of this kind of activity, which was essential to astronomy, if the subject were not to stagnate.

The zīj of al-Battānī, was written in a new and refreshing style, not slavishly repeating everything that had been set down in earlier works, but concentrating on recent developments, such as his newly derived figure for the obliquity of the ecliptic (23;35° as against Ptolemy's poor 23;51,20°), or a new direction for the Sun's apogee, or new formulas in spherical trigonometry. He introduced material on instruments—a sundial with seasonal hours, a new type of armillary, a mural quadrant (that is, a quadrant for mounting on a wall), and a triquetrum (Ptolemy's observing instrument of three hinged rods). There is much that is new, implicit in his extensive tables, and yet at the same time, some of his explanations of planetary theory are hastily thrown together, and even erroneous. We may excuse them as no more than the marks of a talented astronomer in a hurry.

He was not the first with a new figure for the obliquity: a century earlier, al-Ma'mūn's astronomers had found 23;31°, and others had proposed 23;33°. Nor was he the first to detect changes in the solar apogee—Thābit ibn Qurra (or perhaps the Banū Mūsā, two wealthy brothers) had by good fortune previously found what we now know to be a marginally better figure for it. What characterizes al-Battānī's work is his meticulous description of his substantially new methods, a description that allows the reader to assess the *quality* of the result, given the accuracy of the observation. As for this, it was generally good, so that in deriving a new figure for the eccentricity of the Sun's orbit (2;04,45 parts, about 3 percent too high for the time) he improved considerably on Ptolemy's excessively high value. He improved greatly on the figures for precession and the tropical year.

His zīj was influential in select circles in medieval Europe. It was translated in the middle of the twelfth century by Robert of Chester, who also had the distinction of being the first man to translate the Koran into Latin. Another version of his zīj, the only one presently known to survive, was done by Plato of Tivoli at about the same time. The Hebrew tables of Abraham bar Hiyya are a reelaboration of al-Battānī's, and under Alfonso X in the thirteenth century there was a Castilian translation of al-Battānī's canons. These various works were not circulated in great numbers, but those astronomers who, like al-Bīrūnī, thought the zīj worthy of praise were often astronomers of the first rank, scholars such as Abraham ibn Ezra, Richard of Wallingford, Levi ben Gerson, Regiomontanus, Peurbach, and Copernicus. That was praise indeed.

FOUR ASTRONOMERS, FOUR ASPECTS OF ISLAMIC ASTRONOMY

If evidence is needed of the newfound vitality of astronomy, it appears plainly enough in the unprecedented numbers of astronomers with well-deserved and lasting reputations. Consider four men of this class, whose work illustrates different aspects of the subject: al-Ṣūfī (903–986), Abū'l Wafā (940 to 997 or 998), Ibn Yūnus (d. 1009), and Ibn al-Haytham (965 to ca. 1040).

Al-Ṣūfī and Abū'l Wafā were contemporaries who worked for a time in Baghdad, but their contributions were very different. Abū'l Wafā's achievements were mostly mathematical, and it is impossible to do justice to them here, but in brief we may say that he cut down reliance on the theorem of Menelaus—the mathematical workhorse theorem that had been used over and over again in the *Almagest* of Ptolemy and the works of his successors—and introduced a number of new theorems of his own. Unfortunately, however, only the most discriminating of writers took note of his reforms, and his reputation was generally achieved at second and third hand.

Al-Ṣūfī struck a more easily appreciated note. In his *Book of the Fixed Stars*, he made it his mission to integrate Ptolemy's star catalog with the Arab star tradition and terminology and set about defining definite boundaries for the constellations. The constellation drawings associated with his catalog soon became canonical, in Europe too, although there they were regularly supplemented. (Figure 84 shows Eastern and Western examples from the unadulterated Ptolemaic list. Cultural differences fail to hide what are still structurally similar figures, the main difference being in the way the figures are facing.) So well established were the Latinized Arabic star names in Europe by the time Johann Bayer wrote his *Uranometria* in 1603 that it would have been a foolhardy act to try to reform them. Bayer did not, and even though there have been very many revisions in

84. Four constellation figures from *The Book of the Fixed Stars*, the work of tenth-century Muslim astronomer al-Ṣūfī, modeled on a St. Petersburg manuscript (left).

The same, from a fourteenth-century Latin manuscript now in Brussels (right). In order, top to bottom, they are Hercules, Cassiopeia, Gemini, and Navis.

terminology since his time, many of the common names for the stars in use today come indirectly from al-Ṣūfī's work.

Ibn Yūnus had talents of a very different order. Although he earned an enviable reputation as a poet and was certainly concerned with astrology, in many respects his astronomical work has a modern appearance. In his youth he saw the Fatimid conquest of Egypt, and he served two caliphs of the dynasty, making astronomical observations for them between 977 and 1003. To the second, al-Ḥākim, he dedicated his zīj, which is unusual in that it records large numbers of observations, many of them taken from previous observers. He has been associated with instruments in Cairo, instruments of very large dimensions—for instance, with an armillary with rings large enough for a horseman to pass through and an astrolabe three cubits across, although both associations are uncertain. What is certain is that many of the parameters he used in his zīj, on the basis of observations about which he is very vague, were much superior to those of his predecessors. His figure of 23;35° for the obliquity was much quoted, although that value had been found long before by al-Battānī. In the nineteenth century, Simon Newcomb used some of his eclipses to determine the secular acceleration of the Moon.

Like many Islamic astronomers, Ibn Yūnus gave much time to a far from simple problem in spherical trigonometry, that of determining the *qibla*—the direction of Mecca—from the Sun's altitude. The earliest of Muḥammad's followers knew that, when he was in Medina, north of Mecca, he had faced south to pray. Worshippers were thus enjoined to face Mecca in prayer; and more precisely, to face the Ka'ba, the shrine located near the center of the Great Mosque in Mecca. Of course, from distant places there was no significant difference, and how were people to judge the direction of Mecca in any case? Islamic legal scholars had an answer that was far from exact: one must pray in the direction one would stand if facing that part of the Ka'ba which was associated with one's own region. (Even before Islam, its four corners had been associated with four regions: Syria, Iraq, Yemen, and the West.) The Ka'ba, which contains a meteoritic stone that had no doubt inspired wonder when it was found, had a long pagan history antedating Islam, intimately bound up with observation of the heavens in ways common to many prehistoric peoples. It is commonly said to have been orientated so that its southwest face was toward sunset at winter solstice, its northeast face to sunrise at summer solstice, its southeast face toward the rising point of the star Canopus, and its northwest face to the setting point of the handle of the constellation the Plough. This system of alignments—somewhat imperfect, if it was intended as explained here—was much tampered with in later history, when on several occasions it was demolished and rebuilt. Muslim astronomers were occasionally consulted when this was done, but that was a local matter, and a much easier one than the problem of facing in the right direction when praying, or building mosques, far away.

In investigating that theoretical problem of the *qibla*, Ibn Yūnus was continuing an established mathematical tradition. The initial aim had been to represent graphically the *qibla* directions of every town in the Muslim world. To this end, new techniques of map projection were devised, using instruments dedicated to the task. They went back at least as early as Ḥabash al-Hasib in ninth-century Baghdad, and were destined to be improved upon for several centuries thereafter. Ibn Yūnus applied astronomy to Muslim rituals in other ways. He tabulated the five times of daily prayer in relation to the Sun's daily motion, a difficult task twice over, for in addition to the purely astronomical problem there was the fact that prayer times had been defined from the outset in the homely terms of shadow lengths, rules that were not applicable everywhere, and mentioning atmospheric phenomena such as twilight, about which no two observers can ever quite agree. Here, Ibn Yūnus did his utmost to produce an exact science of prayer times, with his meticulous calculations taking into account the atmospheric refraction of the solar ray at the horizon. His figure of forty minutes of arc for the angle between the observed and "true" (level) horizon is perhaps the earliest specific figure recorded for this quantity.

Finally, an important set of tables that Ibn Yūnus based on the parameters of the *Ḥākimī zīj*, give the "equations" of the Sun and Moon, but in a new way, cutting out one stage in the computation of the Moon's longitude. As a result, however, the tables were required to be very extensive. (They are "double entry tables.")

Ibn Yūnus died in 1009—having predicted, it is said, while still in good health, that he would die in seven days time. He saw to his affairs, locked himself away, and recited the Koran until he died on the appointed day. Faith in the truths of astrology does not come in stronger forms than this.

The caliph al-Ḥākim was patron not only of Ibn Yūnus but also of Ibn al-Haytham, one of the greatest of scientific writers of the Middle Ages. (He was usually known in the West as Alhazen, after a part of his full name.) Ibn al-Haytham seems to have come to Egypt from Basra, in Iraq. His most renowned contribution to science was his treatise on optics, which was destined to be of much greater influence in the West than in Islam. (Its influence held until well into the seventeenth century.) He wrote a score of short works on astronomy, however, mostly dealing with specific technical problems, but often also touching on optical matters. Two of his astronomical works merit special attention, for they remind us that there is more to the subject than mathematical abstraction. In *Doubts Concerning Ptolemy*, Ibn al-Haytham covers the *Almagest* at length, the *Planetary Hypotheses*, and even a little of the *Optics*. The work involves a number of misunderstandings, but its main drift is more important than treatment of specific detail. It is directed against Ptolemy's abstract handling of astronomy—in the *Almagest* in particular—which Ibn al-Haytham considers to be a denial of Ptolemy's own physical principles. Poor Ptolemy is in a no-win situation, for when Ibn al-Haytham comes to the *Planetary Hypotheses*, where Ptolemy does take a more physical approach, the complaint is that the work ignores some of the finer mathematical detail in the planetary models of the *Almagest*. How, under these circumstances, could Ptolemy have believed that he had found the true system of the world?

An earlier work than Ibn al-Haytham's *Doubts Concerning Ptolemy* was his *On the Configuration of the World*, which was also critical of Ptolemy's ideas but composed at a level between that of sophisticated mathematical treatises and that of popular and purely descriptive astronomy. Widely known in Arabic, it is especially interesting for the way in which it traveled westward. It was translated into Castilian in the thirteenth century, from Castilian into Latin soon after, and then around the end of that century from the original Arabic into Hebrew. Yet another Hebrew translation followed in 1322, and another Latin version, this time from the earlier Hebrew, in the sixteenth century. Through this multiplicity of versions it became influential in the West—indeed, it was the only complete astronomical work by Ibn al-Haytham that was known in medieval Europe. It addressed the worries of natural philosophers rather than the needs of astronomers, but it is important to remember that the former had the higher status of the two in the western university world of the Middle Ages.

As in his later *Doubts*, the *Almagest* is presented in this early work as a piece of abstract geometry, its imaginary geometrical points, lines, and circles bearing the burden of explaining the movements of real objects in the heavens. Such theories, he had already decided, needed to be clothed in physical reality. (It is possible that Ibn al-Haytham did not know the *Planetary Hypotheses*, except perhaps indirectly, when he wrote the *Configuration*, since there he uses a different type of vocabulary—language apart—and somewhat different concepts.) The theoretical principles by which he said this was to be achieved were all traditional: there was to be no empty space in the universe; celestial bodies were to move with uniform, constant, and circular motions; a natural body could only be supposed to have one natural motion; and to each motion introduced into the *Almagest*, some single spherical body must be made to correspond. These last ideas had repercussions on European criticism of Ptolemy in the fourteenth century.

Ibn al-Haytham later became more conscious of the difficulties of carrying through a thoroughly physicalist program. A work by him that has not survived but that is known through a defense he made of it that survives in Arabic shows him in the guise of constructive critic of the *Planetary Hypotheses*. A serious problem with any physical representation of the Ptolemaic planetary models, as he saw, is that since the epicycle revolves around the equant point rather than the axis of the deferent circle, the motion around the axis of the principal planetary sphere will not be uniform. The accepted physical ("philosophical") guidelines for natural heavenly motion seemed thus to have been broken. He needed a new model in place of the old, and the one he found had an odd resemblance to that of Eudoxus.

In the theory of planetary latitudes as given in the *Almagest*, as we have seen, the epicycles were taken to be inclined to the plane of the ecliptic. Following the physical principles already set forth, each epicycle had to be associated with a single sphere. Modifying Ptolemy's physical model in the *Planetary Hypotheses*, with Earth-centered shells inside the thickness of which are found the eccentrics, equants and epicycles, Ibn al-Haytham found a way of representing the inclined epicycle physically.

The poles of the epicyclic sphere he placed at a distance from the furthest and nearest points of the epicycle (apogee and perigee) equal to Ptolemy's maximum inclination of the appropriate epicycle diameter, and he gave the sphere a rotary motion equal to that of the rotating diameter of the epicycle in *Almagest*. The epicyclic sphere is of course carried round the planet's main (deferent) sphere, but Ibn al-Haytham now suggested adding another, between the main sphere and the epicycle, with a motion equal to the first but in the opposite direction. The resulting motion resembles that produced by the homocentric theory of Eudoxus. Without describing the arrangement in detail, we can say that, as seen from the Earth, the end of the rotating epicycle diameter will trace out a hippopede, a figure-of-eight that is carried round the sky and that it was hoped could account for the planet's motion in latitude.

Ibn al-Haytham wrote much more along the same lines, not so much changing the predictive consequences of conventional theory as giving his audience the impression that he might have found the true physical configuration of the universe, whereas Ptolemy had not. Where he objected to Ptolemy's lunar theory, the pattern of his argument is instructive. He claimed first to have found the only two possible models equivalent to the lunar theory, and then it was a short step to dismissing them with the remark that it was not possible to assume a physical body with the properties of either. Oddly enough, he makes no mention of the great variation in the Moon's distance on Ptolemy's theory, a point on which the lunar model seems blatantly unacceptable.

Dismissive physical arguments of much the same character as Ibn al-Haytham's achieved a certain currency in the West, at first without any great astronomical repercussions. In Eastern Islam, however, in the thirteenth century, the great Naṣīr al-Dīn al-Ṭūsī was stimulated by what he read in Ibn al-Haytham to make yet further criticism of Ptolemy, and to offer an alternative theory of planetary motion.

NAṢĪR AL-DĪN AL-ṬŪSĪ AND HIS FOLLOWERS

Naṣīr al-Dīn al-Ṭūsī (1201–74) was one of those figures in history who combined an acute intellect, an abundance of energy, and the good fortune to occupy a key social position at the center of the Islamic stage.

This made it easy for others to appreciate his worth, and his intellectual influence was probably greater than that of any other single medieval astronomer. Born into a warring continent in Tus in Persia—hence his name—he was educated first at home by his father, one of a long line of Shīʿite scholars, and in several institutions—notably in Nīshāpūr, an important center of learning. He was well trained in virtually all branches of Islamic learning, and he rightly considered himself the heir to Hellenistic science and philosophy. He eventually found security in the service of the Ismāʿīlī ruler al-Alamut, Grand Master of the Assassins—"the Old Man of the Mountains" as he became known in the West. He moved with the court between mountain strongholds, until in 1256 the Īlkhānīd conqueror Hūlāgū, grandson of Genghis Khan, ended Ismāʿīlī rule in northern Persia. The astronomer's fame now ensured him a place in Hūlāgū's entourage. He was at the conquest of Baghdad in 1258, and a year later persuaded Hūlāgū Khān to begin the construction of an observatory at Marāgha in the north-west corner of modern Persia (80 kilometers south of Tabriz). Hūlāgū's brother Moengke, who ruled over a vast area of China, had set in motion plans to build an observatory in Beijing, but this was not completed in his lifetime.

The Marāgha Observatory was in many ways the first research institution on a large scale with a recognizably modern administrative structure. It had an extensive scientific library with a permanent librarian, and a staff of astronomers numbering at least ten, among whom there was at least one Chinese scholar, Fao Mun-ji, and there were very probably more. It was equipped with numerous expensive instruments—a large mural quadrant, parallactic rules, an armillary, and quadrants adjustable in azimuth. The Īlkhānī astronomical tables were completed there in 1272 under the rule of Hūlāgū's successor Abāqā.

Naṣīr al-Dīn al-Ṭūsī wrote a long series of important works on logic, philosophy, mathematics, and theology. He was responsible for a revival of the doctrines of Ibn Sīnā, the great physician and philosopher from Bukhara, who had died nearly two centuries earlier. Al-Ṭūsī was an exceptional geometer, and it is not surprising to find him applying his geometrical skills to the problems in natural philosophy that Ibn al-Haytham had brought to his notice. Composing his own criticism of Ptolemaic astronomy in a work called *Tadhkira* (*The Treasury of Astronomy*), he made it quite clear that the work was to be seen as a summary for non-specialists, and was to include no difficult mathematical proofs. It dealt with the external aspects of earthly as well as celestial bodies. He added to his critique of Ptolemy a positive contribution in the form of some new planetary models. One of the most interesting of these relies on the following theorem:

> If one circle rolls inside the circumference of another with a radius twice as great as that of the first, then any point on the first describes a straight line (a diameter of the fixed circle).

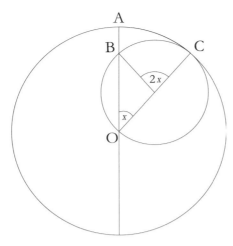

85. The Ṭūsī couple, producing a linear motion by means of one circle rolling inside another. The arcs *BC* and *AC* are equal in length; therefore, if the smaller circle is rolling without slipping inside the larger, the point *B* (regarded as fixed to its circle) must once have been at *A*. This moving point will always lie on *AO*.

This arrangement, today often called a "Ṭūsī couple," is illustrated in figure 85. The theorem is very easily proved: the lengths along the circumferences of the two circles that have been in contact have to be equal, and each is the product of a radius and an angle. For the fixed circle, the angle is only half as great, but the radius is twice as great, as for the rolling circle.

To produce a straight-line motion from a double circular motion was an intriguing thing in itself, but we notice how the rolling circle (or sphere) lends itself to a physical interpretation of the very sort Ptolemy, Ibn al-Haytham, and others had been seeking. Providing the theorem with the clothing of geometry is all very well, as long as we do not lose sight of this fact. Several Marāgha astronomers in addition to al-Ṭūsī were eager to make use of his ideas, and to provide physically viable models more or less equivalent to Ptolemy's, and so answer the criticisms of Ibn al-Haytham, al-Ṭūsī, and others. They included his colleague al-'Urdi (d. 1266), who built the observatory, his student al-Shīrāzī (1236–1311), and an astronomer who lived a century later, Ibn al-Shāṭir (1304–1375).

Al-Ṭūsī generalized the model to three dimensions. By taking the planes of the two circles to be inclined at a small angle, he found that the oscillatory motion approximated to an arc of a great circle. This notion he used in the theory of planetary latitude. What he did is doubly interesting, by virtue of the fact that Copernicus made repeated use of precisely the same device, as well as of other principles from al-Ṭūsī and his followers, so that it is hardly possible to doubt that Copernicus was aware of some text or other in which they were to be found. Greek and Latin materials that made use of al-Ṭūsī's device were circulating in Italy at about the time Copernicus studied there, and this is a suitable place in which to raise the question of the indebtedness of the young Polish scholar.

The evidence is in his greatest work, the *De revolutionibus*, although this was published in the year of his death, 1543. There Copernicus used the al-Ṭūsī device in his model for the variable rate of precession and variation in the obliquity of the ecliptic. In that same book, as well as in his earlier *Commentariolus*, when he was developing his theory of planetary latitudes he used the al-Ṭūsī device to achieve an oscillation in the orbital planes of the planets. In the *Commentariolus*, he used the simpler plane model to achieve a variation in the radius of Mercury's orbit. He did the same tacitly in his *De revolutionibus*. In the *Commentariolus* he based his models for planetary longitude on the models developed by al-'Urdi and Ibn al-Shāṭir, although erroneously in the case of the inferior planets, while in *De revolutionibus*, his models were related to those, and to others by al-'Urdi and al-Shīrāzī. In both works, the lunar model is more or less the same as Ibn al-Shāṭir's—of which more will be said shortly.

There is another example of potential influence on Western Renaissance astronomy from an Islamic source, and one which might have a bearing on Copernicus' procedures, although we shall not pursue that possibility here. The evidence is in a passage of Regiomontanus's *Epitome of the Almagest*, a work published in 1496, twenty years after its author's death. In this, Regiomontanus shows that Ptolemy's epicyclic models for the inner planets can be turned into eccentric models. (Ptolemy had shown this only for the superior planets, and for some unexplained reason denied that it could be done for the others.) The demonstration in question, which might well have helped Copernicus on his way, bears a strong resemblance to a proof—even down to the key diagram—given by the fifteenth-century astronomer 'Alī Qushji, whose career began in Samarqand and ended in Constantinople. He was not unconnected with al-Ṭūsī, on whose theological writings he wrote a commentary. It has been suggested that the manuscript link between East and West in this case was Cardinal Bessarion, who—under circumstances mentioned in chapter 10—had originally suggested to Regiomontanus that he write the *Epitome*.

To return to our sources: rather than enter into the details of the various new planetary models, we may take as an example the way one of them functions. The problem to be solved is that of replacing a motion on an eccentric (say a deferent circle) with a combination of circular motions, the main center being the Earth, the center of the universe. Although we shall use epicycles—the two that make up a Ṭūsī couple—it should be stressed that, to get a model of Ptolemaic type, we need to add yet another epicycle. Here we concern ourselves only with the replacement of the eccentric.

If we take a Ṭūsī couple of carefully chosen dimensions and attach it rigidly, as it were, at the end of a rotating radius passing through the center (*C*) of the deferent circle, then by the properties of the Ṭūsī couple, the oscillating point on the small (rolling) circle will always lie on the rotating radial line through *C* (fig. 86). Several representative positions are shown in the figure, on the assumption that the speed of the "rolling" motion is chosen so that the point on the smaller circle holds the rolling circle's radius in a constant direction. This makes the places where the point is at its greatest and least separations from *C* exactly opposite one another (apogee and perigee, respectively, at the top and bottom of our figure). It will be

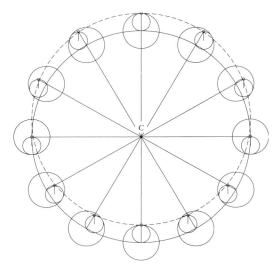

86. The means of replacing an eccentric motion by Ṭūsī-style devices.

seen that an eccentric circle through those opposite points, the broken line in the figure, is almost perfectly accounted for by the three-circle model.

Such models obviously allow great scope for generalization. Inspired by al-Ṭūsī's technique, Ibn al-Shāṭir went further in removing the eccentricity of the deferent and also the equant, using yet additional epicycles. To the Sun he assigned a secondary epicycle riding on a standard epicycle on the deferent, the latter being Earth-centered now. The Moon likewise had a double epicycle, but the proportions and motions were of course different. The lunar model went some way toward correcting the chief blemish on Ptolemy's model, namely the enormous variation in the lunar distance—which the simplest observations show to be an illusion. The planets had nothing less than *triple* epicycles. The bare essentials of the construction for a superior planet are shown in figure 87.

After Ibn al-Shāṭir, the Muslim fashion for designing "philosophically acceptable" non-Ptolemaic schemes seems to have declined somewhat, although there is evidence for it at least as late as the sixteenth century. It had entered medieval Europe by then. It was a strange and long-lived prejudice, but we should be careful to distinguish its varieties. Suppose that we were to observe an oscillatory motion on a straight line. Using some mathematical formula for the variation in position with time, we might succeed in explaining it—simply as a motion on a straight line. A "philosopher," however, might maintain that it should be explained in terms of a Ṭūsī couple and circular motions. Another might go further, and say that—whether it is explained or not—it "really is" a pair of circular motions. This last person needs an independent argument for "what really is." Here, the astronomers called on Aristotle, whose philosophical arguments had begun from very simple observations.

What might seem a less controversial but weaker philosophical position is where we insist that we are giving only a working explanation in

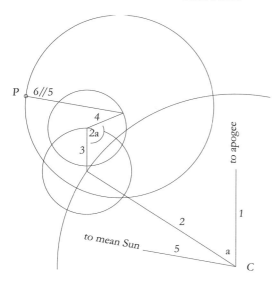

87. Ibn al-Shāṭir's removal of the eccentric deferent and equant. *C* is the Earth, and *P* the planet, the position of which is determined by a configuration that is arrived at as shown. The numerals are added to show the sequence of steps to be followed in arriving at the planet's position. Lines 3 and 6 are parallel to line 1 and 5, respectively. All four basic circular motions on the figure are counterclockwise.

terms of circles. In similar cases today, it is not uncommon to hear an appeal being made, not to an insight into what really exists, but to the aesthetics of the resulting theory, or its simplicity—which often amounts to the same thing. In the past, the less controversial alternative was usually taken by those who themselves accepted the Aristotelian view of what exists, and wished to avoid conflict with it.

THE OBSERVATORY IN ISLAM

Such a remarkable observatory as that at Marāgha, founded by Hūlāgū Khān at al-Ṭūsī's request, could not have arisen without the prior development of some sort of tradition of observation. To be more specific, we might list a hundred instances. There were, for example, the solar and lunar observations made in Dayr Murran near Damascus—on Mount Qasiyūn. There were solstice observations made by Yaḥyā ibn Abi Manṣūr n Baghdad in 829. We know the district of Baghdad where the observations were made, but not, however, whether there was anything that we should call an observatory. There is talk of observation with a "circle," however, in a way suggesting an instrument of some size. In Damascus, we know of a sundial with iron gnomon about five meters high, and a marble mural instrument, perhaps a quadrant, with the same interior radius. These instruments were made on the orders of al-Ma'mūn, the famous 'Abbāsid caliph who had patronized solar and lunar observations in Baghdad. It is not necessary to suppose that these observatories were meant to be permanent institutions, for their aims might have been considered fulfilled as soon as the basic information needed for zījes was obtained. Is there not more to life than collecting a few basic parameters? The information actually gleaned was by no means trivial. The solar parameters found, for instance, were superior to Ptolemy's. Planetary and lunar measurements were somewhat inferior. One important measurement, that in principle needed to be made only

once for the needs of the local faithful, was that of the difference in longitude between one's town and Mecca. This was done on the basis of simultaneous observations of lunar eclipses, and the Baghdad *qibla*, the direction
of Mecca for the purposes of prayer, was no doubt more accurately calculated than most.

An indication of the growing importance attached to this kind of empirical activity is given by the emergence of private observatories. A notable example was set by the brothers Muḥammad, two of the three famous
sons of Mūsā ibn Shakir—the three being always known as the Banū Mūsā,
the "sons of Mūsā," all three of them gifted mathematicians. Muḥammad
and Aḥmad were wealthy patrons of other scholars—their wealth is no
mystery, for their father had been a bandit turned astrologer. The two are
known to have observed the Sun and fixed stars systematically, with the assistance of their protégés, between 840 and 869—mostly in Baghdad, but
also in Sāmarrāʾ and Nīshāpūr.

Official patronage of astronomy waned for a time as a consequence of
political troubles, but it was then renewed from a rather surprising direction. The Buwayhids were a Shīʿite tribal confederation from Daylaman,
a region south of the Caspian Sea. They rose rapidly to prominence in the
tenth and early eleventh centuries, with Buwayhid dynasties taking power
in one center after another: in Jibal, Kerman, Fars, Baghdad—with most
of Iraq—and finally in Rayy. In the mid-eleventh century they fell to the
Seljuks and their allies, but in their heyday they patronized astronomy in
no uncertain way. In Baghdad we hear of the Sharaf al-Dawla, who ruled
from 982 to 989, ordering the astronomer Abū Sahl al-Qūhī to observe
the planets and providing the astronomer with large instruments and a
domed observatory in the palace gardens. The dome is said to have been
pierced with an aperture through which the Sun's rays came, falling on a
surface of 12.5 meters radius. (The surface has been claimed as hemispherical, but is seems more likely to have been an arc of a circle in the meridian. Unfortunately this all fell into disuse with the death of his royal patron.) The Buwayhid ruler of al-Rayy—a town east of modern Teheran in
Iran—subsidized a large-scale instrument of some sort, with which solar
observations were made in 950 and after. It is said that observations were
made in Iṣfahān by al-Ṣūfī, as the basis of his star catalog, but it is not unlikely that he was simply added a correction for precession to older values
of longitudes. He worked under the patronage of two or even three members of the Buwayhi dynasty. The most significant royal patronage from
this dynasty was provided not in Iṣfahān but in al-Rayy, by Fakhr al-Dawla
(d. 997), who supported the activities of Abū Maḥmūd al-Khujandī, an excellent mathematician and a highly pragmatic astronomer.

Fakhrī al-Dawla's name is of some interest, since it was given to a type
of instrument, an example of which he had built for al-Khujandī. This was
a colossal meridian sextant with a radius of 80 cubits (about 20 meters),
constructed in stone. The scale was on a brass strip trapped between two

parallel walls in the meridian plane. The solar image, as cast through some sort of aperture, fell on the brass scale. A movable disk, on which was a cross formed by two diameters, was used to mark exactly the Sun's angular position—in fact its distance from the zenith, the complement of its altitude. The conception of this "Fakhrī sextant" was grander than its execution, for as al-Khujandī confessed to our informant al-Bīrūnī, its sheer weight had shifted the sextant's pivot by a span (say, 10 centimeters). He found a respectable value for the obliquity of the ecliptic, even so, quoting it as 23;32,19°. (A better figure for the year 1000, the year of his death, would have been 23;34,10°.)

We mentioned earlier the colossal instruments seemingly available to Ibn Yūnus, a close contemporary of al-Khujandī, but one working in far-away Egypt. He is said to have had a lavishly equipped observatory, supported by the Fatimid caliph al-Ḥākim, but this is doubtful. In any event, since he records observations from several nearby places, he is likely to have had portable instruments, and so to have been working in a rather different way from his contemporaries in the Near East. It is worth noticing here how eastern Islamic astronomy placed great emphasis on scale, as a prerequisite of accuracy, largely oblivious to those mechanical factors that so often annul the advantages of scale. Al-Khujandī was all too aware of the problem, as we have seen, but in chapter 7 we also saw how Indian astronomers were still constructing large stone instruments in the eighteenth century. Large instruments tend not to survive as well as smaller and more elegant instruments of metal do, and for our knowledge of many of them we are more dependent on ancient report.

Large observatories continued to be built in the Islamic world after the main initiative in theoretical astronomy had passed to Europe. Important instances are those of Samarqand (1420/1421) and Istanbul (1574/1575). The former, housed in a great three-story building, was founded as part of an important research institute by Ulugh Beg (1394–1449), grandson of the famous Timur—better known in English literature as Tamerlane. Ulugh Beg's interest in astronomy was aroused when, as a boy, he visited the remains of the observatory at Marāgha, and perceived its past glory. In due course, the Marāgha observatory set an example for several small-scale copies, but nothing could be strictly compared with it before Ulugh Beg set to work to emulate it in Samarqand. He made its chief instrument an enormous stone Fakhrī sextant, faced in marble. This lay in the plane of the meridian, of course, but was given greater stability than its al-Rayy and Marāgha precursors, by being bedded in a vast trench with a radius of 40 meters, cut into the hillside (fig. 88). Its remains were found in 1908. In 1941, Ulugh Beg's tomb was located in the mausoleum of Tamerlane in Samarqand. He had been killed by an assassin hired by his son, 'Abd al Latif, and the skeleton gave clear evidence that he had met with a violent death.

The staff of Ulugh Beg's observatory included the Persian mathematician and astronomer Jamshīd al-Kāshī, best known today as the author of

88. Ulugh Beg (born Muḥammad Taragai; 1394–1449), as portrayed on a Soviet postage stamp from 1987, together with a cross-section of his observatory. Note the massive "Fakhrī sextant," of masonry lined with marble, at the center of its scale. The occasion for the stamp was the 550th anniversary of Ulugh Beg's completion of his star catalog (1437), an important work for which Edmund Halley still found a use in the seventeenth century, as did John Flamsteed in the eighteenth and Francis Baily in the nineteenth.

the finest Eastern treatise on arithmetic written in the Middle Ages (1427), a work that presented, among other things, the theory of decimal fractions. Among his many remarkable mathematical achievements were his calculations of the values of the sine of 1° and of 2π, both to sixteen decimal places. Needless to say, these two quantities were of great value in astronomy, since they underpin so very many astronomical tables. Ulugh Beg's astronomers produced for him an important zīj under his name, which included some excellent sine and tangent tables, as well as improved planetary parameters and star positions. An unusually large number of these were based on original observations, rather than on a mere updating of Ptolemy or al-Ṣūfī. The star catalog later aroused much interest in Europe, especially in the early days of serious Arabic studies, in the seventeenth century.

Al-Kāshī's contributions to astronomy included the design of a new type of equatorium, an instrument that offered a relatively painless alternative to computing with a zīj. In the simplest kinds of equatorium, the geometrical models for computing planetary positions were simulated by mechanical analogues—circles became graduated disks of metal, radii became rods or threads, and so forth. The disks, when correctly positioned—usually with the help of simple ancillary tables—could yield planetary positions in longitude in a fraction of the time required to grind through ordinary planetary tables of mean motions and equations, as found in zījes. The accuracy of the equatorium could of course never compare with that of a zīj, and although the accuracy of the latter was often quite illusory, the true professional continued to use tables for serious computation. It should not be supposed that al-Kāshī himself was anything but expert in the use of planetary tables. He had a reputation for skill in rapid calculation, and early in life he produced a revision of the zīj of Naṣīr al-Dīn al-Ṭūsī. In this, known as the *Khāqānī Zīj*, in addition to all the usual types of material he included calendars from across the eastern world as far as China; and it is in this work that he reveals himself as one of the very few medieval astronomers to have tried to improve on the complex theory of planetary latitude in Ptolemy's *Almagest*. (Earlier attempts to do so by Ibn al-Haytham and al-Ṭūsī made use of models reminiscent of Eudoxus's nested spheres.) This paragon appears to have been the closest of all the collaborators of Ulugh Beg in the establishment of his observatory-cum-research institute. His ruler's handsome praise was qualified only with remarks about his uncourtly behavior—which was duly excused.

The Istanbul Observatory is interesting because it was so close in time to the great observatory set up by Tycho Brahe at Uraniborg on the Baltic island of Hven. As in the Samarqand Observatory and the eighteenth-century observatories set up at Delhi, Jaipur, Madras, and Benares (the last as shown in fig. 82), by Jai Singh II in the eighteenth century, yet again much use was made of large-scale masonry instruments. More attention was given now to the foundations of these truly monumental instruments, and to their graduation, but their usefulness was limited to a very few types of observation, chiefly of the Sun's position. Even then, the confusion in the solar image cast on the scale (the aperture used to limit the solar image has both an umbra and penumbra) introduced sizable uncertainties into the angles measured.

As precision instruments, armillaries can be discounted, since they are so difficult to make mechanically perfect. A well-known illustration from a sixteenth-century Ottoman Turkish manuscript, now in Istanbul, shows a large bronze armillary instrument—for observation, not demonstration—with a supporting frame constructed entirely in wood. As drawn, the complete instrument would have been nearly five times the height of the men shown using it. One can only hope that this was a case of artistic license.

Instruments for fundamental research were of course always relatively rare. Astrolabes, armillaries, and globes became the symbols of the astronomer, and standard parts of the instrument-maker's repertoire. As in the manufacture of other instruments, Ḥarrān became an important center, and some of the finest astrolabes ever made came from Iran, even as late as the seventeenth century. The fine astrolabe illustrated in chapter 4, figure 68, made for the Safavid ruler Shāh Abbas II, tells us much about the dual purpose of such instruments, which could be as important in the search for patronage as in the search for cosmic truth. The Shāh is described on the front as "The supreme prince, the sultan, the most just, the most great, lord of the centers of command, remover of the causes of tyranny and rebellion, king of the kings of the age." On the back, the message is marginally more devout: "May God Almighty perpetuate his kingdom and his empire and cause his justice and his benefits to spread over the worlds while the spheres revolve and the planets continue in their courses."

A modicum of mathematical skill was called for in making an astro-labe, but skill in casting a sphere was perhaps more difficult to come by. Those who, like al-Ṣūfī, expected accuracy in their globes did not find it. Not all globes were made of metal, but those that were of metal were usually an expensive luxury for the courts of princes and for the rich. For more accurate calculation, the planispheric astrolabe, or some derivative of it, was used, but again, the highest accuracy was vouchsafed only to those able to work with pen, ink, and a zīj. A typical celestial globe had only the main astronomical circles engraved, with perhaps twenty or thirty bright stars. The richest globes would have on them a more substantial fraction of the 1,022 stars that were to be found in such catalogs as al-Ṣūfī's duly plotted by their coordinates. In such cases, the constellation images were usually added, following in the style of the drawings that were often to be found in the catalogs themselves. Together, manuscripts and artifacts helped to hand down Eastern styles of picturing the constellations, styles of surprising constancy, in view of the long routes traveled. Those styles entered Europe, so supplementing the separate traditions that had come through Rome and Byzantium. All have left traces in the modern world, and especially in the demimonde of astrology.

1. Stonehenge, in a view from the south-southwest of the surrounding ditch and bank, allowing the tallest and most finely worked stone to be seen. This was originally the upright of a trilithon, through the middle of which the Sun was to be observed from the Heel Stone at the winter solstice. Its partner, with the lintel they supported, collapsed at some date before 1575.

2. A late but fine Babylonian copy (ca. 500 BC) of the astronomical compendium MUL.APIN. The earliest known copy of this work dates from 687 BC, but it is widely accepted that the observational material recorded on it dates from 1200 BC to 1000 BC. Some have even claimed that the material should be dated a full millennium earlier than this.

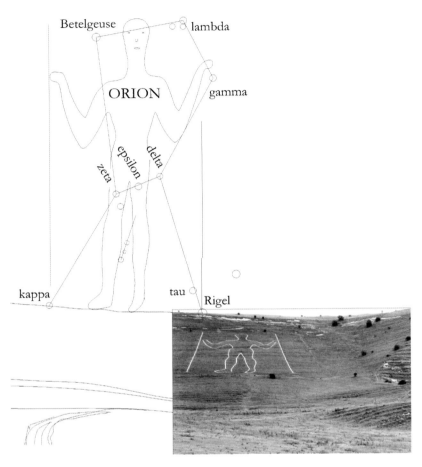

3. Wilmington Man, a chalk figure cut into the downs at Wilmington, in East Sussex, about 8 kilometers from the English Channel. It was possible to locate a point from which the elongated figure had reasonable human proportions, and from that same point the natural curvature of the southern horizon was such that the constellation of Orion, turning with the daily motion of the stars, would have seemed to walk along the ridge. By changing the viewing position, this behavior would have been evident, and reasonably precise, for millennia—from the Neolithic to the modern period. (A priory was established nearby in the Middle Ages.)

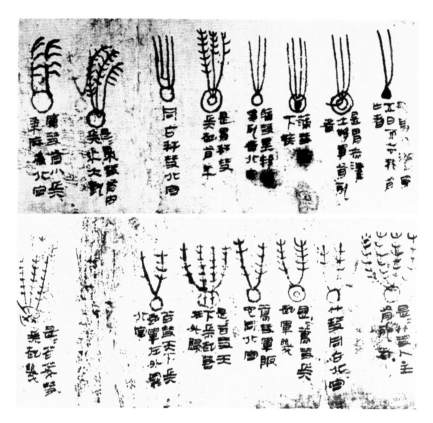

4. Beginning in 1972, a series of tombs dating from the early Han dynasty have been excavated in Mawangdui (in Hunan province, China). Among the many spectacular grave goods, unique texts were found, written on silk. The silk book from which these illustrations are taken (dating from around 205 BC) has much astrological and astronomical content. It records different types of cometary forms, classified by reference to head and tail, with names for the types, and the various disasters associated with them. As far as is known, there is no comparable classification from any other early culture. Chinese astronomers kept systematic records not only of the shapes but also of the appearances, paths, and disappearances of comets. From the same tomb came a book on silk entitled *Wuxingzhan* (Dealing with the five planets), which contains tables of cycles of conjunctions and oppositions of Jupiter, Saturn, and Venus for the period 246 BC–177 BC.

5. The constellation Leo, from al-Ṣūfī's *Suwar al-kawākib* (*Book of Constellation Figures*). The constellations are based on those in Ptolemy's *Almagest*, covering over a thousand stars, with positions duly updated. In this version the figures are drawn twice over, as seen in the sky and as drawn on a celestial globe.

6. Astronomers at work in the short-lived sixteenth-century Istanbul observatory. This was built by Mūrad III for Taqī al-Dīn, clockmaker and astrologer. On the far side of the table, four common instruments are in use: a (wrongly drawn) triquetrum, dividers, a portable quadrant, and an astrolabe. Note the Western-style clock on the table. The terrestrial globe in the foreground was quite up to date. (From an Ottoman manuscript of the *Shāh-nāma*.)

7. *El Caracol* (the Snail) at Chichén Itzá, a Mayan building with Toltec additions of the ninth or tenth centuries A D. Around A D 1200, the city was abandoned in favor of nearby Mayapan, which became the capital of the Yucatan for about two hundred years. The "snail" in the Spanish name refers to the monument's internal structure of concentric circular walls. Horizontal shafts at the top, allowing only narrow views of the horizon, are said to be astronomically aligned on the Sun, and possibly on Venus.

8. Richard of Wallingford in his abbot's study—note his miter on the floor—shown dividing a circular disk. This was probably meant to be his albion. Hanging in the alcove is a quadrant, and the books on the floor perhaps allude to his many writings.

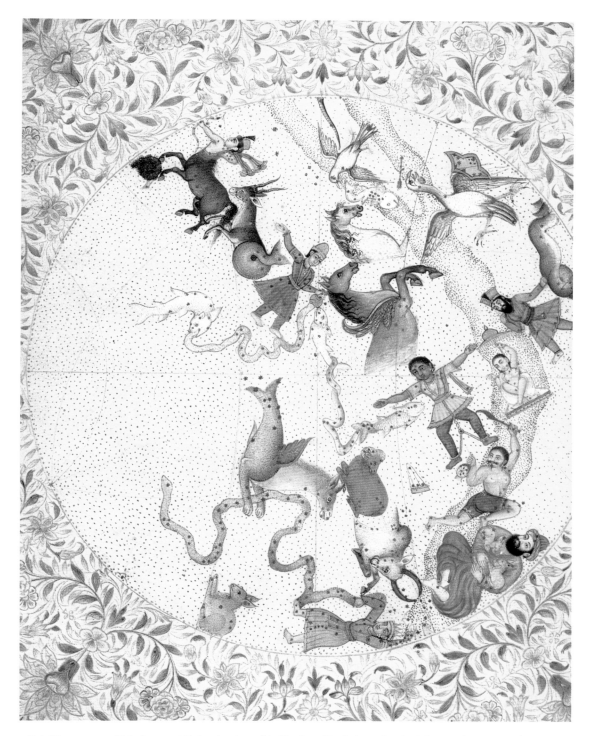

9. Detail from a page with Indian constellation drawings of the Northern Hemisphere, showing influences from many cultures. From a manuscript work by Durgansankara Pathaka, *Sarvasiddhantatattvacudamani* (*The Jewel of the Essence of All Sciences*; 1840). This hemisphere is centered on the vernal point (note Pisces at the center, and Aries and Taurus below). The equator is vertical. The Milky Way, its path marked with dots, runs down the right. The other celestial hemisphere faces this in the original manuscript.

10. *The Ambassadors* by Hans Holbein the Younger. Jean de Dinteville stands on our left, and Georges de Selve on our right. The instruments on the upper shelf of the table are (left to right) a celestial globe, a pillar sundial, a solar instrument designed by Kratzer with a wooden quadrant half hidden behind it, a polyhedral sundial, and a torquetum. On the lower shelf are a handheld terrestrial globe, an arithmetic with set square and compasses, a lute, various flutes, and a hymnal.

✳ 9 ✳

Western Islam and Christian Spain

AL-MAJRĪTĪ AND THE ARRIVAL OF ASTRONOMY IN AL-ANDALUS

Before the end of the tenth century a new impetus was given to astronomy at the western extreme of the civilized world. From the eighth to the fifteenth centuries, Spain was to a greater or lesser extent under Muslim control, and so it became one of the two main channels for the transmission of Arabic science in general to Christian Europe. The other principal route was through Sicily, but from an astronomical point of view this was of lesser importance.

Spain had a tradition of learning long before the arrival of the Muslim conquerors. Few scholars had greater general influence in early medieval Europe than Isidore of Seville (ca. 560–636), an encyclopedic writer who drew mainly on Roman texts. He wrote on broadly cosmological matters, such as atomism, the disposition of the four elements, and so forth, but his knowledge of Greek ideas was almost entirely indirect. The "Isidorian" period in Spanish learning survived the first conquests—the end of the Visigothic kingdom in Spain came in 711—but even then astronomy was still little more than the manipulation of the cycles of the solar and lunar calendars, supplemented later by simple rules for prayer toward Mecca and the orientation of mosques.

In Muslim Spain—al-Andalus—original Greek works, and Arabic and Hebrew commentaries on them, began to arrive on the scene in increasing numbers at a time when a measure of anarchy was descending on Europe generally, in the wake of a decline in the fortunes of the descendents of Charlemagne. With 'Abd al-Raḥmān III (912–961) there began the Caliphate in Córdoba that, culturally speaking, was to outshine the Caliphate of the 'Abbāsids. The second caliph—his son, al-Hakam II—had agents sending him books from as far afield as Baghdad, Damascus, and Cairo. At Córdoba, in the second half of the tenth century, there were scholars studying mathematics, astronomy, and other sciences—scholars who acquired, commented on, and expanded the material coming to them from the East. Whether there were many schools, in the sense of chains of masters

and pupils, is unclear. There was at least one, namely that of Maslama ibn
Aḥmad al-Majrītī, a man active in the last twenty years of the tenth century
who died there around the year 1007. (He is often called simply Maslama.)
As have already seen, among his many achievements was an adaptation of
the tables of al-Khwārizmī to the meridian of Córdoba and the Muslim
calendar (that is, with dates following the era of the Hijra). One or two observations are ascribed to him, but he probably had access to Ma'mūnī observations and built on them.

The example set in Córdoba was later copied by Arab rulers in other
places on the peninsula. Toledo became the main center of astronomical
activity, and Saragossa probably saw some, although it is not well documented. Seville and Valencia are sometimes mentioned in this connection,
although they are more dubious. Astronomy was no mere literary activity:
while it would be wrong to exaggerate their importance, some useful new
observations were made, and important new instruments of calculation
and observation were developed.

One of the reasons for al-Majrītī's historical importance is that he had a
number of pupils who spread his knowledge of astronomy over the whole
of al-Andalus and beyond. The more renowned of them were al-Kirmani
(d. 1066), who worked in Zaragoza, Abū'l-Qāsim Aḥmad (d. 1034; also
known as Ibn al-Ṣaffār), Abū Muslim ibn Khaldūn of Seville, and Ibn
al-Khayyat (d. 1055). Another, Ibn al-Samḥ of Granada (d. 1035; one of
several writers known in Latin texts under the name of Abulcasim), wrote
extensive treatises on the astrolabe and equatorium, and composed a zīj
that again made use of al-Khwārizmī. If the impression has been given
that such things were commonplace, we should not forget that in the rest
of Europe they were as yet an almost total mystery.

CÓRDOBA AND EUROPE

Córdoba's cultural influences quickly spread to the Christian states in the
north of the peninsula, and Latin translations of many scientific works were
produced. Within Islamic Spain there were numerous contacts between
groups professing the three main religions of the area, Christianity, Islam,
and Judaism. Potential translators included, for example, the many bilingual Jews and Mozarabs—that is, Christians whose culture was in many respects the same as that of the Muslims. Even in the tenth century, and long
before the general movement to translate scientific material from Arabic
into Latin, a collection of treatises on arithmetic, geometry, astronomy,
and calendar computation, was produced in Catalonia. One scholarly location with a possible connection to this work was the Christian monastery
of Santa Maria di Ripoll, a Benedictine house at the foot of the Pyrenees,
but Barcelona was very probably a more important center. The one name
we can attach to this activity was a certain Lupitus (called Barchinonensis
or Seniofredus), who was archdeacon of the cathedral there. Such translations as these spread rapidly throughout Europe, helped by the network of

Christian monasteries and the movement of scholars between them. For a long time, it was a mystery how Hermann the Lame (1013–1054), a monk in the monastery of Reichenau, on an island in Lake Constance (bordered now by Switzerland and Germany), could have composed a treatise on the construction and use of the astrolabe. The answer is simply that he had somehow seen a copy of one of the newly translated texts.

One influential scholar who helped to disseminate Spanish science in the tenth century was Gerbert of Aurillac (d. 1003). While still a novice, he had been put in the charge of the Count of Barcelona to receive instruction in the liberal arts. He was taught mathematics, arithmetic, and music by the Bishop of Vich and was perhaps the first scholar of note to carry the new learning across the Pyrenees. In Rome, where he astonished the pope with his learning, he ascended the ecclesiastical hierarchy rapidly: he became abbot of Bobbio and archbishop first of Rheims and then of Ravenna, before becoming the first French pope—as Sylvester II. He did not so much advance scientific knowledge as spread the feeling that it was of real importance. His influence was mainly through cathedral and monastic schools, beginning in Lorraine (the Duchy of Lotharingia, today in eastern France). European knowledge of the astrolabe owed much to him, and he in turn owed it to a text from Ripoll or Barcelona. The West took over some of the Arabic terminology of the astrolabe at this time, although much of it was gradually replaced by Latin equivalents. Other early works on the instrument stem directly or indirectly from his source. They were works written by scholars with independent reputations for learning: Fulbert of Chartres, Hermann the Lame, and Walcher of Malvern.

This was a time when Hindu-Arabic numerals were slowly finding favor among astronomers in Europe. The change was far from sudden. For various reasons, European commerce retained Roman numeration until the sixteenth century and even after, and yet by the late thirteenth century, the best astronomers had begun in earnest to use the new numerals and their associated arithmetical techniques—on which al-Khwārizmī himself had written a standard treatise. Here again it was Spain that provided the stimulus and the point of contact between East and West.

Scholars from other parts of Europe, wanting to learn more of these matters, soon began to travel to the source. The movement was helped along by publicity offered by Petrus Alfonsi, a Spanish Jew (Moshe Sefardi) who had converted to Christianity and who visited the court of Henry I of England around 1110. There he found an interest in astronomical matters, and among others met the Lorraine scholar Walcher, who, after traveling in Italy, had been made prior of Malvern in England. One can judge the sense of intellectual excitement in this new movement, from a work adapted by Walcher from a text by Petrus on eclipse calculation. The work has many weaknesses, and Walcher was often out of his depth, but he wanted desperately to master the new astronomy. Petrus was offering a simple and explicit message: abandon your old and primitive methods of calculation and learn the new techniques from the East. The techniques he was introducing, however, were those of al-Khwārizmī. When Adelard of Bath translated and partially adapted eastern tables, it was al-Khwārizmī's he chose, helping to introduce a consequent confusion into European astronomy that lingered for centuries, for as already explained, al-Khwārizmī's techniques were not purely Ptolemaic, but were mingled with incompatible Hindu theories.

Another zīj—which has since been lost—reported as having been compiled about this time (ca. 1020), was the work of Ibn Mu'ādh al-Jayyānī of the town of Jaén (in Castilian and Latin he was sometimes called Abenmoat). It was put into Latin by the greatest of Latin translators, Gerard of Cremona (ca. 1114–1187), who worked for over forty years in Toledo. The influence of al-Khwārizmī is again clear, although his rules are occasionally found to have been altered—for example, that on first visibility of the lunar crescent. Not the least interesting section of the entire work concerns the division of the astrological "houses," that is, the partitioning of the zodiac into twelve unequal parts, needed when casting a horoscope. There are several mathematical methods of doing this known to history, and here we find a method in use that is traditionally—but mistakenly—ascribed to the Renaissance scholar Regiomontanus. He once owned the manuscript from which the Jaén treatise was later put into print, in the fifteenth century. (This subject will be touched on again in chapter 10.)

IBN AL-ZARQELLU, ACCESS AND RECESS, AND THE TOLEDAN TABLES

In the second half of the eleventh century—a golden age in Andalusian learning—an important group of astronomers formed something approximating a school in Toledo. They included Ibn Sa'id and Ibn al-Zarqellu. (The name of the latter scholar occurs in many different forms, including al-Zarqālī, Ibn al-Zarqiyal or Zarqallu and—in Latin texts—Arzachel.) Ibn al-Zarqellu was to become before long one of the most frequently cited of all astronomical authorities, although in truth, the reference to his name was usually to a set of tables for which he had only a secondary responsibility. This is not to deny his importance. He was a trained artisan who entered the service of Qāḍī Sa'id as a maker of instruments and water clocks. He remained in that city until some time between 1078 and 1080, when the discomfiture of repeated invasion by the Castilians persuaded him to move to Córdoba, and there he lived until his death in 1100.

Ibn al-Zarqellu's true intellectual qualities are manifest in a series of writings. He has been credited with the discovery of the fact that the solar apogee moves—something which we have already seen was known to Thābit and al-Battānī—but he certainly took much trouble to assign a value to the movement: he held it to be (over and above the ordinary precessional motion) 1° in 279 years, on the basis of 25 years of observations.

Ibn al-Zarqellu wrote two works on the "motion of the stars of the eighth sphere," which for brevity's sake we may call precession. In one he described a model for an oscillatory movement of the eighth sphere, an idea that has gone down in history as having originated with Thābit ibn Qurra, although there is no Arabic source attributing the model to him. An early Latin version of the model is quoted at length by John of Spain, who attributed it to Ibn al-Zarqellu. There was a Latin text on the problem, supposedly by Thābit, that might date from only about 1080, a year when observations of the Sun were being made in Toledo and Córdoba. The question is admittedly not closed, however, for Thābit's grandson Ibrāhīm is known to have proposed a complex model of much the same sort, and this was supposed to explain changes in the obliquity of the ecliptic, as well as the increasing longitude of the solar apogee.

It is worth noting that the latter was not taken up by the Iberian or Parisian astronomers who later drafted the Alfonsine Tables in the late thirteenth an early fourteenth centuries, of which we shall soon have more to say. It was accepted by several Iberian astronomers of that later period, including an expert Jewish astronomer Isaac Ibn al-Hadib and might have become a part of established astronomical tradition had anti-Jewish riots in Spain not caused an exodus of Jewish scholars.

The model devised to explain the slow motion of the eighth sphere of fixed stars was one that was easier to draw than to develop mathematically. I shall here do neither, but give only a description in words to convey a rough idea of the complexity of the procedure of determining the movement of the equinoxes, on which the apparent places of the stars depend. The model involves two small circles (with radii of 10.75°), at opposite sides of the celestial sphere, both centered on points on the equator—the mean equinoctial points. Around each of these small circles moves a point, the two moving points being diametrically opposed points on the celestial sphere. The points carry between them a moving ecliptic. Yet other points, where that ecliptic meets the equator, are of course the equinoxes for the time in question. It should be clear that the equinoxes so arrived at shuttle backward and forward along the equator, but since the circuit of the small circles was supposed to take more than four thousand years to complete, the variation in the precessional drift was very slight.

This model of "access and recess," as it is known from its Latin name, is often dismissed today as lamentable nonsense, as something akin to the idea that snakes hatch out of stones. Any assessment of precessional drift, however, required data from the distant past, and since at this time the evidence of earlier authorities seemed to point unequivocally to a variable motion, it would seem more just to regard the model as a triumph of ingenuity. At all events, it became a standard part of astronomical dogma. Petrus Alfonsi reported an early variant: he had tables for an epoch in 1116 in which different parameters were given; and various alternatives appeared over the following centuries, notably one developed by Copernicus.

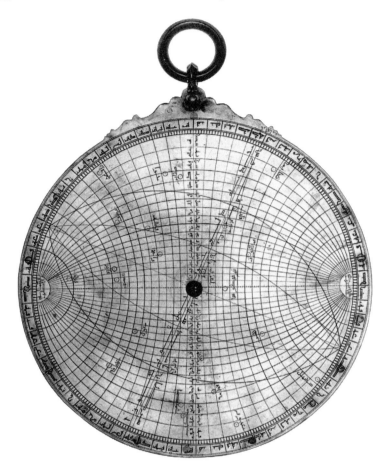

89. Ibn al-Zarqellu's universal astrolabe plate (*al-Shakkāziyya* or *saphea Azarchelis*). This Andalusian (or at least Maghribi, Western) example is engraved on the back of a thirteenth-century
astrolabe with a fairly conventional face. One of its four plates is original, and is for North African latitudes. The others from a later period reach up to the latitudes of central France. It is not
difficult to pick out the two main coordinate systems on this saphea, one equatorial and the other
ecliptic. A full explanation of their use would occupy a short treatise, but it might be helpful to
start by imagining the rotation of the celestial sphere to take place around the horizontal diameter, the lines through the poles giving a measure of time well as right ascension.

Ibn al-Zarqellu wrote original treatises on an equatorium and on
the universal astrolabe. (For the latter, see figure 89.) This last type of
instrument, known in the West as the *saphea Arzachelis* and in al-Andalus
as *al-Shakkāziyya*, was alluded to at the end of chapter 4. It was easier to
make than to understand, and has a complicated history. Robert of Chester, for example, writing in London in 1147, claimed to be translating from
the Arabic when he produced a treatise "by Ptolemy" on the universal
astrolabe, but the work was almost certainly a purely Andalusian fabri-
cation. It is important to remind ourselves here of how thinly spread this
sort of knowledge was, and continued to be. A scholar who, like William
the Englishman in 1231, tried to reproduce the universal astrolabe from

the written source only, could hardly avoid mistakes. The saphea was not well understood in the West until long after Ibn Tibbon's translation of Ibn al-Zarqellu's text in 1263.

More influential by far than these writings by Ibn al-Zarqellu were his canons to the Toledan Tables, which were destined to eclipse all others, in Spain and Europe generally, until the early fourteenth century. It is as well to speak of the evolution of the Toledan Tables, rather than of their composition, for, like most zījes before and after them, they were a compilation of often inconsistent elements. Roughly speaking, the chief debts in this case were to al-Khwārizmī and al-Battānī, but it is quite possible that some effort was put into a program of observation, to verify those older materials.

The original Arabic version even of the canons to the Toledan Tables is lost, but three versions of that introductory material have survived in Latin, together with many variants on the assortment of tables that follow. Those for the solar, lunar, and planetary inequalities (equations) are largely from al-Battānī, with one from al-Khwārizmī. Tables for planetary latitude come from al-Khwārizmī—although a few manuscripts have Ptolemy's added. Others for planetary visibility, stations and retrogradations are from al-Battānī and Ptolemy. There is a considerable debt, in other words, to eastern Islam two or three centuries earlier, but there are some new things. The parameter for the solar mean motion is seemingly new, which might be explained by Ibn al-Zarqellu's interest in the topic; and using this parameter, together with al-Battānī's mean motions, all the other mean motions have been recalculated. There is some astrological material that is possibly new, as far as its appearance in table-form is concerned, but that is not new in any other sense. Star coordinates were updated for precession, of course—and in Latin recensions they were often updated again at the time of copying. The basic epoch is that of the Islamic Hijra, and the meridian of Toledo was used. This meant that all mean motions quoted at epoch had to be recalculated for the time difference between Toledo and Khwārizmī's meridian. Although the difference was not accurately known, the error was not great.

The Toledan Tables influenced Western Islam directly, but also indirectly through at least one of three compilations by Ibn Kammād (ca. 1130). Long afterwards, an astronomer from Seville, Ibn al-Hā'im, issued a zīj (ca. 1205) in which he claimed to be correcting Ibn al-Kammād's's errors. Why should we even take the trouble to put such a claim on record? Are they of any greater importance than a correction to the king's kitchen accounts would have been? Ibn al-Hā'im's was one of the last distinctive zījes to have been produced in al-Andalus. In European terms, it left no mark, but it speaks for an intellectual fever that had taken hold of the Andalusian world and that would soon prove itself contagious. Ibn al-Hā'im's zīj was inevitably overshadowed by the most famous Spanish zīj of all, namely the Alfonsine Tables of the 1270s. These tables, produced under the patronage of the Christian king Alfonso X of León and Castile, will be the subject of a later section.

THE TOLEDAN AND KHWĀRIZMĪAN
TABLES IN COMPETITION

While the old Toledan Tables, in Latin translation, ultimately carried Islamic astronomy well and truly to the heart of Europe, they did so alongside the al-Khwārizmī zīj, and it was long before they became the dominant force. They then remained so until the Alfonsine Tables took their place in the early fourteenth century—and in some places, even later.

Translation from Arabic into Latin and the Castilian vernacular was an activity with a long history in Spain, and we have already seen examples of it. There were at various times important schools of translators in the Pyrenees, the Ebro Valley, and Toledo, for instance. To understand the reason for this activity one must appreciate Christian scholars' realization that they and their schools were seriously deficient in astronomical learning. We have already mentioned the classic example of Walcher, Prior of Malvern Abbey in England, who had been avid to learn from Petrus Alfonsi, at the English court. The zīj to which Petrus introduced him (written ca. 1116) was a rather untidy patchwork of materials, closely related to that of al-Khwārizmī. The planetary mean motions were adapted to the Christian calendar, for instance, in a singularly awkward way. When Adelard of Bath chose to translate a zīj, perhaps with the help of Petrus, it was a version of the al-Khwārizmī Tables, not the Toledan.

The Toledan Tables undoubtedly helped to satisfy a growing European need, but it is hard to see by what criteria they were at length preferred to al-Khwārizmī's. It would be pleasant to think that they were judged on the grounds of predictive accuracy. At all events, tables for Marseilles (ca. 1140) by Raymond of Marseilles, were based on them, and had a certain currency in France. Robert of Chester, who went to Spain in 1141, and who even became Archdeacon of Pamplona before moving to London in 1147, adapted not them but al-Khwārizmī's tables to the meridian of London; but it is significant that he introduced some Toledan material too.

Spanish learning was carried abroad in other texts, and very often by Jewish scholars. A notable example is Abraham bar Hiyya (known as Savasorda, who flourished from 1110 to 1135) from Barcelona, who composed astronomical *Tables of the Prince*, which mix Jewish, Islamic, and Ptolemaic material. At least one Latin manuscript survives. The epoch is the equivalent of 21 September 1104, and the year is not Jewish but the old Egyptian-Ptolemaic-Battānī year of 365 days. Abraham reveals his source, in that he uses al-Battānī's meridian (al-Raqqa)—a strange convention for a Spaniard. Another Spanish Jew was Abraham ibn Ezra of Tudela (ca. 1090 to ca. 1164). He wrote in Latin as well as Hebrew, and traveled widely through Italy, France and England. Abraham wrote tables for Pisa around 1143, following, as he tells us, the zīj of al-Ṣūfī. A London zīj of about

1150 seems to be related to the Pisan Tables, and might also be due to Abraham.

Such intellectual commerce continued throughout most of the Middle Ages. Links with Jewish communities in southern France were strong, but influences spread further afield. This process of exchange lasted to the very time of the expulsions from Spain. Abraham Zacut, for example, born around 1452 in Salamanca, whose name appears on an oft-printed perpetual planetary almanac and tables that were used by Columbus and Vasco da Gama, was himself expelled from the country and died around 1522 in Damascus. Like any other story into which Columbus's name can be woven, it has to be told, although it has recently been shown that all is not quite as simple as was once thought. The canons to the tables are not translated from Zacut's Hebrew works, but probably come from a Portuguese Jew, one José Vizinho. That is much less interesting, however, than the fact that Arabic versions of the tables themselves were made in the sixteenth and seventeenth centuries and were still in use in the nineteenth. In other words, the underlying Alfonsine material had worked its way back into the Arab world, from research done in Southern France by fourteenth-century Jewish astronomers—Jacob ben David Bonjorn and, indirectly, Levi ben Gerson—and for good measure had been on an Atlantic voyage as well.

Versions of the Toledan Tables soon came to hold the European field alone, and continued to do so in some places long after they had been superseded. Well over a hundred manuscripts of them survive to this day, an unusually large number for a genre of text that was of virtually no use once it had been replaced. Scholars in many towns adapted them to their local meridians. The Toulouse Tables were even dignified with the name of "the Christian calendar" in some quarters. One fourteenth-century (Latin) version of the Toledan work was even translated into Greek. The cultural circle was complete.

THE ASTROLABE AND ASTROLABE-QUADRANT

There are at least forty or fifty Western treatises on the traditional form of astrolabe written before the end of the sixteenth century, and yet, including the group already mentioned, they fall into only three families. All stem from the Arabic culture in Spain. As already explained, the general drift of western Europe's first knowledge of the astrolabe was evidently northward and eastward from Catalonia, in the late tenth century. This first move into monastic Europe seems to have petered out by the middle of the eleventh century.

Next came far more powerful intellectual influences, based on more complete texts, that were better argued and better understood. Of central importance was one by the eleventh-century Spanish-Arab astronomer Ibn al-Ṣaffār, and this was translated into Latin twice—first by John of Spain, and then by Plato of Tivoli. The oldest dated western astrolabe—although not the oldest—carries the date 417 of the Hijra, that is, AD 1026/1027, and it is a curious fact that the maker's name shows it to have been made by Ibn al-Ṣaffār's brother.

The third family of texts stems from the school of writers in Castilian at the court of Alfonso x, in the late thirteenth century.

The most influential of the three families was undoubtedly the second. John of Spain's version was used by Raymond of Marseille, and by a writer whose real identity is unknown, but whose work was ascribed to the Jew from Basra and Baghdad, Māshā'allāh. This second family of works introduced further Arabic terminology—including the words "zenith" and "azimuth," for instance—into European astronomy.

As we enter the thirteenth and fourteenth centuries, the Iberian influence in Europe remained just as real, but became less obvious, as astronomers began to blend their sources. Some of the better known European writers were Raymond of Marseille in the twelfth century, and Sacrobosco and Pierre de Maricourt (Petrus Peregrinus) in the thirteenth. With the fourteenth century we enter an age of writing in the vernacular. In French, Pèlerin de Prusse wrote a slender astrolabe treatise for the dauphin, the future Charles v, who was to be crowned king of France in 1364. The English poet Geoffrey Chaucer made a notable contribution, with his treatise in English on the astrolabe—subtitled "Bread and Milk for Children" in one manuscript, presumably by an ironical scribe. This remained the only satisfactory work in English on the instrument before modern times, but it, too, derived from the work of the pseudo-Māshā'allāh.

In the late thirteenth century, a new and inexpensive alternative to the simple astrolabe was found. Imagine an astrolabe with its rete *fixed* in position. (To follow this brief account it might be helpful to consider figure 53 or the astrolabe illustrated in figure 68, both in chapter 4.) A thread through the center carrying a small sliding bead as marker (a seed pearl was the standard marker in the Middle Ages) would serve to locate any point on the rete, such as a star. Stretching the thread over the star, one would simply mark its distance from the center using the bead and note the angle registered on the outer rim. Rotating the thread through a given angle would move the marker to a new position, exactly as if one had rotated the rete. The arrangement would be less intuitive, but mechanically simpler, and easier and cheaper to construct than a complete astrolabe. Suppose now that we take matters one step further, and mentally fold the composite diagram (that is, made up of lines appropriate to the rete and the plate of an astrolabe). We fold it first along a main axis, and then fold it again along the axis at right angles to the first. We shall have a mass of lines, points, and graduated scales, many of them double, but if we have a good grasp of how an astrolabe functions, exactly the same problems can be solved on this "astrolabe-quadrant" as on the astrolabe itself.

To measure altitudes, it was usual to fit small sighting vanes with pinholes to one (radial) edge of the quadrant. In normal use, the quadrant was tilted until a star or other object was seen through the two holes. The

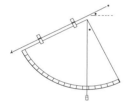

90. Detail of a woodcut by Oronce Fine (1542), in which the
artist is being instructed by Urania, the muse of astronomy. He
holds an astrolabe, its back toward us. Below it is a Profatius
quadrant, and on the ground to our right is a folding sundial.

vertical was settled by means of a plumb bob on a thread hanging from the
geometrical center of the peripheral scale, and the altitude was then read
off from that scale. (This method of finding altitudes was common to most
quadrants, and was not peculiar to the astrolabe-quadrant.)

The earliest known description of an astrolabe-quadrant was written
between 1288 and 1293 in Hebrew by Jacob ben Machir ibn Tibbon, a
writer known in Latin as Profatius Judaeus (ca. 1236–1305). Although he
lived in Provence, he was born in Marseille and died in Montpellier. His
family had come from Granada, a place his grandfather left because of
civil unrest. Both his father and grandfather made reputations as translators from Arabic into Hebrew, and some have assumed that Ibn Tibbon's
so-called new quadrant must have been based on an Islamic prototype.
The fact that he called it the "Israeli quadrant" is hard to reconcile with
that idea. The work was quickly translated into Latin (in 1299 by Armen-
gaud), and the Latin was expanded by the Danish astronomer Peter Nightingale. Through this last treatise, the instrument became well known to
Latin astronomers, and it long remained so. (Figure 90, from a later period,

includes an example.) Its popularity was the result of the extreme ease with which it could be made: a careful drawing on parchment or paper glued to a wooden quadrant was something within every scholar's economic means, although sorting out the mixture of scales and overlapping star-markers might not have been so. For the former reason, no doubt, it became very popular in the Ottoman Empire—usually in the form of a lacquered wooden quadrant—from the fifteenth century onward. It survived in Turkey even into the twentieth century, valued by those who used traditional astronomical methods for ordering religious life.

MACHINERY OF THE HEAVENS

We have seen that in late antiquity, water was used to drive automata, in particular time-measuring devices. Hero of Alexandria (flourished AD 62)—who has often been treated by historians as no more than an ingenious technician, but who has several important mathematical results to his credit—devised numerous machines, including many that were worked by water, steam, and air pressure. A work he wrote on water clocks is now lost. Like his predecessor Archimedes, Hero seems to have influenced the Arabs, and in the tenth century al-Khwārizmī hinted strongly that the old techniques had not been lost.

There are clear signs of Greek influence in the early-thirteenth-century writer al-Jazari. He worked in the service of Naṣīr al-Dīn, the Turkoman king of Diyār Bakr, for whom he made a variety of ingenious machines and gadgets. He acknowledged that the idea for a water-clock he made came from another work, one that he ascribed to Archimedes. Other writers, including the great al-Bīrūnī, mention clocks driven by water or sand, and it is clear that there was a continuing interest in the problem, not simply of turning a wheel to keep time in some abstract way, but of turning a representation of the heavens, be it a plane image or a globe, with a twenty-four hour motion.

Such devices appear in Spain in the eleventh and early twelfth centuries, although there is a much earlier reference to a water clock in a poem written around 887 for the fourth Umayyad emir of Córdoba, ʿAbd al-Raḥmān II. Water clocks are known to have been made in the eleventh century, for instance, by a certain Aḥmad (or perhaps Muḥammad) ibn Khalaf al-Murādī, and later by Abū al-Fath ʿAbd al-Raḥmān al-Manṣūr al-Khazīnī. Little is known of the details of their work. One eleventh-century work describes in detail an elaborate mechanism for operating automata, so continuing a practice that had gone on without break from ancient times, an important source of ideas, although not overtly astronomical. Another author describes a celestial globe turning with the daily motion, the motive power being provided by a falling weight floating on a bed of sand, the level of which fell as sand escaped through an orifice.

Water clocks from fourteenth-century Fez (Morocco) included astrolabe dials. Whether by that time western European knowledge was being conducted back into the Islamic world of northern Africa is hard to decide, and is best left an open question, but certainly, by the time the Fez clock was set up in the Qarawiyyin Mosque there, the European astrolabe clock with a fully mechanical drive was not uncommon. And that supremely important invention was a direct product of the desire to reproduce a moving image of the cosmos, as will be seen in chapter 10.

THE ALFONSINE TABLES IN SPAIN AND PARIS

Paris, beginning in the 1320s, was the single most important point of diffusion of the Alfonsine Tables, which soon thereafter replaced the old Toledan Tables. There are those who would argue that these tables are not truly Alfonsine at all, but are essentially a Parisian creation. No copy of the original tables is known to exist, although the explanatory canons in Castilian survive and have recently been newly edited and provided with a commentary that allows us to appreciate the range and character of the tables. We can reconstruct some of the parameters in the lost tables with a fair degree of confidence. The origins of much of the material are of course to be traced to the various leading zījes that we have already discussed in general terms. Origins and diffusion notwithstanding, it seems reasonable to say that without Alfonso x of León and Castile, "Alfonso the Wise," they would never have existed.

The Alfonsine Tables are only one aspect of the important scholarly activity of that thirteenth-century ruler, who encouraged the translation from Arabic into Castilian of many philosophical and scientific writings, a task already begun under the patronage of his father San Fernando, and one that reminds us strongly of the great patrons of astronomy in Eastern and Western Islam. The very introduction to the canons uses phrases that present the king in this light. The tables make use of an Alfonsine epoch of noon, 31 May 1252, the eve of his coronation, although they were assembled between 1263 and 1272, as we are informed in the prologue. This epoch of a Christian king is then related to the old Spanish era, the Islamic Hijra, and the (Persian) era of Yazdijīrd. The cultural continuity, notwithstanding the great religious divide, symbolizes continuity in the astronomical contents of the collection.

Translating from Arabic into Latin was well established at Toledo at the courts of the archbishops there. (Toledo was the ancient Visigoth capital. During the Moorish period [712–1085], it had been the home of a large Mozarab community. After its capture by King Alfonso vi in 1085, it became the most important political and social center of Castile, and remained so until the capital moved to Madrid in 1560.) Alfonso x established a school that included Christian and Jewish savants, as well as a Muslim convert to Christianity. He even presided over this group in some sense, revising their work and writing parts of the introductions to it. The names of fifteen collaborators are known from the complete collection of Alfonsine books, which includes important treatises on the universal astrolabe in various guises, the spherical astrolabe, a water clock and a mercury clock, the simple quadrant (the "old quadrant," as it is now called), sundials, and equatoria. This rich and encyclopedic collection also includes an astrological text, *The Book of the Crosses*. Most of the Arabic sources used were Spanish-Arabic, and it should not be thought that the drift of ideas from Spain was only in the direction of Europe. One curious fact, that probably has something to do with contacts with Muslim astronomers working under Mongol patronage, is the emergence in the following century of exactly the same value for the obliquity of the ecliptic as appears in two of the Alfonsine books (23;32,30°). This occurs in a zīj by a certain al-Sanjufīnī which was dedicated to the Mongol viceroy of Tibet. The rule that knowledge has always traveled farther afield than we think likely is not a bad rule, but it has to be applied with caution. The text in question is in Arabic, but there are Chinese inscriptions on the first page. Modern scholars found them puzzling, until it was discovered that they were simply shelf marks for use by the Chinese staff responsible for shelving oriental manuscripts in the Bibliothèque Nationale in Paris, the manuscript's modern home.

In view of discrepancies between new observations and predictions based on the old Toledan Tables, Alfonso commanded instruments to be constructed, and observations made, at Toledo. Two Jewish scholars are given as the compilers of the new tables, namely Jehuda ben Moses Cohen and Isaac ben Sid. They made observations at Toledo for more than a year, but the king moved his court often, and much work was no doubt done at Burgos and Seville too.

In the intricate history of the links between Spain and the Parisian tables that went under the name "Alfonsine," there are various key documents. Historically the most significant is one by John of Murs, written in 1321, his *Exposition of the Meaning of King Alfonso in Regard to His Tables*. In 1322, John of Lignères—teacher of both John of Murs and John of Saxony—wrote a work heavily dependent on the canons to the old Toledan Tables, but with clear hints of the "Alfonsine" Tables to come: he used the solar eccentricity that characterizes at least the Parisian Alfonsine Tables—and very probably the originals too. He also assumed that precession (calculated using a theory with access and recess) applied to the positions of the planetary apogees; he used twelve signs of 30° (Aries, Taurus, Gemini, etc) rather than six of 60°; and he took other steps that seem to reflect the usage of the Spanish canons.

I remarked earlier on inconsistencies in the sexagesimal system. The use of zodiacal signs of 30° has ancient roots, as we know, but that usage too is strictly inconsistent, and it is worth noting that later Parisians made use of "physical signs" of 60°, which have certain arithmetical advantages. The astrologers were never very happy with them.

At some time between then and 1327, John of Lignères and his pupils—it is quite certain that they were aware of each other's work—assembled the essential ingredients of what was to become the most popular version of the tables, that is, an edition composed by John of Saxony in 1327. John of Lignères wrote his own canons for the tables between 1322 and 1327, and some time after 1320 he improved them greatly by incorporating tables of combined planetary equations. This was an important step. We recall that planetary equations are terms that are to be applied to the steadily increasing and easily calculated mean motions, to give the final position of a planet at a particular time. There are two main equations to be calculated on each occasion, one of them partly dependent on the other. With the new arrangement, for each planet there was only one table, giving one combined equation. It depended on two parameters, the "mean center" and "mean argument," with which one entered the table. The resulting tables were well described as The Large Tables. In fact a similar step had been taken, probably unbeknown to the Parisians, by Ibn Yūnus three centuries earlier.

There is no evidence for any awareness of the original Alfonsine Tables east of the Pyrenees at a much earlier date than 1321. A remark by Andaló di Negro, made in 1323, suggests that he knew the original canons explaining their use, although John of Murs evidently did not. His work of 1321 was essentially a reconstruction. Andaló is well known to history as an astronomer on whom the great Italian storyteller Boccaccio pours excessive praise. As it happens, three Italian scholars are known to have been engaged in the Alfonsine enterprise, namely John of Messina, John of Cremona, and Egidio de Tebaldis of Parma. The involvement of Italians is not surprising: until 1275, Alfonso regarded himself as a candidate for the crown of Holy Roman Emperor, and regularly exchanged embassies with the Italian states.

Assuming that the tables were brought from Spain to Paris, we cannot say exactly how this happened. There are several possibilities. John of Murs tells us that he knew someone with knowledge of them, but who was keeping the knowledge to himself. This suggests contact with someone schooled in Spain. A Parisian scholar in the 1340s mentions a Spanish version of the Alfonsine book of the fixed stars, "taken from the king's bookcase." The writer also said that he had seen a globe that had been made for Alfonso, with stars appropriately marked. Whether these things were seen in Spain, or only after being transported to Paris, is not clear. Perhaps one day documentation will emerge that serves to fill in the gap in our knowledge of the missing fifty years, and the route of transmission.

John of Murs showed considerable determination and skill in extracting several parameters from his material. There was much hard calculating to be done, and it is not without significance that he addressed a sexagesimal multiplication table to a friend in the very year of his *Exposition*. He does not claim to have done anything original, but he thought his account to be right, because the tables then fitted well with an observation he had made. His loyalty to the Castilian tables seems clear enough from the tone of his writing. A figure for the maximum equation of the Sun $(2;10°)$ seems to be expressly calculated from the Spanish tables. The Parisian planetary parameters would have been much harder to extract, but they surely came from the same source. The hardest part of all would have concerned the Alfonsine theory of precession. This, in brief, combined *two* motions, a *steady* (secular) motion like that proposed by Hipparchus and most early followers, and a long-term *oscillatory* motion in the style of the theory of access and recess erroneously ascribed to Thābit ibn Qurra. The steady component in Alfonsine precession is at the rate of 360° per 49,000 Julian years. This makes use of a very peculiar idea, indeed, for it makes a movement of the stars dependent on a purely arbitrary decision as to how many days we should take for our calendar year. (To take 365 ¼ days is an arbitrary decision—nobody would by this time have claimed that it was exact. The tropical year, according to these astronomers, was about 10;44 minutes less than 365¼ days. In 49,000 years the discrepancy amounts to 365.23 days, hence the choice of that large period.)

In tracing the history of mathematical astronomy it is tempting to concentrate on the grand issues, such as those that converge on Copernicus and Galileo, at the cost of subtler themes. No blockbuster film on Copernicus is ever likely to mention access and recess, for example, even though he flirted with the idea, and even though it well illustrates an old weakness for the veneration of one's ancestors. Historical clues are often concealed in apparently innocuous, even trivial, data. Consider for instance the star catalog that occurs in the Alfonsine *Books of the Eighth Sphere*, as well as in some printed editions of the Alfonsine Tables. This has Ptolemy's star longitudes increased systematically by 17;08°. The first version of this work (1256), essentially based on the Arabic al-Ṣūfī intermediary star list (AD 964), was revised by two Jews, Judah ben Moses and Samuel ha-Levi, and two Christians, John of Messina and John of Cremona, in 1276. The figure of 17;08° does not fit with the double theory of precession, calculating across the time-interval between Ptolemy and 1276. It does, however, fit with a calculation for the interval stretching from AD 16 (which happens to be the starting point for the Alfonsine oscillatory model of precession) to 1252 (the coronation of Alfonso). We know that some Arab astronomers thought that Ptolemy was drawing on a lost work by Menelaus, from the late first century. We now also know that Ptolemy's longitudes were on average rather more than a degree too small in his own day. It is therefore at least conceivable that Alfonso's astronomers, who knew Ptolemy's approximate dates, knew too that his longitudes were systematically in error for his own time, and ascribed them to an astronomer working over a century earlier, in AD 16.

In the following chapter we shall return to the later history of the Alfonsine Tables, which helped to mold European astronomy for well over two centuries. After their first Parisian decade, they continued to be developed in important ways in England. J. L. E. Dreyer thought that he had found the "lost" Spanish tables in their original form in England, in the tables of William Rede, but in fact they can be explained entirely in terms of Parisian intermediaries.

The Spanish role in the revival of Western astronomy was far more important than the impetus coming more directly from the direction of Byzantium. The latter was in some respects potentially stronger, since Byzantine scholars had access to much polished eastern Islamic material, but this they never fully exploited. Byzantine astronomy before the sack of Constantinople by the Venetians in 1204 had relied heavily on Alexandrian sources, and it was to those same sources that Byzantine scholars returned when a great renaissance occurred after the revival of Byzantine political fortunes in the late thirteenth century. The key political figure in the revival was Michael VIII Palaeologus, who reigned as Byzantine emperor from 1259 to 1282, founding the Palaeologan dynasty that would rule the Byzantine empire until the Fall of Constantinople to the Muslims in 1453. Just as astronomy had been fashionable from time to time at the old Byzantine court, so it was again after the renewal of Byzantine fortunes, but its renewed popularity was this time more consistent.

In the second decade of the fourteenth century, Theodorus Metochites—who was already in his mid-forties, and a busy minister of the emperor—wrote what may be counted as the first substantial Greek treatise on Ptolemaic astronomy after the time of Theon. Theodore's pupil Nicephorus Gregoras carried the torch of Ptolemaic learning still higher, writing also on the astrolabe, composing calendars, deriving the constant of precession by observation, and much else besides; but he was still seemingly oblivious to what Byzantine astronomers had learned from Arabic zījes and other sources, three centuries earlier. At this stage, however, there came a new influx of learning from a surprising eastern source, namely Persia. New collections of several Persian tables were made available, in translations by Gregory Chioniades, who had apparently found the originals, having traveled overland to Persia from Trebizond, on the Black Sea. The Persian material was much studied in Constantinople. In due course, Byzantine scholars began to draw on Western sources too, and the result of this eclectic activity was that they seem to have lost sight of the wood for the trees, of the fundamental theory for the morass of incoherent parameters. There was plenty of activity in the closing years of the empire, before the Turkish conquest of 1453, and it has its own history, but it came too late to be of much influence in the world outside Byzantium. There was indeed influence on the West from that quarter, but it came through the classical Greek texts that were brought westward by refugees such as Cardinal Bessarion, and by the book-buying agents of Western humanists. Byzantine influence was ultimately of a literary rather than a scientific kind, and cannot be said to have affected the course of advanced astronomy in the way the many talented astronomers of al-Andalus had done.

✳ 10 ✳

Medieval and Early Renaissance Europe

Since a deep concern with the patterns of movement in the heavens is evident in Europe as far back as Neolithic times, it is not surprising to find traces of it in the earliest historical documents available to us. In his account of his wars in Gaul, Julius Caesar ascribes astronomical knowledge to the Druids, and the Roman historian Pliny the Elder adds a few details. It seems that the peoples of northern Europe began each month, year, and cycle of thirty years with a new moon. Days, in the sense of day-with-night, were counted beginning with the night (note the English word "fortnight"). Fragments of a bronze calendar were found in 1897 at Coligny, near Lyons in France (fig. 91), along with fragments of a statue of a god. Slighter fragments of another calendar had been found in the Lake of Antre near Moirans (in Jura), in 1802. They seem to date from the second century, and to show that a lunisolar calendar was then in use, with months of 29 and 30 days, and that two months of 30 days each were intercalated every five years. The Celts of Britain and Ireland grouped their days in threes and nines. (The week of seven days was introduced only with Christianity.) The Coligny calendar seems to divide months into halves. It shows a complex pattern of marking the months and days, possibly distinguishing them as favorable and unfavorable. A similar distinction—if this is indeed what was intended—was common in calendars throughout the Middle Ages, using a system of "Egyptian days." Early Irish literature shows that births were sometimes delayed to ensure that they took place on a lucky day.

An earlier division of time, done by seasons as marked in particular by the winter and summer solstices, has left traces in folk custom throughout that same Celtic world—especially in Brittany, Ireland, Wales, the Isle of Man, and the Highlands of Scotland. It has often been said that the festivals of the Christian Church were cleverly timed to coincide with the older pagan festivals, the feast of St. John the Baptist on midsummer day, for instance, replacing the feast of Beltane. The story goes that this was under the influence of St. Patrick himself (Patrick did once allude to worship of

91. Fragments of the Coligny calendar. In 1897, beside the old Roman road from Lyon to Stras-
bourg, a deposit was found of 550 fragments of bronze, most of them from a statue of some divinity, but 150 deriving from a Gaulish calendar. The latter is now the richest single source of
the ancient language, with its 2,000 words in sixteen columns. Dating perhaps from the second
century AD, the calendar is luni-solar, covering five years of twelve months, the months being
29 or 30 days, with intercalations of two months in five years to bring the Moon and the Sun
into line. A word found at the middle of each month, *Atenoux*, might indicate the full moon.
Other words may be for feast days. The lettering is Roman, but the calendar ignores the Roman
Julian calendar completely.

the Sun in Ireland). There are pre-Christian pagan elements even in the
Christian festivals, however, and the near-universality of festivals of the
Sun—most of them pagan by definition—makes it difficult to say whether
the imported festival was markedly less pagan than the native versions.

Ancient Germanic and Scandinavian traditions of a similar kind are
if anything harder to discern. There are rock-markings from Sweden suggesting Sun cults, and Sun and Moon, day and night, summer and winter,
are all personified in the poems of the older Edda. We know from the Latin
writer Procopius that when in places north of the Arctic circle the Sun disappeared for (as he says) forty days in the winter, the days were counted

off until it was time to send observers into the mountains to watch for the rising Sun, and so give five days warning of its return to the people below, who then prepared for their greatest festival. There are comparable Sun cults among all ancient Baltic peoples and, naturally, also in Iceland, but whenever among them are found elements resembling the more systematic parts of astronomy from elsewhere, cultural influences can usually be traced. A good example is in the so-called Golden Horns of Gallehus, two large horn-shaped artifacts of gold found near the Danish village of Gallehus, but at different times. They were stolen from the royal treasury in Copenhagen in 1802 and have never been found, but minute drawings of them show that they were covered in human and animal forms that W. Hartner interpreted as stylized inscriptions (written in runes) relating to an eclipse of the Sun on 16 April 413. Hartner found in that symbolism traces of Hellenistic and Oriental astronomy.

THE EARLY CHRISTIAN CHURCH
AND ITS PRIMITIVE COSMOLOGY

The Christian scriptures have been considered by some to have favored the pursuit of astronomy as a science, and by others to have been hostile to it. They certainly make use of many primitive analogies between the universe and the furniture of the everyday world. The tabernacle that Moses constructed in the wilderness was the world; the seven-branched lamp was the Sun, Moon, and planets; the six-winged golden figures were perhaps the Greater and Lesser Bears. This sort of thing was not antagonistic to science until it began to give rise to a large body of mystical commentary that was felt to be in need of defending for the good of the faith. Often, this mysticism was blended with a crude and untutored common sense. Lactantius (ca. 240 to ca. 320) is a good example of this. This "Christian Cicero," one of the most popular of the church fathers, and a man who became tutor to the son of Emperor Constantine, nevertheless preached against Aristotle and in favor of the flatness of the Earth.

It is a common myth—perpetuated by many teachers of young children—that Columbus discovered that the Earth is round. Of course, different people believed different things at different times, and the gap between the educated classes and the uneducated was greater at the time of Columbus than it is now, but the myth can at best have reference to the psychology of those who sailed into the unknown. That the Earth was round was the teaching of the Greeks and their intellectual successors. There are many echoes of Lactantius's hostility to the idea in the literature of late antiquity and the early Middle Ages. Even those who could accept the sphericity of the Earth often had problems with the existence of people at the antipodes. Frequently, those who grudgingly admitted this last possibility were inclined to add that such beings would not be descended from Adam, would be beyond redemption, and by walking with their heads below their feet would be incapable of rational thought. The thousand years that we cram into "the Middle Ages" cannot be treated as a static or uniform whole, but such skepticism was not typical of the views of educated people throughout the millennium.

Some of the church fathers did their utmost to reconcile the scriptures to Greek philosophy: here Ambrosius, bishop of Milan in the fourth century, deserves a prize for his remark that a house may be spherical inside and square when seen from the outside. The book of Genesis raised difficult problems—for example, on the place of the waters suspended somehow above the firmament. St. Augustine (354–430) transformed Latin Christianity—to which he was a late convert—by the Neoplatonic and generally pagan ideas in which he had been educated, and he was quite capable of discussing the question of sphericity, but chose not to give his final judgment on the question. Not for the last time did reverence for the Bible stand in the way of good argument. By comparison with these authorities, John Philoponus (ca. 490 to ca.570) makes a refreshing change. He, above all others, was responsible for turning the Alexandrian school into a Christian one, and his commentaries on Aristotle, including criticism of Aristotle on the eternity and the substance of the heavens ("the fifth essence"), are of a high intellectual standard. He would have deserved a place in history for his *On the Construction of the World* had he written nothing else, for this is a devastating attack on Theodore of Mopsuestia's use of scripture as a scientific text, one that was supposedly capable of proving that the heavens are not spherical and that the stars are moved by angels.

Such early Christian debate had later repercussions on Europe as a whole. Great changes came about with the collapse of the western Roman Empire. Pressures on Rome from the north had been growing for a century when in 476, Odoacer became the first barbarian king of Italy. For a time, the use of Latin alongside local languages throughout the former empire fell into abeyance, and the ties of civilization were loosened as a result. A compensating change in northern Europe came about, however, with the eventual conversion of many of its peoples to Christianity. This helped to restore many of the loosened ties, for the Church brought with it books and learning, and the only way into them was through the Latin language. The Church of course also brought with it a belief in a single God as the creator of heaven and earth, so slowly replacing the idea that a multitude of gods shared responsibility for governing events in the natural world. While it is true that much of the best science of antiquity was devised by people who believed in many gods rather than one, there is a good case for insisting that science progresses more freely, the fewer godly fingers there are in the pie of nature. We might almost view the *Etymologies* of the seventh-century scholar Isidore as arguing for this thesis: he there tried to demonstrate that even the language needed to account for natural and human affairs had no place for mythology.

I mentioned Isidore earlier, in connection with events in Spain. He was a great synthesizer standing in the Roman encyclopedic tradition, rather

92. The four sides of the Bewcastle Cross. The pillar is about 4.4 meters high. Dating from the seventh or early eighth century, it is still standing outside St. Cuthbert's Church, Bewcastle (Cumbria, northern England). The head of the cross has been lost. Dating the cross depends largely on runic inscriptions with personal names. A nobleman holding a hawk seems to have the inscription "Alcfrith . . . Pray for my soul"; but other names, one of a seventh-century Northumbrian subking, are debatable. It has a crude semicircular sundial of a once common type (a "scratch dial"), drawn here a third of the way down the second detail (southern face).

than a great intellect: his technique for staying out of difficulties was to quote Greek authorities—in Latin translation—without any major criticism of them. Isidore's case does not well illustrate a dramatic mixing of cultures that was taking place in his day, even at the outposts of Europe. Here the Venerable Bede, his near-contemporary (672–735), is a better example. Monasticism in Britain was not new—it had begun around the year 430—but Bede was brought up in the Benedictine monasteries of Wearmouth and Jarrow (near Newcastle) that had been founded only a few years earlier by a man who had brought Mediterranean learning in his baggage. This was his abbot, Benedict Biscop, who had studied at the very famous school of Lérins (an island near Cannes) and also at Rome. Benedict brought with him from Rome two scholars from even farther afield: Archbishop Theodore was ultimately from Tarsus in Asia Minor, and Abbot Hadrian was from North Africa and Naples. Later in life Bede himself brought in other scholars to join him from Ireland and the Continent.

Bede was fortunate. The dozen volumes of his many published writings show a mastery of most branches of conventional Christian learning, and a glimmering of good astronomy. Fundamental astronomy was not strongly represented, but what was there had been kept alive in large part simply because it was needed for monastic timekeeping—for the hour of the day and for the calendar. It was to the rituals of monastic life what grammar was to the study of the Bible. The hours of the day, and in particular the "canonical hours" of prayer, were at first registered only by primitive means. Use was made of naive sundials that would have made Vitruvius smile (fig. 92), and eventually also by water clocks of simple kinds. The old Julian calendar remained in use for civil purposes and for the fixed feasts of the Church, but the movable feasts—not only Easter, but feasts on dates at a conventional distances from Easter—were settled by reference to a scheme which had its roots in the lunar calendar of the pre-Christian Jews. As had happened in so many ancient cultures, and as was happening at the very same time in Islam, religion was here setting a value on basic astronomy.

THE CALENDAR AS A REPOSITORY OF SIMPLE ASTRONOMY

From as early as the third century, Christian scholars had begun to devote much attention to calculating the date of Easter. For centuries they struggled to achieve agreement among themselves, and especially as between those in Alexandria and those in Rome. Most were quite out of their depth, astronomically speaking, but the situation improved in AD 457, when Victorius of Aquitaine (also called Victorinus) reviewed the three most commonly used systems, and drew up tables introducing the Roman Church to some of the best of Alexandrian practices. In particular, he made use of the extremely useful astronomical period of 532 years, after which Easter dates repeat. It seems that he was led to this period by a convoluted argument starting from his belief about the date of the creation of the world, and that his discovery of its Easter properties was quite accidental. The period combined the solar cycle of 28 years with the "Metonic" 19-year cycle. In 28 years, the odd quarter of a day in each full year (365¼ days) adds up to a week of days, and so synchronises weekdays and date within the year. We recall that the 19-year cycle brings lunar and solar cycles into harmony, that is, brings full moon back to the same calendar date. A period of 532 years, having both periods as factors, therefore achieves both objectives, and was an important element in what Christians needed for the cycle of the date of Easter Sunday.

This was finally settled as the Sunday following the fourteenth day of the paschal moon, counting from the day of new moon, inclusive. The paschal moon was meant to be grounded in reality, but inevitably it was a theoretical construct, and it embodied some very shaky astronomy. It was defined as the calendar moon whose fourteenth day falls on, or is the next following, the spring equinox—not the true equinox, which was drifting through the Julian calendar, but the conventional equinox of 21 March. This is not the place to enter into a discussion of the intricacies of a subject that took on a life of its own, far removed from the astronomy which set it in motion.

Victorius actually calculated Easter dates for a full cycle of 532 years, and he might have relieved generations of bishops from worry about the problem of calculating it themselves, had all of his dates been acceptable. Unfortunately, some fell outside the traditional limits of Easter in the Roman Church, so that his system was not universally accepted. Not until the sixth century was a general consensus reached, and that was when Pope John I asked that the problem be addressed once more. The matter was referred to a certain Dionysius Exiguus (Denis the Little). Previous calculation had been referred to the Roman calendar, counting from the era of the Emperor Diocletian—a notorious persecutor of Christians. Dionysius found such a chronological system obnoxious, and shifted the grouping of 19-year cycles so that they could be counted from the year of Christ's birth. It is now acknowledged that he was a few years in error over that date. This cannot alter the fact that, by a long historical process, he had arrived at scheme for numbering years that was eventually adopted by almost all Christians, and that in due course would be accepted for civil use throughout most of the world, by people of all faiths.

That the new way of numbering years became widely used throughout the Christian world owed much to the fact that Bede introduced it into two influential treatises on timekeeping, which he had written for the monks of the Jarrow monastery. His first work evidently turned out to be too difficult for most of them, and he was persuaded to write another, at greater length, which he completed in 725. The resulting *De temporum ratione* (*The Reckoning of Time*) was used throughout the Christian world for more than seven centuries, and for good reason. It is technically sound, clearly written, and historically well informed. No finer piece of even moderately original scientific writing would be produced in the Latin world for almost another five centuries, and for most of that time Bede's work offered one of the few ways in which an ordinary scholar could make contact with any science that used numerical methods. It brought the 532-year cycle back into prominence—which Dionysius had not done, for want of true understanding. One thing that Bede's work did not achieve, however, was to establish a uniform convention as to when the new year should begin. It had anciently been assumed to begin at the winter solstice. Pliny, for example, helped to spread the idea that the winter solstice was on 25 December in the Julian calendar. Throughout the Middle Ages, the year was considered by some to begin on Christmas Day, by others on the first day of January, by some on the first of March, by others on the twenty-fifth of that month, while for a time there were even those who accepted a variable starting date, namely Easter. Even today there are countries which start the financial year on 24 March in the old Julian calendar—which is to say 6 April in the Gregorian. One such country is Bede's own.

It is not irrelevant to the course of astronomical history to explain how it came about that Christmas was placed on 25 December, considered to be the date of the solstice. The Gospels say nothing of the date of Christ's birth, and it was not celebrated as a feast day until after the Christians in Egypt took the date of the Nativity to be 6 January. This custom spread throughout the Eastern Church, and would no doubt have been adopted in the Western Church had its members taken that as the date of the solstice. Instead, in the course of the fourth century, they gradually accepted the feast and transferred it to their own solstice. A Syrian writer adds an additional explanation, saying that at the winter solstice it was a custom of the heathen to celebrate the birthday of the Sun, kindling lights in token of the event; and that the doctors of the Church, recognizing that the festival was liked by the populace, made a virtue of necessity. One particularly important group of "heathens" were the followers of Mithraism, which for centuries remained a serious rival to Christianity. Some Roman sources mark 25 December as *Natalis Solis Invicti*, "the birth of the invincible [or unvanquished] Sun." The Sun has been weakening, the days have lengthened, but now, at the solstice, it is *reborn*. What more appropriate day could be wished for, when celebrating the *birth* of Christ? And all this suggests something more. Modern scholars have often seen those prehistoric monuments that are directed to the (winter) solstitial positions of the Sun as celebrating the death of the Sun. Mithraic and early Christian practices rather suggest that this might be a grave mistake, and that they might have had the Sun's rebirth in mind.

To return to Bede: one of his achievements was to stabilize the very language of the computists—a language largely forgotten, now that our calendars are presented to us by anonymous authorities. One important technical term worth mentioning, however, stemming from a later writer of great influence, Alexander of Villedieu, is *golden number*. If the year is AD *Y*, then the golden number is the remainder when $Y + 1$ is divided by 19. This is a concept explained by Alexander, in a work written in 1200. While intimately bound up with the doctrine set out by Bede, it simplified the computist's task in several respects, and golden numbers were usually written against every day in the year in most later ecclesiastical calendars—for reasons explained, together with a page from such a calendar, in figure 93. Every student of the arts in the medieval university, every educated priest, friar, or monk, was expected to know the principles of the computus. And just as Indian and Arab users of *zījes* calculated with the help of their fingers, so the generally less-numerate Western clerk made use of the joints of his hand to assist in calendar reckoning (fig. 94).

Calendars and related texts assumed a great variety of forms. They ranged from sumptuously illuminated and expensive books at one extreme to wooden staves carved with runic characters at the other (fig. 95). Such runic calendars, which in some places survived in regular use even into the nineteenth century, contained elements pointing back to the preChristian period. A very fine early manuscript with a passage on timereckoning by Abbess Herrad of Hohenbourg (1167–1195), who was writing for her nuns in Alsace, contained a full 532-year cycle. This also made use of simple symbols, some of them for numbers and some for feast days, as can be seen in the old transcript of part of the lost manuscript in figure 96. Such symbolism is very commonly found in monastic calendars, and speaks volumes for the low level of arithmetical skill among those who used them.

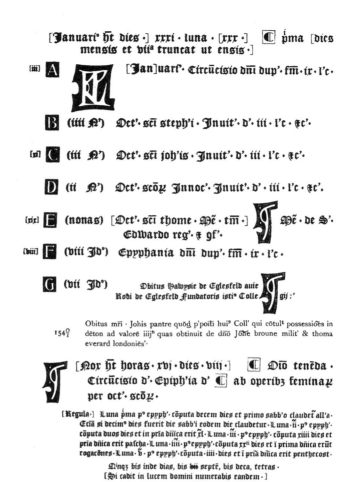

93. A modern printing of the opening page for January of a typical ecclesiastical calendar of the later Middle Ages. For an explanation, see opposite.

COSMOLOGY FROM LATIN SOURCES

Bede's scholastic texts enjoyed an extraordinary reputation and kept alive certain simple general principles of solar and lunar astronomy at a time when many in the Church were deeply suspicious of it. Bede wrote on the tides, and their lunar involvement, largely from first-hand experience, but most of his cosmological beliefs—as in his book *On the Nature of Things*—were inherited from ancient sources. He purveyed a motley collection of ideas on the stars, thunder, earthquakes, the parts of the Earth, and so forth, drawing on Scripture, Isidore, and Pliny the Elder. He taught the sphericity of the Earth, and such basic astronomy as was needed to explain the inequalities of day and night, the variation of both with latitude, and even the rough general pattern of observed planetary motions. Since Bede lived on the periphery of European civilization, his writings might easily have sunk into obscurity had it not been for a small number of key scholars who were able to advertise their merits. Perhaps the most crucial was Alcuin, the great Anglo-Norman scholar from York, who in 781 was invited by Charlemagne to Aachen (Aix-la-Chapelle). Charlemagne, king of the Franks—who in 800 was to become Holy Roman Emperor—was beginning to bring together leading Irish, English, and Italian scholars. The English learning that Alcuin introduced into the Frankish schools included much of Bede, but Alcuin also wrote his own computus; and so, after Alcuin's death in 804, did Hrabanus Maurus, who proceeded to carry his methods deeper into the German-speaking world.

During the eleventh century, while Bede's work on time-reckoning continued in use, his simple cosmological essays were largely put aside, while others—notably older works by Boethius—became fashionable. Boethius (480–524), a Roman aristocrat now best known for his *Consolation of Philosophy*, had written on almost every aspect of scholastic learning except astronomy, a fact which probably reflects the scarcity of good astronomical texts in Rome in his day. When the king of Burgundy asked

The opposite page preserves the general appearance of a fourteenth-century calendar belonging to The Queen's College, Oxford. Published in 1910 by J. R. Magrath, it keeps the medieval abbreviation of the Latin but helps out with a few words in brackets that the reader would have taken as read. The page carries much information. Beginning with the number of days in the calendar month (31) and the lunar month (here 30, alternating with 29 to give an average of 29½) it goes on to identify the first and twenty-fifth days of the month as "Egyptian" (unlucky) days. The twenty-fifth is specified as "seven from the end," using typical inclusive counting. Down the left-hand margin are the golden numbers. Thus iii against the kalends (first day) of the month shows that the new moon falls on that day in the third year of the 19-year cycle. The seven boxed letters of the alphabet from A to G repeat throughout the calendar. They have a simple use for associating a day of the week with a date, still not quite forgotten: if 5 January, for example, prefixed by E, is a Tuesday, then every day prefixed by E will also be a Tuesday. (A simple rule is needed to allow for the extra day in a leap year.) Years could be characterized for certain calendar purposes by these letters, taking the letter opposite the first Sunday in the year. (The custom explains why the letters were called dominical, or Sunday, letters.) The column under the large letter K (for "kalends," the first of the month) has the designation of the days according to the old Roman calendar. The ides of the month (not on this page) was a date near midmonth, and the nones eight days earlier. Days were reckoned backwards, counting inclusively, hence the sequence which in English we should read thus: kalends . . . four days before nones . . . three days before nones . . . and so on. The main script gives the Christian festivals, with some notes on the services, vigils, and obits—here there are also notes for commemorating family members of the college founder. Opposite the lowest large paragraph mark we find the number of hours in the day and night—as astronomically determined, for the local latitude, in this week. Here we also find mention of feasts which everyone was obliged to keep, and those on which women must refrain from their "special work," such as weaving, beating flax, and shearing wool. Where appropriate (not in early January) there are the days on which the Sun enters a new sign of the zodiac, equinoxes and solstices, general chronology, and occasional jottings added by owners relating to such events as eclipses. In rare cases, calendars tabulated eclipses systematically. The page ends with notes and rules which would only be understood by a person with a good training in computus.

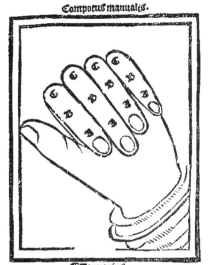

Compotus manualis.

C Januarius.

aredi. dki/   phenie   lari. lici epi/ri abbati/celli.pieii.
Cicio/Janus/epi/lucianus/et hil/fe/mau/mar/iul
pia.   pia.   le.fe.

reu ttm franti/f ianli neti/ citit   ianti
Phil/mul/fab/ag/bin/pete/pau/li/iul/ag/ne/batpl
non bil. poa le. le.fe. noia.   po le.p.

94. A page for the beginning of January, from a short *Computus manualis* (1519–1520), a printed work continuing an old tradition of teaching Oxford students calendar calculation on the joints of the fingers. The verses below, beginning "Cisio Ianus," are the beginning of mnemonics for feast days, university affairs, golden numbers, and other calendrical data—all meant to be transferred to the fingers for counting purposes.

95. In northern Europe, especially Scandinavia, information from the ecclesiastical calendar was often recorded on a wooden staff or other common object. Letters of the runic alphabet, easily carved with mostly straight lines, were used for dates and golden numbers, and easily recognized symbols were used for feast days. Such staves were often more than a meter long, and could include the "Cisio Ianus" verses (see fig. 94). The drawing reproduces a Swedish stave of the 1560s from Dalarne, approximately 60 centimeters at its longest and 2 centimeters thick. Each side covers half of the year. Its runic inscription includes the statement "Mas Jonsun carved me."

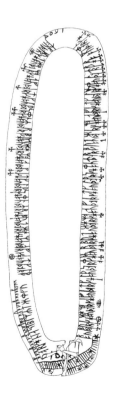

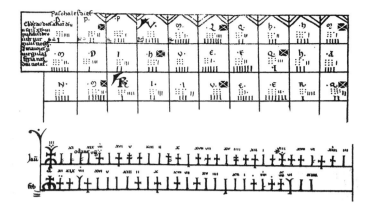

96. Part of Herrad of Hohenbourg's Easter cycle, from her richly illustrated *Hortus deliciarum* (*Garden of Delights*), transcribed in the nineteenth century before the manuscript was destroyed in the Franco-Prussian War. Below are representations of the days of January and February, the importance of any feasts being indicated by appropriate symbols. The Roman numerals there are golden numbers. Above, each square contains data relating to Easter for a single year, beginning 1175. Easter is always six weeks after the first Sunday in Lent (which fell on 2 March in 1175). The period up to the latter is shown by nine dots for weeks and four lines for residual days—bearing in mind that Herrad began her year at Christmas.

Theodoric, the Ostrogoth king of Italy, for the gift of a water clock and sundial, Boethius certainly knew enough astronomy to advise on its construction—but not enough, it seems, to predict that Theodoric would one day order his execution. Boethius's enormous contribution to Western learning did not rest on any technical expertise he might have had, but was a product of his attempts to reconcile the writings of Plato and Aristotle. In the course of doing so, he propagated a simple cosmology that was more or less Aristotle's. Its virtue was its simplicity rather than its subtlety. He was not in the same rank as Aristotle, of course, but he strengthened respect for Aristotle's idea of a universe governed by chains of cause and effect; and he prepared the way for the more serious Aristotelian onslaught of the thirteenth century. Boethian Aristotelianism bore fruit, albeit very slowly, in bringing physical cosmology and theoretical astronomy closer together.

There were two ancient writers with a greater impact than Boethius on the medieval understanding of the simple theoretical principles of astronomy—other than those of time-reckoning. The two were Martianus Capella and Macrobius. Both of these Latin authors were seemingly from North Africa. Martianus—who died around the year 440—composed one of the most popular of all Latin textbooks of the Middle Ages, *On the Marriage of Philology and Mercury*. In this, each of seven bridesmaids presents one of the seven liberal arts, first the "trivial" three, namely the *trivium* of grammar, rhetoric, and dialectic, and then the more advanced four, the *quadrivium* of the sciences of geometry, arithmetic, astronomy, and music. (The structuring of the medieval curriculum around these seven subjects owes more to the writings of Boethius than to any other single person.) The astronomy in Martianus seems to our eyes very basic, and after the arrival of Islamic texts must suddenly have seemed insufferably vague to those who wanted to get to grips with those exciting new materials, but no doubt the feeling was a rare one. The *Marriage* certainly had the makings of good cinema. Astronomy's bridesmaid arrives at the wedding party in a hollow sphere of heavenly light, filled with a transparent fire, and gently rotating. She carries a pair of dividers and a globe, simple objects that were long used as symbols of astronomy. Much attention is given to the guests at the wedding, who behave as wedding guests are prone to do. Astronomy begins her lecture with an account of the numerous circles whose names the student must know, and expresses a certain annoyance with astronomers who build armillary spheres. As she explains, the heavenly circles are idealizations. Martianus achieved an extra touch of fame after Copernicus praised him excessively for saying that Venus and Mercury were centered on the Sun. Of course he was cashing in on the fact that Martianus was already a well-liked figure; but whether the ordinary reader would have taken Copernicus' point is open to doubt. One might imagine a schoolmaster trying to get the point across with the help of one of those hateful armillary spheres. Would the student, unaided, have known what Astronomy meant when she said that the Sun has 183 circles? (It goes round the sky 183 times in passing from solstice to solstice.) Or that while the Sun traverses them, Mars describes twice as many? (Mars covers its orbit in roughly two years.) At least Astronomy stresses the simple point that the Earth is not at the center of the Sun's orbit, as "all men have believed until now." All the men at the wedding, perhaps.

That ninth-century Latin scholars were giving some thought to alternative planetary arrangements, and to heliocentric arrangements of Mercury and Venus in particular, is not in doubt. By the third quarter of the century, we find an utterly unhistorical interpretation of Plato being offered, according to which he believed that Venus and Mercury move in concentric circular paths centered on the Sun. Martianus was supposed to vary this slightly, making both go round the Sun, but in intersecting paths. That idea is illustrated alongside two even stranger alternatives, in a manuscript of the time, now in Paris. This includes a page of sketches, one of which is redrawn here as figure 97, with heliocentric arrangements as supposedly promoted by Pliny, Martianus, and Bede. There is no evidence that Pliny or Bede ever touched on the question.

Macrobius was another writer from much the same period, who gave dramatic form to his simple astronomical account. He presented his story as a commentary on Cicero's *The Dream of Scipio*, a work in which the dreamer had been portrayed as undertaking a journey through the spheres, a journey that allowed him at length to look down on the entire universe. Macrobius's literary style of exposition was less colorful than Martianus's, and owed more to the style of Cicero and Plato. Like Martianus, Macrobius

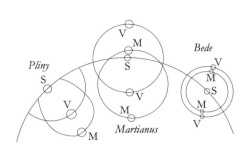

97. Three different interpretations of heliocentrism as discussed by Martianus Capella and attributed in a ninth-century manuscript (from which this is redrawn) to Pliny, Martianus, and Bede. Here *S* is the Sun, *V* is Venus, and *M* is Mercury. The reading of Pliny is very strange: the two planets seem to be denied access to the far side of the Sun. The Mercury oval might just possibly suggest a misunderstanding of the Ptolemaic oval deferent. (The original is in Paris, BN, ms lat. 8671.)

owed more to minor handbooks on astronomy than to the leading authorities of antiquity. An impatient modern reader whose interests are primarily astronomical will find even less of interest in Macrobius than in Martianus, although again there is an interesting ambiguity about the order of the planets and the placing of the Sun. Macrobius has been consistently misinterpreted since the eleventh century on this question, although in the ninth it seems that he was not yet on the list of heliocentrists. His real interests were in Neoplatonic philosophy, which he helped to make popular in medieval Christendom. He was responsible for much of the Pythagorean number-mysticism of the Middle Ages, and writers who dwelt on the theme of the harmony of the heavenly spheres often turn out to have been reading him. He was certainly influential, providing as he did the two great poets Dante and Chaucer with a model for the heavenly journeys they describe. Dante's *Divine Comedy* undoubtedly helped to promote the popularity of Martianus and Macrobius, as well as of Aristotle, in the later Middle Ages.

Astronomy around the Carolingian period was in many ways weak, even by the lights of Roman astronomy in late antiquity, but it was weak for want of texts, not enthusiasm. There are several rather surprising items of astronomical material that found their way into computus collections from the late eighth century onward, when it seems that scholars were trying to discover the fundamental principles underpinning the calendar. Some of the material was drawn from Pliny's *Natural History*, and we even know the circumstances that led to its inclusion, namely a conference of ecclesiastics in the year 809, where the ignorance of those present led the leaders of the meeting to order the compilation of matter from Pliny and from Bede's *On the Nature of Things*. There were passages on the order of the planets from the Earth to the stars, on the harmonic intervals that supposedly separated them, on the planetary apsides, and on the fluctuations in planetary latitudes. The last topic is interesting, for it occasioned some valiant exercises in the graphical representation of latitude changes. Some writers record a wavy line round the circular orbit in an ordinary planetary diagram, but

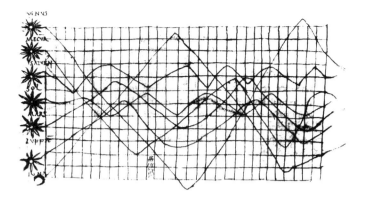

98. Planetary latitudes for the planets, Sun, and Moon, on a rectangular grid. From a tenth-century manuscript. Working upwards, the lines belong to the Moon, Jupiter, Mars, the Sun (which joins the planetary dance in a nonsensical way, for a body that should stick to the ecliptic), Saturn, Mercury, and Venus.

some transfer the pattern of fluctuations to a rectangular grid, covering the cases of Sun, Moon, and all planets (fig. 98).

Another surprising discovery is that the principles of stereographic projection—if that is not too grand a name for the techniques of constructing the circles on an astrolabe, or on an anaphoric clock dial—had either been transmitted from antiquity or had been newly acquired in some roundabout way. The reason for saying this is that the circles are drawn as the construction lines underlying a constellation map in a manuscript now in Munich (fig. 99). The artist shows some hesitation when constructing the equator, but the other main circles are confidently drawn—not only the ecliptic (and the zodiacal band in which it lies), but the tropics of Cancer and Capricorn and even the Milky Way, which is rarely found on later astrolabes, although astrolabe clock dials often include it.

DANTE AND CHAUCER

It is no accident that there, in the greatest of all medieval allegories, Dante provided a moral theme within a framework of Aristotelian cosmology. His *Divine Comedy* may be read at various levels, but taken most literally it is an account of his vision of a journey to Hell, Purgatory, and Paradise. Hell is described as a stepped conical pit with successive circles to which various classes of sinners are assigned. Purgatory is seen as a mountain rising in a succession of circular ledges on which are the several classes of repentant sinners. In Hell and Purgatory, Dante's guide is the poet Virgil, and in both places he sees and converses with his former friends and enemies. Paradise, by contrast, is a region of light and beauty in which his guide is Beatrice, now an angel. This is a vision of what he conceived to be the true state of the world beyond the experience

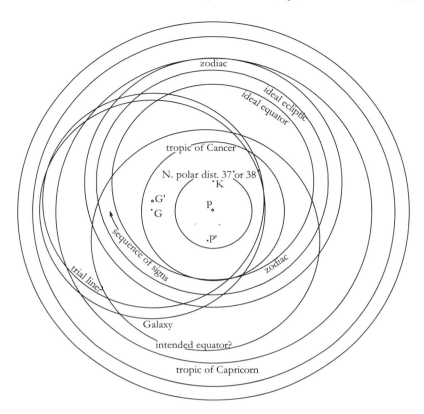

99. A tenth-century stereographic projection from the construction lines for a constellation map. The projection is from the *South* Pole (as on an astrolabe) but the stars (not shown in this redrawing) are as seen, not reversed as on an astrolabe. The tropics do not follow our convention, but touch the zodiac band, not the ecliptic. Lettered points are actual circle centers. The innermost circle might possibly be an arctic circle for latitude 36° (perhaps Rhodes?).

of the living. Allegorically—as he tells us himself—its subject is man's liability to just rewards or just punishment in accordance with the exercise of free choice. Literally understood—and it must not be supposed that this is more than a surface to the poem—his descriptions are more or less those he had been taught by the classical astronomical authorities in use in his day. His universe is a modified Aristotelian world with harmonies of various sorts built into it, and running through his verse is the view that human fortune is intimately linked with those harmonies. (For a sixteenth-century rendering of his universe, see figure 100, which shows an extended version of the perfectly standard Ptolemaic cosmos of contemporary text books, such as that of figure 101.) Beatrice acts as a tutor to him, explaining to him how the innate *forms* of things work, how fire is drawn up to the Moon, how the Earth is bound as one, how the Primum Mobile, the First Mover, moves creatures irrational and

100. One of numerous illustrations from the many printed editions of Dante's *Divine Comedy*, in this case, Bernardino Daniello's 1568 edition. It works outward from the standard Aristotelian picture, with the four elements, the planetary sphere, the eighth sphere of fixed stars, and the Primum Mobile; above the last is added the Empyrean, with its superstructure provided by Christian theology.

rational alike. She explains numerous astronomical technicalities in general terms: the spots on the Earth, the eclipse of the Sun, epicycles and their effects on retrogradation, the pointing of the magnetic needle to the Pole Star, and so forth. How does it come about, Dante asks her, that he can fly through the spheres? If he, Dante, being free from constraint, was to stay on the ground in this Aristotelian world, she implies, that really would be a cause for surprise.

In the twenty-eighth Canto of *Paradiso*, Dante looks into Beatrice's eyes and sees God as an infinitely small but brilliant point of light encircled by nine radiant rings. These Beatrice relates to the movements of the heavens—the Aristotelian or simplified Ptolemaic system, of course—but also to the three hierarchies, or nine orders of angels as described by Dionysius. He, the author of a popular Neoplatonic work called the *Celestial Hierarchy*, was not, as Dante thought, the eminent Athenian Dionysius the Areopagite whom St. Paul converted to Christianity. But pedantry aside, the pedantic Beatrice could not refrain from pointing out that a different opinion had been held by Gregory. Elsewhere, Dante even adopts a third system. The details need not concern us here, but there are points of great interest in all this mixture of crude astronomical and theological elements. One is that the sources of the Dionysian "theological"

101. A version of the "Ptolemaic" world picture, from Gregorius Reisch's very popular *Margarita philosophica* (1503), which went into at least nine editions before 1550. This standard picture does only rough justice to Ptolemy and Aristotle, of course. It ends in the Empyrean, beyond which artistic license takes over. The central figure is Atlas, an unusual choice, although he was sometimes credited with having invented the science of astronomy.

cosmology were apocalyptic writings of Jewish origin, and earlier still, Persian and Babylonian literature that had an astronomical origin of sorts. History had indeed come full circle. Allegorically, the angels were taken to represent the operations of Divine Providence, through which God's love sustains the spiritual order of the universe. There is here perhaps the slightest hint of a way theologians had of avoiding astrological determinism: it was to make the planets agents for God's will. When so Christianized, it was thought that astrology would be above reproach. It is not without interest that the souls of the blessed, when encountered in the different heavens, turn out to be placed in accordance with astrological principles.

There is another sort of cosmic symmetry in Dante's poem, not unrelated to the first, a form of number mysticism involving especially the number nine. It is found in the structure of Hell, for instance, which mirrors that of Paradise in a peculiar way. Dante also manages to work the numbers of lines and cantos into the symmetries of his story, and numbers from the calendar too. The whole story was related to a very specific Easter—it is generally agreed that this was Easter of the year 1300—and some have argued that he wove planetary positions into it. He certainly seems to have added a personal astrological touch: ascending from the heaven of Saturn to the heaven of fixed stars he finds that he has entered his native sign of Gemini, the Twins. He speaks as though the stars in Gemini "poured forth virtue" at his birth. "To you," he says, addressing the stars, "I owe such genius as does in me lie."

This is a story of Divine Love, but there are other passages tinged with astrological belief. In the opening canto of *Purgatorio*, Dante describes Venus in some of the most striking lines in the whole poem:

> The radiant planet, love's own comforter,
> was setting all a-laugh the eastern sky,
> veiling the Fishes that escorted her.

How real was all this meant to be? Was Venus rising with the stars of Pisces at dawn as he set out? At Easter in the year 1301 it was, while at Easter in 1300 it was not. Those anxious to prove a point have been able to produce a manuscript of the period containing a faulty almanac by none other than Profatius, that he might have been using, and that would put Venus in the right place in 1300. True or not, we can say with some confidence that Dante was rather more interested in the formal, architectonic symmetries of his art than in specific astronomical detail. The fact remains that astronomy was far more intimately bound up with the worldview of the Middle Ages than is the case today, even though the average educated person now knows far more about it.

Chaucer, toward the end of the same century, the fourteenth, took a very different approach, when time and again he wove astronomical allegory into his art. As the author of treatises on the astrolabe and equatorium, he was not only a competent but an expert calculator, in possession of the best available edition of the Alfonsine Tables, and capable of working with it to a high level of accuracy, even in his verse. Many of his stories in the *Canterbury Tales* have beneath the surface a strain of astronomical allegory based on quite specific events in the heavens in his own lifetime, indeed almost certainly around the time of composition. There are so many instances, and they are of such different types, that it is hard to know where to begin, but since it is relatively well known, we may take briefly some of the elements in "The Nun's Priest's Tale."

This is the story of a cock, Chauntecleer, and his narrow escape from a fox. By his nature, the cock knew that when the altitude of the Sun was 41° and the solar longitude a little over 21° of Taurus, the time was nine o'clock in the morning. (The poet does not of course set out these facts so baldly, but they are plain enough.) The data we are given fit perfectly with a specific date, Friday, 3 May 1392, following astronomical tables by the friar Nicholas of Lynn, whom Chaucer knew and mentioned elsewhere. The characters in the poem have simple analogs in the heavens: Chauntecleer corresponds to the Sun, and his wives to the stars in the Pleiades, which we can show the Sun to have been actually passing on the day of the story. The fox is Saturn. The story hinges on four different arrangements of the heavens during the fateful day, and it is quite conceivable that these were explained subsequent to the telling of the story with the help of the astrolabe on which Chaucer was so expert. Finally, as a check to skepticism, if not a complete cure for it: it turns out that medieval star lore in many European countries made out the Pleiades to be seven chickens. The Sun was passing the Pleiades. When the action of the tale begins, Chaucer tells us explicitly that Chauntecleer was walking by the side of his seven wives. Astronomy has much to answer for.

THE UNIVERSITIES AND PARISIAN ASTRONOMY

The framework of formal medieval Western education was based on the seven liberal arts. They provided the staple curriculum of the universities and were a great source of strength for those characteristically European institutions. The universities derived some of their patterns of organization from Islamic schools of learning, but what was new was their transcontinental recognition, protected and given privileges as they were by local rulers and by the Pope. The ultimate purpose of the universities was to provide the Church with an educated clergy, but this is not to say that they shared the religious preoccupations of the older cathedral schools. The medieval university—the common description of it as a *studium generale* hints at this fact—bound together different specialized studies into a single institution, organized in a hierarchy of difficulty. The faculty of arts was basic. It took its name from the several liberal arts, the *trivium* and *quadrivium*. The astronomy taught to young students in arts did not reach to a very high technical level, but it gave all university students access to at least the vocabulary of astronomy. Entry to the higher university faculties of law and medicine, and the highest of all, theology, required a long and arduous course of study, and was the privilege of only a small fraction of the university population. Students who advanced to the faculty of medicine were few in number but were deeply committed to learning, and they were expected to have a still greater knowledge of astronomy, since such practices as bloodletting required a knowledge of the phases of the Moon, and since astrological prognostication was an important part of the physician's repertoire. A large number of the surviving manuscript copies of treatises containing more advanced astronomy turn out to have been first owned by medical practitioners.

The universities provided an elite with the knowledge needed to serve church and state. The first meriting the name, in order of foundation, were those set up in Bologna, Paris, and Oxford. Dates of foundation were already a matter of dispute in the Middle Ages, and do not concern us here, but no one would deny that the beginning of the thirteenth century saw a very marked rise in their social and intellectual importance. New introductory texts were needed, and so it was that in the early decades of that century John of Sacrobosco wrote what was to become one of the most widely studied astronomical books of all time, *On the Sphere*. That work, by a man who was perhaps an Oxford master and who was certainly a teacher in Paris, was much ornamented with quotation from classical poets, and it dealt with only elementary spherical astronomy and geography, and hardly anything on planetary theory; but a start had been made. The same writer produced other popular texts on arithmetic and the computus (calendar calculation). At this point, no longer were students listening to bridesmaids and dreams of journeys through space. Sacrobosco's literary ornament is there to sugar the pill, but his is a text to be understood, not merely to be parroted.

Other works at the same level followed—for instance by Robert Grosseteste, Oxford teacher, bishop of Lincoln, and in time effectively chancellor of Oxford University. Grosseteste was one of the greatest of medieval scientists, an early enthusiast for Aristotelian science, but not deeply engaged in astronomy. Books like Sacrobosco's and his needed to be supplemented with planetary theory, and this was done by a type of book known by the generic title *Theorica planetarum* (*Theory of the Planets*). One good example of this type of work, used in the schools from the twelfth century onward, was John of Seville's translation of al-Farghānī's treatise; and Roger of Hereford wrote another. The most famous western example, however, was by an unknown author, and today, as in the Middle Ages, it is simply referred to by its opening Latin words ("*Circulus eccentricus vel egresse cuspidis . . .*"). Its author was guilty of a few technical misunderstandings, and it was sadly uninformative on the question of planetary parameters, but it greatly helped to stabilize Latin astronomical vocabulary and good copies were well illustrated. At the end, it included a smattering of astrology.

With the help of diagrams, this book taught the student the essential elements of the solar, lunar, and planetary models, unlike the canons to tables, which usually taught only the rules of procedure, without justification. One gets the impression, even so, that the writer had such canons in mind as he wrote, for interspersed with general descriptions of the basic Ptolemaic models there are instructions on how to use astronomical tables. At times, those instructions are barely intelligible, and if they were understood by most students it could only have been with the help of a master. The work gave no understanding of how planetary theories had been arrived at in the first place. Ptolemy's *Almagest*, which of course lay behind this type of treatise, could have done that, for it was twice translated into Latin in the twelfth century, once from Greek and once from Arabic. Indeed, the humanist cult of the pure Greek text led to another translation being done in 1451. The *Almagest*, however, was much too long and sophisticated—and costly—for general use, and had been replaced by astronomical digests even in Islam. One of the best known of these, that by al-Farghānī, introduced to Europe the idea that the spheres are nested in such a way as to leave no empty spaces. We have already alluded to the way in which the dimensions of the entire universe may be related to each other, and ultimately to the lunar distance, using this truly Ptolemaic model, a model that goes back to the *Planetary Hypotheses*, and not to the *Almagest*.

Aristotle's *On the Heavens* was studied in the universities for its cosmological content, and many commentaries were written on it. It was kept alive by the fact that it was interwoven with the rest of natural philosophy and metaphysics, which were also part of the staple university diet. Its dual physics, with a celestial region where natural motions are circular, and a terrestrial region where they are straight up or down, was rarely challenged. The level of medieval philosophical discussion, so often derided in later history by those with no time for logic, was very high. All philosophical discussion, including natural philosophy, was intimately bound up with theology. This precondition of all debate was not always as detrimental to freedom of thought as it might appear. Even when the Church forbade the teaching of certain Aristotelian ideas, as happened notoriously in Paris and Oxford in 1277, this could lead to a search for new alternatives, and so breed a new and critical attitude toward the old ideas. To take a single example, with important repercussions lasting for half a millennium: Thomas Bradwardine—who died in the Black Death of 1349, shortly after being appointed archbishop of Canterbury—rejected Aristotle's doctrine that there could be no vacuum outside the world. That he considered the void outside the world to be of a divine nature might not now be considered an insight of much scientific value, but that he spoke of it as infinite was significant. It is clear that cracks were opening up in the old cosmology, with its closed and finite universe. In Paris, Bradwardine's contemporary Jean Buridan, and later Buridan's pupil Nicole Oresme, showed themselves unafraid of this kind of language. These were short steps, but they were steps on a road that would end with cosmologists accepting the idea that the material universe might indeed be infinite.

Where technical astronomy prospered, as it did especially in Paris and Oxford from the late thirteenth century onward, there was a tendency among its practitioners to give ever less attention to Aristotelian commentary, and so gradually to cease paying lip service to the type of naive homocentric planetary astronomy that was assumed by the Aristotelians. As we have seen, in planetary theory this had been superseded long before Ptolemy was born, and the miracle is that it survived for so long. That is did so can be partly explained by the fact that it belonged to a tightly-knit and coherent overall system, namely Aristotle's. As we have often pointed out, even Ptolemy ended up by trying to preserve Aristotle's physical system.

At first sight, the very character of the medieval university might seem to have been unfavorable to the growth of astronomy, regarded as a science related to the observed world. Medieval attitudes to knowledge were strongly influenced by the techniques used in discussing Holy Scripture, that is, as an inheritance, to be purified and restored to its original form, then analyzed and commented upon before being transmitted to later generations. Fortunately, as better material became available, and as scholars learned something of the intellectual pleasures their subject could give, not to mention the promise it seemed to offer of astrological prognostication, a new type of European astronomer emerged. In this way the tempo of intellectual life increased, only to be checked occasionally by such disasters as war, political unrest, and plague—especially the Black Death of 1348–1349. These very dangers, however, helped the cause of learning in one important respect: they affected the movement of scholars. With the rapidly changing social and intellectual order, many more universities were founded, especially in the fourteenth century and after.

PRACTICING ASTRONOMERS

There was no one moment of enlightenment. Roger Bacon (ca. 1219–1292) introduces a mildly empirical note into his writings, but he was no astronomer. The most renowned medieval philosopher, Thomas Aquinas (1224–1274), lent his reputation to the idea that revelation must be supplemented by reason—and by Aristotle—in the pursuit of truth; but neither was he an astronomer at heart. For the real signs of change we should look to a more modest scholar like William of Saint-Cloud, who flourished at the end of the thirteenth century. About his life we know little beyond the fact that he was somehow connected with the French court. In 1285 he recorded an observation of a conjunction of Saturn and Jupiter. He compiled an accurate "almanach," giving the calculated positions of the Sun, Moon, and planets at regular intervals between 1292 and 1312, and introduced it with an account of the observations and planetary tables—those of Toledo and Toulouse—on which the almanac was based, as well as the corrections he found it necessary to make to them.

In connection with his work on almanacs, with their references to solar and lunar eclipses, William considered the projection of the Sun's image on a screen through a pinhole aperture. This would avoid damaging the eyes, he said, as had happened in so many cases at the eclipse of 4 June 1285. Roger of Hereford had mentioned the same technique in the twelfth century, and following a suggestion by William, Levi ben Gerson actually used pinhole images in 1334 to yield information leading to a figure for the eccentricity of the Sun's orbit. Levi observed the Sun at summer and winter solstices, using a combination of the camera obscura and the "Jacob's staff," an instrument of his own invention. Kepler observed a solar eclipse in 1600 in almost the same way. One way of deriving the eccentricity depends on the fact that the diameter of the image is in inverse proportion to the Sun's distance, so that the connection between the quantities observed and the geometry of the eccentric circular orbit is very direct.

The practical side of astronomy had begun to grow rapidly at the end of the thirteenth century, and it continued to do so thereafter in Europe without any real break. A phrase like "astronomical practice," however, can mean many things and cannot be neatly contrasted with "astronomical theory" in a modern sense for the simple reason that there was then a much greater stock of uncritically accepted presupposition. Take, for example, the case of comets. According to Aristotle, they were meteorological in character. He supposed that when the Sun's rays fall on dry land they give rise to dry exhalations, which rise like smoke, and which catch fire as they enter the celestial regions. This sort of speculation was hard to refute, since it fitted so well into Aristotle's comprehensive physical cosmology, which was so widely accepted in most of its detail. His ideas on the nature of comets were seldom challenged—the Roman statesman and writer Seneca, of the first century AD, was a rare critic—and those who observed comets simply took Aristotle as their starting point. When, in the fifteenth century, comets were fairly systematically observed—for instance by Toscanelli (from 1433), and Regiomontanus and Walther (in the 1470s)—there was considerable interest in the outward forms that comets take. Those forms were often interpreted in terms of wild analogy mixed with fairly standard astrological doctrine. That is not what most of us mean by practical astronomy. There was, on the other hand, a serious concern for their precise coordinates, and ultimately for their orbits in space, for which planetary astronomy had created clear precedents. Before the late sixteenth century, following Aristotle, the space to which they belonged was nevertheless sublunar space. As will be seen in chapter 12, not until Tycho Brahe disproved that assumption, in consequence of his observations of the comet of 1577, did Western astronomers as a whole do the same.

There are earlier examples of astronomers taking cometary positions seriously. Peter of Limoges measured the coordinates of the head of the comet of 1299, using a torquetum, an instrument discussed on p. 259, below. Geoffrey of Meaux took the coordinates of the comet of 1315 with reference to neighboring stars. Jacobus Angelus found the longitude of the comet of 1402 from that of the Moon. All three men were highly placed physicians with training in astrology, as indeed was Toscanelli. This particular social group is very well represented when we look down the lists of scholars with a demonstrable interest in the making of astronomical instruments.

Much attention was given to the design of instruments. William of Saint-Cloud, for example, also wrote an unusual work on a dial ("directorium") fitted with a magnetic compass, by which it was to be set. But as illustrating another very practical matter, he composed a new ecclesiastical calendar, commencing in 1292, which was notable for the care bestowed on its astronomical basis. All university students in arts were obliged to learn the rudiments of the ecclesiastical calendar, but they usually proceeded by rote, without any real understanding of the underlying astronomy. Those with a deeper insight into its astronomical basis, such as Grosseteste and Bacon, had complained about the inadequacies of the existing calendar long before William's new version, and others regularly followed suit. Not only had the Julian calendar slipped out of phase with the seasons, but it failed to satisfy the Christian requirements for an Easter canon. Scholarly discontent eventually led to the Gregorian calendar reform of 1582, and the replacement of the old calendar, especially in Catholic countries. William of Saint-Cloud had perhaps been naive in trying to present both ideal and workable alternatives together. That the reform was nearly four centuries in the making, however, was not so much a question of Church conservatism as of the fact that Church councils had more pressing political business to attend to.

Most of Italy, Spain, Portugal, Poland, France, and the Catholic Netherlands followed suit almost immediately before the end of 1582. For reasons of religious pride, Protestant countries were often very slow to follow the lead of the Catholic Church, and even in England, where the reform had been so long advocated, this was the case. Opposition on other than religious grounds had always been widespread, however, especially among the peasantry, whose calendar lore, weather proverbs, and feast day celebrations were thrown out of joint by the initial Gregorian ten-day shift. By and large, the calendar was in the hands of scholars. In England, a number of learned astronomers, including John Dee, Thomas Digges and Henry Savile, reported favorably on the Gregorian proposals, but the English bishops remembered the excommunication of Queen Elizabeth by the pope's predecessor and stymied the reform. German protestants spoke in much less temperate language: the reform, according to one German source, was the work of the Devil. Scotland adopted the change in 1600, but not until 1752 did England follow suit, by which time most European countries had long since fallen in line. By act of Parliament, eleven days were omitted from the English calendar in October of that year, five centuries after the death of the would-be reformer Grosseteste. There was much unrest—"Give us back our eleven days!" became an election campaign slogan, as one of Hogarth's prints reminds us—but in the end, reason prevailed. In the words of a sermon preached by a certain Rev. Peirson Lloyd, had England stayed with the old system, "in Process of Time, the two Festivals of *Christmas* and *Easter* would have been observed on one and the same Day." He refrained from telling his flock how many thousands of years would pass before this happened.

To leave the question of the calendar and return to technical astronomy at the end of the thirteenth century: by observation, William of Saint-Cloud found that the positions of the stars implied that the theory ascribed to Thābit was about a degree in error. For this reason William favored a *steady* precessional motion. All told, his approach to astronomy was unusually critical and constructive, but in his use of fresh observation he set an example which few were yet ready to follow. It seems likely—although we cannot prove as much—that by his example he encouraged John of Lignères and his pupils in the work that culminated with their various editions of the Alfonsine Tables. These, as we saw in the last chapter, date from the 1320s.

Another astronomer of this same generation, with a practical bent and a strong Parisian association, was Peter Nightingale, for some time a canon in Roskilde cathedral in Denmark. He had been teaching astronomy and astrology in Bologna when in 1292 he moved to Paris. There he stayed for perhaps a decade, before returning to Roskilde. Like William, he too composed a calendar, more orthodox in this case, and one that was to become extremely popular. In fact, by his example, Peter Nightingale introduces us to another highly consequential aspect of medieval astronomy, the invention and improvement of instruments for calculation. In his case, when in Paris around 1293, he developed a simple equatorium, as well as other devices for calculating eclipses. As in the case of the equatorium devised by Campanus of Novara, another scholar who had studied in Paris, but in his case in the 1260s, these instruments may be broadly described as moving Ptolemaic diagrams, the circles being graduated, made in metal. The poor student would have made them in wood or parchment. (We introduced the idea in connection with al-Kāshī on p. 212 and have since mentioned other examples.) They continued to be made in those relatively obvious forms—obvious, at least, to all who had studied Ptolemaic astronomy—well into the seventeenth century. An example from a magnificent printed book by Peter Apian is shown in figure 102.

Paris was at this time the most important European center of astronomical activity, and one might be forgiven for thinking that a condition of entry to the study of astronomy at the university there was to have the name of "John." There was a John of Sicily, of Lignères, of Murs, of Saxony, of Speyer, and of Montfort, all in the space of two or three decades. Each of them left his mark on astronomy, and each took a strong interest in the refinement and reorganization of the Alfonsine and related tables. Here it is worth noting the emphasis placed on ease and speed of calculation. As one of many examples of this, we have John of Murs's tables for the conjunctions and oppositions of the Sun and Moon (for 1321–1396), an aspect of ecclesiastical calendar reckoning. Not for nothing did Pope Clement VI invite him to Rome, with Firmin de Bellaval, to advise on calendar reform in 1344 and 1345.

All of these scholars interested themselves in the design of instruments of observation and calculation. In the first category we may mention the use of a quadrant fixed on a wall in the meridian (the mural quadrant), and the parallactic rules used by Ptolemy—there is much to be said for an instrument requiring no circular scales, if workshop practice is not equal to the task of making them accurately. Records of observations are ephemeral things, and the fact that we have so few does not mean that these Parisians were only calculators. We know that this was not so. A manuscript by John of Murs includes records of observations made at five different places between 1321 and 1324.

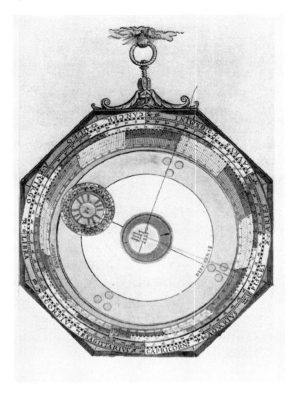

102. A printed equatorium from Peter Apian's *Astronomicum Caesareum* (1540), in this case for the planet Mercury. The magnificent large folio volume, with its colored woodcuts and movable parts, was dedicated to the Emperor Charles v and his brother Ferdinand. The numerous paper volvelles in the volume are pivoted by means of glued paper disks, these simplifying the problem of eccentric positioning.

Falling into the category of instruments of calculation, we have many more examples. To take only John of Lignères: we have treatises by him on a new type of armillary, on the saphea (the universal astrolabe projection), on the Campanus-type equatorium, and on a "directorium"—a calculating instrument related to the astrolabe, but specifically for applying an astrological doctrine, the doctrine of "directions." Most astronomers had their price. The few biographical details we have of many of these Parisians are known to us only because they were in the service of princes and high ecclesiastics, the very classes that provided the richest market for astrology.

RICHARD OF WALLINGFORD

The University of Oxford had a long scientific tradition, especially in the teaching of Aristotelian natural philosophy, and Oxford astronomy during the thirteenth century had been largely concerned with the production of simple teaching tracts for the sphere and the calendar, together with cosmological questions arising out of Aristotle's *On the Heavens*. The Toledan Tables were then in use, and the canons to them helped to consolidate astronomical learning and to fix vocabulary, but not until the early years of the fourteenth century were there signs of much originality. A set of tables drawn up by the Merton college astronomer John Maudith during the period 1310–1316 focussed the attentions of various scholars on the trigonometry underlying spherical astronomy, and provided the first exercises in this subject for one of the most remarkable astronomers of the Middle Ages, Richard of Wallingford (ca. 1292–1336).

It is customary to rank early astronomers according to their originality in devising new planetary systems, but this is to lose sight of what was felt to be most needed in the fourteenth century, namely, methods of rapid calculation and representation. In his efforts to provide these, Richard of Wallingford provided a number of original ideas that reveal great qualities of mind, and that had significant, but largely hidden, repercussions. He was a Benedictine monk, educated at Oxford, where he also taught until 1327. In that year he returned as abbot to his monastery at St. Albans, England's premier monastery (plate 8). Having visited Avignon—then the seat of the papacy—to obtain papal confirmation in his new office, he arrived back in England to discover that he had contracted leprosy. Far from shunning him, his monks were so proud of his achievements that they kept him as their abbot until his death.

Richard of Wallingford's *Quadripartitum* was the first comprehensive treatise on spherical trigonometry to be written in Christian Europe. It was developed on the basis of the *Almagest*, the Toledan canons, and a short treatise perhaps by Campanus of Novara. Whilst he was abbot he found time to revise it, taking into account a work by the twelfth-century Sevillian astronomer Jābir ibn Aflaḥ —one of two scholars known in the West as Geber, the other being more famous as an alchemist. Before leaving Oxford, Richard wrote three other works and some minor pieces. His *Exafrenon* was a treatise on astrological meteorology, a restrained but unoriginal work. He wrote also on an instrument he had designed, the "rectangulus." The third dealt with his equatorium, which he called "albion."

The *Quadripartitum* offered exact solutions of problems in the geometry of the sphere, for instance involving spherical triangles, but such calculation was very tedious. The armillary sphere could give approximate answers, but was difficult to construct accurately. The torquetum, possibly invented by Franco of Polonia in the thirteenth century, while at first sight very different from the armillary sphere, has much in common with it. (For a sixteenth-century example, see figure 103.) The geometrical problem facing all serious astronomers is that of combining vectors in three dimensions. The mechanical problem is one of pivoting a sighting rod around different axes simultaneously, notably those at right angles to the equator and ecliptic. (One is reminded here of the ingenious construction of a Rubik's cube or its spherical counterpart.) Astronomers may offset the axes with impunity, in view of the great distances of what they observe in the heavens. With his rectangulus, Richard of Wallingford took the process of simplification a stage further than the torquetum. His new instrument incorporated a system of seven straight rods, with no circular scales at all (fig. 104). He could not pivot his seven rods in different planes around a single point, but he invented a system of offset pivots, which were all that he needed. The rectangulus was in principle usable for observation, and

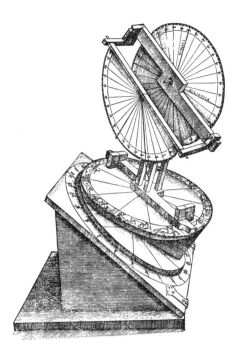

103. A torquetum, from Peter Apian's *Introductio geographica* (1533). Apian reused the woodcut in other works. The mounting of telescopes at a later period of history owed much to the torquetum tradition (see fig. 163 on p. 389 for a very obvious example).

104. Richard of Wallingford's rectangulus. Each of the three pairs of arms opens like a pair of scissors, while the upper two pairs and the sighting arm on top pivot at right angles to the plane of the pair below. Plumb lines hang from four of the arms as shown. The surrounding figures are details of the main structure. The strip with scales shows how the arms are to be graduated so as to allow the angles between them to be measured with the help of the threads, in many cases pulled so as to fall at right angles to the partner arm.

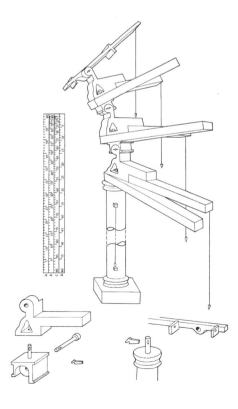

provided coordinates more or less directly. The fact that the rods of the rectangulus were straight meant that great accuracy of construction and graduation was possible. There were some disadvantages, but the design shows great powers of intuition.

His most important finished work was his *Treatise on the Albion*. The albion ("all by one") is in many respects the most notable of the entire genre of medieval planetary equatoria. It is harder to understand than most, because it did not directly simulate the motions of the planetary circles, giving each a metal equivalent, as most of its predecessors had done. Instead, mirroring the use of tables, disks were used to calculate the planetary equations, which were then added to (or subtracted from) the mean motions by rotating the disks through the appropriate angles. This entailed the use of nonuniformly graduated scales; and to lengthen the peripheral scales on the disks, Richard made some of them in spiral form. Any number of turns to the spiral was, in principle, possible, but thirty was not unusual. As a cursor, a thread was drawn out through the center. The whole thing is very reminiscent of circular slide rules—instruments that passed into history, more or less, with the advent of the electronic calculator in the 1970s. The albion had more than sixty scales in all, some of them ovals. Their complexity was not entirely apparent, but was concealed in the various methods of graduation. The instrument incorporated two different types of astrolabe, one a "saphea," but they were not essential to it. There was almost no problem of classical astronomy that could not be solved using the albion. As subsidiary instruments over and above those for planetary positions, it gave parallax, velocities, conjunctions, oppositions, and eclipses of the Sun and Moon.

The other type of equatorium, the simple analog of the planetary models, was easier to understand. For that reason it remained more popular, but was much more limited in its possibilities. The versatility of the albion earned it great esteem, first in England and later in southern Europe, and it remained in vogue, in various anonymous forms, until the sixteenth century. At least seven treatises were derived from it, and astronomers began to abstract its subsidiary instruments, especially its parallax and eclipse instruments. In or around 1430, the Viennese John of Gmunden produced the one that was perhaps the most often copied. Regiomontanus drew from Richard's treatise, and produced a rather careless edition, and John Schöner abstracted its eclipse instrument. The most striking printed work to use it was the *Astronomicum Caesareum* (1540) of Peter Apian of Ingolstadt, introduced earlier (see fig. 102, above) as a repository of relatively straightforward printed equatoria.

These writers took certain underlying principles from Richard—they cannot be explained here, but they concern the graphing of functional dependences—and extended them in ways that had significant repercussions in later history. In France, from 1526 onward, the mathematician and cosmographer Oronce Fine wrote several treatises on equatoria that made use of similar principles; at about the same time in Aragon Francisco

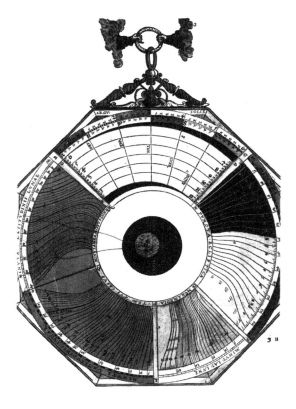

105. One of many instruments from Peter Apian's *Astronomicum Caesareum* (1540), by which one astronomical quantity might be calculated as a function of two or more others. Devices that can be described in this way have a long history in astronomy.

Sarzosa did the same; and there were several other instances of similarly sophisticated techniques that were in later centuries to be given the name of "nomography." Some idea of the procedures involved may perhaps be gleaned from figure 105, taken from Apian's *Astronomicum*. Apian's painstaking print work later came in for passing criticism by Kepler, who called it a waste of time and ingenuity, but its purpose was not to please the likes of the serious Kepler, who was in any case not above seeking easy routes to calculation, or, for that matter, princely patronage.

THE CLOCK AND THE UNIVERSE

At St. Albans, a wealthy monastery, Richard of Wallingford was able to dispose of extremely large sums of money on the building of a mechanical clock. We know from a reference in a commentary on *The Sphere of Sacrobosco*, written by Robert the Englishman in 1271, that astronomers were then working—as yet unsuccessfully—on the problem of controlling a wheel in its rotation, so as to provide it with the daily motion. From many references to the building of expensive church clocks in the 1280s and after, we know that the key invention, the mechanical escapement, had arrived—that is, more than forty years before Richard began his work. Despite this fact, the disordered pile of documents he left at his death, with several drafts for the design, contain the oldest surviving description of any mechanical clock, and also—this is one of the paradoxes of history—what was mechanically the most sophisticated of the Middle Ages. The clock was unfortunately lost to history after the dissolution of the monasteries in the time of Henry VIII. The antiquarian John Leland reported that it showed planetary movements and the changing tides. (These would have been at London Bridge, calculated automatically from Moon positions in a standard medieval way.) The whole thing was built in iron: Richard's own father had been a blacksmith. It was built on an enormous scale, with its frame the height of a man and two or three meters across, seated on ledge on the wall of the southern transept of the abbey church.

This mechanism is highly relevant to the history of astronomy. Not only was it a moving replica of the universe as it was known to the medieval astronomer, but almost every aspect of the design was inspired by astronomical practice, even down to the method of calculating gear ratios and tabulating the spacing of gear teeth. It had hour striking on a 24-hour system—17 strokes of the bell at 17 o'clock, and so on—and it struck at the equal hours of the astronomer, rather than at the common man's seasonal hours. It had spiral gears and an oval wheel to give a carefully calculated variable velocity for the Moon's motion around an astrolabe dial. (This left a theoretical error of only seven parts per million.) It had differential gears for a mechanism showing lunar phases and eclipses. The clock, completed after Richard's premature death, had a single face, and in this was totally unlike the somewhat later astronomical clock, or "astrarium," built by Giovanni de Dondi between 1364 and 1380. The son of an astronomer-physician who had designed a clock for Padua in 1344, Dondi became physician to Emperor Charles IV. His astrarium was in 1381 acquired by one of the Visconti, the dukes of Pavia. It was seen by Regiomontanus in 1463, for whom it was copied, but it was beyond repair by 1530, when Emperor Charles V had it copied.

The Dondi mechanism had a seven-sided frame, with a dial for each of the planets, the Sun, and the Moon. There was a clever digital calendar mechanism. Each planetary mechanism was essentially a geared Ptolemaic diagram. Dondi used sliding rods, and eccentric wheels that were oval only to allow them to mesh at varying distances, and a purist engineer would no doubt consider his mechanical virtuosity less than Richard of Wallingford's, but his astrarium, embodied in brass, was doubtless more appealing close at hand. One important astronomical point that should not escape notice is that, like the simpler type of equatorium that simulated the Ptolemaic models, and indeed like the planetary diagrams in manuscripts of the *Almagest*, it failed to represent the universe as a single system. Richard of Wallingford's clock extended the ancient tradition of the anaphoric astronomical clocks of antiquity, depicting the universe in a single display, but if the planets were present, then they too were no doubt on subsidiary dials.

Most early church clocks had no display, and merely sounded a bell on the hour, but in the course of time most great cathedrals and monastic churches had mechanical clocks built, with some sort of astronomical symbolism in a single display—often supplemented by moving human figures and other automata. Here the water clocks of Islam had helped to show the way. The end result, an astronomical display, tends to hide something far more important, however, which is that the desire to create such a display helped to bring about one of the great turning points in the social and economic history of the world. The mechanical clock imposed its will on all future generations—with or without an astronomical dial.

OXFORD AND THE ALFONSINE TABLES

Richard of Wallingford was at the center of Oxford astronomy when he wrote his *Albion* in 1327, and yet he made no mention of the Alfonsine Tables in that work. He used a version of them, however, around 1330. Around 1340, William Rede of Merton College, Oxford, took a Parisian version of the tables, with sexagesimal time and angle divisions, and converted them to tables for the Oxford meridian in the more familiar "Toledan" form—with signs of 30°, not 60°, for example. The fact that there are surviving versions for other English towns, some even from the 1320s (for Leicester and Northampton), and others for Colchester, Cambridge, York, and London, reminds us that much astronomy was done in religious institutions outside the universities, although no doubt by men with a university education.

These were all relatively simple adaptations. Two much more radical revisions were made in Oxford, the first by an unknown man around 1348, perhaps William Batecombe, the second by John Killingworth in the following century. The 1348 tables go much further than John of Lignères' labor-saving Large Tables, double-entry tables that yielded a single equation. The 1348 tables allowed the planetary longitudes to be extracted more or less directly, apart from a small adjustment for precession. These tables were extensive, and could be used to carry information about the direct motions, stations, retrogradations of the planets, and other matters that were of an astrological importance. Yet again we have evidence that astrology provided an important motive for the intensive study of astronomy. Of course 1348 was the year of the onset of the Black Death in Oxford, and as scholars of the time noted, this turned their thoughts to God. No doubt it occasionally turned them to astrology too.

The 1348 tables were seized upon by scholars in other parts of Europe. There are early manuscripts from Silesia and Prague with adaptations of them. Henry Arnaut of Zwolle, in the northern Netherlands, used them, referring to them as "the English tables." A fifteenth-century Italian Hebrew translation was done by M. Finzi, assisted by an anonymous Christian from Mantua. Giovanni Bianchini of Ferrara, the most notable Italian astronomer of the mid-fifteenth century, was influenced by them in producing a similar set of tables, much used by such leading contemporaries as Peurbach and Regiomontanus. Through this intermediary, and a fourteenth-century eastern European version known as the *Tabulae resolutae*—which was studied by Copernicus when he a student at Kraków—they provided the core of

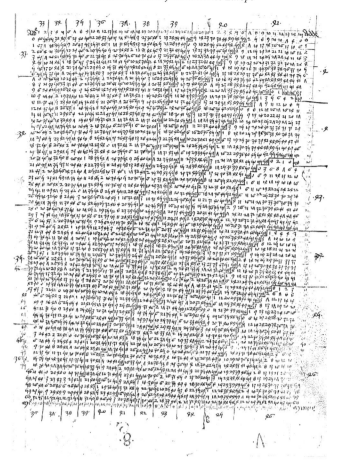

106. A page from the remarkable tables by John Killingworth, which were aimed at reducing the work needed to generate an "almanac," an ephemeris for all the planets, giving their positions at daily intervals according to Alfonsine principles.

John Schöner's tables of the same name (printed in 1536 and 1542). These in turn were widely used for many decades, but their origins had been long forgotten. It was the story of the albion repeating itself; but it was a characteristic of the Middle Ages that truth was God's truth, and that authorship was not something to be contested like territory.

After the 1348 tables, one further set of Oxford "Alfonsine" tables of great merit was produced, by Merton College astronomer John Killingworth (ca. 1410–1445). These were meant to be used to calculate a full planetary almanac (ephemeris). One copy made for Humphrey, Duke of Gloucester, was of great beauty and was heavily interlined with gold leaf. It is not clear whether or not this was meant to reflect the genius of their author, but this was certainly of a very unusual order. For a specimen page from a more workaday copy, see figure 106. Implicit in the tables is a piece of theorizing that he does not write down for us, but that few could reproduce today without recourse to the differential calculus.

PHILOSOPHERS AND THE COSMOS

By 1380, Europe had about thirty active universities, most of them small and recent but eagerly competing with one another. Apart from Prague and Vienna, there was none of importance in eastern or German-speaking Europe. By 1500, there were nearly fifty more foundations to be added to the list. Many of these were of minor importance, but more than a dozen of the new foundations were in German-speaking areas. From there, and places still further east, came a new wave of enthusiasm for astronomy, that for more than a century seems to have outstripped that in the older centers.

This movement was not unrelated to the religious changes taking place at the same time, although the relationship of the two is not a simple one. Although Vienna, for example, was an important center of learning long before the university was founded there in 1365, its rise to importance came only with the schism in the Catholic Church (between 1378 and 1417 there were two, and later three, rival popes, each with his own College of Cardinals and administration). When that began, Vienna provided a natural home to those central European masters and students whose life in Paris had been made difficult—they were unable to endorse the pro-French pope. Later, the university was swelled by those Bohemians driven from Prague, after clashing with the ever more assertive German majority there.

Leipzig University was to be founded by secession from Prague in the stormy days of John Hus, the follower of the English reformer, John Wycliffe; and there are many similar stories that relate to the growing sense of nationhood, in scholarship and religion alike. The tragic history of John Hus and Jerome of Prague, both executed as heretics (in 1415 and 1416, respectively) in the course of the struggle for ecclesiastical reform, is well known. There followed a systematic harassment and execution of Jews—240 persons were burned at the stake in Vienna on a single day in 1421. We should not forget that many of the scholars whose astronomical ideas are being considered here were in some way bound up in the academic discussions that justified such actions. They were contributing, however innocently, to an overall Christian philosophical and theological view of the world that could make such things possible. Their lives as astronomers were not compartmentalized, as are those of many of their successors today. They were usually theologians, concerned with what they saw as deeper questions of truth, and the question of who had the right to decide it; and the roots of many of those questions were emphatically bedded in medieval astronomy. After Copernicus, as will be seen in chapter 12, this discussion was transformed into a less dangerous one, concerning the nature of the knowledge provided by a scientific theory.

One of the most notable of the masters displaced from Paris to Vienna was Henry of Langenstein (ca. 1325–1397), also known as Henry of Hesse.

He was one of the most notorious of Ptolemy's critics. Like several other western astronomers of the later Middle Ages, he followed the example set by Eastern and Andalusian scholars in using physical arguments to criticize the planetary schemes found in the *Almagest*. After leaving Paris, Henry spent the last years of his life in Vienna and played an important part in the reorganization of the university there. He was at heart, however, a man of the schoolroom, whose *Treatise Refuting Eccentrics and Epicycles*, written in 1364 in Paris, is a rather crabby book, much of it is given over to petty academic criticism of the standard university text, *Theorica planetarum*. One of his more serious aims was that of showing that the circles of Ptolemaic astronomy cannot be considered as physically real mechanisms existing in the heavens.

Richard of Wallingford had made the same point earlier in the century, but in such a casual way that it must have been a commonplace, no doubt helped along by Martianus Capella's text, and the Latin version of Ibn al-Haytham's work on optics. For Henry of Hesse, the circles of astronomy were mere mathematical constructions, justified only by the predictions based upon them. Henry was dissatisfied with Ptolemy's account of planetary distances and sizes; he disliked the equant; and he disliked the irregularities introduced into the theory of planetary longitude by the theory of planetary latitude. In a word, he wanted a universe that ran on simpler lines than Ptolemy's. Alas, few philosophers have ever understood how complex is the notion of simplicity.

Henry's arguments are at first sight complicated, but they rest on a number of assumptions about the nature of motion that would not now be regarded as acceptable. As a specimen argument from the *Treatise*: "If epicycles existed, the same simple body would be moving at one and the same time with different motions," and this cannot be, "since one and the same cause cannot produce different effects on the same body simultaneously." Although some scholars were beginning to grasp the point, most shared the difficulty of understanding how two different motions can be combined into one—as when a ship is blown off course by the wind—and conversely, of resolving one motion into two. And why? Because they thought of a motion as a real, unique, and ultimate thing. This is ironic, since the aim was to show that the Ptolemaic circles were not at all real, but purely hypothetical. They had been taught to think in this way through reading Aristotelian physics, and Ibn al-Haytham's influential work *On the Configuration of the World*.

A certain Master Julmann wrote in the same style in 1377, drawing much from Henry, and when he added material of his own, again he revealed the difficulty he had with the idea of a sharing of motions in one body. This problem seemed particularly pressing when it came to visualizing a "Ptolemaic" model of the spheres, of the kind found in Ptolemy's *Planetary Hypotheses* and al-Farghānī's *Theory of the Planets*. The model for longitudes was complicated enough, but add to it the idea of real epicyclic spheres that explain also the movements of the planets in latitude, and one can soon develop a certain sympathy with the writers in question.

An academic politician of much greater genius was Nicholas of Cusa (ca. 1401–1464), who had been educated first in the Netherlands by members of a devotional sect, and then at the universities of Heidelberg and Padua. He is best remembered as a Platonist philosopher, but his interests took in astronomy too, with some curious results. At Padua, with his friend— later a famous geographer—Paolo Toscanelli, he attended lectures by Prosdocimo Beldomandi on astrology. After his ordination to the priesthood, Nicholas became a friend of the humanist Eneo Silvio Piccolomini, one of two members of this noble family to become pope, although not until 1458. Nicholas was fortunate in his friends: after performing a number of diplomatic services for the papal court, he was made cardinal in 1446. Nicholas of Cusa was one of those whose wish for wide reform of the Church was coupled with a wish to reform the calendar, although there he made no great progress. As cardinal, he had the means of buying some very fine astronomical instruments, which fortunately still survive. His fame as a thinker, however, has virtually nothing to do with precision astronomy: he was a philosopher whose scientific imagination, unrestrained by much more than broad analogies, led him to ideas about the place of the Earth in the universe. After his death, his ideas took on a prophetic appearance.

The most influential of his works was one he completed in 1440, *De docta ignorantia* (*On Learned Ignorance*). In this he made much use of a "principle of the coincidence of opposites," a principle at the shrine of which Hegelians and Marxists have since been happy to worship. The general idea was that all problems can be resolved by its use. Apparent contradictions are united at infinity. Each entity is present in every other; the largest number coincides with the smallest ("the maximum of smallness"); the point coincides with the infinite sphere in the same way; and so on. Now, all of this would hardly be worthy of mention were it not that, from the last principle, Nicholas drew the conclusion that since a point includes (or mirrors) the whole universe, there can be neither fixed center nor fixed periphery to the universe. In particular, the Earth cannot be said to occupy the center of the universe. So much for its *place*. As for its *motion*, this hinges on a principle of relativity: the position of anything is relative to the observer. From this it follows that the Earth may be said to move. And having seemingly dislodged it from its traditional place, he was led to speculate that it might not be the only body on which there are living creatures.

Some of these are ancient ideas, held, for instance, by Hermetic philosophers—followers of the mythical Hermes Trismegistus. They had little impact until after Copernicus, and even after Giordano Bruno, that is, after the end of the sixteenth century. It was then that Nicholas of Cusa began to be cited as though he had been a precursor of Copernicus. Descartes quoted him as having proposed the infinity of the world, and his reputation for cosmological sagacity has grown with the centuries.

It is hardly merited. When Nicholas wrote on astronomical matters, he took a traditional line on the centrality of the Earth and the ordering of the planetary spheres. History, though, is mostly made of reputations, not of merit.

It is true that Nicholas played down the inferiority of the Earth with respect to the regions above the Moon, and in fact he took a most unusual step in detracting from the relative perfection of the Sun. He speculated on the possibility that within its bright envelope there may be a layer of watery vapor and pure air, within which in turn there might be a central earth. It may be hard to believe that this view could have been entertained in the fifteenth century, and yet in the eighteenth and nineteenth centuries, such excellent astronomers as Alexander Wilson and the William Herschel were not above making somewhat similar conjectures. The main difference was that they avoided the charge of heresy. Nicholas's political rivals accused him of pantheism, and it was against this charge that he defended himself in his book *Apologia doctae ignorantiae* (1449), in which he quoted the church fathers and Christian Neoplatonist philosophers—from whom he had taken some of his ideas. Oddly enough, in view of the Platonist representation of the world as mathematical, he seems to have been led to a very different idea. From the observation that nothing in our experience is mathematically exact—no object is truly straight, the Earth is not truly spherical, and so on—he concluded that a mathematical treatment of nature is impossible. It is not easy to make a scientific hero out of such a philosopher, although many have tried to do so.

PEURBACH, REGIOMONTANUS, AND THE PRINTED BOOK

Astronomy is frequently represented as having undergone a sudden revival in the middle of the fifteenth century, as though only then, with the recovery of large numbers of Greek texts, were astronomers capable of enlarging their Alexandrian inheritance. This is an illusion created by the invention of printing, which suddenly made the multiplication of books a relatively easy operation. As a direct consequence of this, the reputations of two men in particular rapidly eclipsed those of the scholars whose textbooks they rewrote. Georg Peurbach and Johann Müller were both of them influential scholars in the new literary movement, but their own astronomical writings continued, rather than overturned, the medieval tradition.

Peurbach was an Austrian scholar who inherited the mantle of an almost equally influential astronomer, the first professor in the subject at the university of Vienna, John of Gmunden. John had died in 1442, before his arrival there, but had assembled a large number of invaluable manuscripts and instruments, which he had bequeathed to the university. (He edited the albion text himself, for instance, and owned an example of the instrument.) Peurbach received a master's degree at Vienna in 1453, but both before and after this time he traveled throughout France, Germany, and Italy. He became court astrologer first to Ladislaus V, king of Hungary, and then to the king's uncle, Emperor Frederick III. At Vienna he taught the classics in the new humanist style, and there he also completed the text-book by which he eventually became known to generations of students, the *Theoricae novae planetarum* (*New Theories of the Planets*).

Johann Müller is better known as Regiomontanus, the Latin name for Königsberg in Franconia, where he was born in 1436. A child prodigy, he entered the University of Leipzig at the age of only eleven. Migrating to the University of Vienna in 1450, Regiomontanus took his bachelor's degree there in January 1454. Peurbach, while appreciably older, had begun to lecture as regent master in Vienna in that same year, and after two more years Regiomontanus began to study with him. Together they began a program of observation—of the planets, eclipses, and comets—and carefully noted the astrological implications of what they saw. In 1460, their careers took a new direction with the arrival in Vienna of Cardinal Bessarion, papal legate to the Holy Roman Empire. Greek by birth, his mission was to placate the emperor in a dispute with his brother, and enlist support for the recovery of Constantinople, captured by the Turks in 1453. He was also anxious to accelerate the western intellectual movement to master the Greek classics, and he persuaded Peurbach—who knew no Greek—to produce an improved abridgement of the *Almagest*. George of Trebizond had translated it into Latin out of the Greek in 1451, but his was inferior to the twelfth-century Latin version, done from the Arabic by Gerard of Cremona. Peurbach relied on the latter, which Regiomontanus said he knew "almost by heart," and he was about half way through his abridgement when he died in 1461.

Within two years, Regiomontanus completed the task, although the resulting *Epitome of the Almagest* was destined to be printed only in 1496, twenty years after his own early death. The work depends heavily on a widely used medieval version titled *Abbreviated Almagest*, but his was more comprehensive. In its finished form it was simply the best available commentary on Ptolemy, and remained so until modern times. It is amusing to see with what joy the humanists vilified earlier medieval digests, which had supposedly lost the purity of the original.

Peurbach's works included minor treatises on earlier instruments, but more influential by far was his *Theoricae novae planetarum*, a work based on the dissertation he had presented for the master's degree. Its undoubted importance in education during the next century or two has led many to ascribe great originality to it. It was certainly well written, but it is best seen as a revised, improved, and supplemented version of the old *Theorica planetarum*, the standard text of the medieval university student in arts. In all strictness, there had been several closely similar works going under this generic name, but one of them, by an unknown author of the early thirteenth century, was copied more than any other university textbook in astronomy with the exception of Sacrobosco's *On the Sphere*. There were at least four different printed editions of texts of the old *Theorica* type before 1500, and others later, so that it evidently rivaled Peurbach's in popularity, although in time his became the clear favorite. Used in manuscript copies for its first twenty years, his treatise was first printed by Regiomontanus in 1474, going into nearly sixty editions before it fell into disuse in the seventeenth century. Such was the power of the press. Figure 107 (in two parts) show two of the typically clear illustrations in the new work. The book made popular the type of solid spheres introduced by Ptolemy in his *Planetary Hypotheses* and found in al-Farghānī, Ibn al-Haytham, and others. It contained material on the theories of access and recess (oscillatory precession) attributed to Thābit ibn Qurra, and the somewhat similar Alfonsine theory. What it clearly reveals is how strong was the influence coming from the direction of the various sorts of Alfonsine Tables, in particular those—the 1348 tables, and Giovanni Bianchini's—that were making the calculation of planetary positions so much easier, and therefore discrepancies with observation more readily apparent.

In the late 1450s, Peurbach completed what must have been his most laborious work, Tables of Eclipses (printed first in 1514), calculated first for the meridian of Vienna, and in another version for the town of Oradea, Hungary (*Tabulae Waradienses*).

The first meeting of Regiomontanus and Bessarion in Vienna in 1460 was not the last. At Peurbach's request, the two traveled to Rome together, although only after Peurbach had died. By this time Bessarion had been appointed papal legate to the Venetian Republic, which explains why, in July 1463, he was able to persuade Regiomontanus to settle for a time in the neighborhood of Venice and Padua. Padua was the home of a highly esteemed university with a strong tradition in astronomy, and the young scholar had much to add to the reputation of the place. In that same summer, he began a correspondence with Bianchini, a skilled astronomer whose chief occupation was that of a lawyer in the service of the noble family of Este, in Ferrara. As was not unusual, such exchanges of letters amounted to a polite trial of mathematical strength, and Regiomontanus far outshone the much older man. Regiomontanus was able to point out many of the deficiencies of the Alfonsine Tables, and to formulate a program of reform, mathematical, astronomical, and astrological. He did not remain in Italy, however. In 1467 he moved to Hungary to take up a chair in astronomy at the newly created University of Pressburg in Bratislava. In fact, Regiomontanus was made responsible for selecting an astrologically propitious moment for the actual foundation of the university. (This was not an uncommon procedure. The wonder is that it has not been revived.) In Hungary he collaborated with the royal astronomer Martin Bylica and dedicated a work on trigonometry to the king, Matthias I. This noted patron of humanist scholarship, perhaps better known by his nickname Corvinus, was instituting a program of reform of the anarchic Hungarian state. The would-be reformer of astronomy, Regiomontanus, must have viewed him as a kindred spirit.

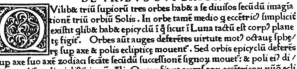

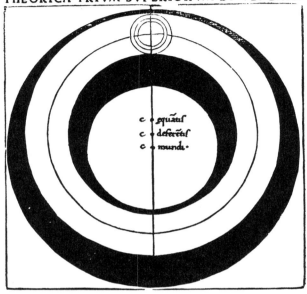

107. Woodcuts illustrating the physical model of the spheres moving a superior planet (above), and the orientation of their axes and their poles (opposite). The three named centers are those of the equant, the deferent, and the world. The sphere carrying the epicyclic sphere has its axis through the center of the deferent. The woodcuts come from the most successful early printed text book of Ptolemaic planetary astronomy, Georg Peurbach's *Theoricae novae planetarum* (*New Theories of the Planets* [1473]; printed by his former pupil Regiomontanus). The work simplified, but also augmented, Ptolemy's *Almagest*, and by introducing (as here) a quasi-physical model of nested solid spheres from Muslim sources, he was unwittingly propagating the ideas contained in Ptolemy's own *Hypotheses planetarum*.

Whether the political instability of Hungary led him to quit Pressburg for Nuremberg, or whether he did so for more material reasons, he made his move in 1471. Nuremberg was the leading commercial center of central Europe, a crossroads between Italy and the states of northern Germany. Its commercial position made communication with other scholars all the easier. There he could obtain fine instruments, and there he could set up a fine printing press, notable for its Latinate typeface—as opposed to the German Gothic forms—no less than for the admirable judgment of its owner. His first publication was the *New Theory of the Planets* by the lamented Peurbach. He followed it with a prospectus of more than forty items that he planned to publish, ranging from the works of Euclid, Apollonius, and Archimedes, to his own compositions. As matters turned out, he did not live

&o in linea augis tantum a centro buiÐ orbis quantum hoc centrum a centro
mundi diſtat elongato: regulariter moueatur. Vnde & punctuſ ille centrum
ꝗuantis dicit̃. & circulus ſup eo ad ꝗntitatem deferentis ſecum in eadẽ ſupſi
cie imaginatus eccentricus ꝗuans appellatur . Neceſſario igit̃ oppoſitum ei
ꝗd ĩ Luna fiebat accidit in iſtis ut ſcilicet centrũ epicycli ꝗnto uiciniÐ augi defe
rẽtis fuerit tanto tardiÐ: ꝗnto uero ꝓpinquiÐ oppoſito tãto uelociuſ moueat̃.

Epicyclus uero duos habẽ motÐ quoꝗ unus eſt ĩ lõgitudinẽ: alt̃ ĩ latitudi
nẽ. De ſecũdo dicẽdũ erit poſtea.   MotÐ aũt eiÐ in lõgitudinẽ eſt quo mouet̃
circa centrũ ſuũ corpus planetꝭ ſibi infixũ ĩ pte ſupiori ſecũdum ſucceſſionẽ:

THEORICA AXIVM ET POLORVM.

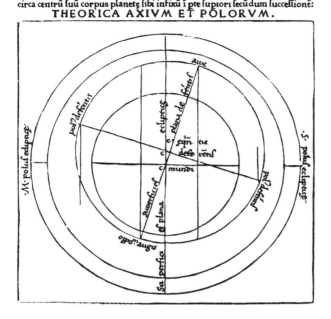

long enough to print more than a few of these titles, but some were issued
in his name after his death. In 1474 he printed a calendar, in both Latin
and German editions, with a simple paper volvelle, diagrams of eclipses,
and other astronomical data based on his own calculations. Before the
century was out, nine editions of this *Calendarium* appeared from presses
in Nuremberg, Venice, and Augsburg, one of which—Ratdolt's edition of
1476—had an important place in the history of printing: it introduced the
idea of the title page, as against the old custom of putting a colophon on
the last page. Also in 1474, Regiomontanus published his own planetary
ephemeris (almanac) for the period 1475–1506. (For a typical opening,
see figure 108.) This was not, of course, the first almanac, but it was the
first *printed* work to exploit a vast potential astrological market for pre-
computed planetary positions. Its popularity can be judged from the fact
that it was reissued at least thirteen times before it became outdated.

Regiomontanus died on a visit to Rome in 1476. One hollow story has
it that he was poisoned by the sons of George of Trebizond, a man he probably never knew personally, but whose translation of, and commentary
on, the *Almagest* of Ptolemy he had openly criticized in abusive terms. It
has been said that he learned Greek from Trebizond, but his knowledge of
the language—which he certainly knew well enough to correct either the
Greek text or Latin translations of Archimedes—is something he seems to

Ianuarius	☉	☽	♄	♃	♂	♀	☿	☊		1475	☉	♄	♃	♂	♀	☿

The pages for January 1475, from Regiomontanus's Ephemerides *(1474).*

108. The pages for January 1475, from Regiomontanus's *Ephemerides* (1474). He names only seven feast days in the month. Having supplied all the dominical letters A, he assumes that the reader can supply the rest (compare fig. 93). The positions of the planets are given within the zodiacal signs (see the second row, and note marks in the table where the sign changes). The planetary order across the top is: Sun, Moon, Saturn, Jupiter, Mars, Venus, Mercury, and finally, the ascending lunar node ("head of the Dragon"). On the right-hand page are columns with "aspects of the Moon to the Sun and planets," followed by miscellaneous aspects of the Sun and planets among themselves. This information was of astrological interest for reasons both meteorological and human. Lunar aspects, for instance, were used for guidance in bloodletting, a very common activity.

have acquired largely on his own, although perhaps with help a satellite of Bessarion. A program of astronomical observation Regiomontanus had begun was continued by an able colleague, Bernhard Walther, over the period 1475 to 1504—an early example of a fairly continuous set of systematic observations. For many of these observations, a brass version of Ptolemy's parallactic rule was used (fig. 109). Walther's observations were eventually published by John Schöner in 1544, in a fine compendium which included works and records of observations by Peurbach and Regiomontanus. Even before that, some of Walther's observations were used by Copernicus for the orbit of Mercury; and at a later date, Tycho Brahe and Kepler took many other observational data from that same Schöner volume.

Regiomontanus's reputation by the time of his death was considerable, and justifiably so, not only for his powerful connections and his enterprise as a scholar-printer, but for his tireless and generally clear way of setting out existing astronomical knowledge, especially its mathematical foundations. In the 1460s he spent much effort on improving the presentation of spherical trigonometry. His work *On Triangles*, first published in 1533,

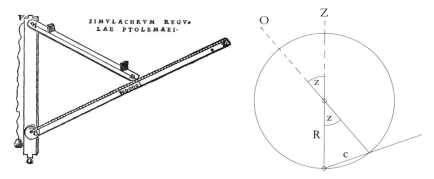

109. The essential parts of the parallactic instrument, much used by Regiomontanus and Walther. Here called "Ptolemy's rule," it follows the description in Book v of Ptolemy's *Almagest*. From John Schöner's *Scripta clarissima mathematici M. Johannis Regiomontani* (*Writings of the Famous Mathematician Master John Regiomontanus*, 1544). In the Middle Ages, this instrument was known as the *triquetrum* (three-limbed), not to be confused with the *torquetum*. For normal use, the fixed upright was placed so that the two movable rods, one with the scale and the other carrying sighting vanes, swung in a vertical plane. The instrument was relatively easy to build and graduate, the simple scale, uniformly divided, registering the chord (c) of the zenith distance (the angle z) of the object (O) under observation. Schöner's woodcut has the rods in the special position where the angle z is 60°, showing us that chords were being measured for a standard radius of 100,000—the norm in Regiomontanus's tables. Ptolemy's standard radius was of 60 units, further divided into sixtieths, and so on, where necessary.

made use of the so-called cosine law, as well as the sine law for spherical triangles. Their originality has been much exaggerated by those unaware of the works of Jābir ibn Aflah, Richard of Wallingford, and others. Even Jerome Cardan, in the same century, noted with some justice Regiomontanus's many debts to earlier medieval writers. His work in trigonometry nevertheless has a somewhat more modern look than theirs, and he supplemented it with something of even greater practical value, namely improved tables of trigonometrical functions. Here he was following in the footsteps of Peurbach, who wrote a work on the techniques required. Regiomontanus broke with traditional practice and no longer used a sexagesimal division of the standard radius (that is, one which made the sine of 90° equal to 60 parts). He gradually changed his own practice, first taking 60,000 parts to the radius—as Peurbach had done—then 6,000,000, and later, 10,000,000. The excellence of his tables of sines and tangents helped to bring about the supremacy of the decimal system.

Regiomontanus was not alone in exploiting the newfound power of the printing press for the furtherance of astronomy. One of the very first pieces of Mainz printing was an unusual partial ephemeris that at the very least covered the period from January to April 1448. It gives dates and times in hours and minutes of new and full moons, with longitudes of the Sun, Moon, and planets, to integer degrees for those times, all calculated using the Alfonsine Tables. This table, printed on parchment and known only from a piece of bookbinder's packing, would have had an astrological use.

110. The northern constellations in Albrecht Dürer's woodcut *Imagines coeli Septentrionales cum duodecim imaginibus zodiaci* (1515). The corner portraits are of four astronomers famous for their accounts of star positions: Aratus, Manilius, Ptolemy, and al-Ṣūfī.

(It probably dates from 1447, although some historians of printing would prefer it to be of a decade later.) Sacrobosco's *Sphere* and the most popular *Theorica planetarum* treatise, the staple of university students in arts, were both printed in 1472. Broadly speaking, the first scientific publications were for the serious-minded. Astrology, including Arabic astrology in translation, began to move into the market place in a significant way only in the sixteenth century. It often did so under the aegis of ephemeral calendars, a valuable source of income for printers, and especially attractive to the populace when replete with woodcut illustrations. For a similar reason, star maps, preferably with constellation images, were put into print. Such maps were often laid out in stereographic projection, as on the retes of astrolabes, but of course with much larger numbers of stars. Such maps, with their origins in antiquity, had been known in European manuscripts as early as the ninth century, and there are notable late-medieval instances, but with Albrecht Dürer's printed star charts of 1515 this type of map had a new lease of life (fig. 110). His woodcuts were copied repeatedly—for instance, by Peter Apian in 1540 and Johannes Honter in

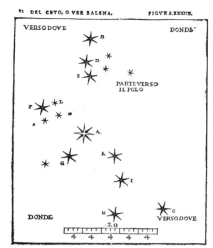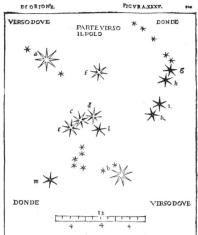

111. Facing pages from Alessandro Piccolomini's atlas, *De le stelle fisse* (*Of the Fixed Stars*, 1540).
He distinguished between stars of four magnitudes, all of them to be seen in these plots of the
constellations of Cetus (left) and Orion (right). Note the scales of degrees, and his lettering of
the stars. His system is a precursor of Johann Bayer's, the basis of our own. Bayer (*Uranometria*,
1603) used Greek letters within each constellation, generally but not invariably starting with α
for the brightest star, and turned to the Latin alphabet after the twenty-four letters of the Greek
alphabet were exhausted.

1541—and the style has never entirely disappeared. In 1540, however, it
was joined by a printed star atlas in a very different style, when Alessandro
Piccolomini published a series of constellation maps. These were lacking
in pictorial content, but in them the positions of the stars were plotted as
accurately as possible, and were marked by symbols showing their magnitudes (fig. 111). A work of this type could never rival a pictorial map
of the Dürer type in popularity. Giovanni Galucci no doubt realized this,
when in 1588 he published his own well-named *Theatrum mundi* (*Theater
of the World*), in which line drawings were added to the bald plots of position and magnitude.

ASTRONOMY AND NAVIGATION

The esteem in which Regiomontanus's printed almanac was held no doubt
helps to explain why Columbus took a copy on his fourth transatlantic
voyage. It is said that he used it to astonish the Jamaican Indians, through
his foreknowledge of the lunar eclipse of 29 February 1504. Whether or
not this happened as reported, Columbus's concern with astronomy raises
the question of its value to the navigator. Extravagant claims are frequently
made in this connection. Even at this date, astronomical methods of ocean
navigation were still rudimentary. When used, the two main principles
were, first, that of sailing east-west at a constant latitude, latitude being
determined by the altitude of the Sun or the Pole Star, and second, that
of sailing along a meridian line (north-south) with the aid of the Sun or

a magnetic compass, a tally of position being kept through the simple measurement of latitude. Most mariners who tried to estimate longitude would have done so by "dead reckoning." This they could use even when not simply following lines of latitude or longitude. The idea was to find the distance traveled and the course—the log and a sand glass together for distance, and a compass for direction. The resulting information was plotted on a "traverse board," a peg board on which was painted a compass rose. Pegs were typically inserted every half hour of a four-hour watch, to record the course and estimated distance covered, and the information was duly interpreted by the pilot and transferred to his portolan chart. We may be reasonably certain that when scholarly guides expatiated on deeper astronomical methods of determining terrestrial longitude, then, as far as most mariners were concerned, they might as well have been whistling into the wind. Even the great explorers of the fourteenth and fifteenth centuries were, for the most part, no more than pilots, that is, they set known courses and found their way when within sight of land by recognizing objects on land. Slowly but surely, however, Sacrobosco's astronomy, with its simple lessons relating to the spheres of the Earth and stars, was beginning to find its way into books for navigators, notably in Portugal.

Astronomers were rarely driven by mariners' needs, except when they had hope of gain, moral or financial, from solving points of theory that would help in the problem of finding terrestrial longitude. That particular problem was not new to astronomers, especially those who wished to adapt astronomical tables to new meridians; but for all who wished to navigate great oceans without placing undue reliance on chance, it was of crucial importance. Fourteenth-century Italian charts had shown the Canaries, Madeira, Porto Santo, the Azores, and the nearest coasts, but with scant respect for the distances between them all. The Portuguese began to colonize Madeira and Porto Santo in earnest early in the fifteenth century, under Prince Henry, "the Navigator." The difficulty of regularly locating these islands, which obviously lacked long coastlines that a navigator could follow, made navigational technique a matter of urgency. They, and the west African coastline, which was being explored at the same period, were soon placed far more accurately on Portuguese charts. This was a local initiative, and the influx of Jewish astronomers expelled from Spain was one important factor. It is a mistake to overestimate the influence of Regiomontanus or other German astronomers here, as was once fashionable.

Portugal long remained a center of navigational study. As early as 1484, King John II set up a commission to improve techniques, especially urgent in the Southern Hemisphere, from which the Pole Star is not visible. Simplified solar tables were consequently drawn up on the basis of some by Zacut of Salamanca, a Jewish astronomer whom we met in the last chapter (on p. 223). They were tested in 1485 off Guinea and gave rise to a type of literature known as a "Regiment of the Sun," a set of simple rules for finding latitude through the Sun's meridian altitude. The mariner's astrolabe,

112. A mariner's astrolabe of perfectly typical form, not to be confused with the far more complex planispheric astrolabe used by astronomers. The seaman's instrument has a simple scale for altitudes, which might have been good for half a degree. (A degree of latitude corresponds to nearly 111 kilometers, or, 60 nautical miles, although the precise length varies.) With the instrument hanging from the thumb, altitudes are taken using the alidade, which is usually (as here) equipped with pinhole sights. When taking the Sun's altitude, the shadow of one vane is made to fall on the other. The closeness of the vanes makes for poor accuracy. The back of the instrument is blank. The heavy casting of the wheel, its ballast to give it a low center of gravity, and piercing to reduce the wind effect, all speak loudly for the conditions under which it was used. Known examples were very few until the rise of underwater archaeology in the second half of the twentieth century, when scores of such instruments were turned up. (This is a copy, dated 1602, with a radius of 17 centimeters and a weight of 2.38 kilograms.)

by which altitudes were usually found, was crude—essentially no more than a heavy scale fitted with a rule on which were sighting vanes (fig. 112). On a pitching ship, it was virtually useless, and the fortunate sailor made a landfall before using it. Its accuracy was rarely better than half a degree. Techniques and instruments generally improved steadily, especially toward the end of the sixteenth century, when the problem of finding a ship's true position at sea—that is, supplementing its latitude by its longitude— was much discussed by astronomers. Unfortunately, few of them had much practical experience of making observations from a pitching and yawing vessel. The problem was often solved to astronomers' satisfaction—to Galileo's, for instance—by proposing that the observer be seated in gimbals. (This device is often called a Cardan suspension, after Jerome Cardan, although it had been used as a lamp holder in ancient China.) No truly acceptable solution to the longitude problem was found before the eighteenth century.

It is perhaps worth adding here that gimbals are often confused with the somewhat similar arrangement found in a universal joint for transmitting power, an arrangement successfully adapted to astronomical use by Robert Hooke in 1674. He used it in a handle for controlling remotely the movement of an instrument for solar observation. A "Hooke's joint"—given various other names when used for power transmission in automobiles—allows a twisting motion in one shaft to be passed on to another, no matter how the two shafts are inclined one to another. This device became very common for telescope control and is still widely used.

CHARTS, RUTTERS, AND SEAGOING INSTRUMENTS

Cartography at its most basic requires the transfer of the coordinates of places and terrestrial features to a suitable grid. By the fifteenth century, both in Islam and the West, there were long lists available, with the coordinates of many hundreds of places, but that left open the question of a suitable grid. There were ready-made theories, grounded in the work of Ptolemy and those later astrolabists who had supplemented the ordinary astrolabe projection. Such theories were the province of scholars, and were not aimed at the needs of mariners. At first, Ptolemy's projection was favored. On a scale of magnificence that put it far outside the practical world of the navigator, but that symbolises reverence for the great Alexandrian, is the extraordinary wall map of the world issued in 1507 by Martin Waldsee-müller (fig. 113 shows a small detail from it). Nearly three square meters in size, it is a horizontally elongated version of the map projection devised by Ptolemy. Numerous early editions of Ptolemy's *Geography* had made his the dominant model, at the very time that knowledge of the world's surface had begun to expand rapidly, as a result of voyages by such men as Dias, Da Gama, Columbus, Cabot, Vespucci, and Magellan. Waldsee-müller's map, of an unprecedented size and complexity, included the latest discoveries of the period, including the full outline of the African continent and the eastern coast of the newly found "fourth" continent. South America he famously named "America" in honor of Amerigo Vespucci, its supposed discoverer. When Waldseemüller eventually discovered his mistake, he tried to correct it and omitted the name from his numerous later maps. As we know, he was too late, and "America" the continent remains.

The mariner could add detail to grandiose Ptolemaic world maps, but he could not make much use of them, nor was his most valued book especially astronomical. It was what the English called a rutter, the French a *routier*, and the Italians a *portolano*, and was originally little more than a notebook in which a captain would record the magnetic compass courses between ports and capes, the distances between them, times of high water at new or full moon, estimates of how far away one was when first sighting the coastline, depths of water, and so forth. (A portolan chart is simply a chart with the courses drawn on it.) This type of literature was highly valued when it was first printed—the earliest was printed in Italy in 1490. As the sixteenth century progressed, more and more theoretical material was added. By their very nature, the great voyages of discovery could not be guided by traditional rutters, but the needs of those who sailed the great European trade routes—for example those between England and Bordeaux

113. Martin Waldseemüller was a scholar active at the court of René II, Duke of Lorraine. In his brief *Cosmographiae introductio*, twice issued in 1507, he announced that there was an accompanying globe and map. The only known copy of the enormous map was discovered at the beginning of the twentieth century at Wolfegg Castle, Württemburg. This detail shows less than a twentieth of the whole sheet. (The main map, with America and the West Indies, is below it.) We here see Ptolemy, holding a new quadrant and flanked by a rendering of his map of the Old World.

or Cadiz for wine—had to be catered for. Miscellanies of simple hints and tricks and secret information from mariners' documents were avidly copied, and each trade route brought with it new publications. Gradually, the North Sea and Baltic trade overtook the wine routes in importance. The Dutch, with their difficult coastline, and the seamen of Antwerp, a city which rose to commercial supremacy early in the sixteenth century, provided the best examples of rutters from that period; and the English were glad to copy them too. In due course, a whole string of English rutters appeared, often unashamed translations from the Dutch—although usually acknowledging their sources. And embedded in all of this were rough and ready rules for using the heavens for navigation.

Then, in the last year of the sixteenth century, there appeared an English text which established English supremacy in the theory and practice of navigation for many decades. Unlike most of the previous literature, it did not come about haphazardly. Edward Wright's *Certaine Errors in Navigation* was the work of a Norfolk man with a good knowledge of mathematics, who from 1587 was fellow of a Cambridge college. In 1589 he was summoned by Queen Elizabeth I to improve English navigation. Elizabeth was in part still apprehensive about the intentions of Philip II of Spain, whose great invading fleet, the Armada, had been defeated in 1588, but she also had an offensive in mind. Wright was sent with the Earl of Cumberland on a raiding mission to the Azores, and in the course of that voyage he made the friendship of John Davis, writer of another well known text and designer of the Davis quadrant. The book Wright eventually compiled, his *Certaine Errors*, was a masterpiece of revisionist writing, by a man with mathematical knowledge who was nevertheless in touch with the practicalities of the sea. Not the least of the "errors" he identified were those in the design and misuse of such instruments as the cross-staff, which could mislead the mariner by as much as a degree in latitude, the equivalent of 60 nautical miles. There was scarcely any branch of the subject, as it then stood, which Wright did not improve. This is not the place to enter into detail, but one subject in particular that deserves mention was cartography, to which again he applied his mathematical knowledge.

NAVIGATION, CHART-MAKING, AND THE INSTRUMENT TRADE

The history of the astronomical side of navigation is one in which the most advanced of all the exact sciences reached down to help solve a practical problem. There is a myth prevalent in some quarters that the debt was in the opposite direction, a debt owed by astronomy, which was supposedly driven and refined in response to the practical needs of navigation. This is nonsense. The main problem before the middle of the sixteenth century was no more than that of educating seamen in the most elementary parts of spherical astronomy and of teaching them to use very simple instruments at sea—not all of which were needed by astronomers comfortably situated on firm ground. One instrument peculiar to the seaman was for finding the true pole by the "rule of the North Star." This came into use in the sixteenth century, and was needed when using the Pole Star for latitude, since this was then more than 3° distant from the true pole. The altitude of the true pole was to be obtained by making a correction to that of the star, its value depending on stars in the constellation of Ursa Minor known as the "guards" (fig. 114). By and large, the family of instruments for measuring angle at sea—and so, ultimately, of measuring longitude—produced nothing of merit that was fundamentally new. The various sorts of cross-staff (or *radius astronomicus*), quadrant, backstaff, and sextant, for instance, were no more than serviceable copies of the more refined devices used in land-based astronomy (fig. 115). The one thing the mariner could use, and the landlubber not, was the distant horizon. Instruments to view this simultaneously with a celestial object were a seaman's desideratum, and several such were developed.

The need for strategic bases overseas, rather than the acquisition of territory for its own sake, followed a growing European demand for precious metals and for such luxury goods as tobacco, spices, and drugs. To satisfy these demands, global economic networks were created—by Portugal, Spain, and the Netherlands, in the first instance—which greatly stimulated the twin trades of instrument-making and chart-making. These last two activities were complementary, and it is no accident that they were practiced together in the Low Countries, from which some of the best work came in the sixteenth and seventeenth centuries. Amsterdam and Antwerp became leading centers for engravers, who could work for printing houses, including those producing maps and sea charts, and who in some cases could work simultaneously as instrument makers. The list of

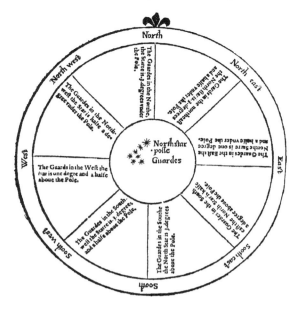

114. A rule for mariners using the guards of the Pole Star to correct the observed altitude of the latter, and so to yield the altitude of the pole—and hence the terrestrial latitude. From William Bourne, *A Regiment for the Sea* (1574), a work that saw ten printings and translation into Dutch before 1631. This type of rule was first given prominence in a printed Portuguese manual of around 1495. It was later embodied in simple instruments, in which a horn-shaped pointer was used to measure the orientation of the guards. Bourne's simpler diagram brings home the rough nature of the exercise. Thus "The Guardes in the South-west the starre is 3.degrees and a halfe aboue the Pole." Strangely enough, mariners were using a more accurate figure for the polar distance than that quoted by the astronomer Johann Werner in 1541 (4°9´).

names of those with a foot in both worlds is impressive, and is not restricted to artisans. One famous instance of a scholar-engraver is Gerard Mercator, whose map projection (1569) should require no introduction. Mercator received a sound university education at Louvain, and after graduating he studied astronomy with Gemma Frisius, who taught medicine at the university—although he is now far better known as an astronomer and designer of instruments. By 1536, Mercator was to be found making a terrestrial globe and embarking on a cartographic career. His influence on calligraphy and printing-type design was of comparable importance, and we cannot really separate these various arts and crafts. A little treatise he published in 1540 on italic lettering was printed from woodblocks he cut himself. By his example he had a revolutionary influence on mapping, book production, picture engraving, and scientific instrument making.

We see tokens of Mercator's great influence in the instruments of the nephew of Gemma Frisius, Gualterus Arsenius, and in the work of an even more influential figure, Thomas Gemini. Gemini (originally Lambert or Lamprechts) was a native of the Liège district, who learned his trade in Louvain before moving to England in the 1540s. He established himself in London as a publisher and editor, earning renown as engraver of the

115. Two simple instruments for use at sea (taken from *The Seaman's Secret* [1607], despite the countryside in the background). The cross-staff (above), invented by Levi ben Gerson in the fourteenth century, was a simple device for measuring angles, not necessarily from the horizon, as here. The mariner had one advantage over the ordinary astronomer, namely, a well-defined horizon. Angles were read from the position of the adjustable cross-piece by reference to graduations on the staff. Some instruments had a set of three or four cross-pieces, for different angle ranges. Davis's quadrant (below) was chiefly used for solar altitudes, when the observer wanted to avoid looking into the Sun. The upright was adjusted so that its shadow just touched a mark on the foresight. This type of back-staff was devised by the English sea captain John Davis in or around 1594. The name "quadrant" refers not to its shape but to the fact that it can measure an arc of 90°.

elaborate plates for his own printing of the *Anatomy* of Andreas Vesalius, which he issued in 1545. (That masterpiece of book production earned him an annuity from King Henry VIII.) If any one person can be considered the founder of the new and highly successful trade in scientific instrument-making in England, it is Gemini. He set up in the Blackfriars district of London as map engraver and mathematical instrument maker, and as a publisher could advertise his wares in his own publications. The excellence of his work can be nowhere better judged than from two magnificent astrolabes by him, and from the eminence of their owners. One is now in the collection of the Museums of Art and History in Brussels. Dated 1552, it bears the arms of King Edward VI and the Duke of Northumberland. The other was made for Queen Elizabeth I and bears her name and arms, with the date 1559. Thomas Gemini's great influence on the London trade in instrument making, which flourished throughout the century that followed, was largely exerted through the work of an even greater sixteenth-century

116. This astrolabe by Humphrey Cole, inscribed with a statement that it was completed on 21 May 1575, should perhaps count as the finest example extant of the work an English astrolabist. Weighing nearly 15 kilograms (even though it now lacks at least one plate), and spanning 60 centimeters, the superb quality of its engraving cannot be judged from a small illustration. (its finest peripheral divisions are spaced at under 0.9 millimeters). Inside the mother (*mater*), there is a nautical square engraved, with points of the compass and the Greek names for the corresponding winds, all based on a diagram published by Gemma Frisius in 1556. In 1679, another London maker, John Marke, made a new plate for this astrolabe, for the latitude of the Palace of Scone, in Scotland. It has been in the possession of the University of St. Andrews for most of the time since then.

practitioner, Humfrey Cole. Cole's metallurgical expertise was enhanced by the fact that he also worked at the Royal Mint, but it was his skill as an engraver that gave him precedence over all other English practitioners (see fig. 116). The range of instruments he offered for sale was enormous, and not the least of his debts to Gemini was his appreciation of the need to advertise them widely. Another factor in the rise of the London trade was the existence of a ready-made guild system to which it could attach itself. Strangely enough, the livery company with which most instrument-makers were associated was the Grocers Company.

There were, of course, comparable situations in many other European centers at this period of history and even earlier. Broadly speaking, there are three main stages in this key segment of astronomical history. In the first, instruments were made by scholars for their own use, with a minimum of outside help. This was usually the case before the fifteenth century, and remained so long afterwards outside the main centers of commerce. Naturally,

there were people of wealth—princes, churchmen, and rich merchants—
who could commission fine instruments as they might commission jewelry,
but this they would have done with scholarly help, and the artisan responsible would have worked on other, unrelated, projects. Next came a period
during which craftsmen in dedicated workshops acted under scholarly
direction—a situation that in many ways still survives. Last came a period
of artisans' independence, where they had enough astronomical knowledge
to work on their own, and even to devise new instruments unaided. The
three stages could obviously coexist, and steps from one to the next were
taken at widely separated times in different centers. From the late fifteenth
century onward, southern Germany became especially rich in those workshop traditions that helped accelerate the evolution of the instrument trade.
Nuremberg, Augsburg, Ingolstadt, and Ulm were all important centers in
which the production of sundials and other relatively simple instruments
for use—or for show—by the rich burgher helped to enhance the skills of
those who created instruments of yet finer quality, destined for the courts
of princes and their scholarly servants. Salzburg, Vienna, Prague, Cologne,
and Brunswick, were among the many towns supporting skilled makers,
and the same was true of Florence, Rome, and Paris, from all of which
places fine instruments survive. By and large, they were not aimed at the few
scholars who would have considered themselves astronomers.

There were dozens of lesser centers, with hundreds of makers whose
products have survived to the present day—sundials, nocturnals, the-
odolites, sectors, armillaries, astrolabes, equatoria, and astronomical
compendia. It is clearly impossible to do justice to this rich European scene
in a short space, in either scientific or economic terms, but some broad gen-
eralizations are perhaps in order. Almost all early makers added artistic
flourishes to their instruments in ways that would today be regarded as irrelevant to their purpose. This tendency, more obvious in some centers than
in others, often masques scientific quality. In the late sixteenth century, for
example, Erasmus Habermel made astrolabes which were ostensibly of the
very finest quality, not heavily decorated, but embodying misunderstandings which seriously limited their astronomical value. At the other extreme,
we find German and Parisian makers satisfying their customers' demand
for Baroque embellishment, obscuring the fact that underneath all the fid-
dledee there is often a finely graduated instrument to be found. The London makers, generally speaking, were more austere than their continental
counterparts, and their concentration on accuracy paid dividends in the
seventeenth century; and in the eighteenth even more so, by which time the
London trade in precision instruments had become preeminent.

As had been the case earlier in Nuremberg and other German centers, the London trade in accurate clocks ran a parallel course with that in
scientific instruments. Astronomy, after all, was still the arbiter of timekeeping; but conversely, there were many occasions when clock-making came
to the aid of astronomy. If we may leave the realm of trade for that of ideas,

117. The design for a clock-driven equatorially mounted sextant. The heavy falling weight (below the clock mechanism on the left) drives the polar axle by means of a screw, while a pendulum escapement (recently invented by Christiaan Huygens) controls the motion. The figure was published by Robert Hooke, in his *Animadversions on the first part of the Machina cœlestis of . . . Johannes Hevelius* (1674), but the design was never implemented by him.

there is one important instance worth mentioning in passing. No user of a telescope—or indeed of any other sighting and measuring instrument— could fail to wish for a means of driving the instrument in such a way that it followed the heavens in their daily motion. It seems that Robert Hooke was the first to produce a plausible design for such a "clock drive," in his case for a sextant with an equatorial mount. (This he published in 1674; see fig. 117.) John Flamsteed, the first Astronomer Royal, shortly afterwards set up the equatorial part of the scheme at Greenwich, but operated it manually, without the clock drive. In time, of course, the clock drive became an almost universal accessory for large telescopes, although it seems not to have been successfully implemented until James Bradley did so in 1719.

THE MATHEMATICS OF HOROSCOPES: ASTROLOGY AS MOTIVE

To return to Regiomontanus, a scholar who did not disdain the connection between the real world and astronomical theory: his Tables of Directions, produced in 1467 during his time in Hungary, were not purely astronomical. They included tables for astrological purposes, in particular for calculating the end-points of the twelve "houses." This term "house" is often used for the 30° signs of the zodiac—Aries, Taurus, Gemini, and so on—in which the planets are supposed by the astrologer to have their homes, their domiciles. It was also used for another type of division of the zodiac, dependent on the time of day and the place from which the sky is observed. The division in question is something about which Ptolemy was strangely silent, although at least five different methods of performing the division have been (wrongly) ascribed to him at one time or another. The division usually—but not always—started from the ascendant, the point of the zodiac (ecliptic) where it crossed the eastern horizon. The houses were then numbered in the direction of increasing longitude, that is, with the first six falling under the horizon. The details do not concern us here, but we must emphasize that many of the eight or more quite different methods for effecting the division were mathematically difficult to apply, in the pre-electronic age. The astrolabe could make the astronomer's life easier, but was not precise enough for the fastidious. Rather as in the case of calculating planetary positions for astrological purposes, the sheer time and energy involved prove the seriousness of purpose of the astrologer. Astrology was not primarily a cynical way of making money or attaining power, but once having presented an intellectual challenge, it took on a life of its own.

Certain names were often attached to the various methods of division, and again the printed word suppressed the admittedly hazy awareness of which astronomer had invented which method. The name of Regiomontanus, for example, became attached to a method that is at least as old as al-Jayyānī and al-Ghāfiqī. The astrolabe in figure 116 is actually inscribed for the use of this method. (The division is made using the lines radiating from a point on the horizon, directly below the center.) German astronomers followed Regiomontanus's lead in calling it the "rational" method, and there were complaints about the way the French astronomer Oronce Fine treated "our Regiomontanus" uncivilly when discussing it. Cardan, an Italian, accused him of plagiarizing Abraham ibn Ezra. The Italians tended to favor "the method of Campanus of Novara," but this too involved a historical misunderstanding. The history of these difficult techniques is a tangle of mistaken ascriptions. What is notable is how much it mattered to astronomers of the sixteenth century, into whose scholarship a kind of nationalism was creeping that was quite alien to the centuries before it. Once again, however, the power of the printed book was such that "the method of Regiomontanus" gradually acquired a large international following.

ASTROLOGY

Astrology, which helped to motivate so many of the best astronomers, had entered a strange limbo in the Latin West. On the one hand, its inherited doctrines, many of them mutually inconsistent, were often accorded the sort of uncritical reverence usually reserved for religious works. On the other hand, the training in Aristotelian philosophy in the Western universities of the later Middle Ages encouraged scholars to rationalize astrological writings. The old texts survived—Manilius, Vettius Valens (who despite his Latin name came from Antioch), and Ptolemy. They were written

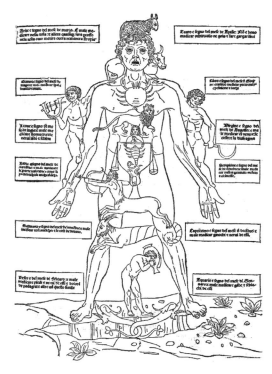

118. Woodcut of the common
type of drawing known as
Zodiac Man, relating the parts
of the body to the signs of the
zodiac, to assist in medical
diagnosis and cure. From a
translation into Italian by
Sebastiano Manilio of Johann
von Kirchheim's "little book of
medicine," *Fasciculus medicinae*
(1493).

before the subject had begun to harden its boundaries, and Ptolemy in particular took astrology to be part of the total rational account of the physical
world. Mixed in with the old half-rationalized magic, there are signs that
it could still be viewed as an empirical science. The Hippocratic writings
in medicine, for instance, contain many passages of astrological medicine
(*iatromathematica*), where the human body is regarded as under the influence and protection of the different parts of the zodiac, and of the planets
(fig. 118). Here the logic of the argument, where it can be detected at all,
usually rested on analogies. Ptolemy, for instance, describes Saturn as having a cooling and drying quality because he is furthest from the warmth of
the Sun and the moisture exhaled from the Earth. There are many similar
analogies in his *Tetrabiblos*, and a few of the scholastics tried to follow in
this "philosophical" tradition.

We have already seen how the Arabs of the eighth and ninth centuries
began the process of collecting Greek, Persian, Syrian, and Indian astrological materials. This was not without opposition from many quarters. The
great Islamic philosopher-theologians, al-Fārābī, Avicenna, Averroes, and
Ibn Khaldūn, all opposed astrology in some measure, often reproducing
arguments—for instance on fate, determinism, and responsibility for one's
actions—reminiscent of those used by their opposite numbers in the early
Christian Church. On the other hand, Māshā'allāh, the encyclopedic writer
al-Kindī, his pupil Abū Ma'shar, and al-Qabīṣī, were all good astronomers
who wrote on astrology in an authoritative way that was very much to the
Western scholastic taste.

The subject was perceived to be spiritually dangerous chiefly when it spilled over into magic and demonology. Here too, however, al-Kindī had tried to introduce a rational physical basis, and his book *On Rays* was on the very fringe of Christian respectability. It was much copied and read in the West. The West even learned something of Aristotelian physics from Abū Ma'shar's astrology in the twelfth century, before the arrival of the full body of Aristotle's writings. An influential astrologer in that same period was Abraham ibn Ezra, whom we have already encountered, and whose works quickly became available in Hebrew, Latin, Catalan, and French. The treatises of Henry Bate, written around the 1280s, continued and reinforced the popular Hebrew-Latin tradition.

The twelfth century saw a flood of Arabic astrological texts reach Latin readers, and this was soon turned into more familiar literary forms. John of Seville not only translated the Arabic, but in 1142 produced a much esteemed summary, an *Epitome of the Whole of Astrology*. The same type of summary continued to be written, often at enormous length: two famous examples were those by Guido Bonatti (some time after 1261) and John Ashenden (1347–1348). An English translation of Bonatti from 1676—only ten years before Newton's *Principia*—and frequent references made to Ashenden at the end of the same century, show the esteem in which such works might be held. A medical astrology by William the Englishman, and a general digest by Leopold of Austria, belong to the best-known works of the thirteenth century. In the fourteenth, we find the subject spawning a vast new literature in which the old ideas were given more local color. Italians such as Pietro of Abano, Cecco of Ascoli, and Andalo di Negro, were much admired, but the Arabic authors remained "classical" for most scholars in the universities. Cultural distance clearly lent enchantment to the view.

Astrology developed a strong meteorological association. There were many texts on the astrological "prognostication of times" (compare the French word *temps*, meaning "weather"), and we have already mentioned Richard of Wallingford's. When the Oxford scholar William Merle added to the list, he incorporated many other sorts of consideration, and his empirical outlook is well and truly proved by his journal of weather observations covering the period from January 1337 to January 1344. Of course, not all such records of observations were made for the reasons for which we now tend to value them. The fourteenth-century scholar was correlating the weather with astronomical events, but we should not be too condescending. It was the weather itself that was the thing of final concern, and a comparable empiricism was hard to find in most of the sciences of the day.

We have already seen that comets, according to the prevailing Aristotelian view, were meteorological in character. Many careful records that survive to us from the Middle Ages are of observations of comets, with classifications of position and color, and so forth. Again, the focus of interest was not on the comets as such, but on the disasters they portended.

To see how deeply entrenched were the presuppositions underlying such a question as "Why comets signify the death of potentates and wars," consider the answer given by so rational a thinker as Albertus Magnus: comets are associated with Mars, and Mars is the cause of war and the destruction of peoples. There was little logic in that sort of reasoning.

The Great Plague—the Black Death—of the late 1340s, the long wars between England and France, the ever present fears of the arrival of the Antichrist, the Hussite heresies, and later the protestant splits in the Church, all helped to turn the thoughts of astronomers to the kind of astrology that predicts the rise and fall of kingdoms and religious sects on the basis of certain patterns in the behavior of conjunctions of the planets Saturn and Jupiter, and possibly also Mars. (For a brief explanation, see figure 136 in chapter 12.) Chaucer worked this theme into his greatest poem, *Troilus and Criseyde*, once more adding a hidden dimension of meaning to his poetry, with carefully calculated astronomical and astrological allusion. Many other English poets tried to copy his style, but none had his arcane skill. On a more obvious level, the whole of European literature was gradually becoming colored by cosmic metaphor and allusion. Some writers continued in the old Cicero-Macrobius "Dream of Scipio" tradition, and after Copernicus some used that literary form to argue the case for a Sun-centered world. Less adventurous writers, anxious to exploit more traditional astronomy and astrology for color, continued to please their audiences by doing so. The products of sixteenth-century France—especially Jacques Peletier and Pierre de Ronsard—are noteworthy examples.

Shakespeare's allusions to the heavens are widely recognized, even if their astrological antecedents are not always understood. Romeo and Juliet, that "pair of star-cross'd lovers," suffered the misfortune of incompatible nativities. There are many other comparable allusions in his writing, although whether or not they lack any deep structural importance is a matter for debate. Was Cassius not speaking for Shakespeare when he exclaimed that "The fault, dear Brutus, is not in our stars, But in ourselves . . ."? Whatever the answer, echoes of astrology in high literature were growing fainter by Shakespeare's time. Later in the seventeenth century, John Milton's *Paradise Lost* seems to play something like the old game, but his is not astrology, and on a closer reading it becomes clear that Milton was trying his hand at modifying the standard astronomical system of his time. In due course, the astrologer in literature became a figure of fun, and even the California of the 1960s failed to redress the situation. The art of turning either astronomy or astrology to poetic advantage is now virtually lost, although as readers of Algernon Charles Swinburne will know, "astrolabe" is one of the few words in the English language that rhyme with "babe."

The three great questions facing those who thought deeply about astrology during the Middle Ages and Renaissance were, first, whether the influences it claimed to describe were real; second, whether human beings could ever reduce it to a workable science; and third, whether such a science would be licit. It has to be said that few scholars would have hesitated to answer Yes to the first two questions, and most would have passed by the third as quickly as possible.

Of those who spoke against astrology from a position of scientific strength, Nicole Oresme is one of the most interesting. Born in Normandy in 1320 and trained in Paris, he became a confidant of the future king Charles v while Charles was still dauphin of France. Nicole died as bishop of Lisieux. His view of the cosmos was in a sense mechanistic—for instance he often used a metaphor likening it to a mechanical clock—and yet he was not ready to break with the Aristotelian division of the cosmos into two regions, one above and one below the Moon. He continued to speak of the spheres as moved by intelligences, in the Aristotelian style. (His master in the university, John Buridan, had said that the spheres could have been given their impetus by God at the time of Creation, so providing them with indefinitely persistent motion.) When Oresme wrote against astrology, it was because he regarded it as incapable of explaining events here on Earth. They arise, he said, from immediate and natural causes, and not from celestial influence. But he was too deeply immersed in the culture of his age to be ruthless in his criticism. The qualities of the stars, signs, degrees, and so forth, could in principle be known, he thought. Predictions from great conjunctions can be made, but only in general terms, not in detail. As for the weather, the same is true—but farmers and sailors he thought likely to be more reliable. Medical prognostication he considered relatively safer from the Sun and Moon than from the planets. Here, however, we are dealing with what for him was a question of Nature. It was at the three final divisions of astrology, those dealing with Fortune, that he drew the line. The casting of birth horoscopes, asking whether a thing will happen, or deciding on propitious times for action—all these concern the freedom of human will, and are to be avoided.

Oresme's view of the situation was shared by many of the more thoughtful intellectuals of his time. It was not particularly new, but it made clear a distinction that is frequently misrepresented. We often read that astrology and astronomy were one and the same thing in the Middle Ages and earlier. For some, however, there were *three* astrologies, one mathematical, which some were prepared to call astronomy, one natural and akin to physics, and one spiritual. Oresme steered a careful course between them.

ALMANACS

Oresme shines out as a beacon of rationality in the later Middle Ages, but it would be misleading to suppose that he was typical of his time. A far truer guide to ordinary opinion about the heavens, popular or even scholarly, can be had by following the fortunes of those types of astrological literature that were most commonly copied or printed. With the advent of printing, the floodgates opened, not just to supply the growing demand for astrology, but to help create it. As soon as almanacs could be printed, they began to play a part in the lives of the semi-educated European populace, and continued to do so well into the twentieth century. Almanacs came in many forms, and had their origin in the twin traditions of zīj and ecclesiastical calendar, as described in earlier chapters. The word "almanac," which was used in Latin and most vernacular languages, is not standard Arabic, despite its appearance, but might have originated in Muslim Spain. It was often used of calendars when they incorporated astronomical data, whether simply lunar phases or something much more elaborate. With the addition of important anniversaries, astronomical and astrological information, and eventually astrological forecasts, especially of the weather, almanacs grew rapidly in popularity in the sixteenth century. The broad framework was set for the future by the *Calendarium* and *Ephemerides* of Regiomontanus, which were mentioned earlier (p. 273). They included his computations of planetary positions and other astronomical events. Continued after his death, they helped to pass on medieval calendar styles to later ages.

Almanacs were at first especially popular in Germany and the Low Countries, some as broadsides, but most in booklet form. The first printed English almanacs date from around 1500, but they soon became very numerous, and until the end of the century most of them were translations of European examples. In 1603, James I granted a monopoly of almanac, prayer book, and Psalter printing to a joint stock company that went on to make very great profits, despite piracy by competitors. In the later seventeenth century, almanacs are said to have sold 400,000 copies annually in England alone. For the barely literate, they were often the only printed material bought, and they helped to propagate simple medical knowledge, as well as religious and political propaganda during the revolutionary period. What never needed to be disguised was the extent to which they were filled with astrological material—accounts of blazing stars, comets, prophecies and predictions, freaks of nature, and diseases and plagues, as well as basic astronomical data. The great majority of them contained, before or after the main calendar, the diagram known as "zodiac man" (such as that in fig. 118), connecting the parts of the human anatomy with the signs of the zodiac. It was commonly used to show what parts of the body should *not* be bled when the Moon was in the corresponding sign. Here, at a less mundane level and in symbolic form for all to see, was the Greek doctrine of a parallelism between microcosm and macrocosm, mankind and universe.

The seventeenth century saw a gradual separation of non-scientific material from the strictly astronomical, such as is still found in modern ephemerides. The *Nautical Almanac* was a title first used of his annual publication by the Astronomer Royal Nevil Maskelyne, for tables that would be of use to astronomers as well as to navigators. It first appeared in 1767. As a result of increasing cooperation between Greenwich and the

United States, from 1960 onwards, this and the *American Ephemeris* have been combined under the new title *The Astronomical Ephemeris*. Tables of a more permanent character—that is, not changing annually—and explanations of the tables generally, are published from time to time in a supplement. Material was eventually pooled with Paris (the *Connaissance des Temps* being the equivalent publication there), with Heidelberg (which has its *Astronomisch-geodätisches Jahrbuch*), and with several other national observatories.

The old pseudo-astrological forms linger on, with such familiar names as *Farmers' Almanac* and *Old Moore's Almanac*, and even zodiac man is still going strong in some of them. The highly respected *Whitaker's Almanack* in Britain still carries vestiges of its forebears in the calendrical material, with its zodiacal decoration. As for the old ecclesiastical calendar, even this was preserved for the future, by diary publishers. Their mistakes in the past were legion. In 1851, the mathematician Augustus de Morgan—of the "godless" University College, London—performed a great service to that community and to historians by drafting a set of all thirty-five possible almanacs of the simple type. De Morgan's scheme actually embodies elements pointing back to a certain "lunar-letter table" devised by the Venerable Bede. As Bede had said, it was meant for those who were too idle or too dull to calculate for themselves. Today, there are computer programs that perform a comparable service. And for the sake of completeness, we might add that even Bede's table can be traced back to Willibrord's calendar of no later than the year 717. It is not easy to shake off the past.

A RETURN TO THE GREEKS?

Renaissance humanism, the intellectual movement in which Regiomontanus found himself being swept along, had from its beginnings in fourteenth-century Italy been generally unfavorable to the natural sciences. The Italian scholar Petrarch, who is often treated as its founding father, had poured scorn on the Oxford logicians who were then becoming fashionable throughout Europe, and in whose writings there are so many of the seeds of the scientific revolution to come. There was a pedantic literary spirit abroad that refused to accept medieval versions of classical treatises. Humanism, however, was concerned with the place of man in history and in nature. Many of the humanists had flirted with astrology, even if they did not remain constant in their loyalty to her. This explains in part why the rebirth of classical studies could easily accommodate astronomy, highly esteemed by many who had no real understanding of it—relations between Bessarion and Regiomontanus illustrate the point. Many humanists, however, grew openly hostile to astrology, the two most famous being Marsilio Ficino and his younger contemporary—but not strictly his pupil—Giovanni Pico della Mirandola.

Marsilio Ficino was the son of the physician of Cosimo I de' Medici, Duke of Florence and Grand Duke of Tuscany, by whose family were

Marsilio's patrons for three generations. A man respected as philosopher, theologian, astrologer, and magician, he was made head of the Platonic Academy in Florence by Cosmo in 1462. He had fallen under the spell of Hermeticism in the 1460s and was fully familiar with traditional astrology, having practiced it seriously, when he developed his own private brand of the subject. It is not easy to summarize the thought of a man who wrote at such length as did he, but we may say that he believed that God somehow communicates with us at a spiritual level through the heavens. His new astrology was not exactly replete with rules, but he tried his hand at applying it, most famously to various great events in his life, notably the publication of his translations of Plato, which were certainly important, and the simultaneous visit of Pico della Mirandola to Florence. On this occasion, Ficino was persuaded to translate Plotinus, which was undoubtedly an important event in the history of scholarship. He famously interpreted that event in terms of Saturn-Jupiter symbolism, in an exercise strongly resembling activities he criticized so severely in others. This is just one of many kinds of inconsistency in his writings, which have long been an embarrassment to his admirers. He changed his mind repeatedly. Ficino was well read not only in astrology but in its standard opponents, from Augustine onward. He came under the influence of Savonarola—the notorious Dominican friar, preacher, and martyr—in the 1470s, and was ordained as a priest in 1477, becoming a canon of the cathedral of Florence. It was in that year that he wrote a work strongly critical of astrological practice, attacking its reliance on metaphor and simile, and its sheer arbitrariness; and yet, only a year later we find him making political predictions on an astrological basis, in a letter to Pope Sixtus IV. Again, in a work of 1489, he put forward his case for medical "astrology," but when in 1494 he found himself confronted with a massive treatise by Pico della Mirandola against astrology, he turned his coat once more, expressed agreement with Pico, and made a feeble attempt to reconcile the views he had expressed in his earlier writings.

Already in the sixteenth century, attempts were being made to excuse his erratic behavior in terms of his mental state, and the number of alternative explanations provided by students of humanism grows steadily. There is no doubt that he was anxious about his religious orthodoxy, but while he sailed near to the wind, Giovanni Pico della Mirandola sailed even closer. Pico, born in 1463, was scion of a noble Lombard family, and even as a youth was distinguished by his erudition. Educated in the newly lauded Greek tradition, he was also put to learn Hebrew and Arabic under Jewish teachers, so Jewish philosophy and such Arabic authors as Averroes were open to him. In 1486 he created a stir with his publication of a collection of nine hundred theses, which he boasted he was prepared to defend against all-comers in public. Several of these were quickly judged heretical by a papal commission, and he fled to France. After his arrest there and return to Italy under the protection of powerful supporters, Pico settled in Florence under the protection of Lorenzo de' Medici. There, among other things, he wrote his *Disputations*

Against Astrology in twelve books, his most extensive work. What is now widely seen as central to all his writing was the theme of the dignity and freedom of man, and this theme had its place in the *Disputations*. He acknowledged the influence of the stars through such physical effects as light and heat, but professed to dismiss occult influences. He was an ardent Hermeticist, however, and defender of Jewish kabbalism, believing that there is such a thing as a primitive fund of divine wisdom. We cannot, he thought, as free spirits, be influenced by the stars, which are bodies of a lower nature. This was conventional enough, but it helped to preserve him in a dangerous climate. He died young, in 1494, not living to see the high point of Savonarola's career, or his execution by hanging and burning in 1498.

Pico's arguments against astrology were numerous and lengthy, but—as in the case of Ficino's—there was very little in them that was deeply original. Neither man started from astronomical premises, and indeed, neither was particularly well versed in astronomy as such. Both could accept celestial influences of sorts, and their reasons for rejecting astrology are not ours. Pico gave his wish to defend the Church as his chief motive for writing, and this may help to explain why his attack, like Ficino's, was much heeded for many decades, even centuries. Perhaps a more potent reason was simply the fame of the two men as philosophers. One way or another, both influenced attitudes within the learned world to astronomy, through the very fact that they were known to have opposed astrology, even though they were far more often cited than read. Gian Domenico Cassini, the first of the great family of astronomers, tells us that it was Pico's work that turned him against astrology and to astronomy. Astrologers treated them warily. When Tommaso Campanella, a Dominican friar, wrote what many regarded as a classic work of astrology, his *Astrologica*, he was careful to point out that it concerned only what Pico had called "physical astrology," and that it had been purged of "the superstitious astrology of the Arabs and Jews." The same sort of distinction was commonly drawn between "natural magic," under which were included topics that we would acknowledge as physical science, and its occult, demonic, and spiritually unacceptable alternatives—to which we now generally apply the name "magic" exclusively.

The debate on magic and astrology that was taking place at this same period occasionally had recourse to some interesting notions of a psychological sort, many of them drawn from the writings of Avicenna, the most renowned philosopher of medieval Islam. Natural magic was then explained as obtaining its very real but seemingly miraculous effects through the medium of faith and imagination. This is where the stars could come into play. As Pico said, "To work magic is to do nothing other than to marry the universe," by which he meant some sort of psychic union. Many Arabic texts had put forward the same idea, but not always ending in the same conclusion. Some, such as Ibn Khaldūn, thought that our innermost thoughts are beyond the reach of the stars. The stance people took depended on the conclusion they wished to draw. In about 1490, Galeotto

Marzio da Narni defended divinatory astrology against attacks by Averroes, and used Avicenna in his defense. The human soul, said Avicenna, has the power to change things through violent longing. Strong personal desires, he thought, could be an agency operating on the lives of people who were on Avicenna's authority affected by the stars at the times of their births. Faith can move mountains, and Galeotto insisted that the actions of a powerful faith are more effective than all instruments and medicines. Others went so far as to treat speech itself as an even more innocent tool of natural magic. Although a measure of skepticism was abroad, it is fair to say that most scholars were happiest when they could find such new ways of defending old orthodoxies. Astrology was reluctant to die.

Renaissance humanism began to spread abroad rapidly from Italy at about this time. From an astronomical point of view, the importance of the new translations from the Greek can be easily exaggerated. Some, such as Thomas Linacre's translation of excerpts from Geminus on the celestial sphere, were only of historical or literary interest. (Printed in 1499, Linacre mistakenly thought the author was Proclus.) Fashions, on the other hand, can be more powerful than reason. When Corpus Christi College was founded specifically to favor such studies in Oxford, provision was made to have astronomy taught there. There was no clearly qualified humanist for the job. The choice fell on Nicolaus Kratzer, who—like his friend and compatriot, the artist Hans Holbein—served at the court of Henry VIII, and who had for long rubbed shoulders with humanist scholars.

Courtly fashion has very often favored the foreign and the exotic, and when the king's father, Henry VII, had employed an astrologer, he had opted for an Italian, William Parron. The distinguished English humanist Sir Thomas More, Henry VIII's chancellor, often expressed his distaste for astrology, but was Kratzer's friend and onetime patron. By engaging scholars from abroad—and from Italy in particular—the tone of any subject at this period was somehow thought to have been elevated, and a Greek connection raised it further. Thus, in a letter to Erasmus from a pupil, Kratzer was in 1517 described as a skilled mathematician, bringing with him astrolabes and armillary spheres and a Greek book. Such was his passport, although he does not seem to have known any Greek himself. When Kratzer built sundials in Oxford, they were duly honored by a colleague's verses in fine neoclassical Latin. Medieval astronomy was being dressed up, but only superficially changed. Despite his many humanist friends, Kratzer's outlook on astronomy involved no real break with the Middle Ages. He had considerable expertise in fundamental spherical astronomy, as we can judge from Holbein's most remarkable painting, *The Ambassadors* (plate 10). There are so many concealed astronomical and astrological gambits of a calculated nature present in the painting that Holbein could not possibly have worked out its plan by himself. It was painted for the young French ambassador at the court of Henry VIII, Jean de Dinteville, whom it depicts, with his young friend Georges de Selve, bishop of Lavaur. When properly understood, it is an allegory suited to Good Friday, 1533. Its most obvious oddity is a distorted skull in the foreground. If one looks down at the skull from a suitable standpoint, one may see a corrected image of the skull. Looking up at the same angle as one previously looked down—an angle which happens to be the angle of the Sun at the moment registered in the painting—one looks toward a small, half-hidden, crucified Christ.

There are many other hidden schemes in the painting, but it is equally memorable for the fact that at its center there is a double-shelved table on which are placed various astronomical, geographical, arithmetical, and musical objects. On the upper shelf, we see a celestial globe, a chilindrum (a pillar sundial), a solar instrument of a rare type, but almost certainly designed by Kratzer, and a wooden quadrant half hidden behind it. There is a polyhedral sundial (with different dials on its several faces), and also a torquetum. On the lower shelf are a hand-held terrestrial globe, an arithmetic with set square and compasses, a lute, various flutes, and a hymnal. These artifacts, in short, illustrate four out of the seven liberal arts, as taught in the arts faculties of all European universities of the time. The painting can serve as a reminder that the arts curriculum of the universities was supremely important in keeping the flame of astronomy alive at a time when it was tending to fade elsewhere.

HOMOCENTRIC SPHERES, THEIR CHARACTER, AND COMETS

One very strange effect of the new literary fashion for all things Greek was a revival of interest in the doctrine of homocentric spheres. It used to be assumed that Eudoxus's homocentric model for planetary motion, transmitted to us through Aristotle's *Metaphysics* and Simplicius's commentary on Aristotle's *De caelo*, was first correctly reconstructed in the nineteenth century by Schiaparelli. In fact it was known, and many of its properties were understood, by Muslim astronomers at least as early as Ibrāhīm ibn Sinan in the tenth century, and was treated at some length in the *Astronomy* of Levi ben Gerson in the fourteenth. In the Latin West, however, apart from a few short passages in translation from Muslim writers, it was not examined closely before Regiomontanus in the fifteenth century, and Amico and Fracastoro in the sixteenth. There is nothing in the *Epitome* to suggest that Regiomontanus wanted a homocentric system, but in his correspondence with Giovanni Bianchini, and in several other writings of the 1460s, he makes it quite clear that he has been trying assiduously to work out a similar non-epicyclic system. He says that he has not yet prepared a scheme for the planets, but intends to do so. Regiomontanus reveals himself in all this as a man with deep cosmological interests, a natural philosopher, and not merely the geometer whom we think we are meeting in his better-known writings.

Girolamo Fracastoro was cast in what was ostensibly a very different mold—a physician and philosopher with a predilection for astrology. He taught at the University of Padua, where in 1501 he made the acquaintance of Copernicus, at that time enrolled in medicine at Padua. Fracastoro's fame stems from a very long narrative poem, written in classical Latin of great elegance on an unlikely subject, *Syphilis, or the Gallic Disease.* (The first draft was finished in 1521.) The disease was supposedly visited on Sifilo, a young shepherd, by the Sun god, to whom Sifilo had been unfaithful. Fracastoro believed in the standard doctrine of the dangers attendant upon a triple conjunction of Saturn, Jupiter, and Mars, and he argued that the spread of the disease was due to the corruption of the air under the malign influence of that event. He later wrote works on medical astrology of a more prosaic sort, but it was in his *Homocentric [Spheres]*, or *Concerning the Stars* (1538) that he set a new astronomical fashion, albeit one of short duration. At Padua he had been a friend of three brothers, of whom one, his teacher Giovan Battista della Torre, had tried to revive the homocentric idea. On his deathbed in 1534, della Torre asked Fracastoro to complete his work. The resulting book was dedicated to Pope Paul III—as was Copernicus's great work of 1543. The obscurity of Fracastoro's might explain its lack of success, although there were instrument makers who thought its schemes worthy of being modeled in metal—no doubt responding to the wishes of those who commissioned them.

It is hard to believe that Fracastoro's ideas on homocentrics were entirely independent of similar ideas presented in a slender work by Giovanni Battista Amico. The fact that this was printed three times in four years (in Venice in 1536 and 1537, and in Paris in 1540) says something about the ambitions of the Aristotelians, who were anxious to find a system in keeping with their ideas. Amico was murdered in Padua in the year of publication of Fracastoro's book, "by the hand of an unknown assassin, it is thought, out of envy of his learning and virtue."

Where the older man took only the special case of Eudoxan spheres, pivoted with axes at right angles, Amico started with that assumption, and then moved on to the general case of arbitrary inclination. His models are not very coherent, but they make extensive use of the theoretical devices which we have already said would be found in Copernicus, devices that we know had been used earlier by Naṣīr al-Dīn al-Ṭūsī. Amico repeatedly applied the Ṭūsī couple—without acknowledging any source—in his theory of planetary longitude and latitude. He also tried his hand at celestial physics, and—as Fracastoro was to do—considered the passage of planets through some medium of varying density ("vapors"), by which he thought their changing brightness could be explained. This, of course, was a serious problem for anyone who proposed a model that made the distances of the planets constant. Amico introduced precession into his model in a way that makes clear his debt to the Alfonsine Tables.

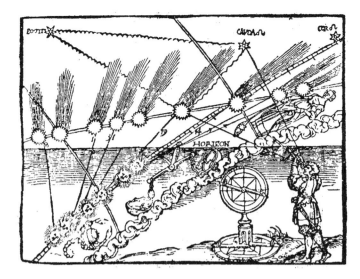

119. Peter Apian's woodcut summarizing his observations of the comet of 1532, showing that the tail of the comet continued to point away from the Sun.

Fracastoro's system, in which he made use of 77 (or perhaps 79) spheres in all, was astronomically much inferior to Ptolemy's, from which a number of ideas were in any case borrowed. He seems to have thought that he was using Eudoxus's own ideas precisely, and one cannot help wondering whether other Paduans of the day had been trying to reconstruct Eudoxus's system, and had hit on the Arabs' constructions in the course of doing so.

At one place in Fracastoro's account, he likened the effect of varying planetary brightness to the changing size and clarity with which we see something through different depths of water, or through a double lens, as opposed to a single lens. (He was not, as has sometimes been quite anachronistically claimed, alluding here to the invention of the telescope.) His remarks on the nature of the celestial spheres are in many respects as interesting as anything on the subject before him. He observed comets carefully enough to realize that their tails always point away from the Sun. Or did he see a depiction of the comet of 1532 printed by Apian, illustrating the same point (fig. 119)? He thought it obvious that comets are nearer than the Moon, for how else could they move freely through the spheres? Tycho Brahe later reversed the argument, and said that since comets were demonstrably more distant than the Moon—by his own measurements— it was the solidity of the spheres that was to be doubted.

Fracastoro's discovery of the direction of cometary tails was published in 1538, seven years after Peter Apian's publication of a diagram showing the same phenomenon. (The general principle is said to have been known in China seven centuries earlier.) The two sixteenth-century discoveries are usually assumed to have been independent, but since Apian's explanation was that comets are spherical lenses, one cannot rule out borrowing.

Apian thought that the comet's tail is simply a fan-shaped pencil of rays passing through the cometary lens. Gemma Frisius and Jerome Cardano later gave publicity to this ingenious idea. Tycho Brahe was another of those who adopted the lens principle, in a work of 1588 concerning the comet of 1577: he decided that in this case the source of the light was not the Sun, but Venus. Even Kepler for long accepted the idea of the solar lens; and despite the fact that from 1618 he offered several criticisms of the idea, Descartes took over a version of it. Newton criticized it once again, later in the century, but it died an astonishingly slow death. At any event, it served a very useful purpose, to the extent that it led astronomers to forsake the Aristotelian view of comets as burning fires, that is, the burning of rising hot and dry terrestrial exhalations in the layer of fire, below the celestial sphere of the Moon. From this point onward, their location—sublunar or celestial—became an open question. That particular problem was one that Tycho was to solve.

All told, while Amico's early death robbed astronomy of an imaginative talent, and Fracastoro's writings had a certain modish quality that made them appealing to contemporaries, both men were attempting to turn the clock back nineteen centuries. Tycho Brahe was later scathing in his criticism of Fracastoro's planetary absurdities. The future lay with other planetary schemes, mathematical and physical. There was still much life left in the fourteen-centuries-old Ptolemy, but the time was near when astronomers would have to contend with a new system, that of Copernicus.

Copernicus'
Planetary Theory

CHANGING BELIEF

If an ability to change a belief that is held almost universally is a measure of greatness, then Copernicus was one of the greatest of astronomers. Historically, he presents us with three distinct problems. One is that of explaining what he said, a second is that of locating the social and intellectual circumstances that led him to his conclusion, and a third is that of accounting for the ways in which his views were transmitted to others. At first his opinion was accepted by only a select few; after a century or two the majority of people educated in the Western tradition were convinced of its truth; and eventually the world at large became convinced.

What he said, roughly expressed, was that the Sun and planets do not travel round the Earth, but that the Earth and all the planets travel round the Sun. The Earth having been freed in that respect, it is freed in another: it spins on its axis. In other words, it is not the heavens that turn around the Earth's axis, but the Earth that spins around a moving axis. The loose way in which people speak of this simplified account as "the Copernican system" is reminiscent of the way in which "Aristotle's cosmology" is remembered, through a simple diagram that omits all of the fine detail. In the case of Copernicus, the best known illustration of the general character of his scheme is that from Book I of his greatest work, his *De revolutionibus*, published in the year of his death (fig. 120). It was meant only as a summary illustration, and does not even hint at the spinning of the Earth on its axis, but many later writers, when they were introducing Copernicus to their readers, thought that nothing more was necessary. There was of course much more to the man than astronomical iconoclasm, as we shall see when we look beyond the simple cosmology of Book I. Perhaps he knew that most readers would not get beyond the first book. For them, he would have achieved something if he had shown them how the planets fall into two natural groups, of those with orbits above the Earth's and of those with orbits below. Readers of Book I would at least be able to see how easily explained were the differences between the risings and settings of the two

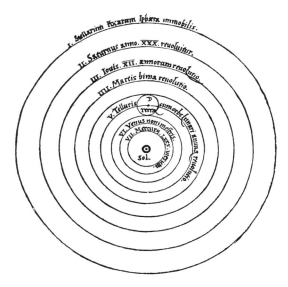

120. The solar system according to Copernicus, from Book I of the first edition of his *De revolutionibus* (1543). *Sol*, the Sun, is at the center, and the outermost sphere is still where the stars are to be found. The fifth sphere is that of the Earth (*terra* or *tellus*—*telluris* meaning "of the Earth"). The Moon still circles the Earth. Note the approximate periods, such as Saturn's of 30 years.

groups, Venus and Mercury being seen easily only early or late in the day, and the others often visible throughout the night, high or low in the sky.

This was something that could be appreciated qualitatively, as was the reason for the frequent retrogradations of the planetary motions—now seen to be a simple consequence of the motion of the Earth from which they were observed. Most of his readers would be on board still. There were a few simple quantitative exercises within the reach of most of those readers who had been inclined to pick up the book in the first place. Introducing the orbital period of the Earth, one year, into the known apparent periodicities of Mercury and Venus, for example, it was not perhaps too hard to see that their "true periods," those round the Sun, were about eighty days and nine months, respectively. (These very approximate results were improved later in the volume.) Ordered now simply by period, the planets fell into a natural sequence, with the Earth between Venus and Mars. As we shall see shortly, the Copernican scheme would make it surprisingly easy to order the planets by distance, happily confirming their ordering by period. The agreement between the two methods was something that seems to have moved Copernicus greatly. He had earlier included in his text a passage praising the Sun, as a lantern placed fittingly at the center of our beautiful world, and called a "visible god" by Hermes Trismegistus. In Book I still, before finally passing on to the basic mathematics he would need later, in his manuscript he added a long encomium, referring to many of those ancient authors his readers were likely to appreciate, but for some reason this did not appear in the printed edition. Perhaps he or his editor were giving up hope of holding the attention of unmathematical readers.

With the advantage of hindsight, a few recent historians have been tempted to represent the changes in planetary astronomy wrought by Copernicus as self-evident. Even if this were true—which it is not—there would remain his undoubted impact on later history. So powerful has this been that the so-called Copernican Revolution has often been taken as a paradigm, a model for many different kinds of intellectual change. Has its importance been exaggerated? Those who from childhood have been taught that the Earth moves are perhaps not the best judges of the difficulty with which the idea was made a part of the realm of platitude. When the question of the Earth's motion was raised in antiquity by Aristarchus, the charge of impiety was brought against him. When Copernicus came to that same conclusion, the situation was, if anything, more dangerous, for he was confronting three weighty authorities: the Church, the Aristotelian orthodoxy of the universities, and the astronomers, all of whom were working in a broadly Ptolemaic tradition. His courage in daring to criticize two of these three pillars of Western ideology was certainly greater than that of many of his latter-day critics.

It is usual to begin a discussion of the sources of the new heliocentrism by reference to such ancient precedents as had been set by Aristarchus, and the ambiguous evidence concerning Martianus and others, which we saw being discussed even in Carolingian times. Some writers attach greater importance to the fourteenth century, and to Oresme in particular, and a suggestion he made that the Earth may not be fixed at the center of the universe, although its center of gravity strives toward that place. He considered carefully the whole question of the movement of the Earth, and discussing this point at all has often been seen as a revolutionary step. Aristotle had done so, however, and in view of the medieval method of writing commentaries on Aristotle, scores of scholastic authors did so as a matter of course. They almost invariably came round to the orthodox opinion, leaving the Earth at the center. In his *Questions on the Heavens* and *Questions on the Sphere*, Oresme emphasized the relativity of motion: the phenomena we observe daily could just as well be explained by the daily rotation of the Earth as by that of the heavens, he argued. In the end, he too opted for the traditional view that the heavens rotate—a matter for persuasion rather than demonstration. Oresme died in 1382, and nearly a century would pass before the birth of Copernicus, who was to be persuaded otherwise—by very different arguments from Oresme's. There were many printings of fourteenth-century commentators on Aristotle's *On the Heavens* during Copernicus' lifetime, so it is not altogether surprising to find him addressing the heliocentric issue. The difference between him and his medieval predecessors was that he was bold enough to break free from tradition and they were not.

COPERNICUS' CAREER

Nicholas Copernicus came across such materials, and no doubt to those Arabic sources mentioned on p. 206 of chapter 8, in the course of a privileged education in Poland and Italy. He owed much to the Establishment that his writings later helped to unsettle. He was born in 1473 into a prosperous merchant family with roots in Silesia, a family now settled in Torun. This was less than a decade after Poland had reached a relatively stable political equilibrium under the rule of Casimir IV. Casimir, grand duke of Lithuania and king of Poland, had won a bloody victory over the Teutonic Knights at Puck in 1462. With papal support, by the second Treaty of Torun (1466), all of western Prussia—"Royal Prussia"—was ceded to Poland. The nationality of Copernicus has been disputed among modern scholars almost as violently as the territory in which he lived has been, down the ages. In brief, we may describe him as a subject of the Polish king, drawing heavily on German culture, but also greatly influenced by Italy and the Church.

Nicholas's father died when he was ten, but with the support of his maternal uncle, Lucas Watzenrode, he was able to attend the university of Kraków between 1491 and 1495. After the uncle became bishop and feudal lord of Varmia (also called Ermeland), a diocese bordering on the territory of the Teutonic order, Nicholas was found a salaried position as canon of the cathedral chapter of Frombork (or, Frauenburg). In 1496, he enrolled in Bologna as a student of canon law, but astronomy seems to have become his main pursuit. In Bologna he became closely associated with the Ferrara-born Domenico Maria de Novara, who taught astronomy in the university there until his death in 1504. In a later period of leave from his chapter to study in Padua—medicine, this time—Copernicus came back in 1503 with a qualification in canon law, from the University of Ferrara. If a hint of willfulness was already showing itself, the remainder of Copernicus' personal life did not conform to it, for he lived out his last forty years in the service of the Varmia cathedral chapter. His comfortable situation allowed him to build an observing tower, from which he used three principal instruments: an armillary, a quadrant, and—following the example of Regiomontanus and Walther—Ptolemy's parallactic instrument.

Copernicus' books at Kraków show that he there learned the use of the Alfonsine Tables, as well as others deriving from them, and also some astrology. He read Peurbach on eclipse calculation. In Italy, we know that he not only assisted Domenico Maria de Novara in Bologna but also gave lectures on mathematics or astronomy in Rome. An important acquisition in this the Italian period was mastery of the Greek language, which enabled him to read classical authors. (Seldom remembered today is a little book that Copernicus published in 1509, containing his Latin translations of some Byzantine Greek letters, wrongly thought to be by the seventh-century historian Theophylact Simocatta.) It is only a matter for speculation, but it was probably also at this time that met with the planetary theories developed in Marāgha, which he somehow acquired, as we noted in chapter 8. A page of notes in his copy of the Alfonsine Tables shows him at work on his own new astronomical system, but the first extensive draft we have of it is to be found in his *Commentariolus* (*Brief Commentary*).

This work of unknown date was almost lost to history. A Kraków library catalog entry of May 1514 that refers to a work "maintaining that the Earth moves while the Sun is at rest" almost certainly refers to it. We have no more definite clue as to its date, but at a guess this might be in the range 1510–1512. Copernicus did not allude to it in his *De revolutionibus*, and his own manuscript copy is lost. The work was no more than a rough outline of his new ideas, and perhaps he was ashamed of mistakes he made in it as a result of hasty composition. It seems to have been circulated anonymously in some quarters, and we are fortunate that eventually Tycho Brahe obtained a copy. This is itself now lost, but three early manuscript copies made from Tycho's do survive.

Copernicus' greatest work, *De revolutionibus orbium caelestium* (*On the Revolutions of the Celestial Spheres*), was his definitive statement. It was completed at the very end of his life (1543) and differed in several technical respects from the *Commentariolus*. He had put that earlier sketch on one side, revisiting it occasionally in the 1520s and again during 1537–1538—judging by notes on his planetary observations in those years. The sketch contained a statement that the calculations in it were reduced to the meridian of Kraków, which he took to be the same as that of Frombork. In arguing that the Earth revolves around the Sun like any other planet, he introduced the names of authorities that he knew would carry weight in his own time, namely the Pythagoreans. In his last work, he named them specifically as Philolaus and Ecphantus, but we do not have to take this reference as more than a canvassing of historical support. In the manuscript of his final work he actually mentioned Aristarchus in the same connection, but then deleted the passage before publication, perhaps concerned lest he be associated with someone with a reputation for impiety (fig. 121). (He mentioned the name elsewhere, in connection with a figure for the obliquity of the ecliptic, but did so then in error for Eratosthenes.)

In his *Commentariolus*, Copernicus had set out his basic assumptions without offering them detailed support, although it is clear that he considered the strongest arguments for his system to be that it conformed to appearances while at the same time being more pleasing to the mind than Ptolemy's. By this he meant, among other things, that he had adhered religiously to the principle of uniform motion on a circle. His models, broadly speaking, followed the pattern of Ibn al-Shāṭir's—so seeming to avoid the equant, although an equant of sorts remains in a geometrically concealed form. (On this point, see fig. 125, below.) What mattered most was that his models were geometrically transformed so as to bring the center of each planet's model to a common point. This common center was not quite at the Sun, but not far away, at the center of the Earth's orbit. A theory of planetary latitude was added. The parameters on which the models were based in his early account were more or less those of the Alfonsine Tables.

121. The deleted passage in Copernicus' manuscript of the *De revolutionibus* where he makes reference to Aristarchus. The deletion covers about two pages, and was omitted from the first four printed editions (1543, 1566, 1617, and 1854), but was included in that of 1873. It contains these words: "Philolaus believed in the Earth's motion . . . . This is plausible, because Aristarchus of Samos also held to that view, according to some people, who were not moved by the argument put forward and then rejected by Aristotle." (Jagiellonian University Library, Kraków, Poland.)

When he decided to write a more thorough account, he found that those parameters gave reasonable results for longitudes, but were not of the precision needed for one of the easiest observations of all to make, namely the timing of conjunctions of the Sun, Moon, and planets. (The uncertainties in the positions of the Sun and Moon enter the calculation; and since their relative speed may be very small, the uncertainty in the time of their meeting will be correspondingly great.) Copernicus therefore set to work to amass the sort of observations he would need to make his schemes more accurate, more acceptable than those they were meant to supersede. He amassed a substantial body of observations, especially during the period 1512 to 1529, and from them derived a set of improved parameters, following the methods set out in Ptolemy's *Almagest* and the *Epitome* of Regiomontanus. He was never, of course, a professional astronomer—at most, a handful of university teachers and court astrologers would have been described as such, in his day.

Throughout this period of his life, Copernicus was busy with his ecclesiastical and medical duties—notice that in his portraits he is shown holding a bunch of lily of the valley, a mark of his medical standing (fig. 122). For long he even had some responsibility for the defense of his frequently beleaguered cathedral and city. For most of his life, Poland was threatened from all sides—by the Hapsburgs, by Moscow, and even still by the Teutonic Order. Battles, treaties, and dynastic marriages created an extraordinarily complex political situation, but one highly significant event at this period was the conversion to Lutheranism of the grand master of the Teutonic Knights, Albert Hohenzollern. In 1525, isolated from the Holy Roman Empire and the papacy, Albert made an act of homage as vassal of the king of Poland. Even after 1525, however, the old threats from Prussia, Brandenburg, and Moscow did not

122. Nicholas Copernicus (1472–1542),
from a sixteenth-century Italian engraving,
thought to derive from a self-portrait.

disappear. Lesser conflicts continued to flare up, threatening Polish access
to the Baltic Sea, but there was stability of a sort, and after 1525 Copernicus found yet another public role. He helped to improve the greatly debased
coinage, and on that subject he wrote an economic tract. He also gave guidance in the regulation of the price of bread—a far more important economic
matter than the well-fed modern reader can easily appreciate.

Finally, from bread to housekeeper: it is customary in biographies
of Copernicus to allude to the fact that he had living with him, in that
capacity, a younger divorced woman, Anna Schillings. In 1539, she and
certain women in the service of two other canons were ordered by the
bishop to leave. One of those canons was later harassed and imprisoned
for the Lutheran heresy. It was in May of the same year that Copernicus
acquired his most vociferous young disciple, Georg Joachim Rheticus; and
as yet another reminder that this was anything but a tranquil age, we note
that when Rheticus was fourteen, his father was executed for fraud.

Between his two chief astronomical works, apart from an almanac,
Copernicus wrote only one other of importance. This was an attack on a
treatise by a competent Nuremberg mathematician, Johannes Werner, "On
the Motion of the Eighth Sphere" (1522). Even so, Copernicus was not
above using some of Werner's ideas in his *De revolutionibus*. We know that
the papal secretary Johann Albrecht Widmanstadt was presented in 1533
with a valuable Greek manuscript as a reward for explaining Copernicus'
ideas on the motion of the Earth to pope Clement VII. In 1536 Cardinal
Nicholas Schoenberg heard of Copernicus' system, and the Cardinal in
turn asked the Polish astronomer to make his ideas public. Copernicus
printed the letter in his *De revolutionibus*, a work he had in any case dedicated to the new pope-a doubly useful defensive tactic.

The young Rheticus, a teacher of mathematics at the University of Wittenberg, had set about designing his own fame by traveling from one distinguished scholar to another, beginning with Johann Schöner, who was then in possession of most of the manuscripts of Regiomontanus, Walther, and Werner. When Rheticus arrived at Copernicus' door, it was with a gift of books, many printed by Schöner's friend, the great Nuremberg printer Johann Petreius. Copernicus was not reluctant to share his latest ideas with his young visitor, who passed them on to the learned world in the form of a long communication addressed to Schöner. This elegant synopsis, *Narratio prima* (*First Account*), was quickly printed twice, in Danzig (also called Gdansk) in 1540 and in Basel in 1541. By this time it was clear that Copernicus would soon be releasing his own magnum opus, that he was working hard on the numerical revision it entailed, and that he was worried about the reception it would have from the natural philosophers. In July 1540 he wrote on this point to Andreas Osiander, a well-connected Lutheran theologian, and in April 1541 received a reply to the effect that astronomical hypotheses and theories were not articles of Christian faith but were simply "bases for computation," devices for representing observed phenomena. Their truth or falsity, he suggested, was of no importance, "as long as they yield the phenomena of motion exactly." Osiander alluded to the well known equivalence of the eccentric and epicyclic models for the motion of the Sun—different models, in his estimation, that produce the same observed results. He advised Copernicus to touch on these matters in his book, in order to placate the Aristotelians and theologians, "whose opposition you fear." (It should not be forgotten that these are the words of a leading Protestant theologian to a canon of the Roman Church.) On the same day, he addressed a similar letter to Rheticus.

Osiander's views have a long philosophical history, and one that continues to provoke discussion. Whether or not they were or are acceptable, they introduced a psychological element into the discussion. There can be no doubt that Copernicus and Rheticus thought the new system to be *true*, physically true, even if unprovable, and they had no wish to water down their claims for it. In May 1542, Rheticus brought the fair copy of Copernicus' *De revolutionibus* manuscript to Petreius in Nuremberg. Printing began soon after, and Rheticus corrected the proofs, until in October he left to take up a professorship in Leipzig. Osiander now took over the supervision of the printing, and added his own anonymous preface to the work. In this, while praising Copernicus, he expressed himself even more strongly on the questions raised in their earlier correspondence. "These hypotheses need not be true, nor even probable," as long as they fit with the observations, he explained, and he pointed out some apparent absurdities concerning the changing apparent sizes of the planets, saying in effect that they were of no importance to a theory of longitude. The new hypotheses were to be taken alongside the ancient ones, "which are no more probable." "Let no one expect anything certain from astronomy, which cannot supply it, lest he accept ideas as true that were conceived for another purpose, and so leave this study a greater fool than when he entered it."

When the book appeared in March 1543, Rheticus and Tiedemann Giese were angry at Petreius and Osiander for this betrayal. An unsuccessful action was brought before the Nuremberg City Council to have the printer replace the original with a corrected edition. It is unclear whether Copernicus himself ever realized what had happened, for in December 1542 he suffered a stroke that left him paralyzed. Even without a severe loss of his mental powers, he would probably not have seen the offending preface before he saw a copy of the work in its entirety—since prefaces are usually printed last. He first saw a copy, according to Giese, on the very day of his death, 24 May 1543.

That Osiander, and not Copernicus, was the author of the anonymous preface was not generally appreciated by most early readers. Kepler discovered the fact from a copy of the book in which the information had been penned by a Nuremberg insider long before. Kepler gave the discovery prominence on the back of a title page of his book on Mars, but this was only in 1609. For three generations, most astronomers had regarded Copernicus as having spent a lifetime in developing a theory that he did not regard as physically true.

THE EVOLUTION OF THE COPERNICAN SYSTEM

By introducing the Sun (or, strictly speaking, a point near the Sun) into the theory of motion of every planet, Copernicus made it possible to represent all in a single system. When Eudoxus, Callippus, and Aristotle, established their systems, it was on the basis of an Earth that provided a center for all motions. The sizes of the spheres were left as an arbitrary question. When Ptolemy achieved a single system, the sizes of the shells accommodating maximum and minimum planetary distances were settled on the principle that there must be no void, no wasted space, between them. The Copernican system followed from a very different observation, namely, that in each of the separate Ptolemaic models for the planetary motions there was a certain line that represented the same real thing, this being (roughly speaking) the Earth-Sun line. To those not irredeemably steeped in Aristotelian philosophy, this in itself must have seemed a more plausible principle. At least it took the mystery out of something that had been half-appreciated since Ptolemy, namely why the mean Sun plays an important role in the motions of the Moon and planets. Unfortunately, this is not how the changes effected by Copernicus were generally viewed. For most people he was simply the man who set the Earth in motion.

The Earth's motion was *not* an inevitable consequence of the single system. Later astronomers were quick to find ways of avoiding it while keeping the system intact. We cannot now, of course, expect to follow his earlier thinking through all its stages, but it seems likely that Copernicus himself gave some thought to an idea favored by Tycho Brahe and others later in the century, according to which the Sun moves in an orbit round the Earth, while the planets move in orbits round the moving Sun. Why did Copernicus not rest content with that idea? His data made the orbit of Mars only about half as big again as that of the Sun, so that the two orbits would have intersected. The Tychonic diagram (fig. 134 in chapter 12) illustrates this point. For Tycho, who had ceased to believe in the reality of the spheres, this intersection was no great problem. Copernicus, however, believed in the reality of the planetary spheres—Kepler recognized that he had done so-and could not have accepted the interpenetration of the spheres of the Sun and Mars.

With his two innovations, the motion of the Earth on its axis and its motion in an orbit around the Sun, Copernicus introduced into astronomy the most far-reaching changes since ancient times, and yet he was in almost every respect a product of the Ptolemaic tradition. His *De revolutionibus* is divided into six books. The first gives a general survey of his system, and ends with two chapters on plane and spherical triangles. The second is a useful textbook on spherical astronomy, but is not in itself revolutionary. The third concerns precession and the motion of the Earth. The fourth deals with the Moon, the fifth with the planets in longitude, and the last with their latitudes. Apart from the actual points of division, the pattern of Copernicus' book is at every stage very closely modeled on that of the *Almagest* of Ptolemy.

Copernicus was a skillful propagandist for his theory. Many of the old arguments for the superiority of circular motion are here, for instance, that it makes endless repetition possible. To Ptolemy's fear that a rotating Earth would imply motions so violent that the Earth would fragment and be dispersed throughout the heavens, Copernicus replied that we should fear more for the stability of the heavenly sphere. As for the supposedly obvious character of the Earth's centrality, he thought that this had been misconceived. It is universally conceded that the distances of the planets from the Earth vary, as do their motions with respect to the Earth. What sort of basis is this, asked Copernicus, for a theory with the Earth at the center of all motions? To the Aristotelians, who might have felt the lack of the idea of an Earth that is surrounded with a spherical region of terrestrial force (up to the lunar sphere), he considered the hypothesis that there might be many such centers, and not just ours. Gravity might be a natural tendency inherent in all particles to unite themselves into a whole in the shape of a sphere, but not necessarily at the center of the Earth. This was one of the first hints of dissatisfaction with the old idea that gravitation was always toward the center of the universe, but Copernicus did not tackle the problem of why matter should congregate in a relatively small number of places—the Sun, Moon, and planets. Not until Newton did so, was universal gravitation to be turned into a truly coherent theory.

Copernicus paid much attention to arguments for the general arrangement of the planets, knowing, no doubt, that most readers would not get much further than this part of the book. Even the order of the planets had never been conclusively established. There was general agreement that the Moon, having the shortest period, was nearest the Earth, and that Saturn, with the longest (29½ years), was most distant. Jupiter and Mars, being explicable by the same model as Saturn, were presumed to come next below it, again basing the order on their periods—nearly 12 and nearly 2 years, respectively. Mercury and Venus were problematical, however. Plato put them above the Sun, Ptolemy put them below, and some Arab astronomers put Venus above and Mercury below. Copernicus simply presented his heliocentric system in a very general way, with the order of planets around the Sun being Mercury—Venus—Earth—Mars—Jupiter—Saturn; and he showed how easily and naturally it explains the relative sizes of the retrograde arcs, Venus's being greater than Mercury's, Mars's greater than Jupiter's, and this greater than Saturn's. His system also explains why the outer planets are brightest in opposition. But not the least of its merits, in his eyes, was the position of the Sun at the center, the lamp at the center of the beautiful temple of the world. Some call the Sun the light of the world, others call it the guide or soul. Hermes Trismegistus called it the visible god . . . And so Copernicus goes on, not averse to employing rhetoric of a pagan tint, if he can thereby persuade the reader of the system's inherent plausibility.

These were only qualitative opening shots. The fine detail was still to come, and this required precise parameters, better than those from the Alfonsine Tables, which he had accepted in his early career. Only about thirty of Copernicus' own observations are mentioned in his final work, but he makes it clear that they were the distillation of a much larger number. They were not of the highest quality, compared with others made in the late Middle Ages, but they were well chosen for his own special purposes. They include oppositions of planets and places close to opposition (for finding the ratio of the radius of the Earth's sphere to the planets' sphere), positions of the Sun (including equinoxes) and Moon, lunar and solar eclipses (which amount to much the same thing), and zenith distances (or altitudes, their complements) of various bodies. The point that quickly emerges, when one sifts through all this material, is that here we have one of the very few astronomers after Ptolemy who appreciated how to build up a planetary model from first principles, rather than by patching up the work of his predecessors.

Of course he did not begin from nothing, any more than did Ptolemy. He assumed certain general principles—under which heading we must place even the assumption of uniform circular component motions. He accepted some inherited prejudices with a tenacity that is hard to justify, even by the lights of his own time—as in the case of his cyclical theory of precession, to which we alluded on a previous occasion. Copernicus finally settled on a precessional theory requiring a double circular motion. These motions amount to oscillations of the Earth's equator at right angles to one another and can be regarded as equivalent to two oscillations of the Earth's

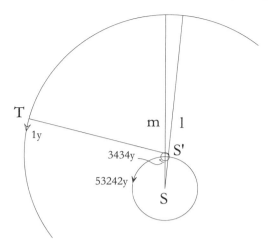

123. The Earth in the Copernican system, in relation to the mean sun.

axis of rotation, one tangent to the Earth's precessional motion (this produced a variability in the precession) and a perpendicular motion varying the obliquity. He assumed that the second period was exactly twice the first, which he set at 1717 years of 365 days; and he decided that the obliquity varied between 23;52° (at some time before Ptolemy) and 23;28° (which was not then reached). As for the steady precessional term, he somehow derived a figure of 360° in 25816 years of 365 days, that is, one degree in 71.66 Julian years, an excellent result. Copernicus measures longitudes from a star (γ Arietis) rather than from the true (fluctuating) equinox.

For the motion of the Earth round the Sun, Copernicus did not strictly need to add much to the simple Ptolemaic model, that is, the simple eccentric (or concentric with epicycle). His observations indicated that for AD 1515 the eccentricity was 0.0323 times the orbit's radius, and that the longitude of the apogee was 96;40°. However, he chose to complicate the final model by making the center of the Earth's orbit, the "mean Sun" (S′ in fig. 123), move in relation to the true Sun (S in the same figure). T is the Earth. S′ was in fact taken to move round a small circle in the same period as that of the obliquity, with both maxima occurring at the same time (65 BC).

The figure, as drawn here, is to scale in all but the orbit of the Earth, which is about a sixth of its correct size. The lines m and 1 are apse lines, 1 giving the direction of aphelion at the given time, and m its direction for maximum and minimum eccentricity. The long periods quoted are those for motions in the two central circles.

Why the peculiar complication of a variable eccentricity? Here we have a second example—the first was his theory of precession—of Copernicus wishing to preserve as far as possible the work of his predecessors, notably Ptolemy, whose figure for the eccentricity was badly flawed. Copernicus made the maximum eccentricity on his model 0.0417 (417 parts of the standard radius of the Earth's orbit of 10,000 parts), only slightly larger than Ptolemy's. Taken together with his own eccentricity, this implies a radius of 48 parts for the small circle.

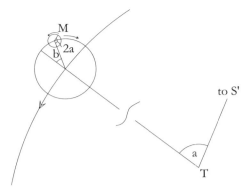

124. The Copernican lunar model.

Copernicus' generosity toward his predecessors resulted in a curious astrological doctrine, put forward by Rheticus in the *Narratio prima*, possibly with Copernicus' cognizance. It was traditional in astrology to suppose that a planet's strength was increased at its apogee and diminished at its perigee. This happened often—even in the case of Saturn it happened once every thirty years or so. There was also the standard astrological doctrine relating to the rise and fall of sects and religions, based on so-called great, major, and maximum conjunctions. (For a brief explanation of these, see figure 136 in chapter 12.) Rheticus saw that the various very long periods inherent in Copernicus' theories lent themselves to being combined with these traditional ideas: he pronounced dogmatically that when the eccentricity was a maximum, the Roman republic was tending toward a monarchy; that, as the eccentricity declined, so did the Roman empire; that, as it reached a mean value, there came the rise of Islam; but that the collapse of the Islamic empire could be expected in the seventeenth century, to coincide with the minimum. The second coming of Christ could be expected with the next following mean value. And all this was no more than a consequence of the fact that Copernicus wished to preserve the truth of Ptolemy.

In the case of the Moon, the new system could be much simpler. Ptolemy-or more precisely "our predecessors"—could be refuted, Copernicus thought, "on the grounds of reason and of sense." "Reason" was violated by the fact that while the epicycle center had been held to move uniformly with reference to the center of the Earth, it had been made to move irregularly on its own eccentric orbit. "Sense," on the other hand, told Copernicus that Ptolemy's model had produced much too great a variation in lunar distance. Copernicus' own lunar model (fig. 124) is identical to Ibn al-Shāṭir's, and he had already used it in the *Commentariolus*. (The radius in the figure as drawn here is broken so that the epicycles can be drawn to scale.) He selected parameters that fitted with the Alfonsine Tables, and that were indeed of much earlier Indian origins, had he but known it; but of course at the root of all this there was the brilliant work of Ptolemy, in his derivation of the second lunar inequality. Copernicus' discussion of the distance, parallax, and apparent diameter of the Sun and Moon, was

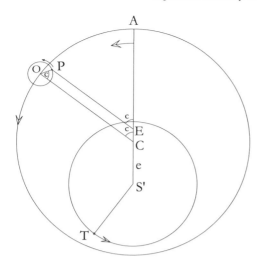

125. The later Copernican model (that of *De revolutionibus*) for a superior planet (*P*). *T* is the Earth, *S′* the center of its orbit, *C* the center of the planetary deferent, and *O* the center of its epicycle. The angles marked *c* increase at a constant rate, and the point *E* has the character of an equant center around which the planet moves uniformly.

somewhat blemished, although inevitably it was much better than Ptolemy's. When showing how to compute eclipses, too, he improved greatly on all of his predecessors.

The fifth book of *De revolutionibus*, dealing with the longitudes of the superior planets, includes some of Copernicus' best work. His is still essentially a rewriting of the Ptolemaic circles, but we should not underestimate the painstaking work of calculating and recalculating the elements of the orbits—by an iterative process, involving many hundreds of calculations. In his planetary theories he had the advantage over Ptolemy that he had only the first inequality to consider, accounting for the revolution of the planet with respect to the stars. Ptolemy had needed one inequality to take care of the movement of our Earth around the Sun. As we should by now have learned to expect, each Copernican model is related not to the true Sun but to the center of the Earth's orbit (*S′* in fig. 125).

The figure is for any of the three superior planets, although not to scale for any of them. He has now found an alternative to the scheme in *Commentariolus*, where he made use of an epicycle on an epicycle. Now he uses a single epicycle. In all three cases, the radius of the epicycle *OP* is more or less equal to a third of the eccentricity *CS′*. The equivalent to Ptolemy's eccentricity of the equant can be shown to be now *OP* + *CS′*, and whereas Ptolemy made the eccentricity of the deferent just half of that total, here it is three-quarters. Note the equal angles marked *c*. It is instructive to analyze the planet's path with respect to *S′*: as Copernicus of course realized, it is not a circle, but of course neither is it a Keplerian ellipse.

Copernicus' handling of the inferior planets is less praiseworthy, partly because he could not obtain the observations he needed. The models he now adopted were different from those in the *Commentariolus*, inasmuch as he shifted the two epicycles from the periphery of the model to the center, where they form a dual eccentricity. An idea of the arrangement may be had from figure 126, where *Q* marks the planet, and where the apparatus

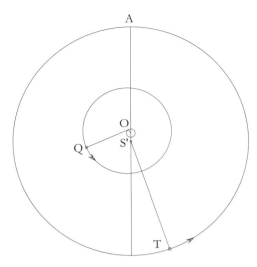

126. The later Copernican model
for an inferior planet.

for fixing the (counterclockwise) moving center *O* of its deferent circle in relation to the center of the Earth's orbit *S′* is shown. The angle at *S′* is half the angle at the center of the small circle. The models now correspond to Ibn al-Shāṭir's, duly inverted for a Sun-centered theory. The actual values of the new parameters are obtained by a process that seems to have relied partly on Ptolemy and partly on unexplained intuition.

In this account, we have said nothing of the motions of the apse lines (lines to apogee and perigee) that Copernicus derived. These motions were of course very slow, and although they left much room for improvement— he was unfortunate in deciding to measure them from the mean Sun *S′*, rather than the true Sun—we have here again a rare instance of an astronomer capable of deriving the elements of a largely *new* theory from fundamental observations.

We saw in chapter 3 how each planet's model contains a common element, namely the mean radius of the Earth's orbit (*TE*), and how that fact implies that the scale of the entire planetary system may be expressed in terms of the common unit, the "astronomical unit." The mean distances from the Sun, derivable from the Copernican parameters of scale, are as follows (with modern figures in parentheses).

Mercury 0.3763 AU (0.3871 AU)
Venus 0.7193 AU (0.7233 AU)
Mars 1.5198 AU (1.5237 AU)
Jupiter 5.2192 AU (5.2028 AU)
Saturn 9.1743 AU (9.5388 AU)

In a sense, these distances came as a free gift with the model, and indeed almost equally good results can be found from Ptolemy's models—those of the *Almagsest*, not the nested spheres of the *Planetary Hypotheses*—by anyone who recognizes the place tacitly occupied by the Sun in those models.

The close agreement between the two sets of figures, ancient and modern, is surprising, when we consider how it depends entirely on geometrical proportions that had been locked away for fourteen centuries in Ptolemaic astronomy, appreciated only for their importance in predicting motions. It is perhaps even more surprising that the geometrical proportions in turn rested wholly on motions, that had been derived from centuries of earlier observations.

As for *absolute* distances, Copernicus made only minor adjustments to the quite inadequate solar parallax found by Hipparchus. Copernicus gave the mean parallax as 0;03,31° and the mean solar distance as 1142 Earth radii.

In view of the new way of obtaining distances within the solar system, it should be added that the older way of scaling the geocentric system, with a nesting of spheres that allowed for no spaces between them, continued to appeal. It appealed even to Copernicans, so immersed were they in the Aristotelian dislike of the notion of a vacuum. Copernicus had, so to say, created large empty spaces—for instance between Venus and Mars, and between Jupiter and Saturn—into which some astronomers—notably Michael Mästlin and Tycho Brahe—felt themselves entitled to try to fit comets, while Kepler once even speculated that there might be hitherto unrecognized planets there.

In Book VI of *De revolutionibus*, Copernicus considered the latitudes of the planets, and here he made little advance over the theory in the *Commentariolus*. Again he penalized himself by using Ptolemy's extreme latitudes to derive the inclinations of the planetary orbits. He gave his planetary planes variable inclinations, using Ptolemy's parameters, and seems to have been almost oblivious to the inherent superiority of the heliocentric hypothesis on this admittedly difficult point. Part of his difficulty stemmed from the fact that his theory was not heliocentric enough, so to speak. He made his planetary orbits lie in planes passing though the center of the Earth's orbit (our point S') and not the physical Sun, as Newton's dynamics would later show to be the case. Copernicus' was a unified system, in the sense that all the subsidiary planetary models were superimposable, and in the sense that the Earth's motion explained away the second anomalies in the older models. His system was not without its difficulties, and these were partly psychological—if that is a reasonable word to use of a mentality bred by a close reading of Aristotle. If the Earth moves round the Sun, then this must affect the apparent directions of the stars over the course of the year—the phenomenon of annual parallax (for the meaning of which see p. 103 above). Since such annual shifts had not been detected, Copernicus was obliged to say that the stars were at distances far more immense than had previously been supposed. The distances in Ptolemy's *Planetary Hypotheses*, for instance, taken together with the Copernican data for the Earth's orbit, imply an annual shift of about 7°. Even a shift of a minute of arc would have demanded a vast empty space between Saturn and the stars. When Tycho Brahe later came to ponder these various consequences, he decided—and surely most of his contemporaries would have agreed— that such vastnesses in God's creation would have been pointless. The idea seemed not only theologically absurd but simply unbelievable. In time, that view would change.

Copernicus' was a unified system but still a geometrical one, rather than a physical system that explained appearances in terms of physical laws. His geometrical transformation of "Ptolemaic" models did not produce any significantly different observational predictions. Kepler later said of Copernicus that he did not know how rich he was and that he had been trying harder to interpret Ptolemy than Nature. This much-quoted statement hits on an important point, but it can also be misleading. Copernicus was trying first and foremost to represent Nature. He often fell back on previous theories, but only because they had already come close to the truth. With his unified scheme, he had come still closer.

A PERIOD OF TRANSFORMATION

In the estimation of leading European astronomers, Copernicus quickly took his place in the hall of fame alongside Ptolemy, and yet the ordinary practitioners of astronomy continued to work with the system—and the tables—that they already knew. Like Rheticus, Erasmus Reinhold taught mathematics at Wittenberg. In 1551 he published new "Copernican" tables in place of the Alfonsine. They were known as the Prutenic Tables (Prussian Tables) in honor of Duke Albrecht. They make use of some newly evaluated parameters as well as those of Copernicus, whom Reinhold praises without referring to the heliocentric character of the underlying hypothesis. There is good reason for thinking that he wished to replace it with an Earth-centered alternative along the lines later adopted by Tycho Brahe.

Reinhold's tables were widely diffused. In England, John Feild used them to prepare an ephemeris for 1557, in the preface to which the scholar and magus John Dee expressed qualified admiration for the Copernican system. Feild, however, seems to have accepted the new system without hesitation when calculating his ephemeris. The writer of mathematics texts, Robert Recorde, a graduate of both Oxford and Cambridge, had previously introduced Copernican ideas briefly into his elementary textbook, *The Castle of Knowledge* (1556), but not until Thomas Digges, a pupil of Dee, did anyone in England give a simple—albeit flawed—exposition of the system. In 1576, responding to a criticism of him by Tycho, he published a book that had been left unpublished by his father Leonard Digges, and appended an English translation of part of Book I of *De revolutionibus*. This work contains a diagram that advocates the infinite extent of the universe of stars—not an entirely new idea, but new to the context of the new astronomy.

Such fragments signifying interest in Copernicanism can be roughly paralleled in most parts of Europe and would perhaps pass unnoticed did we not value the historical phenomenon itself. An English writer whose influence on later astronomy was more substantial, indeed throughout Europe, was William Gilbert. His major work *On the Magnet* (1600) is first and foremost a work on the physics of magnets. It aims to prove that the Earth is itself a spherical magnet (a sphere of lodestone), but the book also contains much of cosmological concern—it propagates the idea of an infinite universe, for instance. Gilbert compares Ptolemy and Copernicus in a general way and is clearly inclined to believe the views of the latter, but Gilbert's greatest influence was the writings of Galileo, who discussed *On the Magnet* at length in his *Two World Systems*. Kepler too helped to publicize Gilbert's book and used it as a basis for some of his own cosmological ideas, as will be seen when I discuss his *New Astronomy*.

Michael Mästlin was Kepler's teacher at Tübingen. He showed Kepler— as Kepler tells us—that the comet of 1577 moved constantly with respect to the motion of Venus as set down by Copernicus. From his failure to measure its distance, which he realized must be much greater than the Moon's, he decided that it was moved by the same sphere as Venus. (A roughly similar conclusion had been reached by Abū Ma'shar in the ninth century.) Here, and even earlier when he formed the same opinion as to the position of the new star of 1572, Mästlin was creating an atmosphere that would eventually help to destroy Aristotelian physics, even though he was perhaps not yet fully committed to Copernicanism. As it turned out, his analysis of the comet was quickly overshadowed by the far more thorough work of Tycho Brahe, but it was Mästlin who first persuaded Kepler to follow the Copernican lead.

As we shall see in the following chapter, Tycho placed the comet of 1577 outside the sphere of Venus "as if it were a fortuitous and extraordinary planet." He praised Mästlin, but emphasized the point that there are no real spheres in the heavens, a point on which he said Mästlin would disagree—note his belief that both the comet and Venus were on the same sphere, an odd thing to say in view of the fact that the comet came and went. While Tycho's astronomy differs from Copernicus', there is no doubt that it developed under the influence of the latter. Throughout his published writings, his letters, and his observational records, we find Tycho making comparisons between what he himself recorded and Copernican (or Prutenic) predictions. As a young student at Leipzig university he found the Saturn-Jupiter conjunction of August 1563 to be better accounted for by the Prutenic Tables than by the Alfonsine. He found a date from the latter which was no less than a month off in error, but before long he became disenchanted with both. He later spoke of this as the turning point in his career, the point at which he turned his thoughts from the study of law to astronomy.

Tycho's extraordinary thoroughness is characteristic of his entire career in astronomy. It is already evident in those Saturn-Jupiter observations he made as a student, using a simple cross-staff. We see it long afterwards, as when, for example, he insists on revising even the latitude of Frombork, as determined by Copernicus. In 1584 he sent an assistant, Elias Olsen Morsing, to Frombork and Königsberg, equipped with a fine sextant that was believed to be accurate within a quarter of a minute of arc. Tycho himself worked out the latitude of Nuremberg, using Walther's observations of solar altitudes and the Landgrave's observations made at Kassel, which he took to be 2° north of Nuremberg. From Frombork, Morsing brought back one of Copernicus' instruments, a parallactic instrument, or triquetrum, which the dean of Frombork had presented to him. It was of pine, about two and a half meters long, and Tycho considered it relatively crude, but it was given a place of honor in Tycho's castle, alongside a portrait of Copernicus. Indeed, so excited was Tycho on receiving the instrument that he immediately composed a heroic Latin poem to the Polish astronomer. Astronomy was on the verge of a sea change in observational technique, and Tycho's poem could as well be regarded as an epitaph to the age that was ending. Using an excusable mixed metaphor, he wrote of a man who "by means of these puny cudgels" had "climbed the lofty Olympus." It was by his own example, however, that Tycho flattered Copernicus most effectively, for it was from Copernicus that he took the geometrical apparatus of planetary astronomy, before reconverting it to an Earth-centered scheme.

The New Empiricism

TYCHO BRAHE AND THE RESURGENCE
OF EMPIRICAL ASTRONOMY

Tyge Brahe was born in 1546 into an aristocratic Danish family at their castle at Knudstrup in Skåne, now the southernmost province of Sweden, but which at that time was ruled by Denmark. Tycho—to give him the Latinized form of his name, and the one by which he is usually known—was well connected on both sides of his family. His mother's family seems to have had a near monopoly of the hierarchy of the Lutheran Church in Denmark. His father, a member of the Royal Council, was eventually given responsibility for the castle of Helsingborg, across the water from Elsinore, at the strategically important entrance to the Baltic. For most of his life, Tycho was in a strong political and social position, although his fortunes did not endure. His upbringing followed more or less the expected pattern: he was brought up at the castle of an uncle, but unlike his brothers, who were likewise sent as squires away from home, he was academically fortunate. At thirteen he attended the Lutheran University of Copenhagen, where he made the acquaintance of a medical professor, Johannes Pratensis, who encouraged his interest in astronomy. He was not yet fourteen when in 1560 a solar eclipse turned his interests to practical questions of observing, and he obtained a copy of the ephemerides of Stadius, which were based on the Copernican Prutenic Tables.

Tycho moved from one university to another—Leipzig, Wittenberg, Rostock, Basel, Augsburg—ostensibly at first to study law, and he did not resettle in his native Denmark until 1570. In 1566, in a friendly duel with another Danish nobleman, he was unfortunate enough to have his opponent's sword slice into the bridge of his nose. He was obliged to go through life with a metal plate over the wound, about which he was—not unnaturally—very sensitive. His interest in alchemical experiments might have been connected with his wish to find a suitable alloy for it: he finally settled on one of gold, silver, and copper.

Tycho's concern with astronomy was growing steadily. Frederick II, the Danish king, was gathering around him a group of talented scholars, and

127. Tycho Brahe (1546–1601), from a sketch by Tobias Gemperle (ca. 1578), as reproduced by the Danish Royal Society in 1901.

had already begun to express interest in the scientific activities of Tycho and his maternal uncle Steen Bille at Herrevad Abbey, when suddenly they acquired new and unexpected intellectual status. After sunset on 11 November 1572, returning for supper from his alchemical laboratory, Tycho discovered a new star in the constellation of Cassiopeia. He knew the constellations well enough to be able to say that it had not previously been noticeable, but in any case, it was brighter than any star he had ever seen before, even brighter than Venus. His instruments were still relatively simple—a sextant and a cross-staff were the ones he used—but they allowed him to measure the star's position relative to neighboring stars and to convince himself that it was not moving, and was therefore not a comet. He kept up these measurements, and recorded the star's changing size (as he thought), brightness, and color, until, in the following March, this new star—*stella nova*—finally faded from view.

Tycho's "nova," at its brightest, was visible in daylight. His careful observations of the star's changes of brightness allowed Walter Baade in 1945 to classify it as a type I supernova. Supernovae are so rare that perhaps only five visible to the naked eye have been seen over the last thousand years. The most brilliant and longest lasting ever recorded was one observed in Europe, the Middle East, and China in AD 1006. Tycho's, carefully measured in position by him, by Michael Mästlin in Germany, and by Thomas Digges in England, left remnants which have been identified with the radio source 3C 10, and have been much studied with the advantage of the early reports.

Tycho's careful program of observation of the new star was eventually seen to be of great significance for the cosmology of his own time. Not only was the star not moving, as a comet would have done, but he was in a position to say that its parallax ruled out its being closer to the Earth than the Moon. It twinkled like a star, it did not have a tail like a comet, and its stability seemed to rule out the idea that it was an exhalation from the Earth's atmosphere. The eventual implications for standard Aristotelian

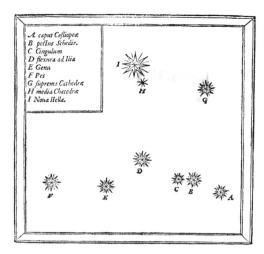

128. The position of the new star (*stella nova*, later classified as a supernova) of 1572 in the constellation of Cassiopeia. Tycho observed it from Herrevad and published this image in the following year.

cosmological doctrine—even he did not appreciate them at first—were serious: the space beyond the Moon, occupied by the planetary spheres, was supposed by the Aristotelians to be unchanging, and yet here was evidence to the contrary. By the beginning of 1573, Tycho was convinced of the reliability of his observations, and he published a short astrological and meteorological treatise into which he wove an account of the star's appearance. (His drawing of it in figure 128 may be compared with that by Hagecius in figure 129). He wrote in a rhetorical style owing much to humanist models such as Ficino and Paracelsus, but there was more to his writing than a superficial deference to fashion. Tycho was daring to cast doubt on none other than Aristotle. He hinted that perhaps comets, like the star, might also turn out to be above the Moon, and so might contradict the Aristotelian view of them.

In 1577 he had a chance to put his speculations to the test. A comet appeared then, as bright as Venus, and with a tail 22° long. Tycho's observations of the comet—which he first noticed as he was catching fish for dinner—were exemplary, as was the use he made of them. He was the first to derive a comet's trajectory in both equatorial and ecliptic coordinates, conducting a large number of trials rather than resting content with a single determination of position. In other words, he deliberately planned to derive an average path on the basis of data in which there was much redundancy. The modernity of that approach is as striking to us as were the forebodings of danger to the more perceptive Aristotelians. Observing the comet between November 1577 and the following January, he found no evidence of any parallax, which confirmed him in his belief that the comet was above the sphere of the Moon. It was moving in a path at 29° 15′ to the ecliptic—much steeper than any of the planetary paths—and in a retrograde direction. He eventually decided that it fell within the sphere of Venus, although a decade later, looking again at his records, he concluded that it went far beyond, and through the sphere of the Sun. By this later time, perhaps prompted by Christoph Rothmann, it

DIALEXIS

IMAGO CASSIOPEÆ.

129. A later record of the supernova of 1572, reaching a wider readership. Published in Frankfurt in 1574 by Thaddaeus Hagecius (Hayck), physician to Rudolph II.

had begun to dawn upon him that the "spheres" so beloved of Aristotelian cosmology could not possibly be solid.

Between observing nova and comet, Tycho had continued to move around freely. He had given lectures at the University of Copenhagen—lectures with a clear Copernican bias—and in 1575 had visited the Landgrave of Hesse-Kassell, William IV, in Kassel. The landgrave was an astronomer himself, with a fine collection of instruments, and the two men made systematic observations for over a week, Tycho using semi-portable instruments that were carried around with him. This was the beginning of a friendship that produced a long and weighty correspondence, which Tycho published twenty years later. After Kassel, he continued on his grand tour. He went on to Frankfurt, Basel, Venice, Augsburg, Regensburg, Nuremberg, and Wittenberg, making contacts with astronomers and collecting ideas for instruments wherever he went. Back in Denmark, there is no doubt that the king and his courtiers were conscious of the political value of his work, for neither he nor they were oblivious to its astrological implications, especially those ascribed to the new star. On several occasions, King Frederick II offered him fiefdoms and a chance to serve the Danish crown as his ancestors had done, but the young astronomer had no wish to do anything of the sort. Indeed, he was tempted to move to Basel, when in February 1576, a few months after his return to Denmark, the king made a new offer. Perhaps as the result of a recommendation by the landgrave, Frederick offered Tycho the island of Hven in the Danish Sound, as well as lavish funds, and asked him to set up an observatory there.

Tycho accepted the king's offer, and for more than twenty years he worked on Hven in what was to become the world's finest astronomical observatory up to that time, Uraniborg. Its name can be translated "Castle of Urania," Urania being one of the nine muses, the muse of astronomy. (Uranus, *Ouranos* in Greek, was god of the sky, but Tycho spoke of the place as his "museum," and as a place consecrated to Urania. Every well-read student knew Urania from reading Martianus Capella's book on the seven liberal arts.) In its heyday, Uraniborg was equipped with a full complement of instruments, not to mention a windmill and paper mill, a printing office, farms and fishponds, and the domestic staff needed for its support. This is to say nothing of the many talented scholar-astronomers Tycho gathered round him. This well-connected nobleman had much personal experience of life in the houses of patrons of learning. One visit in particular must have influenced his outlook; namely, that visit he made to the court of the land-grave, a man of learning who kept a close eye on appointments, and even on the curriculum, at his father's newly created University of Marburg. Here was a notable patron of those who could provide him with the most beautiful of scientific instruments, and of those who could use them. As we shall see later, one of the instruments the landgrave commissioned would prove to be an innocent violation of the Dane's intellectual property.

Tycho was undoubtedly fond of princely society, but his ambitions were not those of a mere astronomer-aesthete. His buildings were care-fully planned, plumbed for water, equipped with kitchens, a library, labo-ratory, and eight rooms for assistants. Around 1584 an adjacent additional observatory was built: Stjerneborg ("Castle of the Stars") had additional instruments on secure foundations in subterranean rooms. The painting on its ceiling depicted Tycho's own astronomical system, and its walls were hung with portraits of six great astronomers of the past, from Timocharis to Copernicus, together with two others, one of Tycho himself, and one of Tychonides, his as-yet unborn descendant. He commissioned several por-traits of himself, and if they have a certain sameness, that is perhaps because they were all done with an eye on a portrait by the Augsburg painter Tobias Gemperle, with which Tycho is said to have been well pleased (fig. 127, above. He had a goodly share of vanity, and was pleased to find that his out-of-the-way foundation was becoming a place of pilgrimage. Among the many socially distinguished visitors to Uraniborg was James VI of Scotland, later to become James I of England.

Tycho's was a research institution in the best traditions of astron-omy, but in its excellent instrumentation it outstripped all before it. For an idea of its scale, see figures 130 and 131, and for a specimen instru-ment, figure 132. His instruments included Ptolemy's rulers, armillaries, sextants, octants, and azimuthal quadrants, some of wood and some of brass. He owned celestial globes, one of them a meter and a half across. On a wall in the plane of the meridian was his finest instrument, a quadrant of radius about 1.8 meters, the scale marked with transversal points to permit easier measurement of its subdivisions of angle. This mural quadrant too was decorated with Tycho's portrait. Tycho made good use of his assistants when making observations. One observer viewed the object through pin-nules on the sighting rule, another entered results in a ledger, and a third noted the time on two clocks beating seconds, clocks that were unreliable but that were checked repeatedly against the heavens (fig. 133). Tycho

130. The white-painted castle of Uraniborg, and its grounds, ca. 1591. (From Tycho's *Astronomiae instauratae mechanica* [1598]).

ORTHOGRAPHIA PRÆCIPVÆ DOMVS

ARCIS VRANIBVRGI IN INSV-
SIA VULGO HVENNA,
RANDÆ GRATIA CIRCA AN-
EXÆDI-

LA PORTHMI DANICI VENV-
ASTRONOMIÆ INSTAV-
NVM 1580 A TYCHONE BRAHE
FICATÆ.

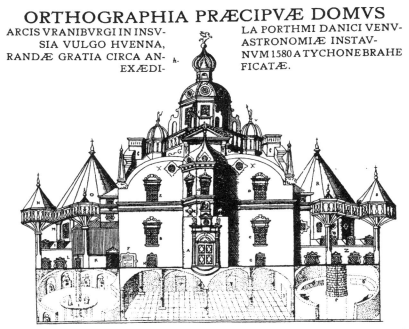

131. The eastern facade of Uraniborg. The upper levels all contained instruments, some of them very fine—such as the equatorial armillary, the azimuth semicircle, and the trigonal sextant—and some that were kept for nostalgia's sake. In the cellars were sixteen furnaces for various purposes—chemical, technical, and household. There were also larders and storage rooms there. (From Tycho's *Astronomiae instauratae mechanica* [1598]).

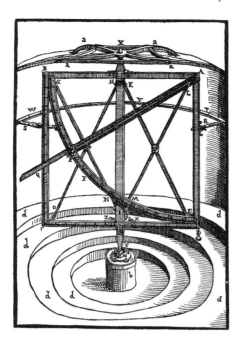

132. Tycho's largest azimuth
quadrant, of steel, rebuilt and
installed in Stjerneborg in 1588.
(From Tycho's *Astronomiae
instauratae mechanica* [1598].)

introduced checks for instrumental error, and cross-checked by comparing results with different instruments.

When using altitudes to obtain the latitude of Hven, he noticed that the
Pole Star—duly corrected, of course, for its true polar distance—gave different results from those he obtained using the Sun. He realized that the cause
was atmospheric refraction. As we know, he was not the first to appreciate
the point, but he studied the phenomenon more intensively than anyone
before him had ever done, and he saw that seasonal and temperature effects were important. He tabulated his results for use with the Sun, and included (faulty) solar parallax in the table. His resulting data were, overall,
of unprecedented accuracy. Much of his early work is reliable to three or
four minutes of arc, and his later accuracy is often better than a minute of
arc for star positions, and hardly much less for those of the planets. This
was better, by a factor of five or even ten, than the level of accuracy of the
best Eastern astronomers, even than that of Ulugh Beg's observatory in
Samarqand. Tycho's precision gave him a sure touch when he came to calculate the precessional motions of the stars, finding it to be $51''$ per annum.
He thought it was probably not constant, but that the quoted value would
be a reasonable one over long periods. In fact he intended to investigate
the question, along with all other potential long-term changes affecting solar theory, but he never fulfilled this ambition. He was certainly anxious to
discover a suitable model for the precession of the equinoxes, but an outline account of one such model that he based on his lunar latitude model,
included in a letter to Scaliger, is unworkable, and does not speak for his
theoretical powers.

133. An engraving of Tycho's great mural quadrant of 1582, from his *Astronomiae instauratae mechanica* (1598). The portrait of Tycho painted within the quadrant is dated by the cartouche at 1587, when he was 40. He is shown pointing to the quadrant's foresight, in the wall. The instrument gave meridian altitude directly, and so declination (the equatorial coordinate) almost as directly. Note the graduation of the quadrant by divided transversals, capable in principle of being read to an accuracy of ten seconds of arc. The clocks would have given short intervals reasonably accurately, but Tycho found them inadequate for timing meridian transits. Timed transits would have given the other equatorial coordinate (hour angle, or right ascension) fairly directly, but he was driven to use his armillaries or sextants for that purpose. The great mural was painted by Hans Knieper, Hans van Steenwinckel, and Tobias Gemperle. The cross-section of his observatory shows men making observations on the top level, the library (with large globe) below, and alchemical experiments on the lowest level. The dog is plainly bored by it all.

Of the excellent assistants with whom he surrounded himself, several—Willem Blaeu the cartographer, Christian Sørensen Longomontanus, Paul Wittich, and Johannes Kepler, for example—became renowned in their own right. Wittich, from Wittenberg, had a fertile intellect, but he stayed for only a few months, and Tycho came to consider him untrustworthy. He was thought to have overstepped the mark in the kind of technical information he passed across to the Landgrave of Hesse, for example. While Tycho's relations with the landgrave were good, there were clearly limits to what he could be expected to share. Wittich was an independent spirit who found it hard to play second fiddle to Tycho. He had already sketched geocentric schemes, broadly resembling a later Tychonic system, in his annotated copy of Copernicus' *De revolutionibus*. He brought with him a useful computational technique for lightening the heavy labor of trigonometrical calculation—replacing multiplications and divisions with additions and subtractions. Tycho seems to have regarded this as something to which he, Tycho, had an exclusive right, and he was angered by its inclusion in a book published in 1588 by his rival Nicholas Reymers Baer. Baer—often known by the Latin equivalent "Ursus," meaning Bear—was the imperial mathematician. He had visited Uraniborg in 1584, and according to Tycho then stole other intellectual property. We shall later return to this charge. A rare printed pamphlet—only one copy is known—dating from around 1599, which was intended as a defense against Tycho's charges and was seemingly written by Ursus, was designed to show that geoheliocentric hypotheses were not new, but could be traced back to Apollonius of Perga and Martianus Capella in antiquity, and were even to be found in Copernicus.

CELESTIAL SPHERES AND PLANETARY SYSTEMS

Tycho's full reasoning about the comet of 1577 was reserved for a Latin work of 1588, published from his own press. Appearing under the title *De mundi aetherii recentioribus phaenomenis* (*Concerning Recent Phenomena of the Aetherial World*), this was the second part of a trilogy he planned but never completed. A short German tract earlier announced its implications to a wider public. His observations of the nova and comet had finally led him to discard the idea that the Aristotelian spheres were real in any strong sense: at least they did not seem to impede the motion of a comet above the Moon, or the generation and decay of a star. This conclusion did not come easily to him: in 1578, he was still a believer in hard and impenetrable solid spheres, and his conversion might have been as late as 1586.

There has been much discussion of the degree of belief that earlier astronomers had in the solidity and hardness of the celestial spheres—the doctrine that Tycho was for long considered to have been the first to undermine. Consulting the writings of medieval commentators on Aristotle is of little help, since Aristotle himself had nothing much to say about the nature of the celestial spheres, their hardness or fluidity. Medieval writers have occasionally been misunderstood by those who have been unaware that "solid" (*solidus* in Latin) usually meant simply "three-dimensional," regardless of hardness, or solidity in the modern sense of admitting no empty space. Thus, an armillary sphere, which is nothing but rings and thin air, was in the Middle Ages a "solid sphere." It is hard to say exactly when the word "solid" and its cognates took on its later meaning, but this had happened before the sixteenth century in some quarters. Whatever the answer, the question of the solidity of the spheres was under discussion well before Tycho's discovery. The latter half of the sixteenth century saw a number of writers who were prepared to consider the planets to be moving either through a liquid, as fishes through water, or even through air. Several reached this conclusion on philosophical or common-sense grounds. Christoph Rothmann used astronomical evidence for his views. Having studied the comet of 1585, he described it in a treatise of 1585–1586, and there he substituted a fluid medium for the solid spheres. The treatise was not published until 1619, but Tycho acknowledged his awareness of Rothmann's views in a letter to him dated 1587. Even before Rothmann, astronomical reasons for rejecting hard celestial spheres had been proposed, for instance in a 1573 treatise by Johann Hardeg, rector of Wittenberg University, on the recent nova (but interpreted as a comet). Tycho was revising his own attitude to the spheres at much the same time as Rothmann, and we may give him the benefit of the doubt when he implied, in his letter to Rothmann, that he had come to the same conclusion independently. Tycho threw in the less than honest remark, however, that he had held these views for some years. He had not.

Putting aside the niceties of the question of priority, there is something of far greater importance here. As long as the planets were believed to be fixed on hard, quasi-material spheres, their movements were simply a consequence of the movements of those spheres. A planetary physics seemed to most people to demand nothing more. As soon as the spheres had disappeared, or at least have lost their rigidity, a new problem had to be faced. What keeps the planets in the motions we observe them to have? The new challenge was to explain the dynamics of the solar system, in the sense of a theory of planetary forces. In this sense, Tycho helped to set the scene for astronomy in the seventeenth century.

Tycho's conclusion might also have seemed destined to clear the way for Copernicanism, but this was not his intention. As early as 1574, he had been lecturing on the mathematical absurdity of Ptolemy's equant and the physical absurdity of Copernicus' moving Earth. Even before this time, Erasmus Reinhold and Gemma Frisius had drawn attention to the ease with which Earth in the Copernican system could be made fixed, allowing all else to turn around it while preserving the geometrical relationships of the system. By 1578 Tycho had been led to the idea that the inferior planets move round the Sun; and by 1584, that the superior planets move likewise. The main objection was that, since Mars's orbit seems to cross that of the

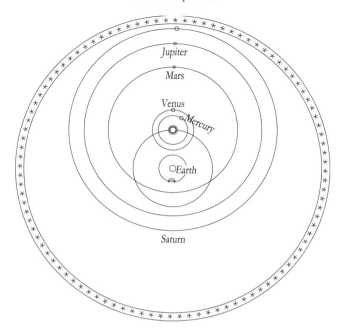

Jupiter

Mars

Venus

Mercury

○*Earth*

Saturn

134. The best-known variant of the Tychonic system, redrawn from Tycho's book on the comet of 1577 (published 1588).

Sun, the intersection of the spheres would require matter to interpenetrate matter. He spent much time in the period 1582–1584 in trying to decide the question by the measurement of the parallax of Mars, with conflicting results. The message of the nova and comet, however, albeit a message that was slow to strike him forcibly, was that the intersection is not a serious problem, since the spheres are not solid.

The Tychonic system was first published in a chapter hastily added to his work of 1588. (Figure 134 is redrawn from Tycho's illustration.) In the eyes of many of his contemporaries, this was his greatest achievement: man was once more at the center of the world. As new evidence accumulated throughout this and the following century that Ptolemy's system was unacceptable, many took refuge in Tycho's, or in related schemes, which were psychologically appealing. There were variant schemes, for example, by Nicholas Baer (Ursus), and by Tycho's most loyal and able assistant Longomontanus. In the following century some Italian astronomers even turned back to Martianus Capella (taking Mercury and Venus to orbit the Sun, with the Sun and the other planets orbiting the Earth, as explained in chapter 10). In 1651, the Jesuit Giambattista Riccioli published a variant of this, in which Mars too went round the Sun. Once the ease of geometrical transformation had been learned, finding such alternatives became an easy intellectual pastime.

Since the stars, even to Tycho's fine instruments, revealed no parallax, there was in principle little to choose between alternatives except on grounds of simplicity—for which there were in any case different criteria.

Tycho regarded the stars as lying in a shell, centered on the Earth, but not all at quite the same distance. They were taken to lie only a little beyond Saturn. Like so many of his predecessors, Tycho did not believe that God would have created and wasted empty space. For all his innovations, he was a traditionalist in many other ways. His physical description of the universe owed much to Aristotle—he got rid of the sphere of fire, but merely to extend the sphere of air—and even owed a little to traditional astrology. He was convinced of the plausibility of astro-meteorology, and recorded the weather daily for fifteen years to prove his point. His study of the horoscopes of the famous led him nowhere, unless it was to ask whether astronomy itself might not be the weak link in the chain of argument.

In astronomy he made some unfortunate decisions, for instance, in accepting an ancient value of 3 minutes for the solar parallax—more than twenty times too great, and presumably taken from Ptolemy. This in turn introduced errors into the obliquity of the ecliptic, and detracted from what even then remained the finest theory of solar motion yet devised. From 1578 to the mid-1590s he recorded noon (meridian) altitudes of the Sun for approximately one day in three, with different instruments. This habit of his was one he never dropped. His night-time observing grew ever more intensive, and multiple sightings with multiple instruments became the order of the day. With lunar eclipses, for instance, he worked with three separate teams of observers. He would gladly spend a decade on a single task, as when he prepared his star catalog. This, the finest ever prepared with non-telescopic instruments, was of course largely the work of assistants, both in observation and calculation, and it has many blemishes in the second respect, but its conception was sound enough. By 1588, Tycho had the position of his base star α Arietis to within 15″ of its true longitude. He knew twenty additional reference stars almost as accurately, and the greater part of the catalog was calculated by reference to them.

TYCHO AND LUNAR THEORY

If the accuracy of his observational work is Tycho's most easily remembered achievement, his most lasting single contribution to theoretical astronomy was a certain aspect of his lunar theory. This was left until a late stage in his overall publishing project, which was ostensibly devoted to the new star, but it had repercussions on most of astronomy as then practiced. He made lunar observations systematically from 1581 onward. He first suspected that something was wrong with traditional theories when he tried to estimate the times of eclipses. He would observe a couple of days before, and so estimate the precise time. In fact he found that in 1590 he missed the first hour of the eclipse, and in 1594 was again late in his estimate: the Moon seemed to have accelerated as it went into opposition. It must, he said, be slowing down on average elsewhere. But where? He settled on the octants, mid-way between syzygies and quadratures, and observed the Moon carefully through these points.

In this way he discovered what is now known as "variation," the first wholly new astronomical inequality to have been discovered since the time of Ptolemy. With this discovery, Tycho made possible a drastic reduction in the residual errors in lunar longitude.

In 1595 he was led to look more deeply into the Moon's motion, in latitude as well as longitude. Before long, Tycho had discovered a slow change in the inclination of the Moon's orbit. He was quick to realize that the old round figure of 5° holds only for syzygy, and that in quadrature 5;15° is more accurate. This led him to a model in which the pole of the Moon's orbit moves round a small circle twice every (synodic) month, and before long he saw that there were similarities with Copernicus' model for the ecliptic, and that the nodes of the Moon's orbit must oscillate around their mean positions. This produced another "correction" in his theory of latitudes. Tycho was no great mathematician, but here we see very clearly how his powerful intuitive feeling for such matters could compensate.

Before 1598, Tycho added yet another correction to the Moon's motion, when he introduced the year, a solar quantity of course, into the reckoning. The maximum size of this term is only 11 minutes of arc, and yet he somehow found an accurate figure for it. (He was less fortunate with the timing of its maxima.)

All of these new effects he wove into the existing Copernican lunar model, and he was in the middle of printing the result, as the capstone to his monumental book, when—greatly helped by Longomontanus—he decided to correct one of that model's shortcomings. Although much better than Ptolemy's, in regard to the variation in the Moon's distance (and hence apparent size), it was still far from satisfactory. Tycho had learned the Copernican lesson of double epicycles. Where Copernicus had one for the first inequality, he now took two. Choosing the parameters of size and velocity carefully, he was able to get a marginally improved, but still far from perfect, model of lunar distance. Its imperfections are in a sense less important than the fact that he saw the desirability of correctly representing the lunar path in space.

The second lunar inequality (known today as "evection") had been explained by Ptolemy in terms of the small circle at the center of the lunar model. Copernicus had taken the mechanism away from the center, but Tycho had two epicycles already, so he moved it back there. The details of the resulting model were more complicated than in Ptolemy's case, and in fact Tycho never managed to fit all of his discoveries into a satisfying lunar model. (He omitted most of the annual equation, for instance.) What he did for lunar theory was nevertheless of immense importance, and it was seen to be so as soon as Kepler succeeded in integrating it into his own, more penetrating, theory.

Tycho worked on a new theory for the planets at much the same time as he struggled with lunar theory, but it was left to Longomontanus to carry out this part of his program. Longomontanus, who had been at Hven from

1589 to 1597, and who had joined Tycho in Prague in 1600 at the end of his patron's life, finally published an *Astronomica Danica* (*Danish Astronomy*) in 1622. Replete with double epicycles, it purveyed a system that combined the Tychonic and Copernican traditions.

That Tycho did not complete the task himself had much to do with external circumstances. After the death of his patron the king, and the ending of a regency in which Tycho's brothers played a part, a young new king, Christian IV, ascended the throne. Tycho quarreled with all and sundry—including his tenants, a pupil who was engaged to his own daughter, and even the king himself. He soon lost his favored status, and had to look elsewhere for patronage. He finally left Hven in 1597—with an enormous baggage train—first for Hamburg. In Rantzov's castle there, but on his own press, Tycho in 1598 finally managed to publish his *Astronomiae instauratae mechanica* (*Instruments of the Restored Astronomy*), an illustrated description of his observatory and instruments. It proved to be extremely influential, and was reprinted again in 1602, in Nuremberg. In time, this book propagated Tycho's standards of excellence in instrumentation right across Europe. Indeed, it took the Dane's designs as far afield as China, as we saw in chapter 5, and Brazil, where the first European observatory in the New World was built. It was established there in the 1640s by Georg Markgraf, a German astronomer in the service of Prince Maurits van Nassau, head of the Dutch colony in Brazil.

In a carefully calculated gesture, Tycho dedicated his book of 1598 to Emperor Rudolf II in Prague. After considering settling in various other parts of Europe, he eventually accepted Rudolf's patronage in June 1599. The move to Bohemia meant that he had to ship his more easily transported instruments, reinstall them, and reorganize his group of assistants. Ensconced in the castle of Benatky, where he began anew to set up an observatory, he spent much of his time putting his earlier observations in order. He had only a year's use of his instruments in Prague, for he died there without much warning in October 1601. Benatky survives, although the instruments do not. As for Uraniborg, virtually all that remains of it are its foundations, duly excavated by archaeologists, but best seen today from the air.

In Prague, Tycho took into his service Kepler, his most famous assistant of all. Kepler was assigned to work on the planet Mars, and it was he who finally saw Tycho's great work into print, under the title *Astronomiae instauratae progymnasmata* (*First Exercises in a Restored Astronomy*). This was in 1602. Another book Kepler was commissioned to write was a defense of Tycho against Baer. Baer died in 1600, and Kepler was only too happy to leave his defense unpublished. It finally appeared in print in the nineteenth century. On his deathbed, Tycho asked Kepler to explain his planetary theory in three ways, Ptolemaic, Copernican, and Tychonic, and this Kepler did, to some extent, in his *Astronomia nova*—setting an example of sorts to those many later textbook writers afraid to make a definite choice. Emperor Rudolf ii was anxious that Kepler complete Tycho's astronomical tables, but the Rudolphine Tables did not appear until 1627, fifteen years after Rudolf's death. In an elaborate frontispiece (which, indeed, is the frontispiece for this book, as well), Kepler paid his respects to history, although Tycho's heirs were unhappy with an early draft of it. As for the tables, Tycho had meant them to be computed following his own principles. While the observational basis for the final version was indeed essentially Tycho's, the underlying theory was not. It was Kepler's own.

HYPOTHESIS OR TRUTH, ASTRONOMY OR PHYSICS?

The controversy between Tycho and Baer well illustrates a well-worn philosophical theme that has owed more to astronomy than to any other science. Do scientific theories carry with them any implications for the reality of the things they describe? The theme had already been given wide publicity through the scurrilous Osiander preface to Copernicus. Baer was a skeptic in much the same mold as Osiander. Copernicus, Tycho, and Kepler, inclined to the view that there was a true, and in some sense real, system waiting to be found, and that a sound astronomical theory did more than merely permit the computation of phenomena in advance. Baer knew full well that accurate prediction does not guarantee the validity of a theory, since true conclusions may follow from false premises—a simple logical point that had often been made in the Middle Ages. This nettled the astronomers, but to understand Tycho's irritation we must bear in mind the colorful stories of how his planetary system was plagiarised by Baer, who had supposedly become apprised of its contents by underhand means.

Nicholas Baer was by origin a farm boy from Ditmarsch, in the southern part of the Jutland Peninsula. For that reason alone, the aristocratic Tycho thought he should know his place. The story went that, while Baer was on a visit to Uraniborg in 1584, as he was sleeping, his pockets were searched by a certain Andreas, a student of Tycho's. Incriminating papers were supposedly found there. Evidence later adduced against him was that, whereas in Tycho's system Mars's orbit intersected the Sun's, Mars's orbit in Baer's system enclosed the Sun's, as in an old diagram by Tycho that had been wrongly drawn.

In the spring of 1586, Baer traveled to the court in Kassel of Landgrave William iv, and there took the opportunity of describing his new planetary system. So impressed was the landgrave that he commissioned Joost Bürgi, one of the finest instrument makers of his day, to construct a mechanical model of it, a planetarium. At much the same time, the landgrave wrote to Tycho on the matter of the comet of 1585, and received a friendly reply, carried by one of Tycho's assistants. This was the beginning of a certain degree of scientific cooperation between the two centers, not all of it entirely voluntary. (We have seen how Wittich helped to transfer many of Tycho's designs for instruments to Kassel, where versions were built.) What was certainly never Tycho's intention was that a model of his planetary system should be built under a rival's name.

Whatever the truth of the supposed theft, Baer's system differed from Tycho's in one important respect: he gave the Earth a daily rotation on its own axis, half-loosening it from its old bonds, so to speak. Baer's is sometimes called the "semi-Tychonic system," although Baer would certainly have disapproved, and the name could just as well be applied to alternatives mentioned earlier. It has to be said that if we disregard this distinction, then there were at least half a dozen writers who claimed to have invented the system independently. Tycho was not pleased by Kepler's remark that it was an obvious step from Copernicus. We have already mentioned a pamphlet from around 1599, seemingly by Ursus, which made the same point.

The loss of the papers by Baer is said to have begun a mental disturbance that led to his loss of imperial patronage. But what was the nature of the stolen ideas? There were complaints of the appropriation of constructions, inventions, formulas, and tables, but not of a fully articulated world system. Kepler, when later given the task of defending Tycho, saw what was at stake more deeply than either man, namely, an integrated viewpoint, a complete system of hypotheses answerable to observations.

Before leaving this controversy it is as well to remind ourselves of the dangers of giving it too modern a color. Sixteenth-century writers did not—as is now so often maintained—normally insist that astronomical theories were "mere fictions." Some were inclined to follow the line taken by Philipp Melanchthon and various astronomers from the Lutheran University of Wittenberg and to accept certain of Copernicus' mathematical techniques without embracing his theory as a whole. Like so many others, they were not truly Copernicans, and yet this did not make them "fictionalists" in a strong modern sense. Most of those who considered the philosophical question seriously firmly believed that astronomy could not pretend to purvey real physical knowledge. It is true that, after Copernicus, there was a growing sense of freedom in regard to discovering the true cosmological system. Indeed, the old certainties were being undermined even before the publication of the *De revolutionibus*. There were, for instance, men like Girolamo Fracastoro, the author of *Homocentrica* in 1538, who aimed to reinstate the homocentrics of Eudoxus. Once again, however, the freedom they felt they had does not imply that they regarded the right choice as anything but a true choice.

It must by now be clear enough that, for many reasons, the road from Copernicus to Kepler was far from smooth. Parts of the road were paved with the pedantry that so often accompanies academic dissent. Baer accused Tycho, Christoph Rothmann, and the Alsatian writer Helisaeus Roeslin, that they had either failed to read Copernicus, or, if they had read him, had failed to understand him. Kepler made a similar criticism of Baer. Vanity played its part. The last decade or so of the sixteenth century saw a cluster of books which a first sight seem to be concerned as much with priority in the formulation of a geo-heliocentric world system as with its truth. There were differences in the method of proving priority, ranging from the legalistic approach of Tycho to the satirical use of invective by Baer, who was no mean rhetorician. Around 1599, for example, Baer wrote a little tract with a long title announcing that it would prove that the system of Apollonius was explicitly described in the works of Martianus Capella and Copernicus. Without mentioning Tycho here, the implication was clear: for Tycho to accuse Baer of plagiarism, when his hypothesis had an ancient pedigree, was absurd. More interesting than all such strategies of one-upmanship, however, are the different opinions as to how astronomical truth should be judged. Tycho, of course, put great weight on his own observations, and here no one could compete. Most, by this time, had become conscious of the need to preserve consistency with physical theory—although not all, of course, agreed on what was good physics. Scriptural authority was important to most. It mattered to Tycho, but it played a larger part in the widely read *De opere Dei creationis* (*On God's Work of Creation*, 1597) by Helisaeus Roeslin. Some placed great weight on the textual criticism of ancient sources. This was not always a mark of obstructive conservatism, for Kepler was one who did so. Had he done nothing more, however, we should not now be inclined to pay much attention to him.

Kepler was, in a sense, on weaker philosophical ground than his opponents, arguing as he did for the final truth of certain astronomical propositions—we may perhaps call them points of view, but he would not have thanked us for calling them hypotheses. Unlike most of those who, throughout history, have been dogmatic in asserting their particular claims, he did not rest his case only on observation and established theory. While Kepler was strong in rhetoric, the arguments he gave for preferring Copernicus to Ptolemy depended heavily on aesthetic considerations of simplicity, harmony, elegance, and the like. He had another very potent argument in regard to his elliptical paths. Unlike the systems with multiple circles, his ellipses showed *the final paths in space*, that is, what would be seen by an observer viewing the system over a long period from far out in space. The ellipses had a sort of reality that bundles of circles lacked.

That, at least, seems to have been the idea that Kepler had in mind when he said—and he said it often—that his system was the first to avoid hypotheses. (Others before him, such as Toscanelli and Apian, had charted the paths of comets, but there was then no involvement in any deep theory of cometary orbits, so that the charted path was the path, pure and simple.) Baer had said, in effect, that if two hypotheses do the job equally well, it did not matter which one took. Kepler insisted that the two may be equivalent, but only in certain limited respects, and not in all; and that physical respects should not be brushed aside. He was not the first to insist on this point, but he saw more clearly than anyone before him something of great importance to the actual growth of scientific theory: he saw how urgent was the need to integrate mathematical astronomy into physics and natural philosophy. He sought for the causes of planetary motion, and not only for the geometry of that motion. It could be argued that this traditional distinction is an erroneous one, but the historical fact remains that established physical modes of thinking were capable of supplementing very potently the old geometrical modes.

Kepler is sometimes treated as the first modern astronomer on the strength of his perception of the need to combine physics with astronomy. Others before him—not least of whom were Aristotle and Ptolemy—plainly had similar ambitions. Kepler, however, lived at a time when conventional (Aristotelian) physics was under attack, for instance, from the writers of several new treatises on comets. This went along with a theological questioning of authority, especially in Protestant countries. Michael Mästlin, for instance, Kepler's teacher, had failed to find any parallax for the comet of 1580 and so openly attacked the old physics. Aristotle was well ensconced in the schools of the time, and remained so for well over a century in some places, but by slow degrees astronomy was removing what some considered the most basic foundations of his cosmology. Kepler's role in all this was crucial, and it is ironical that he was largely inspired by types of reasoning that belong as much to astrology as to astronomy, as we now perceive it.

KEPLER AND PLANETARY ASPECTS

Johannes Kepler was born in Weil der Stadt, near Stuttgart, in 1571, the grandson of a mayor of the town, but the son of a "criminally inclined and quarrelsome" mercenary soldier who eventually abandoned his "garrulous and bad-tempered" wife. The descriptions were Kepler's own, written when he was comparing their characters with their horoscopes. Between 1617 and 1620, Kepler was destined to defend his mother when she was tried for witchcraft.

He eventually attended the University of Tübingen, where he came under the influence of the Copernican adept, Michael Mästlin. After taking his master's degree, Kepler embarked on a study of theology, but then Georg Stadius died, and Tübingen was asked to recommend a replacement for him as mathematician at the Lutheran school in Graz. Kepler was recommended, and in 1594 he left Tübingen to take up his new post. It was just over a year later that he happened by chance upon what he thought was the secret of the structure of the universe.

We recall the curious astrological doctrine put forward by Rheticus in the *Narratio prima*: following an established tradition in astrology, he had supposed that a planet's strength is increased at its apogee and diminished at its perigee (see p. 000 above). Kepler was adept at astrology. Not only did he compile a collection of horoscopes for members of his family in 1596, by which they are compared scientifically in character, but before this—in his capacity as district mathematician—he had published a calendar and prognostication for 1595. The peasant uprisings and invasions by the Turks

135. Johannes Kepler (1571–1630), from an engraving by Jakob von Heyden, based on a painting by an unknown artist and given by the astronomer to a friend in 1620. Most later portraits derive from the one or the other.

that he predicted were perhaps to be expected, but the excessively cold winter which he predicted was not. All of those predictions came to pass. His star was in the ascendant. With a gap of only three years, he published annual prognostications until 1606. He took up the work again between 1618 and 1624, this time more cynically, to compensate for nonpayment of salary. Kepler made some often-quoted early remarks to the effect that astrologers only get it right by luck, or that astrology is the foolish daughter of astronomy, but they do not really tell us more about his early astrological belief than that he had been disappointed in his experience.

We have 800 horoscopes from Kepler's pen that could be put down to a need for ready cash, but this does not explain why he drew up so many horoscopes for himself. He thought long and deeply about the question, which he took seriously for most of his life, perhaps for all. He tells us that it was after he met Tycho in Prague that he abandoned *a part* of astrology, but later he said quite explicitly that there was much good in the subject. In 1598, he wrote to Mästlin that he was "a Lutheran astrologer, throwing out the chaff and keeping the grain." He made the standard protest against a forbidden celestial *magic*, not so much against astrology but against the use of spiritual or demonic magic. In a little work on the "more certain foundations of astrology" he lists three sorts of reasoning underlying astrological prediction: the theories of (1) physical causes, (2) metaphysical or psychological causes, and (3) signs. The first two are valid, the third is not. This is the man of science talking, the physicist-cum-psychologist, the man who no longer believes in the sharp Aristotelian distinction between the regions above and below the sphere of the Moon.

He toyed with the idea that light might be the physical cause he was seeking. Until the telescope had been available for a couple of years, it seems that he shared a common enough belief that the planets shine by their own light; and whereas older astrologers had associated the planets with the ancient deities, and had deduced their properties accordingly, Kepler thought that the colors of the lights they sent to Earth were what determined their astrological characters. Here is an example of Kepler the creative astrologer, who added new doctrine as Rheticus had done. The irony here is that the two leading Copernicans of the century, Rheticus and Kepler, both added to existing astrological doctrine.

One theme runs right through his life, and that is the idea of "imprints" on the human soul, received from celestial configurations when man begins a life independent of his mother. In other words, he approved of the doctrine of aspects—more or less synonymous with configurations—and regarded the question as one for experience to decide. He showed his allegiance to planetary aspects by his example, again and again, and they are in at the very birth of *Mysterium cosmographicum* (*Cosmographic Mystery*), his first important book.

His great experience at Graz was one he had while struggling with the reasons why the numbers, sizes, and motions of the orbs were what they were. In July 1595 he was using a standard diagram to explain to his pupils a commonplace of astrological theory concerning the great conjunctions of Saturn and Jupiter. What no one else had previously noticed was that the lines joining the points of the zodiac circle where the conjunctions occur enclose a second circle, and that the ratio of the diameters of the two circles was very close to that between the orbits of Saturn and Jupiter, now known from Copernican principles (fig. 136).

Kepler could not find similar relationships in the orbits of the other planets until he started looking for analogies in three dimensions, but when he did so he found an extraordinary geometrical scheme which seemed almost perfect. In this, each orbit was circumscribed by one of the five regular polyhedrons, and inscribed in another. He was familiar with Euclid's proof that there can be five—and only five—regular polyhedra, that is, polyhedra with all faces congruent: the cube, the tetrahedron, the octahedron, the dodecahedron, and the icosahedron. These were all he needed, although he did not use them in this order. His finished system fitted the Copernican distances to within about one part in twenty, except in the case of Jupiter (figs. 137 and 138). In an attempt to improve the fit, he enlisted the help of Mästlin for much of the heavy work of calculating distances, eccentricities, and lines of apsides. He tried taking distances not from the center of the Earth's orbit—namely the "mean sun" of Copernican theory—but from the true Sun. This emphasis on the importance of the true Sun would eventually pay great dividends, for as we know, thanks to his later work, the Sun itself is at the heart of our planetary system. In this particular case, taking distances from it did not produce the corrections he had hoped for, but he nevertheless remained convinced that he had discovered the architecture of the universe. As he had written to Mästlin, in the first flush of excitement at his discovery: "I wanted to become a theologian, and for long I was uneasy, but now, see how through my efforts

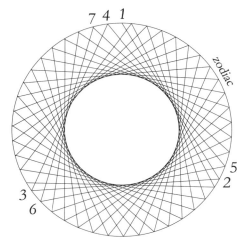

136. The pattern derivable from the conjunctions of Saturn and Jupiter, as hit upon by Kepler and described in his *Mysterium cosmographicum* (*Cosmographic Mystery*). A few numbers are added to this idealized figure to indicate the sequence of places in the zodiac where these "great conjunctions" occur. The ratio of the outer circle to the circle touching the inner envelope of lines in the above figure is 1.93:1. The ratio of the mean distances of Saturn and Jupiter from the Sun derived from Copernican theory was about 1.74:1 (Kepler). A modern value would be 1.83:1. Kepler was conscious of the great astrological importance of Saturn-Jupiter conjunctions, which may be briefly explained. They occur, roughly speaking, every twenty years. Those separated by sixty years occur typically seven or eight degrees apart. If a great conjunction numbered 0 is at the head of Aries, then conjunctions 3, 6, 9 and 12 will all probably lie in Aries; and after 260 years, number 13 will probably have moved into Taurus. If so, the great conjunctions will have moved into a new "triplicity" or "trigon"—the name for a trio of signs of the zodiac separated by 60° (in this case, Taurus, Virgo, and Capricorn). There they might recur for another 240 years, before spilling over into the next. If a conjunction such as the one here numbered 13 was the first to occur in the new triplicity, it was called a "major conjunction." The conjunction starting a new cycle, after about 960 years, was called a "maximum conjunction." Great, major, and maximum conjunctions were all considered to have deep historical significance, especially in connection with the rise and fall of religious sects. A key text on the subject was by Albumasar.

God is being celebrated in astronomy." The closeness of fit of his scheme with the measured distances has even now the power to astonish, as long as we prepared to overlook the five percent error and the existence of planets beyond Saturn.

It is tempting to represent all of this as a scientific triumph with an astrological starting-point, but in Kepler's own psyche the influence must have worked in both directions. Having found himself on the track of the correct geometrical scheme of the heavens, as he then thought, he must have been persuaded that the doctrine of aspects was justified by the astronomy that had provided the distances long before. It is not surprising, therefore, that he scrutinized and eventually added to the standard astrological doctrine of aspects. These aspects, he thought, act on the soul, but only when it has a harmonious instinct. The soul, by means of this instinct, reacts to certain harmonious proportions, subdivisions of the zodiac. It

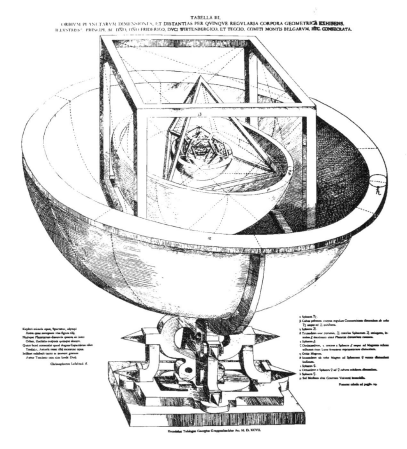

137. The virtuoso plate—not the surrounding print—was cut in 1597 by Georg Gruppen-
bach for Kepler's *Mysterium cosmographicum*. It depicts the astronomer's three-dimensional
scheme for the Copernican planetary system, based on inscribed and circumscribed "Platonic" solids (see also fig. 138). Working inwards, Saturn's sphere circumscribes a cube, which
circumscribes Jupiter's sphere, which circumscribes a tetrahedron, which circumscribes the
sphere of Mars, which circumscribes a dodecahedron, which circumscribes the "great sphere"
of the Earth's orbit. Within this is an icosahedron circumscribing the sphere of Venus, which
circumscribes an octahedron, this circumscribing the sphere of Mercury. At the center of it
all is the Sun.

is not necessary to consider how he added to the collection of traditional
aspects, angular separations of the planets, such as quartile (90°, hostile)
and trine (120°, friendly). The important point is that this was not an aberration of his youth. He discussed these matters in the *Harmonice Mundi*
(*Harmonies of the World*), and there he reminded his readers of another
theme that was for long in his mind, the theory of musical consonances.
Until about 1610, he thought he could link this with his astrological harmonies, but he gradually separated the two. Another link he tried to forge
was between astrology and alchemy, which might have endeared him to
Tycho, but in the end led nowhere.

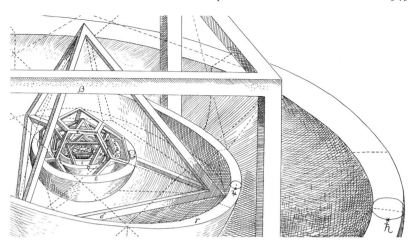

138. A detail of the previous figure. Note the relation between the thickness of the each sphere and the corresponding epicycle.

It is worth noting here that quoting from the *Harmonice Mundi* as evidence for Kepler's intellectual progress is a hazardous procedure, since the book was twenty years in gestating. It was Kepler's favorite of all his writings. Prompted by a controversy with Robert Fludd—an English mystic of a very different sort—Kepler started writing it around 1599. He worked at it intermittently, at the same time as he was developing his astronomical theories and proceeding with the Rudolphine Tables. It was finally published only in 1619.

The strange idea of nested regular solids, which had provided the inspiration for Kepler's *Mysterium cosmographicum* of 1596, reappeared in the *Harmonice*. He tried to remain faithful to that idea all his life, and it has certainly been of great influence down the centuries. In due course it was replaced by other, simpler laws that seemed to fit the planetary distances. One cannot say that they were plucked out of thin air, although they often give that impression. They are framed with some principle of harmony in mind, and with half an eye on the "observed" distances, but usually overlooking the fact that those distances were originally derived from theories incompatible with the supposed harmonies. The Titius-Bode law of planetary distances is an example of such a law, which is scarcely a better fit to observation than Kepler's, but it is still often quoted as a mystery that begs to be explained. Presented by J. E. Bode in 1772 as having been found by J. D. Titius, this gives the orbits of the planets from Venus outward as $0.4 + 0.3 \times 2^n$ AUs (0.7, 1.0, 1.6, and so on, with the minor planets 2.8 and eventually the newly discovered Uranus 19.6). For Mercury (0.4), we simply drop the second term in the expression—not the most elegant of mathematical procedures. Later, in a dissertation of 1801, the "dialectical" philosopher G. W. F. Hegel criticized the arguments that were currently being used to suggest that the gap between Mars and Jupiter must be occupied by another planet. He gave another formula involving simple numbers that fitted roughly the distances as far as Uranus and that did not fill the gap in the same way. What Hegel did not realize was that he had backed a loser and had done so after it had lost, for on the first day of 1801 Giuseppe Piazzi had located an asteroid (or "minor planet") in the gap. Hegel's point was not very clear, but it was not as foolish as it is often presented. He did not "try to prove that there can only be seven planets," as is often said. He said, in effect, "I can find an alternative law which casts doubt on your way of adding more planets." The point he missed is that we need more than a few casual fits with observation before accepting a law in isolation from other parts of a science, whether that law is the Titius-Bode law or his own alternative.

There were to be many similarly mystical and "Platonic" studies and interpretations of Kepler in the nineteenth century, especially in Tübingen, by scholars looking for quick returns, laws without affinities with other parts of science. Kepler would have disabused them of the possibility of quick returns. He had a clear physical vision of the astrological workings of the planets: he seems to have thought of them as somehow disturbing a finely balanced "equilibrium" situation—he suggested, for example, that with a spring atmosphere of saturated water vapors, the right planetary aspect may be strong enough to trigger off showers of rain. He thought that such things would be confirmed by experience, and made very many observations of the weather to do so, for instance, to confirm the idea that a conjunction of Saturn and the Sun causes cold weather. He repeatedly claimed that the efficacy of his new set of astrological aspects was supported by observation of the weather.

It is now usual to describe such matters as instances of Pythagoreanism, a searching for harmonies of a sort beloved in the Renaissance. There is no doubt that he was strongly influenced by his reading of Plato's *Timaeus*, which he viewed as almost a Pythagorean interpretation of the Bible's book of Genesis, but Kepler was doing something far more important: he was matching his geometry against measured quantities. If his harmonies do not quite please us—perhaps we are dissatisfied with that maximum error of one part in twenty—that does not affect the fact that they were rooted in the observed and measured world. This point is easily lost if we allow ourselves to be shocked by his strange remarks about the Earth's soul, but here too he is thoroughly empirical in his outlook. The argument goes like this: an aspect (the angle separating two planets, viewed from the Earth) is a purely geometrical relationship, even when light travels inward along the directions that define it, say, two rays 120° apart. The aspect affects human beings because they may perceive it, or their soul may somehow receive it. How does it affect the weather? The answer is that the Earth has a soul too, a soul that stretches up to the Moon. Kepler, in this way, gives a new look to the sublunary realm. Of course, for him, good Copernican that he was, the more important World Soul is situated in the Sun. In the course of criticizing the notion of a world soul as put forth by the Oxford medical Rosicrucian, Robert Fludd, Kepler set out his own vision of the universe as a living body, with the Sun its heart and the Earth its liver or spleen. As for the harmonies and aspects we perceive, they are appropriate to our earthly position; those relative to the Sun are what decide the velocities of the planets and their orbits.

Regarded in this way, Kepler might no longer seem to be quite the "first modern astronomer" of traditional history, but a word of warning is needed: in the end, it was an abstract doctrine marrying geometry and physics that he accepted, and not the full panoply of traditional astrology, about which he was highly skeptical. As we have already pointed out, he had a strong aversion to astrology considered as a theory of signs, and helped to diminish its popularity in scientific circles. The fact that a great conjunction occurred in what was traditionally "the fiery trigon" (triplicity) meant for most astrologers that it would be accompanied by fire of some sort—for instance, the fires of war or of drought. Kepler saw that the characters of the celestial indicators had been chosen on the grounds of a series of feeble and random human analogies, and these in the end he entirely rejected. He spoke against the zodiacal signs and the mundane houses (the houses into which the zodiac is divided in horoscopes), considered to be the residences of the planets, and yet, as we have seen, he somehow hung on to enough doctrine to practice as an astrologer. In 1608 he drew up a horoscope for an anonymous subject, who later turned out to be the great military leader Albrecht von Wallenstein. A story that was used to underscore his deep insight is that he correctly guessed that Wallenstein was the subject. He cast the horoscope according to the method of computation that was widely ascribed to Regiomontanus. He later "purified" his astrological principles, but the fact remains, had he not been drawn to astrology he would very probably have failed to produce his planetary astronomy in the form we have it.

OPENING SHOTS IN THE BATTLE WITH MARS

Kepler's *Mysterium*, printed in 1597, did much in the following decade to aid the Copernican cause. Galileo acknowledged receipt of a copy, saying that, at the time of writing, he had read only the preface. He admitted, however, that he accepted the Copernican worldview, finding it compatible with natural phenomena that could not be reconciled with traditional alternatives; but that he feared the ridicule that would follow if he published his views. Kepler's reply urged him to proclaim his belief openly, but to little avail. Tycho expressed admiration for Kepler's book, but this he could do with impunity, for although Copernican in character, it was written in a very different style from his own empirical astronomy. Tycho valued many of the calculations, for instance, those of the eccentricities, and realized that Kepler's was a talent worth recruiting. In his search for harmonies, however, Kepler went further than merely wanting to know the harmonies of mathematical scale in the universe. His was a search for *causes*, and he decided that the central position of the Sun must be the key to understanding the causes of the planetary motions. He knew that the planetary periods lengthened with distance from the Sun. He tried to find the relationship between the two quantities, and in the *Mysterium* he settled on a law which made the period (T) proportional to the square of the radius of the orbit (a). The correct power is ³⁄₂, as he was later to discover.

On 28 September 1598, without clear warning, the Catholic authorities—a commission of the Counter-Reformation—ordered all Lutheran teachers to leave Graz at once. Kepler was treated more generously than most, and was allowed to reenter the city, but he needed employment elsewhere, and in February 1600 he arrived in Prague, having been invited to work for Tycho Brahe. Overcoming initial irritation at Tycho's paternalism in astronomical matters, Kepler settled down in Tycho's employ, and was assigned to the theory of Mars. He could not have guessed that his "war on Mars"—as he later called it—would occupy the next six years of his life.

Tycho's assistant Longomontanus had already been engaged on the Mars problem, constrained of course by the geocentric principle that ruled in the Tycho camp. Using a double epicycle, he had trimmed the theory to account very well indeed for a series of oppositions to the Sun, but elsewhere in Mars's orbit the theory was a failure. If we ignore Mercury, with its complicated Ptolemaic theory and the difficulty of compiling a good set of observational data, Mars was the planet with the most eccentric orbit, and therefore the most difficult to deal with in terms of circles. (If we anticipate Kepler's discovery, and think in terms of ellipses, Mercury's eccentricity is more than double that of Mars, but the latter is nearly fourteen times that of Venus.) Kepler, unlike Longomontanus, was helped by his Copernicanism, but also by his discovery that the position of Mars was better referred to the *true* Sun than to the center of the Earth's orbit, Copernicus' choice. He was very gratified to discover that the line of nodes of the planetary orbit (the line of intersection of the orbital planes of the Earth and Mars) passed through the true Sun. This intimation of the fact that the orbital planes of all the planets pass through the true Sun, and that indeed they do not fluctuate significantly, was of great importance in the unfolding of a comprehensive theory of our *solar* system, one in which it was reasonable to seek for solar forces. Among other things, it was the most important step ever taken on the road to a theory of planetary latitude, which for a geocentric theory such as Ptolemy's presented an extremely complex problem. We recall Kepler's remark that Copernicus had not appreciated his own riches: Copernicus had missed by a whisker the advantage inherent in his system, vis-à-vis latitudes, since he had worked from the mean sun, not the true Sun. How pleasant it would be, thought Kepler, if only he could produce a single planetary model to cover motions in latitude as well as longitude.

Already in the *Mysterium*, Kepler had played with the idea that an equant might be a useful way of reconciling the Earth's observed motion with his slowly emerging physical ideas about how the planet's distance from the Sun might affect the motion. (We recall that an equant is simply a point around which a motion, such as that of the center of an epicycle, is uniform. It is distinguished by the name when it is not one of the standard points around which motion had traditionally been taken to be uniform, namely the Earth, or the center of an eccentric deferent circle.) He now made good use of the idea of an equant for Mars, no longer one that was positioned with strict reference to the eccentricity—sometimes rather grandly called "the bisected eccentricity hypothesis"—but an equant that he could vary in position until he found the best fit with observations. The Tychonic data allowed him to obtain a rather accurate orbit for Mars, but only after a long period of difficult calculation, using repeated iterations to approach ever more closely to the observed longitudes. His equant model, which he called his "vicarious hypothesis," matched the observations far better than any before it. The longitudes of oppositions were predicted to better than two minutes of arc. The theory was still unsatisfactory for latitudes, however, and this suggested to him that his distances were wrong.

Tycho's death had left Kepler with heavy obligations—the completion of the Rudolphine Tables, for instance—but it meant that he had access to observational records that Tycho had guarded jealously, and it meant that his loyalty to the Tychonic system could no longer be morally enforced. He continued his struggle with the orbit of Mars. Getting the right distances meant adjusting the relative placement of his equant and the deferent center. He felt himself being pushed in the direction of Ptolemy's "bisected eccentricity" after all; and yet this upset the accuracy of his longitudes, which were now in error by six or eight minutes in the octants. (The octants are the directions at 45° and 135° to that of Mars's opposition to the Sun.) Where others would have let matters rest there, Kepler's confidence in Tycho's observations made him persevere. Those eight minutes of arc were the irritant that stimulated Kepler to bring about a complete revolution in planetary astronomy, as we shall see.

In 1606, Kepler published a work with a certain popular appeal, *De stella nova* (*On the New Star*), which included musings over a wide range of physics and cosmology, but it was his *Astronomia nova* (*New Astronomy*) that marked him out as a new intellectual force to be reckoned with. This work was eventually published in 1609, after he had managed to settle disagreements with Tycho's heirs about his right to use the observations in a non-Tychonic manner. In this period of his life, he was being harassed by other circumstances, public and private. His wife had wealth tied up in estates near Tübingen, and took less kindly to his moves than he. She had little regard for astronomy, no doubt with excellent reason. Obtaining his due salary was one of Kepler's constant headaches. By 1611, spurred by the

Tübingen professoriate for saying that the Calvinists should be treated as Christian brothers, he found promise of employment in Linz. By moving there he hoped to please his wife, but she died of typhus before leaving Prague. Prague was in a state of war, and chaos reigned there, leading to Rudolf II's abdication. Even in Linz, Kepler was involved in endless religious turmoil. This, which at one stage resulted in his being denied access to his own books, and the illness and death of children from his first and second marriages, weighed heavily on him.

He wrote on other subjects: in 1604 and 1611 he published two of his most important works, both on optics, regarded as a necessary part of astronomy. In neither his "Optical part of Astronomy" nor his "Dioptrics" did he have the sine law of refraction. This, "Snell's Law" as it is usually called in the English-speaking world, was first found by Thomas Harriot, no later than 1601. In 1606, Harriot sent Kepler a table of angles of refraction for many different substances, but did not share the sine formula with him. In the second of the two optical treatises, Kepler gave a thorough mathematical treatment of the formation of images by lenses, and the arrangement of two converging lenses into what we now call the "Keplerian" or "astronomical" telescope, one that produces an inverted image. By 1611, the "Dutch" or "Galilean" telescope was a couple of years or so old.

While he wrote a number of minor works, for a time Kepler's prodigious energies were diverted from astronomy. Then, in installments published in 1618 and 1621, came what was to be for many decades the most widely used treatise of advanced theoretical astronomy, his *Epitome astronomiae copernicanae* (*Summary of Copernican Astronomy*). His *Harmonice mundi*, a work we have already mentioned as having been written in the tradition of his *Mysterium cosmographicum*, finally appeared in print only in 1619. It can almost be read as an essay in philosophical relaxation, in which he indulged between long sessions of computing. The same is true of a defense of Tycho against Scipione Chiaramonti, who was trying to uphold Aristotle in the matter of the interpreting of cometary appearances. This minor exercise in Keplerian natural philosophy appeared in 1625. The *Epitome*, however, was his most polished work of astronomy. It was a spirited defense of the Copernican world, much modified by his own ideas. Its great complexity might lead us to ask why it was almost immediately placed on the Catholic Church's *Index of Prohibited Books* (fig. 139). Part of the reason was the stir being created at this time by Copernicanism in Italy, part was Kepler's revisionist anti-Aristotelian physics, and no doubt part was his having introduced analogies between the arrangement of the world and the Holy Trinity. It was fortunate that the Catholic *Index* was powerless to hold Kepler's ideas in check. As a Lutheran, while he suffered from the religious wars, at least he had a measure of intellectual freedom that Catholic astronomers might well have envied.

INDEX
AVCTORVM,
ET LIBRORVM,
QVI AB OFFICIO
S. Rom. & vniuerſalis inquiſi-
tionis cauęri ab omnibus & ſin-
gulis in vniuerſa Chriſtiana Re
publica mandantur, ſub cenſu-
ris contra legentes, vel tenen-
tes libros prohibitos in bulla,
quæ lecta eſt in cœna Do-
mini, expreſſis & ſub
alÿs pœnis in de
creto eiuſdem
ſacri officÿ
conten .
tis .

ROMAE.
EX OFFICINA
Saluiana. XV.
Menſ. Feb.
1 5 5 9.

139. A specimen title page (1559) of the index of books which the Catholic Church forbade its members to read, as they were likely to contaminate faith or to corrupt morals. There were nineteen editions of the *Index* between 1559 and 1966, when it was abolished. Copernicus' *De revolutionibus* was placed on the index in 1616, at the time of the investigation of Galileo, "until it be corrected." Kepler, whose *Epitome* naturally suffered a similar fate, said the phrase should read "until it be explained." Copernicus was removed in 1757. Galileo's *Dialogue* was put on the *Index* in 1633 and removed in 1823.

ATTEMPTS AT A PHYSICS OF PLANETARY MOTION

Accounting for the eight minutes of arc by which the Mars theory was in error at the octants led Kepler to the three laws for which he is now chiefly remembered. Since those laws did not all appear neatly formulated together in one place in his writings, and since the path by which he reached them was extraordinarily convoluted, it will perhaps help us to keep them in mind if we state them briefly here by their conventional numbers, and in our own words: (1) Every planet describes an ellipse with the Sun in one focus; (2) the areas described by radii drawn from the Sun to a planet are proportional to the times in which they are described; and (3) the squares of the periodic times are proportional to the cubes of the mean distances of the planets from the Sun. As we shall see, the second was found before the first; and in due course Newton would show from his law of gravitation that Kepler's first law is not exactly true, and that even if we ignore all other planets and masses in the system, what is at the focus of the ellipse is the center of gravity of the Sun and planet.

An important key to Kepler's success in arriving at these laws was that he began by attempting to reduce the problem to one of physics, and in particular, of magnetism. His interest in books on the magnet by Jean Taisnier (*Opusculum . . . de naturae magnetis*, 1562) and William Gilbert (*De magnete*, 1600) gave him the idea that magnetic forces emanating from the Sun might explain the planetary motions. Kepler was by no means alone in pursuing that idea, for there was a widespread belief that Gilbert had proved the motion of the Earth magnetically. In 1608, the Dutch scientist Simon Stevin published a work written some years earlier, *De Hemelloop* (*The Course of the Heavens*), in which a complex cosmology was based on a theory of cosmic magnetism deriving from the fixed stars. Kepler's work was independent of this. It changed in character and emphasis with time and was well provided with simple analogies, some of which aimed at explaining the theory to the reader, rather than exactly purveying the truth (one favorite analogy, which will be discussed later, dealt with the movement of a boat with the help of an oar). Kepler's more serious theoretical principles made use not only of magnetism but of cosmic intelligences, and also of a mechanism of his own devising that owed something to certain traditional theories of the propagation of light. Just as the Sun was supposed to send out an image of itself in light, so too its corporeal body, Kepler thought, could send out a motive power (*virtus motrix*). Penetrating the planets and spreading out like spokes on a wheel (but in all directions), this power might sweep the planets round as the Sun rotated. When he began his study of Mars, there was no evidence for the rotation of the body of the Sun. Before long, it would be detected through the telescopic observation of sunspots, but we may put it aside for the time being, and concentrate on the broad pattern of Kepler's physical argument. This was lengthy and intensely conjectural. It did not have the survival value of his kinematical work, and until recently, for that reason, tended to be brushed aside by historians of astronomy. It mattered greatly to Kepler, however, and in a sense marks a watershed between planetary astronomy as a largely geometrical pursuit, and as a subject for the new physics of the early modern period.

The two great publications in which Kepler's struggle with Mars is most fully recorded are the *Astronomia nova* (1609) and his *Epitome astronomiae copernicanae* (1618–1621). The first was as much a diary of his thought processes as it was a statement of his final beliefs, but each book tells us much about the evolution of his thought. Where Kepler might seem to be handling a common theme in the two works, on closer examination, his two accounts will often be found to differ fundamentally. Their general character owed something to the excitement of the chase, but it was also shaped by the fact that Kepler's work was so frequently interrupted, as he moved from place to place, at a time of great political uncertainty and danger. There was not a year in his life which does not mark some stage or other in the long-running religious conflict within the German states. In

1609, for instance, the catholic German princes confederated at Würzburg under Maximilian, Duke of Bavaria; and 1618 marks nothing less than the commencement of the Thirty Years War, almost as though it was considered the best way of celebrating the centenary of the German Reformation. Over and above this situation, there were his personal problems. The printing of the second installment of the *Epitome* was taking place at the same time as Kepler's mother was being tried for witchcraft and was published at the time of her acquittal. It is small wonder that Kepler's two great works, the second ostensibly an application of the new type of astronomy demonstrated in the first, are of a fluid and unsettling character. If they were held together by any one thing, however, it was Kepler's faith in the astronomy as a doctrine of the physical world.

Kepler knew much, from previous theories and from his reductions of Tycho's observations, about the geometrical properties of the planetary orbits. How could they be produced in the real world? In chapter 39 of *Astronomia nova*, he says quite explicitly that the planet is inclined by nature to stay at rest where it is put. That is a far cry from Newton's first law of motion, which it might be thought to resemble. According to Kepler, the planet may be moved by a power in the Sun *and also* by a power inherent in the planet itself. He asks what sort of powers would be needed to get it to move on a circular path; and, at a later stage, he will ask what powers are needed to provide an elliptical path. His presentation is convoluted, but may be briefly described in words. First, the geometry of a simple case: he considers a planet moving around the Sun, on a circular path eccentric to the Sun. As it moves, its distance from the Sun shortens and increases alternately, within fixed limits. By a simple geometrical construction, Kepler reduces this oscillatory motion to a motion back and forth along a diameter of a certain circle. (To describe the shuttling motion he uses the Latin equivalent of the word *librates*, referring to the swing of a balance. The word is avoided here, since the concept of libration is firmly associated with a certain property of the Moon.) Now to the physics: what is it that causes the oscillatory motion, the component of the planet's motion along a radius from the Sun? In the *Astronomia nova*, Kepler insists that it is "fitting" that the planet should perform this component of the motion—not the complete motion—of its own accord. He therefore considers the planet's mental processes at considerable length, and suggests that the planet might be able to judge its distance from the Sun by taking note of the Sun's apparent size. He dismisses a few counter-arguments, as well as the ideas that the Sun does the pushing and pulling along the line to the planet, or that the planet has an inherent force. He would later change his mind on some of these questions.

This Platonic Kepler is certainly not the man we envisage today when contemplating his three laws of planetary motion. His model-theoretic approach has always been more easily recognized, perhaps because we are inclined to think of his as the century of mechanistic approaches to nature. We have seen examples of his love of argument by analogy. He always placed a much higher value on geometrical than on arithmetical harmonies. Analogies based purely on numbers correspond to no archetype in the soul of man or mind of God, whereas geometric analogies do so correspond. When he realized, for example, that the force moving the planets falls off with distance from the Sun, geometry persuaded him that the force must be corporeal rather than spiritual. He stands at the crossroads between old and new ways of looking at the macrocosm, the universe. In a letter to Herwart von Hohenburg (dated 10 February 1605) he explains that he plans to show that the celestial machine is not in the nature of a divine living being. It is a kind of clockwork, insofar as the multiplicity of its motions depends on a single driving force. At that time he thought it was magnetic but corporeal, and so comparable with the driving weight of a clock. The planetary system is like a clock—it is the image of a clock. The comparison was not new. We have seen how Oresme used it in the fourteenth century, and it has earlier precedents, but it is now used in a new way, for it is more than an idle comparison. Kepler is more or less insisting that physical causes be given in mechanical terms, with the added qualification that they are subject to mathematics.

This does not mean that he has any wish to drop his earlier adherence to mysticism, or that he abandons the semimystical analogy between microcosm and macrocosm. His *Harmonice mundi* still deals with cosmic harmony, as its title announces. It is instructive to see how, while Kepler entered into a controversy with the mystical Robert Fludd, his own mysticism was of a very different sort. In 1617, Fludd published the first part of a massive work which must surely be counted as the most thorough exposition up of that time of the microcosm-macrocosm analogy. Kepler's response to Fludd tells us much about what had to be rejected before his own brand of astronomy could flourish. He dismissed Fludd as a Hermeticist playing with symbols, occult ideas, and words, and by contrast presents himself as a realist. While at first sight there is little to choose between the styles of their analogical presentations, Kepler believes that Fludd's methods are utterly incapable of delivering a true and accurate account of the observable world, an account that is truly geometrical.

In the *Astronomia nova* he provides us with a rough mechanical model by way of explaining the oscillatory component, and then tries to improve his account, explaining the slow oscillation in planetary distance by use of a magnetic model. For his first model, he asks us to think of the planet as a boat, carried by the steady current around a circular river. The boatman has a single oar. This, like the forward current of the river, influences the boat's motion, bringing it now nearer the Sun, now farther from it (fig. 140). He was evidently proud of this model, although he admits that it has certain shortcomings, and again resorts to an argument from what is "fitting." It is unfitting that the planets should be thought of as having corporeal oars!

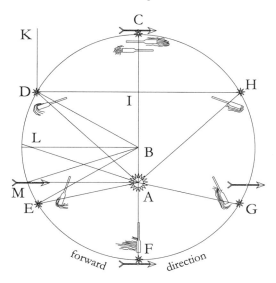

140. A planetary model of which Kepler was fond, and by which—in broad, qualitative, terms—
he explained the force needed to cause the distance of the planet from the Sun (*A*) to fluctuate
in a correct way. (Redrawn from the woodcut in *Astronomia nova*, the faults in which do not
relieve Kepler's obscurity. He knew more about stars than about boats.) He considered a boatman with a single oar, the oar being moved through a right angle as the boat is carried from
aphelion (*C*) to perihelion (*F*), and then through another right angle on the other side of the
boat. The oar cannot be such as a gondolier uses, or a simple scull at the stern; on the contrary,
it starts with blade to the bow. It is pivoted freely so that can be turned steadily through a complete circle. The stream is responsible for the chief movement of the boat, and the oar for forc-
ing the boat on to the correct eccentric path. In doing so it turns the boat through 180° in one
full orbit, interchanging bow and stern in alternate orbits, an admitted blemish on the model.
Added to Kepler's figure are symbols for magnets, later used for his explanation of the fluctuating solar distance in terms of magnetic attractions and repulsions—here, yet again, as corrections to the forced general circulation.

It was William Gilbert's magnetic philosophy, in which the magnetic
character of the Earth was demonstrated, which gave Kepler what he
wanted to explain the toing and froing in planetary distance. He decided
that there were magnetic fibers in the planets, attracted or repelled by magnetic fibers in the Sun. He had no laws of force at his disposal, but he was
not frightened off by the manifest complications in the case. The fibers were
taken to circulate round the Sun along parallels of solar latitude. Those who
are familiar with Newton's study of gravitation will know how, by summing
the forces between a small mass and all individual parts of a large spherical
body, which the Sun may be assumed to be, Newton proved at length that
the total gravitational force could be treated as though the Sun's mass was
concentrated at its center. Kepler also appreciated that he had to add components of magnetic force from all points of the Sun "visible" to the planet;
but he could only take the matter further in a half-qualitative way. We can
see how forbidding the problem must have seemed to him by picturing the
"view" of the magnetic Sun (fig. 141) from a planet, especially when it is not

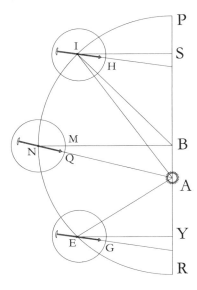

141. In the *Epitome*, Kepler ascribed the deflection of the planetary magnetic fibers to the action of the Sun, which attracted one end and repelled the other. When at aphelion (*P*) they start at right angles to the apse line (*PR*). When the planet reaches *I*, the fibers will be deflected only slightly, in direction *IH*. Kepler postulated that they would head for the Sun (direction *NQ*) after a quarter of the orbit was completed, after which they would be pulled in the opposite direction until eventually coming back to the starting direction half way through the orbit. Magnetism, operating in this way, was supposed to *direct* the planet, and not to force it along its path, or to act as a central force of gravitational type.

on the solar equatorial plane. (The planets have different orbital planes and cannot all share the equatorial plane.) He fell back on an imprecise verbal presentation, but one that followed the lines of his river analogy. He postulated a steady solar force and an adjusting force, corresponding to the oar: the planet is swept along by the Sun; a natural force holds the planet's "axis of power" always in a fixed direction (a slight adjustment is needed for precession); and the Sun-planet magnetic interaction steers the planet on the right course. The magnets, "correctly arranged," will look after the oscillatory motion in planetary distance.

Kepler's bold claim in his *Astronomia nova* that here was a "most beautiful geometrical demonstration" of the toing and froing of the magnetic planet, expresses a pious hope rather than a real achievement. Again, he speaks of planetary minds. A planet's behavior is made dependent on its knowledge of a certain angle—the one he needs for his geometrical account to be correct. He insists that the planet will not be able to obtain knowledge of a certain different angle—one that his demonstration does not require! He never gave up hope of obtaining a sound physical theory, however, and in the *Epitome* we still find him struggling to make the magnetic model fit the known geometry. There is a slight change in the model: whereas previously the planet was taken to move toward the Sun, now it is assumed to be pulled. The twisting of the planetary boat, so to speak, is now done by the Sun, through what we should be inclined to call a magnetic couple (a force couple). The planet's magnetic fibers are pulled at one end and pushed at the other, twisting them from their base position (fig. 142). As Kepler explains, in the lower quadrants the Sun is nearer the planet, and so should affect it more; but the planet is moving faster, making its influence felt less over the time it acts. To turn this into a mathematical physics, he would have needed the methods of the differential and integral calculus, and of a more powerful system of dynamics than was then available. He had to

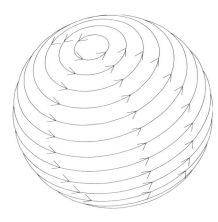

142. The pattern of magnetic fibers that would have been presented to the "view" of a planet not in the Sun's equatorial plane. These fibers, according to Kepler, provided the magnetic force needed to push round the planet through solar rotation and— through attractions and repulsions—keep its fluctuating distance from the Sun correct. The impossibly complex problem in magnetism presented to Kepler did not prevent him from arguing in qualitative terms that the arrangement of fibers could explain how all planets are fairly close to the same (ecliptic) plane.

resort to largely qualitative language, but the miracle is that, embedded in his largely intuitive discussion, there were those remarkable laws that go under his name.

KEPLER'S LAWS OF PLANETARY MOTION

Like so many before him, Kepler assumed that force would be proportional to motion. The power of the rotating Sun that drives the planets spreads out in three dimensions, so that its effect could be expected to diminish in proportion to the square of the distance, but this was an idea that Kepler could not reconcile with the known velocities of the planets. What if the forces were not three-dimensional, but were somehow confined to the plane of the orbit or very close to it? Might their effect on the planet's velocity not then diminish in proportion to the distance? In testing this assumption, instead of formulating his argument in terms of velocities he spoke of "delays," the short interval of time to cover a small arc. If the delay is then proportional to the distance for a small arc, the delay over a large arc should be somehow found by adding all the distances for the constituent small arcs. How to break down the orbit into a series of small increments of arc was something Kepler had learned from Archimedes' practice in geometry. Summing the distances over an interval of arc, in order to find the time taken by the planet to cover the arc, required him to choose an arbitrary small arc interval, and seemed to lead to nonsensical infinite sums. He saw, however, that the area of each component sector swept out by the radius vector from the Sun was roughly proportional to the mean distance of the tiny orbital arc from the Sun, and that the approximation became closer, the smaller the arc. The total time in a sector of any size should therefore be proportional to its area. In this way, he was led to what is now known as the "area law," or Kepler's second law of planetary motion. Otherwise expressed: *the radius vector from the Sun to the planet sweeps out equal areas in equal times* (fig. 143). Strangely enough, he does not appear to have felt very pleased with his demonstration in the *Astronomia nova* and gave the law no prominent place in his writings until he wrote his *Epitome*.

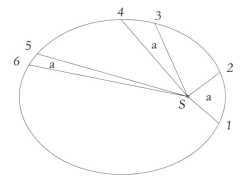

143. Kepler's second law of planetary motion, the "law of areas." Equal areas (such as the three here marked *a*) are swept out in equal times by the line joining the Sun (*S*) to the planet. Isaac Newton later saw that a minor modification was needed, and that in a simple two-body system, *S* should be taken as the center of mass of the Sun and the planet.

By itself, the area law was not enough to eliminate that error of eight minutes of arc. We all know that he had first to arrive at his law of elliptical orbits, but it would be a slight to his industry and his genius to say only that he discovered that next, and in so doing brought to an end the dominance of the circle in planetary theory. He had found the law of areas long before publishing the *Astronomia nova*, and had begun to doubt its truth, before eventually being forced to conclude that he must dispense instead with the assumption that Mars has a circular orbit. He tried an epicycle, its center obeying the area law. This could be made to generate an oval deferent, and he made innumerable calculations with different sorts of oval. Everything depended on the precise form of the orbit, the extent of its deviation from a circle. He knew that if it were an ellipse, an entire branch of familiar geometry would be available to him—the geometry of conic sections of Archimedes and Apollonius. At one stage he used an ellipse as an approximation to the orbit, without apparently viewing it as more than a device for calculating areas. He seems to have thought that fate was unlikely to be so kind as to make this a representation of reality, however. By 1605, with hundreds of trial calculations behind him, and with fifty-one chapters of the *Astronomia nova* completed, he still gave the ellipse only an auxiliary role in his theory. He wrote ten chapters of the book, covering his investigation of ovals, and the errors implied by each at the octants. At one point he tried to reconcile the ellipse to the magnetic hypothesis and failed. The ellipse seemed to give planetary directions correctly, but not the right pattern of fluctuating distances. Then, seven chapters later, he decided that an ellipse could answer both requirements. What we now call Kepler's first law was born, according to which *a planet travels in an elliptical orbit with the Sun at one focus of the ellipse*. It was some time before Kepler was able to transfer the area law to the ellipse, but this he had done before publishing the *Epitome*.

In the course of presenting his account in the *Epitome*, Kepler introduced an equation that has retained, in its own right, a central place in the solution of planetary orbits. He writes as though it was a consequence of his physical theory of the deflection of the magnetic fibers, as indeed it might have been, although he might equally well have arranged that

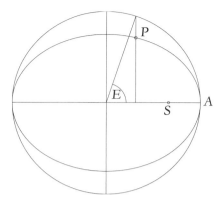

144. Diagram illustrating what has become known as "Kepler's equation."

physical theory to give an elliptical orbit. The following summary account of "Kepler's equation" does not describe its genesis, but is focused rather on its importance for the astronomer—not least for Kepler—who wishes to compute planetary positions. The equation relates two angles, known as the eccentric anomaly (E) and the mean anomaly (M). The former is shown in figure 144, in which P is the planet, S is the Sun, and A is perihelion, the planet's point of nearest approach to the Sun. The angle M is not shown, nor does it need to be shown, if one is prepared to think of these things in analytical terms. It is simply an angle described by a radius vector moving with constant angular speed. We could rewrite our explanation, turning everything into talk of areas, but it should be easy enough to picture M as the angle between SA and a line through S that rotates at the same average rate as the planet, coinciding with it every time it passes through A. Then, if e is the eccentricity of the ellipse, Kepler's equation is simply

$$E - e \sin E = M,$$

where angles are in radian measure.

It should be obvious that M is easily found if E is known, but that the problem facing a calculator of planetary positions will begin with a solution of the inverse problem: to find E, given the time (or given the angle M, steadily increasing with time). There is no simple and exact explicit solution, as Newton and others later proved.

The solution of the equation by different approximation techniques has a long history. As early as the ninth century, Ḥabash al-Hasib had found an approximate solution of this type of problem, although for another astronomical purpose. Before modern computing methods took away much of the importance of finding a mathematically elegant solution to the equation, tables for it were refined from time to time. J. J. Astrand's of 1890 and J. Bauschinger's of 1901 have been much used, but Kepler himself produced some very creditable iteration and interpolation procedures, for use in his Rudolphine Tables.

Before these tables were ready, he had the good fortune to discover John Napier's invention of logarithms (1614), from a work by another author. Appreciating the underlying principle but not having formulas for their construction, he prepared logarithms of his own to a somewhat different convention. The logarithms much lightened the labor of calculation, as they were to do in astronomy thenceforward—and indeed until long after mechanical calculators began to displace them in the nineteenth century. The practical subject of computation is one that may lack the glamour of theoretical astronomy, but it is of course of great practical importance, even at the stage of refining a theory, as Kepler would have been the first to maintain. Logarithms, it should be noted, were usually for decimally expressed numbers, while astronomers traditionally expressed angles sexagesimally. A little-known fact is that in due course massive volumes of logarithms of sexagesimal numbers were produced, especially for navigation at sea using astronomical methods.

The Rudolphine Tables were far more reliable than their predecessors. In the case of Mars, for instance, where errors might approach 5°, they were now less than a thirtieth of that figure. Kepler was able to predict, for the first time in history, transits of Mercury and Venus across the Sun's disk. He died in 1630, a year before this unusual pair was to be seen, but his forecast Mercury transit was indeed seen by Pierre Gassendi in Paris, by Johnann Baptist Cysat in Ingolstadt, and by Remus Quietanus in Ruffach. (The Venus transit was invisible from Europe.)

As we have seen, Kepler's astronomy was founded in physics and geometry. For all his respect for natural philosophy, his love of geometric harmonies never left him. As we have already seen, while his *Harmonice mundi* was a product of the style of scientific thinking of his early years, it was not published until 1619. By the time it appeared in print, he had begun to focus his attention on harmonies that might involve planetary velocities as seen from the Sun. In much the same vein as his earlier account of the Earth's perception of planetary aspects, he now wrote as though there was some sort of sight or instinct and mind in the Sun, capable of perceiving the angular motions of the planets and harmonies of their movements. He looked for new types of harmony, whether for a single planet, or as between adjacent planets; and he wove God's Providence into his account of the harmonies revealed by the planetary eccentricities. (He was not the first to take this approach. In 1531, Philip Melanchthon had written a preface to a new edition of *Sacrobosco*, which became widely known by virtue of the widespread university use of the basic text. Melanchthon wrote of the paths and laws of the planets as having been put in place to help mankind to attain to a knowledge of God and of God's Providence.) Kepler had almost finished writing the *Harmonice mundi* when he hit upon what we now speak of as his "third law." According to this law, *the periodic time of a planet is proportional to the 3/2 power of the semi-major axis of its elliptical orbit.* (The latter distance is the planet's mean distance from the Sun. For a circular orbit, this is simply its radius.) He announced the law almost without comment in that work and offered a justification only in the second installment of the *Epitome*. In his explanation he attempted to quantify the idea of the solar *species*, responsible for pushing the planet round, and the magnetic matter in the planet, which depended on its volume. It is quite clear that this ad hoc explanation was not the route to his discovery.

The third law, as Kepler recognized, is extremely valuable, for it allows the relative sizes of two orbits to be gauged from a reasonably easy comparison of their periods. The semi-major axis of the Earth's orbit is our common unit of distance (the AU, or, astronomical unit), and of course the Earth's periodic time is our common unit of time, namely one year. To take an example: the sidereal period of Mars is easily found as 1.881 (tropical) years, and therefore its mean distance from the Sun is 1.881 to the power of ⅔, that is, 1.523 astronomical units, according to Kepler's third law. Newton, in his *Principia*, showed that the law could be derived from his inverse square law of gravitation, which therefore came to be regarded as justified by Kepler's "observational" law. In fact, as pointed out earlier, Newtonian principles require Kepler's law to be modified very slightly, taking into account the acceleration of the Sun produced by the planet. Since even the most massive planet (Jupiter) is less than a thousandth the mass of the Sun, the correction is very small, and Kepler's third law is still often quoted as though it were exact.

Kepler saw himself as more than an astronomer, in the ordinary sense. He wished to fathom the harmonies of the universe, using highly personal blend of mathematics, physics, philosophy, and pure mysticism. The best astronomers following him all refrained from adopting his eclectic style, but they eventually learned how to exploit his laws for purposes of computation—precisely the purposes for which he had been employed by Tycho. The burden of computing the Rudolphine Tables was a bittersweet affair, running in parallel with his investigation of the highly complex motion of the Moon. The tables were not ephemerides, but were—like most of those before them—the means of calculating single positions or ephemerides. They gave positions far more accurate than anything before them, a few minutes of arc being exchanged for the old errors of a few degrees, in some cases. Kepler hoped to use the Rudolphine Tables to recoup unpaid salary, but printing was interrupted when Counter-Reformation forces occupied Linz. He and his Lutheran printers were allowed to stay, but only in 1627, with Kepler now settled in Ulm, was the printing finally completed there. In view of the fact that his life was almost to its end interrupted by the wars of religion, it is easy to appreciate the flights of fantasy that took him above it all, when revising his *Somnium seu astronomia lunari* (*Dream, or Lunar Astronomy*) in 1609. This science-fiction story of a voyage to the Moon was a literary device—along Ciceronian lines—used to argue for a Copernican arrangement of the planets, which an observer on the Moon is well fitted to appreciate. It was finally seen through the press in 1634, by his son-in-law Jakob Bartsch. When Kepler died, he was still owed more than 12,000 guldens by the state in whose service he had spent most of his life.

THE FIRST TELESCOPES:
THE MAPPING OF THINGS SEEN

Of all astronomical instruments, none has had a more dramatic immediate effect on the course of astronomy than the telescope. Kepler's design was eventually to occupy the center of the astronomical stage for more than a century, but it was not the first, and it did not come into its own immediately. The idea of the telescope certainly long preceded its realization. Simple lenses, as we might now call them, had been known from very early times, not merely as drops of water and ice, but created inadvertently in the process of polishing transparent gems, rock crystal, and, in due course, glass. It requires no great genius to observe that in certain circumstances lenses may magnify, reduce, or invert an image. By the end of the thirteenth century, converging (convex) lenses were in use for reading spectacles (the Latin word *spectaculum* was used for a single lens at the beginning of the same century). There are numerous imprecise references in medieval literature to the possibility of seeing distant objects clearly, as though they were near at hand. By the seventeenth century there was a well-established trade in spectacle lenses, and in some ways it is surprising that the discovery of a method of combining them into a telescope—and later into a compound microscope—was so long in coming. No doubt this has something to do with the relative scarcity of weak converging lenses— lenses of long focal length—and of strong diverging lenses.

It is even more surprising to find that the first well-attested telescopes were "Galilean," that is, that they used a diverging eye lens and a converging object lens, as in the modern opera glass and cheap telescopes. This combination has the advantage that it gives an upright image, unlike the "Keplerian" telescope, with its converging eyepiece and objective; but a magnified image is a magnified image, and the astronomer can work easily enough with an inverted view of the heavens.

Claims for prior invention have been made on behalf of various sixteenth-century scholars, such as John Dee, Leonard and Thomas Digges, and Giambattista della Porta, but they are without foundation and rest on an excessively generous reading of ambiguous texts. Some confusion has been created by the existence of medieval illustrations of philosophers looking at the heavens through tubes. Aristotle himself referred to the power of the tube to improve vision—it can improve contrast by cutting down extraneous light—but the tubes in question were always without lenses. The sixteenth century gave much attention to the theory of "perspective"—the science of direct vision, reflection, and refraction— and "perspective glasses" were recommended to astronomers and military men alike. Kepler, for instance, observed the comet of 1607 (Halley's, as

it happens) *per perspicilla*, probably an eyeglass, and certainly not a telescope. There was much experimenting with lenses at the time, and one
suspects that published accounts were even then misconstrued, so that
an inventor might have mistakenly believed that he was reproducing the
work of another.

Whatever the truth of the case, the first unambiguous evidence that
an effective telescope had been made appears in the form of a letter dated
25 September 1608. This letter is from a Committee of Councilors in the
Province of Zeeland in the Netherlands to their delegation at the States-
General in The Hague. In the letter it is said that the bearer "claims to have
a certain device by means of which all things at a very great distance can be
seen as if they were nearby, by looking through glasses, and this he claims
to be a new invention."

Matters then moved very quickly indeed. A week later Hans Lippershey, a native of Wesel who worked as a spectacle maker in Middelburg,
Zeeland, applied for a patent on his invention. The States-General entered
into negotiations with Lippershey at once, and although he was not granted
his patent he was given a lucrative commission for several binocular telescopes. He was denied the patent on the advice of unknown parties, who
held that the instrument would be all too easy to duplicate. We know that
at least two other men were in fact making the instrument within three
weeks of the original letter—Sacharias Janssen of Middelburg and Jacob
Adriaenszoon of Alkmaar, in the province of Holland. There is a possible fourth reference from the same period, but now the stream of history
is muddied by the fact that the writer, Simon Mayr, had earlier quarreled
disastrously with Galileo, and was to do so again, on questions of scientific
priority relating to the telescope.

Simon Mayr (also known as Marius), was a competent German astronomer who, in his *Mundus Jovialis* of 1614, tells us that his patron had
been offered one of the Dutch instruments at the Frankfurt Book Fair in
1608, but that he would not pay the asking price, especially as one of the
lenses was cracked.

The Spaniards, the old occupying force in the Low Countries, were at
this time in The Hague negotiating a peace settlement, and various other
foreign delegations were also there in force, thus ensuring that the spread
of the telescope's fame throughout Europe was extraordinarily fast. The
Spanish commander-in-chief, Ambrogio Spinola, was evidently alarmed
when he was shown the instrument before the end of September, but is
said to have been assured by the gentlemanly Prince Frederick Hendrick
that, although the Spanish forces would now be visible from afar, the Dutch
would not take advantage of the fact. Lippershey refused to make telescopes
for the French, but the secret was out, and the instrument was on sale in
Paris by April 1609, in Milan by May, and in Venice and Naples by July.
Galileo learned of it by report in the middle of July 1609, made it his own,
and greatly improved its performance. Its value to a maritime power like

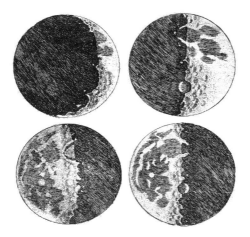

145. Four of the lunar drawings
published in Galileo's *Sidereus
nuncius* (1610).

Venice was very great, and his improvements earned him lifetime tenure of
his chair in mathematics at Padua and an unusually high salary.

These simple telescopes had at first magnifications of 2× or 3×, but
by improving his grinding and polishing techniques Galileo made objectives with ever longer focal lengths, and by 1610 his magnifications were
20× and 30×. He made various other technical improvements, but above
all he recognized the great scientific potential of his new possession. With
it, it was a short step to observing the mountainous character of the surface of the Moon—although we should not suppose that this was evident
immediately to every owner of a telescope. Sir William Lower, writing to
Thomas Harriot at midsummer, 1610, noted that he had seen the "strange
spottednesse al over, but had no conceite that anie partie thereof mighte
be shadowes" before (by implication) reading Galileo. Soon afterwards,
Galileo discovered the four satellites of Jupiter, and the starry constitution
of parts of the Milky Way that could not be resolved by the naked eye (see
figs. 145, 146, and 147). Without warning, astronomy had begun to move
in a completely new direction.

Galileo was not the first to turn the instrument to the sky. A report
of the visit of an embassy from the king of Siam to Prince Maurits in The
Hague—the source of the Spinola story—also noted that the new invention revealed stars that were invisible to the naked eye. That report related
to 10 September 1608. In England, Thomas Harriot was drawing what he
saw of the Moon through his telescope as early as 5 August 1609 (on the
Gregorian calendar), before Galileo's serious studies had begun (fig. 148).
What characterized Galileo, however, was the sheer energy he threw into
the whole enterprise, and the attention he drew to the broad cosmological implications of what was to be seen through the telescope. He rushed
into print with his first telescopic discoveries: his *Sidereus nuncius* (*The
Sidereal Messenger*) came off the press in March 1610. The excitement it
aroused led to a second edition, from Frankfurt, before the year was out.
The book purveyed astronomy of a new kind, a kind that could be easily

[Handwritten facsimile of Galileo's journal notes, dated January 1610, recording observations of Jupiter's satellites]

146. Part of a page of Galileo's journal, based on earlier sheets, with his records of observations of the satellites of Jupiter, which he later called the "Medicean planets." We see that on 7 January 1610, he saw three little stars that he thought fixed, *stelle fisse*, sketched on the second line. The following day they had changed position, as sketched on the third line; and he soon found that they changed systematically, "wandering about Jupiter as Venus and Mercury about the Sun."

147. Stars in the belt and sword of Orion, as seen by Galileo through his telescope. This is one of several drawings of asterisms in his *Sidereus nuncius* (1610). Stars visible to the naked eye that were previously recorded are drawn larger, and distinguished by a central dot. This drawing is worth comparing with what was included in drawings based on naked eye observations, such as Piccolomini's (see fig. 111).

comprehended by the general reader, and yet such that no expert astronomer could fail to be stirred by it. Kepler was quick to publish two short books on the new discoveries, the first before he had even observed these things for himself.

How much Galileo knew of similar activities in other parts of Europe we do not know, but in England, Thomas Harriot had also turned his telescope to the sky, and it was not only in his lunar drawing that he preempted Galileo, as we shall see. Harriot did not publish his findings, but they were not entirely unknown, for he was well connected, socially and intellectually. As a young man, he had been tutor to Sir Walter Raleigh, and in 1585 had been on Raleigh's second expedition to Virginia, but now he was in the circle of the unfortunate Henry Percy, ninth Earl of Northumberland.

148. The first drawing of the Moon obtained telescopically. Thomas Harriot dated this drawing 1609 July 26 (England was still using the Julian calendar) at 9 p.m., noting that the Moon was then five days old. It is not hard to recognize his shaded patches on the modern photograph of the Moon at five days.

The earl was for many years imprisoned in the Tower of London for political reasons, but Harriot visited him there, received a generous pension from him, and was given access to his London houses. He was close to the hub of political life and current international affairs. He observed Jupiter's satellites systematically as soon as he heard of Galileo's observations. In France, Nicolas Claude Fabri de Peiresc did likewise. Peiresc, who was descended from an Italian family, and who had met Galileo at the turn of the century, observed the satellites with the astronomer Joseph Gaultier, and it is worth noting that their patron, Guillaume du Vair, had acquired their telescope even before they read the *Sidereus nuncius* in 1610. It was presumably of a good quality, since Peiresc discovered the nebula in the sword of Orion with its aid. There were many others who were eager to see such things for themselves, but telescopes of good quality were not readily available, and some of the best of astronomers, such as Christoph Clavius, had to wait months for the experience.

Harriot's first lunar drawing (dated 1609) had been a relatively trivial affair, done with a 6× telescope (fig. 148). This he soon cast aside for others with magnifications of 10× and 20×. The lunar drawings made with these have been much misunderstood: they owe nothing of any importance to Galileo's *Sidereus nuncius*. They, like the sketches on which they were based, are artistically attractive but not accurate. In them, however, Galileo did pay great attention to the changing shadows of lunar features, and he later made valiant attempts to calculate the heights of lunar mountains on the basis of the lengths of shadows. Harriot's drawings are of different sorts. The majority were items in a regular series, and were not meant to record everything that could be seen on the lunar surface on the days when they were drawn. They were records of what was to be seen in the neighborhood of the terminator, the moving boundary between light and

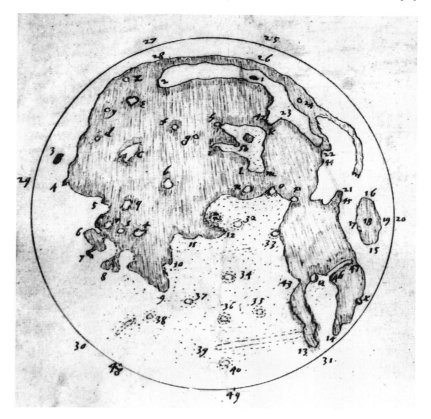

149. (Right) A collage of some of Harriot's lunar drawings made between July and September 1610, with a final example from 1612. Note how he paid most attention to the region of the terminator and was so able to build up maps of the whole surface. Two of these survive, one of them shown here (above). The other is similar, but unannotated. There is no sign here of Galileo's grossly exaggerated crater. Contrast Harriot's, too, with the much inferior Scheiner drawing (fig. 154).

dark. Knowing the day of the lunar month, he could draw the position of the terminator on his blank lunar disk, and then fill in the detail adjacent to it. In this way, as the month went by, he was able to build up a picture of the visible hemisphere, and so was finally in a position to draw two complete lunar maps. One of these, annotated with 72 letters and numbers, is shown in figure 149.

One of Harriot's finest pieces of work was a systematic study of sunspots, which he first observed telescopically in December 1610. Ancient Chinese and early medieval European annals had already recorded spots that could be seen on the Sun with the naked eye. One medieval source speaks of a transit of Mercury lasting eight days, which is nonsensical, and two independent Muslim records also speak of impossibly long transits, one of Venus and one of Mercury. As late as 1607, using a camera obscura, Kepler saw what he took to be a passage of Mercury across the face of the Sun, and in reality that too was a sunspot. The large number and true

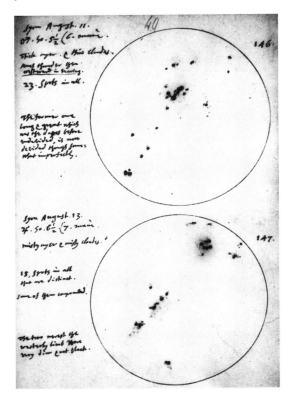

150. Two of Harriot's
drawings of sunspots,
for 11 and 13 August 1612,
numbered 146 and 147
in his long series. These
were done in the early
morning at Syon. He
remarks on changes in
sunspot numbers (23 and 15
distinct), their clustering,
and intensity; and on the
atmosphere ("thick ayer &
thin clouds, much thunder,"
and so forth).

nature of sunspots, however, was Harriot's discovery, and his approach to
the study of the phenomenon was not only earlier, but more systematic,
than that of his contemporaries. Even now we have his notes on 450 separate observations. (For specimen drawings, see fig. 150.) In studies dating from December 1610, he observed their growth, decay, and change of
relative position. He learned the art of tracking a single, changing spot,
and carefully studied the appearance of spots near the limb of the Sun.
He did not appraise the inclination of the solar axis as carefully as Galileo
was to do, but he gave more attention to the solar rotation. His measurements give a sunspot period of 26.87 days at the equator (synodic, as seen
form the Earth) increasing to 28.45 days at solar latitude 30°. These and
other data are very close to those published in 1863 by R. C. Carrington,
and are of a higher order than Galileo's "about one lunar month." From
his papers we can see that Harriot was almost certainly trying to find a
periodicity in sunspot totals. By the end of the eighteenth century, other
astronomers—notably the Dane Christian Horrebow—were beginning to
suspect a regular cycle of change, but not until Heinrich Schwabe, in 1843,
was the sunspot period found to be a question of years rather than months.
(Schwabe opted for ten years. As will be seen in chapter 16, the half cycle
of 11.11 years was found by Rudolf Wolf in 1852.)

Galileo's own study of sunspots was made only in 1612, but it was published in 1613, in the form of a series of letters on the subject that he had

written to the wealthy Augsburg magistrate Mark Welser. Welser, who was also banker to the Jesuits, had in 1612 published two series of letters on sunspots from the Jesuit Christoph Scheiner, protecting the writer under the pseudonym Apelles. In the first series, published in January 1612 and dating from November 1611, sunspots were said to be outside the Sun, floating "like the Cyanean islands in the Euxine sea." An earlier series of observations had been made by the East Frisian Johannes Fabricius, possibly dating from March 1611, and published in June of that year. He realized that they were on the solar surface, as Scheiner did in his second letters, and as Harriot had seen in 1610.

Harriot's work was known to many others, but he seems to have been temperamentally indifferent to its publication. Had his patron not languished in the Tower, things might have gone differently. There was a new excitement in the air, and a letter to Harriot from a friend, William Lower, written in June 1610, caught the mood perfectly. Lower explained how they were "so on fire with these things," and asked for as many "cylinders" as would be needed for such observations. The telescopes used at this early period were given many names, and it would be tedious to list them all, but commonplace were "glass," "instrument," "trunk," "perspective cylinder," "organum," "instrumentum," "perspicillum" (those three in Latin), and "occhiale"—Galileo's favorite Italian word, although he also referred to his "cannone." The first recorded use of the word "telescopio" was apparently by the Greek poet and theologian John Demisiani, at a banquet held in honor of Galileo in April 1611.

The typical reaction to news of Galileo's discoveries was to send for, or try to make, one of the new Dutch telescopes and to attempt to reproduce Galileo's experiences. Kepler, no less excited than the common run of astronomers, chose to make an intensive new study of the theory of optical systems. He had become an expert in the subject in 1600, his practical experience beginning with a pinhole camera he erected at that time in the market square at Graz, intended for the observation—after the manner of Tycho Brahe—of a partial solar eclipse. This experience had led him to study the eye as an optical instrument, and then to compose the first of the two works in optics we mentioned earlier.

Kepler, by this time a recognized authority and mathematician to Emperor Rudolf II, was sent a copy of the *Sidereus nuncius* at Galileo's request, through the Tuscan ambassador. Galileo asked for his opinion. (We recall an earlier exchange when Kepler was less important.) Kepler replied at length, and in 1610 published his views as *Dissertatio cum Nuncio sidereo*. This *Conversation with the Starry Messenger* was both a generous affirmation of faith in Galileo and a summary advertisement for his own views, on matters as diverse as the finiteness of the universe, possible life on the Moon, the planetary orbs in relation to the Platonic solids, and his writings on optics. Galileo was content, and even more so when the imperial mathematician, having in the meantime borrowed a telescope from a noble acquaintance, the Elector Ernst of Cologne, followed with a short tract on the satellites of Jupiter (*Narratio de Jovis satellitibus*, 1611).

As soon as he received the *Sidereus nuncius*, in the space of two months—August and September 1610—Kepler composed his second optical tract. The counterintuitive, inverted image produced by a telescope made to his design, developed as an alternative to Galileo's, worked against it at first. It had the advantage, however, of producing a real image, one that could be focused on a screen beyond the eyepiece. Within a decade or so, it was in regular use for projecting images of the Sun on paper, for instance, for the charting of sunspots.

GALILEO AND THE COSMOLOGICAL IMPLICATIONS OF THE TELESCOPE

To suppose that the excitement caused by the invention of the telescope was entirely the excitement of a peep show would be to underestimate the seriousness of certain questions that the instrument seemed to answer. Aristotelian cosmology was being seriously questioned. The cometary observations worked against Aristotle. As things happened historically, they worked in favor of Copernicus, but that was primarily because he and Aristotle were two leading authorities who conflicted on so many points that a reverse for one was seen as an advance for the other. The rational astronomer could in principle reject Copernicus—indeed Tycho did so—and at the same time accept the cometary evidence. Kepler added to the list of troublesome phenomena. He mentioned a haze in the sky seen in 1547, and the halo round the Sun—the solar corona—which had been seen during the eclipse of 12 October 1605. He mentioned too the new star, the *stella nova*, of 1604, which reinforced in no uncertain manner the cosmological doubts induced by the new star of 1572. That of 1604, which appeared in the constellation of Ophiuchus, gave rise to innumerable university and public debates on the incorruptibility question, and in them Galileo played an active part.

(The new star of 1604, now often called "Kepler's star," was exceptionally bright. Like Tycho's, it is now classed as a supernova. Disregarding the different types of supernova for the time being, we can say that a single example may outshine an entire galaxy of billions of stars. For the remnant of Kepler's star, visible as well as a source of radio emissions, see plate 11. A nova, in modern jargon, is much more modest: starting as a binary star, one component of which is a white dwarf, it increases in brightness by about ten magnitudes, and then usually fades in a matter of months.)

The name of Galileo is certainly better known than that of Kepler, and it was so in his own time. No doubt the fact that Galileo more or less avoided attempting to improve on the intricacies of traditional mathematical planetary astronomy during his entire career did him no harm in this respect: his talents were of a different sort and were more easily appreciated by scholars and the world at large than were Kepler's.

151. Mezzotint after a portrait of Galileo (1564–1642) by Justus Sustermans (1636) in Trinity College, Cambridge.

Born in Pisa in 1564, seven years before Kepler, Galileo Galilei outlived him by eleven years, dying in Arcetri in 1642—the year of Isaac Newton's birth. A copy of an early portrait of the great Tuscan astronomer (fig. 151) hung in Newton's college, which is hardly surprising, since Galileo was from an early date regarded as the figurehead of a new kind of science. It would be hard to name any one person who better symbolizes the change in attitude toward the empirical sciences that characterizes the seventeenth century. He was sent to the university of Pisa by his father Vincenzio—a knowledgeable writer on musical theory from whom Galileo learned much about the art of matching theory to experiment—to study medicine. His interests turned to mathematics, and in 1585 he left Pisa without a degree. He began to study Euclid and Archimedes privately in Florence, while teaching mathematics there and in Siena. An unfinished but revealing Latin dialogue from the period 1586–1587 can be described as physical cosmology, to the extent that it deals with the motions of bodies in general, and the disposition of the elements in the world. The work shows him as a critic of Aristotle, but also as having learned much from Archimedes. In particular, he treats the problem of the cosmological order of the elements along the lines of Archimedes' treatment of floating bodies. Galileo, in other words, was turning his back on Aristotle's teleological account of the elements, and searching for an account in terms of mechanical forces; but still he was waiting in the wings for a university position. In 1588 he was passed over for a chair in mathematics in Bologna, this going to the more experienced astronomer Giovanni Antonio Magini, but a year later he was given the vacant chair at Pisa.

When Galileo is considered only in the narrow context of the history of astronomy, there is a tendency to treat him purely as hero in the struggle to establish the truth of Copernicanism. This treatment, however, oversimplifies the case. Three of his scholastic essays that survived in manuscript form but were not published before modern times were, until the late 1960s, treated as juvenilia belonging to his days as a student in Pisa. They have now been shown to belong to a later period—in one case, from after 1597. They are heavily dependent on Jesuit scholastic authors and illustrate the fact that Galileo could be evenhanded when it came to scholastic writers from whom he felt he could learn. Clavius in particular remained an important influence on him to the end of his life. In his early writings Galileo argued against the Earth's motion, borrowing words from Clavius's widely read *Sphaera*. Even after he had eventually he changed his mind, Galileo's strategy of argument owed much to what he had learned from Clavius.

When teaching in Pisa, Galileo began to openly criticize many parts of Aristotelian natural philosophy as it was then taught. His laws of falling bodies—usually associated with the mythical demonstration at the Leaning Tower of Pisa—are the best known symbol of his dissatisfaction. After his three-year contract at Pisa ended, he needed a new position, and was fortunate enough to obtain the much more important chair at Padua. (This time he defeated Magini in the contest, and in doing so gained an enemy for life.) Padua was the leading Italian university of its day and a focal point for European scholarship. He taught and wrote much on mechanics at this time, and his moderate support for the Copernican theory, expressed in correspondence with Kepler, shows that he valued it for its physical implications—it fitted well with his own ideas on the tides, for instance. He was not an out-and-out Copernican of the Kepler type, for his intellectual career had not brought him to this point from considerations of mathematical astronomy. When he learned of the new star of October 1604, he saw it as an opportunity to lecture to large Paduan audiences on the difficulties inherent in Aristotelian ideas about the incorruptibility of the heavens.

Galileo was nothing if not a great self-publicist, and he reveled in controversy. Those lectures spawned much acrimony, and even more followed the publication of a pseudonymous attack on the Paduan professoriate in 1605. Written in a Paduan country dialect and bearing the name of "Cecco di Ronchitti," there is no conclusive evidence that it was from Galileo's pen, but that does seem likely; and if it was not, then it was certainly from a sympathizer. It did not endear him to his colleagues. It took the form of an amusing dialogue in which two down-to-earth rustics dismantled arguments that a professor of philosophy had directed against Galileo's conclusions concerning the new star of 1604. (What had the substance of the new star to do with its location? For all the mathematicians cared, it might be made of polenta! Land surveyors know more than philosophers about measurement. Such, in paraphrase, were the rustic barbs.) The book sold well enough for it to be reprinted, after a few months, with some changes that suggest—assuming the work to be Galileo's—some hesitancy on his part in adopting the Copernican system. In 1606 another book appeared, this time in more polished Italian but with the same general tendency: it was directed against an amateur Florentine philosopher who later became one of his chief opponents there. Written under the name "Alimberto Mauri," this tract might also have been by Galileo, although the probability here is generally reckoned to be smaller. Both works are amusing indications of discontent with the scholastic thinking and conservatism of the universities, and their plain common sense adds a refreshing dimension to the discussions they contain concerning the borderline between fact and theory in science.

Galileo's talent for controversy was exercised yet again in 1606, when he brought a charge of plagiarism against Simon Mayr, then at Padua, and against one of Mayr's pupils, Baldassar Capra. This case concerned the so-called proportional compass, a calculating instrument Galileo had devised in 1597, and from the sale of which he made a small income (as it happens, the instrument was merely an improvement on an earlier one by Guidobaldo del Monte). Mayr returned to Germany, and Galileo succeeded in having Capra expelled from the university. In this way was the ground prepared for future intellectual disputes. Mayr would soon be claiming to have seen three of Jupiter's satellites before Galileo. The first printed reference to Mayr's claim, however, is in the introduction to Kepler's *Dioptrice* of 1611, and so appeared long after Galileo's *Sidereus nuncius*. Incidentally, Kepler seems to have been the first to use the Latin word *satelles* of them, this being the Latin word for an attendant or courtier, the court in question being Jupiter's. Galileo, with an earthly patron in mind, had lost no time in naming them the "Medicean planets."

The many discoveries made by Galileo, some of which were described earlier in this chapter, affected his own outlook on the world very deeply. The material advantages of a new post as mathematician and philosopher to the grand duke of Tuscany were ostensibly the reason for his return to Florence in the summer of 1610, but there is no doubt that he found the idea of continuing to teach the old Aristotelian philosophy uncongenial. It is true that his first telescopic discoveries did not undermine the foundations of the old ways of thinking: they could all be explained, more or less, in Aristotelian terms. That there are mountains on the Moon, and that the Milky Way is an association of separate stars, are facts that have no strong bearing on the fundamental principle of Aristotelian cosmology that had been brought into question by the new stars of 1572 and 1604, and by the absence of parallax in comets. That Jupiter has satellites was more problematical, for it suggested at least that there were other centers of rotation in the universe than the Earth. Before the end of 1610, however, Galileo was able to announce further spectacular discoveries. He found that Saturn appeared noncircular, as if it were a globe with two handles. He could not resolve what were later found to be rings round Saturn, and assumed that he was looking at satellites very close to the planet. At first he refrained from publishing his findings, except as an anagram that many tried unsuccessfully to resolve. It spelled out, in Latin, the message "I observed the furthest planet to be triple." Soon the appendages disappeared for a time—the rings, as we know, were at the time turned edgeways to the observer—and this greatly puzzled all concerned. "Does Saturn devour his children?" Galileo asked in a letter to Welser. In his book *Il Saggiatore*

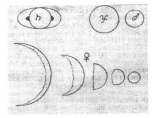

152. Galileo's drawing of Saturn's appearance (upper left), as published in his *Il Saggiatore* (1623). Also on the upper row are the relative sizes of the disks of Jupiter and Mars. The lower part of the illustration shows a sequence of the phases of Venus, as the planet reduces in size with increasing distance from the Earth.

(*The Assayer*; 1623), he still drew Saturn as though it had cup handles (fig. 152). As we shall see later that it was not until the 1650s that the young Christiaan Huygens found the right explanation for Saturn's changing appearance.

Of more immediate importance to the wider cosmological debate were Galileo's discoveries of the seemingly endless aggregate of point-like fixed stars and the phases of Venus. The former did not offer any grave difficulties, for as Clavius had noted, the Bible speaks of innumerable stars. What was seen through a telescope in this respect offered one problem, however, for the few who were prepared to face it: the more stars there are, the more difficult it is to house them in a standard world of Aristotelian type. The phases of Venus were perceived to present greater problems. The various possible arrangements of the Sun (S), Venus (V), and the Earth (E), when they are more or less in line, are limited. According to *all* of the three principal planetary systems in play in the early seventeenth century—the Ptolemaic, the Copernican, and the Tychonic—they may lie in the order EVS, that is, with Venus lying between us and the Sun. (Of course, the orbital planes of Venus and the Earth very rarely coincide. Usually they are not near enough to produce a transit of Venus across the Sun's face, as seen from the Earth.) Venus will in these circumstances not be visible to the naked eye, since it will be lost in the Sun's rays, but the case will be closely analogous to that of the new moon. The case of the fully illuminated Venus, the analog of the full moon, is more interesting. This is ESV, and the Ptolemaic model as traditionally interpreted would not have produced it, since the orbit of Venus was considered to lie entirely within the orbit of the Sun.

One might imagine that there is another possibility of a "full" Venus on the ancient model, namely SEV, but the Ptolemaic motions are so designed that the center of the Venus epicycle is more or less in line with the Sun, so that the planet cannot be in opposition to the Sun—and everyone knew that this was the case in reality.

The upshot of this is that *all* models produce a set of phases of Venus, but the Ptolemaic model *does not have the full set*. For traditional astronomers, Venus should at best show a crescent shape. The opponents of Copernicus had pointed out that the variation in Venus's appearance was not enough to support the idea of a full set of phases. Galileo, however, saw a full set with the help of his telescope, and regarded what he saw as a proof of Copernican astronomy. Later, Kepler, in an appendix to his *Hyperaspistes*

(1625), which was his defense of Tycho's work on comets against the views of the Aristotelian Scipione Chiaramonti, pointed out quite correctly that the Tychonic system could explain the phases of Venus just as well as could the Copernican. By then, however, the battle lines had long been drawn, and Copernicanism was slowly but visibly threatening to topple tradition. And Kepler, of course, was an ally who made his remarks—in the context of criticisms of Galileo's somewhat ill-advised views on comets—more out of loyalty to his old master than in support of the Tychonic world view.

In 1611, Galileo demonstrated his telescopic discoveries to the Jesuits in the Collegio Romano in Rome. The views of the Jesuits were ambivalent, but at last he won most of them over. A key figure in this episode was Roberto Bellarmine, theological adviser to the pope. Bellarmine, who was then nearly seventy, asked the astronomers in the college—of which he was the former rector—to verify Galileo's findings. They did so, and Bellarmine responded warmly to Galileo, although without accepting Copernicanism. He later argued that until it could be rigorously demonstrated, scriptural texts should continue to be accepted at their face value. Perhaps the greatest of all compliments from the Jesuits at this time, however, was the remark by the great scholar Christoph Clavius in the last edition of his commentary on *Sacrobosco* (this was in 1611; Clavius died in the following year). Astronomers, he said, would need to find a new system in agreement with the new discoveries, for the old would serve them no longer. And this after a life of opposition to Copernicanism.

The Jesuits were certainly not converted as a whole to Galileo's way of thinking. One idea many of them seized upon, after word was spread around that the Sun had spots on its face, was that those spots were of the nature of groups of unchangeable stars or planets. When, in a public debate in the college in 1612, a Dominican noted that stars and planets were round, or at least regular in shape, and that sunspots were not, the Jesuits countered with the observation that a group of fifty or so might create an irregular appearance. Galileo later said that he found it hard to accept that fifty stars, like fifty boats joining together at various speeds, would stay together. Besides, he had observed these things with great care—their changes in shape and size, their growth and disappearance, their evident opacity and shading. He had one trump card in such a debate. What was the value of endless citations of traditional authors when—however great their intellects had been—they could not have had access to the new sort of observations?

Back in Florence, Galileo became embroiled in yet more of those running controversies for which he had such a talent. The first concerned floating bodies, another question where the authority of Aristotle was at stake, but now the alternative doctrine was that of Archimedes. While working on a book on the subject, for which he was attacked by at least four Aristotelian professors in Pisa and Florence, Galileo received a copy of a pseudonymous work on sunspots, sent to him for his opinion by the author, Marcus Welser of Augsburg. We have already mentioned Welser, and his relations with

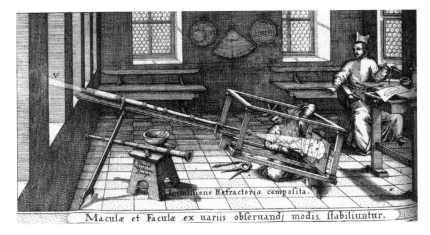

153. The technique used by the Ingolstadt Jesuit Christoph Scheiner for observing sunspots. He projected the solar image on a white screen at the external focus of a suitably adjusted telescope of Keplerian type. From his *Rosa ursina sive sol* (Bracciano, 1626–1630). The technique was widely adopted and was later used for observing the rare transits of Mercury and Venus across the face of the Sun.

154. Christoph Scheiner's drawing of the lunar surface, as it appeared at first quarter. From his *Disquisitiones mathematicae* (1614; written jointly with J. G. Locher).

Christoph Scheiner, one of the principal authors of the idea that the spots were small planets. Galileo replied to Welser in three long letters, claiming priority of discovery and attacking Scheiner's ideas. In this way, he added another to a growing list of enemies. Scheiner was in many ways conservative, but he was an intelligent scholar who taught Hebrew and mathematics at the University of Ingolstadt, and he was a good practical astronomer. His technique for drawing the Sun was by projecting its image through a Keplerian telescope (see fig. 153). The operation was carefully planned and executed: Scheiner's strong mechanical intuition is clear from another of his inventions, the pantograph, a set of hinged rods of a sort still occasionally used for scaling drawings. His lunar drawing of 1614, it must be said, was very undistinguished, as can be seen in figure 154.

Galileo's most influential telescopic work was done. He was becoming slowly entangled in a theological net. On 26 February 1616, he was ordered by a papal commission in Rome to abandon the view of the Earth's motion, and not to defend it. A week later, Copernicus' *De revolutionibus* was put on the Catholic Church's *Index of Prohibited Books*, where it stayed until 1835.

155. Johannes Hevelius and his wife Elisabeth, observing at the 6-foot brass sextant. As Tycho Brahe had done before him, Hevelius published comprehensive and finely illustrated books describing his many instruments. This is taken from his *Machinae Coelestis, Pars Prior* (1673).

Galileo's response was to turn to more practical matters, in fact, to the determination of longitude at sea through the use of Jupiter's satellites as a universal clock. For this he recommended an instrument he had designed—the Jovilabium, a sort of satellite equatorium—and tables of eclipses of the satellites, and their motions generally. Galileo applied to the States-General in the Netherlands for a prize they had offered for a solution to the navigational problem, but all he received was the offer of a gold necklace, which he declined.

He wrote on mechanics and on comets, and in his classic polemic *Il Saggiatore* he even threw a sop to the Aristotelians in the form of a possible defense against the anti-Aristotelian argument from the negligible parallax of comets. One cannot discuss their parallax, he said, unless one is sure that comets are not purely optical in nature, such as they would be if they were formed by refractions in clouds of vapor. (At least he held to Tycho's belief that the new stars were celestial.) That comets are far beyond the Moon would only be well and truly settled after more accurate measurements had been made by the great Danzig astronomer, Johannes Hevelius (fig. 155). Good observational evidence for the idea that they follow parabolic paths, with the Sun at the focus of each path, was first provided by Georg Dörffel in 1680.

A somewhat chastened Galileo now turned to the writing of perhaps his finest literary creation, the *Dialogue Concerning the Two Chief World Systems, the Copernican and the Ptolemaic.* The book is presented as a series of discussions between three men, by name Salviati, Sagredo, and Simplicio. Salviati represents the author, Sagredo, an intelligent listener, and Simplicio, an obtuse Aristotelian. (While Galileo could always say that he had the great sixth-century Aristotle commentator Simplicius in mind, there was a clear double meaning in the name.) Galileo spent six years in writing the book, but he sailed too close to the wind, and after its publication in Florence in 1632 he was finally ordered to come to Rome to account for his actions to the Inquisition. After much prevarication, he finally did so in 1633.

There followed one of the most notorious trials in history. Lest it be thought that this was no more than an intellectual exercise, it is as well to remind ourselves that at one stage the examination was held under threat of torture, and that although there was probably no intention of carrying out the threat, Galileo is unlikely to have taken much consolation in that thought. He continued to maintain that, after it had been condemned by the Congregation of the Index, he had never held the Copernican theory. He was finally condemned to life imprisonment, however, and to certain penances. The sentence was signed by seven cardinals, and while it was not formally ratified by Pope Urban VIII, the ruling of the Holy Office was final as long as he did not object. There is a legend according to which Galileo, rising from his knees after repeating the formula renouncing his supposed offense, stamped on the ground and declared "Eppur si muove!" (And yet it [the Earth] moves!). This story, however, is truer to Galileo's character than to history.

The sentence was immediately reduced to one of permanent house arrest. He was nearly seventy, and lived for another eleven years—blind for the last four of them. Henceforth in Italy, Copernicanism reached the printed page only as the object of criticism. An excellent example of the ease with which a highly talented individual could fall into line with orthodoxy is Giambattista Riccioli: in his exceptionally scholarly survey of the entire history of astronomy—the finest then made—he came down in favor of a modified Tychonic system.

Galileo was not one of the great mathematical astronomers. He remained loyal to the circles used by Copernicus in planetary theory, and even in his *Dialogue Concerning the Two Chief World Systems* he made no reference to the elliptical orbits introduced by Kepler, although he certainly knew of Kepler's work. His mathematical talents were not insignificant, however, and his decision was not a question of incomprehension. He argued long and effectively for the importance of mathematics in natural philosophy, a thesis that was by no means universally acknowledged in his time, although several contemporary Jesuits were of the same mind. One of his greatest strengths had little to do with the mathematization of nature: it was his ability to demolish the insupportable nonsense put forward by so many of his opponents, and of course not only in the name of Aristotle.

THE FIRST TELESCOPIC AGE

Almost within the lifetime of a single human being, the telescope had completely transformed the nature of planetary and stellar astronomy. The subject had at last been opened up to research that required little or no mathematical skill. A new style of publication evolved, representing celestial appearances, not by schematic diagrams but pictorially. The Moon lent itself to this treatment above all other visible bodies, and numerous drawings of it were published. At first, the quality was indifferent, but matters improved greatly after Pierre Gassendi and Nicolas Claude Fabri de Peiresc joined forces in 1624 to try to determine the difference in longitude between Aix-en-Provence and Paris. The method they devised was to watch the timing of small lunar features entering and leaving the Earth's shadow during a lunar eclipse. For this they needed a good map of the Moon, and so it was that in 1634 they engaged the engraver Claude Mellan to prepare one. Mellan used a telescope made from parts provided by Galileo himself, and his map was a very fine one. There were very many other attempts at lunar mapping, rarely as successful as Mellan's, and soon a new type of activity was added: that of adding the names of illustrious men to the lunar features. Michael Florent van Langren added such names to an excellent map he published in 1645, but it was never widely known. He was anxious to do for the Moon what the Spanish Crown had done for territories in the New World, by making claims through naming. Van Langren's Moon was Catholic and Hapsburg, and in the colophon to his map he even mentioned penalties that would follow for those who changed his names for lunar features. Giambattista Riccioli and Johannes Hevelius, the leading lunar cartographers of the day, took no notice, if indeed they ever saw the map. Hevelius avoided the controversial act of naming lunar features after famous people, preferring the safety of classical geographical names. Riccioli used the names of famous scholars and astronomers, and was not ungenerous to Copernicans, with whom he disagreed. By and large, in the end, Riccioli won the day, helped by the hostility of Hooke for Hevelius, and the great influence of Cassini in Paris, whose preference was for Riccioli's shorter names.

Nomenclature apart, Hevelius was the astronomer who achieved greatest fame for his lunar mapping at this time. His magnificent copper engravings of the Moon were at the heart of his weighty and impressive *Selenographia* (1647). The observations on which he based the three large maps in this volume followed, like Harriot's, the changing Moon's appearance as the terminator moved across its surface. His excellent work is not always quite as accurate as it was often thought to be, but he achieved

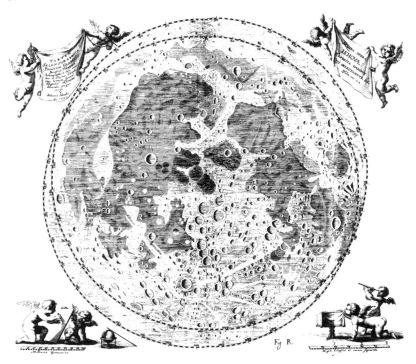

156. This depiction of the topography of the lunar surface, dated 1645, is one of three large plates engraved by Hevelius himself for his monumental folio volume *Selenographia* (1647). The plate covers facing pages. Hevelius had learned much about printing while a student in Leiden, and his work was printed at his own press. The measuring scale at lower left covers 100 German miles; that on the right is three "eclipse digits" long. (It is traditional to divide the Moon's diameter into twelve digits.) The lunules at the edge are those parts of the surface that are visible periodically through libration. Hevelius's skill in engraving what he had observed meant that he could quickly send copies of his drawings to others. When the book itself was complete he sent copies to libraries and friends in Rome, Paris, Oxford, Cambridge, and elsewhere. The pope remarked that it would have been without parallel had it not been the work of a heretic.

something which was scientifically far more important. He charted the libration of the Moon. While the Moon keeps more or less the same face toward the Earth, we can at various times see nearly six-tenths of the total surface area from Earth. Galileo, Harriot, and others had known this but had not described the changing lunar appearance thoroughly. The Moon's rotation on its axis has to remain in step with its direction from the observer, if it is always to keep the same face toward the observer. Its rotation presents a complex dynamical problem, and, in any case, the Moon—as we now know—moves at nonconstant speed around an elliptical orbit. That orbit—as was known from antiquity—is inclined to the ecliptic. Added to all this, our point of view changes slightly in the course of a day, as we are carried on our rotating Earth. The result of these several factors was not something that Hevelius or his contemporary Giambattista Riccioli could explain theoretically, but they could at least appreciate librations in lunar latitude and longitude observationally, and include them on their maps.

For a map by Hevelius, see figure 156. Riccioli's fine lunar maps were published in his *Almagestum novum* (1651).

Since the telescope was so quickly and widely disseminated, it is not surprising to find that numerous priority disputes arose. We have already met with Simon Mayr, whom Galileo had long before forced to leave Padua. He claimed priority on the sighting of three of the satellites of Jupiter (December 1609; Galileo saw four in the same month), and also in the matter of the production of accurate tables of the satellite motions. His measurements do not seem to have been plagiarized, since they are more accurate than any Galileo had then published, and he eventually tried to take movement in latitude into the reckoning, but Mayr's work as a whole is certainly inferior to Galileo's. Not until the publication in 1623 of *Il Saggiatore* did Galileo launch an attack on Mayr in print on these questions. One suspects that what irritated Galileo above all else was Mayr's having named the satellites the "Brandenburg stars," after his own patrons, thus jeopardizing Galileo's standing with the Medici family, after whom he had named the stars.

Since in some human beings the urge to prove priority of discovery seems to take second place to no other, it is amusing to find claims now being made to the effect that the Medicean planets, the satellites of Jupiter, were first found by Chinese astronomers two thousand years before Galileo. *The Kaiyuan Treatise on Astrology*, compiled by Qutan Xida between 718 and 726, refers back to some observations made in the fourth century BC by the astronomer Gan De. He is quoted as having said that in the year of chan yan, Jupiter was seen in the zodiacal division Zi, rising and setting with certain named lunar mansions (Xunu, Xu, and Wei). "It was very large and bright," he added, and "apparently there was a small reddish star appended to its side." This observation, claimed by some to have been of either Ganymede or Callisto, has been dated to 364 BC. On the whole, it seems unlikely that it was an observation of either of those Jovian satellites.

Thomas Harriot had a copy of *Sidereus nuncius* by June of 1610 and began systematic observation of the satellites in October. At first he could only see one, but by December of that year he was charting the motions of four, and for more than a year he worked at analyzing their motions, combining his own observations with Galileo's, and wisely preferring to establish sidereal rather than synodic periods, as the other had done. Harriot derived a very good figure for the first satellite, 42.4353 hours, comparable with the modern figure of 42.4582 hours.

Many other astronomers in plenty made sporadic observations, but the work on the satellites of Jupiter mentioned thus far was the best before Gioanbattista Odierna's. His highly professional *Medicaeorum ephemerides* was published in 1656 in Palermo, in his native Sicily, and remained a rare book. Odierna injected a certain theoretical stiffening into the analysis, using three sorts of periodic inequality, by analogy with existing planetary theory. Nine years later, the Florentine Giovanni Alfonso Borelli tried to draw similar analogies with Keplerian theory, when the Medici acquired a very fine telescope, and asked Borelli to improve on Galileo's tables. The analysis was beyond his powers, and a book he published in 1666 is disappointing.

One thing that is abundantly clear is that instruments alone were not enough. Even the satellite motions derived by Johannes Hevelius, and published in his *Selenographia*, were decidedly inferior to those derived by Galileo, Mayr, and Harriot more than thirty years before. Isolated extensions of the earlier work inevitably followed, however. The phases of Mercury were perhaps first recorded by the Jesuit Ionnes Zupo in 1639. There were several unusually good mid-century Italian telescope makers so that, for example, the Neapolitan Francisco Fontana had an instrument capable of revealing the phases of Mars, which others simply could not make out. The finest work of the century on the Jovian satellites was done with the help of what were surely the finest telescopes then available anywhere: around 1664, Gian Domenico Cassini obtained excellent telescopes from the great telescope makers Giuseppe Campani and Eustachio Divini. Within weeks of obtaining them, and while using one with a length of about 1.5 meters, he was observing the shadows of the second and third satellites on Jupiter's disk when he noticed an uncharted spot. Some days later he saw two or three movable dark spots, which he took to be clouds, and some bright marks, which he thought were volcanoes. This was the beginning of an extended analogy with the solar system and with the theory of planetary motions. The possibilities excited Cassini, as they did others, when he discussed them on a visit to the newly created Académie Royale des Sciences in Paris two years later. Cassini had just published the most accurate ephemerides of the satellites to date, and he was offered such generous terms of employment in Paris that he never returned permanently to Italy. Gian Domenico was the founder of an important Cassini dynasty in French astronomy.

The accuracy of Cassini's work was so great by the time he was issuing tables in 1693 that we can understand Edmond Halley's excitement at the prospects raised. The problem of finding the geographical longitude of distant places—on land at least, where the telescope could be held rigid—was solved. To compare the longitude of a distant place with that of Paris, say, one needed to know the times at which a given and recognizable event was recorded in both places. Jupiter's satellites, duly tabulated against Paris time, could be timed locally against the local noon transit of the Sun. The satellites could be regarded as a universal clock, its hands visible—under certain circumstances—everywhere. At its best, this "clock" was now accurate to better than a minute of time, so that longitudes could be calculated to better than 15 minutes of arc. The astronomer was meant to observe the first satellite (preferably) as it entered or left the shadow of the planet, and a whole series of such observations would obviously improve

CHRISTIANUS HUGENIUS
natus 14 Aprilis 1629.
denatus 8 Junii 1695.

Lugd. Bat. Apud Janssonios Van der Aa, Bibliopolas.

157. Christiaan Huygens (1629–1695),
from the frontispiece to a collection of
his writings (Leiden, 1724).

the results greatly. This method remained a standard one for the positions
of land bases well into the eighteenth century. The practicalities of the
method explain in part, no doubt, the fact that so many attempts were
made at the time to improve upon Galileo's Jovilabium: Fabri de Peiresc,
Odierna, John Flamsteed, Cassini, William Whiston, and Joseph-Jérôme
de Lalande were among those who devised variants on the basically simple device to take into account improvements in the understanding of satellite motions. In due course, this meant taking into account the velocity
of light, which, in 1676, Ole Rømer proved finite. He measured it on the
basis of satellite observations.

Rømer's ideas were of course destined to be of great consequence to
future large-scale astronomy, as well as to fundamental physics. Acceptance of them was rapid and widespread. To quote Francis Roberts, an
early writer who had taken the message to heart, "Light takes up more time
in Travelling from the Stars to us, than we in making a West-India Voyage
(which is ordinarily performed in six Weeks)."

After the first wave of telescopic discovery, the first discovery of a phenomenon of an entirely new kind came with observations of the planet Saturn by Christiaan Huygens (fig. 157), culminating in his realization that
the planet is at the center of a flat ring. We have already seen that Galileo
had studied Saturn in 1610, and had seen "not one but three bodies." Just
as did Scheiner in 1616, he surmised that there were satellites close to the
planet. Twenty years later, a number of skilled astronomers studied the
planet intensively through improved telescopes, but still they could make
little sense of what they saw. Gassendi, Hevelius, Divini, and Odierna all
made drawings which we, with the advantage of hindsight, can interpret
as evidence of a ring, but even by the late 1640s the ring-structure was not

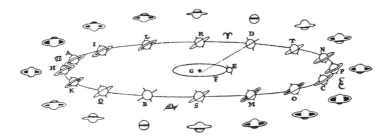

158. Christiaan Huygens's illustration of the how the changing appearances of Saturn (shown on the periphery of the diagram) are a consequence of the inclination of a ring to the plane of Saturn's orbit round the Sun (the outer curve). This plane is close to the plane of the ecliptic (the inner curve, with the Earth shown at *E*). From his *Systema Saturnium* (1659).

understood. In 1655, using a 50× telescope with a length of 3.5 meters, Huygens found that Saturn had a satellite, but this was quite distinct from the strange "cup handles" of the planet. (The satellite is the one we call Titan. Hevelius and Christopher Wren had seen it earlier, but had thought it to be an ordinary star.) What of the cup handles?

In 1657, Christopher Wren proposed an elaborate theory of an elliptical "corona" somehow attached to the planet and spinning slowly around the long axis of the ellipse. What Wren did not realize was that more than a year before he put forward this ingenious hypothesis—for which he had a mechanical model constructed—Huygens had found the correct interpretation of what was to be seen. It was, he said, simply a thin flat ring surrounding the planet and not touching it. Its changing appearance and occasional disappearance were to be explained by the inclination of the ring to the plane of the ecliptic (fig. 158). Huygens had used telescopes first of 4 meters, then of 7, and later of 37 meters, but insight was as important as sight in his discovery. In the 1660s and 1670s, Giuseppe Campani and Gian Domenico Cassini found that they could see shadows cast by the ring on the planet's surface, and in this way they helped to convince the few remaining skeptics of the truth of Huygens's theory. In 1675, Cassini discovered a gap in the ring, so providing yet more evidence of the rapid advance in observing technique and instrumental power. (Looking into the future, photographs taken from the spacecraft *Voyager* 1 in 1980 revealed hundreds of subdivisions, not to mention a rope-like braiding of the outer rings.) Cassini was fortunate in his splendid instruments, made by Giuseppe Campani of Rome: with them he found surface markings on Mars and Jupiter, allowing him to derive rotation periods for both. After moving to Paris he discovered Jupiter's oblate form, the belts on Saturn's surface, and four more Saturnian satellites. He found Rhea and Iapetus in 1671–1672, and Tethys and Dione in 1684. The *Voyager* probes brought the total number of satellites to twenty-one; and among the spectacular images taken by the *Cassini* spacecraft there is one of a recent crater on Rhea

159. A rayed (and therefore relatively young) crater on Saturn's moon Rhea, an image taken in visible light with the *Cassini* spacecraft's narrow-angle camera in 2005. The same crater was observed at a much lower resolution from Cassini in 2004. Rhea's diameter is 1,528 kilometers, comparable with that of our own Moon (1,738 kilometers).

(fig. 159). In the twenty-first century, hardly a month goes by without the addition of new detail to what is known, but the more that is known, the less is its power to astonish. The first chapter in telescopic astonishment had, in a sense, reached its climax before the century was out. The time had come for serious thinking about theories that could bind the new astronomy together, but there were, as always, those with different tastes.

TELESCOPES AND DREAMS

The face of cosmology had been changed by phenomena that were—given the right apparatus and a little intellectual guidance—plainly observable to people with no great knowledge of the technicalities of astronomy. Democratization was an illusion, but it was one that encouraged a revival in what might be seen as the Cicero-Macrobius "Dream of Scipio" tradition in literature. Ludovico Ariosto, the Italian Renaissance poet best known for his epic poem *Orlando furioso* (1516), had contributed to this revival long before. In this poem he tells of a traveler to the Moon and interprets what he sees as being analogous to earthly affairs. Galileo was extremely fond of the poem, although that is not to say that it played any obvious part in his own drawing of parallels between Earth and the Moon. As we have seen, Kepler's *Somnium* made use of a similar framework to argue for Copernicanism, so combining a long literary tradition with the products of the Galileo's newly announced telescopic observations and his own imagination. Perhaps the most influential work in this genre, however, was that by Bernard Le Bovier, sieur de Fontenelle, whose clever dialogues of 1686 in his *Entretiens sur la pluralité des mondes* (*Discussions of the Plurality of Worlds*) were similarly aimed at securing acceptance of the Copernican system. Other writers began to turn the argument around. Accepting the Tychonic system, in 1656 the Jesuit scholar-astronomer Athanasius Kircher used a heavenly journey as the framework for a journey through cosmology. Inevitably, he is confirmed in his belief in the fixity of the Earth. Kircher's is a surprisingly undisciplined fantasy, heavily dependent on astrological tradition. In it, for example, the sweet waters of Venus prompt the question as to whether a Jew or heathen could be duly and rightly baptised in them. Kircher often tied himself in knots when he was led to the idea of multiple planetary systems resembling our own. He published his musings under the title *Itinerarium exstaticum* (Ecstatic journey), following them a year later with a supplement describing the subterranean world. Twice reprinted, the book was never translated out of Latin, and for obvious reasons was frowned upon by members of his Church.

One reader who was plainly influenced by it was Christiaan Huygens, who likewise started by accepting the truth of a cosmological system, but now of course it was that of Copernicanism. Huygens used analogies similar to Kircher's in the hope of proving that the fixed stars must have planetary systems that are inhabited like our own. He carried the argument to great lengths in his last work, *Cosmotheoros*, finished in 1695 but published posthumously in 1698. Here he went on to try to show that there must be a great variety of plants and animals on the planets, some of the animals being rational and with powers of sensation like ours. These, he believed, must be social creatures that will have studied astronomy "and all its subservient arts," and have reached at least to our degree of expertise in these subjects. Huygens rested much of his case on the axiom that, since all of God's creations are for the best, there will be resemblances between them. He often stretched his analogies to breaking point. A typical argument, more bold than persuasive, is the one he offered for hands, rather than elephant-like trunks, on the inhabitants: he could not accept the idea that Nature was kinder not only to us, but to squirrels and monkeys, than to them. While here he was not exactly paving the way for the science fiction to come, he did at least acknowledge that creatures on distant planets might be exceedingly ugly to look at.

Cosmotheoros was first published in Latin, but was quickly issued in an English translation, and within twenty years went into others in Dutch, French, German, and Russian. It is a specimen of a literary genre that acknowledges a debt to astronomy, but only as long as the astronomy is plausible. After looking briefly but disdainfully at Kircher's book on a similar theme, Huygens felt he had to offer something more solid, and the last part of his book surveys the latest results of telescopic astronomy. This gives him an opportunity to speak of his own, and Cassini's, observations. He was able to explain, for example, how the ring of Saturn will appear, over time, to the inhabitants of Jupiter. He was able to introduce an important and original technique for estimating the distance of the star Sirius. For this he admitted light from the Sun through a tiny hole in a plate at the end of his 12-foot telescope, the hole having a tiny lens against it. Changing the arrangement until the light to be seen was comparable with that from Sirius, the brightest of all stars, and using a simple optical calculation, he concluded that the star's distance from us is 27,644 times the Sun's. This was a remarkable result, although only roughly a twentieth of the correct figure. To bring home the vastness of this distance to his readers, he noted that it would take a bullet from a cannon seven hundred thousand years to reach Sirius.

Cosmotheoros was seen through the press by Christiaan's brother, Constantijn Huygens, Dutch secretary to Stadholder William III—the "William of Orange" who had also been king of England, Scotland, and Ireland since the revolution of 1688. Among the many topical asides on Huygenian astronomy in the volume, there is a piece of advice offered to Constantijn. He was told that he might discover more satellites about Saturn, during its northern circuit, if he would but make use of his "two Telescopes of 170 and 210 foot long; the longest, and the best I believe now in the World." These enormous "aerial" telescopes followed designs in a book published by Christiaan in 1684. The object glasses for them had been prepared by Constantijn Huygens in June and July 1686, but ten years later he was preoccupied by politics, shuttling back and forth between London and The Hague. Like the king, both lenses—one plano-convex, one bi-convex—moved to London. They were acquired from the heirs of the brothers in the 1720s by the Royal Society, where they were joined by a 123-foot object glass that Constantijn had donated in 1691. They are there still.

NEW TELESCOPES

There was a clear limit to what was mechanically reasonable in the way of lengthening telescopes to reduce aberrations. Fine as the Huygens lenses were, even by the 1660s refracting telescopes had more or less reached the limits of what was reasonable, in the absence of achromatic lenses. William Derham, for instance, borrowed the 123-foot objective and found it entirely impractical. A number of people competent in mathematical optics had addressed the problem of false color in telescopic images at one time or another without making much headway. It was widely supposed that Isaac Newton had proved the impossibility of combining lenses in such a way so as to avoid spurious coloration. His reputation eventually came to carry so much weight that this belief—whether true or false is immaterial—helped to prevent the development of the achromatic lens. Some argued that the human eye, wrongly supposed to be perfect in this respect, was living proof that an achromatic lens was possible. An English landowner, lawyer, and amateur optician, Chester Moor Hall found (through a series of experiments that he did not publish) that different sorts of glass (crown and flint) could be combined to produce an achromatic combination. He worked out details in the period 1729–1733 and contracted out the work of making the first achromatic lens to professionals. George Bass made the first in 1733, and Hall had at least two achromatic telescopes made at about the same time. He told the well-known English instrument maker John Dollond of his success in the mid-1750s. Euler too saw the possibility and wrote to Dollond to ask him to experiment.

The Swedish physicist Samuel Klingenstierna published his own theoretical account in 1754 and sent it to Dollond but was not mentioned by

Dollond in an account of his own investigations published in 1758. Klingenstierna, nettled by this, published in 1760 by far the most thorough account then available of lens systems avoiding chromatic and spherical distortions, a work of great value to later makers of large astronomical telescopes. The commercial production of achromatic lenses was begun, however, by Dollond. Denmark was a distant land, and a patent was a patent. Dollond later suffered litigation from rival tradesmen, who argued that Hall had been first, but Hall stayed well outside the controversy, and the judge ruled that those who first brought inventions forth for the public good should profit from them. The contrast with the experience of Hans Lippershey is enlightening.

To achieve greater magnification without spherical distortion or spurious coloring of the image, in the refractors of the seventeenth century, apertures had been kept small and focal lengths had been increased, making for extremely long, often tubeless, telescopes. Hevelius built a 46-meter —example in Danzig (fig. 160). As mentioned in connection with his study of Saturn, Huygens had a 37-meter instrument. In this he replaced the tube with a length of wire that connected the objective to the eyepiece. When taut, the wire pulled the objective—hinged on a universal joint at the top of a high pole—in the right direction. Huygens soon discovered an unpleasant fact about all open telescopes: the quality of the image was greatly distorted by air currents. The work of Hall, Klingenstierna, and Dollond made these highly unsatisfactory "aerial telescopes" a thing of the past.

In France, Alexis Clairault simplified the achromatic lens, reducing Dollond's three-component lens to a two-component lens, with the two components in perfect contact. This was made up by several Paris opticians in 1763 and became the most common type. Dollond's son Peter lent their first achromatic telescope to Maskelyne, the Astronomer Royal, in 1765, and results were so impressive that the Dollond's fortune was assured. It was found that other aberrations could be reduced by getting the lenses to share in the refraction—avoiding, for instance, an arrangement where light meets a plane surface as it enters the telescope. The improved quality of object lenses guaranteed refractors in general a long history, and compound eyepieces helped further, by greatly improving magnification. Light-gathering power, on the other hand, depends on aperture, and lenses were fast approaching the limits set by the technology of the day. Something new was needed, and that came in the form of the reflecting telescope.

The Scottish mathematician and astronomer James Gregory had designed such a telescope, and he published details of it as early as 1663. In its practical form, it made use of a parabolic mirror that reflected rays from a distant object back to a small secondary concave mirror outside its focal point. From there the rays were again reflected in a narrow pencil that passed through a hole in the center of the main mirror, and on to a plano-convex eye lens, and so into the eye. The original scheme was geometrically more ingenious: the Gregorian telescope, as originally conceived,

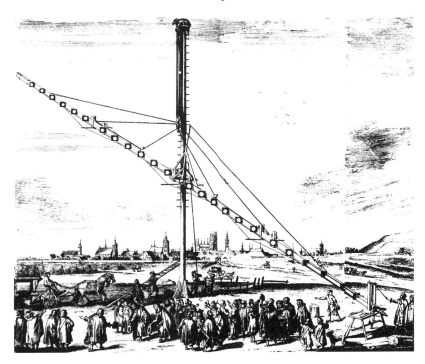

160. The 150-foot (46-meter) telescope built by Hevelius, as illustrated in his *Machina coelestis pars prior* (1673). This was the largest of his telescopes, aimed at eliminating aberration of the image. He built others of the type, but many of them were destroyed in a fire in Stellaburgum, his observatory, in 1679. The Danzig skyline is in the background. The presence of the crowd suggests that erecting the telescope was a public event, but a precarious one—note the workman at the top of the pole. The telescope was a temperamental object of limited value, bending and quivering as it did in the wind. Halley saw it and described it as useless.

uses not a spherical but an *ellipsoidal* secondary mirror with two focal points. The plan was for the main mirror to bring the first image to one focus of the ellipsoid, whereupon the secondary mirror places the second image at its other focus, this coinciding with the focal point of the eye lens. (Throughout, we are assuming the receipt of parallel rays of light from a distant object; see figure 161.) In fact, this idea simply made too great a demand on the workshop skills of the period. Gregory commissioned a London optician, Richard Reeve, to make a telescope with a six-foot focal length to his design. Reeve had long before tried to produce non-spherical lenses, when he worked with the mathematician John Pell in an attempt to produce hyperbolic lenses, as advocated by Descartes. (Johann Wiesel, in Germany, was said to have been successful, but with no clear practical gain.) In all of these cases, however, Reeve failed to produce what was demanded. Gregory was dissatisfied with a substitute spherical secondary mirror, and abandoned his attempts. When Isaac Newton looked into the problem in or around 1668, he settled on a much less ambitious design, using a plane mirror just within the first focus to reflect the converging

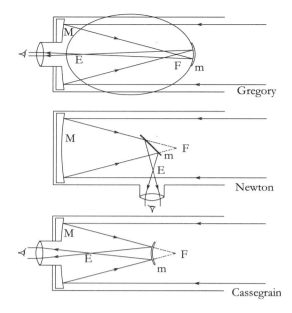

161. The three classic seventeenth-century designs for reflecting telescopes, by Gregory, Newton, and Cassegrain. *M* is the main mirror and *m* the secondary mirror in each case, while the focal point of the main mirror is *F*, and the focal point of the eyepiece is *E*. The name "Gregorian" is often used when the secondary mirror is merely spherical, but Gregory wished it to be ellipsoidal (the ellipsoid is drawn in here, its foci being at *E* and *F*). In time, compound eyepieces were substituted for single lenses.

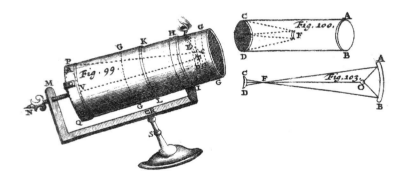

162. Contemporary drawings published by the Royal Society showing Newton's reflector and Cassegrain's alternative design.

rays out of the side of the main tube, and so into the eyepiece (fig. 162). A small reflector to Newton's design, said to be the one he presented to the Royal Society in 1672, is still in the possession of the Society. Its compact form—it is only about 30 centimeters long—and its freedom from the aberrations to which refracting telescopes were then susceptible, were chief among its merits. Its main demerit was the low reflectance of light from the mirrors of the period. Newton worked hard to develop a highly reflective

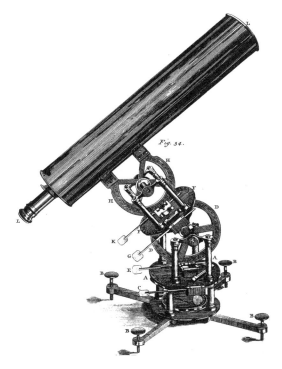

163. James Short made three portable equatorials to this design, as shown and described to the Royal Society in 1749. He knew of earlier uses of the mount with multiple pivots, but thought that adding so large a telescope to it was "somewhat new." It had too many movements to be stable enough for very much larger telescopes, and was too expensive for most amateur observers, but, like the torquetum, it cut down the number of stages in converting between coordinate systems (altazimuthal, equatorial, and ecliptic).

alloy, speculum metal, for his own mirrors, and this alloy became widely used over the next two centuries.

A very similar design to Gregory's, but now using a convex mirror just inside the prime focus, was announced in 1672 by a Frenchman called Cassegrain—a man about whom almost nothing is known. The different designs were rapidly circulated throughout Europe, for example in the Royal Society's *Philosophical Transactions* (figs. 161 and 162). The Royal Society took a special interest in the perfection of the telescope and other scientific instruments, and long continued to do so, with the result that London soon became the center of a large trade in such material. One notable maker was James Short, an Edinburgh man settled in London. He was an astronomer of minor repute, as well as a specialist in making large and very accurately figured mirrors. First these were of glass, but later of speculum metal. Short made no fewer than 1,370 reflectors, and well over a hundred of them still survive. His instruments were in great demand across Europe. A portable equatorial instrument he made for the Prince of Orange (fig. 163) was not only the sort of thing that every gentleman with intellectual pretensions wished to own, but it was also interesting in its own right, a combination of a Gregorian reflector with the kind of mounting used for a torquetum (fig. 103 in chapter 10). He himself observed with an instrument of about 1.5 meters in focal length. Most of his reflectors were appreciably smaller, but the largest was of about 3.6 meters in focal length.

The fourth classical type of reflector was simplicity itself. It was an arrangement used by William Herschel, and involves tilting the main mirror slightly and looking at it directly, through an eyepiece at the edge of the open end of the tube. This eliminates the need for an intermediate reflection, and so reduces light loss at the secondary mirror. It was with such an arrangement that Herschel discovered two satellites of Uranus—Titania and Oberon—in 1787.

After Dollond's manufacture of achromatic lenses, the reflector tended to pass out of favor with astronomers for a time. The speculum mirrors were fragile, they tarnished easily, and accuracy in grinding and polishing them was found to be a more critical matter than in the case of lens surfaces. The spectacular discoveries made by Herschel with his reflectors, however, brought such instruments back into favor with those astronomers whose main concern was magnification and light-gathering power, as opposed to precision measurement.

THE TELESCOPE WITH MICROMETER

The telescope has always been more than a means to obtaining more detailed views of objects in the heavens. It eventually found a serious use as a sighting instrument that could be attached to graduated quadrants and circles, such as those by which Tycho and others had surveyed the stars. Before this happened, it was used for angular measurement in entirely different ways. One method made use of telescopic projection. Using his newly computed Rudolphine Tables to draw up an ephemeris—it appeared in parts, eventually covering the period 1617–1636—Kepler realized that 1631 was a most unusual year, for it would see the face of the Sun crossed by Mercury on 7 November and by Venus on 6 December. As mentioned earlier, Kepler died—on 15 November 1530—before either happened. A printed pamphlet by Bartsch drew attention to Kepler's advice to astronomers to observe those coming transits with great care, since they would assist in settling the thorny question of the size of the planetary orbits in the solar system as a whole, as well as the angular sizes of Mercury and Venus. Several astronomers seem to have been inspired to make the necessary measurements on Mercury. Johann Cysat of Innsbruck and Johannes Quietanus of Rouffach did so, for example, but Pierre Gassendi's measurements on that planet were the most widely disseminated and were the cause of some surprise. He estimated the angular diameter of Mercury at about 20″, appreciably smaller than the estimate made by Kepler. Using Kepler's relative distances of the planets, Gassendi's figure implied that Mercury subtended an angle of about 28″ at the Sun. The timing of the transit showed an error of several hours in Kepler's tables; and when Gassendi tried to observe the transit of Venus, he could not see it at all. Unbeknownst to him, it took place not by day, as the tables predicted, but during the night.

Gassendi's figure for Mercury carried another element of surprise, which was only appreciated after Jeremiah Horrocks had found a figure for the planet Venus. Horrocks was a young and talented English astronomer from a village near Liverpool in Lancashire, northern England. He had been fired with enthusiasm for astronomy from an early age, and he benefited from study at Cambridge, although circumstances forced him to leave in 1635 before he was able to take his degree. Returning home, he continued to study astronomy intensively, and soon discovered a number of shortcomings in contemporary tables, especially in the much admired tables of Philip van Lansberge. In due course, he also found appreciable errors in the Rudolphine Tables. Like another amateur Lancashire astronomer, William Crabtree, a clothier from Broughton near Manchester, Horrocks made a long series of careful astronomical observations, and the two men regularly compared their findings. Horrocks had a good command of the necessary mathematics, and after intensive study of Kepler's writings was able to improve on the great astronomer's theory of the Moon's motion—indeed to such effect that his revisions were later to be adopted by Isaac Newton, for his own theory of lunar motion.

Before making himself familiar with Kepler's writings, and in the course of casting a critical eye over other tables for Venus, Horrocks had reached the surprising conclusion that a hitherto unforeseen transit was due in the autumn of 1639. Since Kepler had calculated that there would be none between 1631 and 1761, there was no widespread study of that in 1639, of which Horrocks and Crabtree took careful measurements. Like Gassendi, they projected the Sun's image on a white screen in a camera obscura, and watched the planet as it moved across the Sun's face. (It was the technique that had been used earlier by Scheiner, Galileo, Fabricius, and others, for studying sunspots. See figure 153, above.) In this way, Horrocks measured the apparent angular size of the disk of Venus, relating it to the known angular size of the Sun. His result was $76'' \pm 4''$ (Crabtree found $63''$), from which it could be deduced by careful calculation that an angle of $29.1''$ would be the angle subtended by the disk of Venus at the Sun—close to the correct value. As a result of Horrocks's untimely death in 1641, the findings of the two men remained unpublished until 1662, when *Venus in sole visa*, a draft of his treatise on the transit of Venus, was published in an annotated edition by Hevelius in Danzig.

Many of Horrocks's unpublished papers survive, including some of his remarkable materials on lunar theory, although many are lost. He became an ardent admirer of Kepler's mathematical theories of planetary motion, but of other parts of Kepler's work as well. He tried to modify Kepler's magnetic theories in the light of Galileo's mechanics, but did not live long enough to make much progress in this direction. He absorbed the spirit of Kepler's doctrine of harmonies, and here he believed that his study of the transit of

Venus was pointing him in the direction of yet another mystical harmony. As pointed out earlier, combining Gassendi's observation of the Mercury transit with Kepler's figures for the relative sizes of the orbits of Mercury and the Earth gave the angular size of Mercury, as seen from the Sun, as 28″. The closeness of this to the figure he had found for Venus convinced him that more than mere coincidence was involved; and taking matters one stage further, he conjectured that *all* planetary disks, including that of the Earth, might subtend the same angle at the Sun. The sizes of Mercury, Venus, and the Earth, do in fact increase, but and if we skip Mars (only half the diameter of the Earth), Jupiter continues the trend; but then Saturn breaks the rule again. The Horrocks rule had one great attraction, however: it immediately implies that the solar parallax (see p. 103) is about 14″, a much better result than Kepler's (60″), but not as good as Cassini's of 1672 (9.5″). As is so often the case in astronomy, wishful thinking stood proxy for hard evidence. In 1672–1673, the Oxford mathematician John Wallis put the tract containing this conjecture into print, along with other parts of Horrocks's writings, in a posthumous edition. The item in question was entitled *Astronomia Kepleriana, defensa et promota* (*Keplerian Astronomy, Put Forward and Defended*).

Such examples as these, of the measurement of very small angles by telescopic means, were not the first, but what had gone before was often much less accurate than astronomers realized. The field of view to be seen through the earliest Dutch (Galilean) instruments was very small—less than the diameter of the Moon, in most important cases. As we have seen, the usual way to improve magnification was to increase the focal length of the objective. For lenses of a given diameter, this has the effect of restricting the field even further. Angles were at first then simply gauged in relation to the field of view. Galileo explained how to make various object-glass diaphragms, and calibrate them. Harriot gave much attention to such angular measurement. Alas, the method is not reliable, for a reason that has rarely been properly understood. With this type of telescope, the result depends on the size of the pupil of the eye, which is effectively the stop in this type of telescope, setting the field of view. Unfortunately, the pupil size may change appreciably, as the overall illumination changes.

With Crabtree and Horrocks, William Gascoigne made up a trio of north country astronomers who kept in close contact with one another, although Gascoigne was in Leeds, on the Yorkshire side of the Pennines. In the late 1630s, he hit on the idea of measuring apparent planetary diameters with an astronomical (Keplerian) telescope fitted with a micrometer. By putting telescopic sights on mural quadrants and sextants, he increased the potential accuracy of measurement, from minutes of arc to seconds. He was by no means the first to use screws as a means of adjusting instruments by small amounts. Regiomontanus made use of the technique, as did Tycho Brahe; and in 1609 Lucas Brunn, then of the University of Altdorf, used a screw as a device for measuring small angular changes in the

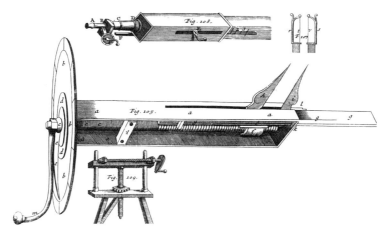

164. The drawings published by the Royal Society in 1667 to accompany an account by Richard Towneley and a description by Robert Hooke of William Gascoigne's micrometer. The upper figure shows how it was fitted to the eyepiece, and the lower shows its construction in more detail. The screw is turned by the handle, the position of which against the circular scale gives an indication of the angular separation of the two pointers, or rather fiducial edges. They could be replaced by two fiducial wires, as drawn above the pointers, hence the name "bifilar micrometer."

setting of an astronomical sector. Brunn, or the maker of his instrument, Christoph Treschler, also added a scale to the screw, to measure the fractions of its turns. The instrument in question was unfortunately destroyed in the bombing of Dresden in 1945, but its details are well recorded. Even if they were not, the instrument was clearly in a very different category from Gascoigne's. In the latter, the cursors (or crosswires) were to be set in the focal plane of a telescope eyepiece, their separation being adjustable by means of a screw. Unlike the Dutch "Galilean" telescope, the Keplerian type has a real focal plane for its eyepiece, and an object in that focal plane can be seen in focus at the same time as the distant object. This fact was not discovered by Gascoigne on theoretical grounds, but by chance, when a spider chose the focal plane for threads in its web.

Gascoigne died in the English Civil War, in the massive defeat of royalist forces at the battle of Marston Moor (1644) in his native Yorkshire. Richard Towneley preserved his papers and letters, however, as well as his micrometer (fig. 164). Use was made of this evidence by the Royal Society, in a priority dispute with the French, after the publication of a letter from Adrien Auzout to the Secretary of the society, Henry Oldenburg. This, the great century of priority disputes, immediately saw others, including Hooke and Huygens, lay claim to the invention—an invention which to modern eyes seems a relatively simple affair. Gascoigne had just two crosswires with a separation changed by the screw. Auzot and Jean Picard had developed the micrometer into a superior instrument for fine angular measurement, using two separate banks of crosswires. To some extent they were indebted to a simpler design by Huygens. In the meantime,

Eustachio Divini had used a gridiron micrometer for lunar mapping, although this was a relatively trivial affair. At all events, by the late 1660s, the direct measurement of apparent planetary diameters and other very small angles had become reasonably reliable. Subsequently, astronomers and instrument makers made dozens of minor modifications to the basic design of the "filar micrometer," and squabbles about originality gradually died away. One important later improvement was Bradley's introduction of a spring to eliminate backlash in the screw, a device which after the 1730s helped to set new standards for accurate measurement. In 1700, Philip de la Hire ruled multiple lines on a glass plate, using a diamond. Placed in the focal plane, as before, this "ocular micrometer" could be used with or without a screw, although the accurate separation of the lines ultimately depended on a screw in the ruling machine.

THE SOLAR DISTANCE

By the mid-seventeenth century, on the basis of Keplerian astronomy, the relative distances of the planets at any arbitrary time could be stated with high precision. From the apparent sizes of their disks, therefore, the relative (actual) sizes of the planets could be derived, with the exception of the Earth. Some tried to make use of a Horrocks-type principle, according to which the sizes of the planets are proportional to their distances from the Sun. As we have seen, the basic principle was a very lame one, and had Horrocks lived longer, he might well have admitted as much. The fact that Mars is far smaller than the principle implies was unfortunate for those who worked from the sizes derived for Mars and Venus, and who then simply put the Earth somewhere in size between them. Doing this led to an astonishingly large distance for the Sun—say twenty or thirty thousand Earth-radii. Huygens suggested 25,086, implying a solar parallax of 8.2″. As it happens, that figure is reasonably accurate, but its accuracy is spurious, for it is based on reasoning that is unacceptable several times over.

It was now obvious that one of the most pressing of tasks for positional astronomy was to discover the solar distance—or equivalently, its parallax—and so the linear scale of the solar system. We might here briefly review the steady reduction of solar parallax, as proposed at successive periods of history. To mention a few values, all quoted for convenience in arc seconds: Hipparchus had 420″, Copernicus, 211″, Tycho Brahe, 180″, Kepler, 60″, and Horrocks, 14″. The perceived size of the solar system was plainly growing, and yet every one of these figures was known to be heavily dependent on conjecture of one kind or another.

The key figure in the next stage of this great enterprise was Gian Domenico Cassini. Even before he left Bologna for Paris, he had taken a strong interest in solar theory, when he had been asked to restore the "gnomon" in the church of San Petronio. This was not a gnomon in the usual sense of a rod or column. Egnatio Danti had arranged for a small hole to be placed high in the church, so that when the Sun crossed the meridian, an image of it was thrown on a calibrated strip, itself in the meridian plane, carefully leveled and bedded in the church floor. Francesco Bianchini, an important astronomer in his own right, later described Cassini's renewed meridian line in the church as "the largest and most exact astronomical facility that exists in Europe." Allowing a small discount for national pride, this was not an unreasonable comment. The arrangement was capable of giving very accurate solar altitudes, and hence the obliquity of the ecliptic and the latitude of the place. Cassini found that the latter did not square with what was found from systematic observations of the Pole Star—which of course he knew moved round the true pole. He knew that parallax and the effects of refraction were significant, and tried to allow for them, but even then he could only explain his observations on the assumption that the solar parallax was less than 12″, and this went so strongly against received opinion that he put the matter aside, only to take it up again after 1669, when he had moved to Paris.

Cassini had earned much fame from his tables of the motions of the satellites of Jupiter. In 1665, to add luster to the newly founded Royal Academy of Sciences in Paris, Colbert had invited a number of renowned foreigners there—Huygens and Cassini among them. Almost from the moment of his arrival, Cassini participated actively in the affairs of the Academy—sometimes to the displeasure of other members. Under his eventual leadership, the observatory—which was associated with the Academy—for a time took the lead in European astronomy. With its lavishly funded instruments, many of them on a large scale and soon to be equipped with micrometers, it was the first to outstrip unequivocally the observatories of Tycho Brahe.

It was clear that accurate measurements required a detailed knowledge of atmospheric refraction and solar parallax, and an expedition was called for to a place on the Earth's surface from which the Sun could be observed at high altitudes, where refraction would matter less—and not at all if the Sun was in the zenith. An expedition under Jean Richer was therefore sent to the French colony in Cayenne, South America (latitude about 5° N).

A movement had already begun to find accurate figures for the spacing of leading observatories, especially now Uraniborg and Paris, and the size of the Earth itself. Jean Picard had attempted the first measurement, and he had also measured the length of a degree on the Earth's surface, from observations in northern France. One way of determining the solar distance (or parallax) was to measure the related parallax of Mars—we have already seen the pattern of the interrelations between these quantities. A long base-line is needed for any parallax measurement. A "diurnal parallax" is a change in position as seen from a single place on the Earth's surface, as between one time of day and another. (That is to say, it has a value less than the maximum parallax. See the right-hand side of figure 172 in chapter 14, which can of course be regarded as applying to any planet.) Of course there are many calculations, adjustments and corrections needed, but essentially the idea is that the observer is carried from one end of the base-line to the other by the rotating Earth. Tycho had found the diurnal parallax of Mars, when the planet was at its nearest approach to the Sun (perihelion) in 1582. John Flamsteed, before becoming Astronomer Royal, did the same in 1672, deriving 10″ for the solar parallax. Cassini found rather similar values: for long he favored 9.5″.

Without giving every twist and turn in the long search for the correct value of solar parallax, for purposes of comparison we note that the figure now accepted by the International Astronomical Union is 8.794158″, with possible error in the last place of decimals. Lalande came close in 1771, putting the value between the limits of 8.55″ and 8.63″. A year later, Pingré gave 8.80″, and over the next two centuries the values given by about a dozen leading authorities all fell between 8.49″ and 8.84″. Between Lalande and Newcomb, most values were obtained from transits of Venus.

When Richer's expedition returned in 1673, it was with an enormous stock of new observational data. It had become clear, for instance, that the period of swing of a pendulum was different as between Cayenne and Paris—a fact explained in due course by the oblate shape of the Earth, which bulges at the equator, like a squashed ball. The expedition found that the solar altitudes measured could be reconciled with parallax and refraction corrections only if the solar parallax were less than 12″. The implications of this were that the obliquity too needed revision (23;29° was now favored). The figure derived from Mars was therefore roughly confirmed.

It was to be improved yet further as a result of work by Edmond Halley—a well-connected scholar and astronomer of wide experience who by the age of thirty had visited Hevelius in Danzig, assisted Flamsteed, cataloged stars on his own initiative off the coast of West Africa, seen Newton's *Principia* into print, and written a paper that is a classic in the history of geophysics, on solar heating as a cause of trade winds and monsoons. In 1663, James Gregory had drawn attention to a method of finding the solar parallax by noting the timing and manner of Venus's crossing the Sun's disk, during any of the rare occurrences of that event. The details were then to be compared with results obtained by another observer at a different latitude. Halley had tried out the scheme with Mercury in 1677, but he saw that Venus, being much nearer to us at its closest, would give more reliable results. In three separate publications (1691–1716) he calculated the details of what would be seen after his death if the transit of Venus were to be observed. In the event, this influenced the French astronomer and geographer Joseph-Nicolas Delisle to lead an attack on the problem, using the transit of Venus. (He also made several attempts using Mercury's transits.) Delisle coordinated a network of world-wide observation on an unprecedented scale—sixty-two stations were in use altogether, many equipped with the new achromatic telescopes. A year after Delisle's death, the 1769 transit yielded further results from sixty-three stations—many of them using reflectors made by James Short. (Short died in 1768, but he had helped to organize the Royal Society's participation in this venture.)

The human history of this magnificent venture is rich in incident. There were several cases in which the astronomers concerned were accused of fabricating evidence to prove a point. The reputation of the Viennese Jesuit astronomer Maximilian Hell, who had observed the 1769 transit in northern Norway, suffered greatly when Lalande hinted that he had manipulated his observations to fit those of other observers. A successor, Karl von Littrow, claimed that he had found proof in the form of ink of different tints. Not until 1883 did Newcomb disprove these allegations—afterwards also discovering that Littrow had been color blind. Less easily disproved was one account of the Rev. Nevil Maskelyne, who had been sent by the Royal Society to St. Helena for the 1761 transit: his personal account for liquors topped £141 out of a total expenditure of less than £292.

Britain and France could collaborate in matters scientific and yet be engaged in the Seven Years War that lasted from 1756 to 1763 (better known in the United States as the French and Indian War, although it involved all the great powers of Europe). The Royal Society dispatched Charles Mason and George Dixon to Bengkulu in Sumatra, but their expedition got off to a bad start when their ship was attacked in the English Channel by a French frigate, which killed eleven of their crew. Threatened by the Royal Society with disgrace, not to say legal action, if they called off their expedition, they eventually made their observations from the Cape of Good Hope.

The French Academy organized several expeditions to observe the 1761 transit. César-François Cassini de Thury was sent to Vienna, capital of France's new ally Austria. Alexandre-Guy Pingré was sent to the reasonably safe island of Rodrigue, in the Indian Ocean—not that the region was entirely safe. Guillaume le Gentil obtained private funding for an expedition to Pondicherry (near Madras). He arrived there to find that the town had been taken by the British. He waited in the area eight years, descending to work as a trader for a time, before setting up his instruments in Pondicherry for the transit of 1769. Sad to say, the Sun was obscured by cloud—neither before nor after, but only at the time of the transit.

Jean-Baptiste d'Auteroche, to take the example of the leader of another French party, was more successful. Sent in 1761 to Tobolsk, in France's ally Russia, he there collaborated with astronomers from the St. Petersburg Academy. Again in 1769, after trekking across Mexico to a base in what is today southern California, he obtained good results. Alas, on this occasion, he and two other members of his party of four astronomers were stricken with disease. They died almost immediately after the Venus transit, and the sole survivor set out for home alone, on another hazardous journey, with the precious records.

Many other illustrations could be given of the difficulties encountered by the numerous groups engaged in obtaining this essentially rare sort of information—rare in the sense that, after 1769, not until 1874 would there be a similar opportunity to obtain it. One of the saddest discoveries of all was that the image of Venus, at the onset and end of transit, was indistinct. This was inevitable, and was the combined effect of the atmosphere of the planet and the corona of the Sun. Many other useful lessons were learned in 1761, in time for the 1769 transit. More than 120 sets of data had been obtained on that earlier occasion, from places stretching round the globe from St. Johns, Newfoundland—where John Winthrop of Harvard observed—to Beijing. Despite the resulting superfluity of data, there was a wide spread in the values derived for the solar parallax, ranging from 8.28″ to 10.60″. The 1769 transit gave more consistency between far fewer observations, with values ranging from 8.43″ to 8.80″. For a time, the astronomical community settled on Lalande's 8.60″, to which Laplace gave his blessing, starting from his theory of the Moon. Early in the nineteenth century, Johann Franz Encke spent much effort, during more than a decade, in trying to refine the old data using new mathematical methods, in particular, the mathematician Carl Friedrich Gauss's method of least squares. In announcing his final result as 8.57116″±0.0371″, Encke was far too ambitious, and later nineteenth-century transit observations did not improve matters greatly. Eventually, astronomers turned the clock back, and remeasured the parallax of Mars, using this to produce the solar parallax. In 1930–1931, when the asteroid Eros was near opposition to the Sun, its parallax was used to derive the value of 8.794″, which is the figure presently accepted by the International Astronomical Union. Seldom has an astronomical parameter been so hard won.

The Rise of Physical Astronomy

The sound physical basis Kepler had tried in vain to find for his planetary astronomy was eventually provided by Newton, with the theoretical mechanics he built on foundations laid by his predecessors and a theory of gravitation that he had every right to claim as his own. So overwhelming have the implications of Newton's work for astronomy been that the feverish nature of activity in the intervening period has often been overlooked. To understand its character we must first appreciate how inconspicuous Kepler's laws then were, and likewise, the astronomical tables that rested on them. Except in the cases of Mercury and Mars, the known planetary orbits did not diverge greatly from circular paths. The law of areas could be assessed only very indirectly, and was not much heeded by astronomers at first. The third law—relating the periods of the planets to the dimensions of their orbits—was more easily seen, and Jeremiah Horrocks and Thomas Streete made use of it to derive the (relative) dimensions from the easily measured periods. The real test of Kepler, for most of those who were not so prejudiced that they could not even consider accepting his Copernican ideas, was the unprecedented accuracy of the Rudolphine Tables, and yet, even a Copernican like the respected Belgian astronomer Philip van Lansberge, with his follower Martinus Hortensius, could fail to appreciate Kepler's merits. Van Lansberge prepared tables that were as inaccurate as they would have been had Kepler never lived, and yet they were republished twice, were modified slightly by others, and remained in use across Europe for longer than thirty years. They were torn apart mercilessly in a brilliantly devastating little book by the Frisian adherent to Keplerian principles, Jan Fokkens Holwarda of Franeker. Writing under a Latinized form of his name, Johan Phocylides Holwarda, he published his attack first in 1640 and again in another book in 1642. It has to be said that few astronomers seem to have taken much notice of either.

Holwarda will emerge again in chapter 16, in connection with his measurement of the period of variation in the brightness of the star Mira. This

165. Mezzotint after a portrait of
Isaac Newton (1642–1727) by John
Vanderbank (1725), done for the Royal
Society.

was an important discovery, although the study of variable stars made little
real theoretical progress before the alliance of astronomy with spectroscopy in the nineteenth century.

Horrocks's criticisms, not to mention his generally positive support for
Kepler in matters of planetary theory, seem to have had rather more repercussions. One of the outstanding difficulties left by Kepler was in regard
to his equation for the eccentric anomaly (discussed on p. 391 above). As
already explained, there could be found no exact solution that astronomers
could readily use. Horrocks, like Bonaventura Cavalieri at about the same
time, tried repeatedly, and seemingly independently, they found similar
approximate formulas. Horrocks argued against Kepler's magnetic theories and proposed the conical pendulum—a plumb bob swinging in an
oval orbit—as an alternative model. In the 1660s, Robert Hooke, secretary
of the Royal Society in London, revived the analogy. Had Horrocks not
died in 1641 at such an early age, his ideas might well have led to something more valuable.

Among the leading astronomers of Europe, many were dissatisfied
with Kepler's work for reasons often having little to do with the prediction
of planetary positions. Ismael Boulliau, a devoted admirer of Kepler, was a
noteworthy example. A French Calvinist from Loudun who converted to
Catholicism and became an ordained priest, he entered Parisian astronomical circles around 1633 at the very time of the Galilean crisis in the Catholic Church. This did not prevent him from joining forces with his friend
Gassendi in speaking up in Galileo's favor. He accepted Kepler's elliptical
orbits, and in 1645 published tables based on them, but with a different law
of motion from Kepler's.

Kepler had attempted to use the inverse square law of illumination to
support an analogy between light and power that would account for his

mathematical laws of planetary motion. Boulliau took this over in a work of 1638, before modifying it. A printed account appeared in a work called *Philolaus* (1639), named after Philolaus of Tarentum, the supposed author of the Pythagorean astronomy that had displaced the Earth from the center of the universe. (In the ancient world even Plato had been accused of plagiarizing Philolaus, when writing his *Timaeus*.) It was in his more polished *Astronomia philolaïca* (1645), however, that Boulliau brought to wide public notice Kepler's attempts to discover the moving power of the planets. He struggled to reinstate Kepler's analogy with light and made several sharp criticisms of Kepler's special pleading when trying to make the analogy work. His admiration for the mathematical side of Kepler's work was never in question.

Boulliau's own planetary theory was kinematic, that is, force-free and descriptive, in the way that all before Kepler's had been. It is extremely complex, and it contains many mathematical errors, some of which were pointed out at the time by Paul Neile and by Seth Ward, the Savilian professor of astronomy at Oxford. In his *Astronomia philolaïca*, Boulliau had used certain principles that were at best arrived at intuitively. As an example, he assumed that in moving 90° in mean motion from aphelion, the planet reaches the average of its true speeds at aphelion and perihelion. He decided that only a path that was a section of a cone—he took an oblique cone—satisfied this rule. Of course an ellipse is a conic section, and this is what he accepted as the path. Like Kepler, he thus had an ellipse, but his laws of motion, that is, the laws for calculating the eccentric anomaly, were very different from Kepler's. As Ward showed, however, they did not even follow from Boulliau's own principles.

Ward showed that from these principles the planet should move uniformly around the empty focus of the ellipse, the focus not occupied by the Sun. In a strange act of deference, Ward now adopted this principle, which makes the empty focus an *equant* point, as it were, for the motion. That different way of calculating the planetary motion was published in Ward's *Astronomia geometrica* (1656), which is actually dedicated to Boulliau, among others. Boulliau counterattacked with another book in 1657. This acknowledged some errors and scored off Ward chiefly by noting the impracticality of his proposals for deriving planetary parameters. The whole episode proved only one thing: astronomy was a theoretical science working at the very limits of the observational techniques by which it was to be justified or rejected.

Many other, lesser minds, failed to see that Boulliau's results were inferior to Kepler's, and it is hard to avoid the conclusion that this was because Boulliau was a more blatant adherent of astrology, whose works were therefore respected uncritically by others of the same persuasion. Jeremy Shakerley, John Newton, and Vincent Wing all produced works and astronomical tables for London that were greatly influenced by Boulliau's. Isaac Newton remarked on the accuracy of Boulliau's tables, perhaps surprisingly in view of their known shortcomings. Boulliau was elected a member of the Royal Society but was never elected to the Paris Academy, and this is consistent with the expressed dissatisfaction of Huygens and Picard with his tables, which they regarded as still inferior to the Rudolphine Tables. They also knew from the work of Horrocks, which by this time had been published posthumously, that Boulliau's figure for the mean solar parallax (141″) was far too large.

One of the more regrettable habits of seventeenth-century astronomers—one of which medieval astronomers had been guilty, but with more excuse—was that of drafting planetary tables using parameters that had been drawn from inconsistent theories. In drawing up his *Astronomia Carolina* (1661), the London-based Irishman Thomas Streete made use of parameters from Kepler, Boulliau, Horrocks, and others in a careful blend, but one that could only be justified *a posteriori*, that is, by virtue of the fact that Streete's planetary tables made possible more accurate predictions than most rivals. No truly systematic comparison with the heavens was ever made, and it is doubtful whether Flamsteed, for instance, who praised them greatly, realized how best to go about doing this. They were reissued in 1689 with slight amendments by Nicholas Greenwood, put into Latin by Johann Gabriel Doppelmayer in 1705, and between 1710 and 1728 were reissued five times in all by Edmond Halley and William Whiston—in Whiston's case, with his own book. It was from Streete's *Astronomia Carolina* that Newton learned Kepler's first and third laws of planetary motion. The second law he probably learned from Nicholas Mercator's *Institutiones astronomicae* (1676).

It was Mercator—who was born in Denmark as Niklaus Kauffman but settled in England—who effectively put an end to the Boulliau-Ward hypothesis of an ellipse with equant at the empty focus. He showed that Cassini's method of determining the line of apsides of a planetary orbit was dependent on it, and that therefore that too must go. Mercator was more important as a mathematician than as an astronomer, but in astronomy he helped to clear away some of the last traces of mist that drifted between Kepler and later theoretical astronomers.

Despite the inaccuracies of his predictions, astronomers were long in improving on the lunar theories of Tycho, perhaps because his ideas were so complex that most found them impenetrable. An important breakthrough came in 1672, with Flamsteed's publication of Horrocks's lunar theory. Incomplete though the theory was, it revealed—with much intermediate reliance on Kepler's account—both a variation in the eccentricity of the orbit, and an oscillatory motion in the line of apsides. Horrocks's manuscripts had been scattered, and some were lost, but Hevelius had printed those on the Venus transit, and the Royal Society published most of the remainder in 1672–1673. (For more on Horrocks's concern for Keplerian ideas, see p. 391 above.) Flamsteed added tables to this edition, making some improvements in the process—as well as a serious mistake that Halley later corrected. It is to Flamsteed's great credit that he recognized the talents of Horrocks, who had died thirty years before.

CARTESIAN COSMOLOGY AND THE VORTEX THEORY OF PLANETARY MOTION

When Tycho Brahe and those of like mind dismissed the solid spheres of the Aristotelian heavens, they left an intellectual vacuum. Most scholars were unhappy with idea of banishing material from the heavens entirely, and few were truly satisfied with the magnetic cosmology that Kepler had introduced. The attempts at improving on Kepler's planetary theory took relatively little account of physical arguments: Kepler was a rare example of a scholar who could marry the two streams of thought. When the French philosopher René Descartes came on the scene with an elaborate substitute for the Aristotelian theory of matter, he gave cosmologists exactly what most of them wanted, a universe in which movements took place under the action of matter on matter. Alas, for Descartes, some of the best mathematical minds in Europe were unable to turn his imaginative cosmology into a theoretical scheme to rival Kepler's, and certainly not to rival Newton's, later in the century. Endless attempts to do so were made, nonetheless, and in some quarters the struggle persisted well into the eighteenth century. As with Aristotle's physics before it, Descartes' was psychologically satisfying, and it had a large following, but in the end it proved to be a superfluous psychological luxury.

René Descartes was born in La Haye in Touraine, France, in 1596. A member of a family of minor nobility and of independent means, he was well educated, even in the latest scientific matters, at the Jesuit college at La Flèche. Having graduated in law from the University of Poitiers, where again there was a strong Jesuit presence, he became not a priest but a volunteer in the army of Prince Maurits of Nassau—hereditary stadholder of the United Provinces of the Netherlands. In 1618, while stationed in the town of Breda, Descartes had the good fortune to meet Isaac Beeckman, assistant headmaster of a school on the island of Walcheren.

Despite his less than illustrious position, Beeckman was an important man of science in his own right. He had a lively interest in the natural sciences, and was at that time making astronomical observations with Philip van Lansberge. He introduced Descartes to a number of recent problems in mechanics, and some of Descartes' most important insights into algebraic geometry—Cartesian geometry, as we know it—stem from this time. The philosophical career for which he is more often remembered did not really begin in earnest for another ten years, during which time he traveled widely. From 1628 to 1649, he lived for the most part in the Netherlands. Persuaded then to serve as philosopher to Queen Christina of Sweden, he died in Sweden in 1650, victim of a cold climate and a Utrecht physician.

Descartes' best known work, his *Discourse on Method*, was published in 1637, together with the three treatises *Meteors*, *Dioptrics*, and *Geometry*, in a single volume. In *Dioptrics*, he extended the theory of lenses set out by Kepler, but now with the sine law of refraction, and this treatise was of great importance for the later development of a subject on which astronomy was becoming increasingly dependent. In cosmological matters, his influence was of a different sort. He argued, in ways not unlike Aristotle's, against the existence of a vacuum, and he insisted on the idea that mechanical effects be explained by the action of matter on matter.

Descartes treated motion in a straight line as a *state*, just as *rest* is a state; and since a cause was needed to change a state of rest, so it followed that a cause was needed to change a state of motion in a straight line. This, a form of the law of inertia, was to assume great importance when adapted by Newton.

The same is true of Descartes' law of the conservation of quantity of motion (a product of the separate magnitudes and velocities of bodies in a closed system). The steps by which this was transformed into—as we should say—a law of the conservation of momentum are important, but are not our main concern. In this connection, Descartes developed a theory of colliding bodies, but it was highly unsatisfactory, and Huygens greatly improved on it in the 1650s.

Of more immediate importance for any theory of planetary motion was Huygens's theory of centrifugal force, developed by him in the late 1650s, but not then connected with any idea of a gravitational force of the Newtonian type. The law expressing the acceleration of a rotating body toward the center of its path, in terms of its velocity and the radius of its path, does not require the Newtonian laws of motion, but those are needed to turn that simple law into a law of centrifugal force. In the strictest sense, therefore, the latter cannot be counted as Huygens's.

In the years 1629–1633, Descartes developed a system of the world that rested on a theory of celestial vortices, whirlpools of subtle matter. He explained terrestrial gravitation as an effect of these vortices. A treatise he wrote on the subject, *Le Monde, ou traité de la lumière* (*The World, or, Treatise on Light*) was ready for publication, but hearing of the condemnation of Galileo he decided not to publish, and it was first printed only posthumously in 1664. As he gradually realized, a universe of vortices is one in which every natural body may be at rest with respect to the local matter and yet moving with respect to distant bodies. This seemed to him to be an answer to the problem of placating both the Copernicans and those who took the Earth to be stationary. He could say that, in a sense, they were both right. Emboldened by this insight, he made public his vortex theory in his *Principia philosophiae* (*Principles of Philosophy*) in 1644. The book that was soon translated into French and became highly influential.

His cosmology made use of the idea that there are three different forms or elements: luminous, transparent, and opaque. The first was the finest, with particles moving at great speeds, and it made up the Sun and stars. The Earth and planets were of the coarse third element, and the second element, filling the spaces between these different sorts of bodies, was of globules in rapid motion. Celestial matter was supposedly capable of penetrating the pores of terrestrial matter. It was no easy task to rationalize the vortex motions in three dimensions. Each vortex has an equator and poles, and it is no easy matter to explain how they are fitted together. He developed a theory of the movement of matter, from the equator of one vortex to a pole of another, for instance, with collisions modifying the shapes of particles. The shapes were supposedly so designed as to facilitate the passage of particles through gaps between others. Magnetism was seen as evidence for the vortices, and so magnetism was worked into the cosmic scheme, as were sunspots—elements of the third type, floating on the Sun for a while, in the course of the vortical process. (For a specimen of Descartes' highly complex cosmological theorizing—one might even say temporizing—using the vortex principle, see fig. 166.) The comets too were found a place, as were the satellites around planets, and the Moon of course, and the Earth's diurnal motion. This was a theory truly meant to solve all physicists' problems. Gravitation toward the Earth's center was seen as analogous to the tendency of floating bodies to move to the centers of whirlpools on water. It was all very ingenious, but almost entirely qualitative, and in several places, inconsistent. It is not known whether Descartes even knew of Kepler's laws of planetary motion. If he did, he seems to have made no attempt to explain them.

The Cartesian ideas were welcomed at first, with other parts of his philosophy, in the universities of the Netherlands, and later in informal discussions held among scientists and scholars in Paris. At the beginning of the seventeenth century, atomism, of the type taught by the Greeks, was widely regarded as spiritually dangerous, and was linked in many scholars' minds with atheism. Descartes was anxious to avoid being associated the atomists, but eventually the link was made, and it was made all the more easily since Pierre Gassendi—in fact an opponent of Descartes—had sweetened the pill by making spirits, or minds internal to them, the causes of atomic movements. The popularity of Cartesianism nevertheless snowballed, supported by such disciples as Henricus Regius in Utrecht, and Jacques Rohault, Pierre Sylvain Régis, and Nicolas Malebranche in Paris. They added new phenomena to the list of those treated by Descartes, but their treatment was still qualitative, and like Descartes, none of them ever made a serious attempt to explain Kepler's laws—which indeed received only passing mention by Cartesians even as late as the eighteenth century. It was as though astronomers and natural philosophers inhabited two different worlds.

Huygens was that rare phenomenon, an early Cartesian capable of making use of quantitative argument. It is significant, however, that by the time he and others—for example, Gottfried Wilhelm Leibniz—were beginning to achieve a measure of success in explaining gravitation in the solar system on Cartesian principles, they were tacitly making use of

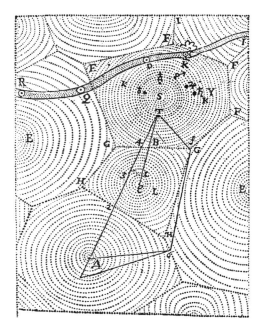

166. The vortex structure of the solar system described by Descartes in *The World*, composed in the period 1629–1633. The Sun is at the center of one of the many vortices that fill up the Universe completely. The Sun comprises the smallest filings of matter, the first element. Larger spherical pieces of matter, the second element, tend in their rotation to be flung outwards. Those small spheres carry round the planets, made of larger pieces still, the third element. All vortex centers are like *S*, in having planets around them. The small spheres at the periphery of the vortex spin fastest, but the slowing as we move inwards suddenly changes on reaching Saturn. Thereafter, the spheres must become smaller, since they revolve more rapidly. The reversal was forced on Descartes by the shortening of the planetary periods, moving inwards (Saturn nearly thirty years, Mercury about three months). He concluded that the Sun, being innermost, must turn faster than any planet. When revising the scheme for *The Principles of Philosophy*, however, he learned of the recent discovery of sunspots, and of their slower rotational speed than the planets. He therefore made another ad hoc proposal, saying that the Sun has an atmosphere, extending as far as Mercury, that slows the spots down. (This is no more than one stage in his long and involved, but entirely qualitative, cosmology.)

certain principles developed by Newton, and so were working within a very different intellectual system. Unfortunately, they were the exceptions to the general rule. Most of those who wrote in the Cartesian tradition seem to have been overwhelmed by the notion that cosmological argument was within the reach of all who had a clear and distinct idea of change that comes about by the action of matter pushing matter.

THE NATURE OF COMETS

The discovery by Apian, and later Fracastoro, of the fact that cometary tails always point more or less away from the Sun had an unforeseen consequence, in that it introduced the Sun into the discussion of the nature of comets and their tails. We have already mentioned one new line of discussion based on the principle that comets are spherical lenses. Another shift in opinion came as a consequence of Tycho Brahe's evaluation of their distances, starting with the comet of 1577. He stirred up the debate again a decade later, putting forward the idea that comets are made of a hitherto unrecognized celestial material, neither pure and transparent nor opaque. Comets, he decided, were capable of entrapping solar rays, as well as the rays from planets—for we recall that he gave Venus responsibility for the comet of 1577. This entrapment explained their visibility, but he also suggested that they are partly porous, and so let some rays escape, forming a tail. Almost every major astronomer after him addressed this problem, although the conclusions they reached were inevitably of a speculative nature. For Galileo, comets were reflections of solar radiation from terrestrial exhalations—a very Aristotelian solution. For Kepler, they were of a matter denser than the ether, but not solid, and their tails were expelled from the nucleus by the impact of solar radiation. That master of speculative cosmology, Descartes, fitted them neatly into his vortex theory in a very different way, making them out to be optical phenomena. They are, he said, in the outer reaches of vortices, each being well beyond the planets belonging to its vortex. Indeed, they could change their vortex. The tail of each was taken to be a purely visual effect of reflection at the comet's head, although Descartes explained how tails take on their appearance—referring the reader to an inordinately complex, wholly qualitative, theory of the deflection of light from particles of different sizes. Flamsteed and Hooke both proposed ingenious variants of Kepler's material theory of tail formation. There are many more variants to be found in the astronomical literature of the time, but toward the end of the seventeenth century the situation changed when comets finally became recognized as remote members of the solar system.

It is easy to appreciate how natural it was for Isaac Newton to assume that comets were planetary in character, for in doing so he brought them into his gravitational world system. He had other reasons, however. He calculated the heat they received from the Sun and concluded that it was too great to allow them to survive unless they were highly compact. He decided that each was bathed in a sea of vapors expelled from the nucleus, and that a comet's tail was vaporous, set in motion in some way by the Sun. He gave much thought to this spreading of cometary vapor, assigning to it responsibility for nurturing the planets and replenishing the stars, redressing their loss of matter by radiation. But how did he suppose that the vapor was made to stream into a cometary tail? He took the analogy of smoke in a chimney, in which particles of soot are carried away from the gravitational center (the Earth) with the rarefied hot air. In the case of a comet, the gravitational center is the Sun, and the analog of the hot air in the chimney is ether that has been warmed by the Sun. This warming process he supposed to be indirect, with the Sun heating the comet's atmosphere, which then heated the adjacent ether. Newton's analysis was plainly of a higher order than earlier accounts, grounded as it was in terrestrial physics. There remained, however, the thorny question of the mechanical ether, and the workings of this were something that Newton made several

different attempts to explain. His explanations fall into two groups. In the
1670s he supposed the ether to be a subtle medium, penetrating matter,
but capable of acting by impact on matter—indeed, capable of giving rise
to what we call gravitation. After 1710, or thereabouts, he suggested that
small particles that repelled one another and that were repelled by gross
matter comprised the ether. This hypothesis was put to work in an ingenious way: ether particles were supposedly rarer in matter—as in the stars,
planets, and comets—and the consequent reduction in repulsive force explained why dense matter moves toward dense matter. Newton proposed
several variants, but they need not be pursued here, for they were not specifically concerned with comets.

Newton's theory of cometary tail formation attracted much attention,
especially through the influence of textbooks by Willem 's Gravesande.
This Dutch scholar's Latin works brought the first comprehensive accounts
of Newtonian natural philosophy to continental Europe—from whence
they returned to the English-speaking world in translation. His familiarity
with Newton's work, and with Newton himself, came about from his having spent a year in England (1715–1716) as secretary to a Dutch embassy,
sent to congratulate George I on his accession to the throne. He became
actively involved in the affairs of the Royal Society, corresponding with
several of its leading members, but his later career was marked by a whole
series of excellent textbooks, including his *Mathematical Elements of Natural Philosophy* (in Latin and in English, both in 1720, with many later editions). A similar work first published in French, but almost immediately
put into English, was Voltaire's *Elements of the Philosophy of Newton* (both
versions were published in London in 1738). Voltaire made the acquaintance of his subject while in exile in England, in 1726, but 's Gravesande
was an important source for him, and Voltaire actually visited Leiden to
have the Dutch scholar give his approval for the contents of his own work.
Notwithstanding this fact, Voltaire's account misrepresents Newton on the
nature of the "smoke" of cometary tails.

Much the same could be said for several English writers of the time,
for example, William Whiston, David Gregory, and Henry Pemberton, although most of them would all have insisted that they were aiming to go
beyond, and improve, their source. Whiston and Pemberton, for instance,
said more about cometary atmospheres, noting that these could not be like
planetary atmospheres, or planets would also have tails. Gregory delved
into the nature of light pressure, as a means of propelling the tail particles.
All three men were well known to Newton. Pemberton was even the editor
of the third edition (1727) of his *Principia*, and yet no hint is to be found
in that work of its author's having changed his views. As the eighteenth
century progressed, dozens of astronomers put forward theories of their
own, usually modifying only slightly the several versions mentioned here,
but not one was given high scientific standing. It is true that they were easily understandable, unlike Newton's theory of gravitation, by comparison

with which they seemed to be almost trivial. Whether or not one could understand it, there was a world of difference between the arguments of the cometary theorizers, for the most part lacking in an empirical basis, and the well-grounded planetary theories, with two thousand years of history behind them. For the time being, the positions of celestial objects in general, and of comets in particular, were measurable, whereas their natures were only guessable. This situation would only change in the nineteenth century, with the introduction of astronomical spectroscopy.

ISAAC NEWTON AND UNIVERSAL GRAVITATION

The Cartesian search for a plausible theory of gravitation had come to nothing. Another largely qualitative style of cosmological thinking that paralleled Cartesianism, and occasionally overlapped with it, was an extension of the magnetic philosophy of Gilbert and Kepler. This survived, and was long actively pursued, in England, where it was kept alive especially in discussion groups centered at first on Gresham College in London, and later the Royal Society. John Wilkins and Christopher Wren were both spokesmen for Gilbert. In 1640 Wilkins published two easily comprehensible books concerning the possibility of a voyage to the Moon, both affording some publicity to Gilbert's and Kepler's ideas, namely, *Discourse Concerning a New Planet* and *Discovery of a World in the Moone*. In 1654 Walter Charleton, a follower of Wilkins, published a blend of these ideas with Gassendi's atomism. Even so, it is hard to believe that without Wren, Edmond Halley, and Robert Hooke, the English discussion would have remained focused, as it did, on the laws governing the precise forms of the orbits, and their relation with laws of mechanics—such as a law of inertia and some sort of law of a central (Sun-directed) force of attraction.

The discussion was stimulated by the appearance of a comet in 1664. Wren was at this time Savilian Professor of Astronomy at Oxford, and John Wallis was Savilian Professor of Geometry. Wallis was advancing the cometary theory of Horrocks, who had analyzed the motion of the comet of 1577 into a straight line motion, modified by the Sun's magnetic action. Wren now attempted to derive the new comet's path on the basis of four observations, assuming with Kepler that it followed a straight line at constant speed.

Within months of the first, a second comet appeared, which served to stir flagging interests in a problem that was evidently beyond the powers of all concerned. Hooke tried the hypothesis of circular motion, but tended to favor the rectilinear alternative, with some sort of solar attractive power. He suggested that there might be an all-pervading ether that was vibrating, with the vibrations diminishing as the distance from the Sun increased. Here was a mechanism that might have owed something to Descartes' idea of a circulatory ether, but that was much more in tune with the law of a central attractive force that was in the thoughts of Hooke and his friends at the Royal Society.

Hooke, curator of experiments there, was beginning at this time to conduct a long series of experiments on gravitation. Ten years later, he had made no appreciable theoretical progress, if we are to judge by the theories of Newton that were soon to emerge, but we may use him as a measure of changing attitudes toward the old Aristotelian universe. For Hooke, as he claimed in a lecture of 1670 (first published in 1674), *all* celestial bodies have an attraction or a gravitating power toward their own centers, which binds their parts together and also acts on other heavenly bodies that fall "within the sphere of their activity." This last qualification suggests that he thought the force to fall off to zero at some finite distance. He admitted that he had not yet verified the law of force—but he did not say what he thought it to be.

In or around 1677, Newton discussed these matters with Wren and took it that Wren was assuming a law that made the force fall off as the inverse-square of the distance between the attracting bodies. His own announcement of the inverse-square law was first published in his *Principia mathematica philosophiae naturalis* (*Mathematical Principles of Natural Philosophy*, but always known simply as the *Principia*; 1687). It seems likely that he had become convinced of its truth only about three years earlier, but the law was only one piece in a complex jigsaw puzzle, and for the rest we must look to earlier stages in his career.

Isaac Newton was born on Christmas Day 1642 in Woolsthorpe, Lincolnshire. His father died before he was born, and his early years were spent with his grandmother, as his mother had married a clergyman for whom Isaac seems to have had no great affection. After school in Grantham, he entered Trinity College, Cambridge, in 1661, and four years later he returned home when the university was closed because of an outbreak of the plague. He was already developing an interest in current mathematical and scientific affairs. He read widely—he learned much mathematics from the works of Descartes and Wallis—and the interlude during the plague gave him time to make a number of original discoveries of his own. By 1669 his qualities were such that he succeeded Isaac Barrow as Lucasian Professor of Mathematics at Cambridge. He went often to meetings of the Royal Society in London, but left Cambridge only in 1696, when he was made warden of the Mint. He died in 1727, a national figure with an unrivaled international scientific reputation. He wrote much on religion and on a variety of other subjects peripheral to those for which he is best known, namely mathematics, physics (especially optics), and theoretical mechanics. And it is through these that he influenced the course of astronomy most materially.

One of Newton's student notebooks, begun in 1661, shows that he was then aware of Kepler's third law, and of some of Horrocks's observational records, and that he had studied the methods for finding planetary position given in Thomas Streete's *Astronomia Carolina*. By 1664 he had improved on Descartes' work on the "conservation of motion"—he realized that the directions of the "motions" had to be taken into account—and he had developed a theory of centrifugal force—which Huygens announced independently only in 1673. Combining this with Kepler's third law, he applied it to the case of the Moon, which is about sixty Earth-radii distant, and to an object at the Earth's surface—the famous falling apple, for instance. Doing this, he confirmed a previous conjecture that the force acting on such bodies diminished inversely as the square of the distance.

His data were poor and did not confirm the idea as well as he could have wished, and it is for this reason that he is usually supposed to have set the question aside for twenty years—in fact until 1685, when he was writing the *Principia*. In the meantime we know that he read Borelli, who in 1666 had written of the planets that their curved orbits implied a centrifugal force which could be regarded as equal and opposite to a force of attraction by the central body. This passage would certainly have meshed very well with his own ideas, but it is not clear that it was new to him.

It was at the very end of the 1670s, or even after 1680, that Newton, having gained a mastery of his newly developed dynamical principles, encountered Kepler's law of areas. This, combined with an exchange of correspondence with Hooke, set Newton on what turned out to be a very fruitful course. Hooke wanted to know the law of central force that would turn the straight-line motion of a planet into an ellipse of the sort Kepler said the planets followed. Newton was in possession of precisely the tools needed—namely, the methods of his own infinitesimal calculus and his dynamical principles—to show that the law of areas implies that the force is indeed directed to a single center, and that it is as the inverse square of the distance. An essential step in the proof was that a homogeneous material sphere exerts a gravitational force exactly as if all its mass is concentrated at its center.

In December 1684 Newton asked Flamsteed for data—distances and periods—on Jupiter's satellites, and Flamsteed replied, in effect, that they are in agreement with Kepler's third law, the law relating period to orbital size. Newton also asked him whether there was anything to suggest that he was right in a hunch that Jupiter might perturb the orbit of Saturn. Flamsteed explained some errors he had detected in Kepler's parameters for these planets. Both answers pleased Newton, for in the first case the implication was that the effect of the Sun on the satellites could be ignored. The second answer meant that Kepler's data were not above reproach, and freed him, as he thought, from an obligation to take other types of force than gravitational into account. They could not, at least, be proven on the strength of Kepler's data.

It was in 1684 that the learned world began to learn of the intellectual riches that Newton had been amassing. Halley visited him in Cambridge to ask what path a planet would follow under the action of a force proportional to the inverse square of the distance. He explained that Wren, Hooke, and he had failed to solve the problem. Newton's answer was that it was an ellipse, and that although he could not find the proof, he would send it on to Halley. (After correspondence with Flamsteed, he had decided that cometary orbits were parabolic, a consequence of the inverse-square law. This was no less important a conclusion.) This led Halley, after seeing some of Newton's remarkable writings on mechanics, to press him to publish. The *Principia* was the result, put together in a remarkably short time.

No sooner was the work published than a number of disputes as to priority arose. Hooke was loud in protesting that he had priority in applying the inverse-square law to the problem. He had certainly mused at length on Boulliau's criticisms of Kepler on this subject. He had rejected Boulliau's idea that light is neither body nor substance, but a sort of geometrical medium between the two. On the contrary, said Hooke, it is wholly corporeal, a motion of the parts of the luminous body caused by the source. He wrote much along these descriptive and impressionistic lines. In 1685 he penned some speculations about how gravity could be explained as a "continual impulse" from the center of the Earth, with a power reducing as the inverse square of the distance. His model was the old one, with light thinning out, as it were, in inverse proportion to the base of the light cone, and until he came to assert his own priority, he seems to have regarded that as common knowledge, at least since Kepler. So much for the Earth's gravity; but when he came to apply similar ideas to the Sun, he offered instead an inverse fourth-power law. And hard as he tried, he never came close to an explanation of Kepler's results. In a letter to Halley of 1686, Newton was not boorish enough to draw attention to Hooke's mathematical inadequacies, but gently pointed out that what he himself had proved was with Hooke no more than a hypothesis. He went further and said that even Kepler had only conjectured that his ovals were ellipses. Newton, however, could now settle the matter at issue, using the "correct" law of force.

Newton did not mention Kepler's name in his *Principia* until the third book, but there was no thought of concealing a debt. As Halley noted in his review of the *Principia*, his first eleven propositions were in full conformity with the "Phenomena of the Celestial Motions as found by the great Sagacity and Diligence of Kepler." Book III of the *Principia* is called "The System of the World," and there is a strong sense in which it is the very first complete explanation of material movement in all parts of the universe under the action of a single set of physical laws. The motions of the planets and their satellites, of the comets, of the Earth and the tides in its seas—all are explained in terms of a *universal* gravitation. The planet attracts the Sun as the Sun attracts the planet. All matter attracts all other matter, and the force is independent of the type of matter. Only the "quantity of matter" and the separation were significant. He performed experiments with pendulums made of different materials, and found no consequent difference in their mechanical behavior.

As we saw from his correspondence with Flamsteed, he realized that while the gravitational forces between the central body and the planets are considerably greater than those between the planets themselves, the latter cannot be ignored, especially where planets make close approaches to one another. The perturbing action of the Sun on the Moon is another important non-central force that cannot be ignored. The important theory of planetary perturbations thus entered Newtonian celestial mechanics at its very birth.

With his powerful dynamics and theory of gravitation at his disposal, Newton was in a position to explain the flattened shape of the Earth, and how the Sun's force of attraction for the near side of the bulge is very slightly greater than that for the far side. This discrepancy produces a turning effect (couple) on the axis of the Earth. This couple, as he could then show, leads to a precessional, conical motion of the Earth's axis, equivalent to a precession of the equinoxes. For the first time in history, this phenomenon had been explained in terms of physical laws.

There was much in the *Principia* concerning comets, and their virtually parabolic or elliptical orbits. Newton realized that comets shine by the reflected light of the Sun, and that the space through which they move could not be offering significant resistance to their motions. Newton is often disregarded as an astronomical observer, and of course that was not his profession, but he was quite capable of supplementing the observations of others—collected from wherever he thought worthy—with some of his own. In *Principia* Book III he showed how, from three given observations, one could determine the orbit of a comet moving in a parabola. The mathematical underpinning of his semigraphical solution to this "problem of very great difficulty" cannot be summarized briefly or simply, but when taking the comet of 1680–1681 as an example, after tabulating fourteen observations by Flamsteed he added seven excellent observations of his own, made with a 7-foot telescope fitted with a filar micrometer. He found his theoretical solution just in time for the first edition of 1686. In the second edition, printed in 1713, he gave Halley's recalculation of some of the data; and in the third, in 1726, he gave more calculations by Halley, based on the idea that the comet was periodic, and identical with the comets of 44 BC, AD 531, and AD 1106. That particular identification was disproved in the nineteenth century, but by then the periodicity of comets had been proved brilliantly by Halley, in connection with another comet entirely, as we shall see.

With the help of his own observations, Newton set to work on his analysis of the comet's path. First he improved the star positions in Perseus, the background to the comet's observed path. His analysis was not only of the path but of the direction of the tail, which he discussed with far more insight than anyone before him. It seems that he made a large scale drawing of the path, on which the Earth-Sun distance was about 41.5 centimeters. (This is the distance of *GH* from *D*, in fig. 167.) Assessing the accuracy of his results in the 1920s, A. N. Kriloff found that on such a scale it corresponded to better than 0.05 millimeters, far greater than could have been

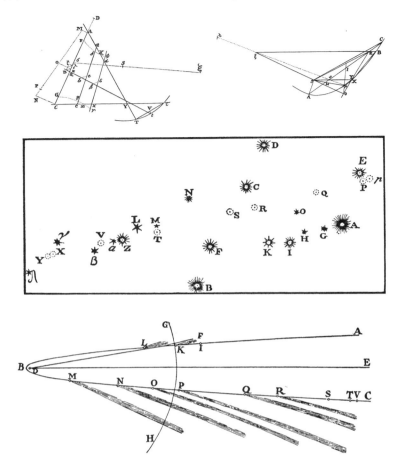

167. In *Principia* Book III, Newton showed how, from three given observations, to determine the orbit of a comet moving in a parabola. The two uppermost diagrams, which were used in this connection, should give a feeling for the complexity of the problem. Applying his method to the comet of 1680–1681, with the help of his own observations he first improved the star positions in Perseus (middle figure), relative to which he placed the succession of comet positions (of which here note *PQRSTV*). Finally, he plotted the orbit and tail-directions (lowest figure, with corresponding points *PQRSTV* and other comet positions). Some editions had a finer and larger drawing, not as plainly labeled, but with points given dates and great attention paid to the tails. The arc *GH* represents the sphere of the Earth's orbit. The Sun is at *D*, the focus of the parabola.

achieved using the method he described. Whatever the explanation, there is no doubt that he made good use of observations made in very many places. In addition to Flamsteed's and his own, he used some by Hooke and James Pound in London, others from Italy, Nuremberg, the East Indies, Jamaica, and Boston (in the American colonies), and yet others by Gottfried Kirch in Saxony. No previous comet had been subjected to such powerful and effective scrutiny. (It is perhaps worth noting that Gottfried Kirch—member of a family that included four good astronomers—was in his day not only known for his prolific correspondence with scientists throughout Europe, but for his having discovered several comets.)

Newton's analysis did not end, however, with his plot of "a true representation of the orbit which this comet described, and of the tail which it emitted in several places." He went on to investigate its physical properties, concluding that comets must be "solid, compact, fixed, and durable," or they could not survive their passage close to the Sun. He gave ten closely argued pages to his discussion of the constitution of comets and the properties of their tails, yet again drawing on reports from elsewhere—indeed, not only from Brazil but now from such historical sources as Aristotle, the *Saxon Chronicle*, Matthew Paris, and Simeon, monk of Durham. When discussing the physical properties of comets, he was not speaking impressionistically, for he had experimented with the heating effect of summer sunshine on dry soil, and so had calculated that the comet at perihelion must have received heat "about 2,000 times greater than the heat of red-hot iron." Having next raised the question of the comet's ability to retain its heat, again he reduced the problem to numbers. While he left these physical problems open, for future experimenters to investigate, he stated clearly what those problems were. It would be many decades before the challenge was taken up seriously by others.

Among the most compelling parts of Newton's work, at least for the few experts who were capable of reading it with understanding when it first appeared, were those dealing with the Moon's motion. Newton explained in general terms the gravitational causes of the known inequalities in the Moon's motion, the motion of the nodes of the orbit, and the reason for the Moon presenting always the same face to us. In the second and third editions he supplemented his lunar theory, having been asked by Halley to continue with his work on it. By the time he had finished, he had as many as seven "equations" of lunar motion, some of them possibly found from Flamsteed's observations rather than from fundamental gravitational arguments. His data were such that lunar tables based on his work, by such as Flamsteed, Charles Leadbetter, and Halley, were hardly better than those based on Horrocks's methods. It was in its potential that the real merits of Newton's theory lay.

In the 1690s, Newton desperately needed Flamsteed's observational data, but the two men, disagreeing strongly as to whether theory should lead observation or follow its lead, quarreled violently. The fact that there was friction between Halley and Flamsteed did not help, but as Newton had aged he had become increasingly autocratic. As warden of the Mint, he had distributed largesse to Halley, in the form of comptrollership of the Chester Mint (1696). In 1699 matters grew more serious when he told Flamsteed that he was interested in his observations, not his calculations. Flamsteed had always felt that his observations, made with instruments largely financed out of his own pocket, were his own. Newton and Halley took the view that the Astronomer Royal's work was public property, and in 1712 they published a considerable part of his work without his approval. The arrival of a Whig government in 1715 led to an improvement in Flamsteed's fortunes. Through the Lords of the Treasury he obtained three hundred unsold copies of the 1712 *Historia coelestis*. After removing parts that he had personally approved, and keeping a few copies to justify himself in the eyes of his friends, he had the somber pleasure of burning the rest. As he said, he "made a Sacrifice of them to heavenly Truth." Flamsteed was long in preparing to publish his work himself, the three-volume *Historia coelestis Britannica*. (In this *British History of the Heavens*, the word "history" has the sense of "data" in modern parlance.) He died in 1719, before it was ready for the press, but it was eventually published in 1725. The companion atlas of stars, of which we shall say more in chapter 15, was seen into print by his widow, Margaret, and one of his assistants, James Hodgson, in 1729. They judiciously omitted a section Flamsteed had prepared, bitterly describing his dealings with Newton and Halley. Flamsteed's catalog and atlas far surpassed in accuracy everything before it, but his reputation has always been overshadowed by Newton's.

A curious footnote to Newton's theory of the Moon concerns his estimate of its average density in relation to the Earth's, based on the relative tidal effects of the Sun and Moon. In the first edition of the *Principia*, this was overestimated by a factor of three. The fact that the Earth seemed so relatively light led Halley to the conclusion that four-ninths of it must be hollow. He was not the first with the idea, which will be found, for instance, in Thomas Burnet's *Sacred Theory of the Earth* (1681, first in Latin), but there one finds little more than an old tradition of grottos and caves, as found in ancient myth. Halley was making a valiant attempt to find a theory of the Earth's magnetism. From the time of a voyage he made to St. Helena in 1676, he had studied magnetic declination (the difference in angle between true and magnetic north), and by 1683 had concluded that the Earth had four magnetic poles. He now thought that they might best be understood on the hypothesis that the Earth is a system of a sphere within a sphere—perhaps even more spheres were involved—in relative rotation, each carrying magnetic poles. There are slight similarities between this and an earlier model of the Earth put forward by Hooke, but in Halley's case the model had, as it seemed, a double justification, and like his paper on monsoons it gives him an honorable place in the history of geophysics, a science that has always maintained strong links with astronomy. The hollow Earth, however, soon disappeared, when Newton's mistake was corrected. A portrait of Halley as Astronomer Royal in 1736 shows him holding a hollow Earth. William Whiston had by then propagated the idea widely in a book of 1717 and had even provided biblical evidence for the notion that the cavity is inhabited.

Newton's *Principia* is often described as the most important work ever published in the physical sciences. The criteria for such judgments are hard to define and easy to vary, but the work can certainly be seen as marking the end of one historical era and the beginning of another. It gave physical reasons for Kepler's descriptive laws of planetary motion, and in that sense legitimized them, or as Newton would have said, changed them from speculation to fact. And it presented a program for astronomical research that—while augmented—still continues. His demonstrations were not always complete; indeed, in many cases Newton had not yet developed the necessary mathematical techniques to give compelling proofs. As later astronomers were to discover to their surprise, however, he had a remarkable instinct for correct conclusions, even when he had to paper over the cracks in his arguments.

When first published, Newton's work aroused much hostility on philosophical grounds. Leibniz, for instance, objected to Newton's ideas about absolute space and time, and "action at a distance," which he deemed an occult quality. There were many in England who felt uneasy at the idea that gravitation could act through empty space, without action-by-contact, and Cartesian vortices had a wide currency even there. Throughout most of continental Europe, Descartes ruled supreme until after Newton's death. Leibniz and Newton had quarreled over priority in the invention of the calculus. When an embittered Leibniz, through the mediation of Caroline, princess of Wales, began a philosophical exchange with Samuel Clarke, a supporter of Newton, he had every reason to suspect that Newton was colluding with Clarke.

Such scruples as Leibniz's raised the question of whether we should reject a physical theory that seems to work perfectly as such, while yet being philosophically rebarbative. It is a question that has never entirely disappeared from philosophical discussion, but astronomers are usually content to ignore it completely. Does it really matter if Newton's idea of absolute space is illogical? Could Leibniz or his followers quantify a seventh inequality of the Moon? When relativistic arguments surfaced in astronomy at the beginning of the twentieth century, it was often recognized that there was a tenuous connection with the tradition in which Leibniz had argued his case. If the far more crucial link with Newton's theory was less often mentioned, that was because it had so entered the lifeblood of science that it was taken for granted, like an old coat. Even on a cosmological scale, it is not yet worn out.

New Astronomical Problems

By 1685, Newton had written a little work with the title *De mundi systemate* (*On the System of the World*), which was intended as a tailpiece to his *Principia*. It was destined to be published only in 1728, a year after his death. In this he used a technique devised by James Gregory in 1668 to show that the stars lie at much greater distances from the Sun than had previously been supposed. The method was photometric, depending on a comparison of the brightness of the Sun with that of a star, and on the inverse square law of photometry. The comparison could not of course be made directly, but rather was done by considering the sunlight reflected off Saturn. Certain assumptions had to be made, for instance, about the nature of the reflection and the absence of light-loss in space, and it was further assumed that the star considered was equal in brightness to the Sun, but these assumptions all seemed plausible enough. When Newton used the method on Sirius, for instance, he found its distance to be a million times that of the mean distance of the Sun from the Earth (the astronomical unit). The figure is actually too great, but there is a case for counting this as the first acceptable determination of a star's distance.

If the stars were at enormous distances, Newton felt he could assume that their gravitational attractions on one another were minimal. This was a vague conclusion, but it was important to him, for he was perplexed by the fact that the world did not collapse on itself, under gravity. By the time his *Principia* was finished, however, he had developed a test for external forces acting on the solar system: large forces would produce detectable rotations of the apse lines of the planets. Such were not observed at a significant level, so external forces must be negligible, and this fact fitted with the idea that the stars were very distant, indeed.

Late in 1692, Richard Bentley, a brilliant young classical scholar who was then chaplain to the Bishop of Worcester, was giving the first of a series of lectures founded by Robert Boyle with the purpose of defending natural and revealed religion. One of his themes was that "the observed structure of the universe could only have arisen under God's guiding hand." Before going into print, he asked Newton for advice. What would happen if matter were spread uniformly throughout space and were to be allowed to move under gravity? If space is limited, said Newton, it would fall into one large spherical mass, and if infinite, into infinitely many masses. But surely, replied Bentley, if matter is *evenly* spread, there is no sufficient cause for a particle to move one way rather than another. Newton's reply was that this evenness in relation to even a single particle is unlikely, as unlikely as that one could make a needle stand on its point on a looking glass. How much more improbable, then, to find *all* particles so placed. God could have made them so, however, and then they would have stayed in place. But then, said Bentley, consider the universe to be divided by a plane into two parts. A particle in the plane will be pulled by an infinite gravitational force to one side of the plane, and this will be balanced by an infinite force pulling in the opposite direction. Why should the presence of the Sun in the particle's neighborhood have any effect on its behavior? Would its attraction not simply be incorporated in one of the infinite forces, leaving it still infinite? Newton's reply was that not all infinites are equal. A particle in equilibrium will, he said, be moved by an extra force. The two men were getting into deep waters, waters that had proved to be too deep for most philosophers for over two millennia. Bentley sent a summary of his seventh sermon. The universe was not homogeneous, and the conclusion was more or less that if the universe is in equilibrium, it is God who keeps it so.

Newton was engaged around this time in revising his *Principia* for a second edition. It must have seemed that he was glad to drop the subject, but in reality it continued to occupy his thoughts, as his unpublished papers show. He tried to find a geometrical model of the universe in which the stars are distributed in an exactly regular way, so as to be in equilibrium. Of course, even the simplest of observations of our uneven world throw doubt on that idea, but such a cosmological model is meant only as an approximate representation. He tried out the idea that the stars are all on spherical surfaces, each at one unit distance from neighboring stars on its own sphere, the spheres being centered on the Sun. He took the radii of the spheres in question to be one unit, two units, three units, and so forth. The advantage of the scheme is that for large radii the distribution will correspond to a thin uniform shell of matter, and in *Principia* he had shown that the net gravitational attraction on any star at an arbitrary place within such a shell is zero—a very comforting result, bearing in mind the perplexing discussion with Bentley.

Newton investigated the geometrical properties of his model. How many stars can be put at unit distance on a sphere of unit radius? Kepler had examined that problem, and thought that the answer was at most twelve. Newton thought possibly thirteen. Whatever the answer, there will be four times as many on the sphere of radius two units, then nine times as many for radius three, and so on. Newton at first assumed that those on the innermost sphere could be treated as stars of apparent magnitude 1, that the next would then be of magnitude 2, the third of magnitude 3, and so on (a century later, Herschel took more or less the same stance). An observational test is therefore not only possible but simple as well. One merely counts the stars of successive magnitudes in the best catalogs available. In a rough and ready sort of way, the stars of the six traditional visual magnitudes seemed to fit the scheme, although there was a tendency to accumulate faster than in the model, so Newton crossed out magnitudes 5 and 6.

There was another problem remaining: was he not making the Sun the focus of the universe? Newton tried more adjustments to his model. The details are less interesting than a paradox that might be thought to have escaped him—although he was presiding over the Royal Society at a meeting in 1721 when the paradox was raised by Halley. On Newton's model, stars accumulate as the areas of the surfaces of the spheres, with a certain number on the first, four times as many on the second, nine times as many on the third, and so on, as explained. At a distance of two units, however, each star is a quarter as bright as a star at one unit; and at three units a ninth as bright, and so on. In other words, the total light from the stars at any particular distance is constant, so that in an infinite universe the sky should be ablaze with light, the sum of an endless series of constant totals. (We are here assuming point sources and that stars do not stand in the way of light from other stars. If they do, we shall still have an entirely bright sky.)

It is not known who first appreciated this paradox, but Halley said that he had "heard it urged" by someone he did not name. David Gregory has been suggested as a possible candidate, for we know that in 1694 he was engaged in discussions with Newton of the cosmological problems Bentley had raised and later wrote of them in a book of his own. Perhaps a stronger candidate is William Stukeley. Halley's paradox was for long ascribed to W. H. M. Olbers, as we shall see, but Olbers's formulation of it came more than a century later.

HALLEY AND COMETS

The career of Edmond Halley was so intimately bound up with those of Newton, Flamsteed, and other leading figures of the time, that it is easy to make the mistake of treating him only as a satellite to them. He was a man of great originality, and his contributions to astronomy were substantial. His learning was considerable and his interests wide. He produced editions and Latin translations of Apollonius and Menelaus, for example, and wrote on archaeology, Arabic astronomical tables, and geophysics. His best known achievement, however, was his demonstration of the fact that the comets of 1531, 1607, and 1682 were one and the same object, and that it would return—he allowed for perturbation by Jupiter—in December 1758. It did, but of course he did not see it. Only with its return was Halley's achievement brought to the notice of a wide public.

168. Edmond Halley (1656–1743) at the age of eighty. Mezzotint after a portrait in the Royal Society ascribed to Michael Dahl (1736).

It almost goes without saying that Halley's comet is remembered because it was the first to return, as he had predicted it would. We recall that the same prediction he made concerning the comet of 1680–1681, as considered by Newton in Book III of the *Principia*, was not so obliging. In the more confined world of astronomers, however, Halley's comet was valued because it served to underwrite Newtonian science. Drawing on Newton's method for calculating the orbital parameters of planetary bodies from three observation of position, Halley calculated five parameters for the orbit of "comet Halley" (This is an honorific title, since the comet had been seen by at least nine other astronomers before him, the earliest on 15 August 1682. He first observed it on 5 September of that same year.) He assumed a parabolic orbit, following Newton's method, which was reasonable for a comet in the vicinity of the Sun. Had he not done so, the number of parameters would have been six. He did not produce his theory of cometary return immediately, but embarked on an intensive study of cometary orbits. This study resulted in his main publications on the subject only in 1705, first in the *Philosophical Transactions* of the Royal Society, and then in a book that became a classic on the subject: his *Astronomiœ cometicœ synopsis*, which appeared in the same year, as did an English translation, *A synopsis of the astronomy of comets*.

Halley was a calculator of a kind not then uncommon, but now scarcely understood in these days of electronic computers and calculators. He would have considered himself rich by comparison with older generations, having logarithms at his disposal. He was assiduous not only in computation but in his use of libraries, collecting together as much information as he could find on twenty-four comets between the years 1337 and 1698. When he compared the orbits he had deduced from them, he eventually found that in three cases the orbital planes were very close to one another—all at

169. Stretching seventy meters in all, the *Bayeux Tapestry* tells the story of the conquest of England by William of Normandy. On the right in this scene is Harold, the newly crowned English king. On the left, dubious about his claim to the throne, a group of his subjects "marvel at the star." This omen, the periodic comet now named after Halley, appeared between April and June, 1066.

about 18° to the Earth's orbit. He found that the three comets moved in a retrograde sense and came to perihelion in similar directions and at similar distances. It was inconceivable that these were chance coincidences. This is how he was led to his discovery of the periodic character of comet Halley. Later he identified the object with the bright comets of 1305, 1380, and 1456. The periodicity of about 76 years allowed him to predicted a return in the year 1758. Looking further back into history, he added 1456 and 1066 to his list of the same comet's appearances. He worried about these earlier dates, which seemed to imply a lengthening of the period time, but explained the fact away by the probable perturbation of the orbit by a massive planet such as Jupiter or Saturn.

Since comets had from ancient times been associated with the downfall of princes, it is amusing to see that comet Halley appeared to have had this very property in 1066. This is—or at least once was—the best-remembered date in English history, when William of Normandy interpreted the comet as a favorable omen for his conquest of England. The association is famously recorded on the *Bayeux Tapestry* (fig. 169). Such traditions die hard. Abraham Lincoln was described as "The Comet of 1861" on patriotic envelopes supplied to the Union troops. In Italy, the comet of 1861 was widely seen as a sign that the old despotism had been suppressed, and replaced by a constitutional monarchy. There is something about comets that seems to stir the imagination of *homo sapiens*, or at least of news editors. When comet Halley made another appearance in 1910, the world's popular press found no difficulty in whipping up a veritable comet mania. The *New York Times* announced that police reserves had to be called out to calm and clear a crowd that had been terrified by a toy balloon carrying a white light. The paper was faintly dismissive of unrest throughout the city, although it hastened to point out that the more superstitious inhabitants were from sections largely inhabited by foreigners. We read stories of widespread panic, of businessmen refusing to sign contracts before the comet's effects were known, of a New Jersey man who confessed to murder, in fear of the comet—but perhaps he was a foreigner—and a Montreal girl who spontaneously expired for the same reason. The deposed sultan of Turkey, Abdul Hamid, refused to eat for several days running. Clever entrepreneurs sold anti-comet pills, and a German company insured people against the end of the world. Halley's century was less well organized. In 1985–1986, Halley's comet returned, but on this occasion, in relation to expectations, it was disappointing. By this time, Moon landings had helped to take much of the mystery out of the sky.

HALLEY AND PROPER MOTIONS

A practical man, Halley was as much at home writing on gunnery and annuity tables and the properties of thick lenses as on comets. In 1676, at the age of twenty, he had made a voyage to St. Helena, off the African coast, to catalog southern stars. Between 1698 and 1700 he captained a mutinous ship across the Atlantic, and as we have already seen, charted varying magnetic declination, with the result that he developed his strange but ingenious theory of the Earth's structure. He was sixty-four when he finally succeeded Flamsteed as Astronomer Royal, but he promptly set in motion a program for observing the Sun and Moon over an eighteen-year cycle, and he lived to see the cycle through.

   With rare exceptions before the eighteenth century, the stars were regarded as fixed—at least relatively to one another. Hipparchus's discovery of precession merely introduced the need for catalog makers to add to *all* ecliptic longitudes a constant appropriate to the date. Ptolemy's catalog of over a thousand stars was repeatedly revised in this simple way, and even when alternative catalogs were drafted—by Ulugh Beg, Tycho Brahe, Hevelius, Flamsteed, and the rest—there was always the assumption of their internal constancy. The "new star" phenomenon, which helped to make Tycho's reputation, did nothing to change this view. The situation did change, however, with Halley's discovery that some at least of the stars were in relative motion. His scholarly talent assisted him in this achievement—one that in retrospect proved his instincts to have been as important as his statistics. In a paper published in 1718, he explained how he compared modern observations with those of the Greeks. He had been studying star catalogs, especially Ptolemy's, since about 1710, and had come to the conclusion that precession and observational error were not enough to explain the discrepancies. He was convinced that a southerly motion for the bright stars Aldebaran, Arcturus, and Sirius was proven, that there were "proper motions" of the fainter stars—apparent motions across the celestial sphere as a result of their own real motions relative to the solar system—and that these would have been more obvious had the stars not been so distant.

Jacques Cassini confirmed Halley's claim in 1738. In his case he could detect a shift in the position of Arcturus even from as recent a measurement as one made by Jean Richer at Cayenne in 1672, which was of course a much more precise and certain measurement than Ptolemy's. Cassini argued that this was indeed a true proper motion of Arcturus, rather than a consequence of some shift in the ecliptic, since it was not shared by a faint star nearby. (Many years earlier Cassini had expressed irritation with Halley, who had found a mistake in his claim to have measured the parallax of Sirius.)

The discovery of proper motions in the stars opened up for serious discussion a completely new vista in astronomy. Stellar parallaxes were as yet undetected in anything approaching a direct way, but the fact that the stars have different motions, as seen from the Earth, seemed to confirm the conjecture of several writers of the late sixteenth and early seventeenth centuries that the stars were scattered throughout space—whether finite or infinite space was another question. That they were scattered would seem to follow whether the observed motions, which differ as between stars, are due to the Earth's motion or to motions intrinsic to the stars themselves. As for the idea that only the solar system moves, that is, through a system of stars at relative rest, James Bradley, at the same time as he announced nutation in 1748 (see p. 432 below), suggested that it would be many ages before there was evidence to decide between this and the alternatives. In this he was mistaken.

THE EIGHTEENTH CENTURY

Looking back over the history of eighteenth-century astronomy, from the perspective of one living at the end of the nineteenth, Agnes Clerke wrote that it "ran in general an even and logical course." She saw the age of Newton as one lasting almost exactly a hundred years, having ended in 1787 when Laplace explained to the French Academy the cause of a certain acceleration in the Moon's motion. The only anomaly in her description, as she believed, was the rise of William Herschel, whose work did so much to influence the course of later events, but whose starting point was not that of Newtonian dynamics.

There is much to be said for this simple account, but it is too narrowly focused. There were other forces at work, other motives to study astronomy than a wish merely to extend the monumental system of the *Principia*. Many important discoveries were made in the course of practicing astronomy in a perfectly traditional way, but with new instruments—and new intellects. There was an ever-present desire to make new telescopic discoveries, and although here Herschel was preeminent, he was not alone. Astronomy no longer occupied the compulsory place that it had formerly held in the university arts curriculum, but it was a subject in which there was a strong cult interest, creating a demand for traveling lecturers, for example. Without astronomy, no gentleman regarded his children as properly educated, a fact that produced new types of popular literature and educational instruments. There were simple telescopes, of course, and globes—terrestrial and celestial were usually paired—and simple orreries eventually became commonplace. These moving models of the solar system were named after Charles Boyle, Earl of Orrery, who merely happened to commission a particularly fine example. The orrery tradition—one that stretched back through centuries of astronomical clock making to the planetary models of antiquity—was thus merely continued by him at a popular level.

One sort of literature that achieved a new popularity, and that required no special expertise, was natural theology, the attempt to argue from nature, and especially from the harmony of the cosmos, to the existence and attributes of God. One of the most influential works in this genre was the *Astro-Theology* (1714) of William Derham. William Paley was much indebted to it, in his even more influential writings at the end of the century. Paley's best work, his *Natural Theology* (1802), provides an important measure of a change that took place in the intellectual atmosphere in the course of the century. For Newton's contemporaries—for Bentley, for example—the ordered celestial universe gave proof of God's existence. For Paley, it was necessary to introduce also biological considerations, although the universe was a benevolent one still. Some of Paley's writings were required reading at Cambridge when Charles Darwin was an undergraduate, and he took much pleasure from them, but in the middle of the nineteenth century they created a climate that worked against his theory of evolution, in which a benevolent God was conspicuous by his absence. This conflict of intellectual and religious interests was one of the less obvious legacies of centuries of discussion of the cosmic harmonies, a discussion to which Plato, Kepler, Newton, Leibniz, and scores of lesser scholars had contributed.

THE INSTRUMENT MAKERS

On a more pragmatic note, the eighteenth century in European astronomy is characterized by a rapid growth in the number of official observatories, that is, observatories maintained by states, universities, scientific communities, and religious groups. Medicine apart, no other science could boast such large numbers of people professionally engaged on research, even though the research was of a routine nature. Ever greater precision in the recording of celestial coordinates paid great dividends in the long run. It was an expensive business, but it could be justified in terms of its practical utility for navigation, the surveying and mapping of country and empire, and still even for theology, the mapping of Mankind's place within Creation. National observatories became symbols of power and principle. They had to be well provided, however, and it was no longer enough to employ local artisans on what had become highly specialized work.

We have already seen how the face of astronomy was drastically changed by the Paris Observatory and the employment of Cassini there.

The founding of the Greenwich Observatory in England owed much to French influence, but of a highly unusual sort. In late 1674, the master of the king's ordinance was raising promises of money for an observatory, when one of Charles II's mistresses, Louise de Kéroualle—a Breton lady who had recently been made Duchess of Portsmouth for her services to the Crown—recommended to the king a certain Sieur de St. Pierre. He was claiming to be able to find terrestrial longitudes "from easy celestial observations."

As already explained, the longitude problem reduces to that of finding a universal clock that will allow a comparison of local celestial phenomena with what would be seen at a standard meridian, such as Uraniborg, or Paris, or Greenwich. (For example, if the Sun is on the local meridian, what is its position as seen at Greenwich? To know the answer is to know one's relative longitude.) One such universal clock is a transportable time-keeper. This was not available in a reliable enough form until after 1763, when John Harrison—after a long and painful struggle with machines and men—was awarded a first Board of Longitude prize for one of his chronometers. Another "clock" is constituted by the satellites of Jupiter, as we have already seen, for from their relative positions around the central planet the time can be found from tables prepared at one of the great observatories.

St. Pierre kept his method secret at first, but Flamsteed and others guessed correctly that it was to use the rapidly moving Moon as timekeeper (the idea was not original, and he probably took it from Jean Morin). The Royal Society was instructed by the king to collect the necessary lunar data, and the services of John Flamsteed were obtained. His verdict was that neither lunar nor stellar positions were well enough known to make the method reliable. At any event, the king was thus moved to found the observatory, and John Flamsteed was duly appointed his "astronomical observator," with the task of improving tables of motions and star positions for the benefit of both navigation and astronomy. In chapter 15, we shall encounter Flamsteed's important star catalog, which was highly relevant to this program, but it was destined to be published only in 1729. It was in July of 1676 that he moved into the new observatory building, designed by Sir Christopher Wren, and erected on the hill above the port of Greenwich on the river Thames. He was more fortunate in his architect than Cassini had been in Paris; Claude Perrault's building there was more splendid, but less functional by far.

We have seen something of Flamsteed's relations with Newton and Halley. He would not have been pleased to know that, on his death in 1719, he would be succeeded at Greenwich by Halley, although he might have taken consolation in the fact that his heirs kept his instruments—on the grounds that he had paid for most of them himself. Most of the instruments had been in government service for forty-four years when he died, and most were of a high quality. He started out with an iron sextant with a seven-foot radius, mounted on an equatorial axis and fitted with a pair of telescopic sights—one fixed at zero of the scale (for a first observer), the other moving over the scale (for a second, observing simultaneously). He had a meridian quadrant that proved too flimsy, but he spent £120 of his own money on another, again of seven-foot radius, and paired with the sextant. He had longcase pendulum clocks, which he supplemented with a sidereal regulator, the most accurate clock of its day, by the great maker Thomas Tompion. He followed the fashion of the day, with his very-long-focus telescopes, fitted with eyepiece micrometers. Hoping to determine the annual parallax of γ Draconis, a star known to cross the zenith, he sank a 90-foot telescope down a well; but his hopes were in vain, for he failed to appreciate the nature of the measurements he was obtaining, mixed up as they were the aberration of starlight, the very concept of which was still undiscovered. With this array of instruments, he undertook a systematic program of observations, settling his local parameters, latitude, year lengths, and so forth. Flamsteed gave very much attention to the Moon throughout his career, but he was a perfectionist in such fundamental matters as the parameters of the Sun's motion, matters that almost all other astronomers were prepared to consider settled. His measurements of angle were of unprecedented accuracy, his errors being typically a tenth or even a twentieth those of Tycho's measurements. His "British catalogue" of 3,000 stars (in the third volume of his *Historia*) was for many years by far the best available.

The question of observational error is an important one, although to do justice to it here would take us too far afield. Taking a simple average value of many observations wherever possible—a procedure that now seems to be a minimal precautionary requirement—became standard practice only in the eighteenth century, a surprisingly late date. There are instances in Tycho and Kepler where a mean is found between pairs of values. Newton averaged multiple readings in his optical experiments on occasion, as increasing numbers of his contemporaries were doing in laboratory work in the late seventeenth century. Jacques Cassini and Flamsteed did the same with astronomical readings fairly regularly, and there are other instances, but for long there seems to have been a feeling that to consider many measurements of the same thing was a sign of weakness, and certainly not something to which one should confess in published work. In time, finding analytical methods for reducing observations would become almost a subject in itself, the most momentous step forward being taken by the Brunswick mathematician Carl Friedrich Gauss. In 1809, he published methods that he had been using in connection with the gravitational theory of planetary motion since 1795. His powerful analysis connected the so-called postulate of the arithmetic mean—the assumption that a simple average gives the answer that is most probably correct—with the mathematical theory of probability. (That "postulate" is today often seen as derivable from still more fundamental principles.) His most famous achievement was to put

170. The observatory buildings at Greenwich in the time of John Flamsteed. The river Thames is visible in the distance. Compare the outdoor telescope with those of Hevelius, which Halley criticized so strongly (fig. 160).

his method of least squares on a logical foundation. Such ideas were a very far cry from the practice of eighteenth-century astronomers, but Flamsteed's procedures of reduction were by no means without merit, as we shall see shortly (p. 467).

Not only did Flamsteed have to provide most of his own instruments, but he was also often more than a year in arrears in obtaining his modest salary from the king—his history is in some ways reminiscent of Kepler's. In the long term, however, the relatively large investment of the British government in the fabric of the observatory at Greenwich (fig. 170) provided a stimulus for the growth of a profession of instrument making in London that for the best part of the eighteenth century became the supplier to all of Europe. The time was past when professional astronomers could be their own instrument makers, although they continued to have a strong interest and involvement in the art; and the makers themselves were often passable astronomers, capable of designing new instruments from firsthand knowledge.

New designs came from other quarters, too. To take one valuable new idea: Roger Cotes, first Plumian Professor of Astronomy at Cambridge, and editor of the second edition of Newton's *Principia*, sent Newton a design for a heliostat. This allowed a solar image to be reflected from a moving mirror, driven by clockwork, into a static telescope. Hooke designed another, and the principle is still in use. The first satisfactory heliostats on a large scale followed the design of Jean Foucault, while a famous example on a still larger scale was that built at the Mount Wilson Solar Observatory in California and has been used from 1903 onward. A notable recent example is that at Sacramento Peak, New Mexico, while yet another is that at Kitt Peak National Observatory, Arizona. There, by reflecting the image down the polar axis, it is possible to use only a single upper mirror.

George Graham was one of the first great specialist makers to produce a series of designs covering almost every aspect of observatory installation—mural quadrants, transit instruments, zenith sectors, astronomical regulators (precision clocks), and many more. He was of a generation later than Flamsteed, and his first great instrument was in fact a large quadrant made for Halley, in 1725, when he, the new Astronomer Royal, was nearly seventy. Graham was already famous as a watch and clock maker, but he achieved much publicity from the Halley quadrant, the fame of which was spread when it was described and praised by Robert Smith in his widely read textbook of optics. The idea of a quadrant mounted on a solid wall (usually) in the meridian was of course not new, but Graham added a central telescope and an axis in the form of a double cone that made his work superior to everything before it, and the design soon became standard. It was Graham who introduced the two main technical improvements to the longcase clock that made it acceptable as an astronomical regulator—namely, the mercury compensation pendulum and the dead-beat escapement.

BRADLEY'S TWO GREAT DISCOVERIES

Graham's fame benefited further from his association with two of the most important astronomical discoveries of the eighteenth century, both made by James Bradley (fig. 171). The first was made when Bradley was Savilian professor of Astronomy at Oxford, the second after he succeeded Halley as Astronomer Royal upon Halley's death in 1743. It had long been realized that the best hope of detecting the parallactic displacements of stars was to make use of the Earth's shifting position as it moves round the Sun. The maximum displacement measurable would obviously be that between the star's apparent positions when the Earth was at the ends of the major axis of its elliptical orbit. To appreciate the displacement, a very remote set of reference stars was obviously necessary: the Earth's elliptical motion would then be expected to make nearby stars describe tiny ellipses against that remote background. Nothing of the sort had been detected with anything approaching certainty, when in 1725 a wealthy amateur, Samuel Molyneux, tried to improve on Hooke's abortive attempts, made in 1669, to measure the parallax of the star γ Draconis. Molyneux used a very long (24-foot) zenith sector made by Graham, that is, a telescope with a short but very accurate scale arranged to measure angles over a small range in the neighborhood of the zenith (at the zenith, refraction can be effectively ignored). With Bradley's help, he did indeed observe displacements, but they were too large, and not at all in the expected direction, even though it eventually became clear that they followed an annual cycle, as parallactic shifts would do.

Bradley took over the observations, and in 1727 examined other stars, using another, smaller, zenith sector made by Graham. This was set up at the home of his maternal uncle, the Rev. James Pound, at Wanstead in

171. James Bradley (1693–1762), some time after his appointment as Astronomer Royal, following Halley's death in 1742. (Oxford copy of a Royal Society portrait in oils by T. Hudson.)

Essex. Pound was himself no mean astronomer—we recall that Newton used a cometary observation of his from 1681. Long after Bradley's appointment in Oxford, he returned to Wanstead to use his uncle's instruments. It now became clear to him that all stars shared in the same peculiar type of motion, although its size was not constant: it depended in some way on a star's position in relation to the ecliptic. He tried various hypotheses, but the story goes that he hit on the right one only when traveling on the Thames in a pleasure boat, watching the shifts in direction of the pennant at the masthead. He decided that the shifts in a star's position resulted from changes in the combined effect of the Earth's orbital velocity and the large but finite velocity of the incoming light from the star. (The effect, which may be recognized as an application of the "parallelogram of velocities," is explained in figure 172. It is often explained in simple terms by the use of an analogy with the direction at which one is met by falling rain when walking through it.) The aberrational displacement, like that resulting from the (as yet unobserved) annual parallax, also makes the star describe a small ellipse—the former because of the Earth's motion, the latter because of the Earth's (changing) position. The velocity of light was known approximately, following Ole Rømer, and so was the changing velocity of the Earth, as it moved round its orbit. The size of the aberrational ellipse could therefore be derived theoretically; but conversely, by measurement of it, the velocity of light could be deduced. (Doing this, Bradley found a value the equivalent of about 301,000 km/second. The currently accepted figure is 299,792.5 km/second.) The optical explanation he had found fitted excellently with his observations, and in 1728 he wrote to announce his findings to Halley. This discovery of the "aberration of light," as it was later called, was finally announced to the Royal Society in 1729.

Bradley was the first to explain the phenomenon but was not the first to be perplexed by the odd behavior of carefully measured star positions. The

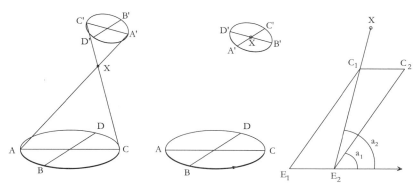

172. Annual parallax, in the first figure, is here distinguished from annual aberration, in the second and third. In all cases, X is the observed star, A and D are extremes of the major axis of the Earth's elliptic orbit, and the corresponding positions of the observed star are at A' and D'. The upper ellipses show the apparent positions of X against the background of the remotest stars. Parallactic shifts are due to the observer's changing *position*, while aberration is an effect of the observer's *velocity*. The effects are very different in magnitude (the maximum parallax of Bradley's γ Draconis is under 0.02″, while its aberration is a thousand times as great), but the most obvious difference is in the directions of the star's displacement (compare the differences in the lettering of the upper ellipses). No attempt is made in the second figure to explain the aberrational shift, for which see the third figure. There C_1 is the center of the telescope's object lens, and E_1 is the eyepiece, both at the moment when light from star X meets C_1. When the light arrives at the eyepiece, the telescope will have moved with the Earth to position C_2E_2. If the Earth were stationary, the star would be seen in the direction E_2C_1, the "true" direction of X (that is, at angle a_2). When the Earth is moving, however, the telescope scale records the angle a_1 when the star is being observed. The difference is the aberration due to the Earth's annual motion. (There is a very much smaller effect caused by the Earth's rotational velocity, ignored entirely here.) It is not difficult to show that the difference between a_2 and a_1 is proportional to the sine of a_2, to a high degree of approximation, so that aberration depends on the star's place in the sky. It reaches a maximum of about 20.5″ (known as the "aberration constant"). Applying aberration to the star's true coordinates, or correcting the observed coordinates to obtain the true, require more complicated calculations.

context was the hunt for parallax. Flamsteed at one stage thought that he had found a parallactic motion of the Pole Star, but then Cassini corrected his mistake, and the search continued. Perhaps the most significant sets of relevant observations were those assembled in Bologna by Eustachio Manfredi, astronomer to the Institute of Sciences there. Some were made after 1705, and some in the early 1720s, but Manfredi's interest in the problem was stirred again late in 1726, after he heard of the studies by Molyneux and Bradley, and so he made more measurements. (The news reached Bologna through a letter written by the English astronomer William Derham to his brother, who was English ambassador in Rome.) Assembling his data in a pamphlet in 1729, he ran into difficulties for a time with the papal physician, who thought that he was paying too much attention to the thesis of the Earth's motion. The pamphlet was finally given the Bologna inquisitor's approval, however, and when it appeared in print it was dedicated to the archbishop of Rimini—a covert Copernican, if we are to judge by a gift he made to the Bologna Observatory in 1726. The gift was an armillary model of the Copernican world, and is still to be seen there.

Nearly a century had passed following the judgment given against Galileo, and Manfredi's predicament illustrates well the difficulties under which Catholic astronomers were still working. It is not easy to decide on Manfredi's innermost feelings on the Copernican question, although he later published an acknowledgement that Bradley's hypothesis was compatible with all of his best observations. He trod a difficult path, surrounded as he was by conformist anti-Copernican churchmen. We have a letter to him from a friend, advising him to publish some inconclusive parallax observations by Horrebow, so as to lift suspicion from his own work. Whatever his views, he made a palpable hit with the simple Latin word *aberratio*, aberration. He used it in the title of his publication of 1729, and eventually it became generally accepted in the world of astronomy, to describe what in Bradley's original paper was simply "A New Discovered Motion."

The argument in Bradley's paper in the *Philosophical Transactions* of the Royal Society was extremely well crafted, for in it he made a very powerful negative statement about the parallaxes of the stars. Had they been as great as a second of arc, he said, he would have been able to detect them. In other words, the stars were clearly very much more distant than had been generally supposed. (The parallax of γ Draconis is in fact less than a fiftieth of a second of arc.) The most impressive of all conclusions to be drawn from his work, however, as far as the world of learning was concerned, was that the Earth is in motion relative to the frame fixed by the distant stars—it is in motion round the Sun, as the usual interpretation would have it. Another strut had been added to the Copernican edifice, the Bologna inquisitor notwithstanding. When the Bolognese-Parisian Francesco Algarotti wrote his highly influential *Newtonianism For Ladies*—important because it carried the new science into educated circles across the length and breadth of Europe—one of those indiscretions that led the Congregation of the *Index* to condemn it was his assertion that aberration proves the reality of the Earth's motion.

In 1727, Bradley had noticed something else of concern to all who wanted an accurate stellar astronomy: the declinations of certain stars seemed to be erratic. Five years later he had found an explanation: the Earth's axis is "nodding" as a result of the Moon's attraction on its equatorial bulge. From this nodding there results an apparent displacement of the stars, such that each seems to describe a tiny ellipse about its true (mean) position, in a period of about 18.6 years—the period of the lunar nodes. Nutation, as he called the effect, was discovered with the same instrument as aberration. This was one of the two great discoveries that earned Bradley international fame, at a time when precise measurement was rated so highly. In fact, Jean-Baptiste Delambre, who had written knowledgeably and at great length about the astronomy of all periods before his own, in 1827—sixty-five years after Bradley's death—said that he considered those two discoveries to have entitled him to "the most distinguished place after that of Hipparchus and Kepler, and above the greatest astronomers of all ages and all countries." Bradley must have known that his work owed as much to his geometrical insight as to his instruments, but he nevertheless expressed his gratitude to Graham in extravagant terms. Helped by Bradley's findings, Graham's flourishing trade in instruments in Europe began with the zenith sector. There had previously been something of a fashion for the French portable quadrant, but now zenith sectors were all the rage, although totally unsuitable for most types of routine observation.

ZENITH SECTORS AND THE LONDON TRADE IN INSTRUMENTS

When Pierre Louis Moreau de Maupertuis was sponsored by the French Academy of Sciences to lead the famous expedition to Lapland in 1736, an expedition that settled the controversy over the shape of the Earth, the principal instrument was again one of Graham's zenith sectors. This was an important episode, inasmuch as it made many French converts to Newtonian principles. Two generations of excellent astronomical observers trained in the Paris school of the Cassini family had formed a nucleus of broadly Cartesian and anti-Newtonian natural philosophy, and the general opinion there was that the Earth was a prolate rather than an oblate spheroid—a rugby ball or an American football, rather than a flattened orange, as we might say. Expeditions were sent to Peru and Lapland to measure a degree of terrestrial longitude in both places for comparison with that in France. After a difficult expedition that involved shipwreck on the return journey, and a long period during which the observations had to be evaluated, Maupertuis pronounced in favor of Newton. Voltaire, one of Newton's few supporters in France, congratulated him on having flattened both the poles and the Cassinis. The precise form of the Earth continued to call forth both practical and theoretical energy of large numbers of scientists, and some of the best mathematicians of the following century—Clairaut, d'Alembert, Legendre, Laplace, Gauss, and Poisson, for instance—gave the problem a central place when developing Newton's theory of gravitation.

In eighteenth-century England, the tradition of fine instrument making was extended, largely on the basis of Graham's designs, by a succession of excellent makers—Jonathan Sisson, John Bird, Jesse Ramsden, John Dollond, and Edward Troughton all earned respect throughout Europe. As a measure of progress: Flamsteed's mural arc was good to about ten seconds of arc, while Graham's work was accurate to five or six, and Bird's to a single arc second. It was John Bird who fitted out the second most important English observatory of the time, the Radcliffe Observatory of Thomas Hornsby at Oxford. Bird wrote an influential account of the method of dividing the scales of instruments. By the mid-1760s, he had made large instruments for Greenwich, Paris, St. Petersburg, Göttingen, and Cadiz. Jonathan Sisson had the advantage of having worked under

173. Jesse Ramsden (1735–1800), the leading scientific instrument maker of his day, with his transit circle behind. (After a painting by Robert Home, in or after 1791.)

Graham's direction. He made instruments for European observatories, one of them indeed loaned for a time by Le Monnier to the Berlin Academy, to allow them to supplement lunar parallax observations made by Lacaille at the Cape of Good Hope.

Later in the century the preeminent maker was Jesse Ramsden (fig. 173)—who supplied Giuseppe Piazzi and Franz Xaver von Zach, for example. Ramsden supplied many European observatories with achromatic telescopes equipped with finely graduated scales, read by micrometer microscopes that he had developed. His most famous instrument was the "Palermo circle" he built for Piazzi, a 1.5-meter refractor in an altazimuth mounting, shown in figure 174. This soon proved to the world that proper motions of the stars were the rule rather than the exception, and at the same time it gave evidence of the extraordinary stability and accuracy that might be achieved in a fine instrument.

It would be foolish to pretend that this trade was economically as important as, for example, the London trade in clocks and watches of the same period, nor should we forget the rapid expansion in the trade in sextants used for astronomical navigation. Ramsden, for instance, with a staff of sixty artisans, had produced a thousand sextants by 1789, quite apart from his other work. Astronomical methods of navigation have a history of their own, but it is worth noting here that in 1756 Bradley had reported to the Board of Admiralty that the new lunar tables by the

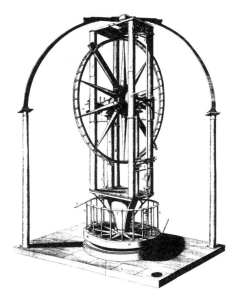

174. The great five-foot vertical circle, commissioned by Giuseppe Piazzi from Jesse Ramsden, during a visit to England in 1788. Piazzi installed it in his new observatory on top of the royal palace in Palermo, northern Sicily, in 1789. He used it to catalog more stars than were in any previous catalog, and this with unprecedented accuracy. With it, in January 1801, he was the first to discover an asteroid (Ceres), so giving rise to a long-running controversy as to whether such objects merit the name of planet.

Göttingen astronomer Johann Tobias Mayer should give terrestrial longitude to an accuracy of half a degree. (Mayer, incidentally, used a Bird quadrant.) After trials at sea, Bradley decided that this was overly ambitious, but after correcting Mayer's tables he decided that an accuracy of better than one degree was attainable.

Most sizable countries had their instrument makers, in one sense or another, but in most cases their influence had been strictly local. The London makers were noteworthy for the way in which they set an international standard and style of practice, and this they did with the encouragement of the Royal Society, to which the best were elected as members. The Royal Observatory at Greenwich provided the third vertex of this fortunate triangle. Delambre was exaggerating, but only mildly, when he wrote that if all other materials of the kind were to be destroyed, the Greenwich records alone would suffice for the restoration of astronomy. There was scarcely an astronomical textbook of importance at this time without either a description or an illustration of the work of the English makers. Lalande's books are good examples. He had an observatory at the École Militaire in Paris equipped with a Bird quadrant superior to anything then to be found at the Paris Observatory, despite the far more lavish funding of the latter.

Bird's friend George Dixon, together with Charles Mason, took some of his instruments on the voyage mentioned in chapter 12, which ended with observations of the Venus transit of 1761 at the Cape of Good Hope. They made observations that turned the Cape into one of the best-surveyed places in the world, until, that is, two years later they surveyed the boundary between Pennsylvania and Maryland—the Mason-Dixon Line—and in the process provided an extremely accurate figure for the size of a degree of terrestrial latitude. Here are a few of the consequences of the new instrument industry that had grown up under the patronage of astronomy; and in return for patronage, the London trade raised astronomical standards, and so supplemented the new wave of excellence in astronomical writing that had followed in the wake of Newton's work. Together they created a strong international sense of astronomical purpose, and it is one that has lasted, more or less, to the present day.

As a postscript, it might be added that William Pitt, the prime minister, and the British government were responsible for almost destroying the London optical industry, with the introduction of a punitive tax, at first on windows and then on glass. Another factor was the invention by Pierre Guinand, a Swiss bell caster and glass maker, of a new technique for stirring molten optical glass to achieve a homogeneous mix. In 1805 he moved to Munich, and was shortly afterwards joined by an assistant, Joseph Fraunhofer. Fraunhofer, the son of a glazier, and trained as such himself, was a man of outstanding technical ability, and although he died at a relatively early age, he was to become a dominant influence on nineteenth-century astronomical practice, as we shall see in chapter 15.

WILLIAM HERSCHEL AND CAROLINE HERSCHEL

After Halley's discovery of proper motions in the stars had been well and truly confirmed by others, astronomers began to make systematic measurement of them. Johann Tobias Mayer, in Göttingen, published the proper motions of eighty stars in 1760, on the basis of a comparison of his own and Lacaille's measurements with Ole Rømer's of 1706. Mayer stated clearly an important consequence of a motion of the solar system through the stars: those in the general direction toward which we are heading, the direction of the "solar apex," will seem to become more widely spaced, and in fact will seem to radiate from the apex. Those in the direction away from which we move (the "solar antapex") will seem to close up on one another, and to draw toward the antapex. No one familiar with *Star Trek* and its imitators can be unaware of these effects.

Mayer could not see any such pattern in his proper motions, but in 1783 William Herschel found exactly what had been expected, by examining a limited number of stars that had been observed by the fifth Astronomer Royal, Nevil Maskelyne. In that same year, the Swiss astronomer Prevost showed that Mayer's own data yielded a similar result. (Herschel learned of this when his attention was drawn to Prevost's article in the *Astronomisches Jahrbuch* for 1786.) Herschel placed the solar apex at a point in the constellation of Hercules (a little to the north of the star λ Herculis).

Prevost's apex was about 30° away, while Georg Simon Klügel in Berlin found an apex only about 4° from Herschel's. Together it seemed that these convergent results provided a remarkable new perspective on the large-scale structure of the universe.

It seemed so at the time, as it does today, and yet, in the first two decades of the nineteenth century a number of leading astronomers—including Jean-Baptiste Biot and Friedrich Wilhelm Bessel—claimed that the evidence justified no such conclusion. Later in the century, however, as more and more data were collected, not only was the result confirmed qualitatively by the best observers in both Northern and Southern Hemispheres, but Herschel's original position for the apex remained respectably close to the newly derived positions. And when data were eventually obtained for the distances of the stars whose proper motions were analyzed, it became possible to quote a figure for the speed of the solar system, as well as its direction. (Otto Struve gave the speed as 154 million miles per annum, that is, about 7.85 kilometers per second.)

Of course it went without saying that this entire exercise was only of statistical value, since there may be—and indeed were eventually proved to be—motions among the stars themselves, regardless of our motion through them. At first this began as pure conjecture, even before Herschel's work. Thomas Wright believed the solar system to be in motion around a central body, and various alternative schemes involving the stars and system of the Milky Way were devised, before Herschel at length tempered the more ardent imaginings with the chill of observation.

If any astronomer of the eighteenth century turned the subject in the direction it was to follow in the nineteenth, that person was William Herschel. In an age when astronomy was becoming on the one hand allied with the most advanced mathematics then practiced, and on the other increasingly institutionalized, it seems paradoxical that an amateur, standing entirely outside both traditions, should have been able to influence almost every corner of the subject. Friedrich Wilhelm Herschel—William, as he later became known—was born in 1738 in Hanover, in the duchy of Braunschweig, a principality that might easily have been absorbed by Brandenburg or Prussia, had it not been for its connections with the English royal family: George I, successor to Queen Anne, was of the House of Hanover. Herschel first visited England at the age of eighteen as an oboist in the Hanoverian Guards—his father's regiment, in English service. A year later, fleeing the French army, which had defeated the Guards at Hastenbeck and captured his father, William moved permanently to England. He began at first to earn his living by copying and teaching music, but made himself acquainted with astronomy and telescope making through Robert Smith's renowned textbook of optics—which was mentioned in connection with Halley and Graham.

Over the years, he became so adept at lens and mirror grinding that by the 1770s he was in possession of telescopes on a par with the best available in the country. What drove Herschel was not a clear vision of what he hoped to discover, but a desire to see what others saw, and more, and to excel in an art that was highly valued by the society of his day. His ambitions were more than fulfilled. In 1782, after a comparison done at Greenwich, Maskelyne acknowledged that Herschel's was a better instrument than any in use at the Royal Observatory there.

In 1772, while he was still an organist at Bath, in the west of England, Herschel brought his younger sister Caroline over from Hanover, and she helped him greatly, but during the first decade less as an astronomer and more as a housekeeper. She attempted to create a career for herself as a singer, but after William earned royal fame with his discovery of a new planet in 1781, they moved house to a quiet village in the neighborhood of Windsor. There Caroline was dismayed to learn from her brother that she was "to be trained for an assistant Astronomer." By the time she died, she was an astronomer in her own right. She began by copying astronomical catalogs and other material that William was able to borrow, and she helped to record and organize his observations, at first without any great understanding of their point. Vast quantities of her immaculate manuscript records of William's observations are witness to her key role in his research. She later helped him with the work of grinding and polishing large mirrors. In 1782, at William's insistence, when she was not assisting him she began to search systematically for double stars, clusters, comets, and nebulae, and several of those she discovered were unknown to Charles Messier—the great comet hunter of the time, who had begun to record other oddities in order not to be misled by them. In 1787, Caroline was granted a salary of £50 a year by the king. William had been similarly rewarded in 1781. If a salary is what makes an astronomer a professional, then she may be regarded as the first professional woman astronomer. Launched on her own career, her personal composure was much disturbed only a year later, when her brother, at the age of nearly fifty, married Mary Pitt, the young widow of a neighbor. Their partnership too was a fruitful one for astronomy, but in a different way, for their son, John Herschel, born in 1792, was destined to be almost as famous as his father.

Between 1786 and 1797, equipped at first with a modest telescope, and later with a grand instrument of 5 feet focal length built for her by her brother, Caroline found at least eight new comets. It was in 1797 that— for reasons we can only guess—she moved out of the cottage next to her brother, and into lodgings. In 1798 she revised Flamsteed's star catalog for publication, but she still occasionally helped her brother in his "sweeping" of the skies for nebulae and clusters of stars. When they ceased this work in 1802, it seems to have been because he, not she, was wearying of the task, and some areas of the sky were still unexamined when the catalog was finally published in the *Philosophical Transactions of the Royal Society*. After William's death in 1822, Caroline edited much of his work. This remarkable

175. William Herschel
(1738–1822), after a pastel by
J. Russell (1794).

woman, who never sought the limelight, lived on until 1848, when she was
nearly ninety-eight. At a later date, her 5-foot instrument was taken to the
Cape of Good Hope by her nephew, John Herschel.

William Herschel, an autodidact in astronomy, had turned forty before
he came into the limelight. Having spent many years preparing himself
with instruments, he had decided to acquaint himself thoroughly with the
entire sky of bright stars and nebulae, almost one by one, and by degrees he
formed the idea of charting their distribution throughout the entire universe. By 1779 he had surveyed the sky down to stars of the fourth magnitude, and he was undertaking a second review—of which more in due
course—when on 13 March 1781 he found an object which he knew was
not a star. At first he supposed that it must be a comet.

It tells us much about his prowess that, after Herschel had announced
his discovery, Maskelyne could not measure its position relative to the
reference-stars Herschel had named, for the stars were so faint that the
light from the cross-wires of his micrometer made them invisible. Thomas
Hornsby, in Oxford's Radcliffe Observatory—England's most beautiful
observatory building, modeled on the Tower of the Winds in Athens—
could not even locate the object. Herschel's quoted diameter of 5 seconds
of arc, was likewise unverifiable, so inferior were other telescopes even
to his modest reflector of 6.2 inches aperture. French astronomers spent
much time calculating the orbit, but the data were difficult to handle, and
it became clear that observations would be needed over a very long period before the matter was settled. The scales were more or less finally
tipped when, in the summer of 1781, Anders Johann Lexell, the Imperial

176. Mezzotint of Caroline
Lucretia Herschel (1750–1848),
after a portrait by Tielemann
(1829).

Astronomer at St. Petersburg, who was then visiting London, calculated
the first broadly acceptable elements of the orbit of the new object.

Most professional astronomers henceforth accepted that it was not the
orbit of a comet, and that in fact Herschel was the first to have discovered
a planet in historic times. The Hanoverian Herschel called it Georgium
Sidus, in honor of the king of England, George III, of the House of Hanover.
It will be noticed that in his 1794 portrait (fig. 175), Herschel holds a drawing labeled "The Georgian Planet with its Satellites." He had discovered
the two satellites Titania and Oberon in 1787, and in 1798 he was to find
that they move with retrograde motion. For the planet itself, various alternative names were proposed. Erik Prosperin of Uppsala even suggested
"Neptune," and Lexell liked the idea of "Neptune de George III," or "Neptune de Grand-Bretagne." Following a proposal by the Berlin astronomer
Johann Bode, who no doubt thought a Hanoverian name too parochial,
astronomers gradually settled on Uranus—in mythology, father of Saturn
and grandfather of Jupiter, for whom the next two planets below Uranus
were named. In France, largely under Lalande's influence, "Herschelium"
was long favored. In fact, three different names for the planet were used
simultaneously for at least sixty years, and as a fossil of the controversy, two
different symbols for it are in use to this day.

Herschel was now famous. He was granted a royal pension—appreciably
less than his salary as organist in Bath—in return for occasional instruction and display to the royal family. It was this that made it possible for him
to move from his old abode to the neighborhood of the royal residence
of Windsor Castle. Later he moved to nearby Slough. He was not entirely

177. William Herschel's 40-foot telescope, completed in 1789 with the financial support of King George III. The entire structure was on wheels, and could be rotated with the help of a geared winding device. Note the platform for the observer at the opening of the tube. (What is still known as a Herschelian reflector dispenses with a secondary mirror or prism, and the observer looks towards the objective, the main mirror.) Elevation of the telescope was by ropes and pulleys.

dependent on his pension; he went into the business of manufacturing telescopes for sale, and at the same time worked at ever larger mirrors. The king financed the largest, a 40-foot telescope with a 48-inch mirror that cost much trouble and energy (fig. 177). The mirror was four times the area of that in his previous telescope, which until then had been the largest in the world. The first two attempts resulted in failure, in one case with molten speculum metal spilling on to the floor and exploding the stones like shrapnel. The final mirror weighed over half a tonne, and when finally in use—as Herschel realized—its great weight made it subject to flexure, affecting its optical performance. A team of twenty-four workmen hired for the grinding and polishing produced poor results, and eventually Herschel built a large machine to finish the work. Completed in August 1789, on the second night of its use it revealed a sixth satellite round Saturn (named Enceladus). Herschel had a paper in press on nebulae at that time, and he wrote to Sir Joseph Banks, President of the Royal Society, modestly asking that he add at the end of it: "P.S. Saturn has six satellites. 40 feet reflector."

The telescope was vast but difficult to maneuver, and on the whole was less useful to Herschel than his 20-foot reflector. It was used until 1815, and when it was finally laid to rest in the garden at Slough, the family held a Requiem Mass inside its 12-meter tube.

In the early 1780s, Herschel's interest in the white patches of light in the sky known as *nebulae* (Latin for "mist" or "clouds") was aroused when he was given a copy of Messier's catalog of a hundred such objects. In many respects this marks the beginning of a completely new phase in cosmological thought. To map the universe in three dimensions, a knowledge of stellar distances was needed, and virtually the only thing known of such distances was that they were too great to yield measurable parallaxes. Under these circumstances, it was necessary to make conjectures, judging the plausibility of these by considerations having little to do with astronomy as such—considerations of symmetry and analogy, for example. Herschel was not the first to enter this territory, which was opened up from two different directions. On the question of the possibly infinite extent of the universe, considered in a very general sense, we have seen Newton and Bentley confronting difficulties with gravitational equilibrium. Halley, and later the Swiss Jean-Philippe Loys de Cheseaux, had noted the paradoxical consequence for the illumination of the sky at night, which on too simple a picture leaves no dark spaces at all. Complementing such abstract approaches were others that began from the actual structure of the system of stars that we see as the Milky Way.

The Greek word for the Milky Way is *galaxias*, hence our word "galaxy." Distant star systems were only called galaxies beginning relatively recently, that is, when there was evidence that they somehow resemble our Milky Way. The path to this knowledge was long and arduous. Before the telescope, most who touched on the subject at all spoke of the Milky Way as a cloud, but some—even as early as Democritus—treated it as a conglomeration of tiny stars that were so close as to be indistinguishable. Even after Galileo's telescope had resolved much of it into stars, the overall structure of the system was not at first a subject of great concern. After Wright, Kant, and Lambert had written on the subject in the mid-eighteenth century, however, it became an open question as to whether there might not even be systems of higher order, that is, systems of milky ways.

NEBULAE AND STAR CLUSTERS BEFORE HERSCHEL

The early history of attitudes to the nebulae has no very profound chapters before the advent of the telescope. Ptolemy used the descriptions "cloud-like" or "misty" for five or six items in his catalog, four of which are now classified as true (galactic) clusters and the others as asterisms. Since most later catalogs of any importance before Tycho Brahe's followed Ptolemy's to a greater or lesser extent, the list of nebulae remained virtually unaltered.

Tycho's half-dozen had only one in common with most of his predecessors, but he missed the Andromeda Nebula, which had been included in al-Ṣūfī's tenth-century list. Simon Mayr recorded it in 1612, and of course he could see it more clearly since he owned a telescope. He described it as being like the flame of a candle shining at night through transparent horn. When in 1656 Huygens found another nebula in the sword of Orion—which was to become coequal in fame to that in Andromeda, although we now know that it is of a completely different character—the area of sky around it was so black that he thought he was looking through a hole in the heavens into the luminous region beyond.

The number of nebulae recorded grew only slowly. Hevelius's and Flamsteed's lists, for instance, included only fourteen and fifteen, respectively. Few astronomers seem to have taken any great interest in individual nebulae for their own sake. Halley was one of those who did so. He spoke of patches of light seen in his telescope that look like stars to the naked eye, but that are in reality light coming from "an extraordinary great Space in the Ether; through which a lucid *Medium* is diffused, that shines with its own proper lustre"—that is, without the aid of an embedded star. He knew of others that seemed to shine because there was a star within them. In these vast and very distant places, he noted, there would be perpetual day. This thought caused him to defend Moses against those who had criticized the biblical account of the creation of the world, saying that Moses was wrong to speak of light before God's creation of the Sun.

Not all speculation was so restrained, and when the 75-year-old clergyman William Derham culled material from various sources, and added a few ideas of his own, he brought to general notice the old idea that stars may be openings in the heavens to a brighter region beyond. Might the same idea not apply rather to the nebulae, as Huygens had suggested of that in Orion's sword? His supposition that these were plainly much more distant than the fixed stars was without foundation, as Maupertuis pointed out in a justly critical rejoinder, which also emphasized our helplessness in discussing luminosities in the absence of firm knowledge as to distance.

As telescopes improved, astronomers began to search for and list nebulae as objects of interest in themselves. Cheseaux listed twenty objects in 1745–1746, at least eight of which were true nebulae or clusters not recorded previously. Guillaume le Gentil added others, including the elliptical companion to the (elliptical) Andromeda Nebula. Lacaille, Messier, and Bode added more, and yet by 1780, a hundred and seventy years of telescopic observation had increased the number of "genuine" clusters and nebulae from nine to only around ninety. By the end of Herschel's program, he had charted 2,500.

Gradually, nebula hunting was becoming a pursuit in its own right, although for long it was held in less esteem than comet hunting. Charles Messier, whose catalog led Herschel in 1783 to launch his twenty-year program of searching for nebulae, was first and foremost a comet hunter, indeed, he was the leading European adept at this sport in the 1760s. He was an observing assistant at the Marine Observatory in the Hôtel de Cluny in Paris, and had more than a dozen new comets to his name. It is said that the need to be at his wife's deathbed cost him another, to his sorrow.

Messier began his catalog of nebulae, like his lists of double stars, simply to make comet searching more reliable, for nebulae and comets were all too easily confused. When, at the end of the century, he referred to two thousand nebulae charted by Herschel, he added coolly that the knowledge gained would not simplify the comet hunter's task. So much for the priorities of the age. He resurveyed the old lists and eliminated much spurious material from them. His last published catalog had 101 distinct items, and for all his lack of interest in them on their own account, these, the most conspicuous nebulae, are still often known by their "Messier numbers." The nebula in Orion's sword is M42, for example, while that in Andromeda is M31, and its companion is M32.

THE MILKY WAY: WRIGHT, KANT, LAMBERT,
AND HERSCHEL

Newton may have started from very general and tenuous cosmological principles, but they were certainly more securely founded than most of those that followed in the eighteenth century. Almost by accident, before the middle of the century, an intellectual lead of sorts was given by a man with very different credentials, Thomas Wright of Durham. Wright was at first apprenticed to a clock maker, but taught himself practical astronomy effectively enough to go on to teach navigation to others and to work as a land surveyor. This last occupation led him to write successful works on English and Irish antiquities, and in the second half of his life he followed the career of architect, making no contribution of intrinsic importance to astronomy. That he is remembered in this connection today is chiefly because his ideas were seized upon and embroidered by others.

In 1742 Wright prepared a "key to the heavens," a volume explaining a very large (2.2 square meters) plan of the universe, as he thought it might be. In 1750, he published his most influential work, *An Original Theory or New Hypothesis of the Universe*. Wright was anxious to give a religious dimension to his schemes. As early as 1734, in a lecture-sermon illustrated by one of his large paper plans, he had identified the divine center of the universe with the gravitational center around which he believed the Sun and stars all move in orbit. This movement, for which Halley's proper motions seemed to provide slight evidence, gave an ingenious explanation of why the universe does not collapse into a single body under gravitation, the problem that had worried Bentley.

The Milky Way, in Wright's early model, was the cross-section of the universe we see when looking in the direction of the grand center. This was not well thought out, and in 1750 he changed the model, placing the stars—including our own Sun—in a thin spherical shell (fig. 178). Looking inward and outward from the shell, we see relatively few stars, but looking along any direction parallel to the tangent plane to the shell, we see stars at a high density. Here was a simple hypothesis that explained appearances, more or less. The Milky Way is of course irregular and of uneven density,

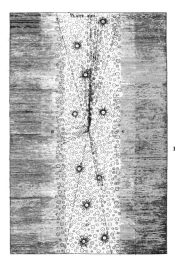 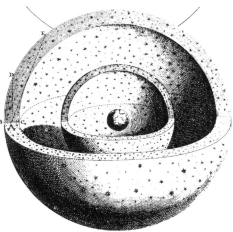

178. Thomas Wright's favored model of the Universe, from his *An Original Theory of the Universe* (1750). On the right (a detail of his Plate xxvii) is the complete model, while on the left (Plate xx) we have a detail, in which on account of the vastness of the sphere, he represents a segment of the shell of stars as trapped between parallel planes. In both figures, an observer at *A* sees relatively few stars when looking towards *B* or *C*, but vast numbers in the direction towards *D* or *E*, so producing the illusion of a Milky Way.

but that could simply be taken to indicate that there are irregularities in the shell of the universe. As he saw, however, there are other possible models. One he considered was a flat ring, a mere slice of the spherical shell, as it were. The Milky Way is explained as the effect of looking along tangents to the ring, and again, looking inward or outward produces a thin spread of stars, although now the variations in the actual Milky Way are slightly better represented. Finally, Wright considered that our universe might contain many such star systems, each with its supernatural center.

A manuscript ("Second Thoughts") survives, showing that he was not altogether content with this scheme. He now proposed an infinite set of concentric shells around the divine center. He supposed that each looks like a fiery sun from the outside, but that from inside it is pierced with volcanoes, which we see as stars, and as the Milky Way. He supposed that divine punishment was achieved by God moving the soul from one shell to a more confined shell.

With these schemes, Wright had shown how numerous are the hypotheses open to a vivid imagination, and how slender was the evidence by which to choose between them. Neither of these considerations seems to have unduly worried the great philosopher Immanuel Kant. Kant learned of Wright's book only through a review of it in a Hamburg newspaper in 1751, and the reviewer had unfortunately misunderstood some essential points. In his *Universal Natural History and Theory of the Heavens*, published in 1755, Kant shows that he did not recognize the supernatural nature of the center in Wright's theory, and therefore took Wright's ring model to be a *disk* model. In view of the shaky foundations of Wright's ideas this was of no great consequence, and Kant did add a number of new speculations to the existing stock.

He thought that the Sun and planets of the solar system might originate by condensation from some thin primordial matter—we are perhaps reminded of the Cartesian theory. He gave a rough qualitative "Newtonian" explanation of how under gravity such diffuse matter could form a disk, before condensation occurred, and he considered that this process is going on throughout the universe. The universe was thus, for Kant, a nonstatic affair. Bodies evolve, suns condense, then heat up to a point when they explode into fine matter within which the process can repeat itself. This process, he thought, goes on throughout an infinite space and over an infinite time—concepts that were destined to cause him a great deal of trouble in his later so-called critical philosophical writings.

When Johann Heinrich Lambert first seriously turned his thoughts to the structure of the Milky Way, around 1749 if we are to accept his own account, he had not heard of the works of Wright and Kant. Rather, as Kant was to do, Lambert took the Milky Way to be a (convex) lens-shaped structure, but he took the Sun and stars in its neighborhood to be a subsystem of the whole, and one of many such. In similar fashion he proposed that the Milky Way is a member of a higher-order system of milky ways. Unlike Wright and Kant, Lambert was a good mathematician. He was aware of the difficulties that Euler—who would one day be his colleague—and others were having with the theory of perturbations of Jupiter and Saturn, and he decided that there must be forces at work within the solar system that come from outside it. (He himself tried later to represent the motions of both planets by empirical equations, and he even anticipated some of the results that Lagrange subsequently obtained on theoretical grounds.) This encouraged his vision of a hierarchical universe, and although purely speculative, it is an early example of an astronomer trying to introduce the large-scale distribution of masses in the universe into an analysis of specific local effects.

Lambert's ideas were published in 1761 as *Cosmologische Briefe* (*Cosmological Letters*) and became popular in Germany and abroad, for they were translated into French, Russian, and English. When Herschel first came across this work is unclear, but there was in any case a widespread discussion of the problem of the nebulae: were they, too, resolvable into stars? His telescopes were better fitted to answering that question than any in the world, when he studied the nebulae in the early 1780s, and he soon found to his delight that many of them—in his enthusiasm he said "most"—were indeed resolvable. In 1790, however, he confirmed a suspicion that there was another class of nebulae, for then he found one that was plainly a cloud of luminous gas with a single central star. This discovery did not alter the truth of the first, although it left a certain ambiguity in the vocabulary being used by others in the discussion.

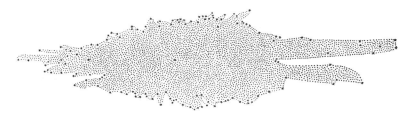

179. A section of the Galaxy, as drawn by William Herschel to illustrate his 1785 paper.

Clusters of stars fitted well with the idea of a process of a drawing-to-gether of matter through gravitation, ending up at places where the stars had originally been somewhat closer together than average. He found clusters too within the Milky Way. In a series of papers published between 1784 and 1789, he presented the evidence, and there announced the preliminary findings of his study of the Milky Way as a whole.

How was he to chart its outline, in the absence of stellar distances? He made two assumptions: first, that his telescope could reach to the farthest limits of the Milky Way, and second, that the stars are distributed regularly within its limits. On these assumptions, his program was one of counting stars within identical segments of the sky of the same (solid) angular size. These star counts he called "gauges." The 20-foot reflector telescope he used had a field of view of about 15 minutes of arc. Occasionally he registered no stars at all within a segment, or just 2 or 3, while in one segment he registered as many as 588. He took specimen gauges from well over 3,000 places in the sky. Of course he knew that the assumptions behind his star gauges were not impeccable, but he hoped that they would give statistically acceptable results, and indeed, were we ignorant of the method by which it was obtained, we should today judge his first chart, that of 1785, to have been a quite remarkable success (fig. 179).

It was in a sense unfortunate that, as his techniques improved and as he made use of the more powerful 40-foot telescope after 1789, his confidence in his early work was shaken. Uniform distribution was unacceptable, and the new telescope introduced so many more stars into the gauges that he could not guarantee that a bigger telescope would not add yet others. He still clung to a belief in condensation and clustering through gravitation, however, for the more he observed, the more examples he found of both—that is, of nebulae having the brightest patch in the middle, and stars in resolvable clusters as the middle was approached. He found regions at opposite sides of the sky that seemed to be unusually densely packed, and this and other items of evidence he produced would have fitted well into what we now know to be the spiral form of our Galaxy. In a paper of 1814, he drew attention to the way clusters of stars seem to favor the plane of the Milky Way—the galactic clusters, as we now realize, are a part of our own galactic system. He saw them as threatening to close in eventually on our galactic center, by gravitation, and so to bring about an end to the Milky Way. This

180. John Herschel's twenty-foot reflecting telescope at Feldhausen, Cape of Good Hope. He made his observations there in the period 1834–1838.

conclusion did not, however, take into account possible dynamical (rotatory) effects. At any event, observational cosmology was at last under way.

Herschel's gauges were continued by others. John Herschel continued his father's work in the Southern Hemisphere—at Feldhausen, an old estate southeast of Cape Town (fig. 180). An astronomical tradition had been inaugurated at the cape with Lacaille's work there between 1751 and 1753, and a British observatory had been established in 1820 with the practical aim of improving navigation in southern waters. At the cape, John Herschel used similar telescopes to his father's, including a 20-foot reflector and the smaller instrument formerly used with such success by his aunt Caroline; and he followed the same general methods as his father had used. John Herschel was an excellent mathematician who would probably have earned distinction in that subject had he not turned to astronomy out of duty to his father. He had learned the art of mirror making from his father, and made several large telescopes of his own. By 1838 he had cataloged 1,707 nebulae and clusters, 2,102 binary stars, and for his gauges had counted 68,948 stars in 2,299 fields of view. He noted that the Southern Hemisphere is richer in stars than the Northern Hemisphere. Otto Wilhelm Struve later confirmed many of his conclusions, in particular, that the Sun is not in the galactic plane but a little to the north of it.

Struve combined Herschel's observations with others by Bessel and Friedrich Wilhelm August Argelander to find a formula expressing the clear tendency of brighter stars to favor the plane of the Milky Way. Struve was a prolific and extraordinarily energetic astronomer, bred in the tradition of systematic star surveys, to which his father—Friedrich Georg Wilhelm Struve, director of the new observatory of Pulkovo, near St. Petersburg in Russia—introduced him. When Pulkovo was founded in 1839, under the patronage of Czar Nicholas I, it was provided with a 38-centimeter (15-inch) refracting telescope that was then the largest in the world. Otto was then twenty, with a couple of years' experience in the Dorpat Observatory already behind him. He was to spend fifty years at Pulkovo and become father to two well-known astronomers and grandfather to two more. More will be said about this dynasty and the Dorpat-Pulkovo link in the next chapter. It is worth noticing here the great influence exerted on Otto Wilhelm Struve's work and methods by William Herschel's example. Struve, having broadly accepted Herschel's views on star formation, was driven to make a series of intensive studies of the Great Nebula in Orion, which seemed to be the most promising star nursery within telescopic reach, and in which, in 1857, he claimed to have detected changes that had taken place during his time studying it. He gave less attention to star systems, and was probably typical of most astronomers of his time, in insisting that the Milky Way is unique, and perhaps even of infinite extent.

JOHN MICHELL, WILLIAM HERSCHEL, AND STELLAR DISTANCE

The concern of both William and John Herschel with double stars—they used a micrometer to gauge their separation and position angle—was connected with the hope that star distances might somehow be found. In 1767, John Michell had published a paper in which he had pointed out that close pairs of stars occur much more frequently than one would expect on the assumption that stars are uniformly distributed in space. The conclusion—an interesting early use of statistical reasoning in astronomy—was that there is a high probability that stars that seem to be close together are in reality so. Michell, who was the rector of a church near Leeds, in the north of England, decided to leave the reader with two alternatives: there is a general law at work, perhaps the law of gravitation, or there is a law of the Creator. Astronomers have tended to ignore the second alternative and to speak of Michell as the man who proved the existence of physical binary stars.

Michell, in whose house William Herschel was often a guest before he moved to Bath, has a second claim to distinction, for in 1784 he offered an argument leading to what would now be considered a plausible distance for a star. Saturn, he said, when at opposition to the Sun, is about as bright as the star Vega. The distance of Saturn is known to be, at such a time, about 9.5 times as distant as the Sun (9.5 AUs), and the angular size of Saturn is about 20 seconds of arc. Assuming that Vega is of the same intrinsic brightness as the Sun, and that Saturn returns to the Earth all (or a known fraction) of the light that falls on it, it is a simple matter to calculate its distance. It is necessary to use the inverse square law of brightness, but this had been established by the excellent work of Pierre Bouguer half a century before. In fact Kepler had stated the law, but Bouguer, royal professor of hydrography in Paris, was the first to turn astronomical photometry into an exact experimental subject.

It is just as well, when one discusses questions of photometry, to be aware of the loose ways in which its terminology is used in astronomical texts, and especially the word "brightness." The brightness of a star, which is effectively a point source, may be defined in terms of the intensity of radiation emitted from it, or the intensity of radiation actually received from it. The brightness of an extended surface, on the other hand, has to be defined differently. Sometimes called "intrinsic brightness," "brilliance," or "luminance," it is usually defined in terms of the amount of light emitted or reemitted by unit area of the surface. In an ideal situation, where emission from the surface is the same in all directions, as it is with a matt surface, it can be shown—paradoxically, perhaps—that the surface will look equally bright when viewed from any angle. This is not the place to launch into photometric theory—in this case into Lambert's cosine law—but it should at least be pointed out that potential ambiguities are obviously important when we are comparing the "brightness" of a star with that of the Moon, or the Sun, or the planets.

The distance Michell derived for Vega was approximately 460,000 astronomical units. In fact Vega is now known to be much more luminous than the Sun, and in 1837, F. G. W. Struve showed from trigonometric measurements that its distance was about four times Michell's estimate, but the only earlier estimate of a star's distance that was remotely accurate was Newton's figure for Sirius.

Michell's treatment of double stars was in stark contrast to another, which indeed set Herschel off on his study of doubles. Galileo and others had hoped to use "optical doubles"—to use our term for stars that are only accidentally in line as seen from the Earth—in the hope that the faint distant component might be used as a standard of reference. It was to be a token stationary point, with reference to which the fluctuating position of the much nearer star would be measurable. The hoped-for fluctuation would be that due to the Earth's orbiting round the Sun, and would, it was thought, provide the star's parallax. After intensive searching, with Caroline's involvement, Herschel published three catalogs of double stars (in 1782, 1785, and 1821) with 848 examples. After the first catalog, Michell drew attention to his own earlier reasoning, and in 1802 Herschel remeasured his doubles, finding to his surprise that many of them showed relative movements that had nothing to do with parallactic shifts—that is, they were not the result of the Earth's annual movement. (On this, see figure 57 in chapter 4.) This was the first step on the road to a proof that there are stars moving around one another in gravitational orbits, and that gravitational attraction is at work far beyond the solar system.

It was Herschel's hope that the relative brightnesses of stars would show their relative distances, applying the inverse square law. For a long time, he followed a common assumption that stellar magnitudes were a guide to relative distance—with third magnitude stars, for instance, three times as distant as first magnitude stars. In 1817, he devised an ingenious procedure for comparing brightnesses: he pointed near-identical telescopes at the two stars and cut down the aperture of that which was pointing to the brighter until the two stars appeared equally bright. His observations of double stars produced some unsettling results. If we are certain that two stars are orbiting around one another, and one seems much brighter than the other, then it is surely intrinsically so. Alas, the data Herschel was now obtaining from his photometric comparisons led him to a distribution of the stars in space that he saw to be unacceptable. He had himself discredited one of the chief assumptions that led to his old plan of the structure of the Milky Way.

We have already seen how the motion of the Sun through the stars was a factor in their proper motions, how the solar apex is the point away from which the resulting tiny motions seem to radiate, and how the antapex is the point to which stars in the other half of the sky seem to converge. There is a simple relationship between the proper motions due to this effect (measured, say, in seconds of arc per century), the speed of the Sun, and the distance of the star. The proper motions can be measured, and the Sun's speed can then be found, if we know the star's distance. At any event, the Sun's speed and direction should have unique values with respect to the system of stars as a whole, no matter which star's proper motion it is derived from—as long as the Sun is the only moving star. If it is not, then some sort of statistical analysis will be needed. Ignoring local motions, all stars should yield the same answer for the solar velocity, if their distances have been correctly assessed. When Herschel derived relative distances from brightness alone, however, he found very discrepant results. The best he could do, therefore, was find a statistically acceptable answer for the solar velocity.

Turning the argument round now: from a given value for the solar velocity, however it is obtained, one might expect to be able to derive stellar distances from the proper motions of the stars, if these are produced solely by our motion (with the Sun) through the star-system. Here Herschel was at first disappointed. There were bright stars—which one would expect to be close—that seemed to show no proper motion, and that should therefore be at a very great distance. At length, he saw the reason: his argument was valid only in a generally static universe. To us, this seems hardly surprising. Why should the Sun be moving and the rest not? Herschel realized that if some stars shared the Sun's motion, then their distances would not show up on the argument previously presented. In this way he received the first intimation of the phenomenon of "star streaming," but it has to be said that his discovery made little impact at the time. Not until Jacobus Kapteyn's announcement of 1904 was this phenomenon properly appreciated.

After Herschel had been using his largest reflector for some years, he learned to distinguish between nebulae that appeared nebulous simply because the stars constituting them had not yet been resolved, and nebulae made out of continuous luminous matter, and usually with one or more stars discernible within it. The more examples he accumulated, the more

certain he was that he possessed visual evidence for the way stars first condense out of diffuse matter, and then cluster under gravitational forces. In
other words, by the mid-1810s he had introduced yet another argument
for change in the remotest parts of the known universe.

Herschel never lost his interest in the solar system. He studied the
surface of Venus, for example and felt obliged to dismiss Johann Hiero-
nymus Schröter's claims—made public in a printed work of 1796—to
have found colossal mountains on that planet, one of them supposedly
43 kilometers high. Schröter was an Erfurt lawyer whose interest in astronomy Herschel had fostered, and to whom Herschel had sold instruments.
Schröter was not the first to read more into his telescopic observations of
that planet than the evidence justified, but he was at least right to suggest
in 1788 that many of the surface markings he detected were atmospheric.
Earlier claims to have seen markings on the Venusian surface had been
made by Francesco Fontana in 1645 and by Francesco Bianchini in 1726.
Bianchini had actually produced a map showing nine maria (seas), eight
straits, and twelve promontories, to all of which he was confident enough
to give names. At length, when these several apparitions finally evaporated,
the names were left high and dry.

Herschel studied Mars and found its rotation period and axis; and he
made measurements of the newly discovered asteroids Ceres (by Piazzi,
1801) and Pallas (by Olbers, 1802). In observing the Sun in 1800, he noticed that the sensation of heat in the image projected by his lens did not
correspond with the light at the same place; and in this way he was led to
experiment with thermometer and prism, and so to discover infrared radiation. (He found no heating effect at the violet end of the spectrum. ultraviolet light was detected soon afterwards by R. W. Ritter, from its blackening
effect on silver chloride—and this long before the advent of photography.)
In short, Herschel was a man of an irrepressible experimental talent that
counterbalanced the brilliance of contemporary theoretical astronomy, especially that in France. He had an instinctive appreciation of experimental
method. Where many would reach a conclusion only to close their eyes to
contrary evidence, Herschel actively sought out potential conflict. To take
a single example, that of the motion to the solar apex: ending his account of
this in 1783, he expressed the hope that future observations would "either
fully establish or overturn the hypothesis of the motion of the whole solar
system," adding that to this end he had "already begun a series of observation upon several zones of double stars." His interests took in the whole of
astronomy, but his most lasting contributions concerned the universe of
the stars and the nebulae. In that last respect, after his death, many decades
passed before his work was substantially extended by others.

MATHEMATICS AND THE SOLAR SYSTEM

At a theoretical level, astronomy after Newton profited greatly from the
very rapid advances being made in mathematics, which ran parallel to

improvements in practical matters. The advances rested heavily on the foundations Newton had laid. One of the worthiest of his early followers was the Scottish mathematician Colin Maclaurin, who investigated the equilibrium of ellipsoids—such as the Earth—and the tides. British mathematicians were unfortunately often too loyal to Newton when sticking to certain of his techniques on which continental mathematicians had improved, and the lead soon passed to the continent. The Basle mathematician Leonhard Euler was one who advanced virtually every branch of the mathematics of his time, pure and applied. Time and again he won prizes from the Paris Academy of Sciences for his work. Without any title to practical experience, he gave astronomy some of its most useful mathematical procedures, for instance, the theory of instrumental errors, ways of finding the solar parallax, and the determination of orbits—whether of planets or comets—from a few appropriate observations. The problem of the lunar perigee was a worthy test of his abilities.

Following Newton's principles, Alexis-Claude Clairaut and Jean Le Rond d'Alembert had derived a value of about eighteen years for the period of revolution of the Moon's perigee, the nearest point on its orbit to the Earth. (This is not to be confused with the period of 18.6 years for the Moon's nodes.) From observation, the figure was known to be only about half as great, and for a long time Euler and others believed that the only remedy was to make adjustments to Newton's law of gravitation. In 1749, Clairaut found a mistake in the method of approximation that all had been adopting. Euler did not at first agree, with the result that he composed a treatise on lunar theory that outshone everything before it. This *Theoria motus lunae exhibens omnes eius inequalitates* (*Theory of the Motion of the Moon, Showing all its Inequalities*, 1753) included a method for an approximate solution to the three-body problem—in this case, the problem of the Sun-Earth-Moon system. The work makes use of a new technique that was destined to prove of enormous value to future mathematical astronomy and physics, the "method of variation of the elements." More immediately comforting was the fact that Clairaut and he had shown that Newtonian gravitation and dynamics passed this stringent test.

Euler devoted much labor to the three-body problem, and the problem of the perturbation of planetary orbits which is essentially similar. His greatest work on lunar theory appeared in 1772, a work that was not properly appreciated for more than a century, when it was rescued from neglect, and was further developed, by the American mathematical astronomer George William Hill. (Hill was the leading person in his field. He seems to have had difficulties precisely the reverse of Kepler's and Flamsteed's, for he insisted on returning his salary to Columbia University.)

One of the most difficult problems confronting mathematical astronomers of the eighteenth century, and one that acted as a constant spur to progress, again concerned an inequality in the Moon's motion. The Moon's mean motion averaged over a reasonably long period—say a century rather than a millennium—is not constant when the averages are examined across much longer tracts of time, but rather, it accelerates. This trend was first suspected by Edmond Halley around 1693 on the basis of a comparison of ancient eclipse records with what the best modern tables gave for the same eclipses. In 1749, Richard Dunthorne revived the subject, and added further ancient data to confirm Halley's suspicions. The acceleration was extremely small, and indeed its small size is a useful pointer to the progressive refinement in astronomical accuracy. Dunthorne fixed it at only 10″ per century, and others later in the eighteenth century, such as Mayer and Lalande, settled on figures between 7″ and 10″ per century. But what was its physical cause? In 1770, the Paris Academy offered a prize for a solution, which Euler took jointly with his son Johann Albrecht. They were under the impression, however, that they had proved that the steady ("secular") acceleration of the Moon could not be explained by Newtonian gravitational forces.

Here once again was something of a crisis in Newtonian science, and the subject was proposed for the Academy's prize of 1772. On this occasion it was awarded to Euler jointly with Lagrange.

Joseph Louis Lagrange was born to an Italian family of French ancestry in Turin, Italy. (His French name represents only the last version of a continuously varying quantity.) Before he was twenty he drew attention to his mathematical talents through his correspondence with Euler. Following Euler's lead in the 1760s, he introduced some brilliantly original methods of his own into a study of the motion of the Moon, and another, of the perturbations of Jupiter and Saturn, which won him the Paris Academy prize for 1766 and much fame. He was found a position in Berlin, thanks to d'Alembert's friendship with Frederick II of Prussia. Euler, who was on the point of leaving a post in Berlin for one in St. Petersburg, did not manage to persuade Lagrange to join him there, but in Berlin he had several stimulating colleagues, including Johann Lambert, whose cosmological ideas we have already encountered. Lagrange's prodigious mathematical talents soon became obvious to all. In 1772 he shared the Academy prize with Euler for an essay on the three-body problem, in this case on lunar motion. Euler, in his essay of 1772, now maintained that gravitation could offer no explanation for the secular acceleration of the Moon, but that there must be some sort of ethereal fluid in space, offering resistance to the motion of the Moon and the Earth. Lagrange offered a new solution to the three-body problem, but did not explain the secular acceleration.

Again in 1774, the Academy prize was offered for a solution, and again Lagrange was successful, with an account of how the shape of the Moon has an effect on its motion—and similarly for the Earth. Still he could find no explanation for the secular acceleration, and in a study of the historical evidence for it, he pronounced it a dubious idea that should be abandoned.

The series of Academy prizes continued to attract work of the highest quality, but Lagrange was tiring of the constraints they placed on his work, preferring to write independent memoirs. His last entry was for the 1780 prize, which he won with an important study of the perturbation of cometary orbits through the action of planets. He contributed several additional memoirs of great value on Newtonian planetary theory. Brought to Paris in 1787, he managed to survive through the turbulent years of the Revolution. He became a member of the Bureau of Longitudes there and was able to assist in the practical needs of astronomy, such as the production of ephemerides—of which he had experience in Berlin. He was honored by Napoleon, and when he died in 1813 his funeral oration was delivered in the Pantheon by Pierre Simon Laplace—who had by this time solved the problem that had eluded Lagrange and others for so long.

Laplace was born in Normandy, where he studied at the University of Caen before leaving for Paris in 1768 with a recommendation to d'Alembert. Within five years, a brilliant series of mathematical papers he had published resulted in his election to the Paris Academy of Sciences. He wrote on the integral calculus, celestial mechanics, and the theory of probability. Successive volumes of his *Mécanique céleste* (*Celestial Mechanics*) appeared between 1799 and 1825, and in them, as in his important writings on physics, he made much use of a number of mathematical techniques that he had himself developed, and that are still widely used with his name attached. Laplace's work was plainly too important for it ever to have gone unnoticed in the English-speaking world, but the translations of the first four volumes by the self-educated mathematician and astronomer Nathaniel Bowditch, of Boston, Massachusetts, certainly did it no harm (these appeared between 1830 and 1839).

Laplace was not above pointing out his own incomparable genius and lost many friends as a result, but he was conscious too of the need to make the mathematical sciences available to a wider audience, and one of his most popular works was his eminently readable *Exposition du système du monde* (*Exposition of the System of the World*, first published in 1796). This dealt with a very wide range of cosmological matters. His work in mathematical astronomy reached its peak during the revolutionary period in France, and he was able to exert very great influence over the structuring of intellectual life in France at every level. During the Empire, he was honored in various ways by Napoleon, with whom he discussed astronomical matters—in one case on the field of battle, or so it is said. It was the returning Bourbons, however, after Napoleon's exile, who eventually made him a Marquis.

When Laplace came to study the possible acceleration in the Moon's motion, he began by dismissing the claims of skeptics, that the historical evidence for it was unreliable. He also dismissed a solution that was being offered, that the effect was no more than an illusion, caused by the slowing down of the Earth's rotation due to friction, possibly caused by the Earth's winds. Why, he asked, in this case, did the planets' mean motions not also increase? There was no answer. As for Euler's notion of an ethereal fluid, he rejected that for want of independent evidence. In short, he faced up squarely to the problem as it had stood three generations before.

Still, however, he was unable to solve it, and so it was that he tried to modify Newton's law of gravitation. It had been generally assumed that the gravitational force exerted by one body on another acted instantaneously, but what if it required a finite time to act? Laplace showed that this could result in a secular acceleration of the Moon, but only if the speed of transmission of gravity were more than eight million times as great as the speed of light. (And he showed that if the secular acceleration could be explained in other ways, then the speed of gravity must be fifty million times that of light for it not to be otherwise in evidence.) He was not happy with his solution, which, like the ether, was not exactly obvious elsewhere; but then in 1787 he found a much more acceptable alternative. He had found that the shape of the Earth's orbit was changing; in fact, he had found that the eccentricity of the ellipse was decreasing, and he was able to connect this effect with the gradual shortening of the length of the month. The analysis was supplemented by his study of the motions of Jupiter's satellites. (Jupiter actually enters the calculation of our own Moon's behavior.) He calculated a theoretical expression for the secular acceleration of the Moon, that for his own day gave a figure of approximately 10.1816 ″, close to the best historical evidence; and he showed that after about 24,000 years the secular change would reverse, and the month would lengthen.

When Lagrange read the paper announcing these findings, he looked over his own earlier work of 1783 and found an omission, which, when put right, yielded almost exactly Laplace's results. Long afterwards, John Couch Adams, "discoverer of Neptune," demonstrated that Laplace's theory could not account for all of the known effect, but Laplace's achievement was unquestionably important, and for long was regarded as the acme of dynamical astronomy.

Laplace had many other results to his credit in gravitational theory. For instance, not only did he find a relationship between the shape of the Earth and certain irregularities in the Moon's motion, but he introduced that and the rotation of the Earth into tidal theory. One of his greatest achievements was to explain certain fluctuations in the orbital speeds of Jupiter and Saturn. He found that they resulted from a curious relationship between the periodic times of the planets, five times Jupiter's period being very nearly twice Saturn's. One of his greatest achievements, however, seemed to touch on a subject of great concern to those who were trying to relate astronomy to natural religion. This was his work on the stability of the solar system. Will it continue forever, without intervention by a divine clock maker? Leibniz had taunted Samuel Clarke with the imperfection of a Newtonian universe, which he said would need winding up by God from time to time, so implying that God was an inferior artisan.

The stability question was a true test of mathematical prowess. Laplace made much use of Lagrange's method of introducing variations into the

181. Pierre-Simon, Marquis de Laplace
(1749–1827).

six elements of a planet's orbit—the eccentricity, direction of aphelion, and
other parameters that define it—and in 1773 he was able to prove that,
even if one planet's elements are perturbed by another planet, its mean
distance from the Sun will not change appreciably, even over millennia.
Over the next few years Laplace followed this with more complex theorems relating the distances, eccentricities, and angles of the orbital planes,
and again these seemed to point in the same direction: the solar system is
highly stable. He showed that there is a certain plane in the solar system
about which the whole system oscillates. In more recent studies, the fric-
tional effects of the tides have been introduced into the account, and again
it has been found necessary to qualify Laplace's claims, but the skeleton
of his analysis remains, a remarkable testimony to the achievements of
Newton's great successors in the century following his death.

LAPLACE'S NEBULAR HYPOTHESIS

Laplace's *Exposition du système du monde* provided a non-mathematical
account of his chief work, as well as many new ideas. Book v of this often-
revised work surveyed subjects which he had not dealt with mathematically
elsewhere, with five highly selective chapters on the history of astronomy,
and a sixth and final chapter expressing his new ideas on broadly cosmological subjects. This last chapter, only eleven pages long, caught the attention of his popular audience, as well as of professional astronomers, in
particular for what is now generally known as Laplace's "nebular hypothesis." It is clear that in allowing his imagination free rein, he was inspired by
some of the ideas of a fellow lecturer, the zoologist Baron Georges Cuvier.
Whereas Cuvier tried to demonstrate how living species are conserved, Laplace wished to show that the laws of nature could guarantee the duration

of the solar system, and by extension that of the whole universe. This was a view which he considered to have been partly sanctioned by his theoretical work on the stability of the solar system—given that there are no extraneous causes.

A generation earlier, the French naturalist Georges-Louis Leclerc de Buffon had speculated about the origin of the planets in our system. He suggested that a comet, striking the Sun, ejected material that cooled and condensed into planets. Laplace's approach to the question was vastly superior. In view of the vastness of our system, he decided that the matter from which the comets were formed must have been an atmosphere of some sort or a fluid expanding to an atmosphere, presumably a solar atmosphere. The planets would then have formed by condensation in the plane of the solar equator, each planet holding on to an atmosphere from which its satellites could be formed. Comets too could be formed, rather as planets are, but they would not all be made to move in a similar way: those with elongated orbits would eventually seem to have been placed in their orbits by chance. Chance, however, was not an option in the formation of the system as a whole, as he demonstrated in a brilliant piece of analysis, easily understandable by his ordinary readers, and occupying less than a page of his last short chapter.

He considered the orbits of the seven planets then known—including Uranus—and their fourteen known satellites, those that had been discovered at the time of publication of first edition of his book in 1796. All revolved in roughly the same plane and—he mistakenly thought—in the same direction, and they shared this direction with the rotation of the Sun, Moon, five planets, and the rings and outer satellites of Saturn. (As mentioned earlier, the retrograde motion of Titania and Oberon, satellites of Uranus, was discovered by Herschel only in 1798.) There were thus 29 known circular motions, and so 2^{29} possible arrangements of senses of rotation. Left to chance, the probability that just one planet out of 29 would move in a retrograde sense, was 2^{-29} to 1, in short, a virtual certainty. (He noted that the probability of this is far, far, superior to that of many historical events, the truth of which we never doubt.) This is to speak in terms of chance; but our experience shows that the world is not as chance would decree, and therefore the world is as it is by virtue of a cause. Other considerations of chance, in particular, the fact that all planetary eccentricities are relatively small, led to the same conclusion.

We have already seen what he considered the cause to be. It was as described by his nebular theory. He must have known that it would be controverted at a later date, but even had it collapsed in his own lifetime, he would have remained proud of his argument for a causal system. In fact, while his nebular hypothesis had many adherents, it displeased many a religious fundamentalist, and in time it was recognized as fitting badly with what was known of the masses and energies of the Sun and planets. The planets have only a thousandth part of the mass of the solar system, but 98 percent of the rotational energy of the whole, far more than Laplace's ideas require. He would no doubt have offered an explanation.

One final word about the words "nebula" and "nebular." Laplace's was a "nebular hypothesis" in the sense that the atmosphere amounted to a cloud, a nebula in that sense, and not in the sense of blurred, or barely resolved, patches of light, as seen through powerful telescopes in the heavens. Laplace tends to speak of "atmospheres," but "nebular hypothesis" is a phrase that has been used so often that one is more or less obliged to keep it. He himself went halfway to applying his hypothesis to the nebulous objects that Herschel had discovered, "nebulae" in which apparent points of condensation had been detected. It seems not unlikely that Herschel's observations were a source of inspiration to this great mathematical astronomer.

Precision and the New Astrophysics

Maps of the stars make up a genre in themselves, but by the seventeenth century the genre was plainly branching into two. Both owed much to the ancient listings of bright stars with their coordinates, as in the tradition of Ptolemy and al-Ṣūfī, but whereas the serious-minded astronomer was more concerned with the recorded coordinates, there was a much larger potential following for the imagery of the constellations. The same ambivalence was being exploited by publishers of terrestrial atlases, with which star atlases were often twinned. By the mid-nineteenth century, professional astronomers had almost entirely abandoned the pictorial aspect of mapping—the wonder is that they nurtured it for so long. By the end of the century, most had lost interest even in producing the very large printed sheets which some German observatories had begun to issue, with all their recorded star positions, magnitudes, names, and other information duly plotted on them. The high period of pictorial mapping covers little more than the seventeenth, eighteenth, and half of the nineteenth centuries, during which time there were dozens of examples produced, although most of them were directly cribbed from a mere handful that towered above the rest.

We saw earlier how Piccolomini's austere maps, in his "book of the fixed stars," *De le stelle fisse* (1540), broke with the pictorial tradition so beautifully exemplified in the printed star charts of Albrecht Dürer (1515) and in works scarcely distinguishable from Dürer's at first glance, such as those of Apian (1540) and Honter (1541). In 1588, Giovanni Gallucci followed Piccolomini's way of setting stars over a coordinate framework of limited coverage, as opposed to coverage of a whole hemisphere, although Gallucci used Copernicus' star coordinates where possible and added crude constellation drawings to his star markers. The pictorial element is what draws the eye of the casual judge of a chart's quality, but beneath the artistic surface, important things were happening. In 1603, Johann Bayer, taking his star positions from the 1602 printed catalog of his contemporary Tycho Brahe, issued the first star atlas to be printed from engraved copper

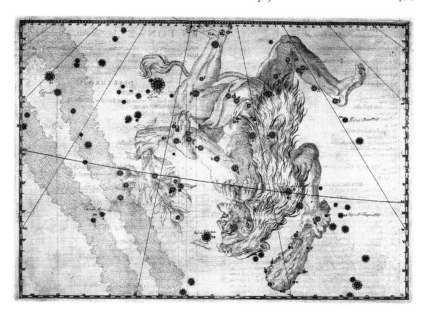

182. The constellation of Hercules, from Johann Bayer, *Uranometria* (1603). The stars are mapped as seen from the Earth. On this much-reduced figure it is hard to see the small Greek letters written against the stars, but for a few of Bayer's star letters, see the next figure. The paper used for the first edition of the book was thin, and print on the verso shows through. In all later editions (beginning in 1624), the versos of the plates were left blank, and the tables were printed separately.

plates as opposed to woodcuts. His *Uranometria, omnium asterismorum continens schemata*, printed in Augsburg, promises in its title that it contains the figures of all asterisms (constellations). The plates were drafted with great precision and artistry by Alexander Mair, a Bavarian artist active in Landshut (for a specimen, see fig. 182). There were no fewer than fifty-one star charts in Bayer's atlas, one for each of the traditional forty-eight Ptolemaic constellations, two planispheres not unlike Dürer's, and a chart of the newly discovered southern sky. This included twelve hitherto uncharted southern constellations, including Phoenix, Toucan, Peacock, and Bird of Paradise. Here Bayer was working from somewhat inaccurate data, which he took from the work of the northern Dutch mariner Pieter Dirckszoon Keyzer, alias Petrus Theodori.

What was strikingly new in Bayer's work was his reformation of the method of designating stars. In labeling every star visible to the naked eye, he assigned to each star in a constellation a letter of the Greek alphabet, generally in order of apparent brightness, using the Latin alphabet after the Greek when it was needed. He did not restrict himself to mapping the heavens, but added enormously valuable lists and commentaries to simplify the task of cross-referring to classical catalogs, many elements of which he carefully amended (fig. 183). His stellar nomenclature, somewhat expanded, is still in use today. *Uranometria* set the pattern for the star atlases of the next two centuries and more. Of course accuracy is not to be

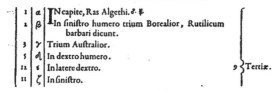

TABVLA SEPTIMA.
HERCVLES.

ΕΝ ΓΟΝΑΣΙ, Ingeniculus, Prociduus in genua,
Incuruatus in genu, Genu flexus, Nixus, Ὀκλάζων, Homero γνὺξ ἐριπὼν. M. T. Ciceroni
Nifus, Vitruuio vt quidam volunt, Neſſus, aliis Saltator, Aper, Cetheus, Theſeus, Alcides,
Ixion, Prometheus, Thamyras, Orpheus, Arato Εἴδωλον ἀπευθές καὶ ἄιςον, nomen anti-
quiſſimum, Imago laboranti ſimilis, Algiethi, veriùs Algethi, Alphonſinis malè
Raſaben, Arab. Elgezialerulxbachei, Perſis
Ternuelles.

DIARTHROSIS.

1	α	IN capite, Ras Algethi. ♂. ♃	
1	β	In ſiniſtro humero trium Borealior, Rutilicum barbari dicunt.	
3	γ	Trium Auſtralior.	
5	δ	In dextro humero.	
11	ε	In latere dextro.	9 } Tertiæ.
11	ζ	In ſiniſtro.	

183. An excerpt from Bayer's listing of the stars in the constellation Hercules. Opposite his Greek letters are some traditional appellations, many of them Ptolemaic, other imported from Arabic—as in the case of Ras Algethi, now α Herculis, the brightest star in the constellation. Like the other stars in this excerpt, the latter is only a third magnitude star, as indicated by the word *tertiae*. One odd characteristic of Bayer's work is that he often views figures from the back, making nonsense of such traditional descriptions as "star in the left foot."

judged instantly by eye alone, but on close examination it turns out that in this respect, too, his work set a quiet revolution in motion. One of the greatest authorities on the subject of star positions in the mid-nineteenth century, the Bonn astronomer Argelander, valued Bayer's accuracy enough to publish a careful star-by-star study of it.

Of those star atlases that were sold primarily on the basis of their pictorial content, the best known was the work of a schoolmaster, Andreas Cellarius. Born in Neuhausen, near Worms, presumably with the name Keller, Cellarius seems to have moved around eastern Europe for some years until he settled in Holland, where eventually—in 1637, at the age of about 41—he was appointed rector of a school in Hoorn, north of Amsterdam. This seems an unpromising start for a man whose colorful celestial imagery is today perhaps the most often reproduced of all—in picture books and magazines, on greetings cards and similar places, where its precise meaning is immaterial. Cellarius had written on the history of Poland and on military machinery, but his best known work was undoubtedly that which he wrote at Hoorn, subtitled *Atlas coelestis*. He would have been saddened to think that his *Macrocosmic Harmony*—its main title was *Harmonia Macrocosmica*—was to be regarded as a mere picture book. Published in 1660 in relatively small print run, it sold out quickly and was reprinted in 1661 with a much larger run. His Amsterdam publisher, Johannes Janssonius, was hoping for a supplement to an ordinary atlas he was printing and was less interested in the Latin commentary that Cellarius added to his plates in the hope that one day his *Harmonia* would serve as the historical section of a grand two-volume treatise on cosmography. This he never completed. When the plates

184. The "first hemisphere of the starry Christian heavens," from Andreas Cellarius, *Harmonia Macrocosmica* (1660), broadly following Julius Schiller's iconography (1627). Most maps of the hemispheres divide the sky between north and south. Cellarius here divides it between east and west. On this western half, the ecliptic constellations (with traditional equivalents) are the saints James (Gemini), John (Cancer), Thomas (Leo), the younger James (Virgo), Philip (Libra), and Bartholomew (Scorpio). What the protestants of Hoorn thought of the prominence given to Pope St. Silvester I (Boötes) is anyone's guess. The enticingly clad woman at the south is Eve (Peacock, Chameleon, and Fly).

were reprinted in Amsterdam in 1708, the Latin commentary was dropped, showing clearly enough what his particular audience valued most.

The *Harmonia* was originally intended to instruct, depicting as it did simple versions of the chief planetary theories and a touch of astrology. He illustrated the sphere of stars by means of several magnificently drafted hemispheres, Northern and Southern according to "ancient tradition"—although the new Toucan, Peacock, and Bird of Paradise in the southern sky were included—and a Christianized sky divided into eastern and western halves. For the latter depiction he was following closely in the footsteps of the Augsburg lawyer Julius Schiller, whose Christianized atlas had appeared in 1627. Schiller claimed to have been spurred on by consulting an edition of the collected works of the Venerable Bede. Judging by sales, he seems to have misjudged his public, but Cellarius's publisher certainly did not. His work would have been extremely useful in the schoolroom, although few schools are likely to have been rich enough to buy it. For an example, see his version of the Christian sky (fig. 184).

After Cellarius had shown the way, a few others followed suit—Johann Doppelmayr was one of the last of much merit, in 1729—but the fashion for such luxury did not last much longer. While the fashion had an even more opulent parallel in the craft of globe making, globes were only for the discriminating rich. The undisputed master in this craft was Vincenzo Coronelli of Venice. Coronelli's enormous creations, of larger than human proportions, used gores he had printed, which were painted only after being applied to the globe. His globes could in principle have provided a useful source of instruction, as well as amusement, but one suspects that the abundance of textual information on their magnificently drafted surfaces was seldom read.

Mapping the heavens continued to have its more serious side, of course. Classification, as well as accuracy, was seen to be a matter of importance. As the seventeenth century wore on, astronomers began to introduce new constellations, some from the southern sky, some to honor patrons, others perhaps to draw attention to themselves. We get Giraffe, Dove, and Unicorn, from Jakob Bartsch, Kepler's son-in-law; and Lynx, Fox, Hunting Dogs, and the Shield of Sobiesci from Hevelius. Jan Sobiesci was his patron, the elective king of Poland who had earned European fame by driving back the Ottoman Turks when they besieged Vienna in 1683. There were many comparable attempts to place patrons in the heavens in the hope of earthly reward, but the agreement of foreign astronomers was as hard to find as local patronage, and such attempts were generally futile. Some astronomers were content to rename existing constellations. William Schickard, for example, used biblical names—Perseus became David, Draco was presented as the dragon in Revelation, and so on. Julius Schiller introduced an interesting variant, assigning northern constellations to the New Testament, and southern to the Old—to be followed by Cellarius, as we saw. In the long term, such practices led nowhere in particular.

The case was very different with the *Uranographia* of Johannes Hevelius (1690) and the *Atlas Coelestis* of John Flamsteed (1729; posthumously published). Hevelius's work was based on a catalog of more than 1,500 stars, more comprehensive and more accurate than Tycho's, making his accompanying star atlas the first to rival Bayer's. That atlas was a tour de force containing fifty-six finely engraved charts, each a double-page spread. It was unusual in one respect: Hevelius chose to depict the constellations as they would appear on a globe, rather than as they appear in the sky. He introduced eleven new constellations, including seven that are still accepted: Scutum Sobiescanum (Sobiesci's Shield), Canes Venatici (Hunting Dogs), Leo minor, Lynx, Sextans, Lacerta (Lizard), and Vulpecula cum Anser (Little Fox with a Goose). For a specimen of his work, see figure 185, and compare the Flamsteed half with figure 186. Like Bayer, Hevelius included southern asterisms. Following Edmond Halley's 1676 voyage to the island of St. Helena in the South Atlantic, which had resulted in a catalog of the positions of 341 southern stars in 1679, Hevelius could make use of that very useful source. Halley corresponded with Hevelius often—we recall that he visited him in Danzig.

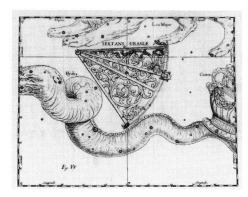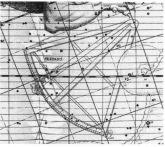

185. *Left,* Hevelius's engraving figuring his newly formed constellation of Sextans Uraniae, from his *Firmamentum Sobiescianum sive Uranographia* (1690). The book, *Sobieski's Firmament, or Uranographia* was a form of advertisement for the Polish king, while the constellation was an advertisement for the highly successful use of such a sextant in Danzig between 1658 and 1679. Hevelius put the constellation between Leo and Hydra, with their fiery association in astrology, to commemorate the destruction of his house and instruments by fire in 1679. *Right,* a detail of the same constellation, from Flamsteed's *Atlas Coelestis* (1729, published posthumously). Note the straightforward borrowing, apart from the baroque network. Flamsteed's stars are as seen in the sky, while Hevelius's are in reverse, as on a globe. Contrast Flamsteed's figure with figure 186.

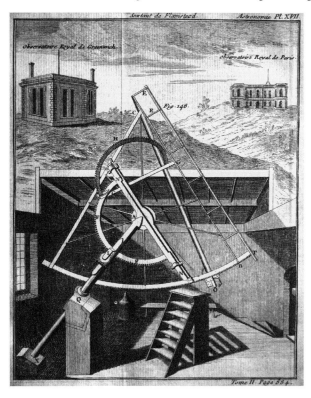

186. Flamsteed's sextant as used at Greenwich. From J. de la Lande, *Astronomie,* vol. 1 (1764), copied in mirror image from Flamsteed's *Atlas Coelestis,* but with images of the Greenwich and Paris observatories added. Notice how different the sextant is from the borrowed drawing of a sextant in the atlas (fig. 185).

Not surprisingly, the atlases that appeared in the following half-century show signs of having been influenced in one respect or another by Hevelius—examples are works by Johann Rost (1723), Corbinianus Thomas (1730), Christoph Semler (1731), and Johann Doppelmayr (1742). While most purchasers of these lavish and expensive volumes were doubtless more interested in artistic splendor than in accuracy, an accurate atlas can contain hidden treasure, and there is no saying in advance what that will be. The one great step forward during this time was not dependent on the tastes of dilettanti, but was rather the publication of Flamsteed's work. He had been dead for a decade by the time it appeared. His long program of stellar observation had begun soon after his installation in the Royal Observatory at Greenwich in 1676. Between then and 1689, he made about twenty thousand observations, accurate to about ten seconds of arc, using a seven-foot sextant presented to him by Sir Jonas Moore. These were all observations of relative—not absolute—distances, and not until his father's death was he able to afford out of his inheritance a large mural sector to provide anchoring points for his earlier readings. From 1689 he made use of two excellent clocks, again presented to him by Moore, to measure time intervals between certain solar and stellar observations, allowing him to establish the positions of forty reference stars. From them, he computed the precise coordinates of the 3,000 stars that were to be included in his finished catalog, using the old sextant measurements.

We have already seen how his quarrels with Newton and Halley delayed the publication of his findings, and how his three-volume *Historia coelestis Britannica* only appeared in 1725, six years after his death; and how his *Atlas coelestis* had to wait until 1729. Flamsteed's situation improved during the last years of his life. He had a devoted wife and good friends, one of whom was a well-known painter, Sir James Thornhill—a man whose painting of the ceiling of the lower hall of the Royal Naval Hospital at Greenwich is proof of his friendship. It portrays Tycho, Copernicus, and Flamsteed, with his assistant Thomas Weston, and with Flamsteed's mural arc, as well as a diagram of the total solar eclipse of 1715. Newton, on the other hand, is there depicted as an ancient philosopher, holding mathematical diagrams. Thornhill had an important part to play in helping Margaret Flamsteed with the atlas, providing some of its fine engravings and coordinating the work of other engravers. A specimen constellation is shown in figure 187. A few charts had been drawn up as early as 1703 by Thomas Weston, ornamented by the artist Paul van Somer—not to be confused with the more famous Jacobean artist of the same name.

Like all of his serious precursors of the seventeenth century, Flamsteed adopted Bayer's Greek-letter notation. His atlas was created from what was by far the fullest star catalog published at that time, but like all such works, the massive endeavor hidden within each and every one of its final coordinates remained hidden from view. With a few notable exceptions, those who had designed earlier star atlases simply copied the coordinates

187. The constellation Andromeda, in a
detail from plate 16 of John Flamsteed's
Atlas coelestis.

from the best available lists. The secret of Flamsteed's quality lies not only
in his own innumerable calculations, but in the fact that, from 1696 onward, he had calculators working for him in Derbyshire and at Greenwich,
so that any errors could be detected by comparing their findings. Independent calculation soon became standard practice, and for the rest of the
century Flamsteed's two posthumous works set a standard for others to
follow. Particularly important was its adoption by leading French astronomers. There were two new French editions, in 1776 and 1795—the second
being presented as though it were the third. For these, Joseph-Jérôme de
Lalande and Pierre Méchain take most of the credit. They added nebulae
from Messier and five new constellations, one of them being The Telescope
of Herschel, between Orion's head and the horns of Taurus. For an example of their handling of another constellation, Hercules, which had stars to
spare, see figure 188.

The atlases of Bayer and Hevelius, and even the more complete atlas
by Flamsteed, showed only stars visible to the naked eye. Johann Elert
Bode's *Uranographia* of 1801 took matters a step further, with a gargantuan atlas that marks a change in emphasis. The artistic element was still
present, indeed revised and refreshed, but its days were numbered. There
was little point in marking invisible stars on a pretty picture. To compare
a specimen of his and Flamsteed's figures, see figure 189. As for the contents of Bode's, which was surely the largest star atlas ever published in an
ostensibly traditional form, it had positions for more than 17,000 stars and
for the 2,500 nebulae that had been discovered and cataloged by William
Herschel. Bode was determined to omit nothing. Virtually every constellation ever named in the Western world was included, filling up most of the
vacant spaces in the traditional sky in a quite ridiculous way. There were
constellations like George's Harp, Brandenburg Sceptre, Sculptor's Carving,
Clock, Chemical Apparatus, Electrical Machine, and Sculptor's Apparatus,
to name but seven examples from a single plate. Bode's work was an early
use of constellation boundaries, which, with constellation names, were in
due course standardized—although not without modification, and greatly

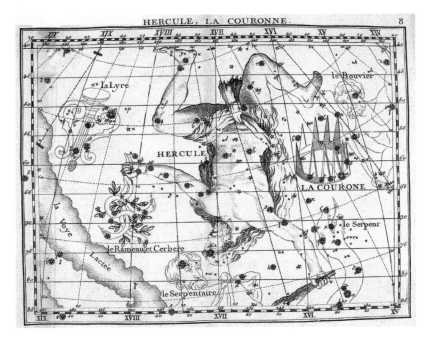

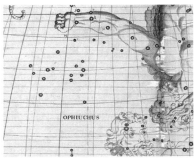

188. The constellations Hercules and Corona Borealis, from the French edition of Flamsteed's *Atlas Coelestis* (1776). Opposite is a detail from the original edition. The French edition has added the Milky Way (La Voye Lactée) and The Branch and Cerberus (Le Rameau et Cerbere). The allusion here is to Cerberus, the dog guarding the Underworld, which could sprout writhing snakes. This strange French addition, like many others, failed to obtain general acceptance.

helped by Friedrich Argelander's *Uranometria Nova* (1843) and Benjamin A. Gould's *Uranometria Argentina* (1877–79).

In time, astronomers began to lose patience with the old style of atlas-cum-catalog. Specifying much fainter stars required new notations. A definitive and manageable list of 88 constellations was established in 1930 by the International Astronomical Union. Its rectilinear constellation boundaries preserve, more or less, the traditional arrangements of naked-eye stars. Before then, however, constellation names were already being ignored by many astronomers, stars being simply listed by right ascension and declination, or some other property. Catalogs of double stars and variables have their own conventions. Ptolemy would have been astonished to

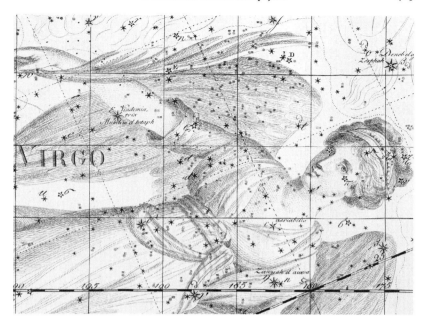

189. Details of the constellation Virgo from (below) Flamsteed's *Atlas Coelestis* (1729) and (above) Bode's *Uranographia* (1801). It is impossible to do justice to the latter here, in view of its enormous scale—it maps more than 17,000 stars and about 2,500 nebulae—but these two illustrations give an idea of relative numbers of stars in the two works.

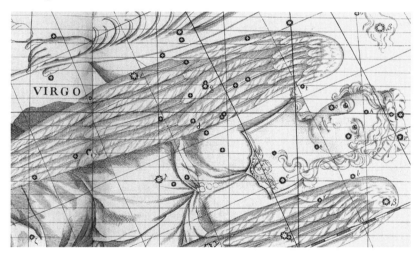

learn that a day would come when the number of printed catalogs would exceed the number of stars listed in his own.

NEW PRECISION: BESSEL AND STELLAR PARALLAX

The desire to follow in the old traditions waned, but a desire for accuracy did not. Astronomy in the first part of the nineteenth century benefited greatly from the industrial advances that had diffused from England and

France to Germany and Europe generally, in the latter part of the previous century. There was now much worthy competition for the London workshops, from Germany especially—for example from the firms of J. G. Repsold (Hamburg, 1802) and G. Reichenbach (Munich, 1804). These firms modified very successfully the English designs produced by such firms as Dollond, Ramsden, and Cary. A strong market for instruments for surveying and navigation continued to provide a technological basis for the grander requirements of new, and often rich, observatories, and Germany was fortunate inasmuch as the leading mathematician of the age, Carl Friedrich Gauss, took an active interest in the practicalities of instrument making.

For exact measurement, the transit instrument replaced the mural quadrant in astronomers' affections. With its refracting telescope mounted on an east-west axis so as to align on the meridian, it was fitted with complete and finely graduated circular scales, that were read by micrometer microscopes. The fact that the circle was complete made centering and other pivoting errors easier to assess. Accurate timekeepers, "regulators," were used for the measurement of certain angles by the timing of stars as they crossed reticles of illuminated wire—or better, of spider web—in the focal plane of the transit telescope. New theories of instrumental errors were developed and, as a consequence, angular measurement became very much more precise. Timing was done electrically, after the chronograph was introduced from America after 1844, although the first electric chronographs were not particularly accurate. As Simon Newcomb said of the first, at the Naval Observatory in Washington, "its only drawbacks were that it would not keep time and had never, so far as I am aware, served any purpose but that of an ornament."

If one name stands out above all others in the question of precision, it is that of Friedrich Wilhelm Bessel. Bessel was apprenticed at the age of fourteen to a large Bremen merchant company, but privately applied himself to astronomical calculation so thoroughly, and with such success, that Wilhelm Olbers took him under his wing. It is curious to find that Bessel had been greatly stimulated by reading about Harriot's 1607 observations of what became known as Halley's comet. With the help of works by Lalande and Olbers, Bessel—at the age of twenty—wrote a masterly account of how he had improved the elements of the comet's orbit. It was after reading a manuscript account of this work that Olbers recognized his mettle. Long afterwards, Olbers modestly described his patronage of Bessel as his greatest service to astronomy. He found Bessel employment in a private observatory near Bremen, where he developed his techniques further before turning to a reduction of Bradley's observations of over 3,000 stars—all this at Olbers's suggestion. So successful was he, in the early stages of all this, that when Friedrich Wilhelm III of Prussia founded a new observatory at Königsberg (ceded to the Soviet Union and renamed Kaliningrad after the Second World War) in 1810, Bessel was appointed director. He remained there for the rest of his life, constantly regretting the climate but refusing invitations to move elsewhere. Königsberg, formerly the capital of the dukes of Prussia, and later the capital of East Prussia, was the home of a distinguished university, where Immanuel Kant taught. The philosopher had died in 1804, his tomb in the cathedral reminding us of his wider interests by its often-quoted inscription: "The starry heavens above me and the moral law within me."

Bradley's published observations were valuable very largely because Bradley himself had determined the errors of his instruments or had made observations from which they could be deduced. For his reduction, Bessel needed accurate values for fundamental astronomical constants such as aberration and refraction. Even a modest correction for refraction amounts to two or three minutes of arc, which might be several thousand times as great as the probable error claimed for the declination coordinate. The greatest difficulty is in allowing for atmospheric temperature and pressure. Bessel began with a study of astronomical refraction, and this to such effect that in 1811 he was presented with a prize by the Institut de France for his tables. With his first task completed, in 1818 he published the results of his reductions in his *Fundamentals of Astronomy*.

This all brought Bessel's name to the attention of an astronomical world that was grateful for a supply of proper motions far better than had ever been available before. Soon after, the Dollond transit instrument and Cary circle with which the observatory was equipped were replaced by a fine new Reichenbach-Ertel meridian circle (1819). This was followed by a Repsold circle in 1841, by which time Bessel's *Tabulae Regiomontae*, the Königsberg Tables, had been in use throughout the world's best observatories for eleven years. Bessel's influence was immense twice over, for by his example he gave to the meridian circle almost the status of a cult instrument. Every new observatory, large or small, had to be equipped with one. Greenwich fell into line, and there the newly appointed Astronomer Royal, George Biddell Airy, acquired one on the German model in 1835. From this period onward, astronomers were regularly quoting star declinations to a hundredth of a second of arc, and right ascensions to a thousandth of a second of time. The measurement of star coordinates became almost an end in itself—rather like cleanliness and godliness—regardless of potential utility, although of course this was often forthcoming.

In 1844, Bessel himself made an important discovery from his measurements of the places of the stars Procyon and Sirius, which are on Maskelyne's list of fundamental stars: he found that their proper motions are variable. He drew the conclusion that each had an invisible companion, massive enough to make the motion of the brighter component around their center of mass visible. He was not the first to make such a claim, and in fact a controversy over the same sort of claim had been simmering since John Pond at Greenwich had made it—in 1825 and 1833—for a large number of stars. Bessel successfully dismissed Pond's argument, arguing that

190. Carl Friedrich Gauss (1777–1855), "Prince of Mathematics." Pictured here on a West German postage stamp celebrating the bicentennial of his birth.

his stars were far too distant for such an effect to manifest itself; but as for Sirius, the companion was actually seen telescopically by Alvan Clark in 1862, when he was testing a new telescope he had made. It was an eighth magnitude star, and yet with a mass—derived later from its orbit—of about half the mass of Sirius. (Its orbit had been analyzed by C. A. F. Peters in 1850, before it was seen.) The dark companion to Procyon was found by John Martin Schaeberle at the Lick telescope, in 1895. It was even fainter— of the thirteenth magnitude. When eventually the distance of Sirius was found, it was possible to deduce the size of its orbit and hence the masses of the components. They were of the order of one and two solar masses; and when similar calculations were applied to other binaries, it became clear— to the great surprise of those engaged in the work—that while the luminosities of the stars may differ by a factor of millions, the masses were rarely as much as ten times greater or smaller than the Sun's mass.

As a postscript, underlining the great value of the raw materials bequeathed by Bradley, we note that more unpublished work of his was used later in the century by Airy, who supplemented it with his own observations at Greenwich. They were in turn taken over in the 1860s by Arthur J. G. F. von Auwers, who had also worked at Königsberg but was by now in Berlin, and von Auwers made yet another reduction of the material, improving on Bessel's. Von Auwers's three-volume catalog (1882–1903) set new standards of accuracy, and with Bessel's catalog of 75,000 stars of brighter than the ninth magnitude, occupies an important place in the trunk of the family tree of later material.

Bessel's work was continued by a man whose work he long encouraged, Friedrich Argelander, who similarly followed the ninth magnitude limit. Their combined work was the basis of the *Bonner Durchmusterung* (*Bonn Review*), which, since its inception in 1859, has continued to serve as a standard of reference, not least in providing a comprehensive system for identifying stars. (The old Greek-Arabic names, even Bayer's Greek letter labels, apply only to a thousand stars or so.) The positions of over a hundred thousand stars were recorded, these being referred to certain fundamental stars whose coordinates had been measured to a very high degree of accuracy—whether the 36 stars Maskelyne had first selected, which Bessel increased to 38, or the 400 and more so-called Nautical Almanac stars, or Bradley's much longer list.

Although far too extensive even to list here, later fundamental tables were assembled in ways that mark out astronomy as the science of international cooperation par excellence. In 1871, the German Astronomical Society organized cooperation between thirteen observatories (later sixteen) in different parts of the world, each of them assigned a zone of declination north or south of the equator, covering all of the sky between both poles. (Thus one might take declination 20° to 25°, and another –35° to –40°.) This was the basis of a continuing program of proper-motion measurements that naturally improves in quality as time passes. When the results of the collaboration began to come in, cross-checks between coordinates assigned to the same stars by different observatories showed up numerous discrepancies, and unsuspected errors. Clocks contributed some of the most serious errors. True to history, astronomers concerned themselves with improvements in methods of timekeeping. The Repsold pendulum, Shortt pendulum, and atomic and quartz clocks are instances of advances made during the last two centuries with astronomical involvement. Greenwich and other national observatories have been the principal guardians of time standards. Here again, Bessel's presence was long felt, for it was he who drew attention to the fact that personal errors in registering time were to be taken seriously, unavoidable, but open to systematic assessment. The irascible Airy might not have dismissed an assistant whose observational records differed from his own, had he recognized the inevitability of a "personal equation," that is, of a physiological delay in registering any reading that varies from one person to another.

Bessel's most memorable achievement was to do what astronomers had tried and failed to do for centuries, namely, measure the distance of a star by relatively straightforward, trigonometrical means. We recall how Bradley had hoped to detect the shift in star positions that results from the movement of the Earth, in particular when it moves from one extreme of its orbit to the diametrically opposite extreme. In this way he was led on to his discovery of the aberration of light, but not of simple parallax. It was Bradley who thus made astronomers aware of how very small were the parallactic shifts they should expect—smaller than half a second of arc, say. John Brinkley, the first Astronomer Royal for Ireland, announced the parallaxes of a handful of bright stars in the period 1808–1814, all in the region of 2″ arc, but John Pond, at Greenwich, disputed his findings over a period of many years. Attention had already turned to measuring star positions relative to much fainter—and therefore, presumably, much more distant—stars. Doing this, as we saw, Herschel had also discovered something other than what he had set out to find, that is, pairs of stars physically related. In this way, he and others gradually began to realize that there were many faint stars with large proper motions.

Bessel now assumed that a large proper motion was a surer sign of proximity to the Earth than brightness, so he focused attention on the star 61 Cygni, which had the largest proper motion then known, 5.2″ per annum. (Actually, this is a double star, with 16″ between the components, so the line joining the pair is a useful pointer to direction in the sky, when measuring.) He observed 61 Cygni for eighteen months, in relation to two much fainter stars nearby, and by the end of 1838 had found the parallactic shift he sought. He found it to be less than a third of a second of arc (0.314″ ± 0.020″; it is still being quoted as at the lower end of this range). In astronomical units, semi-diameters of the Earth's orbit, its distance is about 657,000.

Within a year or two, other parallaxes were found by F. G. W. Struve at Dorpat and by Thomas Henderson at the Cape Observatory. Henderson found that the star α Centauri had a parallax well over double that of 61 Cygni, that is, its distance seemed less than half as great. His measurements had actually been made in 1832–1833, for another purpose, but were not used for a derivation of parallax until after Bessel's announcement.

The instrument used by Bessel to measure the small angular separations of 61 Cygni and the reference stars was the heliometer, as designed by John Dollond, but in this case made by Joseph Fraunhofer. The name of the instrument comes from its use for measuring the Sun's diameter. The principle of the heliometer is simple: the telescope objective is sliced into two semicircular halves, which can slide sideways (by amounts measured on a vernier screw) relatively to one another. An image of the first star is seen through one half lens, and an image of the second through the other. The components of the split lens, having next been adjusted until the two images coincide, the movement needed can be easily translated into an angle.

OPTICS AND THE NEW ASTRONOMY

As a result of the serious reverses suffered by the English optical industry in the late eighteenth century, here again Germany seized the commercial initiative. Joseph Fraunhofer's name became synonymous with the new and vital scientific activity that resulted. In 1806, this glazier's son joined the Munich firm of Utzschneider, Reichenbach, and Liebherr. At first he worked with Guinand, between 1809 and 1813, improving the quality of the different mixes of optical glass. When grinding and polishing the components of achromatic lenses, most previous makers had used trial and error, using whatever disks had been cast, but in 1814 Fraunhofer proceeded more scientifically, determining the optical properties of glasses in advance, using the bright yellow lines in a flame spectrum as his source of light. He made accurate measurements of the angles of light entering and leaving a prism by adapting an ordinary theodolite. In this way he created what is now a standard piece of laboratory apparatus, the spectrometer.

Examining the colors of stars by means of spectroscopes was not new. As early as 1798, William Herschel had compared the spectra of six of the brightest stars in the sky, noting that they differed appreciably in the proportions and intensities of different colors. He was unable, of course, to explain what he saw. Fraunhofer's approach, concentrating on the spectrum of the Sun, proved to be an invaluable gateway into the whole subject. In the course of comparing the flame spectrum with that produced in a prism by the Sun's light, he noticed that the latter was crossed by innumerable dark lines. (Some of these had actually been recorded by William Hyde Wollaston in 1802.) He later realized that some of the dark lines had counterparts in the spectra obtained from flames in the laboratory, and that while there were broad similarities between the spectra of the Sun and bright stars, there were many subtle differences between them. These observations were the basis of research by a number of physicists over the following decades, until in 1859 a revolutionary interpretation of spectra was offered by Kirchhoff and Bunsen. This interpretation marks nothing less than the beginning of a new type of astrophysics—astronomical spectroscopy.

Robert Wilhelm Bunsen was one of the century's leading experimental chemists, while Gustav Robert Kirchhoff was a distinguished theoretical physicist who had followed Bunsen to Heidelberg in 1854. John Herschel and W. H. Fox Talbot, in 1823 and 1826, respectively, had advocated chemical analysis from spectral observation, and by the 1850s the method was widely known, although rarely used. Bunsen was experimenting with such methods for analyzing salts, that is, by the coloration of the flames in which they are burned—as when sodium salts produce a bright yellow flame, a flame that Fraunhofer had associated with the "D-lines" in the Sun's spectrum. Kirchhoff's new and crucial contribution was to introduce an element of precision into Bunsen's work, by measuring his colors on a spectrometer. By 1860, the two men had shown that each metal has a highly characteristic line spectrum. This led Bunsen to analyze alkaline compounds, and so to make the discovery of two new elements, cesium and rubidium.

Already in 1859, Kirchhoff had made the discovery that most affected all future astronomy. He had been surprised to discover that if weak enough sunlight was passed through a sodium flame before entering the spectrometer, the dark spectral D-lines ("Fraunhofer lines") were replaced by bright lines from the flame, but that with bright sunlight the flame could make the dark lines relatively darker. His interpretation of the situation was that a substance capable of emitting a spectral line—that is, light of a particular wavelength—is also able to absorb light of the same wavelength. Ten years earlier, William Stokes had come to a similar conclusion concerning the D-lines in the laboratory flame and the absorption spectrum, and had even given a theoretical interpretation in terms of atomic resonance, but his ideas remained dormant, and it is odd that Kirchhoff later wished to have nothing to do with this fundamental explanation of the phenomenon.

191. Gustav Robert Kirchhoff (1824–1887), from an East German postage stamp celebrating the sesquicentennial of his birth.

Kirchhoff's main conclusion, however, one which he drew within a day of making his observation of line reversals, was dramatic enough. He saw that not only is sodium present in the Sun, but many other identifiable elements are present too, as shown by the appropriate Fraunhofer lines. What many had held to be a perfect example of unattainable knowledge, that is, the chemistry of celestial bodies, was at last seen to be within reach.

Within a few weeks, Kirchhoff had developed a quantitative theory of the emission and absorption of light, not as fundamental as Stokes's, but in many ways more useful for the development of physics. His conclusion was that the ratio of the power of a surface to absorb radiation to its power to emit radiation is the same for all bodies at a given temperature. (Within a given interval of wavelength, the same is true.) This law of radiation became one of the principal pillars of the thermodynamics of radiation. What it gave to astrophysics was an opportunity to add temperature to the list of measurable celestial parameters. In yet another respect, therefore, astronomy was revolutionized by a discovery that seemed at first to have little to do with astronomy as such.

THE ASTEROIDS AND NEPTUNE

There was still room for astronomical discovery of the old familiar sort. It seems that the whole world is excited by the idea of a new planetary neighbor. The two that have been found since Herschel's discovery of Uranus are Neptune, found in 1846, and Pluto, found in 1930. (Whether Pluto is to be regarded as a planet was a question reserved for the twenty-first century.) There are so many points in common in the manner of their discovery that it is instructive to take them together. They were found only after predictions had been made on the basis of Newtonian planetary dynamics, in other words, ostensibly in quite a different way from that by which Herschel discovered Uranus. They were found *after* those predictions, but it was later suggested that in at least one case the prediction was faulted, and the discovery was by chance.

Before they were found, another sort of planetary neighbor had been found, and a very different line of reasoning had been applied to it. The Titius-Bode law of planetary distances—for which see p. 343—had seemed to most astronomers to have been pulled out of thin air, without any real justification, and yet it eventually seemed to be broadly supported by the case of Uranus, discovered by Herschel nine years after the law had first been mooted. The radius of the orbit predicted by the law agreed with the observed radius, within two or three parts in a hundred. This in turn reinforced a belief in the existence of a planet between Mars and Jupiter—something suspected even by Kepler, on different grounds—which corresponded to a value for *n* of 3 in the law. Led by von Zach, a group of German astronomers even went so far as to found a club in 1800, the Lilienthal Detectives, to search for the missing planet. The gap is in fact occupied by the asteroids, the "minor planets," the first of which was found in 1801 by Giuseppe Piazzi of Palermo, Sicily. Piazzi was not searching for a missing planet when he found what he at first took to be simply a faint star. It proved to be moving, first with retrograde and then with direct motion. He confided in his friend Barnaba Oriani in Milan, saying that he had found a new *planet*, but when he wrote to Lalande in Paris and to Bode in Berlin, he was more cautious in saying that he believed he had found a *comet*. The Lilienthal Detectives were nevertheless convinced that here was their missing planet. Whatever it was, Piazzi's object became for a time lost in the glare of the Sun. Soon the young Gauss was able to calculate a new orbit, and on the last night of 1801 the object was found again, where Gauss had predicted. Piazzi chose the name Ceres Ferdinandea after the goddess Ceres and the ruler of Naples and Sicily, Ferdinand IV. When Herschel measured its size, however, he and all concerned were in for a surprise: he estimated a diameter equivalent to only 259 kilometers.

Another surprise came in March 1802, when Olbers, one of the Lilienthal Detectives, found a similar object. This he named Pallas, and again Gauss found for it an orbit placing it between Mars and Jupiter. Herschel estimated its size as about two thirds that of Ceres, and noted that the Titius-Bode law would be overturned if both were to be regarded as planets. Olbers, on the other hand, pointed out that one could save the law if the two asteroids—Herschel's word—were remnants of a single object that had disintegrated in the past. This suggestion led to a minor industry when, later in the century, yet more asteroids were found—Juno in 1804, Vesta in 1807, and so forth. The problem was conceived to be one of calculating the time when all of them would last have been in one place. In 1809, however, before this phase was entered, and prompted by his researches into the asteroids' path, Gauss had published his *Theory of the Motion of Celestial Bodies Orbiting the Sun in Conic Sections*. This was the most exquisite mathematical analysis ever written on the general problem of determining an orbit from any number of observations greater than three. For this purpose, Gauss developed his method of least squares, which is of use today in almost any science that requires a matching of theory and measured observation. (On this question, see also p. 427 above.) In a sense, Gauss's theoretical work leaves the discovery of the asteroids quite in the shade.

Even before the planetary character of Herschel's new discovery was recognized, attempts were made to find the parameters of the orbit, that is,

of the future Uranus. Old sightings of it helped here. Bode found that
Johann Tobias Mayer had seen it—without appreciating that it was a
planet—in 1756, and Flamsteed, in 1690. It is in fact visible to the naked
eye, given good eyesight. Almost a score of similar old "star" sightings that
were really of Uranus were found over the next forty years. These sightings
supplemented current observations and allowed half a dozen astronomers
to specify its orbit and draw up tables of its motion from the year 1788 onward. As the years went by, however, it became clear that Uranus was a
most intractable planet, and even the best of tables—Delambre's of 1790
long held that position—ran quickly into difficulties. After a time, even so,
interest in the planet's orbit seems to have waned. The Napoleonic wars did
not entirely close down scientific communication, as we have seen, but on
the continent of Europe the wars were not conducive to tranquil research.
Piazzi's patron was deposed, and Gauss's died soon after the battle of Jena.

After Napoleon's defeat, Uranus came once more under attack. New
tables were drawn up in 1821 by Alexis Bouvard, a farmer's boy from
the Alps by origin, but one with a genius for computation, who had become an invaluable assistant to Laplace. Within eleven years we find Airy
complaining that the tables were nearly half a minute of arc in error. For
several years—before and after his appointment as Astronomer Royal in
1835—he worked on the problem, and his published transit measurements of the planet allowed astronomers throughout Europe to try their
hand at eliciting the cause of the anomalous readings. It is interesting
to observe how confident astronomers had become of the truth of the
Kepler-Newton complex of laws, and how disturbed they were by the discovery that the planet was misbehaving to the tune of only twenty or
thirty arc seconds, while it had not even been under observation over a
full circuit of its 84-year orbit since the time of its discovery. Various possible explanations of the anomaly presented themselves. Had Uranus been
struck by a comet since its discovery? Did it have an invisible but massive
satellite? Was there an interplanetary fluid, impeding planetary motions?
Did Newton's law of gravitation cease to operate at great distances? Or was
there, perhaps, an invisible planet that was exerting a perturbing effect on
Uranus?

Clairaut had long before put forward this last hypothesis, to explain
oddities in the movement of Halley's comet; and now an English amateur,
T. J. Hussey, made the same suggestion for Uranus. Niccolo Cacciatore in
Palermo thought he had actually seen a planet beyond Uranus in 1835;
Louis François Wartmann in Geneva later said he had seen one in 1831.
Both claims were shown to be misguided, but gradually the belief spread
that some perturbing planet existed. In 1842, the Göttingen Academy of
Sciences offered a prize for the solution of the Uranus problem. Under
these circumstances, the fact that two men came up with a solution quite
independently of one another—often presented as an example of some sort
of mystical synchronicity—is no real cause for surprise.

Urbain Jean Joseph Leverrier, the son of a local government official in Normandy, had distinguished himself in mathematics as a student at the Ecole Polytechnique and had further studied chemistry under Gay-Lussac, before turning to celestial mechanics. He studied the perturbation of cometary orbits, extended Lagrange's general theory of perturbation, and his successes were such that he was seen in some quarters as a successor to Lagrange and Laplace. In 1845 he revised Bouvard's theory of Uranus, having been deputed to work on the problem by François Arago, who was director of the Paris Observatory and also a friend of Airy. As early as June 1845, he had written to Airy, promising to send him a planetary position as soon as he had calculated it, and asking Airy to search for it; but the offer was declined. On 18 September 1846, Leverrier's calculations were complete, whereupon he wrote to Johann Galle at the Berlin Observatory, asking him to search at a specified place in the sky for a new planet. On the night and early morning of 23 and 24 September 1846, Galle and Heinrich d'Arrest found the planet, within a degree of the calculated position. (The name eventually given to it, Neptune, had already been proposed for Uranus, as we saw earlier.) There was much jubilation, not least in France, but this was quickly soured when it was suggested elsewhere that Leverrier's prediction had been anticipated.

John Couch Adams was the son of a Cornish tenant farmer, whose struggle to send him to Cambridge was justified when he headed the class lists in mathematics in 1843. On graduating, he was made a fellow of his college, St. Johns, and at the beginning of the long vacation he set to work on the Uranus problem. By October he had an approximate solution, and in February 1844 he applied through James Challis to Astronomer Royal Airy for more exact data on Uranus. Using Airy's figures he calculated values for mass, heliocentric longitude, and elements of the elliptical orbit of the presumed planet. Adams gave his findings to Challis in September 1845. They were in fair agreement with the later findings of Leverrier, but Leverrier had given a first longitude prediction in June of 1845. The historical record of who predicted what, and when, is incomplete, and would not be considered especially important without the priority dispute that was to come. After twice trying to see Airy in person—on the first occasion, Airy was in Paris—Adams left a copy of his calculations at Greenwich on 21 October 1845. Airy wrote to Challis with a misconceived criticism some weeks later, getting Adams's name wrong, and clearly misconstruing his status—he supposed that Adams was an older clergyman. In November, he wrote to Airy with some constructive criticisms, but Adams did not reply. Airy had half-heartedly asked Challis to begin a search for the planet in July 1845, but had irritated Challis by his language. A year later, beginning on 29 July 1846, Challis began a search of almost unparalleled thoroughness, using what was then the largest British refractor, and being guided by a paper drawn up for him by Adams. Despite the fact that the paper in question seems to rely heavily on some work by Leverrier dating from 1

June of that year, Adams was nevertheless pointing Challis in a direction far removed from the right one, and the Cambridge sweep of the sky therefore produced no new planet. It was Leverrier's later investigation that led to the discovery of Neptune, as we have seen, in September.

John Herschel was the person who first drew the public's attention to Adams's achievement, in a letter to the London-based literary journal *Athenaeum* in October 1846. In strongly rhetorical prose, he referred to an address he had given to the British Association for the Advancement of Science on 10 September, in which he purported to have spoken of the movements of the planet as having been felt "trembling along the far-reaching line of our analysis"—an allusion to perturbation theory—"with a certainty hardly inferior to ocular demonstration." This dubious retrospective by the son of the discoverer of Uranus was plainly seen as claiming the discovery for British astronomy. It is possible that, when first uttered, there was no conscious reference to Adams, or indeed to any one person. Whatever the truth of the matter, a bitter controversy followed, partly over the facts of priority, and partly concerning the behavior of Airy and Challis in the affair. Challis, who had succeeded Airy as Plumian Professor of Astronomy at Cambridge when Airy became Astronomer Royal in 1835, announced that while searching for the planet in 1845, and recording the necessary star positions in the hope that one of them would change over time, one of those he found turned out to have corresponded to the planet. He had not gone through the necessary process of elimination, however, since he was busy with comets, and did not wholeheartedly believe in Adams's prediction. Some took this to amount to a claim for priority, and a claim to have "found" the planet on 12 August, rather than as a confession of negligence. At all events, Leverrier was showered with honors and Adams received very few, although in 1848 the Royal Society was to give him its highest honor, the Copley medal. It had given this same honor immediately after the discovery to Leverrier alone.

In the long term, the two great peacemakers in this unhappy episode were John Herschel in England and Jean-Baptiste Biot in France. Chauvinistic controversy centering on Adams did not end here, however. In the 1850s he discovered a substantial error in Laplace's treatment of the secular acceleration of the Moon's motion. This was decided in Adams's favor, but not until 1861, after yet more French recriminations. As a result, the figure for the acceleration was reduced by about half, from 10.58″ to 5.70″. That same year of 1861 saw Adams succeed Challis as director of the Cambridge Observatory.

LOWELL, MARS, AND PLUTO

For many years the story of Neptune was an object lesson in armchair discovery: one might be too poor to afford a great telescope but not to learn mathematics. The discovery of yet another planet beyond Neptune taught a somewhat different moral. Although it had to wait for the year 1930, the search was certainly encouraged by the earlier efforts of a rich New

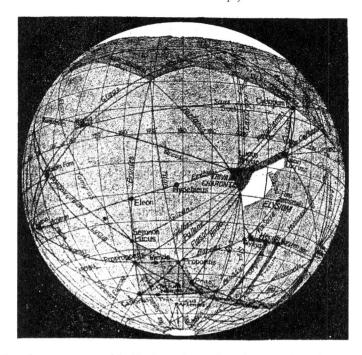

192. One of numerous maps of the Martian surface, as drawn by Percivall Lowell. He provided this for the eleventh edition of the *Encyclopedia Britannica*, where the accompanying article by Simon Newcomb was not ungenerous. The map shows half the sphere, and a second figure showed the other half. They were both based on a globe Lowell made. The labeling of features is highly imaginative, with its use of classical Greek. The most conspicuous is Elysium, the abode of the blessed after death.

England amateur, Percival Lowell. Lowell was convinced that he had a key to the mystery of a missing planet, just as surely as the Lilienthal group had: he believed he knew of certain "resonances" in the motions of the planets.

Lowell had set up an observatory in Flagstaff, Arizona, where the air—or so he had been told—was the most stable in North America. He had earned much publicity for his observations of the surface of Mars, and in 1908 for a book entitled *Mars as the Abode of Life*. The theme of the book had been arousing interest for thirty years or so, especially in the wake of Schiaparelli's use of a fine Merz refractor at the Brera Observatory, by which—in the late 1870s—he studied and mapped the planet of Mars. Drawing a series of fine dark lines across the orange surface of the planet gave it an intriguing character, which he took to be geological, but by using the innocent Italian word *canali* (channels), some of his readers decided that his lines represented artificially constructed canals. Lowell was writing in the wake of the ideas that stemmed from this exciting possibility. His drawings of Martian canals were more reminiscent of the London Underground than the irregular dark markings on Schiaparelli's prototype maps. Lowell was persuaded by his own studies of the Martian surface that the planet experienced seasons, and that he could detect changes in its vegetation (fig. 192).

He suggested that intelligent beings, faced with a shortage of water, had constructed the canals for purposes of irrigation—that they carried water from the melting of the polar snows to their habitations. He thought that what we see are not the canals themselves, but strips of vegetation flanking them. The idea was treated skeptically by most professional astronomers, especially when he interpreted his images of Venus and Mercury as also showing canals, although it held sway in some quarters until the Mariner space missions of the 1960s—especially Mariner 4 in 1965. These put an end to the idea in most astronomers' minds, although one enthusiast, Carl Sagan, refused to give up hope, insisting that plants and animals might survive in moist crevices. His belief, in view of his popular writings and his high profile on American television, helped materially in guaranteeing the funds—well over a billion dollars—necessary for the Viking project that explored the planet in 1976. Again, no evidence for Martian life in any form was found. Hope springs eternal in the Martian quest, however. In 1996 it was suggested that a Martian meteorite found in Antarctica contained bacteria-like spherules within it; but $2 million later, that idea too was abandoned. And then in 2006, a gully on Mars was spotted by the US National Aeronautics and Space Administration's (NASA's) Mars Global Surveyor spacecraft, a gully down which it was suggested water could have flowed during the last few years. (For the creation of NASA, see p. 691 below.) While some preferred the idea of liquid carbon dioxide, others revived the notion that water below the Martian surface might be supporting microbial life. Whatever the situation on Mars, there is still much life in astrobiology.

Lowell's ideas about resonances fell out of favor fairly quickly, but not before they had achieved a remarkable effect. He accepted the hypothesis of Thomas Chamberlin and Forest Ray Moulton, that the planets were formed out of material pulled from the Sun after the near-collision of a star with the Sun. (This theme will be continued in chapter 16.) His idea was that, after the formation of one planet, the next would tend to be formed at a place where its period of revolution was in a simple ratio to that of the former—say 5 to 2, in the case of Saturn and Jupiter. Uranus follows Saturn with a period three times as long; and Neptune follows Uranus with twice its period. (Lowell juggled the numbers slightly by talking of perturbations in a rather vague way.) Should there not be a "Planet X" beyond Neptune, with a period of revolution fitting such a scheme of simple ratios?

He tried to follow the example set by Adams and Leverrier, now adding Neptune to the list of perturbing planets, but he was not their mathematical equal. With the help of C. O. Lampland, who took photographs of the appropriate regions, he searched unsuccessfully for Planet X from 1905 to the end of his life. The search continued as an act of piety at his observatory after his death, with the advantage of increasingly refined calculations, and an object was eventually found by Clyde William Tombaugh in 1930, using a 13-inch telescope with a wide field of view, especially built for the purpose. It was through its movement, not its appearance, that the object, which was to be called the planet Pluto, was detected on 23 and 29 January. It was later found that the planet had been recorded on earlier plates, but had been overlooked; and that Milton L. Humason's 1919 search for a planet beyond Neptune had failed only because, while Pluto registered on two of his plates, in one case its image had fallen on a flaw in the emulsion, and in the other it had been virtually masked by a bright star.

The name had a curious origin. Falconer Madan, a well-known Oxford scholar and librarian, was reading about the discovery in the *Times* (London), which remarked on the need for a name. He mentioned the fact to his eleven-year-old granddaughter, Venetia Burney, who suggested "Pluto," a suggestion that so appealed to Madan that he went at once to call on his friend H. H. Turner, professor of astronomy. Turner was out at a meeting of the Royal Astronomical Society, at which, as it happened, no one had been able to offer an acceptable name. Returning to find a note from Madan, he was so impressed as to telegram the Lowell Observatory at once. A month later, the name was settled. It was highly appropriate, since the planet could be given an astronomical symbol formed from Lowell's initials, P and L. At the age of 87, Venetia Phair, neé Burney, on the grounds of age, was obliged to decline an invitation from NASA to watch the launch of the first space mission to Pluto. She could still recite the planetary distances and had not forgotten that the names "Phobos" and "Deimos" had been first suggested for the moons of Mars by her great uncle, Henry Madan.

Despite the marked theoretical component in the work leading up to the discovery of Pluto, there were flaws in the mathematics, and systematic searching played a greater part. Pluto's mass is now known to be simply too small to produce the supposed perturbations of Neptune and Uranus. Neptune's mass is in fact ten thousand times as great as Pluto's, which underlines the futility of Lowell's attempt to emulate the procedures of Adams and Leverrier; but he was not alone in this. William Henry Pickering, younger brother of the more famous Harvard astronomer Edward Charles Pickering, was Lowell's friend and adviser, and he too had been searching with photographic help for a planet beyond Neptune from 1907 onward, using calculations hardly more meaningful than those from Flagstaff. After Tombaugh's discovery, however, the planet was found on plates taken for Pickering in 1919.

The charge that the discovery was in some sense accidental provides a strange echo of a similar charge that had been brought against Adams and Leverrier by Benjamin Peirce and S. C. Walker in America. After Neptune had been under observation for some months, it became clear that the orbit was very different from what had been predicted, that is, on the assumption of a distance fitting the Titius-Bode law. It was pointed out that several different solutions fitted the known data, and that in the period when Neptune was found, the "real" planet by chance just happened to coincide roughly in position with the calculated planet. Many European astronomers, not least Leverrier, were at pains to rebut the charge, while others thought that it simply proved the unremitting honesty and openness of American science. The American criticism had much to recommend it, but was presented in a misleading way. Its authors seem to have implied that they somehow "knew the real planet" because they knew better values for Neptune's parameters—such as distance—than their predecessors. At the foot of this slippery slope is the conclusion that astronomers can never know the real Neptune, or for that matter, specify any other "real" celestial body.

Pluto is now known to have a twin, with a diameter rather more than half its own. If we describe its partner as a twin, rather than a moon, then the pair comprise the only known planetary double in the solar system. The discovery was made James W. Christy, of the U.S. Naval Observatory, who was studying photographic plates of Pluto, with a view to improving our knowledge of its orbit and position. The new moon or planet was named Charon, after the ferryman who rowed shades of the dead across the River Styx—a lake, rather than a river, in some accounts—to Hades, the underworld realm of Pluto. Greek mythology was still not exhausted. The crucial image of Charon and Pluto was taken on 2 July 1978. In May 2005, observations with NASA's Hubble Space Telescope—about which we shall say more on p. 733 below—showed the presence of two more moons around Pluto, observations that were confirmed in the following February. Scraping the barrel of classical knowledge for new names was becoming increasingly difficult. (The fifth edition of the International Astronomical Union's (IAU's) *Dictionary of Minor Planet Names*, a book which also explains the basis of the names of the very many asteroids between Mars and Jupiter, runs to more than a thousand pages.) The problem will grow, in all likelihood, by the year 2015 or thereabouts, when the NASA mission New Horizons reaches Pluto, which is likely to have neighbors that are as yet undiscovered. The spacecraft, with instruments to probe the composition and structure of the planet's atmosphere, to study dust grains from the Kuiper Belt, and to study effects of the solar wind at that great distance from the Sun, was launched on board an Atlas v rocket from Cape Canaveral, in January 2006.

If all goes well, after Pluto, the spacecraft will travel on to examine an object from the Kuiper Belt. This takes its name from Gerard Pieter Kuiper, a Dutch-born American, and in his time a leading light in planetary astronomy. Kuiper's discoveries included a moon of Jupiter and one of Uranus (Miranda, 1948), as well as an atmosphere on Titan. He argued for the existence of a belt of comet-like debris at the edge of our solar system, which he saw as a potential source of short-period comets. Kuiper died in 1973, twenty years before his ideas about the "Kuiper Belt" were finally confirmed. Even Triton, satellite of Neptune, is now thought to have been a Kuiper Belt object before being captured by Neptune.

However worthy of the title "planet" Pluto may be, and however "accidental" we may judge its discovery, we should not underestimate the difficulties of locating it. Its angular size, as seen from the Earth, is less than a third of an arc second, well below the limit at which any surface detail was visible. There are twenty million stars in our sky that appear as brightly as Pluto, and it was often said that any comparable planet in the solar system would only be systematically found in future using an instrument as potent as the Hubble Space Telescope. (To echo the boasts of the creators of that telescope, it could detect a dimly glowing firefly at a distance of 8,500 miles, as well as the firefly's mate, if they were only ten feet apart.) Just such a comparable object was discovered by the team of Michael Brown, Chad Trujillo, and David Rabinowitz, on 5 January 2005, on images taken on 21 October 2003. The team had been systematically scanning for large bodies in the Kuiper Belt beyond Neptune for several years, and had previously found other examples, although not as large as that new object. Designated 2003 UB313 and provisionally nicknamed Xena, after the protagonist of an American television series, it was eventually given the official name Eris (or 136199 Eris), after the Greek goddess of discord. In view of later events, the *Star Trek* Vorta character of that same name might have been a more suitable role model.

This new "trans-Neptunian object" (TNO) was found to have a diameter somewhat greater than Pluto's—2,400 kilometers, compared with 2,306—and therefore the inevitable question was asked: Should it be counted as the Sun's tenth planet? Two further large TNOs had been announced on the same day in July 2005, and it was clear that in time there were likely to be others, debasing—as some thought—the word "planet." At a general assembly of the IAU held in Prague in August 2006, an attempt was made to establish new guidelines for the use of that title, and the debate was protracted and heated. It was eventually decided to redefine the word in ways that meant reclassifying Pluto, Ceres, and Eris, which were henceforth to be known as "dwarf planets." As a concession, Pluto was taken as the prototype of this new family of trans-Neptunian objects, but to the many—inside and outside the profession—who had become emotionally attached to the idea of Plutonian planethood, that was not enough; and adding Pluto to the list of minor planets, with the number 134340, seemed only to add salt to the wound. An Internet petition sponsored by the Planetary Science Institute and the Southwestern Research Institute in the United States was signed by three hundred professional astronomers who pledged themselves not to use the new definition of a planet. Mike Brown, on the other hand, one of the team that discovered Xena/Eris, came out in favor of the IAU decision. Questions were frequently raised as to whether the IAU had the right to legislate on a matter of such broad cultural concern. If not, then who should decide? The question has not yet been referred to the United Nations Security Council.

In 2005 there had been disagreement of another kind, in connection with the discovery of yet another object in the Kuiper Belt. In quick succession, teams from the Sierra Nevada Observatory in Spain and the California Institute of Technology in the United States made public their observations of a planet-like object with a mass approaching that of Pluto, but with some more remarkable properties. It was found to be spinning once every 3.9 hours and was said to have a shape like that of an American football—a rugby ball, to the rest of the world—squashed sideways. Kuiper Belt Object 2003 EL61, as it is designated, while waiting for a suitable name, has a reflectivity almost that of pure snow, and is thought to be coated in ice. It has at least two moons. The Spanish group eventually withdrew their claim to prior discovery, after some controversy. The object was later found in images dating back to 1955. The old story.

PROFESSIONAL REFRACTORS, AMATEUR REFLECTORS

Bessel's parallax measurements confirmed professional astronomers in their preference for the achromatic refracting telescope over the large and unwieldy reflector, as a tool for precision measurement. A professional whose astronomical reputation came to equal Bessel's was F. G. W. Struve, who—as we pointed out earlier—left Dorpat in 1833 to found a new and glittering Imperial Observatory at Pulkovo, near St. Petersburg. Funded by Czar Nicholas I, Struve was then able to patronize the best instrument makers.

Since there were to be no fewer than six astronomers in the Struve family, several of them distinguished, it will be as well to note their relationship here, for they are often confused. Friedrich Georg Wilhelm was the father of Otto Wilhelm—not to mention his seventeen other children by two wives. Otto's sons included Karl Hermann and Gustav Wilhelm Ludwig, both astronomers. Each of the last had an astronomer son, Georg Otto Hermann and Otto, respectively. The founder of this dynasty had originally been sent from Germany to Russia to avoid military conscription. His son Otto Wilhelm followed him at the Pulkovo Observatory, but on retirement, after fifty years of service at the observatory, he moved to Germany. In the third generation, Gustav took various astronomical posts in Russia, while Karl Hermann remained there until 1895, when he accepted the directorship of the Königsberg observatory. Karl's son's career was consequently in Germany—and of lesser distinction, although he made important studies of Saturn. Otto, youngest of the six, suffered great hardships when his studies were interrupted by the civil war in Russia, for he joined General Denikin's army, only to be driven out of his homeland by the Red Army. In 1921, by way of Turkey, he managed to reach America and the Yerkes observatory (Wisconsin), of which he eventually became director.

When the first Friedrich Struve left Dorpat for St. Petersburg, he took with him the splendid Dorpat refractor made by Fraunhofer (fig. 193). The czar bought for him numerous other fine instruments, by such makers as Ertel, Repsold, Merz, Troughton, Dent, Plössl, and Pistor, and within a decade or so, he had at Pulkovo what was for a time the best-equipped observatory in the world. He obtained for it the biggest refractor then built—the 38-centimeter achromatic lens by Merz and Mahler, in a mounting by

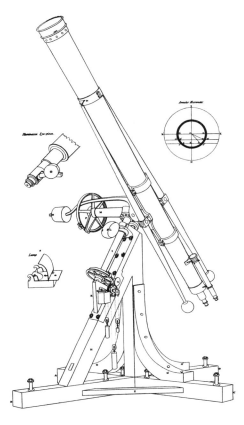

193. The refractor of aperture 26 centimeters (10 inches) built by Joseph Fraunhofer for F. G. W. Struve's observatory in Dorpat (now Tartu, Estonia). This was Fraunhofer's last, largest, and finest telescope. Note the rods carrying the heavy balls near the eyepiece, designed to counteract the bending of the telescope tube under its own weight.

Repsold and Sons of Hamburg. Later, when his son learned of the excellence of the workmanship of the American firm of Alvan Clark and Sons, a 30-inch object lens was commissioned from them. Mounted by the firm of Repsold, this too was for a very brief time the largest in the world (fig. 194). It was marginally outstripped by a Meudon instrument, and finally by the Alvan Clark glasses in the refractors at Lick (36 inches [91 centimeters]) and Yerkes (40 inches [102 centimeters]). The last of these instruments was destined to fall eventually within the domain of another member of the Struve dynasty. More than a century after it was built, it remains the world's largest refracting telescope, although it has to be said that size is not everything. The Yerkes instrument (1898), ten years younger than Lick's, was in a much inferior setting—Wisconsin as against Mount Hamilton, California. When Lick supplemented its refractor with a Crossley 36-inch reflector in that same year of 1898, it could still claim to be the world's premier observatory. It remained so for another ten years, after which the Mount Wilson Observatory, with its 60-inch reflector, took that title.

194. The Pulkovo 30-inch (76-centimeter) refractor of 1885. The mounting was by Repsold and Sons, of Hamburg, and the optics were by Alvan Clark and Sons, of Cambridge, Massachusetts. For two years this was the world's largest refractor, just as the 15-inch Pulkovo refractor of 1839 had been before it. In 1887 another 30-inch was built at Nice, while others marginally larger were built for the Paris Observatory's outpost at Meudon. Soon the giant refractors at Lick (36-inch) and Yerkes (40-inch) took the lead, and kept it, thanks to the new professional preference for reflecting telescopes.

After this aside on the last great age of refractors, let us return to Struve: apart from fulfilling the duties of a state astronomer whose assistance was sorely needed for the charting of a vast terrestrial empire, he set himself the task of continuing William Herschel's study of double stars, and by the time he published his 1827 catalog, he had recorded the positions of 122,000 stars, of which 3,112 were double. By 1847, he and his staff had covered the northern sky, and a catalog of 1852, making comparisons of the positions of 2,874 stars with those recorded by his predecessors—from Bradley to Groombridge—is a monument to nineteenth-century thoroughness comparable with Bessel's. Struve looked further afield, however. He wanted to solve the problems that Herschel had set. Are the stars distributed in a recognizable pattern? Is there a meaningful relationship between the distances of the stars and their magnitudes? He came to the conclusion that the loss of brightness of the stars examined by Herschel, and now by the staff at Pulkovo, was only partly explained by the fact that light falls off following an inverse square law. It was also the partly result, he thought, of the absorption of light in interstellar space. The actual figures he quoted are reasonably close to modern figures in the neighborhood of the Milky Way, the object of his study.

Despite the preference shown by professional observatories for the refracting telescope, especially in precision measurement, it was the reflecting telescope that had all the advantages, where light-gathering power was needed. Advances here came slowly, however, and would never have come at all had everyone accepted the consensus among most telescope makers. According to William Parsons, they all seemed to think that "since Fraunhofer's discoveries, the refractor has entirely superseded the reflector, and that all attempts to improve the latter instrument are useless." When, in 1833, John Herschel took to the Cape his father's favorite reflector, that of 47-centimeter aperture (quoting its length, this was usually known as the "20-foot" telescope), there was no better reflector in the world, and yet it was already half a century old.

The revival in the fortunes of the reflector came about in the 1840s as the result of the work of three amateurs: William Lassell, an English brewer, who mounted his reflectors (of 23 centimeters and 61 centimeters aperture) in equatorial mountings of the Fraunhofer type; the Scot James Hall Nasmyth, one of the greatest engineers of the century, who built several fine instruments (notably those of 25, 33, and 51 centimeters) with excellent mechanical properties and with a new type of viewing arrangement; and the Irish landowner William Parsons, third Earl of Rosse, who achieved his ambition to build a mirror larger than any of Herschel's.

Lassell designed and used machines for grinding and polishing his mirrors of speculum. In 1846, with his larger instrument, he discovered a satellite, Triton, orbiting the newly discovered planet Neptune. Two years later, simultaneously with W. C. Bond of Harvard, he found the eighth of Saturn's satellites, Hyperion; and in 1851 he discovered two hitherto unknown satellites of Uranus—Ariel and Umbriel. He took the instrument to Malta, where he charted over 600 new nebulae, hoping to make further satellite discoveries, but he made none. Nasmyth had been a keen amateur astronomer from his youth, and he designed and built several telescopes. After retirement in 1856, his efforts redoubled, but even before then he had concentrated his attentions on the Moon and Sun to great effect. His drawings of the Moon were outstanding, and he was awarded a gold medal for them at the Great Exhibition of 1851 in London. He was the first to appreciate—and to cause some controversy by announcing, in 1861—that the mottled surface of the Sun is patterned, as he said "like willow leaves." In the 1870s he produced impressive early photographs of "the lunar surface," not strictly of the Moon itself, which was not easily photographed at this period, but of plaster models he had created. In oblique sunlight they cast strong and very plausible shadows. His best remembered contribution to astronomy, however, was a mechanical one.

Nasmyth built a 51-centimeter (20-inch) reflector and mounted it in 1849 on a turntable of his own invention, a design that came back into fashion in the late twentieth century for the class of very large altazimuth instruments. In his design, the eyepiece was mounted at the end of one

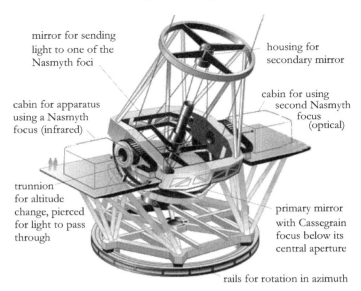

mirror for sending
light to one of the
Nasmyth foci

housing for
secondary mirror

cabin for apparatus
using a Nasmyth
focus (infrared)

cabin for using
second Nasmyth
focus
(optical)

trunnion
for altitude
change, pierced
for light to pass
through

primary mirror
with Cassegrain
focus below its
central aperture

rails for rotation in azimuth

195. A typical modern altazimuth telescope of very large proportions. This uses a telescope of
Cassegrain type, and James Nasmyth's idea of intercepting light returning from the secondary
mirror (top), and reflecting it through the middle of one or other trunnion to a "Nasmyth focus." Nasmyth's own instrument of 1849 broadly resembled a simple cannon, rather like the central tube in the present illustration.

of the hollowed-out trunnions on which the instrument was pivoted, a
mirror inside the main tube having deflected the light returning from the
secondary mirror in a Cassegrain arrangement. The normal Cassegrain
telescope, with an eyepiece behind the main mirror, is especially troublesome when one is looking toward the zenith, but now Nasmyth could
stand upright and view from the side. If we are to judge by earlier accounts
of mishaps with large reflectors, the design must have saved many broken
limbs. Modern altazimuth instruments can carry observing cabins and
large quantities of equipment on platforms at both ends of the trunnion
without having to be tilted. An internal mirror deflects the light down the
appropriate half of the trunnion to its "Nasmyth focus." For a typical modern arrangement, see figure 195 (the Cassegrain principle has already been
illustrated in figure 161 in chapter 12).

After graduating at Dublin and Oxford, William Parsons—who was
Lord Oxmantown until he took the title Earl of Rosse after his father's
death—became a member of the House of Commons. Renowned in his
day as that rare thing, an honest man in Irish politics, his obsessions were
rather with making telescope mirrors of the finest quality. He worked with
speculum metal, ground and polished on a steam-driven machine, and
constantly cooled during the process. The metal is brittle, and Herschel
had included more copper to make it less likely to shatter, but in doing so
had lowered its reflectance. Rosse tried mirrors cast in segments, these being held on a brass backplate with the same coefficient of expansion as the

speculum metal. The fine lines of division in the resulting mosaic were very troublesome, and the image was at first of low quality, but after repeated experiments he found another solution. He saw that the problem in casting was that of controlling loss of heat. He experimented with different sorts of molds, partly of sand and partly of metal, and allowed his 90-centimeter mirror no less than a fortnight to cool in a purpose-built edifice. The mirror was of excellent quality. The weather at his home, Birr Castle, was not, and yet in rare intervals of clear sky he saw much new detail in clusters and nebulae. At the center of the nebula M57, a disk-like object in the constellation Lyra, he found a very faint blue star. This was another of those "planetary nebulae" of the sort that in 1790 had persuaded Herschel of the existence of true nebulosity.

Rosse was so encouraged by his various discoveries that, with his estate workers and others, he now set to work on a mirror of 183 centimeters aperture and nearly 4 tonnes weight. After five attempts and four disasters in five years, the mirror of "the Leviathan of Parsonstown" was successfully cast, ground and polished. The suspension of the massive telescope tube (fig. 196) was extremely difficult: it was elevated in the gap between two bearing walls that stood about 18 meters high, in the meridian, so that stars could only be followed for an hour or so at most. To prevent the vast mirror from flexing under its own weight, it was supported on felt-covered cast-iron platforms.

The Leviathan, now by far the biggest telescope in the world, was brought into use in February 1845. By April, Rosse had made his most momentous discovery, that of the spiral structure of the nebula M51, of which he prepared an excellent drawing. For this, see figure197, and compare it with an image taken by the Hubble Space Telescope (plate 12). He found the same structure hinted at in several other nebulae. Ireland was in the throes of a potato famine, however, and he had to neglect his telescope for some months. (Rosse returned the major part of his rents to alleviate the poverty of his tenants at this time.) When observing resumed, he had a clear program in front of him, to investigate the shapes of the nebulae. He was assisted in this by his eldest son, who continued the program after his death, and by the Rev. Thomas Romney Robinson and various other friends, as well as hired observers.

Rosse was convinced that the spiral structure of nebulae must hold important dynamical information. He held a degree in mathematics, but without much more information he was helpless, and it would be many decades before that problem could be attacked formally. The Birr Castle astronomers therefore began a systematic search for spiral nebulae, and they found many. Several were of a type since known as Seyfert galaxies. (These have highly condensed centers, and it has recently been thought that they are powered by black holes.) The detail in the drawings produced at Birr often continued to be of very great value, even after the rise of astronomical photography. The great reflector continued in regular use in the same place until 1878, when

196. The mirror telescope of 6-foot (1.8-meter) aperture, built at Parsonstown, Ireland, by the Earl of Rosse. This was equipped for use in either a Newtonian or Herschelian mode. On the left is a view from the south, "showing the position of the telescope when a man enters the tube to fix the small speculum [mirror], and remove the cover of the large one, in preparing for the night's work." On the right is the telescope housing as seen from the north.

197. The Earl of Rosse's drawing (published 1850) of the spiral structure which he discovered in the spring of 1845 in the "nebula" Messier 51, using his new giant telescope. He did not fully appreciate what he had found for some time, but spirality is even more obvious in another sketch done by June 1845. He referred to the views of his predecessors: "a double nebula without stars" (Messier); "a bright round nebula, surrounded by a halo or glory at a distance from it" (William Herschel). John Herschel had detected structure in it, but saw it as a divided ring.

the Danish-born J. L. E. Dreyer resigned as Birr astronomer to accept a post at the national observatory at Dunsink. Dreyer, well known as a historian of astronomy, was also later to produce the famous *New General Catalogue of Nebulae and Clusters of Stars*. Its NGC numbers are still often used to identify such objects: Herschel's planetary nebula, for example, is NGC 1514.

Among other things, Rosse's work helped to change fashions in telescope making in favor of large reflectors. Astronomers of course continued to make discoveries with large refractors—as when Asaph Hall, in 1877, discovered satellites around Mars using the Washington instrument with a Clark lens. In addition to examples of refractors mentioned earlier, in the 1880s the Dublin firm of Grubb made a lens of 63 centimeters aperture for Cambridge, while the brothers Paul and Prosper Henry in Paris made refractors of 76 centimeters for Nice and 83 centimeters for Meudon. Ever better lenses were produced—for example in France at the workshop of Mantois, using glass from the factory of St. Gobain, and in Germany at the firm of Carl Zeiss at Jena, founded with the support of a famous theoretician in optics, Ernst Abbe. Astronomy was especially beholden to Carl Zeiss, who had begun as a toymaker. He established his optical business in a small way in 1846, in Jena, and by the 1880s he was working in close collaboration with Otto Schott's nearby glassworks. The combination was unbeatable. With the help of a Prussian state subsidy, the joint firm was soon well on the way to becoming the largest European manufacturer of high-quality optical instruments of all types, a position it held with ease between the two world wars.

The future of telescopes was with reflectors, and this simply because the glass at the center of lenses of the order of a meter in diameter is so thick that the absorption of light there is intolerable. Physical deformation under the glass's weight is also a problem. A new medium was needed, however, to replace the unworkable speculum metal. Glass mirrors were not new, and glass grinding and polishing was a highly developed art, but early methods of silvering the glass were crude. In 1853, the German chemist Justus von Liebig devised a technique for depositing a thin and uniform layer of silver on a clean glass surface, from an aqueous solution of silver nitrate. The technique had been shown at the Great Exhibition of 1851 in London but was seemingly reinvented by Liebig, who brought it to the attention of a wide scientific circle. A few years afterwards, the renowned Munich instrument maker Carl August von Steinheil of Munich, and Jean Bernard Léon Foucault of Paris, both made use of the technique, depositing layers of silver on glass mirrors for use in telescope—at first in fairly small ones. Foucault, who had begun his career in an intellectually unpromising way, was beginning to show his true worth, first with his work on the pendulum, demonstrating the rotation of the Earth, and then with experiments that allowed him to compare the velocity of light in air and in water. In 1853, he was rewarded by Napoleon III with a post at the Paris Observatory. His silvering of mirrors followed in 1857; and a year later he introduced another technologically important innovation, the "Foucault test" for the image quality produced by the mirror.

With his second telescope, that which he took to Malta, Lassell followed the French example, silvering a glass mirror. Soon, glass mirrors became the standard choice for reflectors almost everywhere. The last large

198. The mounting for the 48-inch Cassegrain reflector built by Thomas Grubb of Dublin and erected at Melbourne, Australia. It is shown as it was in 1867, before the erection of a sliding-roof enclosure. The mirror created for it by Grubb was of speculum metal, the silver-on-glass option having been rejected—an unfortunate decision.

mirror to be cast in speculum was a 120-centimeter example for Melbourne, made in 1870 by Thomas Grubb of Dublin, with Royal Society advice. This turned out to be a failure, for reasons both optical and mechanical (fig. 198). The mirror was shipped with a protective coating of shellac. In removing it, the surface was ruined. The director of the observatory had to try to learn the technique of refiguring one of the biggest telescopes in the world in a few months. He failed, of course; but then, Grubb had been doing the work for over thirty years.

With the advent of glass as the material for mirror objectives, reflectors quickly became generally preferred to refractors. The simplest achromatic lens has four optical surfaces to be figured, while a mirror has only one. Improving it, by modifying its spherical surface, making it paraboloidal, was not very difficult. There was another important consideration. With the advent of astronomical photography, objective lenses that had been regarded as achromatic could no longer be treated as such, since photographic emulsions are sensitive to a wider range of light wavelengths than that perceived by the human eye.

While small mirrors were relatively easy to grind and figure—as many thousands of amateurs have proved—the age of the amateur builder of the very largest instruments was fast coming to an end. The Grubb firm eventually redeemed its reputation with a 51-centimeter glass reflector made for

199. M31, the "Great Nebula" (a galaxy) in the constellation Andromeda, photographed by Isaac Roberts on 1 October 1888, using his 20-inch (51-centimeter) silver-on-glass reflector. Built by Sir Howard Grubb in 1885, with astronomical photography in mind, the telescope is now to be seen in the South Kensington Science Museum, London. This image, from a glass negative that required an exposure of three hours, speaks highly for Grubb's clock drive.

Isaac Roberts in 1885. With it, Roberts scored a notable hit with a photograph of M31, the great "nebula" in Andromeda, clearly showing its spiral structure, and settling a number of doubts about a rather ambiguous drawing of M31 by Bond in 1847 (fig. 199). In November 1886, Roberts achieved another success with a photograph of the Orion nebula. It had been photographed earlier by Draper (1880), Janssen (1881), and notably by A. Ainslie Common (1883), but the image obtained by Roberts showed that the nebula was at least six times larger than had previously been suspected, and that an adjacent nebula, cataloged separately by Messier, was joined to it.

By the end of the century, the tide had well and truly turned. George Hale's experience at Yerkes had taught him the limitations of refractors. He was fortunate to have the services of an optical worker of great expertise, George Ritchey. And when a new telescope was begun at Mount Wilson, with the help of a grant from the Carnegie Institution, there was no doubt that it had to be a reflector. Twice in the course of building it, Hale had a change of heart, and plans for a 1.5-meter mirror gave place to the reality of a 2.5-meter mirror-the 100-inch Hooker telescope that produced the countless spectacular astronomical photographs so widely used in the first part of the twentieth century. Completed in 1918, and weighing about 100 tonnes, the open-girder "tube" is held in a yoke polar mounting of great rigidity, although it does not allow the tube to point to stars near the pole (fig. 200). At last, an instrument was available that was worthy to continue the work on the forms and distribution of the nebulae, begun so long before by Herschel and continued by Rosse.

THE PARADOX OF THE DARK SKY

As telescopes increased in power, evidence began to accumulate for the existence of many unexpected and often strange distributions of matter in space. There was also Struve's argument for the importance of interstellar absorption, which in turn had a bearing on the paradox noted by Halley and others. When it was believed that the stars were finite in number—whether or not they were situated on a sphere made little difference—there was no mystery about the general darkness of the night sky, but as Kepler wrote in his *Conversation with the Starry Messenger* in 1610, "in an infinite universe the stars would fill the heavens as they are seen by us." Intuitively considered, no matter what the direction in which one looks, if the stars

200. The 100-inch (2.5-meter) Hooker telescope at Mount Wilson, completed in 1918. Weighing over 100 tonnes, it is carried between mercury floats in drums at the end of the polar axis. The clock drive is powered by a falling weight of 2 tonnes.

are not specially arranged for the particular viewpoint—for instance so as to leave straight avenues of empty space—then in an infinite universe with uniformly distributed stars, sooner or later the line of sight will encounter the surface of a star. The sky should be filled with light. This very simple way of presenting the case does not seem to have been found disturbing to astronomers before Olbers. We have already discussed Halley's more intricate route to a similar conclusion and his possible sources. In 1720 Halley published two papers on the problem of the dark night sky, and his solution to the paradox was that light from distant stars is not indefinitely divisible, and that at great distances it diminishes faster than the inverse square law suggests, so that distant stars are simply too faint to be detected by the human eye.

A generation later, in 1744, Cheseaux again asked why the sky is not filled with light of the average surface brightness of stars. His explanation for the dark night sky was in terms of interstellar absorption. This effect might seem to have the same consequences as Halley's, but as John Herschel was to point out in 1848, absorption alone is no answer, for the interstellar absorbing substance will heat up until it re-emits as much energy as it receives. Cheseaux's was too easy a solution. In 1823. the Bremen physician and astronomer Wilhelm Olbers repeated the explanation in terms of interstellar absorption, and since most of those who discussed the problem in the first half of the twentieth century knew of it through him, the paradox of the dark night sky became known as "Olbers' Paradox." Olbers was an acquaintance of Struve, and this was part of the context of Struve's study of interstellar absorption.

Absorption was not the only way of resolving the paradox, however. In 1861, J. H. Mädler favored another explanation, in terms of a finite age of the universe. The idea was that light from distant stars has simply not had time to reach us, assuming that the time of the light's travel is less than the age of the universe. Since there was no firm and independent astronomical knowledge of many of the distances, or of the processes of creation, this was no more than conjecture, but it foreshadowed future explanations in a curious way, as we shall see. In 1901, William Thomson, Lord Kelvin, took a similar route to the solving of the puzzle. He also considered that the individual stars might have a finite lifetime. His solution was carefully worked out, and can be extended to cover an expanding universe of finite age. Not until later in the twentieth century did this problem of the dark night sky occupy the center of the astronomical stage. Before it could do so, astronomers had to learn to take seriously the physics of the universe as a whole. This they were largely forced to do by scientific developments that were taking place outside astronomy, as it was then generally conceived. We shall return to this subject in chapter 17.

THE PHOTOGRAPHIC REVOLUTION

The introduction of photography into astronomy in the nineteenth century is in many ways reminiscent of the introduction of the telescope in the seventeenth, for both brought to light new and entirely unexpected phenomena. The first photographs of the Sun and Moon showed almost instantly a degree of detail that would have required many hours of careful drafting by hand. As time progressed, and materials became more sensitive, ever fainter objects were recorded on photographic plates, and by the end of the century numerous discoveries had been made that simply could not have been made without photographic aid. We should not forget, however, that even before workable photographic processes were available, it was known that certain materials were sensitive to light invisible to the eye, so that some of the new detail could be explained on that account.

The history of photography was far more protracted than that of the telescope. It has been common knowledge since ancient times that substances may be bleached, and others darkened, by sunlight. Many chemists of the seventeenth and eighteenth centuries—among them Joseph Priestley (1772), Carl Wilhelm Scheele (1777) and Jean Senebier (1782)— investigated the chemical action of light, including light of different colors. In 1802, Humphry Davy published a report of the experiments of Thomas Wedgwood on methods of copying glass paintings or drawings on paper or leather treated by silver nitrate or silver chloride. He had no means of making them permanent. The first real progress came only after decades of experimenting by Joseph Nicéphore Nièpce of Chalon-sur-Saône, France. The camera obscura, or darkened chamber with a pinhole or lens through which an image was cast on a screen within it, was then already in wide-spread use for purposes of display and astonishment, and for assistance in sketching. For long, Nièpce struggled to produce a camera-made image on a plate that would be of use to a printer. His first moderately satisfactory images were produced in 1816, although of course they did not satisfy his original need.

In 1829, still experimenting, he went into formal partnership with a scenic painter and showman of similar interests, Louis Jacques Mandé Daguerre. There is scope for endless controversy as to the precise nature of their numerous processes, who was responsible for them, and whose was the priority of invention; and matters are complicated by the work of William Henry Fox Talbot, who by 1835 was quite independently producing tiny negatives in England. Talbot's friend John Herschel also took a strong interest, and among other things suggested to him the use of hypo (sodium thiosulfate), formerly called hyposulfite—as a fixing agent. (Herschel has to his credit the invention of the photographic terms "positive" and "negative," "snapshot," and even, in 1839, the word "photograph" itself.) Through Daguerre's adoption of hypo, this became the classic fixing agent. This entire period was one of a steady rise in sensitivity, both human and photographic. Fox Talbot was open with his inventions, but patented them and expected royalties, although he often had his claims contested. It was Daguerre, however—never particularly sensitive to the rights of his rivals—who provided the enterprise that brought photography into the consciousness of the whole world, not least because he published a clear handbook of instruction that went into numerous languages and editions, and that advertised the apparatus he sold.

Just as astronomy needed maximum apertures for light-grasp, so too in photography, exposures could be shortened by increasing aperture. (We should refer more strictly to *relative* aperture. Most readers will be aware that more light is admitted to the film, or the charge-coupled device (CCD) in a digital camera, by decreasing "f/number." The latter is the ratio of focal length to the actual aperture.) In 1840, hoping to maximize the light from the subject, and following astronomical practice, Alexander Wolcott opened the world's first portrait studio in New York using a camera with a large concave mirror. A more important turning point came in that same year, when the Viennese mathematician Josef Petzval designed a double lens with what was then an unusually large aperture of f/3.6. Earlier exposures had usually been of the order of minutes. To give an idea of what the new lens made possible: by 1841 a Viennese photographer was able to take a military parade in bright sunlight with an exposure of one second.

A photograph of the Sun itself clearly needed a shorter exposure than this, and indeed, Foucault and Fizeau in Paris in 1845 found it very difficult to make exposures *short* enough. A typical refractor might have a focal ratio of about f/8. Recording the faint images of the stars was for long out of the question, but a few crude daguerreotypes of the Moon were obtained by J. W. Draper of New York in 1840. In Cambridge, Massachusetts, in the later 1840s, William Cranch Bond, the first director of the Harvard College Observatory, together with his son George Phillips Bond—the future director of that same institution—carried out a series of photographic experiments. The 38-centimeter refractor there shared the distinction of being the world's largest refractor. Using that, they obtained a very much finer daguerreotype of the Moon than Draper's, a photograph that obtained much publicity for both sciences. Less dramatic, except to professional astronomers, was their photograph of the star Vega. A decade later, in 1857, the younger Bond showed the merits of the newer wet collodion process, although he was not the first to do so.

Astronomical photography was thus launched, owing as much to amateur as to professional astronomers. Photographs of considerable value were those taken by Warren de la Rue, a wealthy Guernsey-born paper manufacturer in England, whose chemical and technical interests covered a wide range. At the Great Exhibition of 1851 in London—for which he and his father were jury members—he saw daguerreotypes of the Moon taken by the Bonds. He repeated their achievement, although without the advantage of their fine telescope; and then in 1853, still focusing on the Moon, he tried out the wet collodion process. De la Rue's contribution to astronomical photography did not end there. The need to follow celestial objects during the necessarily long exposure-time meant following the object with a telescope accurately driven, mechanically, rather than by hand. A mechanism to do this was one that de la Rue did much to perfect. We shall shortly say more of his pioneering work in solar photography.

The wet collodion process had been devised in March 1851 by Frederick Scott Archer, a sculptor. It involved coating a glass plate with a mixture of potassium iodide and collodion, also known as "gun cotton." The coated glass plate was then sensitized in a bath of silver nitrate. As the adjective "wet" in the name of the process implies, plates were exposed immediately after they had been sensitized, making them difficult to handle, but their higher sensitivity made it possible to record much fainter detail than with daguerreotypes. The first successful attempt to photograph a comet, Donati's comet of 1858, was by William Underwood, using a collodion plate. George Bond had not been successful and was beaten to the post by an amateur astronomer. Underwood was simply described in the Dorking Post Office Directory in Surrey, England, as "portrait painter and photographer." In the 1870s, it was another British amateur who was responsible for improving the more advanced albumen dry-plate process, specifically to advance plans for the British 1874 expedition to observe the transit of Venus.

A rough idea of progress in astronomical photography in the course of the century may be had by considering typical exposure times for a terrestrial subject in bright sunlight. With the original daguerreotype, this was at first of the order of half an hour, and rarely less than a fifth of that time. With the wet collodion plate, the time dropped to about 10 seconds, and 15 seconds with dry collodion. "Dry plates" with rapid gelatin emulsions, available by the end of the century, reduced exposures to about $\frac{1}{15}$ of a second. Gradually, the large photographic companies were happy to play their part in the search for ever faster emulsions, and here C. E. Kenneth Mees of Eastman Kodak deserves special mention for his close collaboration with astronomers in the 1920s and 1930s.

Important milestones in later periods include the development of the red- and blue-sensitive emulsions that were used between 1949 and 1957 to take 1,800 photographs, covering the sky north of declination −30°. In this, the U.S. National Geographic Society collaborated with Palomar Observatory. Forty years later, the task was repeated with still more sensitive emulsions; and in the meantime, the southern sky had been added to the survey. Electronic means were eventually found for intensifying the received image. Television joined forces with photography in the post-war period; and after the 1970s, CCDs were much used, giving way to film photography only where image size was too large for CCDs to handle. Now that there are CCDs of well over a thousand megapixels, film will no doubt soon join cuneiform tablets in museum archives.

PHOTOGRAPHY AND THE SUN

During the period in which photography was first finding its way into astronomy, there were still many unresolved questions concerning solar phenomena, and solutions to them were being offered that may now strike us as very naïve. There were still some who were prepared to entertain William Herschel's strange ideas—in part, theologically inspired—about the Sun as a "lucid planet," a body that was potentially inhabited by people who were protected by a thick cloud-canopy from its intensely illuminated upper atmosphere. (It was J. H. Schröter who gave this the name "photosphere.") A sunspot, in this Herschelian view, was a hole in the photosphere, below which there was a lesser hole in the penumbral cloud, the latter lying between the photosphere and the body of the Sun. The difference in the sizes of the holes was supposed to explain why there seem to be two distinct degrees of shading at the edges of sunspots (fig. 201). Herschel hesitantly suggested volcanic eruptions as a cause.

Herschel's suggestion was slowly dying a natural death when, long afterwards, his son John Herschel explained the spots in a very different way. He regarded them not as signs of volcanic activity coming up from the Sun but as whirling storms, boring their way down through the photosphere and cloud. He drew up a lengthy analogy with the terrestrial atmosphere, its varying temperatures, and trade winds, but his ideas seem to have found little support. Another of John Herschel's suggestions did find an approving audience in some quarters, however, although not immediately, and

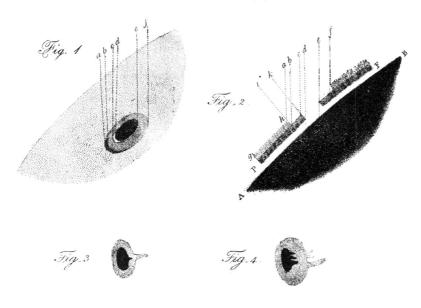

201. Illustrating William Herschel's explanation for the structure of the Sun, based on the shading of sunspots. A selection from his account in the *Philosophical Transactions of the Royal Society* (1801), plate XVIII.

not for long. It was that some of the spots may be due to great meteors falling into the Sun. In the 1860s and 1870s, Norman Lockyer spent much time and thought on a "meteoritic hypothesis" which chimed with Herschel's ideas. Lockyer's was a scheme of cosmic evolution which, while it did not gain much support, foreshadowed several modern ideas. According to Lockyer, all celestial self-luminous bodies are composed either of swarms of meteorites or of masses of meteoritic vapor generated by heat. He devised an involved scheme covering all the stages of cosmic evolution—as he saw it—starting with nebulous swarms of meteors. Small swarms would form comets, while large would amalgamate as stars, these first rising and then falling in temperature, destined in the end to wander through a cold, dead, and aimless universe. Without his strong desire to justify this over-ambitious hypothesis, Lockyer would never have undertaken his lengthy studies of meteoritic spectra, which were indeed valuable and influential.

There were two curiously isolated achievements in solar spectroscopy in the 1840s, and the first of them can almost be described as photographic. It was when John Herschel devised a way of noting the spectrum in the infrared, by moistening black paper in alcohol, and observing the "dark" lines by virtue of the fact that they dried last. Others found his published results unrepeatable, but long afterwards, in the 1880s, they were seen to be real enough when the wavelengths deduced were found to be those corresponding to absorption in the water vapor of the Earth's atmosphere. The second achievement that others could not immediately repeat was in 1842, when Edmond Becquerel used a daguerreotype plate to record the whole Fraunhofer spectrum of the Sun. This he recorded even into the ultraviolet, where the plate is naturally sensitive, and he was able to extend Fraunhofer's system of labeling the dark lines. Becquerel was the son and the father of talented Parisian physicists. His feat seems not to have been repeated for more than thirty years, although in the meantime—in 1852—the ultraviolet spectrum of the Sun had been observed by George Stokes, using its ability to make certain substances fluoresce.

There were more of Warren de la Rue's photographic achievements. His great ingenuity would not let him be satisfied with simple snapshots of the Sun. To provide answers to such questions as were raised by the arguments of the Herschels, he wanted a three-dimensional picture of the Sun. He knew that the profiles of sunspots could be studied as they came to the visible edge of the Sun, with the Sun's rotation. Alexander Wilson of Glasgow had examined a very large spot in this way in 1769 and found perspective effects that led him to study the problem intently. Many others later did the same. By 1866, a completely new kind of evidence would be called upon, when—as we shall see later—Norman Lockyer applied the spectroscope to the study of sunspots themselves. De la Rue's earlier approach to the problem, however, was both ingenious and much easier to understand. Stereoscopic photographs of ordinary scenes were then much in vogue. Two photographs were taken simultaneously by two cameras, spaced more or less at the spacing of the eyes. One then looked through a viewer at the two photographs, each seen by one eye only, whereupon the scene took on a three-dimensional appearance. This was of no use with the Sun, for the views from our two eyes are to all intents and purposes the same. Nevertheless, inspired by the idea, rather than photograph the Sun from two widely separated stations, de la Rue simply took one shot, and waited for the Sun to turn on its axis before taking another, so providing the view that would have been seen from a far-distant viewpoint. He found 26 minutes a suitable time interval. (He also applied a similar technique to photographs of the Moon, relying on its slight rotational movement with respect to the Earth, namely, libration.)

By this type of three-dimensional analysis, de la Rue found that the Sun's bright *faculae* (Latin for "little torches") are high in the solar photosphere (the visible layer), and that the dark part of a sunspot seems lower than its surrounding penumbra, above which the faculae seemed to float. This at once disposed of a host of fanciful ideas about the Sun, not only those of the Derham type, which made out sunspots to be volcanic, but also Lalande's notion that a sunspot is a rocky island in a luminous sea, with the penumbra, so to speak, the sandbank on the shore.

Rather as in the case of the astronomical use of the first telescopes, trying out the new method of observation was far easier than drawing theoretical conclusions from what was seen. De la Rue had to his credit what might well be described as the first contribution by photography to astronomical theory. There were several early attempts to apply photography to the problem of the fiery corona ("crown") of the Sun, seen at the moment of its total eclipse by the Moon, and the small pink prominences in it. These had been described long before—indeed, as early as 1185, in a medieval Russian monastic chronicle, there was mention of flames issuing forth from the eclipsed Sun, as though from red-hot charcoals. But did such flames, the prominences, belong to the Sun? Were they an optical illusion, a mirage, perhaps? Birger Wassenius, who had seen them in 1733 from Gothenburg in Sweden, described them as red clouds in the Moon's atmosphere. The eclipse of 1851, however, also seen from Sweden, was generally thought to have indicated that the place of the prominences was truly the Sun. The total solar eclipse of 18 July 1860 was a heaven-sent opportunity for European astronomers to settle this question once and for all. The central track of the eclipse touched southwest Ireland, the Channel Islands, and northwest France, from where it crossed Europe to Sicily. De la Rue observed it from the upper Ebro valley, while Father Angelo Secchi viewed it from Desierto de las Palmas, 400 kilometers to the southeast of him. The similarity of what they saw, and of what was now recorded for all to see on their photographs, finally convinced the astronomical profession of the solar nature of the prominences.

There were many drawings of the prominences and corona, made at many different viewing sites for total eclipses, in the decades following. In 1879, A. C. Ranyard and W. H. Wesley published more than a hundred earlier drawings and photographs. What surprised many was the wide variety of the forms the corona took, and the fact that those forms could change appreciably in much less than an hour—as observed at different stations. Too late for inclusion in their publication was a photograph of the corona taken by the young German-born spectroscopist Arthur Schuster in Egypt at the total eclipse of 1882, a photograph on which an unknown comet appeared (fig. 202). From 1882 onward, the photographic spectrum of the corona was recorded at every possible eclipse, and the rich assortment of spectral lines found in the ultraviolet—where the human eye cannot follow photographic emulsions—added a completely new dimension to solar research.

The camera was used increasingly in the study of the fine detail of sunspots, as well as the granulation of the Sun's surface. Pierre Jules César Janssen was a pioneer in this respect. By the time he began to use photography, Janssen had already earned much fame from an ingenious device he had constructed, after observing the total solar eclipse of 18 August 1868 from Guntur, near the Bay of Bengal. Pointing the slit of his spectroscope at two great prominences, when the Sun was in total eclipse, he found intense spectral lines appropriate to hydrogen. It occurred to him that if he were to admit only light of this particular wavelength—that is, occupying this position in the spectrum—then by rapid scanning of the Sun with the slit of his spectroscope he should be able to produce a picture of the Sun, and so follow its changes on a regular basis, instead of having to wait for eclipses. He gave his spectroscope a rotary movement, to form the solar image out

202. Photograph of the solar corona, taken by Arthur Schuster in Egypt during the eclipse of 1882. Notice the comet, which was discovered fortuitously through this photograph.

of its slit-like components. This perhaps reminds us of the mechanics of the cinematograph. In fact another invention of his, the photographic revolver, came even closer to it. This was a device designed to take a rapid succession of photographs during the transit of Venus in 1874.

If Janssen's name was known to the French public in his lifetime, it was surely as the man of daring who left the besieged city of Paris by balloon during the Franco-Prussian war to observe the eclipse of 22 December 1870. Lockyer had obtained a permit from the Prussians to allow Janssen to pass through their lines, but honor would not allow him to use it, since he intended to carry war dispatches. Having asked the Academy of Sciences to support his attempt, Janssen was provided with a balloon, the *Volta*. Together with an assistant, he reached an altitude of 2,000 meters and was carried westward by the wind, landing safely with his instruments—and precious dispatches—near the Atlantic coast. He reached Oran (in Algiers) to observe the eclipse, but the weather was less cooperative than the Academy of Sciences and the Prussians had been, and his journey had a symbolic rather than a scientific value.

Janssen's studies, which would now be thought perfectly straightforward astrophysics, were not then perceived as falling squarely under traditional astronomy, and it is worth noting that he found much difficulty in obtaining serious government support. It was fortunate that the minister of education, Victor Duruy, supported him, and strove to equip him with an observatory, eventually turning the tide in his favor. At the end of seven years, in 1876, Janssen was granted a choice of two sites, and he chose Meudon. He

203. Photograph of sunspots, taken by P. J. C. Janssen at Meudon, 1 June 1878.

204. A fine specimen of a drawing of a sunspot (23 December 1873) by Samuel Pierpont Langley, of the Alegheny Observatory, Pittsburgh, Pennsylvania. Langley was the inventor of the spectrobolometer, a highly sensitive device for measuring the heat received via the slit of a spectroscope, through the change in electrical resistance of a very fine wire behind the slit. This allowed him to map the energy of the spectrum. He used it in 1878 on solar and lunar spectra, and in 1894 published an impressive map going far into the infrared. His observatory during this period was the leading American center for solar physics.

was eighty years old before the astronomical staff was increased beyond a total of two; but in the meantime he had conducted a highly important research program, producing an atlas of solar photographs covering the period 1876–1903 (see fig. 203 for an example of his work). And so began one of the world's most distinguished solar research centers.

Early solar photographs were often compared unfavorably with drawings—figure 204 shows a fine example of what could be done by a good draftsman. Complaints were often made about the smudging of parts of the photographic images, but it was wrong to suspect the blemishes as having been always caused by poor collodion plates. Janssen realized that the phenomenon is truly solar, and that there is a network of some kind covering the Sun, which he called the *réseau photosphérique* (photospheric reticulation). He was proud of this discovery of a cellular structure in the solar surface, and claimed it to be the first discovery that simply could not have been found without photography.

There is a doubly interesting parallel between Janssen's career and that of Norman Lockyer in England. Lockyer, a civil servant without university training, had attached a spectroscope to his 16-centimeter refractor in the mid-1860s, and in 1866 conceived the same idea as Janssen's of 1868, that is, of viewing the Sun only in light of the color of the prominences. He obtained a government grant in 1867, but received a suitable high-dispersion spectroscope only a year later, observing the prominence as planned only on 20 October 1868. He tried an oscillating slit with poor results, but helped by William Huggins he came to see that a wide slit was enough, and he waxed eloquent over the strange shapes in the forest-like solar atmosphere. Stranger still was what happened when he communicated his findings to the French Academy of Sciences: his letter, and another from Janssen explaining the same method, arrived within minutes of each other. The sad story of Neptune was only twenty years old. On this occasion, the French government generously celebrated the near-simultaneous discovery by striking a medal carrying the portraits of the two astronomers.

Despite his many maverick achievements in astrophysics, Lockyer never really found a niche in establishment astronomy. A man of enormous vitality who worked with missionary zeal to improve public scientific awareness, he was the founder, and for half a century the editor, of the famous scientific journal *Nature*. He achieved very limited success, even so, in persuading the government to found a national astrophysical observatory. The Solar Physics Observatory was set up in South Kensington, London, but the site was later reclaimed, as it was needed for the Science Museum. At the age of seventy-six, a disappointed man, he built himself a new observatory at Sidmouth, in Devonshire, where he worked until his death eight years later.

It is easy now to be critical of this apparent government lack of concern for the advancement of science, but it is as well to remind ourselves that high science was then regarded as the province of the universities, which in Britain—in contradistinction to many other European countries—were not seen as primarily a government concern. Astronomy was in a borderline position. Airy, a generation earlier, had been the first Astronomer Royal to be able to rest entirely dependent on his official government salary. (Halley had received a naval pension, and all other incumbents of that office were in holy orders, and so could draw small stipends from the Church.) Greenwich, however, was not seen as a research institute, any more than was its American counterpart, the Naval Observatory in Washington, where Lockyer's contemporary Simon Newcomb was pursuing a clear vision of the need to improve planetary and lunar theory. These institutions were founded to serve the state, to supply tables and astronomical constants as a public utility for purely practical ends. Fortunately for astronomy, this did not rule out the lavish funding of observatories from private sources, especially in the United States, where local pride was an important factor. In Britain, since we have already heard much of Warren de la Rue, we might use him as an example of someone who bridged the gap between astronomer and patron. In 1873, having heard from Charles Pritchard that Oxford University was to establish a new observatory, de la Rue presented his telescope and other equipment. In 1887 he went further and bought the university a photographic refractor to enable the observatory to participate in the new international *Carte du ciel* star cataloging project. Britain had changed little, it seems, since the days when Flamsteed bought his own instruments.

Established by the International Astrographic Congress held in Paris in 1887, the aim of the *Carte du ciel* was to produce a photographic chart of the whole sky. It was not completed until 1964, however, and it should not be allowed to eclipse the colossal achievement of David Gill, director of the Royal Observatory at the Cape of Good Hope, who collaborated with Jacobus Kapteyn of Groningen in a more modest photographic survey. Gill's southern star catalog, the *Cape Photographic Durchmusterung*, was completed in 1900. It included the positions and photographic magnitudes of 454,875 stars. Gill's telescope, mounted in 1886, had a 9-inch (22.9-centimeter) object glass by Grubb, and it is worth noting that this was designed for observation by eye, and needed correction for photographic work.

Photography had at last become an instrument of the astronomical establishment. It had become an essential tool, not only of positional astronomy, but of every branch of astrophysics. Most large new telescopes from the last quarter of the nineteenth century and afterwards incorporated a photographic facility. Photographic recording was never without its problems, of course, but it remained the supreme recording tool until the 1970s, when CCDs began to take over more and more of its duties.

STRANGE SPECTRA: REAL AND ILLUSORY
NEW ELEMENTS AND LIFE ELSEWHERE

Huggins, Lockyer's senior by twelve years, resembled him in some ways. Both came to astronomy as amateurs, without any university education, but with wealth enough to buy fine equipment; and both were eventually knighted for their services to science. Lockyer was the speculator and Huggins the generally cautious observer-who at length was rewarded with presidency of the Royal Society between 1900 and 1905. Huggins's reputation was made by his pioneering work in spectroscopy, mostly done in the private observatory at his home, about four miles south of Westminster (fig. 205). Learning of Kirchhoff's findings, he enlisted the help of the chemist William Allen Miller of King's College, London, in investigating stellar spectra. Miller had already obtained much expertise in spectroscopy, and the two men happened to be neighbors. In 1862, they began to investigate the spectra of celestial bodies, including the Moon, Jupiter, Mars, and many of the fixed stars. Their preliminary findings were presented to the Royal Society in February 1863 and included diagrams of the spectra of Sirius, Aldebaran, and Betelgeuse—the three brightest stars in Canis Major, Taurus, and Orion, respectively. A few

205. Huggins's 8-inch Cooke equatorial refractor, with an Alvan Clark objective, used intensively between 1860 and 1869 at his home at Upper Tulse Hill, southwest London. (This French redrawing of a less clear original was printed reversed, but is here corrected.) One of his spectroscopes is mounted in place of a normal eyepiece. To help him to identify spectral lines, he used a comparison spectrum, produced by a spark between dissimilar metals, passed into the field of view by an extra prism. Note the induction coil at lower left, and the switch on the chair arm. Working with his friend W. A. Miller, Huggins's first attempts to use the telescope for photographing stellar spectra were in 1863, but the wet collodion process—faster than dry plates—proved troublesome. A year later, Huggins discovered the gaseous nature of some nebulae using this instrument. With it, in 1868, he measured the velocities of bright stars relative to the Earth and studied cometary spectra.

of the spectral lines were found by measurement, but others were estimated by eye. In the following year, using new and better equipment, they obtained far better results, measuring seventy or eighty lines in each of the brightest stars. (See plate 13 for the spectra of Aldebaran and Betelgeuse.)

The gold medal of the Royal Astronomical Society was presented to Huggins and Miller jointly in 1867 for their meticulous spectral analyses, of which of course there were very many more. Their work was a source of much excitement, and the far-reaching conclusions they drew from it even more so. They speculated about the source of light from the stars and our Sun: their spectra, while crossed by dark absorption lines and bands, were considered to be fundamentally continuous, which they interpreted as indicating "solid or liquid bodies in a state of incandescence." Invoking the nebular hypothesis for the creation of stars, they concluded that the significant differences evident in the constitution of different stars suggested that the chemical elements existed in very different proportions at different points of the nebulous mass. The time was not yet ripe for a theory of chemical transformation within stars. They could at least draw attention to the many points of similarity between distant stars and the Sun, concluding that this is likely to mean that many stars have planetary systems of their own; and that "if matter identical with that upon the earth exists in the stars, the same matter would also probably be present in the planets genetically connected with them, as is the case in our solar system." From here, they went on to argue that living beings were no doubt to be found in those remote planetary systems. This was not the typical wild conjecture but was based on the fact that the elements required by living creatures on Earth—they mentioned hydrogen, sodium, magnesium, and iron—are the very elements they had found widely diffused throughout the stars.

In claiming that their finding amounted to an "experimental basis" for the diffusion of life throughout the universe, Huggins and Miller were touching on a debate that had been vigorously pursued in Britain for half a century and more. There were the old classics in this genre, of course, many deriving from Cicero and the dream of Scipio. Kepler, in his *Conversations with Galileo's Sidereal Messenger* (1610), argued from the moons of Jupiter to the probability of life on that planet—for are they not analogous to our Moon, the purpose of which is to serve us? There had been Fontenelle's witty descriptions of the beings that live on other planets, not to mention the sober extension by Kant of his Categorical Imperative, which he said must govern the morality of planetary beings. Matters suddenly reached a less scholarly tribe, however, with the preaching of the learned Presbyterian divine Thomas Chalmers. He could fill his Glasgow church four hours ahead of his lunchtime sermons, in which he told of how God had filled the universe with intelligent life. Covering that same topic in his *Astronomical Discourses* (1817) no doubt helped to make it into the bestseller it soon became. The most intellectually influential debate of the period stemmed largely from his example. It was one that involved, among many others, the Cambridge mathematician William Whewell, the Scottish publisher and writer Robert Chambers, and the Scottish physicist, Sir David Brewster—an ardent Calvinist and biographer of Newton.

It is impossible to do justice to their arguments in a brief space, but their bearing on Charles Darwin's highly controversial *The Origin of Species* (1859), despite its ostensibly different theme, should not be missed. The temperature rose sharply in 1844, when Chambers published anonymously a tract entitled *Vestiges of the Natural History of Creation*. In this, starting from a Laplacian-style nebular hypothesis, he argued for an evolutionary life cycle throughout the universe, grounded in the world of chemical elements. Whewell, who in 1841 had been elected master of Trinity College, Cambridge, having previously tentatively accepted Chalmers's views, now changed direction and reacted slowly but forcibly to *Vestiges*. On theological grounds he insisted that human life is unique, and that it arrived on Earth entirely providentially, miraculously. Evolution was anathema, and life elsewhere would seem to require it. He too published anonymously, in 1853, but made little effort to hide his authorship. Brewster now entered the fray. An old critic of Whewell, he threw his considerable reputation behind Chalmers—first in an angry review, and in 1854 in a full-length book. In particular, he took the insignificant and peripheral nature of mankind's abode, the Earth, as proof of the grace and favor of God towards us. The debate was taken up by many others, and Darwin's bombshell did nothing to calm it. In 1833, Whewell had used an argument somewhat akin to what today is called the "anthropic principle" (which will be discussed briefly on p. 781), pointing out that if the laws of Nature as imposed by God were different, life would be impossible. There is an intriguing entry in one of Darwin's notebooks: "quote Whewell as profound, because he says length of days adapted to duration of sleep of man!!! Whole universe so adapted!!! And not man to planets-instance of arrogance!!." By the time *The Origin of Species* was published, there were already at least twenty books responding to Whewell alone. The findings of Huggins and Miller were certainly not the sort of thing to go unnoticed, as might be the case today.

Among the many spectral discoveries made by Huggins and Miller was one made in 1864, when two strange green lines were identified in the spectrum of the Great Nebula in the constellation of Orion. Huggins believed that he had found a new element there, unknown on Earth, to which he gave the name "nebulium." Five years later, John Herschel visually confirmed an observation by William Parsons, Lord Rosse, of a continuous spectrum in the light of the Orion nebula. On the basis of this he postulated the existence of a whole series of nebulae, ranging from those with purely continuous spectra to those with pure emission spectra. This did nothing to solve the nebulium problem, but it raised awareness of the much greater complexity of the cosmos than had been generally expected. Astronomers and physicists returned to the nebulium question often without solving the mystery, and only in 1928 did I. S. Bowen show that the lines were the so-called forbidden lines of oxygen and nitrogen. (Forbidden lines are such as are found in the spectra of certain nebulae in the H II regions but not in laboratory spectra, since we cannot rarefy the gases sufficiently on Earth. It was Bowen who first explained this phenomenon, and we shall soon see how very important his explanation was.) The fact that Huggins was mistaken should not be allowed to obscure the great importance of his observation, which proved that the nebula in question was gaseous, and not, for example, a solid or liquid, as was sometimes supposed.

Huggins's pseudo-discovery of nebulium contrasts with the very real discovery of the element helium, found in the Sun, by Norman Lockyer. The discovery of terrestrial helium did not follow until 1895, when it was isolated by Sir William Ramsay. Strangely enough, it was Lockyer who made most speculative capital out of nebulium, for he thought that it confirmed a theory he had of celestial evolution. There was much friction between him and Huggins over the question of whether the green lines corresponded to part of the spectrum of a magnesium spark in the laboratory. Huggins here was right: they do not. Another question, however, was whether Huggins's "green nebulae" were the breeding ground of stars, following the old idea developed by William Herschel, for instance. Again Lockyer seized on the notion, and wove it into his so-called dissociation hypothesis of stellar evolution. Huggins was more circumspect. He knew that the spectra of stars contained signs of many chemical elements, and those of the gaseous nebulae very few. Not for many decades was a theory of stellar evolution forthcoming that could account for these spectral differences.

Huggins seems to have been happy to leave the spectra of bright objects to others and to have concentrated on such faint objects as comets and stars, including the nova of 1866. In 1868 he made one of the most useful of all extensions of spectroscopy to astronomy. The Austrian physicist Christian Doppler had in 1841 given theoretical reasons for a change in the wavelength of a source moving relatively to an observer-a change of pitch in a source of sound, or of color in a light source. A. H. L. Fizeau-better known to physics than astronomy-had seen the possibilities of using the dark Fraunhofer lines as reference colors, and Huggins had enough knowledge of their overall patterns to make comparisons between laboratory sources of light and the feeble spectra of stars. In 1868 he found for the first time a stellar velocity by means of the Doppler effect. He gave the velocity of Sirius as 29.4 miles per second away from the Sun-*away*, since he found a shift of the spectrum toward the *red*, indicating a lowering of frequency, that is, a lengthening of wavelength. This visually acquired figure was on the high side, and he later revised it downward, but the application of the Doppler principle in astronomy was to prove of the utmost importance, especially in cosmology, when it was eventually applied to the light from entire galaxies.

Huggins had been trying to use photography on the spectrum of Sirius since 1863, but at first he obtained mere streaks of light on his plates. In 1872, Henry Draper in New York got a photograph with four lines crossing the spectrum of Vega (α Lyrae); and then, in 1875, Huggins began to obtain better and better results, from such bright stars as Sirius and Vega. He was now pioneering the new "dry-plate" photographic process, with sensitized gelatin in place of the old wet collodion. Four years later, he was able to record ultraviolet spectra. Together with others, some obtained earlier by H. W. Vogel in Berlin, and some later by M. A. Cornu in Paris, these showed him that white stars are abundant in hydrogen. This was an important discovery, marking the beginning of a general awareness of the overwhelming preponderance of hydrogen in the universe.

The year 1875 was an important turning point in Huggins's life, not to say in astrophysics, for it was then that he married Margaret Lindsay Murray of Dublin. Although only half his age, she quickly became an invaluable intellectual partner, both in astronomical observation and in the joint publication of their findings. Together they worked on the spectra of stars: it was no easy matter, working alone, to observe a spectrum while holding the image of a moving star on the spectroscope's slit less than a tenth of a millimeter wide, for an exposure lasting an hour. Together, around 1889, they obtained a photograph of the spectrum of light from the planet Uranus. Angelo Secchi had first observed its spectrum by eye in 1869, and in the intervening decades others had repeated the observation, some suggesting that the spectrum gave signs that Uranus was shining partly by its own light. The Huggins pair scotched this idea, showing that the spectrum was more or less that of light from the Sun, so that there was no reason to discard the assumption that its light came by simple reflection.

From the eclipse of 1882 onward, the spectrum of the solar corona was regularly photographed. The results showed more clearly than ever the spectral lines appropriate to hydrogen that had been found in the prominences by Janssen at the eclipse of 1868. Another spectroscopic discovery, made in the following year, was of a much more mysterious nature, in the sense that there was no known analog in the laboratory. On 7 August 1869, the track of a total solar eclipse crossed North America. The corona was studied by an enthusiastic army of astronomers, the best work being done at Des Moines, Iowa, by William Harkness of the Naval Observatory, Washington. He obtained from the corona a continuous spectrum of the sort expected, but he found that it was crossed by a single green spectral line. The discovery was also made, independently, by Charles Augustus Young of Dartmouth—but shortly to move to the College of New Jersey, later Princeton University. The place of the green line on the spectrum suggested to the physicists Kirchhoff and Anders Ångstrom that the element responsible was iron, although there was much opposition to this identification. In 1876, Young was able to resolve the green line into two components, one due to iron. An unknown element in the coronal gas was postulated to explain the other, as yet unexplained. It had no known terrestrial equivalent, and the supposed element was eventually given the name of coronium.

In the decades that followed, many more spectral lines that were characteristic of the corona were found and measured that were not to be found in any terrestrial source. In 1925, however, some lines in the spectra of nova RR Pictoris seemed to Harold Spencer Jones to hint at a connection with those in the corona; and in 1933 some of those strange lines were definitely found in the outburst of nova RS Ophiuchi. Stimulated by Spencer Jones's claim, Bengt Edlén of Uppsala, Sweden, began a laboratory study of higher spark spectra of the elements, and by 1939 he and I. S Bowen had proved that the novae and solar corona shared certain emission lines belonging to the element iron. They proved to be "forbidden lines," like those already

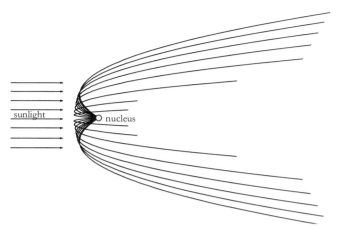

206. Bessel's fountain model of a comet, in which sunlight somehow pushes back material coming out from the nucleus, material that eventually forms a tail. A century later, a much more complex solar wind theory would be found necessary to explain the pattern of material ahead of the comet.

explained by Bowen as being a consequence of very low density in the corona. Almost immediately afterwards, W. Grotrian of the Potsdam Astrophysical Observatory—an important research center southwest of Berlin—suggested other correspondences, still with lines belonging to iron. Within a year, Edlén strengthened his argument, and formulated a list of coronal lines belonging to very highly ionized atoms of iron, calcium, and nickel, suggesting to him a corona at a temperature of about a million degrees Celsius. In short, by 1940 the ghost was laid, and it was plain—at least to those who in wartime had an opportunity to absorb the new information—that coronium was an illusion, and that no new exotic element was called for.

THE SOLAR WIND

The possibility that light may exert pressure had been a source of speculation from the seventeenth century onward. The first significant application of the idea was in Bessel's "fountain theory" of cometary forms, put forward after the German astronomer, in studying Halley's comet on its 1835 reappearance, saw that material thrown out by the nucleus toward the Sun was being pushed back so as to present a fountain-like head (fig. 206). There were many historical drawings available to him, from which the idea of light pressure from the Sun might have been encouraged, and here we may use an engraving by Thomas Wright of Durham to stand proxy for them all (fig. 207). Bessel was not able to say much about the mechanism, but that there was a solar force of some sort seemed undeniable. He tried to calculate the effect of the reactive force on the orbital period of the comet, but was not very successful. Unfortunately, comets are not often seen broadside-on, so that the analysis is difficult, but Bessel, an excellent mathematician, calculated the likely appearances that would be presented to us, given various different laws of solar

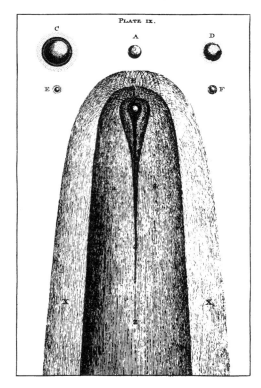

207. Thomas Wright's engraving of various cometary nuclei, and a complete image of the comet of 1680. The "balls" are those of the comets of 1665 (*D*), 1682 (*C*), 1742 (*E*), and 1744 (*F*), and are supposedly in true proportion to the Earth (*A*). For the comet of 1680, Wright distinguishes between its "natural atmosphere" (*aa*), "the denser matter winding itself into the train" (*zzz*), and "the inflamed atmosphere and tail dilated near the Sun" (*xx*).

repulsion. His theory went a long way to explaining the forms of planetary tails. Fountain-like displays are not rare. There was to be a dramatic example in the case of comet Donati in 1858, which continued to throw out fountain-like envelopes for several weeks (fig. 208).

There was a general—but mistaken—belief that, if only it could be demonstrated experimentally, proving that light exerts pressure would validate the corpuscular theory of light. It is amusing to read Jules Verne's novel *From the Earth to the Moon*, published in 1867, in which a central character insists that projectiles would be the vehicles of the future, driven by electricity *or light*. Physicists had no great confidence in the existence of the phenomenon until 1873, when, in his *Treatise on Electricity and Magnetism*, James Clerk Maxwell gave a formal proof that it could be deduced from his theory of electromagnetism, and that a precise value could be set on it. In 1900, the Russian physicist Petr Nikolaevich Lebedev finally demonstrated experimentally the minute pressure that light exerts on bodies. There were many implications of all this for astrophysics in the long term: quite apart from its likely effect on cometary tails, it also meant that stellar radiation pressure on atomic particles in interstellar space could no longer

208. Donati's comet of 1858, as drawn by Otto Wilhelm Struve at Pulkovo (5 October 1858). In 1954, Nicholas Bobrovnikoff calculated the expansion speeds of the halos of this and many other comets, and Fred Whipple of Harvard College Observatory later showed that the highly regular halo production could be explained in terms of the rotation of the nucleus.

be ignored; and that light pressure has a role in restricting the size of the stars—radiation pressure blowing off their surface layers, so to speak.

In 1943, Cuno Hoffmeister—founder of the Sonneberg Observatory in Germany—suggested that a solar radiation of particles, rather than light, was affecting the direction of cometary tails. He gave reasons for this belief, based on accumulated photographic records, but it was his fellow countryman Ludwig Biermann who, in 1951, synthesized theories of cometary tails and radiation in general from the Sun. He found that ions in the cometary tail were propelled by the continuous high-speed streaming of a million-degree plasma of ionized gas from the Sun, rather than simply neutral atoms or molecules. There was still no sound explanation for this particle radiation, however, until 1958, when Eugene Parker of the University of Chicago investigated the equilibrium structure of the corona—the interplay of gravity and the heat-driven velocities of its constituent particles. What were the conditions for what we might view as the "vaporization" of the particles? A model based on the behavior of such atmospheres as that of the Earth was of no use. Parker found that the conduction of heat interfered with the sort of equilibrium normally assumed. He eventually concluded that the uppermost layers of the corona flowed away from the Sun at a velocity comparable to that proposed by Biermann for his radiated particles.

The outward flow was named the "solar wind." It was quickly realized how, during its passage through the solar system, it sweeps up evaporated gases from planets and comets, meteoritic dust, and even affects cosmic radiation from our galaxy. (For the moment we may say briefly that cosmic radiation includes energetic charged particles that reach the Earth from outer space as well as from the Sun. It includes electrons, positrons, ions, alpha particles, and protons.) Looking back into history, there are many earlier findings that can be seen as anticipating the discovery, and they are worth listing, if only to show their diversity. Annibale Riccó of the Palermo Observatory, for example, a specialist in solar physics, had noted as early as 1892 that terrestrial magnetic storms begin forty to forty-five hours after the passage of large groups of sunspots. He decided, therefore, that the storms must be caused by an agent traveling from the Sun at a speed of around 1,000 kilometers per second. In 1896, the Norwegian physicist Olaf Kristian Birkeland suggested there was an electrically charged corpuscular radiation which, drawn in near the pole by the terrestrial magnetic field, gave rise to the northern aurorae (*aurora boreales*, the northern lights). With the advent of communication by radio and telephone, perturbations of the terrestrial magnetic field were found to interfere with transmission, and they too were later found to be correlated with solar eruptions. Around 1930, Sydney Chapman and V. C. A. Ferraro provided a theoretical model for this. They calculated that a cloud of ions ejected by the Sun would move at a speed of 1,000 to 2,000 kilometers per second, reaching Earth in one or two days, behaving as magnetic storms were known to behave. Then, in the late 1940s, Scott Ellsworth Forbush discovered that cosmic rays reaching Earth had a low intensity when the Sun was active, but faded rapidly during magnetic storms. He interpreted this as indicating that a magnetic field, transported by a stream of charged particles from the Sun, was obstructing galactic cosmic rays. Only when Ludwig Biermann's work on cometary tails was published, however, and was supplemented by Eugene Parker's model for the wind's mechanism, did these earlier findings begin to fall into place. And the most convincing confirmation of all was through the measurement of densities and velocities of particles *in situ*, starting in 1959, by instruments aboard the Soviet spaceprobes *Luna* 2 and *Luna* 3, and later by the American *Voyager* 2. Parker's model was then broadly confirmed.

These and subsequent studies showed the solar wind to be of extremely low density by the time it reaches the Earth—between 0.1 and 30 particles per cubic centimeter, far lower than that in the best laboratory "vacuum." Despite this fact, the wind affects the Earth's magnetosphere dramatically and supplies enormous energy to its many processes. (The magnetosphere is a cavity created when the solar wind encounters the Earth's magnetic field. It extends around ten Earth radii in the Sun's direction, and a hundred times as far on the other side. The solar wind cannot enter.) It was agreed that the solar wind includes protons, with helium, oxygen, and other elements, with a mean speed of about 400 kilometers per second, and that it extends far beyond the outermost planets of the solar system. What gradually became clear to those working in this field, however, was that the properties of the solar wind vary between wide limits, and that simple models are not enough. Quite apart from wild density variations, its velocity can vary between 300 and 1,000 kilometers per second. The corona itself has been found to possess a complex structure, with holes in it—revealed by X-rays—corresponding to the places from which come the most rapid fluxes of solar wind.

It should by now be clear how a mere handful of nineteenth-century spectroscopic discoveries were the seeds of enormously rich streams of astronomy in the century following. That those early discoveries were found to be linked with the study of cometary tails was remarkable in itself, but the ways in which they led on to a better understanding of solar structure, and to that of the magnetism of Sun and Earth, and even to much fundamental physics, are all fine instances of serendipity. There were so many paradoxes waiting to be swept away. It was not easy to get used to the idea that the two chief components of the corona, hydrogen and helium, send us very little light. Once their extremely high temperatures were accepted, that became easy to understand: they had lost their electrons, and the tenuous light seen during an eclipse is that provided by the heavier atoms. Even then, atoms of iron might typically have half of their 26 electrons stripped away. This was only the beginning of the chase, however. There are plasma effects to be studied in the terrestrial aurorae, southern and northern; but more significant, on a cosmic scale, is the fact that, since the discovery of the nature of the corona, low-density plasmas at extremely high temperatures have been identified throughout the universe. They have been found in the atmospheres of other stars, in supernova remnants, and in the outer reaches of galaxies. It is because low-density plasmas radiate so little that they can reach such high temperatures. Light quanta at these temperatures include ultraviolet light and soft X-rays. It is by detecting helium absorption and X-ray emission in stars similar to our Sun that coronas have been found not only to exist, but to be commonplace. The idea of coronium was illusory, but it had some very real and astonishing consequences.

THE CORONA NEWLY PHOTOGRAPHED

To return to the nineteenth century: one phenomenon that was then regularly photographed was the spectrum of the chromosphere—the narrow pink layer between the photosphere and the corona—with its strong lines appropriate to calcium. The first photograph of the corona's spectrum was taken by yet another outsider to regular astronomy, Arthur Schuster, a Jewish emigrant from Frankfurt, who held a post in physics at Owens College in Manchester, in the north of England. Schuster's speciality was spectroscopy, and his was the first successful attempt to photograph the spectrum of the corona with an ordinary slit spectroscope. This was at the total solar eclipse of 17 May 1882. His straightforward photographs of the complete corona during that eclipse were hardly bettered before the twentieth century. They showed its great extent, something that had been appreciated visually at the 1878 eclipse, seen from America: parts of it in the neighborhood of its equator were now to be seen around two full diameters from the solar center. (Schuster was even fortunate enough to catch the image of a comet on that occasion.) The eclipse in question marked a useful step forward in international collaboration in astronomy. It was observed from Sohag in Upper Egypt by a large number of astronomers, each having agreed to undertake a particular type of observation during those unrepeatable seventy-four seconds of totality.

Inspired by the 1882 records, Huggins experimented on photographing the corona without an eclipse. His idea of using a restricted part of the spectrum was probably inspired by that set down in Lockyer's original request for government aid. Huggins noticed that Schuster's negatives showed that there was a great concentration of light from the corona in a certain region of the spectrum-a region in the ultraviolet. Could he not photograph the Sun with photographic plates sensitive to just this region? He chose silver chloride as the sensitive material, and after several trials, at about the time of the eclipse of 6 May 1883, obtained results very similar to those obtained during the eclipse itself by standard methods. Unfortunately, it was another three years before the method gave further photographic evidence of the same sort, and there were many who doubted its authenticity. The reason for difficulty in repeating Huggins's results was partly the filtering of light through volcanic matter thrown into the Earth's upper atmosphere by the great Krakatoa eruption in the Straits of Sunda in August 1883. Agreement was long in coming. Other techniques for separating out the light of appropriate color were gradually developed, and Huggins's ideas were well and truly vindicated.

FURTHER APPLICATIONS OF PHOTOGRAPHY

As the century wore on, photography became integrated into advanced astronomical practice, which indeed eventually became unthinkable without it. There were, of course, the ever more spectacular photographs of gaseous and spiral nebulae. (For a fine early photograph of the Great Nebula in Orion, taken in 1883 by A. Ainslie Common, see figure 209, and compare it with the drawing in figure 210.) If the nineteenth century was the golden age of star catalogs, then it was only proper that photography should be applied to their production too. Juan Thomé in Cordoba, Argentina, labored until his death in 1908 to extend the Bonn catalog southward, reaching to latitude 62°S. Not until 1930 did others extend the lists of his *Cordoba Durchmusterung* down to the South Pole. One of the weaknesses of this work was in the estimation of star magnitudes, and here it was that photography, together with some essential theories about image density on the plate, came to the rescue. It was found, paradoxically, that

209. The Great Nebula in Orion, photographed by A. Ainslie Common in January 1883 (exposure 37 minutes, 36-inch silver on glass mirror).

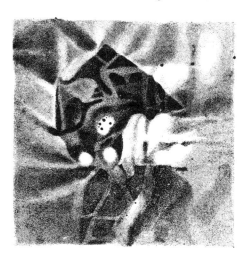

210. A drawing of what the Earl of Rosse called the "Huygenian region" (the central portion) of the Great Nebula in Orion. Compare the photograph in the previous figure. The drawing was made in 1852, using the Parsonstown telescope, by Bindon Stoney, a civil engineer and friend of Rosse, who prized it greatly. He was chary of the way in which professional astronomers allowed their preconceptions to guide their pencils. He and his assistants kept the nebula under observation for many years, in the belief that they could detect changes taking place in it.

poor lenses gave better results than the finest. Techniques were later developed that required a slight defocusing of the image.

Even more useful was the rapidity with which star coordinates could be measured from photographic plates. The idea occurred to the brothers Paul and Prosper Henry of Paris, but neither they nor the members of an international conference convened in Paris in 1887 appreciated all the errors inherent in the technique they developed. Nevertheless, the great *Carte du ciel* project was born, with the aim of mapping the sky photographically down to the fourteenth magnitude—that is, to magnitude 15.0. It was planned that there would be a new catalog of stars, titled the *Astrographic*

Catalogue, of stars down to the eleventh magnitude. Only after several decades was the work put on a really reliable footing. Even before it had begun, however, Jacobus Cornelius Kapteyn was at work, producing one of the century's great monuments of cataloging, using a simple and elegant technique, which was again photographic.

Kapteyn was in some ways fortunate to be at a university—the University of Groningen—reluctant to provide him with a large telescope. (The Netherlands, the United Kingdom, and the Irish Republic redressed the injury to some extent by establishing a "Jacobus Kapteyn telescope" with a 1-meter aperture at La Palma, on the Canary Islands, in the 1970s.) He used instead a set of photographic plates taken by David Gill at the Cape Observatory between 1885 and 1890. By the ingenious use of a theodolite in his laboratory, viewing singly the stars on the plate, which he placed at a distance equal to the focal length of Gill's telescope, he could measure each star's coordinates—right ascension and declination—directly, and even more accurately than had been done for the Bonn catalog. He also found stellar magnitudes by measurements on the star images, so that in the space of thirteen years, and only ten for the measurements, in two small rooms of the physiology laboratory in Groningen, the *Cape Photographic Durchmusterung* was produced, with its 454,875 stars between 18°S and the pole, down to the tenth magnitude.

The Harvard Observatory later went one stage further, and distributed its "atlas" in the form of boxes of plates. It is amusing to compare Kapteyn's with the Harvard tradition of processing measurements, which was to use the services of ladies with arithmetical abilities and time on their hands. Kapteyn, living in a more conservative society, persuaded the governor of the Groningen state prison to lend him the services of selected male guests of that place. Another instance of educated women being used to measure and reduce data for a cataloging project was that at the Paris Observatory, where a group was recruited to work on the *Carte du ciel*. The group was led by a remarkable Californian woman, Dorothea Klumpke, and her career was linked with astronomical photography in more ways than one.

Born in San Francisco, one of five daughters of a failed goldminer who later made a fortune in real estate, Dorothea Klumpke and her sisters were sent to schools in Europe for their education. Having taken her bachelor's degree in mathematics in 1886 at the Sorbonne in Paris, she joined the staff of the Paris Observatory as an assistant. When the *Carte du ciel* project was launched, she was successful—in competition against fifty men—in being appointed to the post of director of the Bureau of Measurements, becoming the first woman to hold this or any comparable scientific position in France. She worked with a number of distinguished astronomers, and with those two experts on all matters instrumental, Paul and Prosper Henry, who were then busy with a project for photographing asteroids with a new 34-centimeter refractor. In 1893, Klumpke became the first woman to proceed to a mathematics doctorate in Paris, with a dissertation on Saturn's

211. Dorothea Klumpke (1861–1942), an American-born
astronomer in Paris, who in 1901 married Isaac Roberts.

rings that brought her yet more local fame; and when Pierre Jules César
Janssen-director of Meudon Observatory, and by this time president of
the French Society of Aerial Navigation—had to choose an astronomer to
be taken up by balloon to observe the Leonid meteor shower of 1889, his
choice fell on her. The Leonids were less obliging, a disappointment not
only to the intrepid astronomer at 480 meters altitude, but to the entire astronomical world.

Dorothea Klumpke's career in astronomical photography had taken a
new turn before this hair-raising experience. On a vessel sailing for Vadsø,
Norway, with a group of astronomers who were hoping to observe the total
solar eclipse of 9 August 1896, she met that pioneer of astronomical photography, Isaac Roberts. They married in 1901, their thirty-one year age
difference (her forty years to his seventy-one) bridged by their passion for
the subject. Roberts died suddenly three years later. His remains were fittingly interred in a granite monument with ancient Egyptian motifs and
a bas-relief based on two of the most famous of his enormous collection
of photographic images, one of the great galactic nebula in Andromeda,
M31, the other of the gaseous nebula NGC 1499. The latter carries a message that is not altogether obvious, for it is the California Nebula. This
dim red object is very extensive, but its faint red form makes it virtually
invisible to the naked eye, and it was discovered as late 1884—by E. E. Barnard, just in time for it to be added to Dreyer's *New General Catalogue*. Its
supposed resemblance to the coastline of Dorothea's native state no doubt
explains her affection for it. Soon she returned to France, and with her, the
collection of photographic plates taken with the 20-inch reflector. Faithful to her husband's memory, she spent twenty-five years measuring those
plates, and in 1929 she published the results of her phenomenal labors in
*The Isaac Roberts Atlas of 52 Regions: A Guide to William Herschel's Fields
of Nebulosity*. In 1934 she returned at last to the land of NGC 1499, where
she remained until her death in 1942.

Hardly a branch of astronomy remained untouched by photography.
The camera was used from several stations, for example, to observe the
transit of Venus in 1874. Planetary photography as such required good atmospheric conditions to obtain a steady image. Mars was photographed by
B. A. Gould from Cordoba, Argentina, in 1879; and in 1890, in a succession

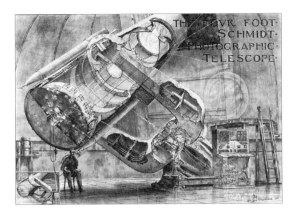

212. The Palomar 48-inch Schmidt reflecting telescope, as drawn by Russell N. Porter in 1941, when it became the largest instrument of its type. The moving parts weighed more than 12 tonnes, the main mirror alone weighing more than a tonne.

of photographs, W. H. Pickering, at Wilson's Peak, California, showed the southern polar cap of the planet. To everyone's surprise, and to the excitement of the many who were then speculating on the possibility that Mars, with its canals, was populated with intelligent beings, it appeared that the polar cap was actually changing in area.

Jupiter was systematically photographed with the great Lick telescope in 1890–1892, when the research observatory was new and enjoying its status as the country's finest. The planet was near opposition, and the "great red spot" on its surface—which had long been a subject of study through the telescope—seemed to be threatening to disappear (for a modern image of the spot, see plate 14). Even asteroids became subjects of the photographic lens. They were now being detected in ever greater numbers by virtue of the simple fact that they often left trails of light on photographs of star fields. The first asteroid so found was discovered by Max Wolf of Heidelberg in 1891. In the fifty years from 1890 to 1940, the number of recognized asteroids increased from fewer than 300 to almost 1,500. Among the more important of those now known are Eros, discovered by Gustav Witt in 1898, and Icarus, which Walter Baade discovered by chance in 1949, as a mere streak on a photographic plate made with the newly completed Schmidt telescope at Palomar (fig. 212). Icarus can approach close to two-thirds of Mercury's distance from the Sun, and yet its aphelion is far outside the orbit of Mars. Eros was the first asteroid known to come within the Earth's orbit, but asteroid Hermes comes closer, and indeed it may come uncomfortably close. In 1937 it was only about twice the Moon's distance.

At the time of this discovery, most astronomers had other missiles in their thoughts, but in the postwar era there was much speculation over the best course of action should an asteroid seem to be heading for planet Earth. The Hawks tended to favor a nuclear attack on it, and the Doves, a rocket engine that would put it off course. The 1979 film *Meteor*, starring

Sean Connery, presented the hawkish alternative and, together with the cult of computer war games, may have much to answer for. In 1991, NASA went so far as to set up the Interceptor Committee. The committee is said to have proposed a battery of laser guns on the Moon and an orbiting fleet of nuclear warheads, and "nuking" a few specimen asteroids for target practice. These proposals and the great meteor crater in Arizona, remind us that the danger need not come from an asteroid at all.

Icarus is of the order of a kilometer in diameter. The object responsible for the Arizona crater was almost certainly less than a tenth of that, in other words, less than a thousandth of its mass. As for the sizes and velocities of the fragments of the unfortunate asteroid, if ever it is hit, they are anyone's guess. And as for the difficulties of detecting such an asteroid, most methods rely on minute changes in the appearance of the sky, that is, movements across the field of view of the detector. The real danger will of course be from an object that is not moving across the field of view.

Of objects that have reached Earth in the past, not all have left such a clear sign of their nature as the Arizona meteor. One especially mysterious visitor from space caused an extremely violent explosion on the morning of 30 June 1908, in central Siberia. It landed in the valley of the river Podkammenaya Tunguska, northwest of Lake Baikal, and witnesses spoke of an enormous fireball, visible in the sky for only a few seconds. Witnesses as far as 60 kilometers away from its area of impact were knocked to the ground by the shock wave it caused. For long, political objections were raised to visits by scientists, who were generally inclined to believe that they would find at Tunguska at least a meteor crater. The first visit by an expert was in 1927, and to the astonishment of all concerned, no crater was found. On the other hand, it was found that trees were laid flat over an area 30 or 40 kilometers across. In 1930, the English astronomer Francis Whipple proposed that a comet, or cometary fragment, was responsible, and there is now wide acceptance that this was so. Analysis of the trajectory suggests that the fragment may have been part of the nucleus of comet Encke. The idea is supported by reports of a persistently bright night sky, lasting nearly two months after the event, and assumed to be caused by cometary dust.

Accidental discoveries have often been made with the camera's help. For example, Edward Emerson Barnard—a man with considerable reputation for finding comets by honest searching—was working at the Lick observatory, photographing stars in the constellation Aquila, in 1892, when he found a cometary trail on the plate. This comet was not the first to be photographed, but was the first to be discovered by photography. Barnard's systematic photography of regions of the Milky Way, and of comets, was of material help in advancing a knowledge of both. As the speed of photographic plates improved, the task became easier, but that simply meant pushing out to more distant regions, where it was less so. And not all advances came with the fastest of cameras. Barnard, who was then a junior member of staff under the autocratic direction of E. S. Holden, produced his fine Milky Way photographs with poor apparatuses requiring exposures of up to six hours each, using a guiding telescope without an illuminated reticle. As the nineteenth century ended, it had in Barnard an excellent example of a new type of astronomer, one who not only grasped at the opportunity to introduce photography wholesale into astronomy, but who was prepared to give it precedence over those methods of observing with which the century had begun.

PHOTOGRAPHY AND THE STORAGE OF DATA

By focusing only on new observational techniques, we are in danger of overlooking developments of comparable importance in the storage of data, and here again, photography played an important part. When positional astronomy became fashionable—indeed, almost an obsession—in the first half of the nineteenth century, large observatories could generally cope with the publication of their own observational records and their reduction to usable information, but that soon became impossible. The first of the old conventions to disappear was that of entering positions on maps. Photographic plates replaced them, up to a point, and as time went on, the photographic record was gradually allowed to go largely uninterpreted, or rather, the plates were shelved until needed. They remained invaluable for the study of novae, variable stars, magnitudes generally, shapes, sizes, and the distribution of matter-but like human beings, astronomy is not what it eats but what it digests. Photographs in the form of spectrograms kept their value less well, in view of rapid improvements in technique, but an astronomer wishing to check on a star in the *Henry Draper Catalogue*—which with its extension covers more than a quarter of a million stars—can still, in principle, consult the corresponding spectrogram in the Harvard photographic archives. The Harvard College Observatory plate vault has more than half a million ordinary photographs, going back to 1885; and a similar story could be told of many other observatories of comparable age and standing.

Positional and other types of measurements on photographic plates were originally made by eye, as we have already seen, using plate-measuring machines, or possibly Kapteyn's theodolite method. After the advent of the electronic computer, it became possible to automate this process up to a point, afterwards keeping the plate and discarding unwanted information. The pace of recent change, however, would have astonished even Kapteyn. To take the example of spectrography: as late as the 1990s, it was usual to obtain the redshift of only a single galaxy from each exposure and placing of the telescope. There are now multi-object spectrographs that can register the spectra of hundreds of galaxies in a single field of view; and there are computer processes for analyzing these spectra automatically, in many different ways. Automation of different kinds, by electronic means, was already being used in the 1960s as a means of directing the telescope, ensuring, for instance, that it would remain locked on to a suitable star at the center of the area being photographed. With radio telescopes, there is nothing to see by eye, and it is natural enough to have data collected by "observers" in a comfortable room, or even in a comfortable bed, while a computer carries out its instructions. This sort of arrangement was extended, before long, to photographic and solid-state data recording, which could be done without an astronomer sitting in the dome and looking through a guide telescope, risking frostbite or pneumonia in the process. The world has become a softer place, but there are certain nonhuman advantages too: arrays of solid-state detectors can be artificially cooled—for instance, using liquid nitrogen—in order to improve results by reducing the electrical "noise" in the system. Recording might last for several days before the results are downloaded; and the person responsible for that might even be in another country, or a continent away. Needless to say, all of the techniques of remote operation have found a use with satellite observatories.

The electronic computer has brought about other changes in astronomical tradition. Not only has it reduced the tedium of reducing data, but in many cases the data downloaded are actually processed before they are received to be put on record. Gone are the days of logarithms and mechanical calculators. Another change that crept in by degrees, especially over the last half of the twentieth century, has been in the allocation of what might still be loosely called "telescope time"—a much easier task in the past, for the director of an observatory who was faced only with those energetic few who could make their way to the dome. Out of the forty or fifty thousand aspiring astronomers who live in the hope of being allocated facilities at one or other of a few hundred research observatories, many will be disappointed; and when committees for allocating telescope time try to please too many, long-term projects tend to suffer. A career option that is open to disappointed applicants with computer skills might be astronomical modeling, a relatively new but rapidly expanding type of research, using computer simulations to answer questions from almost any branch of astronomy. Cosmological modeling has already shed much light on questions of the origins of galaxies, why they take on the shapes they do (spheroidal, elliptical, disk-shaped, etc.), why their sizes seem to be limited, and whether they originated in a golden age of star-forming in the early universe, or in a more leisurely process.

The older techniques of astronomical photography were inevitably exchanged for digital imaging in the last decades of the twentieth century, in line with developments the world over—and often in advance of the rest of the world. That very process has made the dissemination of images easier, in particular via the internet, but the same has been true of information generally, resulting in a drastic change in the tempo of research. This has not been without its disadvantages—cut-and-paste excerpts from provisional drafts, which may be multiplied by hundreds within a matter of weeks, or even days, each version less reliable than the last. These phenomena are somewhat analogous to dark matter in the universe. Another problem relates to the storage and retrieval of images, a problem that was already widely recognized in the 1970s, when technologies suddenly began to change almost monthly. Reading a photograph a hundred years after it was taken is often easier than reading a digital device on which data was stored a decade ago—stored, perhaps, on a device which is now utterly obsolete, if not completely unavailable.

ON THE NATURE OF COMETS

After Newton, cometary orbits were at last rather well understood, and new studies of the appearances of cometary tails, and their directions in relation to the now well-described cometary orbits, gave scope at last for some reasonably sedate speculation. Only after the advent of spectroscopy, however, could astronomers only begin to make well-founded conjectures about the constitution of cometary nuclei, the very hearts of comets. We have seen how modern studies of the direction of cometary tails, for instance by Hoffmeister and Biermann in the 1940s, led to a synthesis of theories relating comets to the solar corona through the solar wind, but the basic knowledge on which that connection was made was available only in the second quarter of the last century. At an earlier stage, appearance was the only guide, supplemented by occasional forays into traditional—and often unreliable—lore, concerning oddities from the past. With telescopes, it could be seen that when comets were near the Sun, they could develop not only a coma of vapor, gases, or dust—quite which, was not certain—but that their conspicuous tails were then most obvious. The common belief that their tails proved them to be some sort of celestial firework was hard to disprove. (The Greek word κομήτης refers to the hairy appearance of the typical cometary *tail*. It is confusing that the word "coma," from the same Greek root, is now used only of the immediate envelope of the *nucleus*.)

It will perhaps be helpful if we begin with a handful of simple but unhistorical points, which may be kept in mind as we look back into the past. The tail of a comet has two chief components. One is dust, often yellowish, which is blown away by radiation from the Sun. Following the same laws of motion as other gravitating bodies, the dust lags behind the head as it streams outward, creating a curve to the tail. As we have seen, Newton drew the subtle curves with great care, but others before him had noticed them. The other component of the tail is a plasma of ionized gases, often bluish in color. Recognition of the distinction between the two components came only in the latter half of the twentieth century. Modern historical accounts often ignore it, when they tell of Apian's "discovery" that cometary tails in space point away from the Sun. That they do so is only roughly true, and even then, only for the first component. The plasma tail is much more difficult to account for, and it is the fact that the different tails may lie in very different directions that makes comets present such curious shapes— shapes that have been imagined as crucifixes, swords, hammers, and all the

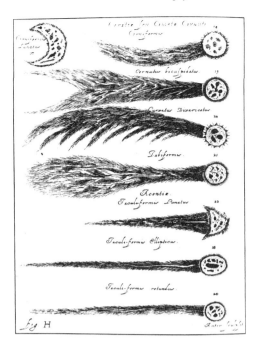

213. Various sixteenth- and seventeenth-century cometary forms, as drawn by Hevelius for his *Cometographia* (here much reduced). Despite their somewhat stylized and artificial character, the drawings record several important properties that other astronomers often missed, such as the curvature and splitting of tails, and the multiple bands of light round cometary nuclei.

rest. For some of the drawings of comets by Hevelius, half imagined and half real, see figure 213. The plasma tail might even break away from the coma. Yet another cause of the peculiar shapes of comets is the effect of dust left along their paths, which may scatter sunlight in peculiar ways.

Does a knowledge of the tail and outermost particles of the comet tell us anything of the nucleus? Is it not obvious that they are formed by the accretion of similar particles? But why should the particles come together at all? Some valuable clues were obtained after Giovanni Schiaparelli proved that there is a connection between meteor showers and comets. It is a curious fact that whereas Aristotle's views of comets as something of a meteorological character was generally rejected after Tycho Brahe's study of the comet of 1577, nevertheless, the idea that shooting stars were meteorological, inflammable, vapors accidentally set alight in our atmosphere lingered on into the eighteenth century. Halley and the German physicist Ernst Chladni had both proposed a cosmic origin for them. In 1794, Chladni suggested that space was filled with atoms that were drawn down to the Earth by gravitation to be ignited by friction in the atmosphere. This was purely speculative, but he suggested to two of his students at the University of Göttingen, Heinrich W. Brandes and Johann F. Benzenberg, that they study the altitudes of shooting stars by simultaneous sightings from distant locations. They were all presumably unaware that a similar procedure had been proposed by the mathematician Abū Sahl al-Kūhī more than eight centuries earlier, although the treatise he wrote suggests that he was more interested in the geometry of the case than the natural philosophy. In 1798, the two Göttingen students found that meteors became visible at an average height of 97 kilometers, and traveled with "planetary velocities" of several kilometers per second, all of which indicated an origin outside Earth's atmosphere. In fact, Edmond Halley had calculated that an extraordinarily bright meteor of 19 March 1719 was at a height of 60 miles (37.5 kilometers) and had a velocity of 300 miles a minute (8 kilometers per second). Several astronomers had made similar estimates in other cases, but the Benzenberg-Brandes study was more thorough.

It is a strange fact that external opposition to these findings led even Chladni to lose faith for a time. Laplace was quite dogmatic, in his insistence that shooting stars (meteors) originate with materials thrown out by immense volcanoes on the Moon. That idea had probably originated with Olbers, writing in 1795. A popular science encyclopedia of 1821, edited by E. Polehampton and J. M. Good, is a useful touchstone of belief at that time. Having devoted seventy pages to meteors, meteorites, and fireballs, incorporating all varieties of expert opinion, they come down in favor of the volcanic hypothesis for meteoritic stones, but not for meteors, "since they sometimes ascend as well as fall." There they favor electricity, or terrestrial exhalations. The situation changed in a doubly dramatic fashion when, on the night and early morning of 12 and 13 November 1833, one of the most spectacular storms of meteors ever seen terrified and excited people in both Europe and North America, in particular over a swathe stretching from Halifax, Nova Scotia, to the Gulf of Mexico (fig. 214). Estimates put the total number seen at nearly a quarter of a million, but it was the properties of the lines of light in the sky, rather than their number, that was so revealing: they all seemed to radiate from the same part of the sky, more or less a point, in the constellation of Leo. Similar properties had been noted in previous showers of meteors, but without anyone reasoning clearly on the basis of the fact. A typical arrangement is shown in figure 215. That point, the "radiant," moved with the stars in the course of the night, thus proving that their source was a point far outside the Earth's atmosphere. While it may be hard to see that their paths through space were indeed parallel, and not, as it were, like the paths of fragments from an exploding firework close at hand, the matter can be easily understood if one imagines a set of long parallel lines—such as several sets of straight parallel railroad lines with multiple overhead conducting wires, all parallel to them—viewed by looking along them from one end. They will seem to diverge from a distant point, but this is simply a property of perspective. It should now be easy to understand why many witnesses—judging by the 1821 summary chapters—could say that "they sometimes ascend as well as fall."

Their interests newly aroused, astronomers began to search early manuscript and printed records for the timing and circumstances of meteor

214. An impression of the great Leonid meteor (shooting star) storm of November 12–13, 1833, drawn after the event by an Adventist minister, Joseph Harvey Waggoner, on his way from Florida to New Orleans. He had not grasped the notion of meteor radiants (*Bible Readings for the Home Circle*, 1889).

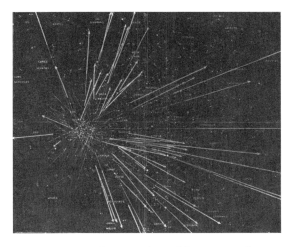

215. Paths of the most striking Leonid meteors observed from Greenwich on 13 November 1866. Here accurately plotted in right ascension and declination, they appear to diverge more or less from a point in the constellation of Leo.

showers, with some success. As half-expected, they found preferences for certain parts of the sky, and for more or less the same date in the calendar. It was clear to them that this meant that each family of meteors was associated with a particular place on the Earth's orbit, which was presumed to be a particular crossing with a ring or other band of particles that burn up as shooting stars when they enter the Earth's atmosphere. The Leonids, as those with a radiant in Leo were now called, were always on the night spanning 12 and 13 November or the following night. The search for historical data led astronomers to study the periodicities of displays. One highly significant result was announced after the Leonid shower of 1866. Here was another dramatic display: in Europe, it was possible to count meteors at rates of rarely less than a thousand per hour, at the peak of activity. Hubert Newton was by no means a pioneer in the study of periodicities, but he was more precise than most, and he concluded that the November meteors come from something which revolves around the Sun in a period of 33.27 years, and moves in an elliptic orbit stretching out between the orbits of the Earth and Uranus. He correctly predicted a return in November 1899, although those people who had hoped for a great firework display in that year were disappointed. After periodicities were found for other showers, the fact that the entire community of interested parties—many of them amateurs—was forewarned, and could prepare for such occasions, led to a marked improvement in knowledge of their nature.

Giovanni Schiaparelli was a graduate of Turin University, but he had also worked at the Berlin and Pulkovo observatories before returning to Italy, and was at the Brera Observatory in Milan in 1860. During the 1860s he spent much time analyzing new and old data with a bearing on the nature of meteors. He summarized his findings in letters to Angelo Secchi, who published them. Through these letters, Schiaparelli put previous work in the shade, with a demonstration that the August meteors—known as Perseids since their radiant was in Perseus—*move in the same orbit as the bright comet of 1862.* (The comet in question is known as Swift-Tuttle, having been discovered by Lewis Swift on 16 July 1862 and independently by Horace Parnell Tuttle three days later. Shortly after his discovery, Tuttle left Harvard to fight in the Civil War, achieving some fame as a combatant and even more as an embezzler of military funds.)

Just as Tycho Brahe had given comets astronomical status, the same was now the case with meteors and meteoritic material generally—it being assumed that these two classes differed only in particle size. Possessed of Schiaparelli's insight, several astronomers set to work to correlate as many meteor showers as possible with all of the computed cometary orbits then known, and this was done with great success, leading to some scores of identifications. Yet again, some astronomers began to look back into history, this time to produce the names of anyone—preferably a fellow countryman—who had anticipated Schiaparelli's great discovery. This tells us much about the level of excitement the discovery created. The names of Kepler, Maskelyne, Chladni, Morstadt, and Kirkwood were all put forward. All were certainly on the brink of the discovery, but it was Schiaparelli who calculated the coincidence fully and argued his case rationally.

The Leonids, the best known of all meteor storms, were highly active in 1799, 1833, 1866, and 1966. Other well known storms are the Perseids (highly active in 1861, 1862, 1990), Lyrids (1982), Ursids (1795, 1945, 1986), Draconids (1933, 1946), and Andromedids (1872, 1885).

After Schiaparelli, while the old problems of the constitution of comets and of meteors were coalesced, astronomers were still far from a solution. Was it true that meteorites—sometimes called aerolites—were of the same general nature as that of whatever it was that caused meteors? Meteorites were objects that landed on the earth, some of them enormous, and often causing a degree of devastation. When found, they tended to be preserved, and it had long been recognized that all the examples that had been preserved had a similar appearance, and that they often contained much iron. Chemical analyses in the first decade of the nineteenth century showed that they also contained nickel, iron oxide, sulfur, silica, magnesium, and manganese. Until Schiaparelli broke the spell, there were still many who were prepared to believe that this cocktail of substances was actually formed in the atmosphere. Now it had become necessary to look further afield.

Could these substances be found in comets? In 1864, Giambattista Donati observed a spectrum from the atmosphere of the comet now named 1864 II Tempel. He found three bright bands, yellow, green, and blue with dark areas between them. This offered the first proof that the light of comets was not—as was widely supposed—entirely of reflected sunlight, but was in part light from a self-luminous gas (a faint component of a continuous solar spectrum was later detected). Donati himself was content to note that what he observed resembled the spectra of metals. It is now recognized that there is much reradiation from the solar ultraviolet light absorbed by the atoms and molecules in the comet. Their vibrational and rotational motions give rise to hundreds of lines that create the bands of light first observed.

Between 1868 and 1880, the spectra of eighteen comets were analyzed, showing signs of hydrocarbons in every case. Then, in 1881, Janssen, Huggins, and Draper all independently photographed the spectrum of the comet of that year, with the same result (figs. 216 and 217). Lines for metals were also found, especially iron, as well as lines for an unstable form (CN) of the poisonous gas cyanogen (C_2N_2). Knowledge of this fact, spread abroad in 1906 by H. G. Wells's novel *In the Days of the Comet*, caused many to stay indoors during the 1910 reappearance of Halley's comet. Hawkers made handsome profits selling "comet pills" to neutralize the supposed danger.

Research into cometary shapes—their heads and tails and curvature—gradually took second place to spectroscopic studies, although there were

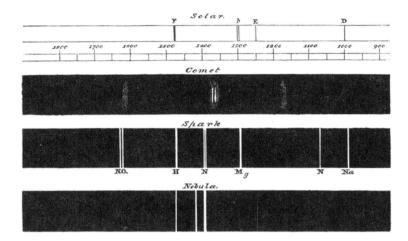

216. A drawing of the spectrum found by Huggins for comet Brorsen (1868). He tried several spectroscopes before obtaining good results. Here he draws as reference spectra part of the solar spectrum above, a spark spectrum below that, and a spectrum of the type he had seen in gaseous nebulae at the bottom. He speculated that this differed from a cometary nebula by virtue of different temperature and molecular state. He noted the resemblance to the diagram given by Donati for Comet 1864 II Tempel.

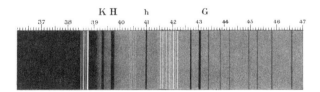

217. The impact of photography on the recording of spectra may be seen by comparing this, a lithograph of the first photograph of a comet's spectrum, with the previous figure, both from Huggins. His photograph was of what is now labeled comet 1881 III Tebbutt. Along the top edge is a scale of wavelength in hundred-Ångstrom units. The continuous part of the spectrum is from sunlight reflected from the comet's dust particles, the dark lines being absorption lines from the same—the Fraunhofer lines that result when light from the Sun passes through a cooler gas. Lines *K* and *H* are of this sort, due to calcium. Line *G* is due to absorption by the CN radical, but bright lines to the left and right of *K* are emission lines from CN and carbon molecules in the cometary gases. This lithograph was prepared for a meeting of the British Association for the Advancement of Science in 1881 and created a huge impression on those present.

still many surprises to come in the twentieth century, in particular, with the understanding of two different types of cometary tail: Type I tails shine entirely by fluorescence of ionized atoms, Type II by sunlight reflected from solid grains. The positive charges on the ions in Type I result in a repulsive force up to a thousand times that of the gravitational attraction of the Sun on the particles. This knowledge was only acquired in the mid-twentieth century, but it is an example of a half-forgotten doctrine that went back to the beginning of the nineteenth century. As early as 1812, Olbers had put forward the view that cometary tails are made up of particles subject

218. Comet 1858 VI Donati (as it is now named), after a drawing by George Bond, director of the Harvard College Observatory. The curved tail is of dust, the two faint components are of ionized gases. The bright star to the left of the head is Arcturus.

to electrical repulsion from the Sun. This idea was by no means ignored. It was taken up by Bessel in Königsberg, W. A. Norton in Yale, C. F. Pape in Altona, Fedor Aleksandrovich Bredikhin in Moscow, and Johann Karl Friedrich Zöllner in Leipzig, each of whom made attempts at a mathematically precise theory. Bredikhin's contribution was especially valuable and surprisingly modern in character. He analyzed numerous comets, having himself studied directly the spectra of many of their heads and tails. Theorizing about the solar electrical repulsion, in 1877 he postulated three types of comet. By 1879 he had distinguished between them on the basis of their constitution: in his Type I he supposed that hydrocarbons predominate; in Type II he thought that light metals, such as sodium, were predominant; and in Type III, iron molecules. At first he supposed the repulsive force to be the same for all, but by 1885, drawing on a study of forty examples, he differentiated between types on the basis of the ratio of repulsion to gravitation: for Type I it was of the order of 14, for Type II, about 1.1, and for Type III, about 0.1. To each type there corresponded a different curvature of the tail.

These ideas were at first dismissed by more astronomers than accepted them. Some found Type I hard to reconcile with spectroscopic findings; some favored explanations in terms of heat-repulsion; others thought cometary tails to be nothing more than immaterial electrical discharges in space. A drawing by Bond, of the comet 1858 VI Donati, shows very clearly the coexistence of three different tails (fig. 218), fitting beautifully into Bredikhin's theory. Discovered on 2 June in the constellation of Leo, a tail first began to appear in mid-August, its triple tail was seen around mid-September. The comet's nucleus was itself complex, at times resembling a batswing burner. No earlier comet had been so intensively studied by astronomers around the world, George Bond at the Harvard College Observatory being one of the most assiduous. The data collected became useful at a later time; for the moment, Bond's claim that the light from it was reflected sunlight was generally accepted. Four years later, Donati inaugurated a new phase in the understanding of cometary light, and—as explained earlier—he found a component of nonsolar light, so that Bond's conclusion needed to be modified.

Spectroscopy brought a celestial form of chemistry into astronomy, but the identification of spectral lines proved to be more difficult than had at first seemed likely. The *relative* intensity of cometary lines often differs radically from what is found in the laboratory. The very placement of lines from a single comet may change with time. In 1941, the Belgian astronomer Polydore Swings advanced considerably the methods used for interpreting cometary spectra when he took into account Doppler shifts in the solar radiation that excites the lines in the cometary spectrum. Another milestone in the analysis of cometary spectra came in 1957, when Swings and Jesse L. Greenstein, working at Palomar, took high-resolution spectrograms of comet 1957 v Mrkos. In them they found evidence for forbidden lines of neutral oxygen, lines that had earlier been found in the solar corona and in nebulae, and that require an extremely rarefied atmosphere. Other properties of comets were added steadily to the list during this postwar period. Spending on astronomy was slowly beginning to recover, especially in the United States, although everywhere the astronomy of the solar system tended to be undervalued, being then considered less fashionable than theories of the universe on a grand scale. A revival of solar astronomy came with the invention of new techniques. Given a tool, people are inclined to invent uses for it. To studies of the nature of dust particles were added studies of their sizes—a few microns at most, in the majority of cases. This last work was done mostly at infrared wavelengths. Observations of a comet in the infrared had been made by Carl O. Lampland at the Lowell Observatory as early as 1927, but did not come into its own until the work of Eric Becklin and James Westphal, beginning in 1965. Observations of comets at radio wavelengths also began, with attempts to observe comet 1957 III Arend-Roland—at wavelengths of 21 centimeters, 50 centimeters, and 11 meters. The first reported detection by radio of a molecule in a comet (methyl cyanide, CH_3CN) was by Bobby L. Ulich and Edward K. Conklin at Green Bank, West Virginia, in December 1973; and others quickly followed. Solar astronomy was back in the limelight, and it was beginning to be obvious that it had a bearing on cosmological issues into the bargain.

The study of the physical composition of comets had also entered a new phase in the postwar period. For many years, cometary nuclei had been seen as swarms of small particles moving together in the same direction. This was still the view of Raymond A. Lyttleton, when he proposed his "sand bank" model. Lyttleton and Fred Hoyle had met in Cambridge in 1939, and thereafter produced a series of highly influential papers on this and other subjects. Some concerned the internal constitution of stars, and others the gravitational accretion of inter-stellar matter by stars—thus pioneering the study of the two-way interaction between stars and the interstellar medium. In 1949 and thereafter, they used the expertise they had gained in those earlier studies to explain cometary accretion. The sandbank theory was based on the idea that the Sun, moving through a cloud of gas and dust, draws both together to follow its path by gravitational attraction. They supposed that the streams of gas and dust would eventually converge, collide, and form a cometary nucleus.

The theory aroused heated debate, as more and more of its shortcomings were pointed out. One of its serious failings was that it could offer no source for the colossal wastage of gas into space that it required. Within a year or two, it had lost ground to a far more plausible account of the nature of cometary matter and formation, the "dirty snowball theory." One of the clues that led up to this was the existence of the radical OH in the heads of comets, and the positive ion OH+ in their tails. Swings and others suggested that the OH might come from water, a parent molecule that would have a very weak spectrum. By a parallel argument, it seemed that similarly stable—but as yet undetected—molecules of methane, ammonia, and carbon dioxide might be present. At this stage, in the late 1940s, Fred Whipple, of the Smithsonian Astrophysical Observatory and Harvard, formulated a theory of cometary nuclei as masses of ice embedded with dust and meteoritic particles. Dormant when distant from the Sun, they would lose molecules by sublimation when in the Sun's neighborhood, at speeds of the order of 300 meters per second—enough to carry away any embedded grains of rock. Gas and grains, and a theory that could explain the longevity of comets, which could remain active for hundreds of revolutions around the Sun—these virtues were very telling, but how could the theory be proved?

One approach to a proof was to test some of the consequences of the basic idea. One such consequence was that sunlight, falling on the icy nucleus, would be active on the side nearest the Sun, and so would produce a jet effect, propelling the comet away from the Sun. If a comet was rotating, however, the direction would be different. Just as in the case of the Earth, the comet will go through a daily cycle of changes in temperature, with a highest temperature an hour or so after true noon. If a comet rotates in the same sense as the Earth, the jet force will be inclined to the line to the Sun, and will push it forward in its orbit, increasing its period of revolution around the Sun. Whipple had studied Encke's comet, spinning with a retrograde motion and known to have a decreasing orbital period. This was exactly what the theory could explain, and when numerical data were fed into the model, the results were even more encouraging. The theory was soon found to fit well with the known motions of other comets, and one early finding was rather surprising: roughly equal numbers of comets were turning with retrograde as with direct motions. The theory could explain meteor streams, as a product of large pieces being blown off the parent body; and it could explain the tidal forces experienced by comets close to the Sun, forces that were able to split a comet into several pieces. (The sungrazing comet 1882 II, for instance, is in four parts.)

The dirty snowball model was quickly accepted by most astronomers, although it created problems in plenty. When and where and how are comets created? The Dutch astronomer Jan Oort favored creation with the asteroids in the inner solar system. Others argued that the low temperatures needed for water, ammonia, and carbon dioxide to freeze, spoke in favor of the outer reaches of the solar system, from which they could be thrown into the Öpik-Oort cloud, which we shall introduce shortly. To help solve this, and more modest problems of detail, new methods of gathering evidence were plainly needed. They came with the space age, and will be touched upon in chapter 19.

WHERE DO COMETS BELONG?

The nuclei of comets are usually only a few kilometers across, and in their orbit around the Sun they are not visible even with a telescope until they come fairly close to the Sun. There are innumerable small asteroids of a similar size, but comets differ from them in two respects. As Whipple and others had decided, a comet's nucleus contains a high proportion of water in the form of ice, and it is this, when vaporised, that produces the surrounding coma, which may be as much as a million kilometers across, making the comet. Another difference is that comets' paths through space are much more eccentric than those of asteroids, so that comets' distances from the Sun vary enormously as they move round their orbits. Their paths are also often greatly inclined to the zodiac, in which the paths of the planets all lie. Only since the time of Newton and Halley has it been known that they are periodic, like the planets; but since some comets have periods of the order of thousands of years, they have been seen only once in historical times. To take an extreme case, for example, it will be another million years before comet Kahoutek reappears.

By the beginning of the twentieth century, comet hunting had been an often frenzied pastime for nearly three centuries. Discoveries were frequently rewarded by monetary prizes. In 1840, for example, the king of Denmark offered a gold medal for every new discovery. In or around 1885, E. E. Barnard, later of the Lick Observatory, was able to build himself a small house out of award money for his several discoveries. The prince of comet hunters, however, was Jean Louis Pons, who in the first three decades of the nineteenth century discovered around thirty comets (his career had begun as doorkeeper of the Marseilles Observatory). Equipment did not have to be high powered, at least at the discovery stage, so this became a sport in which amateurs often excelled. Only in the last two or three decades of the twentieth century did professional discoveries finally put amateur work into the shade.

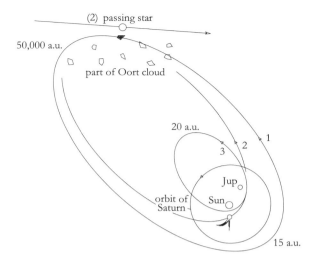

219. The Öpik-Oort cloud. The figure, which of necessity is out of proportion, illustrates three different situations. In the first (numbered 1, ignoring the passing star and Jupiter, and orbits 2 and 3), the millions of nuclei of potential comets are at present in the Öpik-Oort cloud, in eccentric orbits but mostly far from the Sun. It is assumed that they were first created when the planetary system as a whole was formed, say, 4.6 billion years ago, but were then in a circular orbit only about 15 AU from the Sun. They were put into their present orbits through planetary perturbations in the course of the long history of the system. By Kepler's second law, they spend most of their time at distances of perhaps 50,000 or 60,000 AU. A passing star (see 2) can easily perturb an object in the cloud, however, sending it into a long-period orbit going much closer to the Sun. When this happens, it becomes visible. In its passage through the solar system, it runs the risk of being perturbed by a planet, especially by the massive planet Jupiter. It may then be changed into a short-period comet (see 3) with its perihelion much nearer than before (say, 20 AU), as in the case of Halley's comet.

There was more to cometary studies than the hunt. Those who worked in celestial mechanics had problems enough. Laplace, for instance, spent enormous effort on explaining why so many short-period comets cluster near the plane of the orbit of the planet Jupiter. Hubert Anson Newton of Yale, starting from Laplace's suggestion that they were originally randomly organized, calculated their chances of being captured by Jupiter. He decided that the chances were very slight—one capture in a million, perhaps—but that this would be no obstacle, given millions of comets, most of which are lost. If there were or are millions of comets, what of their source? In 1932, Ernst J. Öpik—an Estonian who was then at Harvard—asked himself how remote from our Sun comets could be, and still be held gravitationally by the Sun, without being lost to other stars. He gave an answer of 60,000 astronomical units, within a radius of which we can talk of our cloud of comets. Other astronomers followed with calculations of motions, which must often be affected by planetary proximities, and of origins. The most successful theorist here was Jan Oort.

As we shall see in the following chapter, Oort had already gained considerable experience working on the dynamics of stars in the Galaxy. In

1950, he formulated a theory according to which comets originate from a vast cloud of billions of small bodies orbiting the Sun at a distance of about 60,000 astronomical units. These, he decided, are the constituents of comets, and these are sent into orbits close to the Sun when their original orbits are perturbed by approaching stars. His idea was that a subsequent perturbation by a massive planet may result in a much smaller orbit, with much reduced intervals between returns to the Sun's neighborhood. (For more detail, see fig. 219.) This hypothesis of the "Oort Cloud," or "Öpik-Oort Cloud," soon gained wide acceptance. The steady accumulation of data, and the advent of electronic computing, have made amending the parameters of Oort's model much easier than its initial formulation. Brian G. Marsden and his colleagues at the Smithsonian Astrophysical Observatory in Cambridge, Massachusetts, estimated that out of well-observed comets with periods above 250 years, only about ten per cent came in from outer space, the others being gravitationally bound to the Sun. As for the capture of comets by Jupiter, Laplace's problem, the odds are smaller by a factor of perhaps 500. Indeed, the larger planets were now said to be responsible for actually expelling comets from the solar system, by means of a gravitational slingshot action on them—the sort of "gravity-assist" action (to use the latest jargon) that is now often used to accelerate spacecraft missions on their way to the outer regions of our planetary system. With those spacecraft, set to pursue and even bombard the things they imitate, came the next phase of cometary study.

Galaxies, Stars, and Atoms

Optics was not the only branch of physics to make a welcome intrusion into astronomy in the nineteenth century. In the middle of the century, a new and coherent theory of heat—thermodynamics—was assembled from laws that had been derived in relative isolation—the conservation of energy, the law of entropy, and so forth. No sooner had the relevant laws been established by a succession of physicists—including Sadi Carnot, Julius Mayer, James Joule, Hermann von Helmholtz, Rudolf Clausius, and William Thomson—than they were considered universal and were applied to the heavens, beginning with the Sun. Once it was realized that there is a balance sheet on which all forms of energy must be accounted for— energy of motion, energy of position, heat, electrical, chemical energy, and so on—then it became clear that most of the energy that manifests itself on the Earth is ultimately derived from solar radiation. What is the origin of that solar energy? The Sun's radiant energy could in principle have been transformed from energy of another type, but if so, from which?

In a privately published work of 1848, Julius Robert Mayer suggested that it came from the mechanical energy released with the continuous bombardment of the Sun by meteors. Mayer's work was not widely known, but the same idea was later put forth independently by the much-wronged Scottish scientist John James Waterston, and for a short time the "meteoric hypothesis" attracted much attention. (Waterston's is the sad case of man whose ideas, especially on the kinetic theory of gases, were not properly appreciated until several years after his death.) It was a simple matter to calculate how much mass would need to fall into the Sun from a great distance to produce the heat that it is found by measurement to radiate. John Herschel and the French physicist Claude-Servais-Mathias Pouillet had independently measured the heat received from the Sun fairly accurately and had made estimates of the amount of heat absorbed in the Earth's atmosphere. The annual infall of mass was calculated differently by different authors, but William

Thomson's figure is representative: the annual infall was said to be of the order of a seventeen-millionth of the Sun's mass.

Here is yet another illustration of the growing number of interrelated physical quantities. Celestial mechanics had been brought to such a state of perfection that even this minuscule quantity could immediately be ruled out as much too great. As Thomson showed in 1854, it would imply a shortening of the time of revolution of the Earth by about a couple of seconds a year, an easily detectable quantity over the interval between the Babylonians and the nineteenth century. Hermann von Helmholtz, on the other hand, had a more subtle mechanical theory. Helmholtz suggested that the Sun's heat came from the conversion of gravitational energy to heat energy in the process of condensation of material that began as a vast cloud. (This version of a "nebular hypothesis" was also sometimes confusingly called the "meteoric hypothesis.") The Sun might at present seem to be a well-formed object, but his idea was that the same processes of contraction are still operative, and that the gravitational energy (potential energy) of the Sun's slowly contracting material continues to be converted into the energy that the Sun releases, mainly in the form of heat.

This hypothesis seemed very much preferable to an alternative theory of solar energy supply, which some were then proposing, namely, that chemical reactions in the Sun were the source of its heat. As Thomson pointed out, the most energetic chemical reactions then known would not keep the Sun radiating for more than about 3,000 years. Even the theologians wanted more time than that.

There are many slight variants of the data quoted, and the following calculation is meant only to show the path to an important new conclusion that was then reached. The contraction needed to provide the enormous radiant energy of the Sun would reduce its diameter by only about 75 meters per annum, far too small a quantity to be measured, even over centuries. (The diameter of the Sun is almost 1.4 million kilometers.) Helmholtz's contraction hypothesis was thus safe from criticism on this account. It led on, however, to a conclusion as to the vastness of the time-scale of solar activity that seemed theologically so dangerous that astronomers frequently apologized in advance for it. There is no difficulty in accounting for 20 million years of heat, by the theory. Ten million years, said Thomson, is a minimum requirement, and 50 or even 100 million years would be quite arguable.

This was the first coherent physical argument for the age of the Sun, and of course it provided a lower limit to the age of the universe. Needless to say, those who believed that God created the world around 4,000 years before Christ were displeased by such reasoning. Some geologists were admittedly by this time demanding a much longer period to account for the geological changes that the Earth had seen. Even Buffon, in the mid-eighteenth century, had estimated 75,000 years, on the basis of the rate of the Earth's cooling, but three million years on the basis of the deposition of sediments—although he did not publish the second figure. By the time of Thomson's calculation, however, geologists such as Charles Lyell were quoting figures far in excess of his, and this paradoxical situation remained a thorn in the side of solar astronomers for more than half a century, although they did not find it difficult to expose errors in geologists' standard methods of calculating cooling times for the Earth. How the geologists came to terms with their problem falls outside the scope of this book, but, broadly speaking, there was always one easy way of shortening time scales: they simply needed to introduce some great catastrophe or other into the world's history. One might even say that the Bible had shown them the way.

There was another awkward conclusion to be drawn from thermodynamics. This predicted a running down of the universe, "a state of universal rest and death," as William Thomson called it, in his article "On the age of the Sun's heat, assuming that the universe was "finite and left to obey existing laws." He avoided this by arguing for "an overruling creative power" that was responsible for introducing living creatures to the universe, and that so removed the need for "dispiriting views" as to human destiny. He wrote these words in 1862, less than three years after Charles Darwin had published his theologically controversial *The Origin of Species*, with its biological theory of the evolution of living forms, a theory that had some interesting parallels with the law that Thomson and others found so disturbing—the law of increasing entropy.

This, Clausius's second law of thermodynamics, amounts to saying that natural processes as a whole move in one direction, that the entropy of the world can only increase. Entropy is often casually described as a measure of disorder, but there is something to be said for examining the way in which the concept unfolded historically. In 1847, Thomson pointed out that, while energy cannot be created or destroyed—this being the first law of thermodynamics—it does lose its capacity to do work. This is the case, for instance, when heat energy is transferred from a warm body to a cold one, doing work in the process. The "dissipation" of energy, he said, amounts to a one-way process in nature. Before long, we find Rudolf Clausius and the Scottish engineer Macquorn Rankine formulating this new concept more carefully. In 1865 Clausius gave it the name entropy and formulated the twin laws: one, the constancy of energy in the universe, and the other, the law that its entropy tends to a maximum, when there will be no longer any energy available to do work. Time could therefore be seen as an arrow pointing in the direction of the "heat death" of the universe, when the temperature will be the same throughout it. Some, like Thomson, have found this melancholic idea of a final thermodynamic equilibrium distasteful; and counter-arguments and escape clauses are still being offered at regular intervals, many of them rooted in theology, a few of which are highly emotive.

Another important step in the development of a theory of the Sun came when astrophysicists began to consider the detailed structure of its interior and other possible sources of energy within it. Jonathan Homer Lane was a rather shadowy figure on the edge of American science, for some years an examiner in the U.S. Patents Office in Washington, D.C. In 1869, Lane developed Thomson's arguments further, assuming that there are convection currents in the Sun. He investigated the conditions under which the Sun might remain in equilibrium and found that it could do so if its temperature changed in inverse ratio to the radius. If, when it contracts under gravity, only part of the heat generated by contraction is radiated and the rest can is kept to increase the temperature of the sphere, it is possible for this retained portion to keep it in equilibrium. The Sun may thus lose energy and yet grow hotter.

Although it was soon appreciated that "Lane's Law" breaks down when the contraction eventually produces a gas of very high density, it stimulated others to investigate the Sun's structure, and the structure of stars generally, and it forced Kelvin to reconsider his own argument. A similar argument to Lane's was presented independently by August Ritter in 1872. It is a curious fact that both took meteorological models as their starting point. When, in 1907, the Swiss physicist Robert Emden published what was to become a classic textbook of this branch of astrophysics, he applied his theory of spherical distributions of gases to both cosmological and meteorological problems in the same work.

Another example of interplay between astrophysics and other scientific subjects was when George H. Darwin, the son of Charles Darwin, considered friction in the tides as an agent of cosmic as well as biological evolution. Darwin was a strictly theoretical astronomer and the successor to Challis as Plumian Professor at Cambridge. In a series of studies, beginning in 1879, he projected the motions in the Earth-Moon system back in time and found that there was a time when the daily rotation and the monthly revolution were equal. Going further back still, he was led to a situation when it seemed that the Moon and Earth must have been a single body, destined to split into two. The conclusion fitted closely with theories of the equilibrium of rotating fluid bodies that were developed around 1885 by the great French mathematician Henri Poincaré. Darwin explained how the split might have taken place, and his work led others to investigate the wider questions of the formation of the entire planetary system. The model usually assumed was some variant of the old Laplacean model of condensation from a rotating cloud. There were some serious difficulties here, however. The simple model did not explain why Jupiter has so much rotational momentum (angular momentum); indeed, it has nearly two thirds of that tied up in the solar system as a whole, whereas the mass of Jupiter is only a thousandth of the whole. The Sun, on the other hand, with most of the mass, has only one-fiftieth of the total angular momentum.

In 1898, the American astronomer Forest Ray Moulton, still a graduate student at the University of Chicago, working with the chairman of the geology department there, Thomas Crowder Chamberlin, began a study of the formation of the planets, although he broke with the straightforward Laplacean view. After studying photographs of the solar eclipse of 28 May 1900, they were eventually led to form a hypothesis of "planetesimals," lumps of matter that had supposedly solidified out of the original condensing nebula. By 1906, considerations of the oddities of angular momentum led them to the idea that the planetary system had originated with the close approach of another star to the Sun, which had drawn matter out gravitationally, giving rise to motions of a sort they were able to explain, at least approximately. Their idea was that planetesimals collected near the Sun and formed small planets. Indeed, the energies released in their collisions when coalescing were said to be responsible for the high temperatures inside the planets, known at least from study of the Earth, and assumed to be likely in all planets.

It seemed clear that there were at least three quite different explanations possible for the formation of the planetary system. (The words "cosmogeny" and "cosmogony" were, and are, used indiscriminately here. The first strictly refers to an *origin* of the world or a *beginning*, and the second to a creative act, a *begetting*. Religious belief apart, the second is best used where a process is concerned.) It could have been the result of a rare accident, such as the approach of another star to the Sun, or it could have arisen during the typical evolution of a star from gas and dust in the Laplacean manner, or perhaps in the course of a rarer sort of stellar evolution, whatever that might be. The rarity we ascribe to our own solar system hangs on the choice made. In the first case, planetary systems would be rare; in the second, common; and in the third, presumably somewhere between the two. In the 1920s, the Cambridge mathematical physicist James Jeans expressed a preference for the first alternative, and thought the odds against a star being surrounded by planets to be about a hundred thousand to one. At the same period, Arthur Stanley Eddington—the leading British theoretical astronomer of his day, although rivaled by Jeans—went even further and hazarded the suggestion that our world, with its living beings, might not simply be rare, but unique. This made many commentators uneasy, for no better reason than that it seems to go against the Copernican trend in history. Man began by taking first the Earth, then the solar system, and then our Galaxy, as central to the universe. Uniqueness of character is uncomfortably similar to spatial centrality. More modern theories have tended to start from the observation that clouds of dust are frequently observed around stars in the process of formation, and that in those cases a very small fraction of matter seems to remains behind in the cloud. From this point on, two different sorts of theory sprang up, and are both still in vogue. One supposes that giant protoplanets are formed fairly rapidly out of a fragmenting protoplanetary disk. The other supposes a slow building up of planets out of solid chunks of ever larger size. Current majority opinion inclines perhaps more to the second alternative, for which strong arguments have been marshaled by the Russian astronomer Victor Safronov, although the question is still an open one.

Astronomers gradually became disillusioned with William Thomson's age for the Sun, since it was so very much smaller than the age of the oldest rocks in the Earth's crust, not only by the old criteria, but also following estimates made from rates of radioactive decay. It became clear that there was some other, much more abundant, supply of energy available in the Sun. After the moderately wide acceptance of Einstein's special theory of relativity—which had been put forward in 1905, but which took two or three decades to work its way into establishment physics—astrophysicists began to consider possible processes for the conversion of mass into energy as an explanation for the Sun's energy output. No one appreciated this better than Arthur Eddington, who from 1917 had been working on a theory of the internal constitution of stars—within which, of course, a theory of solar energy was bound to take pride of place. Noting that gravitational contraction could account for only a hundred thousand years or so for giant stars, he repeatedly argued for nuclear processes as an energy source. In 1920, in a presidential address to the British Association—a body with a largely nonspecialist audience—he was more specific, offering the theory that the hydrogen in a star could be transformed to helium, the resultant difference in mass being released as energy radiated by the star. He also offered a suggestion that was farsighted twice over: "If, indeed, the sub-atomic energy in the stars is being freely used to maintain their great furnaces, it seems to bring a little nearer to fulfillment our dream of controlling this latent power for the well-being of the human race—or for its suicide."

Such fundamental ideas as Eddington's did not gain universal acceptance, and one of his most persistent but distinguished critics was James Jeans, who continued to argue for gravitational contraction as the main energy source in stars—until, at length, he too changed direction. Jeans even maintained that he had put forward a similar idea in the journal *Nature* in 1904—before Einstein's 1905 equation of mass with energy—but there was an element of wishful thinking in his claim. He now played with the idea that stellar energy must be the product of some sort of radioactive transformation, involving massive atoms. One by one he began to question doctrines that had been generally assumed, such as that the material in stars obeyed the gas laws of the laboratory physicist. While Jeans continued his search for suitable energy sources in the early 1920s, Francis William Aston was investigating the properties of isotopes. Aston was showing how "uranium lead," "thorium lead," and "ordinary lead," for example, differed in their atomic weights, that is, in one important identity tag, and yet were indistinguishable in their chemical properties. Jeans and others saw that in the development of stars, the transmutation of elements—even between other states than mere

isotopes—may take place on a colossal scale with the corresponding release of very large amounts of energy. Many of his ideas were short-lived, but he and Eddington gave a considerable impetus to this entire new branch of astrophysics.

Another physicist to do so was Jean Perrin, who as early as 1919 perceived that what would now be called "thermonuclear fusion reactions"— reactions in which atomic nuclei join to form heavier elements—might be the source of energy in the stars. This speculative insight, only expressed in qualitative terms, was ultimately to prove correct as an ingredient in the full story. Another two decades passed, however, before the details of such reactions could be given satisfactory form, independently by Carl Friedrich von Weizsäcker and Hans Bethe. Their work, and the studies by Eddington and others leading up to it, will be considered in more detail in chapter 16.

SUNSPOTS AND MAGNETISM

Solar astronomy in the nineteenth century was conducted, as astronomy always has been, at two different levels: the one merely observational, and the other largely theoretical. Far from being directed toward a critical assessment of theory, much observation, especially in amateur circles, was no more than a pious recording of curiosities. Even moderately acceptable theories often followed observation at a considerable distance in time, if at all, and often with a strong element of serendipity. There are many illustrations of this. In 1826, Samuel Heinrich Schwabe, an apothecary from Dessau in Germany, was hoping to discover a planet below Mercury. For this reason—and notice how closely the story parallels that of Messier and the nebulae—he registered sunspot positions, simply in order to eliminate them from his searches. Reviewing his records after twelve years of observations, he began to suspect that sunspot totals fluctuate over a period of about ten years. To make sure, he went on with his work, and in 1843 tentatively published that idea of a ten-year periodicity.

Little attention was paid to this until 1851, when Alexander von Humboldt published Schwabe's table of results with some supplementary data. In 1852, the Swiss Johann Rudolf Wolf, working first in Berne and later in Zurich, assembled all the historical material on sunspots he could discover, and gave the average period as 11.11 years. He continued to publish reports on the numbers of sunspots, almost up to the time of his death. In 1851, John Lamont, a Scottish-born astronomer who had left his native country for Bavaria in 1817, published a discovery he had made, that the Earth's magnetic field also seems to vary with a period of about ten years, and that the alternate periods are weak and strong—in short, that the full cycle is twice that of the sunspots. Immediately, Wolf and the influential Anglo-Irish geophysicist Edward Sabine, noted that, broadly speaking, sunspots do indeed follow magnetic changes, including changes in the aurorae, in all their irregularities. That there was some strange connection between solar phenomena and terrestrial effects could not be doubted, but a century would pass before any plausible explanation could be found. And despite modern advances in our knowledge of the physics of the Sun, there is still no widely accepted explanation for the regular fluctuation in sunspot numbers.

No less elusive was a theory to explain the periodic changes in sunspot behavior discovered by the wealthy English amateur astronomer Richard Christopher Carrington. Carrington was for a time a salaried observer at the newly founded University of Durham, but he resigned in order to be able to set up a private observatory of his own near Reigate in Surrey in order to complete the zone surveys of Bessel and Argelander to within nine degrees of the north celestial pole. For seven and a half years, between 1853 and 1861, he systematically and meticulously observed sunspots by a simple and accurate method, and so discovered that the period of their rotation increases with their distance from the solar equator. This showed that they could not be regarded as fixed to a rigid and solid solar object. He found that when spots were generally most numerous, they tended to approach the equator, and to become extinct around 5° latitude, at the time of sunspot minimum. He found that the first spots of the new cycle start to appear at the same time.

Prompted by such findings as these, the Royal Observatory at Greenwich, under the direction of W. H. M. Christie, at last decided to enter into a program of astrophysical observations, appointing Edward Walter Maunder as photographic and spectroscopic assistant in 1873. Maunder was one of a class of "computers" recruited from the ranks of society that did not have university training, and some of the resulting social tensions at Greenwich are no doubt to be detected in the anecdotes Maunder told of Airy's punctilious nature—for example, that he, with his staff, spent a whole afternoon sticking the label "empty" on boxes that were, indeed, empty. Maunder's professed admiration for Airy was undoubtedly genuine, especially for Airy's vision in founding the Magnetic and Meteorological Department of the observatory in the late 1830s, but Maunder could not resist protesting against Airy's draconian employment of the "computers," the "boys" who were working at the great lunar program from eight in the morning to eight at night, with an hour's rest at midday.

From 1891, under Christie, a "lady computer" in a different class was recruited to assist Maunder at Greenwich. She was Annie S. D. Russell, who had graduated from Cambridge with a degree in mathematics two years earlier. They married in 1895, after his first wife's death. With her mathematical skills, she contributed many new ideas, helping especially with the statistical analysis of the daily photographic records of the Sun for which Maunder had been responsible, records that included the positions, sizes, and movements of sunspots. One important outcome of their work was the famous butterfly-shaped "Maunder diagram," showing the numbers of spots to be found at all latitudes at any time, a diagram that became a standard means of representing the situation graphically. It was first used in 1904, to show the latitude drift of sunspots during the sunspot cycle, when Maunder was busy demonstrating the close correlation of geomagnetic disturbances with the presence of large sunspots on the solar disk.

Maunder was highly active in the Royal Astronomical Society, in which he held several offices, and with the help of his brother he founded the British Astronomical Association, an amateur organization, in 1890. The latter body was not only for amateurs but also for women, who were still not accepted as Fellows of the professional body. In an official capacity, he and his wife traveled the world widely in the pursuit of solar photographs, and in 1898, using a camera of her own design, she obtained a spectacular photograph of a streamer from the Sun's corona, far longer—six solar radii—than any previously observed. Maunder's career was unusual in that he retired before the First World War but came back to the observatory during the war, with his wife, to help out in the absence of many staff; and after the war they made a useful study of historical sources on sunspots. From this a great paucity of sunspots in the second half of the seventeenth century became obvious. During what is now known as the "Maunder minimum," generally reckoned as occurring from 1645 to 1715, a period of thirty years then might have yielded only about fifty spots, whereas a current figure for an equivalent period of time might be a thousand times as great.

The relationship between the Earth's magnetic field and the sunspot cycle was a problem that went deeper into physics than the statistical surveys made by the Maunders. It had been under investigation by several astronomers since Wolf's and Lamont's work of the 1850s. We have already mentioned Birkeland's idea that peaks of geomagnetic activity, "magnetic storms," were caused by beams of electrically charged corpuscles from the Sun. Before the turn of the century, he began a number of laboratory experiments to support his belief, his idea being that these charged particles, drawn in to the Earth's magnetic field near the poles, were what gave rise to the aurorae, the northern and southern lights. Two discoveries of great significance were now made by Maunder, and published in 1904 (this was the context of his Maunder diagram). He found that the greatest geomagnetic storms were associated with large sunspots near the central meridian of the solar disk—on average, they had passed the central meridian about 26 hours earlier. He also found that geomagnetic storms as a whole recur at intervals of about 27 days, which is the Sun's period of rotation with respect to the Earth. His conclusion was that they were due to some streaming effect from localized regions of the Sun, and that this streaming took about a day to reach the Earth. In the case of the great storms, sunspots seemed to be the cause, but lesser storms could occur when the disk was virtually blank. The 27-day cycle seemed to settle the solar origin of all of them, however. That is where matters were left for over two decades, before others—notably W. H. M. Greaves and H. W. Newton—reexamined the relationship. Maunder's results were confirmed and extended, but the question of cause long thereafter remained a mystery.

A breakthrough came at last as a result of work done with the spectrohelio-scope in the 1930s by George Ellery Hale, to whom practical solar astronomy already owed innumerable debts. Hale had invented a photographic form of that instrument in 1889 for photographing the solar prominences in nearly monochromatic light. A number of others had tried to do the same, but had failed to get good practical results. (All were effectively following the lead of Janssen and Lockyer, with their visual methods. Hale was not able to perform the more difficult task of photographing the corona of the non-eclipsed Sun. This was first done by Bernard Lyot in France in 1930.) Hale called his first instrument a spectroheliograph. The Paris astronomer Henri Deslandres designed a similar instrument to Hale's, a few months later, quite independently. The idea is simple: the telescope, which is driven to follow the Sun precisely, forms an image of the Sun on the slit of the spectrograph. Light of a chosen wavelength (that is, of a particular color) emerging from the spectrograph is then brought to a focus on a photo-graphic plate. The Sun is scanned by moving the slit across the solar image, but *only in declination*, at right angles to the celestial equator. As the im-age of the Sun moves across the slit, the photographic plate is moved in harmony by suitable linkages and using an electrical drive for both slit and plate. In this way, a series of slit images on the plate builds up into a final image of the Sun, in light of virtually a single wavelength.

Out of his early success with the spectroheliograph (fig. 220) came Hale's concern with solar research. Having graduated from the Massa-chusetts Institute of Technology, he moved back to his home in Chi-cago, where his father financed a 12-inch (30-centimeter) refractor. With this, in 1892, he achieved excellent results, for instance, photo-graphing bright solar clouds showing calcium lines and prominences round the entire disk. At the end of the century he designed a spectro-heliograph for the great Yerkes refractor, and with it he found the dark hydrogen clouds, as well as investigating the processes of circulation of calcium at various levels.

A man of great drive who was well-placed financially, Hale next per-suaded his father to pay for a 152-centimeter mirror to further his re-search. The University of Chicago would not finance its mounting, and the disk remained unused for twelve years, until in 1908 it was set up on Mount Wilson, above Pasadena in California, as part of what was then the largest reflector in the world—but not for long. Even before this date, he had persuaded a wealthy Los Angeles businessman, John D. Hooker, to pay for a 254-centimeter mirror, and this was eventu-ally made and mounted in a telescope built with funds from the Carn-egie Institution of Washington. The Hooker telescope was completed in

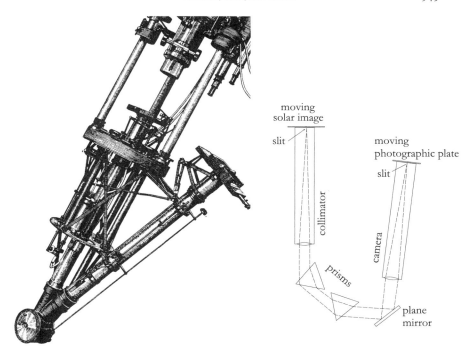

220. G. E. Hale's spectroheliograph, mounted on a refractor bought by his father, together with a schematic plan of the instrument. A solar image falls over the slit of the spectrometer, and a series of linkages shuttles the slit back and forth while doing the same for the exit slit at which the photographic plate is exposed. The number of prisms in a train was often six or more. The result is an image of the Sun in light of a single wavelength. It was found that using different spectral lines (that is, light of different wavelengths) produced substantially different images, for they relate to activity at different depths in the solar atmosphere.

1917. These two telescopes, the "sixty inch" and especially the "hundred inch," became household names throughout the world of popular science in the period following the First World War.

No sooner had the "hundred inch" been proved successful than Hale was planning something more. In 1928 his plans were more or less outlined, and he was able to raise six million dollars from the Rockefeller Foundation for a 200-inch (5-meter) telescope. This was donated to the California Institute of Technology—an institution that already owed much of its greatness to Hale—and was set up on Palomar Mountain in Southern California. Hale died before it was finished, and the Second World War interrupted progress. The mirror blank was cast by 1934 and shipped to Pasadena in 1936, but only finished and installed in 1947. Dedicated in 1948, the instrument was with some justice designated "the Hale telescope." It was an extraordinary engineering project. The tube assembly, for instance, weighs 520 tonnes, and the dome, which is 41 meters high, weighs almost 1,000 tonnes. The telescope was the first to allow the observer to ride at the prime focus. It is worth noticing that the Corning Glass Works, where the Pyrex blank for the mirror was cast, worked their way up to a 200-inch disk through disks of 30, 60, and 120 inches, and that the last of these, used for testing the 200-inch, eventually became the main mirror of the Shane reflector at Lick.

At the beginning of the century, ambitions of other sorts had occupied Hale's horizon. In 1904, he had received $150,000 from the recently founded Carnegie Institution to set up a solar observatory on Mount Wilson. The materials for this new observatory were transported under great difficulties up the mountain, by burro and mule, and the whole enterprise soon became for many a symbol of a type of astronomy done by pioneers, living and working in cabins and bivouacs outside the reach of courts and cities, academies and universities. With the Snow telescope brought from Yerkes Observatory, a solar telescope driven by a coelostat, Hale obtained in 1905 the first photograph of a sunspot spectrum. With colleagues in his mountain laboratory, he found that the spectral lines that are strong in sunspots are those that are strong in laboratory sources at relatively low temperature, so that sunspots must be—as many had long suspected, but not proved—cooler than neighboring regions of the disk.

The Snow telescope suffered from distortions due to solar heating, so in 1908 Hale designed a second telescope, and in 1912 a third, still larger, each in the form of a tower topped by a two-mirror guidance system, which was arranged so as to feed its image into the telescope—of 18 meters, then 45 meters, in focal length—and thence into an underground spectrograph—first 9 meters in length, then 22 meters. With the first he was able to detect vortex motions in the hydrogen clouds (flocculi) near sunspots, and he felt sure that they could be the source of magnetic fields, and that the widening of the lines in sunspot spectra must be due to those magnetic fields. Certain double lines had been seen earlier in sunspot spectra, but misunderstood.

Perhaps Hale's most inspired single discovery was that what he was observing here was an example of the "Zeeman effect," the splitting of spectral lines that occurs when light is passed through an intense magnetic field. The effect had been observed by the Dutch physicist Pieter Zeeman in his laboratory in Leiden in 1897. In Hale we have an early example of an astrophysicist trained in university physics, who was therefore able to make such associations easily. In this case, he followed with a study of the polarity of sunspots, the orientation of their magnetic poles, and so was led to the discovery that at the end of the 11-year sunspot cycle the polarity reverses: the true periodicity could therefore now be stated as one of 22 or 23 years, as others had earlier and more vaguely speculated.

Hale's 1912 instrument was aimed at solving another problem, that of the general magnetic field that the Sun was thought to have, as judged, for instance, by the shape of its corona seen during an eclipse—something to which Frank H. Bigelow drew attention after observing an eclipse in 1889. The difficulty of observing such an effect was very great indeed, for a weak magnetic field would be expected to cause a displacement of lines on the spectrographic plate of less than a thousandth of a millimeter. It would be another sixty years before it was directly measured—in 1952, after Harold and Horace Babcock invented the solar magnetograph at the Hale Solar Laboratory in Pasadena. As for Hale's ambitions, what of Birkeland's hypothesis of an inflow of charged particles from the Sun? Could the Sun's magnetic field drive such particles? Hale, with a number of colleagues, did what they could to measure the field, but so small was the effect that their results varied widely, and they made no real progress. The solar magnetic system is enormously complicated, and we have already seen how a knowledge of its character was gradually acquired by indirect means. An understanding of it was built up gradually, after a series of seemingly disparate problems: fade-outs in radio and telephone communication; Chapman and Ferraro's calculations of the speed and likely effects of ions thrown out by the Sun; Forbush's analysis of a weakening of cosmic radiation during magnetic storms, as though a magnetic field was blocking the path of cosmic rays from the Galaxy and perhaps beyond; Biermann's study of cometary tails; and only in the late 1950s was the first satisfactory covering model arrived at by Parker. It is now known that there is a complex three-dimensional whirlpool structure to the solar magnetic field in which the entire planetary system is immersed. Kepler would have been delighted. The field is dragged, as it were, by plasma ejected from the corona. The electrically charged particles themselves move in tight spirals, like springs wrapped round the lines of force. Hale was simply born too early.

As though to prove that the rocket programs of the superpowers were drafted primarily with astronomy in view, space probes now came opportunely to the aid of solar physics. The first probes on board Soviet spacecraft *Luna* 2 and *Luna* 3 provided confirmation, in 1959, of what had been painstakingly pieced together in the course of more than seventy years, about the outpourings of the Sun. As we shall see in chapter 19, a year before this, the first intimation of the Van Allen belts of charged particles around the Earth had been given by the U. S. vessel *Explorer* 1. Eugene Parker's model was well confirmed in outline by a whole series of tests. Many properties of the "solar wind" became the object of research, for example its relationship with the Galaxy in which it is embedded. The mechanism for heating the corona is, of course, an aspect of the overall model of the active Sun. How, for instance, are the complex patterns of sunspot behavior to be explained? There have been many explanations for the "solar dynamo" that powers the system, but the foundations for them all were laid in the late nineteenth century, by a relatively small number of people with a sound knowledge of fundamental physics.

THE STRUCTURE OF THE SUN AND THE SCHWARZSCHILD RADIUS

In no other branch of astronomy had so many disciplines been called upon as in the study of the Sun in these early years. As the solar scheme was pieced together, it was found that the sunspot cycle coincides with fluctuations in growth in plants, as evidenced by tree rings, radioactive dating by carbon-14 residues, silt deposits, fish stocks in the oceans, and so on. It became evident, on the basis of geological evidence, that the sunspot cycle has persisted for 700 million years, a fact of obvious significance to any theory of the character of the Sun. Astrophysics, which is rarely in a position to experiment with its objects, must always be heavily dependent on laboratory physics, which usually can. On the other hand, there is a reciprocal exchange, since there are situations in the cosmos that are unrepeatable on Earth. As Hale pointed out when canvassing support for the 200-inch reflector, the extremes of mass, density, pressure, and temperature available in the heavens so far transcend those of the laboratory that "many of the most fundamental advances in physics depend upon the utilization of these conditions." In this he was peering into the future. At the end of the nineteenth century, what both physics and astrophysics sorely needed was something nearer at hand, and in particular, a sound theory of radiation.

Such a theory slowly emerged, with a succession of physical laws that are now a standard part of physics, even though they have known limitations: Stefan's law (1879), Wien's law (1893), and—comprehending them all—Planck's formula (1906). Wien's law states that the wavelength at which blackbody radiation reaches maximum intensity is inversely proportional to its absolute temperature. This law allows meaningful statements to be made about the Sun's surface temperature, which had previously been assigned wildly differing values. Planck's formula took more detail into account, relating as it did radiation, temperature, and wavelength. Using these laws, the new figure for the temperature of the photosphere was about 6,000°K (degrees Kelvin can be roughly defined as degrees Celsius from an absolute zero of about −273°). The temperatures of the corona and chromosphere were much more difficult to chart, and in estimating them, little of importance was achieved until techniques were developed for receiving radiations outside the visual range—radio waves, ultraviolet rays, X-rays, and so on. When this was achieved in the second half of the twentieth century, the great surprise was that, in the transition zone between the chromosphere and corona, which is only a few hundred kilometers in thickness, the temperature increases dramatically, from 10,000°K to about 1,000,000°K.

In the absence of an acceptable theory of radiation, astronomers fought shy of speculation about the interior of the Sun. Exceptions to the rule were August Ritter and Robert Emden, mentioned above. Emden's model made use of heat transfer in stars—supposed gaseous—by conduction and convection. The fact that the Sun's surface is granulated had been known since the 1860s, and his interpretation of the granulation as the visible parts of convection elements was essentially correct. The most remarkable studies of the structure of stars made in the early years of the century, however, were those by Karl Schwarzschild, the most talented German astrophysicist of his generation.

Born in 1873 in Frankfurt-am-Main to a Jewish family, Schwarzschild took his doctoral degree *summa cum laude* in 1896, with a dissertation applying Poincaré's theory of stability in rotating bodies to a variety of pressing problems, among them the origin of the solar system. His interests gradually focused on a subject that had never been given the attention it deserved, namely, stellar photometry, the measurement of the radiant energy received from the stars. Apart from the introduction of some rough photographic techniques—which Schwarzschild supplemented and refined—measurement was chiefly done as it had always been done, estimating and comparing brightnesses with the human eye. Working in Vienna, Schwarzschild applied his photographic methods to the magnitudes of 367 stars, some of them variable; and in following one variable star, η Aquilae, he found that the range of change in its magnitude was much greater when estimated photographically than when judged by eye. He realized that this was an indication that its surface temperature fluctuates, an important discovery relating to a type of star—Cepheid variables—that was soon to assume enormous importance in the development of astronomy.

In June 1899, Schwarzschild returned to Munich, and in 1901 he moved on to Göttingen, where he was director of the observatory built and fitted out by Gauss. Here he continued his photometric work, and when visiting Algeria to observe the total solar eclipse of 1905 he obtained a remarkable series of ultraviolet photographs of the solar spectrum—sixteen in thirty seconds—that led him into a study of the transfer of energy in the neighborhood of the Sun's surface. Discarding the old cloud theory of the photosphere, he took the Sun to be layered, and considering the net energy absorbed from below, and that emitted upward, he was led to a series of equations, amounting to what became known as the "Schuster-Schwarzschild model for a gray atmosphere." The two men had arrived at the model independently, Arthur Schuster in 1905 and Schwarzschild a year later. In this model, temperature and the density of solar matter increase with depth. Later practical investigations showed that this did not account well for the flow of energy, and not long after Schwarzschild's death in 1916, it was abandoned in favor of other solar models, first that proposed by E. A. Milne in 1921, and then by A. S. Eddington's of 1923.

In 1909 Schwarzschild became director of the Potsdam Astrophysical Observatory. (He married in the same year, and one of his sons, Martin Schwarzschild, became a well-known American astronomer.) After volunteering for the army and serving in various scientific capacities in Belgium and France, he moved to the Russian front. It was there that he contracted a disease and died in 1916, but it was also from there that he wrote two papers on general relativity that were to become his most lasting monument. The first concerned the gravitational effect of a point mass in empty space, on Einstein's theory (it was the first exact solution of Einstein's "field equations"). The second paper concerns the gravitational field of a uniform sphere of material—not a model true to the Sun, of course, but an important beginning. Again he found an exact solution, and this time it was one with a very surprising property. This concerns a certain distance from the center of the sphere, "the Schwarzschild radius," which is related to the mass of the sphere in a very simple way. If a star collapses under gravitational forces so that its radius becomes smaller than this critical Schwarzschild radius, it becomes incapable of emitting radiation. It becomes a black hole, a concept of which we shall say more in the last chapter. The Sun would constitute a black hole if it shrank to a radius of 2.5 kilometers.

Schwarzschild's result had an important consequence: since the critical radius is proportional to the mass, the horizon around a collapsed star (or black hole) can be on any scale, as long as the condition of compactness is satisfied—say the mass of a mountain in the volume of an atom, the mass of the Earth in an aniseed ball, the mass of the Sun in the volume of an asteroid, the mass of a small galaxy in the space of the solar system, and so on.

The idea of a star that cannot emit radiation and so cannot be seen, by virtue of the high concentration of its mass, has interesting parallels in the eighteenth century—although it would be foolish to pretend that they are in any deep sense anticipations of what came later. In 1772, Joseph Priestley discussed in print unpublished ideas by John Michell, ideas that Michell presented in a paper to the Royal Society in 1783. The central idea was that, just as any ballistic missile can leave the gravitational field of the Earth only if it is projected with a sufficiently high velocity, so too can light only leave the Sun if it has a sufficiently high velocity, a velocity that is easily calculated. In fact, Michell computed that the known velocity of light was 497 times greater than was needed for escape; or that light from a star with a radius 500 times that of the Sun would not escape from it. Again, in 1791, William Herschel suggested that the nebulous character of the matter he could see through his telescope might be explained by the gravitational opposition to light in its attempt to leave or to pass gravitating matter, that is, making its escape difficult. And then, best known of all, came the statement by Laplace in 1796, in his *Exposition du système du monde*, to the effect that any star with the density of our Sun, but with diameter 250 times as great, will be able to recapture all of the light it radiates. In 1799, Laplace published the calculation, which was new only in its details. He dropped the topic from the third edition of his book, perhaps because he appreciated a difficulty it encounters: if light is like any ordinary projectile, it loses speed as it is shot away from a gravitating body. The speed of light, however, was generally supposed to be constant.

These isolated ideas can hardly be said to have led toward anything approaching modern theories of black holes, but—at least after 1844—anyone who thought about the problem of detecting a massive body from which light could not leave did not need to give up hope. It was in that year that Bessel was able to detect the invisible companion to Sirius by its gravitational effects. In this case, admittedly, the companion was directly observed in 1862 by Alvan G. Clark—it is now classified as a white dwarf, Sirius B—but the principle was plain enough.

STELLAR DISTANCES WITHOUT SPECTROSCOPY

The use of the spectroscope to study the Sun slowly yielded knowledge of its structure and composition. In due course, the instrument was applied to the stars, allowing the findings of solar physics to be extended and also providing distances and velocities. In this way the spectroscope provided a key to the structure of the universe at large. I shall first explain how spectroscopic methods were related to other ways of determining distance. We have already encountered Bessel's stellar distances, "annual parallaxes," which were trigonometrically obtained. Unfortunately, the method is of use only with the nearest stars, say, up to 100 parsecs, beyond which the annual displacement is too small to measure. Beyond this limit, the method based on proper motions could be used. Herschel and later astronomers knew the Sun's motion in space, and its direction, both reasonably accurately. This motion gives nearby stars a large proper motion and distant stars a lesser one, judged against very distant stars, just as nearby objects seen from a train appear to move faster than distant objects. Of course the objects in question, like the stars, may have other motions of their own, but if we use suitable averaging procedures, and independent reasoning as to what these other motions might be, we can estimate distances from proper motions, given the Sun's velocity.

This last method, bearing in mind the necessary averaging, is called the method of statistical parallaxes. It was applied by several astronomers before Kapteyn, but at the very beginning of the twentieth century he used it with great effect to ascend one rung higher on the distance ladder. From the proper motions at hand, he analyzed the relative frequency of stellar (absolute, intrinsic) magnitudes in the Sun's neighborhood. (Absolute magnitude is the magnitude a star would be judged to have were it at a certain standard distance. From a star's actual distance, and its apparent magnitude, its absolute magnitude could be found. By convention, the standard distance is ten parsecs, that is, a distance equivalent to a parallax of a tenth of a second of arc.) Assuming that the same proportions would be found elsewhere, Kapteyn could examine groups of distant stars, and give probable values to their brightnesses, and so, statistically, estimate their distances. Even then, however, the method would not take him out to great distances. The key to further expansion came unexpectedly from spectroscopy.

THE NEW SPECTROSCOPY AND STELLAR EVOLUTION

Stellar spectroscopy had proceeded by fits and starts after Fraunhofer first described lines he had seen in 1814 in Sirius, Castor, Pollux, Capella,

Betelgeuse, and Procyon. There were isolated attempts by others to observe the stellar lines—for example, by J. Lamont in the late 1830s, W. Swan in the 1850s, and G. B. Donati in the early 1860s; and then came a flurry of activity in 1862–1863, with the work of Lewis M. Rutherfurd (an American amateur), Airy, Huggins, and Secchi. Out of Airy's work there developed the Greenwich program for measuring star velocities through the Doppler displacements in their spectra. Rutherfurd's and Secchi's publications were of importance in another way, for they turned astronomers' thoughts to the *classification* of stars through their spectra. Rutherfurd's was a simple threefold division: stars with lines and bands like the Sun, white stars like Sirius with very different spectra, and finally white stars with apparently no lines—"perhaps they contain no mineral substance," he wrote in a paper in *Silliman's Journal*, "or are incandescent without flame." Secchi's first classification, published marginally later but introducing more specific spectral criteria, distinguished two classes; in 1866 he decided on three; and in 1868, at a meeting of the British Association for the Advancement of Science, he described four classes. Briefly stated, these were: (1) stars like Sirius, whitish or bluish, with dark bars due to hydrogen, and faint lines due to metals; (2) stars of solar type, like Capella and Arcturus, with spectra strong in the middle (yellow) area, and many dark lines; (3) red, often variable, stars, like Betelgeuse and Mira, with spectra showing broad bands, brighter at the red end, and generally of a fluted appearance; and (4) stars, not numerous, of low brightness, but redder than the third type, which they resembled somewhat. Secchi made further adjustments to these divisions until his death in 1878. Such schemes as his led a number of astronomers to speculate on the possible evolution of stars. In 1865, for example, the Leipzig astronomer Friedrich Zöllner, in an important book on the measurement of stellar brightness, proposed that stars were created with a high temperature, but cooled as they passed through such stages as were described by Secchi, finishing up as red stars.

In the collection of stellar spectroscopic data, especially photographically, Huggins was long preeminent, although increasing numbers of astronomers were entering the field, including many who were excited by his study of the spectrum of the nova of 1866. New spectral types were proposed. The astronomers Charles Joseph Étienne Wolf and Georges Antoine Pons Rayet of the Paris Observatory, for example, announced in 1867 that they had found three very faint (eighth magnitude) stars in the constellation of Cygnus, with several broad emission lines on a continuous background spectrum. Then, and indeed now, this spectrum seemed very peculiar: it is now known that the Wolf-Rayet stars are as much as ten times or twenty times as massive as the Sun, and are surrounded by an envelope of matter, matter that they are ejecting at very great velocities, of the order of a thousand kilometers per second and more. They are young stars at an evolutionary stage that seems likely to be of relatively short duration—a hypothesis that would explain their rarity. Only two or three hundred are known.

New schemes for classifying stellar spectra multiplied, as the century wore on, but what was conspicuous by its absence was any deep, unifying principle, such as might have been provided had more been known of the physics of stars and their production of light. The man who came nearest to this insight was Hermann Carl Vogel, who in 1870 was appointed director of the privately owned Von Bülow Observatory near Kiel in Germany, which then sported the largest refractor in the country. One of his first memorable achievements was to measure the rotation of the Sun—already evident, of course, from sunspot movements—by measuring the Doppler shifts in the spectra of light from its approaching and receding limbs. With his colleague W. O. Lohse, Vogel next began a survey of spectra of visible stars. They tried to determine the radial velocities of the stars, on the Doppler principle, but without clear success. When a glittering new observatory was established at Potsdam, jointly directed by G. R. Kirchhoff, Wilhelm Foerster, and Arthur von Auwers, Vogel was appointed to a post there, and began to apply photometric methods to the spectra of stars. In 1876 he had a striking success, when he was able to show changes taking place in the spectrum of a fading nova. It was then that he began a study of the solar spectrum, improving on earlier mapping of the solar lines by Kirchhoff and Ångstrom, but soon being obliged to bow to the superiority of a similar study by the American physicist H. A. Rowland in the 1880s. (This was the man responsible for the "Rowland diffraction grating" known to all physicists, and which for a wide spectral spread is far superior to prisms.)

At this point Vogel turned back to his studies of stellar spectra, now with a view to classifying the stars, ultimately hoping to understand thereby their evolutionary track. In 1883, he published his observatory's first catalog of stellar spectra, in the course of analyzing which he discovered "spectroscopic binaries," binary systems with components that cannot be resolved visually, but that can be detected from the mingling of two sets of spectral lines, shifting with the Doppler shift. Even when only a single spectrum is seen, from the brightest component, in the case of a binary the spectral lines will move around periodically. Vogel did not only measure the positions of spectral lines, he also devised a clever photometric method; he assessed the brightnesses of lines by comparing results when altering photographic exposure times. By painstaking analysis, he showed spectroscopically that the bright stars Algol (β Persei) and Spica (α Virginis) are eclipsing binaries, and in 1889 was able to announce plausible data for the total mass of each pair and the separations of their components. (In the same year, E. C. Pickering of Harvard College Observatory had noticed spectral shifts in the star Mizar, ζ Ursa Majoris.) And then at last, between 1888 and 1892, Vogel was able to do what he had tried and failed to do at the Von Bülow Observatory: he found reliable Doppler shifts for fifty stars—far more, and more certain, than anyone had found previously.

The spectrographic catalog from Potsdam went from strength to strength, and its detail was exemplary. Its fame in the long term has far outstripped that of Vogel's spectral classification. Even a second version he produced was disappointing, although there he was able to take into account the helium lines that were recognized after Lockyer's discovery of the element. Vogel's schemes are mostly of interest for their emphasis on subtle variations in the character of the hydrogen lines, and later the helium lines, which would eventually be brought into the evolutionary picture in no uncertain way. When further progress came, it was not from Potsdam. It came rapidly, from the land of mass production, by the application of a technique that seems in retrospect to be astonishingly simple.

In 1886 the widow of the New York physician and astronomer Henry Draper had established a fund to support the photography, measurement, and classification of star spectra and the publication of results, as a memorial to her late husband. Edward Charles Pickering, who had been trained as a physicist, had been director of the Harvard College Observatory since 1869, and now funds were allocated by Mrs. Draper to him and his colleagues to photograph, measure, and classify stellar spectra and publish the results as a memorial to her late husband. Pickering now devised his new and simple technique for producing spectrograms of many stars at once: he placed a large narrow prism of low dispersion in front of the object lens of his telescope, so that each star created a tiny spectral strip on the photographic plate instead of a point image. A slight loss of definition was more than compensated for by the mass production of spectra that the arrangement made possible. The spectra still required measurement and analysis, of course, but *The Henry Draper Catalogue* was substantially ready before the end of the century. Publication of the main work, in nine volumes, appeared only between 1918 and 1924, but that was largely because it was being combed meticulously with a view to deriving a classificatory scheme. For an astrophysical, as opposed to an astronomical, catalog, it was on an unprecedented scale, including as it did nearly a quarter of a million stellar spectra.

There were many changes of direction in the technique of classifying the resulting spectra. At first they were classified by the strengths of hydrogen absorption lines, and an alphabetical series of labels (A, B, C . . .) was then used. This sequence seemed to make no sense of the absorption lines for other chemical elements. Pickering was assisted on this project by Williamina P. Fleming, Antonia Maury (Draper's niece), Annie Jump Cannon, and more than a dozen women who helped with the computing. Mrs. Fleming was a Scot who had been taken on as second maid in Pickering's household. Impressed by her accuracy and perseverance, he engaged her to work part-time at the observatory in 1879, and from 1881 until the year of her death in 1911 she was a staff member (fig. 221). Useful though she was, it was Annie Cannon, who joined the team in 1895, who was to find a way of rearranging the spectra—subdividing some, and reclassifying types C and D—so that the progressive changes in the lines seemed orderly over more or less all types. This resulted in a disorderly jumble of the old letters,

221. E. C. Pickering and his "computers" outside the Harvard College Observatory. May, 1913. Annie Jump Cannon is in the back row, fourth from the right.

but it was too late to change them easily, and the sequence that emerged became the basis of the one that is still used. The old sequence had become: O, B, A, F, G, K, M, R, N, S. Attributed to Henry Norris Russell was a useful mnemonic device, one version of which is "*Oh Be A Fine Girl, Kiss Me Right Now, Smack.*" Circumspect astronomers now only speak it inwardly, if at all, and never refer to "Pickering's harem."

Already by 1901, Annie Cannon was able to publish spectra of well over a thousand brighter stars; in the nine later volumes of the *Draper Catalogue* she listed 225,300 spectra, many of them belonging to stars as faint as magnitude 10. She developed rare skills in rapid classification and in spotting peculiarities, but she did not develop any theory to explain the sequence she detected. Several astrophysicists realized very quickly that surface temperature was somehow related to the sequence, with O-type stars the hottest, and Antonia Maury was the first to appreciate that even within a given type there were various possible differences of line width in the spectra. Here were the raw materials for an extremely important discovery, made almost simultaneously by two astronomers working elsewhere, Ejnar Hertzsprung and Henry Norris Russell.

A Danish astronomer who had first been trained in chemical engineering, Hertzsprung was later to work in Germany with Karl Schwarzschild, both at Göttingen and Potsdam. His visit to Mount Wilson in 1912 was to prove particularly important, but it was in two separate papers, published in 1905 and 1907 in a journal devoted to photography and its allied chemistry, *Zeitschrift für Wissenschaftliche Photographie* that he showed how Antonia Maury's line widths could be related to the brightnesses of stars. Using proper motions to determine distance, and thence intrinsic brightness (that is, absolute magnitude, for which see p. 555 above), Hertzsprung showed that her c-type stars, with sharp and deep absorption lines, were more luminous than the others. He thus developed the idea of "spectroscopic parallax," the idea that line width generally could be correlated with absolute magnitude, so that the latter could be found directly from a simple observation of the former. Since Maury types a, b, and c were subdivisions of a given spectral type, all stars of which were taken to be at the same temperature, he drew the conclusion that what made some (c-type) brighter than others was their actual physical size. In effect, he had discovered giant stars, hidden in the data produced and duly analyzed at Harvard.

Hertzsprung went further and decided that stars could be divided into two series, one now known as the "main sequence," and one a sequence of high-luminosity giant stars. His first diagram of this was done in 1906 for the stars of the Pleiades cluster, and was not known in America, where remarkably similar ideas were being developed by Russell.

The "method of spectroscopic parallaxes," as now generally understood, still involves using a form of the Hertzsprung-Russell diagram for stellar distances (or parallaxes), but follows a slightly different procedure. Knowing a star's spectral class (or color index) puts it somewhere on a vertical line (or within a vertical band, if we are to be realistic) on the diagram, but the diagram was eventually found to require the addition of many different branches—covering, for instance, red supergiants, giants, and white dwarfs, as well as main sequence stars. We need to know the star's luminosity class, if we are to choose the correct branch, after which we can read off the appropriate curve (this too is a wide band rather than a neat curve) a rough value for the star's absolute magnitude (M). Knowing its apparent magnitude (m) from measurement, the usual formula $m - M = 5 \log_{10}(d \div 10)$ gives the star's distance (d) in parsecs.

After studying astronomy at Princeton, Russell worked for a time in physical laboratories in London and Cambridge (England), as well as at the Cambridge University Observatory. There he worked with Arthur Hinks on determining stellar parallax photographically, and he continued this work after returning to a post at Princeton in 1905. By 1910 he had assembled a mass of data, allowing him to correlate spectral type with absolute magnitude, as Hertzsprung had done. The branching graph showing the correlation, now known as the Hertzsprung-Russell diagram, was not generally known until Russell presented his results to the Royal Astronomical

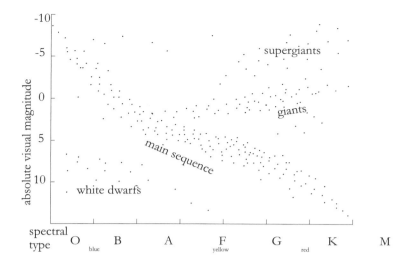

222. A simplified outline of the main elements of a Hertzsprung-Russell diagram, correlating the spectral types of important classes of stars with their absolute magnitudes.

Society in London in 1913. It is conventional to refer to this as the H-R diagram, and we shall generally use the convention here, for the sake of brevity. For versions of it, see figures 222 and 223.

With his version of it, Russell gave an interpretation that differed, as it was to turn out, from Hertzsprung's. Both took the diagram to show stars at different points in a general pattern of evolution. Russell thought that stars started out as red giants that heated up as they contracted to bright blue stars; and that they then cooled without much further change in size. In his address to the Royal Astronomical Society, indeed, he took it as read that "almost everyone will agree that a star contracts as it grows older," from which he concluded that his red giants were at an early stage of evolution. For the next decade, Russell and most other astronomers, but not Hertzspung, tried to establish an evolutionary scheme in which the gravitational contraction of stars was the overriding influence on their development. Not for another thirty or forty years was it widely accepted that stars could expand against gravity, and that giants stars were not young stars on their way to join the "main sequence," but old stars that had come from the main sequence. Hertzsprung at first treated the double graph as indicating two different evolutionary paths. The theoretical study of these questions very soon made great progress, when it passed to a more mathematically en-dowed class of astrophysicists. The most talented of these was Arthur Stan-ley Eddington, who had heard Russell deliver his 1912 paper in London.

Eddington, educated at Manchester and Cambridge, was an excellent mathematician with a sound knowledge of observatory practice that he had obtained when he worked for a time at Greenwich. From 1913 until his death in 1944, he was Plumian Professor of Astronomy at Cambridge, from which place he acted as an incomparable stimulus to world astrophysics.

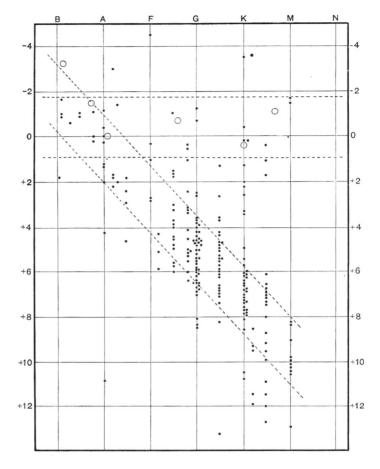

223. Russell's earlier version of the previous figure (1913) as reproduced by Eddington in 1914. Hertzsprung and Russell both realized that stars of widely different luminosity could have the same spectral type, from which came the hypothesis of "giant" and "dwarf" stars—an idea which others later elaborated.

He now took as his theoretical starting point Schwarzschild's theory of the outer atmosphere of a star, which explained how the outward pressure due to radiation could be in equilibrium with the inward pressure due to gravitation. Eddington also took gas pressure into account, and extended the study to all depths within the star. The "Eddington model" turned out to have some unexpected properties. So rapidly did radiation pressure increase with increasing mass that Eddington decided that stars with masses ten times greater than the Sun's would be relatively rare.

The discovery of a relationship between the masses of stars and their luminosities was one of the keys to an understanding of the properties of stars, and it came at an opportune moment. The first to argue that masses are correlated with spectral type, and hence with luminosities, was Jacob Halm, in Edinburgh. A convincing empirical survey of main sequence stars was published by Ejnar Hertzsprung in 1919. He interpreted his empirical relationship as indicating a law according to which luminosity increases with the seventh power of the mass. (Later studies made it a fourth-power relationship.) By 1924, still assuming that stars could be treated as gaseous, rather than liquid, spheres, as Jeans believed, Eddington was able to publish a theoretical relation between the mass and luminosity of a star. By then it was known that dwarf stars could have very high densities, and many agreed with Jeans that they at least could not be gaseous, but they fitted Eddington's model for giant stars so well that he decided otherwise. (It is ironic that his assumption that stars could maintain their perfect gas state throughout their lives rested heavily on Jeans's arguments for a high degree of ionization in stellar interiors.) With data from George Ellery Hale and Walter S. Adams at Mount Wilson, Eddington's model triumphed over widespread skepticism when he applied it to the companion of Sirius, to which he assigned an extraordinary density, 50,000 grams per cubic centimeter. His theory was enlarged upon by Heinrich Vogt in 1926, and in that same year was followed by Ralph H. Fowler's studies of the superdense gas—or plasma, as it would now be called—using ideas drawn from the new branch of physics known as wave mechanics.

In 1926, Eddington presented his ideas in his book *The Internal Constitution of the Stars*, which will be mentioned again later in connection with other ideas of his on stellar evolution, and with the work of his student Chandrasekhar. It was natural enough to do what Russell had done, which was to interpret the H-R diagram as a template of stellar evolution. On some hypotheses, this seemed to imply a time scale for evolution unparalleled in other astronomical theories, a time scale of the order of a million million (or, a trillion) years. It seemed now that a massive star, say of type O or B, would require as long a time as that to bring it down to the mass of a white dwarf. This whole question was one that would later be used to link the theory of stellar evolution with theories of the age of the universe. Eddington, one of the leading exponents of Einstein's theories of relativity, out of which so many new cosmological ideas were springing forth, was no stranger to the equivalence of mass and energy, and as early as 1917 had developed a theory of the subatomic origin of stellar energy (electron-proton annihilation). He later developed alternative explanations (especially electron-positron annihilation), and lived to see them incorporated in a solution to the problem that was later found by Hans Bethe and Carl von Weizsäcker in 1938, namely the CNO cycle (or carbon-nitrogen-oxygen-carbon cycle (for which see p. 609 below). The topic of stellar constitution will be discussed further later in the chapter (p. 592). It is one that now seems to belong to the very heart of astronomy, but in its early stages it appeared to many astronomers as a strange island with a culture of its own.

In the following pages there will be numerous references to the H-R diagram—or rather class of diagrams, since it underwent several changes in style and name. While these will be generally ignored here, the main alternatives are worth mentioning. The original diagrams were seen to display the spectral types of stars on the horizontal axis and their absolute magnitudes on the vertical axis. Since spectral type is open to some ambiguity, it was later generally replaced by a numerical parameter for color, the "color index"—hence the alternative name "color-magnitude diagram." This parameter, however, could also be ambiguous. Originally, "color index" referred to the difference between magnitudes assessed visually and magnitudes assessed photographically (photographic plates were predominantly blue-sensitive). As photometry improved in the mid-twentieth century, color filters were employed to allow comparisons between magnitudes measured at different but fairly well-defined bands of wavelength. Here, a quantity "B–V" is frequently quoted without further comment, B and V being meant to correspond approximately to magnitudes over wavelengths predominating in photographic ("blue") and visual work. The B–V color index (put on the lower axis of the diagram) was could be measured quickly and easily, and was a fairly reliable indicator of spectral type. Another concession to ease and speed is when apparent magnitudes are plotted, rather than absolute magnitudes. This creates a diagram based only on observables, but is usually meant for later adjustment. In another related type of diagram, the temperature of stars is plotted against their luminosities. Effective temperatures are derivable from B–V. Temperature-luminosity diagrams are often used when modeling stellar evolution. Needless to say, the exact transformation between one type of diagram and another is not trivial, and depends on many other (often highly problematical) parameters.

SOME EARLY STUDIES OF VARIABLE STARS

At a time when astronomers so desperately needed techniques for determining distances, yet another came to hand unexpectedly, from work at Harvard on the variability of the brightness of stars. This subject, today so important, hardly existed before the later eighteenth century, although there had of course been the new stars of 1572 and 1604, as well as a handful of relevant discoveries in the seventeenth century. Sporadic though these were, they illustrate well how certain threads of research, which for centuries are considered to be unrelated, can eventually be drawn together to great advantage.

The first significant discoveries by Western astronomers concerning variability in the brightness of stars were made by two men who would have counted themselves Frisians. The first was David Fabricius of Esens, in East Friesland, then an independent county, but now part of Germany. In August 1596 he noticed the long-term variability of a star in the constellation of Cetus (the Whale). Within a few months, he saw that it had faded from a third magnitude star to invisibility. The star in question is not in Ptolemy's catalog, but has been known as o (omicron) Ceti since Johann Bayer's catalog of 1603. Fabricius noted that in 1609 it had resumed third-magnitude status, and assumed that it was some sort of nova, an opinion shared by some astronomers as late as the end of the nineteenth century. The strange star was observed again by A. G. Pingré in 1631, but was generally ignored until its rediscovery in 1638 by Johann Fokkens Holwarda, of Franeker in Friesland—the province of that name in the northern Netherlands. Holwarda was able to announce that it varied in brightness periodically, with a period of eleven months. (Its period is not entirely regular, but is indeed approximately 332 days.) Alas, Fabricius did not live to learn of Holwarda's findings, for in 1617 he had the misfortune to be murdered with a shovel by a local peasant, whom he had accused from the pulpit of stealing geese.

Johannes Hevelius later named this star Mira, "Wonderful," and the word is now used to name a class of long-period variables. (It is in fact a giant of spectral type M, but there are other M stars that are dwarfs.) Its variability was studied by the French Catholic priest and astronomer Ismael Boulliau, who in the 1660s set its period at 333 days, and conjectured that it was a rotating "half sun," a star with extensive dark patches. In view of the rotation of the Sun, and the phenomenon of sunspots, this type of hypothesis still found plenty of support during the next two centuries, as we shall see.

Another Mira type variable, χ Cygni, was found in 1687 by Gottfried Kirch, who had worked briefly for Hevelius in 1674, and whom we met earlier as a dogged comet hunter. The large amplitude of the variation of naked-eye examples of Mira type variables makes them relatively easy to spot, and it is unlikely that more will be found in the future. Telescopic examples are another matter. By 1896, a total of 251 Mira type variables had been discovered, mostly by photography, and by the end of the twentieth century, more than 6,000 had been recorded.

Many claims were made later in the seventeenth century for the discovery of further variable stars, but when a skeptical Edmond Halley reviewed the evidence in 1715, he could add only three more, all in the constellation of Cepheus. He did not include Algol (β Persei), which had been noticed in 1670 by Geminiano Montanari in Bologna. (The argument that its Arabic name means "Demon," and that therefore its variability must have been noticed long before, does not hold water.) What Montanari had not appreciated, however, was the fact that Algol's variability was periodic, and that its period was short—under three days. The way in which its periodicity was discovered reflects in an interesting way on a growing class of relatively wealthy amateurs who—while remote from the great European centers, as were Fabricius and Holwarda—could afford the finest of instruments and had the leisure to use them.

Edward Pigott was not the discoverer of the truly periodic nature of light from Algol, but he played an important part, for he was the mentor of John Goodricke, who was. At the time, both were resident in York, in northern England, and both were comfortably endowed. Pigott's was an old Catholic family with Yorkshire connections, but Edward's mother was a Louvain woman, and he himself was educated in France. The family spent much time wandering through Europe, making influential connections with leading astronomers. Edward's father, Nathaniel, was rich enough to buy astronomical instruments from all the best London makers, and even had time enough on his hands to survey much of the southern Netherlands at his own expense. On returning to York, he had a fine observatory built for him, fitted out with yet more fine instruments, to the great advantage of his son. Edward, duly indoctrinated, made several minor discoveries, including a new comet, and entered into correspondence with William Herschel, at a time when Herschel was still a musician in Bath. His most important connection, however, proved to be with a near neighbor in York, John Goodricke.

Goodricke, Pigott's junior by eleven years, was born in 1764 in Groningen, the Netherlands, to a French mother and a father who was in the British consular service. He clearly had much in common with Pigott, but not by way of health. Goodricke, who since infancy had been deaf and mute, was sent at the age of eight to an Edinburgh school for deaf mutes. He later transferred to an academy for Dissenters in Lancashire, and despite his infirmity was commended there for his knowledge of mathematics. His family moved to York from the Netherlands, and in 1781, Nathaniel and Edward Pigott came to the same neighborhood, and they quickly discovered a common interest in stellar astronomy. Goodricke, who at seventeen was soon to inherit from his grandfather, Sir John Goodricke, had no financial worries. He acquired his own Dollond telescope, and he and Pigott observed regularly in silence together. Both corresponded with Herschel on his new "comet," before it was recognized as a planet—Uranus. And together they searched the literature for data on variable stars.

In November 1782, Goodricke discovered what was already known: that Algol is a variable star. Having observed the star systematically from that date until it was no longer visible by night, he went further and established the short-term periodic nature of its changes. He was still only eighteen when he wrote to the Plumian Professor of Astronomy at Cambridge with his findings. Three days later, his letter was read to the Royal Society, whose members were sufficiently impressed to award him one of the Society's two Copley medals for the year 1783. His figure for the periodic time was $2^d20^h45^m$, just 4 minutes from its currently accepted value; and his explanation for the variation has, in a certain sense, likewise stood the test of time. He proposed a large body revolving around Algol, obscuring it periodically. As we saw earlier in this chapter, not until 1889 did Hermann Vogel show, with the help of his spectroscope, that the spectrum received from Algol also fluctuated. Vogel interpreted this as a Doppler effect, with conjunctions of two component stars coinciding with the times of minimum brightness of what we see by eye. Algol, in short, was one of Vogel's spectroscopic binaries. In fact recent spectroscopic studies have been thought to indicate that yet a third star is present, and that matter is passing between the main star and the other two.

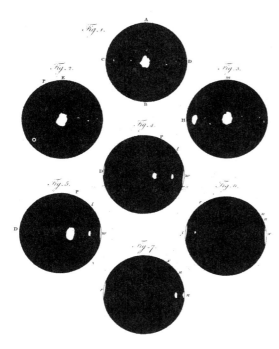

224. One of Edward Pigott's illustrations of his theory of variation in the brightness of short-term variable stars. Such stars are rotating, mostly "unenlightened," but surrounded by a medium with bright patches that are both irregular in size and in movement. From his 1805 paper, published in the *Philosophical Transactions* of the Royal Society.

Goodricke soon discovered several other variables, and some about which he was less certain, but β Lyrae and δ Cephei were firmly established. Each of them later became a prototype for a new class of variable. His notebooks show that he was beginning to appreciate that the levels of minimum brightness of a single variable star are not all identical, as he had previously thought. Pigott, in the meantime, found that η Aquilae was also a variable, and began a long series of studies of this and other variables, studies which he pursued almost until the 1820s. The astronomical community was growing increasingly excited by their work, and in April 1786 Goodricke was elected a fellow of the Royal Society. His good fortune did not last long. Only two weeks later, at the age of only twenty-one, he died, "of a cold from exposure to night air in astronomical observations."

Pigott, now alone, maintained his interest in variable stars, and contributed greatly to what was fast becoming a subject in its own right. As an illustration of his dedication, note his account of no less than five years of observation of the variable R Scuti—"the variable star in Sobiecki's shield," as he called it when he published his findings in the Royal Society's *Philosophical Transactions* in 1805. At a much earlier date he had decided that the causes of apparent change in the short-period variables that he and Goodricke had found were spots on them, combined with the fact that they were rotating. In 1793 he moved house to Bath, Herschel's old place of residence. He read Herschel on the structure of the Sun, and the possibility of life on it. When Herschel told him of his use of physical models to try out his various hypotheses, Pigott followed suit and built models

of his own to help him decide on the causes of variability. He gave his final opinion in his 1805 paper, a work he wrote partly in Bath and partly in
Fontainebleau—where he had been free to visit, after the Treaty of Amiens
in 1802, but where he was detained by the French when hostilities broke
out again. He gave lectures in France on his ideas, and his old scholarly
acquaintances there no doubt helped to ensure his repatriation to England,
war or no war.

Pigott thought that the variable stars are dark, solid, rotating bodies,
and that their rotation is quite regular, but that the medium surrounding
them behaves erratically (fig. 224). The medium, he said, generates and
absorbs luminous particles, "nearly similar to what has been lately so in-
geniously illustrated by the great investigator of the heavens, Dr Herschel,
with regard to the sun's atmosphere" (see fig. 201 in chapter 15). His hypothesis that stars may exist at times in an "unenlightened" state gave him
an added bonus: it could account for the areas of the sky which are relatively empty of stars, to our sight. How numerous are such dark stars? He
wondered whether they might be even as numerous as bright stars; and
he expressed the view that our Sun might itself eventually be reduced to
small patches of light. Such intimations of stellar evolution, based as they
were on an unsustainable hypothesis, are easily dismissed, but there is no
doubt that they helped to wear away the traditional view of constancy and
perfection in the stars.

William Herschel had discovered new variables with his telescopes,
and Pigott was eventually able to draw up a list of thirty-nine in all, a list
that was slow in being augmented in the first half of the nineteenth century.
Part of the problem was that the assessment of brightness was done in
a relatively unsystematic way. There were several useful advances in this
respect. A very simple but intelligent change in practice was Herschel's,
which involved placing the stars within a limited group in order of brightness, doing this for many groups, and relating one group to another by
choosing stars that were shared by those groups. In this way, a whole network could be built up. What was sorely needed, however, was a suitable
photometric instrument, and a step in this direction was taken by John
Herschel. On his visit to the Cape (1837–1838), his interest in photometry
had been aroused by his discovery of the intriguing variability of the star
η Argus, a star seemingly in the middle of the nebula in Argo. He devised
what he called his "astrometer," in which he set up an artificial star with
adjustable brightness, created using moonlight and prisms, so that it could
be made equal in brightness to the natural star being investigated. Using
it, and using the brightness of α Centauri as a unit of magnitude measurement, he drew up a table of 191 stars against which other astronomers
could make their comparisons, without an instrument. When Argelander
set about compiling a catalog of stellar magnitudes a few years later, he
relied on no photometer, but on a system similar to William Herschel's,
arranging stars in magnitude order, and "estimating magnitude steps."

(While the human eye is poor at estimating the brightnesses of stars, it is extremely good at detecting minute differences between the brightnesses of stars seen near together.)

The first astrophotometer to be readily reproducible was not Herschel's, but one designed by Johann Karl Friedrich Zöllner about twenty years later. Zöllner, the son of the owner of a small cotton printing concern, had studied in Basel before setting up a small private observatory in a Berlin suburb. His instrument was designed in the hope of winning a prize offered for competition by the Vienna Academy of Sciences (as it happens, the prize was not awarded to any of the entrants). He had made an intensive study of photometric principles. In order to create an artificial comparison star in his device, he made use of a standard petroleum lamp, the light from which was passed through adjustable Nicol prisms (polarizing prisms). His design was later used in many of the world's leading observatories. It was at his suggestion that the Potsdam Observatory made use of it to produce the first highly accurate photometric catalog of the northern heavens, *Photometrische Durchmusterung des nördlichen Himmels*. E. C. Pickering of Harvard later used an instrument based on the same principles during the period 1879–1882 to produce a photometric catalog of 4,260 stars, while in 1885, Charles Pritchard, the Savilian Professor of Astronomy at Oxford, used a different instrument (a wedge photometer) for a similar catalog of 2,784 stars. Pritchard argued convincingly for the unlikelihood of obtaining close agreement over magnitude measurements, as they were then obtained, and yet over seventy percent of his and Pickering's magnitudes did, in fact, agree to within 0.25. In time, their techniques were superseded by photographic photometry, especially after the first decade of the twentieth century, and this held the field until mid-century, when photoelectric techniques were introduced.

Improved methods of registering magnitudes led inevitably to a growth in the numbers of variable stars known, and of their periodicities. In 1884, a modest list of 190 was published by the Royal Irish Academy; and in a new edition of 1888, the number had increased to 243. The century ended with several leading astronomers making a plea for amateurs to take up the search. This they did with ever-increasing enthusiasm, in Britain and America in particular, and close studies of photographic plates—made mostly by professional astronomers—swelled the numbers still further. By 1903, Edward Pickering could list 701, and that number more than doubled over the next decade.

Numbers alone could not solve the problem of the nature of variable stars, of course. The simplicity of Pigott's hypothesis kept similar explanations in circulation for many decades, and in the mid-nineteenth century Johann Rudolf Wolf turned astronomers' attention to the possibility that we have a variable star on our very doorstep, in the form of the Sun. He believed that he could detect meaningful similarities between his graphs of sunspot frequency and those representing the changing luminosities of many variable stars. There was a lack of constancy about the peaks and troughs; there were instances of double peaks; and both graphs seemed to climb quickly and decay slowly. Attractive as the idea seemed, attention was soon drawn away from it, when the spectroscope was applied to these enigmatic objects.

CEPHEID VARIABLE STARS AND MODELS
OF THE MILKY WAY

As early as 1881, Edward Pickering had proposed a classification of variables into five groups, largely based on the length of the period of variability. Like so many classifications, his would have benefited from a reliable insight into the causes of variation. Were such stars dark, with irregular bright patches, as Pigott had conjectured, guided by Herschel? Were enormous "sunspots" responsible? Secchi was still supporting Rudolf Wolf's idea in 1869. Or were there irregular swarms of meteors surrounding such stars, as Lockyer speculated, in 1887? Goodricke's conjecture that the variability of Algol might be caused by the eclipsing of the bright star by a dark object seemed to be coming true, when Vogel's spectroscope showed Algol to be a binary system. But what reason was there for thinking this to be a common situation? Were there no conceivable *intrinsic* causes of variability? Unfortunately, so little was known of the nature of stellar evolution that an answer to such a fundamental question had to be left for the future. One or two attempts were made to explain how a pulsation might be responsible, and how it could be accounted for. In 1879, for instance, August Ritter tried to apply thermodynamic arguments, but his ideas attracted little attention. In 1913, H. C. Plummer interpreted fluctuation in the spectral lines of Cepheid-type variables as a consequence of their pulsation, but this did not settle the question of the cause of such a pulsation. (Their periods fall within the range of about 1 to 70 days.) This idea was shortly to be provided with some theoretical underpinning by Eddington, although the main cause of the pulsation of Cepheids was not to be discovered until S. A. Zhevakin—in 1953—and John P. Cox and Charles A. Witney—in 1958—traced it to a second stage of helium ionization. This was far in the future in 1913, when an exciting discovery was made by Ejnar Hertzsprung, based on work at Harvard.

It was in 1908, while studying stars of fluctuating brightness in the southern group known as the Small Magellanic Cloud, that Henrietta Swan Leavitt at Harvard noticed a simple property they seemed to have. For the sixteen stars whose light variations she had measured carefully, the longer the period of variation, the brighter the star seemed to be. Four years later she found a simple mathematical relationship that fitted the observations very well: the apparent magnitudes were nearly proportional to the logarithms of the periods. Since all of the stars in that particular group were almost certainly at more or less the same distances from us, the same sort of relationship should hold for the absolute magnitudes of the stars. She saw the potential value of the stars as distance indicators, but could take the technique no further for the time being.

The first of her announcements had attracted little attention, but after reading her second, Ejnar Hertzsprung realized that the pattern of variation in the luminosity of the stars she was studying resembled those of Cepheid variable stars. He saw that since such stars seemed to proclaim their identity by the pattern of their light curves, they should make excellent distance indicators in the universe at large, granted that the Leavitt period-luminosity relationship could be calibrated, that is, could be made to yield a measure of absolute luminosity from the periodic time of fluctuation. This was a very promising idea, but unfortunately there were no known Cepheid-type variables near enough to the Sun for their distances to be measurable by trigonometrical methods, that is, as annual parallaxes. Hertzsprung was able to use proper-motions for the method of "statistical parallax," however, and so proceeded with the necessary calibration of Leavitt's curves. Although there were to be many subsequent revisions of the calibration, this new method opened up a new era in the measurement of distances, for now a single Cepheid-type star, for instance in a distant nebula, could yield a plausible distance from a relatively simple measurement of its periodicity, together with its apparent magnitude. Revision of his calibration curve proved to be necessary, for Hertzsprung had made an arithmetical error that made his distances too small by a factor of ten, but the general principle was established (the distances were so large by the standards of the time that his error went unnoticed and persisted in some quarters for twenty years).

At much the same time, but independently, H. N. Russell was working out the absolute magnitudes of the Cepheids in the Milky Way, and he found values close to Hertzsprung's, but without appreciating the technique that rested on Henrietta Leavitt's discovery. At this time, Harlow Shapley was a young doctoral student, in fact, Russell's first. They saw that the Cepheids could not—as some had inevitably argued—be eclipsing binaries. Like Plummer, Shapley suggested that they might be pulsating. For the time being, it was clear only that they were very bright and very large. In 1914—after a tour of Europe, some months in Princeton to complete his work on eclipsing binaries, and marriage to Martha Betz, who would eventually become an authority on this subject—Harlow Shapley moved to work at Mount Wilson. There he began a study of variable stars in those clusters of tens of thousands of stars whose apparently spherical form has given them the name "globular clusters." (Many globular clusters are easily seen with binoculars. The number associated with our Galaxy is of the order of a hundred—but their status was unknown at that time.) Working with the 60-inch reflector at Mount Wilson, Shapley found a few Cepheids among the variables, and these seemed to follow the pattern of Leavitt's law. Their apparent magnitudes differed from one cluster to another, but this was only to be expected if the clusters were at different distances. In fact here he had a means of determining the *relative* distances of clusters, even without calibrating the Leavitt graph. In 1918 he produced a calibration in the way Hertzsprung had done, and announced that the typical globular cluster was of the order of 50,000 light-years distant. He found that they were not symmetrically arranged, however, and decided that the center of the system of globular clusters was probably identical with the center of the Milky Way system of stars, some tens of thousands of light-years distant from us.

This result, placing the Sun on the edge of the system, was much disliked by many astronomers for a variety of reasons. It was not that many of them expected the Sun to be at the center of things, but that the distances seemed to fit badly with existing, hard-won ideas. In a series of papers, beginning in 1884 and covering two decades, Hugo von Seeliger, having developed a number of new principles of stellar statistics, made counts of stars of different apparent magnitudes in various regions of the sky, and found from them a flattened-disk model of the Milky Way. In overall form this was not unlike Herschel's model, with the Sun not far from the center. In 1901, Kapteyn used his statistics of proper motions, as already explained, to provide a distance scale for Seeliger's work, making the system about 10 kiloparsecs in diameter and 2 kiloparsecs in thickness. (A kiloparsec is about 3,262 light-years.) Kapteyn saw that one unknown quantity, interstellar absorption, might have strongly affected his data, and he made various attempts to measure absorption, but with limited success. By 1918, working in Groningen with his assistant Pieter Johannes van Rhijn, he had concluded from the absence of serious reddening of starlight that the effect was not very significant, and that his basic model was a reasonable representation of the Milky Way.

The model, which circulated widely among his many acquaintances, was not published in detail until 1922, the year of Kapteyn's death. The overall shape of his model was an ellipsoid—a squashed sphere rather than a rugby ball—with its axes in the ratio of about 5 to 1, its major diameter being about 16 kiloparsecs. Having a statistical idea of stellar distances, he could make estimates of the thinning out of stars with distance. He took the Sun to be about 0.65 kiloparsecs from the center and off the central plane, but for convenience often spoke as though the Sun was at the center. He thought that at 8 kiloparsecs distance from the center there were only about a hundredth as many stars in a given volume as in the Sun's neighborhood, while at 4 kiloparsecs the density was about a twentieth that near the Sun. Broadly speaking, his and Van Rhijn's estimates of the pattern of thinning of stars with distance was reasonably correct, for directions well away from the central plane of the Milky Way, but was in serious error in the central plane, where interstellar matter congregates.

That this was so was first conclusively shown by the Swiss-born Robert Julius Trumpler, studying open clusters—clusters less tightly bound than globular clusters—at the Lick Observatory in California, in the late 1920s.

Trumpler's method can be very simply explained. He sorted his clusters into a small number of categories, using criteria that need not concern us. Assuming that clusters within a given category were all of much the same physical size and character, the relative distances of his clusters could be estimated twice over, once on the basis of their faintness, and once on the basis of their apparent overall size. The two methods in practice turned out to give quite different results, and he decided that the brightness criterion was being upset by interstellar absorption. Expressed otherwise: if we accept distances based on the size criterion, then brightness provides a measure of absorption in the direction in question.

In all his work on distances, Kapteyn had shown a preference for proper motions over apparent magnitudes, as distance indicators, since the magnitudes depend on the intrinsic properties of stars, which vary widely. Proper motions are in large part a reflection of the Sun's motion through space, and it was generally assumed that on average they depended on nothing else, that is, that the motions of the stars have a random character, like the molecules of a gas. Kapteyn found, however, that this assumption gave various inconsistent results. It seemed to him at an early stage in his work that stars belong to two different groups, two populations, that are intermingled. His discovery of the two "star streams" was announced at a congress at St. Louis, Missouri in 1904 and caused a great stir in astronomical circles. Karl Schwarzschild, master of stellar statistics, not to be outdone, devised a model in 1907 that explained the measured proper motions on the assumption of a carefully chosen relationship of velocity and position within the Milky Way, avoiding the assumption of intermixed populations. These were the beginnings of a long and continuing study of motions within our Galaxy, that even in Kapteyn's work made it possible to introduce gravitational considerations, and so strengthen the models derived from counts of stars with measured magnitudes and proper motions.

Even before Kapteyn's final model had been made known, a rival was being worked out by Harlow Shapley, after his move to Mount Wilson. As we saw, he had noted that globular clusters, for which he had tentative distances, are not symmetrically arranged in the sky. He noted an idea that had been put forward by Bohlin in 1909 without much supporting evidence. This idea, which was simply that the Sun's position is not central, while it seemed to explain the asymmetry in the clusters, did not fit with Shapley's own estimates of their distance, at least taken together with conventional wisdom on the size of the Milky Way system. By 1916 he had found from Cepheid measurements that the globular cluster Messier 13 was 30 kiloparsecs from the Sun, and therefore far beyond the limits of Kapteyn's Galaxy. A year later, with more evidence from other globular clusters, he returned to Bohlin's idea, and decided that they were indeed all associated with the Galaxy, the Milky Way, and centered on the invisible nucleus of our Galaxy, which was somewhere in the direction of Sagittarius. As many as a third of them occupied only a twentieth part of the sky in that direction. He concluded that the Galaxy must be ten times larger than was generally believed.

Broadly speaking, Shapley had overestimated by a factor of two, partly because he had ignored interstellar absorption. His zero point of the period-luminosity graph was approximately 1.5 magnitudes too low. He made another fundamental mistake, however, for want of knowledge that would not be available for another thirty years. His mistake was to have used a type of Cepheid variable that Walter Baade, in the 1940s, distinguished as Type II. Baade found that stars of this type, with spectra that were typically low in metals, were to be found chiefly in globular clusters, elliptical galaxies, and the galactic bulge of spirals. Stars of Baade's Type I, relatively rich in metals, were younger, and were concentrated in the spiral arms of galaxies. Shapley was doomed to make use of both types indiscriminately, naturally assuming that Cepheids in globular clusters are like those in the Sun's neighborhood. It was only by a stroke of luck that he also underestimated the brightness of the latter by a factor of four, so that these factors canceled out. Even had the extent of absorption been appreciated, he would still, however, have been regarded by most astronomers as wildly wrong.

As a footnote to Baade's discovery: the variables of shortest period and lowest mass, the RR Lyrae stars, were henceforth recalibrated and made the standard of reference for the globular clusters associated with our own Galaxy.

FITTING THE SPIRALS INTO THE ACCOUNT

The ultimate reasons for dismissing Shapley's claims did not rest on Kapteyn's great authority, but had to do with the spiral nebulae, and beliefs about their status. Spectroscopy had greatly enlarged astronomers' knowledge of them. After Huggins's first excitement at finding that they seemed to have the line spectra of a luminous gas, he found ever more with continuous spectra. Distinguishing between the two sorts as "green" and "white" nebulae, astronomers struggled to resolve the white nebulae into stars telescopically. The first success on this score was not until 1924, with the resolution of the "nebula" in Andromeda, M31—called a nebula, even though its spectrum had been shown to resemble that of a cluster of stars by Julius Scheiner, shortly before the end of the previous century.

While it is possible, with the advantage of hindsight, to distinguish between two separate fields of study, that of our own Galaxy and that of the "white," predominantly spiral, nebulae that we now regard as coequal galaxies, such a clear-cut division was not feasible as long as there was a possibility that the second were merely appendages to the first. Kapteyn's model of the first was carefully devised on a statistical basis, but there were earlier alternatives based on impressions received from visual and photographic evidence. John Herschel had noted what appeared to be streams of

225. Photographs of "nebulae" (in fact, spiral galaxies), taken by G. W. Ritchey with the 60-inch telescope at the Mount Wilson Observatory, ca. 1909. On the left is the Whirlpool Nebula, M51 (NGC 5194–5, where it is treated as two objects), in the constellation Canes Venatici; and on the right is H.V. 24 (NGC 4565), in the constellation Coma Berenices, a spiral seen edgeways-on.

stars between us and the main body of the Milky Way. Giovanni Celoria of Milan had in 1879 proposed a model that made the Milky Way equivalent to two more or less concentric rings of stars. Seeliger marshaled evidence against these notions, and yet before long, strong evidence for a much more exciting arrangement was produced by Cornelis Easton, a Dutch amateur astronomer with a longstanding interest in this problem.

Easton had begun to study the problem in 1881, at the age of eighteen, and during more than thirty years he produced a succession of drawings of the distribution of brightness in the Milky Way, based chiefly on published data. He was by profession a journalist, but his drawings of the northern Milky Way were published in Paris in 1893, and so highly did Kapteyn rate them that he persuaded the University of Groningen to grant Easton an honorary doctorate. Easton found that counts of stars down to the ninth magnitude in the *Bonner Durchmusterung* were closely correlated with his drawings, and this decided him to try to chart them in three dimensions. By 1900 he was claiming that the Milky Way had an overall spiral structure, resembling that of several bodies then known, such as M51 and M101, and he seems to have been the first person with this supremely important idea. (For a contemporary Mount Wilson photograph of M51, see fig. 225.) He tried various arrangements, settling on Cygnus as the direction of the galactic center, and summarizing his work in a widely read article in *The Astrophysical Journal*, in 1913 (fig. 226). This direction was eventually found to be in error by about ninety degrees; and still, he thought, the other spirals were not comparable to ours. They were "small eddies in the convolution of the great one." In these matters he was wrong, but his basic idea survived intact.

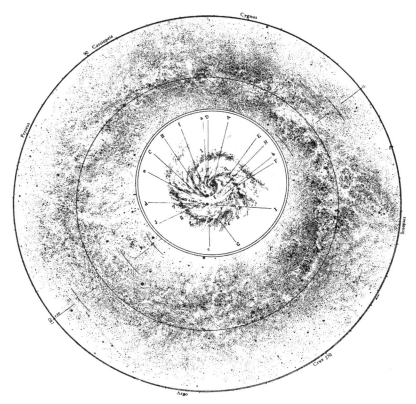

226. One of Cornelis Easton's double diagrams, in support of his contention that the Milky Way has a spiral structure (inner plan). The outer diagram is to be thought of as the Milky Way projected on the inside of a cylindrical surface and then folded out flat. The image is negative, with bright matter printed black. The Sun is at S, the center of the outer circle, and the galactic center in the direction of Cygnus is above it. At this scale it is impossible to appreciate the vast quantities of data that were built into the drawing.

Despite forceful arguments to the contrary, especially by Eddington, there was still a strong feeling in the astronomical community that the Milky Way was a unique focus of the universe. The arguments for this view were broadly five in number. The dimensions of the nebulae were thought to be insignificant by comparison with the size of the Milky Way. Second, the star-like spectra of the white nebulae—those showing Fraunhofer lines—could be explained by starlight reflected from a diffuse nebula, such as V. M. Slipher found in 1912. Third, for those prepared to swallow Shapley's pill, the Milky Way was big enough to include at least some of the white nebulae, according to distances estimated for them at the time. Fourth, the white nebulae seemed to avoid the plane of the Milky Way, suggesting that they were symmetrically arranged around it. And last, new evidence as to their velocities was coming in that likewise suggested symmetry.

The first type of argument seemed at the time to be particularly strong, especially in a version that began from evidence as to the internal motions of the nebulae. Adriaan van Maanen had studied with Kapteyn in

Groningen before taking a post at Mount Wilson in 1912. From the time of its completion in 1914, he used the 60-inch reflector there, and in 1916 was able to publish the results of measurements concerning the rotations of spiral nebulae, deduced from photographs taken over an interval of time. Checked by an experienced colleague, Seth Nicholson, motions outward along the spiral arms of nine nebulae were surprisingly large. The actual velocities of matter along the arms could in some cases be measured by spectroscopy—by the Doppler shift. Taking the two results together for M33, a distance of only 2 kiloparsecs was deduced, putting the object well within a Kapteyn universe.

Van Maanen's results were unfortunately soon to be rejected. Knut Lundmark measured his plates again in 1927, and found movements only a tenth as great as he had said earlier. Even he later halved the movement, but stuck to his general conclusion, and his friend Shapley accepted the same, and argued for nearby nebulae. Edwin Hubble later showed his findings to have been influenced by some sort of systematic error, and indeed, although an excellent practical astronomer, in claiming to measure movements well below the limits of his apparatus, Van Maanen was far too sanguine. He also had a way of brushing aside the findings of other astronomers—C. O. Lampland of the Lowell Observatory, W. J. A. Schouten of Groningen, and H. D. Curtis, for example. Knut Lundmark struggled hard to refute Van Maanen's inferences, and to uphold the idea that the nebulae were "island universes." (Alexander von Humboldt gave this last expression—*Weltinseln* in German—a wide currency in his book *Kosmos*, in 1850.) Van Maanen's findings could not be easily refuted, however, without coming close to questioning his integrity, and many continued to accept them until around 1933.

This entire episode well illustrates, as it happens, how criteria other than the purely scientific may play an important but largely invisible role in the evolution of a science. Here there were clear personal forces working for and against the island universe idea. It was no secret that the socially ebullient bachelor Van Maanen was strongly disliked by Hubble, and that although Shapley and he were from Missouri, during his time as a Rhodes scholar in Oxford, Hubble had acquired mannerisms that impressed the ladies in the vicinity of Mount Wilson more than they impressed Shapley. In his autobiography, Shapley tells of how he heard that Hubble, refereeing a paper he had written for a general readership, had simply scribbled "of no consequence" across the top. The editors told Shapley that his paper had first been accidentally set in type with "Shapley—of no consequence" at its head.

Among those mentioned as having been opposed to Van Maanen's findings was Heber D. Curtis of the Lick Observatory, who in 1914 had begun a study of the spirals, which he already suspected to be "inconceivably distant galaxies of stars or separate stellar universes." But how to measure their distances? One answer was found almost by chance, as a consequence of a discovery made twice over. In 1917, Curtis discovered a nova in one of the spirals he was examining, but did not announce the fact. Then, in the same year, George W. Ritchey, working at Mount Wilson, was in the process of photographing spirals in the hope of finding rotations and proper motions, when in one nebula—NGC 6946—he too found evidence of a nova. This led him to examine old plates, and so immediately to discover a handful of other nebulae with novae that had previously gone unnoticed. Astronomers elsewhere did likewise, and quickly the number reached to more than a dozen. Curtis was delighted with the implications of the apparent brightness of the novae, since they were extremely faint by comparison with novae in the Milky Way. He said that, with two exceptions, the average difference was ten stellar magnitudes, making them a hundred times more distant, and thus far outside the limits of the Galaxy, whether on Kapteyn's or any other model then thought acceptable. The two exceptions would later be recognized as supernovae.

There were two markedly different opinions about the spirals, therefore, at this time. Shapley and Van Maanen placed them so near to the Sun that they could be treated as subsidiary to the Galaxy, the Milky Way. Curtis placed them so far away that their physical sizes, as deduced from the apparent sizes, must be comparable with the Galaxy's. If one were to include the globular clusters in the overall system of the Galaxy, as Shapley did, the case was less clear, but even then Curtis's evidence, if accepted at its face value, would have been difficult to incorporate.

In 1920, it seemed that matters were coming to a head. Curtis and Shapley agreed to debate the question of the scale of the universe at a meeting of the National Academy of Sciences in Washington, D.C. In the event, the meeting seems to have been in a low key, with the two men at odds over what they were meant to be discussing. Shapley took seven out of nineteen pages of his script before reaching the definition of a light-year, a fact conceivably related to the presence at the lecture of those who might choose to appoint him director of the Harvard College Observatory, for which he was then applying. He concentrated on the size of the Galaxy, and said little about the spirals, while Curtis made them his theme. Their differences of opinion made more of an impression on the astronomical world when the two went into print with their views, shortly afterwards. In the light of later developments, we may say that, despite many imperfections in the data available, each was broadly right on his main theme—Shapley on the Galaxy and Curtis on the spirals—but that Shapley was greatly disadvantaged by his acceptance of Van Maanen's work on the spirals.

THE SPIRAL FORM OF THE GALAXY:
DARK MATTER

In 1871 the Swedish astronomer Hugo Gyldén, working in Stockholm, examined large numbers of large proper motions, that is, of stars that were likely to be relatively close to us. He discovered that they are not symmetrically arranged, but that they seemed to be concentrated in one half of the sky, and that there they drift in a roughly similar direction. At right angles to the greatest motions, the stars had negligible motions. He interpreted this as a sign of rotation in the Galaxy. Quite independently of this discovery, when studying stellar motions spectroscopically using the Doppler effect, Benjamin Boss, Walter Adams, and Arnold Kohlschütter found a similar asymmetry in the arrangement of high-velocity stars. In 1914 they announced that they had found that three-quarters of those recorded seemed to be approaching the Sun. (At the same time, they drew attention to an important but subtle difference between the spectra of main-sequence and giant-branch stars of the same spectral type, based on the relative strengths of certain pairs of spectral lines. This was an important aid to placing stars on the H-R diagram.)

Adams continued this study of stellar velocities, and it was also taken up by Jan Hendrik Oort, who had studied with Kapteyn in Groningen, and who completed his doctoral thesis there in 1926 on this very subject, while working with Van Rhijn—after Kapteyn's death. In 1922, however, he published an interim study showing that the Sun was approaching high-velocity stars in one half of the sky—more strictly, between galactic longitudes 310° and 162°—and that it was receding from those in the other half. Oort found another peculiarity: below 62 kilometers per second, the radial velocities were random, while above this value they showed the asymmetry. He attempted a gravitational analysis of the Galaxy, and assumed, mistakenly, that the high-velocity stars had entered the system from outside. He assumed that 62 kilometers per second was the velocity above which a star would be able to escape from the system's gravity. From this he was able to deduce an average stellar mass of 0.65 solar masses. This was hard to reconcile with the dynamical equilibrium of stars within the system, however, which required masses nearly eight times as great.

The general picture Oort drew from the velocities was that the high-velocity stars are moving round the center of the main system, and also with respect to the globular clusters. The Sun he took to be a member of a local system—a cloud of stars—moving in much the same way, but somewhat faster. In all this, Oort was assuming a Kapteyn model of the Galaxy. When he examined the radial velocities of the globular clusters, he found that, not only did they favor one part of the sky, but that they too had their own asymmetry in velocity. A Galaxy arranged along the lines of Kapteyn's model would not be able to hold them gravitationally, and yet they certainly seemed to belong to the Galaxy, judging by their symmetrical arrangement about its principal plane. The explanation, Oort believed, was that the Galaxy is more massive, indeed as much as 200 times more massive, than the visible matter in it—that is, mainly visible stars—would lead us to believe. This was the beginning of a new interest in a difficult but vital part of astronomy, the study of dark matter.

Oort's tentative explanation of the invisible material was that it was obscured by matter in the galactic plane. (Hubble was embarking on a

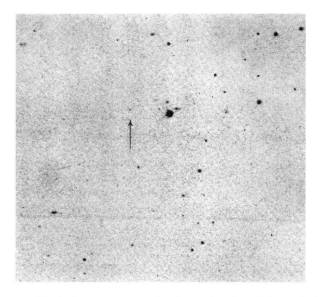

227. However nondescript it may now seem, when it was taken through the 100-inch telescope in March 1934, this (negative) photograph could be counted the most spectacular ever taken of deep space. The reason, while not obvious in a reproduction, was that it showed appreciably more nebulae than resolvable stars. The negative was taken on a special emulsion of extreme sensitivity, developed by Eastman Kodak, and was exposed for 200 minutes.

survey of galactic obscuration at about this time. Figure 227 shows a specimen of what the 100-inch telescope could achieve at this time, while figure 228 illustrates Hubble's work on obscuration.) By 1932, Oort was insisting that we may deduce the existence of two or three times the visible mass, using dynamical arguments. By this time, astronomers were well accustomed to the idea that ionized atoms were to be found in clouds in outer space. Johannes Franz Hartmann had discovered a spectral line for ionized calcium in 1904, by studying the spectrum of δ Orionis, and by the 1920s lines of ionized sodium and titanium were also recognized. In anticipation of much later developments, however, it should be added here that the "missing mass" is not only of dust and small particles but also contains a contribution by stars of very low mass and low luminosity belonging to the halo of the Galaxy. Such stars might be black holes or burned out dwarf stars, of a type to be mentioned again in due course.

We have already encountered early arguments—for instance, Michell's and Laplace's—suggesting that under certain conditions light will not be able to escape from massive bodies. In other words, these bodies will be invisible, and the only way of detecting them might be through their gravitational effect on neighboring visible bodies. "Dark matter," which may be invisible for many reasons, assumed increasing importance over the course of the twentieth century. E. E. Barnard's fine collection of photographs of the Milky Way star clouds, often showing dark areas, had provided evidence of a general sort. The first really penetrating theoretical study of the problem

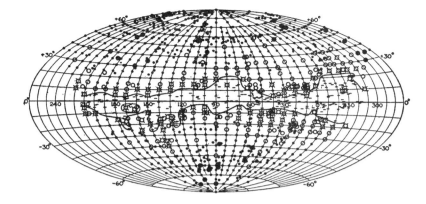

228. In a large program of sampling the nebulae (galaxies) with the help of the 60-inch and 100-inch telescopes, Hubble and his colleagues were able to show the effect of obscuration due to our own Galaxy, the Milky Way. This type of map (1934), showing a central "zone of avoidance," helped those who wished to correct for the obscuration, a necessary step on the way to obtaining an overall picture of the distribution of matter in the Universe.

of the interstellar medium, however, was Eddington's, made in response to important spectrographic studies of calcium lines by the Canadian astronomer John Stanley Plaskett, director of the Dominion Astrophysical Observatory in Victoria, British Columbia. Plaskett had concluded from line displacements that clouds of calcium were not moving together with the stars whose light allowed him to detect both. Eddington agreed, but went further in maintaining that there was a continuous cloud occupying the entire Galaxy, and more or less at rest in relation to it. He calculated the energy density of starlight in space, and so the temperature of the gas, which he found to be high enough for the gas to be doubly ionized. This work of Eddington's, done around 1926, stimulated more intensive observational work, for example by Otto Struve at Yerkes, who soon came to Plaskett's defense on the question of whether the calcium was in discrete clouds, or was all-pervading. Over the next few years, as Struve pursued the problem further, he withdrew his arguments, and with Trumpler's work at Lick on the globular clusters, mentioned previously, the issue was settled in Eddington's favor. (Trumpler's results were published in 1930.)

As we have already seen, the spectrographic route was not the only approach to the problem. Before Oort, Kapteyn had considered the observed mass-to-light ratio in the Sun's neighborhood within the Galaxy and had compared it with what was calculated assuming that the Galaxy is in equilibrium. He had concluded that the mass of dark matter could not be excessive. Oort's work, also using dynamical arguments, had now dispelled that idea.

A new model for the Galaxy was needed that would cover its form, its internal movements, and its relationship with the globular clusters and possibly the spiral nebulae. After Easton, many others had speculated about a spiral structure for the Galaxy itself, but this was singularly difficult to prove. Bart J. Bok, another Groningen astronomer—who went to work with Shapley on his doctoral thesis in 1928 at the Harvard College Observatory and remained until 1955—had applied Kapteyn's numerical methods to the Galaxy, but after attempts spanning more than a decade he failed to produce any conclusive proof of spirality.

After Oort's dissertation of 1926, it became clear that rotation of the Galaxy must be taken into account, but Bertil Lindblad of Sweden had been laying the foundations for this even earlier, in a mathematical treatment of stellar statistics begun in earnest around 1922. Lindblad's was a system of interpenetrating subsystems of stars, all moving in elliptical orbits around the galactic center, as do the planets in our solar system, following Newton's laws. The galactic case is of course very much more complex. We recall that Kapteyn's model, for example, made its form ellipsoidal. Over the years, Lindblad became a leading authority on the dynamics of systems with spiral arms. His theories provided a stimulus to Swedish astronomy, and led many at the observatories of Lund, Uppsala, and Stockholm to specialize in the same branch of study. More immediately, his work was what led Oort to work out a new model for differential galactic rotation, and to estimate the distance and direction of the galactic center. In 1927 he estimated the distance to be about 6.3 kiloparsecs, only a third of Shapley's value, but of the same order of magnitude.

Others joined in the mapping of velocities in the Galaxy. At the beginning of the century, the two great American telescopes, those at Lick and at Yerkes, were manned respectively by William Wallace Campbell and Edwin Brant Frost, who were unequaled in their use of the spectrograph to measure stellar (radial) velocities. Their careers ran along curiously parallel courses, even to the extent that both became blind—Campbell partially and Frost totally. Still more relevant to the Oort problem than the materials they had assembled were data obtained by Plaskett, who was highly experienced in measuring stellar velocities with the spectrograph. Stars of types O and B are intrinsically very bright, and so can be seen at great distances. When Plaskett heard of Oort's work, he had extensive data on stars in these classes, that he and J. A. Pearce had long been collecting, especially of types between O5 and B7. He could therefore immediately test Oort's theory, which had been based on relatively sparse data. Agreement with Oort's parameters was surprisingly close.

Oort's discoveries, of galactic rotation and of the enormous importance of invisible matter, led to considerable new activity in the development of galactic models. On the question of dark matter, progress was slow. Kapteyn's suspicions have already been mentioned. In 1922, James Jeans estimated that there must be three dark stars in the universe to every bright star. The first study that could be called definitive was yet again by Oort, in 1932, and this set a value for the density in space, determined by dynamical arguments. This, known as the "Oort limit," he set at roughly one solar mass per ten cubic parsecs. By 1965 he had increased his estimate by fifty percent, and he then thought forty percent of the total to be invisible stars and gas.

An approach distinct from Oort's was developed by Fritz Zwicky, an astronomer born in Bulgaria of Swiss parentage, who moved to the California Institute of Technology. In 1933 Zwicky found a very surprising result. Taking the Sun as a standard of measurement, he compared the ratio of mass to luminosity in a single galaxy with the figure for a certain cluster of galaxies—the Coma cluster, containing more than a thousand bright galaxies. The second figure, based on the spread of velocities of the component galaxies, was fifty times greater than the first. Although the analysis was crude, it was clear that non-luminous matter on a cosmic scale was of far greater significance than had previously been supposed. In 1936, Sinclair Smith repeated the procedure for members of the rich and nearby Virgo cluster of galaxies, and found there a mass per galaxy a hundred times that implied by the luminosities, working from what was known of individual galaxies. Zwicky believed that, in addition to intergalactic material, there must be intergalactic stars and even dark dwarf galaxies. There was now general agreement that—regardless of the fine detail—the universe had been behaving as a dark horse throughout all of human history.

During the period following the Second World War, evidence was found to suggest that in some systems the mass-to-luminosity ratio might be as much as 1,000, and that even single galaxies might have far more dark matter than had been suspected. As we shall see later, the steady-state cosmology of this time called for a field of dark matter being constantly replenished as the universe expands, a field out of which stars form. The whole issue therefore became in some quarters highly topical. Even so, in singling out this theme for discussion, it has to be said that in the 1950s and 1960s there were many astronomers strongly opposed to the idea that dark matter could be of much overall importance in the universe. In the 1960s, those astronomers who considered the nature of clusters of galaxies, for example were split into two groups, one arguing that galaxies in clusters were bound together by dark matter, the other that this was not so, and that the clusters were relatively short lived, each expanding from some sort of explosion peculiar to its own system. Opinion is still not unanimous, but there seems to be a consensus that mass to luminosity increases steadily with the size of the system considered, and there is no doubt that the hypothesis of dark matter is here to stay.

To return to the form of our Galaxy as it was understood in the two decades after 1930: when further inroads were made into the problem, it was as a consequence of closer studies of spiral galaxies beyond our own, galaxies such as M31 and others nearby. In this work, Walter Baade produced the crucial evidence. We have already come across many striking results obtained by Baade, but some of them will be worth recapitulating here. Born in Westphalia in Germany, he had acquired experience at the Bergedorf Observatory of Hamburg University. A meeting with Shapley in 1920 had led him to a study of globular clusters, for which his telescope was barely adequate, and in 1926–1927 he was able to enlarge his experience when, on a Rockefeller Fellowship, he visited the large Californian telescopes. After his return to Bergedorf, he became a close friend of the telescope builder Bernard Voldemar Schmidt, an eccentric Estonian genius whose best ideas are said to have come to him when in a state of complete intoxication. Their friendship grew on a sea voyage to the Philippines, where they were heading to observe an eclipse, and it resulted in Schmidt's designing a new optical system of the greatest importance for astronomy.

This uses a thin correcting plate across the upper end of the reflecting telescope, and makes possible high relative apertures (low f/ numbers, and shorter photographic exposures). Schmidt was fond of pointing out that he made his telescopes single-handed—in fact he had lost an arm as a boy. Schmidt telescopes would eventually prove to be almost indispensable aids to the survey of the sky that led to the proof of spiral arms in our Galaxy. The best known is perhaps that at Palomar, illustrated in chapter 15 in figure 212.

In 1931 Baade accepted an invitation to join the staff at Mount Wilson. During the Second World War he used the 100-inch telescope to study M31 and its satellite galaxies M32 and NGC 205. From this he was led, as we shall explain, to his discovery that different regions of galaxies are home to stars of distinct "populations," but this discovery, in retrospect, can be seen as having been foreshadowed by an awareness on the part of Shapley and others that stars in galactic clusters differed from those in globular clusters. The differences had been suspected in view of the different situations of the different classes of star on the H-R diagram, but confidence in the brightnesses of stars in globular clusters was small, in view of their great distances. Shapley enlisted help from Cecilia Payne-Gaposchkin and others in the mid-1930s, but the question remained largely unresolved when Baade took advantage of the fact that the skies over Los Angeles were blacked out during wartime. (He also managed to block natural atmospheric emissions with the use of narrow red filters.) This made it possible for him to photograph individual stars in the inner regions of M31. Crudely expressed, he discovered that the brightest stars there were red, and not blue, as in the spiral arms. Stars in the gas and dust-laden disk appeared on his H-R diagram as though they were galactic or open clusters, while stars in the galaxy's nucleus were more like those in globular clusters. The paper in which he announced this discovery gave the names Type I and Type II, respectively, to stars of these types. The types had in effect been hinted at even before Shapley's findings, albeit without any analysis of their internal characteristics, in Kapteyn's two "star streams" in our own Galaxy—types that had in turn been the subject of Oort's work. Baade's Type I was of O- and B-type stars, highly luminous and blue, while Type II were the brightest red stars, also found in globular clusters. Baade did not, as he might have done had he been sympathetic to the parties in the contemporary debate on stellar evolution, reveal any sympathy for the rival schemes on offer there, although he was aware that his populations pointed to some sort of change in circumstance. In time it became clear that the distinction he had identified was one between young stars and old.

In due course an important application to galactic astronomy was found for Baade's populations. In June 1950, at the dedication of a Schmidt telescope in Michigan, Baade expressed his conviction that our own Galaxy is a spiral of a type known as Sb—using Hubble's classification—simply because its nucleus resembled that of M31, which was of that type. Type 1 stars might serve as markers for the spirals, but we can only plot them in three dimensions if we know their distances, and this requires a knowledge of their absolute magnitudes. William W. Morgan of Yerkes and Jason J. Nassau of Warner and Swasey Observatory were at this time studying that very problem, and shortly afterwards announced that, on the basis of 49 estimated distances out of 900 stars of these types they had examined, it seemed that our Sun is on the outer edge of a spiral arm of the Galaxy. Another clue had become available, since O- and B-type stars were by this time known to be associated with large regions of ionized hydrogen (H II). By the end of 1951, they, as well as Stewart Sharpless and Donald Osterbrock, had used H II as a tracer outlining two spiral arms, one through the Sun and the other beyond the galactic center. Tracing the second was a remarkable achievement. To all intents and purposes, the problem that had been attacked for so long using star-counting techniques had yielded at last to quite a different technique.

Only just in time, however. In the same year, 1951, radio emissions were detected from neutral hydrogen, and they very soon supplemented the evidence that had been based on the visible emissions from ionized hydrogen. The discovery was made almost simultaneously in three countries—the United States, the Netherlands, and Australia. The whole of the Galaxy immediately became open to observation. Oort and Bok, and their numerous colleagues and students, quickly threw themselves into a study of the Galaxy's structure as marked out by these radio sources, and within a year the spiral structure of the Galaxy was established beyond all doubt. Oort had by this time been long working in the University of Leiden, and of course supplementary information was needed for those parts of the Milky Way appearing only in the southern sky. Some of the earliest radio work was in fact done in the 1950s by astronomers in Sydney, Australia.

The origin of the spiral structure of the Galaxy was not satisfactorily explained until the 1960s, when a theory of density waves was developed by Lin and Shu. One of the great advantages of the radio measurements was that the intensity of the emissions from the neutral gas gives a measure of the concentrations of gas, and so of the gravitational field. Unfortunately, there are important missing ingredients in all of this, namely, molecules, dust, and gas, that provide a site for star formation. This material was detected through radio emissions in the decades that followed.

From the 1970s onward, it became possible to study the rotation of external galaxies in more detail, using in particular the radio emission (the 21-centimeter line) from the neutral hydrogen in relatively nearby galaxies. In 1970, using evidence of the rotation of the two satellite galaxies NGC 300 and M33, K. C. Freeman deduced from the measured rotations that there must be dark matter in them at least as massive as our Galaxy and that it must be distributed in a way quite different from the visible matter in those galaxies. The measurements were done with a single dish antenna. When data from multiple, synthesizing, radio telescopes became available—first from Owens Valley and more particularly from Westerbork, near Groningen in the Netherlands—it became clear that the total masses of galaxies are roughly proportional to their radii. This was not a result to be expected from a simple spherical or elliptical model of our own Galaxy, for example. Massive invisible halos were deduced from the patterns of the observed rotations. It had been expected that the outermost gas clouds detected by their distinctive radiation would be much slower at the optically visible edge of any galaxy. The fact that their velocities were found to differ little from those of the inner gas clouds showed the existence of the massive invisible halo. Improvements in knowledge of the patterns of rotation in other galaxies provide perhaps the most reliable way of estimating the quantities of dark matter in them; and the great technical improvements made in the early 1980s on such telescopes as that at Westerbork greatly advanced this type of study.

Dark matter is not simply operative in galactic neighborhoods. As Fritz Zwicky noted in the 1930s, long before radio observations of the kind mentioned here were being made, the random motions of galaxies in clusters are so great that the only possible explanation as to why the clusters do not disintegrate is that unknown gravitational forces are at work. In the 1990s it became possible to detect the effect of the total masses of clusters, by using the Hubble Space Telescope to observe the bending of light as it passed them. It was discovered that images of galaxies, seen beyond clusters but far more remote, were distorted, as though smudged and magnified. The explanation was that the cluster and its dark matter were acting as a gigantic lens, the refraction of light being caused by gravitation, in a way predicted by Einstein's general theory of relativity. This discovery pointed the way to a technique for detecting dark matter in the universe, which some in the twenty-first century would judge to be the most important outstanding problem of astronomy. This will be discussed in more detail in chapter 20.

THEORIES OF GALACTIC EVOLUTION

One of Van Maanen's lines of defense, when he produced what purported to be evidence that the nebulae were relatively near, was to show that this idea fitted well with a theory of how the nebulae actually form and evolve. In this he followed a lead given by James Jeans in an important study published in 1919: *Problems of Cosmogony and Stellar Dynamics*. Jeans had retired from university teaching in 1912 on the grounds of ill health, but he continued to make fundamental contributions to theoretical mechanics and physics until, from the late 1920s until his death, he turned to writing for a wider public—which he did with great acclaim. In his 1919 monograph, Jeans explained how a spherical mass of gas would contract under gravity and flatten with spin until it became unstable. It would then throw out filaments of matter from its edge, and they would eventually form themselves into spiral arms. One of the main issues was whether the chief motions observed in the spirals were rotational or—as Jeans thought—outward along the arms, with condensations occurring in the arms and so giving birth to giant stars.

In 1921 Van Maanen supplied what he thought to be evidence for the second view. Jeans was so impressed by the evidence that he went so far as to toy with the idea of modifying Newton's law of gravitation, since the two did not seem to be reconcilable. His enchantment with Van Maanen's data did not last long, however, once Hubble—from 1924 onward—began to produce new evidence as to the vast distances of the spirals, based on the discovery of Cepheids in them. Van Maanen's claimed rotations remained a problem, as explained earlier. The instincts of those who chose to solve the problem by simply ignoring it turned out to be sound.

The crucial positive evidence, like that long afterwards which discredited Van Maanen's measurements, was supplied by Hubble. Born in 1889, Edwin Powell Hubble was the son of a Missouri lawyer, but he grew up in Wheaton, Illinois. Introduced to astronomy at the University of Chicago by G. E. Hale, after taking a degree in mathematics and astronomy he went on to Oxford for a time, where read law before studying Spanish. A successful athlete and boxer—surely the only astronomer to have entered the ring with the great French champion Georges Carpentier—he eventually returned to American and taught in an Indiana high school before practicing law in Kentucky. There he remained until, in 1914, he joined the staff of the Yerkes Observatory of his old university. In 1917, Hubble completed a doctoral dissertation on the photography and classification of nebulae, and Hale soon offered him a post at Mount Wilson. This first attempt at what was essentially a revision of Max Wolf's classification was not particularly memorable, and there were several alternatives in play at the time. He returned to the problem later, and resisted the temptation to pay too much attention to the nebular evolution theory of Jeans. This is ironic, in view of the fact that his later scheme was rejected by the International Astronomical Union in 1925, on the grounds that it employed terms suggestive of a physical theory of evolution. That it did so is hardly surprising, for as Jeans later pointed out, it agreed almost exactly with his own ideas.

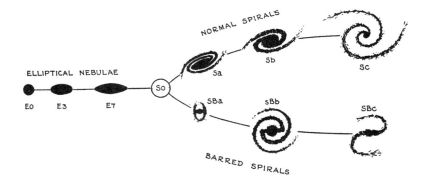

229. The evolution of nebulae, or rather "sequence of nebular types," according to Edwin Hubble (1936). He noted that the transition stage So was hypothetical, and that no nebulae had been recognized between E7 and Sa.

Lundmark thought that it plagiarized his own scheme, and a bitter dispute arose. The reason for the similarities was surely quite simply that Jeans's ideas—and some of the ideas of even earlier writers—were in the thoughts of both. At any event, Hubble's later version of his classification of nebulae, developed in 1923, provided a basis for that which is substantially still the one in use. He distinguished between regular (with a clear nucleus) and irregular (relatively scarce, lacking a nucleus), and he divided the regular nebulae into normal spirals, barred spirals, and ellipticals, each class being represented by subclasses that seemed to be in an evolutionary sequence. The diagram from Hubble's *Realm of the Nebulae* (1936) is reproduced in figure 229.

When the United States eventually entered the First World War in 1917, Hubble enlisted in the infantry, and was not free to move to Mount Wilson until two years later. Working there eventually with the 60-inch telescope, he was at once in a position to record numerous new objects. He noted that diffuse nebulae—as we now know, *within* our Galaxy—are made luminous by nearby blue stars of high surface temperature. (We recall William Herschel's example of the illumination of a nebula by a nearby star.) Hubble gave the phenomenon a theoretical basis, but his reputation was made in a very different way. By 1923, now also using the 100-inch telescope, he was able to resolve the outer regions of the spirals M31 and M33 into stars, and altogether 34 of those stars were found to vary in brightness following the pattern of Cepheid variables. At last it was clear that M31 lies far outside the limits of the Galaxy. By the end of 1924, Hubble was able to quote its distance as 285 kiloparsecs, of the order of ten times the greatest diameter of the Galaxy, as then widely accepted.

During the late 1920s, Hubble made estimates of the distances of further nebulae, and tried to extend the ladder of distances, that is, by finding other criteria than Cepheids, which could then only be seen up to about 4,000 kiloparsecs. He very tentatively proposed using a "brightest star" criterion in late-type spirals, bright objects that seemed to be about 50,000 times as luminous as the Sun. It hardly mattered what the objects were. (In 1958, Allan Rex Sandage, of the Mount Wilson and Palomar observatories, showed that they are hot stars surrounded by clouds of ionized hydrogen.) By averaging the properties of such bright objects visible in galaxies of the great Virgo cluster, all of which galaxies which must be of comparable distance, Hubble found data that he could apply to still more remote galaxies. He so built up an incomparable body of data concerning galactic distances. By 1929 he had distances for eighteen galaxies and four members of the Virgo cluster.

In this way Hubble provided what proved to be one of the essential keys to testing a subject that, for more than a decade, had been theoretically prepared for it. The subject in question was a type of cosmology that had grown, quite unforeseen, out of Einstein's general theory of relativity. By his painstaking deductions of galactic distances, and their correlation with the galactic redshifts, he had the makings of a law of the expansion of the universe, the law now usually associated with his name. The law in question will be a subject for the following chapter.

GIANT OR DWARF? STELLAR INTERFEROMETRY

What twentieth-century theories of stellar structure and evolution owed to the nineteenth was mainly derived from stellar spectroscopy. As we have seen, the first two decades of the new century were witness to the development of especially valuable tools for the study of stellar and galactic structure, in the form of the H-R diagram and the mass-luminosity relationship. The H-R diagram, as explained earlier, has been presented in various guises, not only as a plot of spectral type against absolute (visual) magnitude. Spectral type is usually replaced by a numerical color index, or by effective temperature, the connection being seen from considerations of theoretical physics. Spectral type is chiefly assessed by the presence or absence of specific spectral lines. The important message from physics was that these lines are generally better indicators of differences in effective temperature than of chemical differences. (In the hottest stars, for instance, the hydrogen is almost completely ionized, so that its absorption lines are weak or absent. Hydrogen is nevertheless still present in those stars.) Another complication was that two stars—Rigel (β Orionis) and Regulus (α Leonis) provide good examples—might be of the same spectral type, and yet differ in the breadth of their lines. It was recognized that this difference was a reflection of their different luminosities, and so of their sizes; and line-width therefore soon became viewed as an important parameter.

Astronomers were at first skeptical about the enormous spread of stellar dimensions and luminosities implied by the new interpretations, but most skeptics were eventually convinced. Walter Adams's technique for determining parallaxes spectroscopically (see p. 560 above) soon resulted in a vast body of evidence for the existence of stars at the two extremes, giants and dwarfs. Eddington pleaded desperately for more evidence, in the form of measurements of stellar diameters that were in some sense made directly. Here, astronomers were faced with an enormous practical problem. How could the sizes of stars be measured?

Numerous attempts to settle the angular sizes of stellar disks had been made in the past. Tycho Brahe, for instance, had estimated them at two minutes of arc for a first magnitude star, and a quarter of that for a fifth magnitude star, which is as faint as most people can easily see. (This was a time when a star's "magnitude" was thought to be a measure of its angular size—on which, of course, its brightness was supposed to depend.) Galileo's telescope had shown that these sizes were far too great, and that—ignoring false disks that were clearly spurious—the stars still seemed point-like. He nevertheless tried to measure the angular size of the bright star Vega, using a fine silk thread of measured thickness, and trying to decide on the distance at which the star was barely hidden by it. We now know that his figure of 5 seconds of arc was well over a thousand times too large. Newton's approach was quite different. Asking how distant the Sun would have to be to appear as bright as a first magnitude star, and assuming that stars were of the same (known) physical size, his estimate of approximately two-thousandths of a second of arc was a very reasonable one. It was not, however, a direct measurement of a particular star. The first plausible measurements were made by the use of interferometry, in the nineteenth century, following a suggestion made in 1868 by the physicist Armand-Hippolyte-Louis Fizeau. The idea was to emulate Thomas Young's classic double-slit interference experiment, where light beams from a single-slit source are passed through two other slits, and recombined on a screen. Having traveled by slightly different paths, at points where the two light waves are in phase, they will reinforce to produce a bright image on the screen, where they are out of phase it will be dark, and of course at intermediate points the result will be between the two extremes. In the astronomical case, the double slit is replaced by two apertures in front of the telescope.

The first attempt along these lines was by Édouard Jean Marie Stephan, in 1873, at the Marseilles Observatory. Stephan masked all of the objective of his telescope—an 80-centimeter reflector—except for two small apertures at opposite points on a diameter of the mirror. These played the part of Young's double slit. There is nothing analogous to the first slit in the astronomical case, however. Its purpose in Young's arrangement was to provide a narrow and coherent single source of light, with the true source behind it. Had the star been without significant area, that would have played the part of Young's lamp and first slit, but a more correct analysis must take into account the fringes created by all parts of the star. In brief, the final pattern of fringes, as seen in the focal plane of the telescope, arises from the superposition of two distinct groups of diffraction fringes, one group from each telescope aperture. (We call them a group because they come from all parts of the star.) It can be shown that the superimposed fringes will disappear when the separation of the two apertures takes on a certain value,

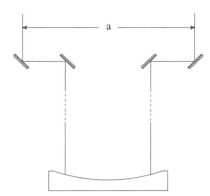

230. Michelson's 1920 interferometer arrangement. The effective aperture is *a*.

a value inversely proportional to the angular size of the disk, and that this can then be easily worked out from the geometry of the case.

Stephan examined several bright stars in this way, and found that in no case could he make the fringes disappear. Whether or not it was correct, his was not a trivial piece of evidence, for it set an upper limit to the angular diameters, which he decided had to be less than 0.158″. There the matter rested until 1891, when the Polish-American astronomer A. A. Michelson applied the method to the case of Jupiter's satellites. He knew, however, that this type of instrument could not be expected to work with the much smaller disks presented by stars, unless it could be used with one of the largest telescopes, and that even then he would need to increase the separation of the analogs of "Young's slits." This was done by a kind of double periscope (fig. 230), and after some hurried preparations, the Mount Wilson 100-inch reflector was adapted to this kind of use (the small outer mirrors in this arrangement were twenty feet apart). At last, in December 1920, F. G. Pease and J. A. Anderson succeeded in measuring the angular diameter of Betelgeuse (α Orionis). Combining their result of 0.047″—close to the 0.051″ predicted by Eddington—with the supposed distance of the star, the astonishing conclusion reached was that the star was larger than the orbit of the Earth around the Sun. (It is actually nearly four times as great, and its diameter is about 800 times as great as the Sun's.) Betelgeuse was a happy choice, when it came to proving a point, for this red supergiant is still one of the largest known.

It has to be said that, after the initial success was repeated for a handful of other giant stars, the method failed for stars on the main sequence, owing to atmospheric turbulence and the practical difficulties of making large configurations of mirrors. Some data were found from eclipsing binaries, but only in 1952, with a new type of interferometry devised by Robert Hanbury Brown and Richard Q. Twiss, was the subject truly revived. Their method of "intensity interferometry" was first used at radio frequencies, but by 1956 it had been adapted by them to optical use, and successfully tested. Using apparatus patched together at low cost—their reflectors, for instance, were old army searchlights with mirrors of 156 centimeter diameter but silvered on the back—their method gave high resolution and largely avoided the problem of atmospheric scintillation that had beset the Stephan-Michelson technique. Their first measurement was of Sirius, made at Jodrell Bank. There they found conditions far from ideal—they had clear night sky for only eighteen hours in five months—and the group eventually looked elsewhere for a clearer sky. Settling on Narrabri in New South Wales, Australia, there they assembled two 6.5-meter reflectors, each comprising some hundreds of small mirrors. Mounted on trolleys on a circular track of 94 meters radius, their separation could be varied from ten meters to almost twenty times this. Between 1965 and 1974, the angular sizes of several hundred bright main sequence stars were measured with this arrangement. What mattered more than numbers, however, in the first instance, was the maximum resolution attained. Their first attempt with Sirius yielded a diameter of 0.0068″ with the probable error a dramatic ±0.0005″. Soon the method of intensity interferometry was being used successfully in many other centers, and over the next two decades, at radio frequencies, using a baseline almost the size of the Earth, star disks of a thousandth of a second of arc were being routinely measured.

The intensity interferometer does not require an image of high quality—hence the old searchlight mirrors—but it has a very poor signal-to-noise ratio compared to the classical arrangement, and Twiss eventually decided to persevere with the latter at the National Physical Laboratory at Teddington in the United Kingdom. He later moved operations to an outpost of the Edinburgh Royal Observatory at Monteporzio Catone, south of Rome. John Davis and Hanbury Brown also later turned their attention to the older type, partly on the grounds of cost; and other interferometric techniques were developed, such as that of speckle interferometry, devised by Antoine Labeyrie. (Here an attempt is made to recreate the telescope image of the object as it was before being corrupted by the atmosphere. The technique is also used in microscopy.)

STELLAR TYPES AND EVOLUTION

Physical size, and not only mass, was an important parameter in the theory of stellar interiors, on which Eddington was working. Michelson's data, which did not cover the enormous spread of stellar dimensions that had been suspected, did at least prove the existence of stars with giant dimensions, and so persuaded those working on problems of stellar structure that they were on the right lines. What then of the mechanisms by which heat is brought to the surface of stars of these vastly different sizes from their interiors? Almost every nineteenth-century account had assumed that convection currents were responsible, since the granular appearance of the Sun's photosphere seemed to speak for convection cells in a boiling liquid. Two lone voices for the predominance of radiation through the solar atmosphere, and for a minimum of convection, were those of Johannes Wilsing of Potsdam, in 1891, and Ralph Allen Sampson of Durham, in

1894. They were heavily criticized in their day, but a decade later, Arthur Schuster and Karl Schwarzschild separately helped to tip the scales in favor of the idea of radiative transfer of the Sun's internal energy. We have already said something of Schwarzschild's achievements. It was with much reliance on his thermodynamic studies that Eddington began to build his theory of radiative energy transfer.

In the early 1920s, Eddington and most of those who were studying stellar constitution seriously, accepted Russell's ideas in broad outline. As we saw earlier (p. 561), they believed that stars began life as enormous, rarefied, red giants. These were then supposed to contract, and to heat up as they did so, moving to the left on the H-R diagram, until reaching the main sequence. (As before, the abbreviation H-R is being used for the Hertzsprung-Russell diagram, or, "Russell's magnitude-spectral type diagram," as some called it at the time. The x-axis on the usual H-R diagram is one of decreasing temperature. See figure 222 on p. 561 above, for instance.) It was assumed that when it reached the main series the star had reached a critical density, and would then begin to evolve by moving down the main sequence to the right. To this theory, Eddington added the idea that—as stars of low density—the giants would obey the ideal-gas laws. He believed that the pressure of radiation from the center of giant stars contended with the pressure of gravitational contraction and that these pressures could be in equilibrium. In that case, roughly expressed, the bigger the mass, the brighter the star. This was not to suppose that gravitational contraction was believed to provide the radiant heat, which had been a common idea, as we have seen, in the previous century, although it was an idea that gave an unacceptably short lifetime for the stars—a few tens of millions of years. There were two new processes under consideration, for the production of radiant heat, electron-proton annihilations, and a building up of hydrogen into heavier elements. In a sense, all that was needed of the mechanism was that temperature and density should increase with contraction in size, and decrease with expansion.

Eddington now tried to apply his theory of giant stars to dwarfs, such as the Sun. He tried replacing the ideal gas laws with the Van der Waals equations for an imperfect gas. (In 1873, Johannes van der Waals had modified the familiar simple gas laws relating pressure, volume, and temperature, replacing them with laws that take into account the sizes of the gas molecules and the weak forces of attraction between them.) He was surprised to find that this implied extremely high temperatures and densities in the cores of stars like our Sun. What was the mechanism by which they held back so much of the expected radiation? In trying to find answers, he encouraged Edward A. Milne, then a young member of the Solar Physics Laboratory in Cambridge, to reexamine his theory of giant stars.

Milne produced a surprising answer: the atoms of most elements would be able to hold no more than two electrons. After further refinement of Milne's theory of ionization—his first equations were modified by increasing the importance of temperature as against density—it became clear to Eddington that dwarf stars, like giants, should obey the ideal-gas equations. The mass-luminosity relation for giants should likewise apply to (most) dwarfs. By a fortunate chance, in 1920, Paul W. Merrill at Mount Wilson had determined the masses and luminosities of two giant stars, and Eddington in 1924 was able to show that they could be fitted extremely well into the mass-luminosity curve for the Sun and thirty-seven other stars. It now seemed clear that atoms in the interiors of dwarfs could behave like an *ideal* gas because they were—stripped of electrons—smaller than terrestrial atoms. The matter in stars should therefore be highly compressible, and Eddington could postulate the existence of stars with densities far in excess of the Sun's. The density of the companion to Sirius, for example, was thought to be as high as 53,000 grams per cubic centimeter, fitting perfectly well with these new ideas. What is more, in 1925, Walter Adams at Mount Wilson Observatory measured the gravitational redshift produced by Sirius's companion, and found a figure for its density reasonably close to Eddington's claim.

The older theories of Lane and Lockyer were by this time being generally ignored, and Eddington's ideas were taken on board by most in his profession. That did not mean an end to the question of whether the H-R main sequence could be considered an evolutionary track. What was the mechanism for producing heat in the interiors of stars? If it was element conversion, then stellar masses would not change greatly, so the main sequence would not be evolutionary. Eddington was inclined to accept the alternative, electron-proton annihilation, with a gradual movement along the curve as mass was lost to the star. Russell continued to work hard to make his giant-to-dwarf theory compatible with Eddington's new findings. It is amusing to see him following Eddington's style as he tried to save his ideas about stellar evolution by postulating two different forms of stellar energy, "giant" and "dwarf" forms, to account for the differences between the two types of star. James Jeans supported the electron-proton process, but attacked Eddington on the question of whether temperature and density played the part in energy generation that Eddington had claimed. The most novel theory of the nature of white dwarfs was Fowler's, using ideas proposed by his student Paul Dirac. The exceptional density of white dwarfs could be explained, he showed, in terms of the new Fermi-Dirac statistics.

This was the situation in 1926, when Eddington produced his *The Internal Constitution of the Stars*, a masterly summary of his ideas on stellar structure. In it he faced up to serious problems with his predictions of the opacity of stellar matter. He returned once more to the question of the mechanism of energy production, but still favored electron-proton annihilation. This was the tentative orthodoxy until the early 1930s, even though he had been obliged to fight off attacks from two old colleagues, Jeans and Milne. As we have seen, he was no stranger to disputes with

Jeans. The controversy was joined, in time, by others—including Bengt Strömgren, Ludwig Biermann, Subrahmanyan Chandrasekhar, and Thomas Cowling—but still Eddington's reputation brought large numbers of astrophysicists on to his side. He was helped by more than his reputation, however. The question of the age of the universe was a very thorny one. It seemed obvious to all concerned that the long time-scale required by Jeans for the age of the stars—trillions, not billions, of years—had to be compatible with what was then being found for the age of the universe. (The value for the latter was being derived from the rate of expansion of the galaxies, a subject for chapter 17.) Before long, expert opinion moved in favor of an intermediate figure for the age of the universe, and so Jeans lost support. Eddington next disposed of Milne by solving the opacity problem in a way not open to this other main rival. In March 1932 he received a letter from Strömgren, who had reached a similar conclusion through a different kind of argument. (Strömgren, who was then only 24, was a Swedish-born Danish astronomer, whose father was Director of the Copenhagen University Observatory.) Hydrogen, he maintained, predominated not only in the surface layers of stars but in their interiors, which had generally been thought to contain a much higher proportion of heavier elements. As we shall see later, this question of the relative abundance of hydrogen was only one aspect of a much larger problem of relative abundances of the chemical elements, a problem not only of what their proportions were, but of how they had been arrived at in the history of the universe.

Around this time, the subject of stellar energy was being studied by a loosely connected cohort. Many of them were professional physicists, of whom members of the more conventional astronomical establishment were only dimly aware. One such person was Robert d'Escourt Atkinson, a Welsh-born physicist who worked at Greenwich and then in Germany, before moving in 1930 to Rutgers University in New Jersey. In the late 1920s, Atkinson argued that the building of elements in stars, mainly the synthesis of helium, was the chief source of energy generation within them. In 1931 he published a revised account, into which he wove an elaborate theory of stellar evolution. Influenced by the ideas of George Gamow, his basic idea was that thermonuclear reactions might be set in motion by proton-proton reactions. (In such a reaction, two protons combine to form a deuteron and a positron.)

Atkinson had the misfortune to be working in an area of nuclear physics that was on the point of rapid transformation, the result of experimental work with particle accelerators. As we shall see later in this chapter, the key to the future was the neutron. As early as 1920, Ernest Rutherford in Cambridge had speculated that protons and electrons could combine to form a neutral particle, a "neutron" that might reside in atomic nuclei. Such particles were eventually identified in 1932 by James Chadwick, a junior colleague of Rutherford's. Here was one of several factors necessitating further revision of Atkinson's work, but Atkinson himself moved into other fields of research, and the revision was left to others. In 1934, for example, his compatriot Thomas George Cowling, a former student of Milne who had lost faith in Milne's approach, came to the conclusion that stars conforming to Atkinson's model would generally be stable if they were not too large, and that those hostile to the idea of thermonuclear reactions as the energy source should not use instability as an argument against it. Other remarkable studies were being made at this time, by several researchers of a rising generation, but one of Cowling's models obtained a loyal following well into the 1940s. (It presupposed a convective core driven by some sort of nuclear energy source, and a radiative envelope.) There was no obvious way of deciding between the "Cowling Model," as Chandrasekhar called it, and Eddington's, although the former was more congenial to the nuclear processes soon to be developed by Bethe and Weizsäcker. Not until 1952 was it shown that even there it applied only to certain stars on the upper main sequence. (The proof required their CNO cycle, which will be introduced later in this chapter.)

THE CHANDRASEKHAR AND
OPPENHEIMER-VOLKOV LIMITS

The structure of the Sun and stars is a theme in which considerations of gravitation—even in the work of Karl Schwarzschild—begged to be supplemented by the rapidly advancing knowledge of the physics of fundamental particles. Here, in the period following the First World War, most astrophysicists would no doubt have put Arthur Eddington in pride of place, but there was one application of quantum mechanics to this particular theme which he failed to appreciate. It was an application due almost entirely to the young Indian astrophysicist, Subrahmanyan Chandrasekhar, and viewed in the light of later history, it represents a crucial step on the road to an adequate formulation of the concept of "black hole," and the search for plausible examples. For the time being, we may take the idea in the form we have seen it take so often before, the idea that there are huge masses in space whose gravitational pull is so strong that not even light escapes them, but that these masses, while cut off from the rest of the universe, leave other sorts of evidence of their existence. Not until the last quarter of the twentieth century did this old idea become more or less universally accepted, but Chandrasekhar's first contribution to the theory was made long before its acceptance, indeed even before he was twenty. As for the phrase "black hole," while we have used it already and shall go on doing so for the sake of continuity, it must be said that it was coined only in 1968, by John Wheeler of Princeton. Before then, writers in English tended to speak of "collapsed stars," and in Russian of "frozen stars"—the latter phrase picturing a collapsing star that freezes only when reaching the Schwarzschild limit. It has been pointed out that the different phrases, which are plainly associated with different mental images, have influenced attitudes to the real existence of the entities in question, even as regards their dynamical, energy-storing, and energy-emitting properties.

Chandrasekhar was born in privileged circumstances in Lahore, which is now in Pakistan, the nephew of the Nobel prizewinning physicist C. V. Raman. In 1929, he began graduate study at Trinity College, Cambridge, having already worked his way through Eddington's *The Internal Constitution of the Stars*. On the sea voyage to England he had struggled with the question of the fate of stars that have used up their supply of nuclear energy. Did such stars merely turn into dead white dwarfs, dense cinders emitting no more light? He saw that as a star contracts, in general there will come about a situation where the inward gravitational attraction is balanced by a certain repulsion, of a type demanded by what is known as the "exclusion principle." (This law of quantum mechanics is now associated with the name of the Austrian physicist Wolfgang Pauli, who formulated it clearly in 1925.) By 1930, Chandra—the name by which Chandrasekhar was, and is, usually known—had decided that it is impossible for a white dwarf star, supported solely by a degenerate gas of electrons, to be stable if its mass is greater than 1.44 times the mass of the Sun. (It may be somewhat more if the star does not completely exhaust its thermonuclear fuel, but the value of the mass, now universally known as the "Chandrasekhar limit," has in any case since been revised downward.)

Following work by Edmund Stoner, codiscoverer of the exclusion principle, work that was further extended by Chandrasekhar, it was then realized that Einstein's general theory of relativity will come into play if the mass of a star is greater than the Chandrasekhar limit of 1.44 solar masses. The load on the upper layers will then be so great that it will collapse catastrophically. In fact if the star ends its nuclear-burning lifetime with a mass greater than the limit, it must become either a neutron star or what would now be called a black hole, and the condition for a black hole was calculated not long afterwards, as we shall see.

Chandra's mentor, Eddington, was a fellow of Trinity College during the time that Chandra was there, and while the two were undoubtedly on good terms, Eddington was reluctant to accept the graduate student's idea. For reasons that seem quite out of character, he dismissed it in a peremptory way at a 1935 meeting of the Royal Astronomical Society, to which it had been presented by Chandra. Chandra was naturally disappointed, and he changed the direction of his work for a time and two years later joined the University of Chicago at Yerkes Observatory. There he remained for the rest of a distinguished career, during which he advanced astrophysics in numerous other ways. The Nobel Prize for Physics, which he shared with William A. Fowler in 1983, however, was for his earliest work, his research into the "origin, evolution, and composition of stars." After half a century, his doctrine had by slow degrees become accepted. In the interval, exciting studies of neutron stars and black holes had been made, and the scientific community was at long last waking up to his role in providing a vital key to their discovery.

Of those professional astronomers who gave much thought to the matter in the early 1930s, it is probably true to say that most sided with Eddington. Indeed, all of those who disliked the notion had an easy, albeit arbitrary, answer to the problem of the awkward theoretical properties of stars over the Chandrasekhar limit: the ejection of mass from them, whether continuous or explosive, could act to bring down the mass to a point where there was no paradox. Chandra was not entirely alone in his conclusions, however. In 1932 another young researcher, the Russian physicist Lev Davidovich Landau, proved similar results to his, but more simply. He showed that there is another possible final state for a star, even smaller than that reached by a white dwarf. If the repulsive force explained by the exclusion principle was the result of protons and neutrons, rather than electrons, the stars might be even smaller and denser. Neutron stars was about to enter the arena.

The neutron star was an entity that was known to theory long before any such thing was identified. That happened, as will be seen in subsequent chapters, only in the 1960s. In 1934, only two years after James Chadwick had identified the neutron, Fritz Zwicky and Walter Baade wrote a paper together titled, "Cosmic rays from Supernovae." They ended with the conjecture that a supernova represents the transition of an ordinary star into one that comprises mainly neutrons. They pointed out that a very small radius and an extremely high density was possible with neutrons, and that the gravitational energy of the very closely packed nuclei in a cold neutron star may become very large indeed. The entire contents might be contained within a radius of 10 kilometers. (An ordinary thimble would then contain around ninety million tonnes of such matter.) The conjecture, made almost casually, seems to have gone generally unnoticed, and neither Baade nor Zwicky seems to have pursued it further.

The most spectacular visible remnant of a supernova was that in the Crab nebula, the debris from the supernova explosion of the year 1054. The object, the nebulous character of which had been discovered by John Bevis of Oxford in 1731, was known to have an unusual star at the center. Baade asked his friend Rudolf Minkowski to obtain its spectrum with the 100-inch telescope, and it was found to be extraordinary in two respects: it was lacking in spectral lines, and it was ten times hotter than any known star. The mystery of why Baade did not match up these findings with the 1934 conjecture is now thought to be best explained by the fact that he and Zwicky had quarreled famously in the meantime.

It cannot be said that such ideas occupied the thoughts of many astronomers, and the Second World War, of course, played a part in delaying recognition of Chandrasekhar's more prominent conclusions, but there was one extremely important discovery made in this connection before the war began. The American physicist J. Robert Oppenheimer had studied in Cambridge, England, as well as with Max Born in Göttingen, before accepting posts at the California Institute of Technology and the University of California at Berkeley. There he led a group working on an extraordinary range of problems in quantum electrodynamics, nuclear physics, and astrophysics, including the problem of the mass limits of stars. In 1939, he and his collaborator G. M. Volkov announced that while stars above the Chandrasekhar limit do not stabilize as white dwarfs, but rather, condense further and become neutron stars; nevertheless, if they are of more than about 3.2 solar masses, they become black holes (as we may call them). This new limit is now usually known as the Oppenheimer-Volkov limit. Oppenheimer and his group would no doubt have continued work along these lines had the war not intervened; but, as is well known, circumstances moved him to the Manhattan Project, with explosive consequences of another kind.

Not until the 1960s was there much further progress on these problems, and then it was as the result of an accumulation of empirical discoveries. Highly energetic phenomena were found, both on the scale of a star and on the scale of a galaxy, that seemed to call for black holes or neutron stars as their explanation. Using satellite telescopes, systems of binary stars were discovered in which one exceedingly compact component, optically invisible, is emitting a very great flux of X-rays while the other is visible. It was suggested that the massive flux of X-radiation comes from the conversion into energy of material—for instance from its companion's atmosphere—falling into a massive star, such as a neutron star or a black hole. By settling the mass of the invisible component, with reference to the Chandrasekhar and Oppenheimer-Volkov limits, a decision is made as to whether a black hole or a neutron star is involved. There is an ever increasing number of candidates for both, within our Galaxy. We shall return to this subject in chapter 20.

On the galactic scale too, there are objects—Seyfert galaxies and quasars, for instance—that emit very much more energy than normal galaxies at all wavelengths. In such cases we may be dealing with black holes of the order of a billion star-masses, each drawing in gas and dust of many star-masses annually. To put this idea to the test will require techniques as yet unproven, the most promising being, perhaps, the detection of gravitational waves, that are theoretically emitted with the infall of material into the black hole.

THE RELATIVE ABUNDANCES OF THE CHEMICAL ELEMENTS

Eddington's belief, expressed in 1920, was that the energy stored in atomic nuclei was providing the Sun's power, and that it was "sufficient . . . to maintain output of heat for 15 billion years." By the end of the decade, it had come to be generally assumed that stars were gaseous spheres in radiative equilibrium, evolving somehow along the H-R diagram, with electron-proton annihilation converting their mass into radiant energy over not tens but perhaps even thousands of billions of years. By the mid-1930s, however, opinion was rapidly changing. Evidence for the extremely high abundances of hydrogen had been accumulating for a decade, and this encouraged a belief in energy-generation by element building. New observational evidence was coming from several directions, and astro-physicists who tried to interpret it were eventually to discover that they had not always understood it well, but false leads in one study often yielded useful results in another.

We have come across many occasions, beginning with the findings of Huggins, Vogel, and Cornu (p. 511), on which astronomers were taken by surprise by the high apparent abundance of hydrogen, as indicated by spectra of the Sun and stars. The problem of the relative abundances of different chemical elements soon took on a life of its own, but it was a problem with obvious links to theories dealing with the chief sources of stellar energy, and those theories were constantly undergoing revision, helped by the rapid evolution of fundamental physics. There was a long period, how-ever, from the mid-1920s to the mid-1940s, when the two lines of research did not combine as they might have done for their mutual advantage. Part of the reason for this lack of engagement was uncertainty about how much could be inferred from what was observed to be happening at the Sun's surface about processes and relative abundances in the Sun's interior. Even this is to oversimplify the problem, for evidence from the Earth, a part of the solar system, could hardly be ignored; and here the sort of difficulty we have with the Sun—that is, of distinguishing between what we know of its interior and its photosphere—reappears.

How does the Earth's accessible crust, and what lands upon it from space, differ from its inaccessible core? To ask that question led further afield still, in the direction of cosmogony and the formation of the Earth itself. It is easy to forget that until gravitational measurements on the Scot-tish mountain of Schiehallion, made by a group of British scientists under the leadership of Nevil Maskelyne in the 1770s, there were many intelli-gent physicists who thought the Earth to be a hollow shell. Edmond Hal-ley was one: he argued that the Earth comprised three hollow concentric spheres. This was at least an advance on Athanasius Kircher's ideas about giants of fire living under the Earth's crust. Nineteenth-century America even had its religious sect of Koreshanity, founded by one Cyrus Teed, ac-cording to whom we are all living *inside* such a hollow sphere with our heads to the center, without appreciating the fact. The ideas of more recent hollow-earthers, however, defy summary treatment. It is easy to smile, but how do we prove the opposite? A much underrated milestone was passed when, in May 1778, on the basis of the Schiehallion measurements, the mathematician Charles Hutton presented the results of massive calcula-tions that led him to the conclusion that the interior of our planet is not only solid but is likely to be composed of metals of more than double the Earth's mean specific gravity, which he set at 4.5. Further light has been shed on this problem of the Earth's constitution ever since, from gravita-tional, meteoritic, seismic, magnetic, and thermal evidence.

The constitution of meteorites is of supreme importance. About ten percent of them are of chiefly iron-nickel alloy. Others are a blend of silicate minerals and nickel-iron alloy in roughly equal amounts. Stony meteorites resemble much of terrestrial rock, especially peridotite, which is the chief component of the Earth's mantle. About nine-tenths of all stony meteorites contain round silicate grains—chondrules, hence the name for such meteorites, "chondrites." Some of these chondrites ("carbonaceous") contain carbon, hydrocarbon compounds, and amino acids. The most famous, which exploded over the Mexican village of Pueblito de Allende in 1969, has been assigned—on the basis of its radioactive isotope content—an age of 4.6 billion years, which is taken to be much the same as the age of the solar system and a third of the age of the universe. Its composition is obviously highly relevant to the abundances issue. We can examine meteorites directly, but to enlarge upon Hutton's claims the pattern of seismic reflections and refractions through the Earth, following on earthquakes and explosions, is used to throw light on the physical properties of the materials of mantle and core, and their distribution—allowance having been made for effects in the known continental and oceanic crust. It is now accepted that the core is mainly of iron, silicon, sulfur, and nickel, and that it has a liquid outer part and a solid inner part. With specific gravity of around 12 or 13, it makes the average figure for the Earth about 5.5. Hutton's intuition was sound, but the modern evidence for an iron-nickel core, while it goes far beyond what was available to him, is still of a similarly inferential nature.

A realization that such terrestrial data on the abundances of the elements might be relevant to the universe as a whole was slow in coming. When it came, in the course of the second and third quarters of the twentieth century, it led to a type of physical cosmology that paralleled, and that could be successfully grafted on to, relativistic cosmological models. As matters turned out, the exercise was valuable but flawed. The notion of a *single-parameter* "cosmic abundance" was widely adopted for a time, but as we shall see later, it was eventually well and truly abandoned. More than half of the atoms in the Earth's crust comprise oxygen. How could the makeup of the Earth be expected to resemble that of the Sun, especially when the Sun seemed to be mostly of hydrogen? What could the Sun be expected to have in common with meteorites, themselves members of the solar system, but with a demonstrably high iron content? Turning to the stars, their spectra often seemed even more disparate. Gradually, however, making allowance for appropriate secondary processes, such as the loss of volatile materials by meteorites in the course of their entry to our atmosphere, ratios of elemental abundances began to appear similar in a surprising number of respects. Taking the relative proportions of the heavier elements in many meteorites and terrestrial rocks, for example, with appropriate adjustment as explained, they turned out to be very similar to the corresponding solar ratios. In due course, interstellar gas, planetary nebulae, and many types of star appeared to approximate to some sort of normal elemental distribution. In retrospect, the first important indications that this might be so can be found in papers by two theoretical chemists, Giuseppe Oddo of the University of Pavia in Italy, in 1914, and William Draper Harkins of the University of Chicago, in 1917. Both showed that the key to understanding abundances lay not in the chemical properties of the elements so much as in their nuclear properties. More particularly, they independently arrived at what is today often called the Oddo-Harkins rule, that elements with even atomic number (such as carbon) are more stable and more abundant than those with odd atomic number (such as nitrogen).

It should be noted here that atomic number originally corresponded only to an ordering of the elements by their chemical properties, an ordering that was seen to agree with ordering by atomic mass, but that appeared to reflect some property other than mass. Largely as the result of work by Bohr and Rutherford, it was becoming clear that the atomic number (Z) corresponds to the electric charge of the nucleus, that is, the number of protons in it—which is how it is defined today.

Harkins was at this time one of the few American scientists working on the structure and reactions of atomic nuclei. From 1915 and into the 1930s, he published long a series of important papers on the synthesis of nuclei from protons, deuterium, tritium—a radioactive isotope of hydrogen— and alpha particles. (Alpha radiation was discovered by Ernest Rutherford in 1899. In 1903 he showed that it was of the character of particles. Their nature was not understood before the discovery of the neutron by James Chadwick in 1932, after Rutherford had predicted its existence. Alpha particles are helium nuclei, with two protons and two neutrons. A tritium nucleus contains one proton and two neutrons.) It was his early work, however, that most materially affected the direction of astrophysical thinking on abundances. He investigated the possibilities, as then understood, for synthesizing higher elements out of hydrogen, noting the relevance of his "hydrogen-helium system" to astronomy. He referred to the astronomers' ordering of the elements in the stars as "first nebulium, hydrogen and helium, then such of the lighter elements as calcium, magnesium, oxygen, and nitrogen, and finally iron, and the other heavy metals," adding that "in the present system it has not been found necessary to include nebulium." (As we saw on p. 510, the nebulium question was one of long standing, and was not to be truly solved until 1928.) Harkins regarded his findings as stronger support for "the theory of the evolution of the heavier atoms from those which are lighter" than anything astronomy could then offer unaided. He calculated the vast amounts of energy produced in the nuclear fusion of hydrogen to produce helium, and was a protagonist of the idea that this was a source of stellar energy. It was he who coined the physicists' phrase "packing effect" for the decrease in mass—which provides the energy—in nuclear synthesis. Most significant, however, was his proof, which was more advanced than Oddo's, that this was less in complex nuclei that were of *even* atomic number than those of odd atomic number.

The theory was confirmed soon afterwards in experiments by Ernest Rutherford in England.

Having decided that elements with even Z were more stable, Harkins went on to show that they are the very elements that turn out to be most plentiful in stars, meteorites, and the Earth's crust; and so he set the pattern for building databases for elemental abundances. He warned against assuming that elemental abundances in the Earth's crust and in the Sun's gaseous envelope represent the overall composition of these bodies, a warning that was frequently ignored. He shed much light on the history of the solar system, insisting that the composition of meteorites is a better guide to the Earth's evolution than the very limited part of Earth's material to which we have access. After making chemical analyses of 318 iron meteorites and 125 stone meteorites, he concluded that the first seven elements in order of abundance consist of Fe, O, Ni, Si, Mg, S, and Ca (with the usual symbols for the elements). Here was support for the theoretically derived "Oddo-Harkins rule," for here were his elements with even atomic numbers, seven elements that he decided could between them account for no less than 98.6 percent of the material of the meteorites. This was not exactly a *cosmic* abundances ratio, but the whole subject was certainly moving in that direction.

A more recognizably astrophysical contribution made at this period was the work of Cecilia Payne, someone who is often treated as a lone pioneer, but who had the good fortune to study in two important centers with some of the leading scientists of the day. As an undergraduate at Cambridge University in England, she heard lectures by J. J. Thomson, Ernest Rutherford—newly returned there in 1919 after periods in Montreal and Manchester, and now at the peak of his fame—and others at the Cavendish Laboratory, including Niels Bohr. It was when hearing Eddington's lecture on the famous 1919 eclipse expedition to test Einstein's general theory, however, that she decided to pursue a career in astronomy. A fortuitous meeting with Harlow Shapley, then newly appointed as director of the Harvard College Observatory, led her in 1922 to move to the other Cambridge, and there Shapley persuaded her to write a doctoral dissertation, although there was as yet no graduate program in astronomy. Her remarkable thesis on stellar atmospheres (1925) would not have been possible without the Harvard collection of stellar spectra, but just as important was her awareness of new developments in physics. In particular, she made use of the so-called Saha ionization equation, and the work of Ludwig Boltzmann, which was already better known.

The Saha equation, bringing together ideas from quantum mechanics and statistical mechanics, had been developed by the young Indian physicist Meghnad Saha in 1920. This was a time of rapid change in physicists' understanding of the structure of the atom. In 1911 Ernest Rutherford had shown that most of the atom's mass, and its positive charge, are in a nucleus at its center, with electrons outside it—the electron having been discovered as early as 1897. The most relevant new advance, however, was due to Niels Bohr, who had been working with Rutherford in Manchester. Bohr saw that any atom could exist only in a discrete set of stable or stationary states, each characterized by a definite energy level. The atomic model he developed and published in 1913 after his return to Denmark allowed him to account with great accuracy for the series of lines observed in the spectrum of light emitted by atomic hydrogen. A central idea in his theory was that an atom does not emit radiation while it is in one of its stable states, but only when it makes a transition between states, so that the atom does not absorb or emit radiation continuously, but only in finite quantum jumps. This idea flew in the face of the usual assumption, that the frequencies of the radiation emitted by an atom were decided by the frequencies with which the electrons moved round within it. Saha was able to make good use of Bohr's ideas, and of other work that rested on Bohr's. He had been studying in Calcutta, but having made visits to Alfred Fowler at Imperial College, London, in 1919, and Hermann Nernst's institute in Berlin, he wrote a fundamental paper on ionization in the Sun's chromosphere.

The equation at the heart of his paper relates the degree of ionization of an atom to temperature and pressure. (More generally expressed, it gives the degree of ionization of a plasma comprising an electron gas mixed with a gas of atomic ions and neutral atoms as a function of the temperature, density, and ionization energies of the atoms.) The intensities of spectral lines, therefore, depending on ionization levels, can be an indicator of pressure and temperature. Before long, Saha had used his equation to correlate spectral types with temperatures, and his equation provided a starting point work by Edward Milne and Ralph Fowler—Cambridge physicist and son-in-law of Rutherford, as it happens. Milne and Fowler demonstrated that the number of atoms (or ions) responsible for a spectral line can be estimated from the line's intensity, after the temperature and pressure of the stellar atmosphere have been determined. Cecilia Payne now proceeded to make good use of this application of the Saha equation.

Accepting the idea that the great variation in stellar absorption lines was a consequence of their different levels of ionization, and so of their different temperatures, was not enough. She made the further assumption that the number of atoms needed to make a star's spectral line be just visible is the same for all lines of all elements. Eddington, in his *The Internal Constitution of the Stars* (1926), commented that this hypothesis "is not so wild as we might suppose at first," and gave his reasons, before tabulating Payne's findings. She had decided that silicon, carbon, and other common elements seen in the Sun were found in about the same relative amounts as on Earth, but that the helium and especially hydrogen in the Sun were very much more abundant—at first it seemed that the factor might be of the order of a million, in the case of hydrogen. It was not uncommon during the 1920s to hear the Sun's composition being estimated at around 65 percent iron and 35 percent hydrogen, while Russell thought hydrogen to be far rarer even than that. Russell famously refused to accept Payne's idea, although over the next two decades he and others of like mind grudgingly admitted the existence of more and more hydrogen as the years went by, starting from less than 10 percent and finally admitting to well over 70. Payne was herself naturally puzzled by the seemingly vast proportion of hydrogen, and—being greatly influenced by Russell—she decided that the intensity of the hydrogen lines must be, not a sign of its abundance, but of some unexplained abnormality in hydrogen's behavior. By 1928, however, Albreht Unsöld had shown by very different means that her first instincts were correct, and that hydrogen is a million times more abundant than any other element in the Sun—a conclusion confirmed a year later by William McCrea.

Payne's career completed, in a sense, the changes begun at Harvard College Observatory under Pickering, whom we have seen offering limited opportunities to women, but beginning with Payne's PhD—the first in astronomy from Harvard—women began to see that more was possible. Not that Harvard has ever acted quickly: only in 1956 was she given the rank of professor, although even then she was only the second to be given that title (by this time she was known by the name of Payne-Gaposchkin, having in 1934 married a Russian-born astronomer, Sergei Gaposchkin). It is hardly surprising that she has now had thrust upon her the character of a woman oppressed by the dominant male, in this case Russell, on whose advice she had come to reject the extremely high abundance of hydrogen as physically inexplicable. Those who use her biography in this way overlook her indebtedness to Russell, who in 1914—even before Harkins—had used the solar spectrum to point out "the apparent similarity in composition between the crust of the earth, the atmosphere of the star, and the meteorites of the stony variety" (her words, published in "The Solar Spectrum and the Earth's Crust"). Her protagonists also have the advantage of hindsight, ignoring the fact that she, like others who were even more familiar with the new physics, had good reasons for doubt. When in 1926, for example, Eddington referred to Saha's theory of stellar spectra, he commented that it "determines the temperature of a layer rather vaguely defined," and Cecilia Payne was not unaware of that vagueness. While noting that her findings did not apply to giant stars, Eddington was grateful for the temperature scale she produced using Saha's work. (He found good agreement with temperatures he had calculated independently.) Commenting on a certain detail, however, he added this cautionary note:

The abundance here determined depends on the ability of the element to rise to the upper part of the photosphere and may not be typical even of the photosphere itself. The heavy elements are likely to be badly handicapped in showing themselves.

We are reminded of the warning issued by Harkins. It was after the findings of Unsöld and McCrea, which eventually convinced Russell of his error, that Bengt Strömgren—as already mentioned—took Eddington's theory of the interiors of stars a stage further, using the hypothesis that the material in the star is thoroughly mixed and chemically uniform. In 1932 he published a paper announcing that Eddington was more or less correct, and that the observed luminosities were consistent with the idea that hydrogen makes up about a third of the star by mass. This did not entirely settle the thorny problem of surface-versus-interior in the minds of all concerned. Fred Hoyle tells us in his autobiography that the results obtained by Payne and Russell were considered to be valid only for abundances in the atmospheres of stars, not in the depths of their interiors, and that it was for this reason that he, Eddington, and many other astronomers continued to believe in an iron-rich model of the Sun until after the Second World War.

The work on abundances discussed up to this point had dealt with the status quo in the universe, based for the most part on observation, rather than on theories that were aimed at explaining those ways in which it might have come about. In the long term, accounting for the present situation in terms of nuclear processes and their products was done in two fundamentally different ways, and while in principle these could be complementary, they were not always seen in that light by those who developed them. The first is often dated from the time when George Gamow proposed to explain the cosmic abundance distribution of the chemical elements as a result of nuclear transformations of various sorts within stars, transformations in which the recently discovered neutron played a part. His loosely sketched proposals did not lead him very far along the road to accounting for relative abundances, and by 1942 he had abandoned this type of theory— sometimes called "equilibrium theory"—for reasons of the so-called heavy element catastrophe. His first route to a theory of relative abundances through nucleosynthesis in stars did not really come into its own until after the Second World War, in a movement led by Fred Hoyle, and this will be a subject for the following chapter, as will the alternative approach to which Gamow was now favoring. The latter was based on the hypothesis that the key to present-day abundances is to be sought in the thermonuclear reactions that took place in a very early and extremely hot dense phase of the evolving universe. Gamow's ideas were not unlike others that had been put forward earlier by Georges Lemaître. In both cases, the context was that of the sort of model universes being discussed at the time in terms of general relativity. As we shall see, after the Second World War Gamow was joined in his search by Ralph Alpher, Robert Herman, and others.

In due course, the two research paths were combined. Anticipating to some extent what we shall discover when we eventually examine the "cosmological" route to abundances more closely, we may say that it had much success with the proportions of the lighter elements, while stellar nucleosynthesis was more successful with the heavier. For the time being it will be enough to give the merest sketch of a combined explanation. It is now widely assumed that material with an initial composition determined by a hot initial phase of the universe—the Big Bang, as it was to be called—condenses into galaxies. The first generations of stars more massive than the Sun are then assumed to undergo nuclear reactions, the products of which are returned to the interstellar medium. Successive generations of stars, including the Sun and its planets, are next taken to be formed from the material that has been so enriched in heavy elements. Finally, theories of galaxy formation and dynamics are added. Attempts at this last extension of the theory were beginning to achieve some success from the early 1960s onward. Before that final phase was entered, there was a long period of competition between the two broad theories of elemental abundances. One of the most striking things about that competition is how it resulted in the wholesale abandonment of many of the conclusions concerning abundances that had been so dearly won in the 1920s and 1930s. Conclusions that had begun to be treated as sacrosanct were, one by one, discovered to be simply false.

NUCLEAR PHYSICS AND THE SOURCES OF STELLAR ENERGY

Gamow's ideas on "nuclear transformations and the origin of the chemical elements," first published in a rather obscure Ohio journal in 1935, was more concerned with the nuclear transformations than with the abundance ratios they produce. For several years this inspired sketch seems to have been largely ignored, and in some astrophysical quarters to have been entirely unknown. George Gamow himself was still better known in physics than in astrophysics, although that situation changed quickly over the next decade. Born Georgiy Antonovich Gamov in 1904 in the Ukraine, then part of the Russian Empire, he had studied at the university of Odessa there, and then in Leningrad, where he learned something of relativistic cosmology under Aleksandr Friedmann—of whom we shall have more to say in the following chapter, in another connection. Gamow was to physics and astrophysics what a bee is to a flower garden: from Leningrad he went on to study quantum theory in Göttingen, and then between 1928 and 1931 he moved to Bohr's institute in Copenhagen. (He had already shown Bohr his talent in nuclear physics, having in 1928 proposed an important new theory for the alpha decay of a nucleus via a process he described as "tunneling.") During a Copenhagen interlude, he went for a time to Rutherford, in Cambridge now, continuing to study the atomic nucleus, but also working on stellar physics (and abundances) with Robert Atkinson and Atkinson's Polish collaborator Fritz Houtermans. Finally, like so many Europeans of his generation, he settled in America.

Gamow's worth was plain enough to those in the United States who had the means of rescuing him from an increasingly repressive, not to say dangerous, existence in the Soviet Union. He had already published (in Oxford) a well-received monograph on atomic physics when he attended a Solvay Conference for physicists in Brussels in 1933. On that occasion he and his wife defected, and in 1934 he was given a position at The George Washington University, where he spent the next twenty years. One of his conditions was that the Hungarian-born refugee Edward Teller, then in Birkbeck College, London, be appointed with him. He had met Teller when in Copenhagen with Bohr, and Teller would later prove enormously helpful to him. Gamow certainly fulfilled his early promise. In 1938 he interpreted the H-R diagram of stellar evolution, and the mass-luminosity relation in stellar theory, in terms of nuclear reactions. He arranged a conference on this theme, and it was partly as a result of this that Hans Albrecht Bethe went on to discover—later in the same year, and independently of Weizsäcker—the CNO cycle, the key to understanding the generation of energy in massive stars.

In his 1935 Ohio paper, Gamow sketched the many different processes that might be expected to take place inside stars, forming and transforming different elements. He left open the question of the relative importance of the various processes and said little of a quantitative kind. He drew attention to the bombarding of nuclei with protons, and to John Cockroft's recent experiments that had been partly based on Gamow's earlier theories. He noted experiments performed by Enrico Fermi, in Rome in 1934, in which heavy elements had been bombarded with neutrons, which transformed them into their heavier isotopes; and also drew attention to the theory of radiative capture of slow neutrons by heavy nuclei, as worked out by Hans Bethe. (Fermi found that neutrons slowed down in passing through light elements became very effective in performing nuclear transformations.) Gamow's hope that he and his colleagues might prove the existence of comparable transformations in stars was great, in view of the extreme thermal velocities there, quite large enough to produce artificial transformations of the lightest elements. In fact, using Gamow's formula for the probability of nuclear transformations by collision, Atkinson and Houtermans had already calculated the probability of protons with thermal velocity penetrating the nuclei of different elements. They had shown that only the lightest elements are easily transformed by proton bombardment under conditions in the interior of stars. Gamow now explained how even the Fermi effect might be significant, with heavy nuclei capturing neutrons ejected from the nuclei of light elements when the latter collided with protons. In short, he conjectured that all of these newly discovered effects might have some role in element formation in the interiors of stars.

As confidence grew in the idea that stellar energy is dependent on nuclear reactions triggered by very high temperatures in the stellar interior, the lifetimes of stars were once again taken to be of the order of tens of billions of years. The precise nature of the nuclear reactions in stellar interiors, however, and their relative importance, presented so many conceptual difficulties that very few people had the courage to become deeply involved. It was fortunate that the few who did so were physicists of the highest caliber. Gamow was a man with a strong intuitive streak, and he and the great calculator Edward Teller made a fine partnership. Like them, Carl Friedrich von Weizsäcker, Lev Landau, Hans Bethe, and Robert Oppenheimer, were all men with independently distinguished careers in nuclear physics. Of the whole cluster, however, Weizsäcker was perhaps closest to the astrophysical tradition.

THE CNO CYCLE

A pupil of Werner Heisenberg, Weizsäcker had studied Eddington's writings, and had been in contact with Atkinson and Strömgren—he met the latter while visiting Niels Bohr's institute in Copenhagen. Using many of Atkinson's and Cowling's ideas, by the late 1930s he too was in a position to take advantage of new knowledge of the neutron, of deuterium (an isotope of hydrogen, having a neutron as well as a proton in the nucleus), and of various nuclear reactions that had been recently reported by physicists. In 1936, Weizsäcker began a serious study of the most probable nuclear reactions occurring in stars, and of how they would influence the relative abundance of elements. He restated and improved on those ideas we met earlier that had been put forward by Atkinson and Cowling. He explained how light elements might be produced by deuterons resulting from proton-proton reactions; how neutrons might result from deuteron-deuteron reactions; and how heavier elements might be synthesised by reactions involving those neutrons, so taking us up the periodic table of elements. His ideas, when published in 1937, were well received, although the theory was still very largely qualitative in character, and it was hard to reconcile with much of current opinion about the abundances of the elements in stellar atmospheres—of helium in particular, but also of the heavier elements.

In 1938, Weizsäcker abandoned his assumption that the stars began as gaseous spheres of hydrogen, and he started afresh from the assumption that all the chemical elements were formed before the formation of stars. (As stated earlier, this approach to the problem of abundances will be postponed to the following chapter.) Freed from a need to concentrate on hydrogen and helium, he was led to his discovery of the so-called CNO chain of reactions. Its name—although it is sometimes called the Bethe-Weizsäcker cycle—is derived from the fact that carbon, nitrogen, and oxygen act as catalysts in the reaction. The details of the two thermonuclear processes will not be considered here. Suffice it to say that for stars at temperatures above 20 million °K, that is, for stars above about 1.5 solar masses, the CNO cycle is predominant, while for stars around or below the solar mass, the proton-proton reaction is dominant.

The paper in which Weizsäcker presented his second thoughts was remarkable for another reason. He noted that in order to account for the observed abundances of the heavier elements, very high nuclear densities and temperatures would be required. He estimated temperatures of the order of $2 \times 10^{11} °$K. This led him to speculate about a "great primeval aggregation of matter, perhaps comprising pure hydrogen." This, collapsing under its own gravity, he thought might be raised in temperature sufficiently for the required nuclear reactions to take place. He conjectured that this "primeval aggregation" might have encompassed not only the Milky Way but the entire universe. This is often cited as foreshadowing the notion of a primeval fireball, or whatever was responsible for the microwave background radiation—again a subject that will be reserved for chapter 17.

Not long before Weizsäcker completed his revised theory, he learned from Gamow that a similar, but now more quantitative, theory was being developed by Bethe. A former student of the well-known theoretical physicist Arnold Sommerfeld, Bethe had taught in Munich and Tübingen before being forced to leave Germany after the Nazi Party came to power in 1933. Moving to the United States, he eventually joined Cornell University, where he concentrated on nuclear physics in general. It was not until 1938, when attending a Washington conference organized by Gamow, that he was first persuaded to turn his attention to the astrophysical problem of stellar energy creation. Helped by Chandrasekhar and Strömgren, his progress was astonishingly rapid. Moving up through the periodic table, he considered how atomic nuclei would interact with protons. Like Weizsäcker, he decided that there was a break in the chain needed to explain the abundances of the elements through a theory of element-building. Both were stymied by the fact that nuclei with mass numbers 5 and 8 were not known to exist, so that the building of elements beyond helium could not take place. (That some other mechanism was needed was what had led Weizsäcker to turn to the idea of a primeval atom.) Like Weizsäcker, Bethe favored the proton-proton reaction chain and the CNO reaction cycle as the most promising candidates for energy-production in main sequence stars, the former being dominant in less massive, cooler, stars, the latter in more massive, hotter, stars. His highly polished work was greeted with instant acclaim by almost all of the leading authorities in the field. As the tempo of astrophysics slowed down with the advent of war, Bethe's findings seemed to many to be opening a tantalizing new chapter in the understanding of stellar affairs. This would have to wait. Bethe became head of the theoretical division at Los Alamos during the development of the first atomic bomb—although after the war he became an untiring campaigner for arms control. When the astrophysical debate opened up again after the war, the two radically different solutions to the problem of relative abundances, cosmological and stellar, were at its storm center, as we shall see.

NEUTRINOS

Neither Bethe nor Weizsäcker explicitly considered the role of neutrinos in either proton-proton or CNO reactions, although they do need to be included in a complete account; but those who followed up their lead, after the war, did consider them, and it soon became apparent that neutrinos could act as our informant about the solar interior, since they have the property of being able to travel through the Sun with ease.

Three types of neutrino are now recognized. The first—now distinguished as the "electron-neutrino"—was proposed in 1931, by the Austrian physicist Wolfgang Pauli. He postulated the existence of a particle of zero mass and zero electrical charge, to carry away the energy that seemed to be lost in the process of beta decay. (This refers not to the decay of electrons, but to their emission in the form of the "beta-rays" that had been detected coming from decaying radioactive nuclei.) Enrico Fermi later developed Pauli's idea, and gave the particle its present name. An electron-neutrino is emitted along with a positron in what is known as positive beta decay, while an electron-antineutrino is emitted with an electron in negative beta decay. Neutrinos are the most penetrating of sub-atomic particles. They react with matter only through the so-called weak interaction, and do not cause ionization, since they are not electrically charged. A neutrino directed so as to pass through the center of the Earth from outside has only a one in ten billion chance of interacting with a proton or neutron as it does so. It is no wonder that the electron-neutrino was not detected until 1956, when—in experiments done by Frederick Reines and Clyde L. Cowan—a beam of anti-neutrinos from a nuclear reactor produced neutrons and positrons by reacting with protons.

The need to take neutrinos into account in many parts of astronomy became increasingly obvious from the 1940s onward. In 1941 Gamow considered neutrino production as a factor in the cooling of the helium cores of stars at a certain stage in their evolution. Cosmologists adhering to the Big Bang theory of the universe—of which more will be said in the next chapter—soon decided that neutrinos left over from its opening moments are likely to be the most abundant particles in the universe, with perhaps as many as a hundred per cubic centimeter (at an effective temperature of $2°K$). They are produced in abundance in supernova explosions, and a surprising opportunity for observing such neutrinos came with the appearance of a supernova in 1987. They are emitted by the Sun and—as explained—mostly pass clean through the Earth. (It is often pointed out that neutrinos from the Sun shine down on us during the day, and shine up on us during the night.) The Sun was in fact being recommended as our most reliable laboratory for a study of the physics of stellar interiors through neutrino activity there, an idea that was first taken seriously in 1955, and discussed in a paper by Raymond Davis.

Davis proposed a way of detecting neutrinos on the basis of the fact that when chlorine absorbs them, argon is produced. In 1967, he and his colleagues constructed their first detector, in the form of a very large tank of cleaning fluid, placed down a South Dakota coalmine. This they used to show that solar neutrinos do indeed exist, but in far fewer numbers than had been predicted by John Bahcall and others, using what then was the standard model for energy production in the Sun. Were their measurements incorrect? Later measurements were not very different. Was the standard solar model wrong? It does correctly predict many other properties of the Sun. Are electron-neutrinos capable of change into the other types (muon neutrinos or tau neutrinos), which are not being detected? (For more on muon neutrinos and tau neutrinos, see p. 749 in chapter 19.)

Strong evidence for this third possibility was reported in 1998 by a team of Japanese and American physicists, in what has become widely known as the Super-Kamiokande experiment. The evidence came from the first two years of data, obtained from a giant $100 million experiment, using a cavity carved out under the Japanese Alps, lined with stainless steel and holding 45,000 tonnes of ultra-pure water. This was observed with the help of 13,000 light detectors. (It is not the detector that is struck, but an atom that then emits a flash of light, which is what is detected.) Briefly expressed, it was found that muon neutrinos can disappear and reappear as they travel through the earth. It was also decided that neutrinos are not without mass, and that those created in the Big Bang may account for a significant fraction of the mass of the universe. These findings created a considerable stir in fundamental physics.

In the case of solar neutrinos, however, the distance of travel (Sun to Earth) does not vary enough to allow measurements of the type made in the Japanese program. It was realized that a more promising way of proving that the electron neutrinos produced in the Sun are switching into muon neutrinos or tau neutrinos would be to look directly for the latter types. This was the aim of a group responsible for a neutrino observatory in a mine near Sudbury, Ontario, Canada. More than 2 kilometers underground, 1,000 tonnes of heavy water (D_2O) are held in an acrylic vessel surrounded by an ordinary water shield and over 9,500 photomultiplier tubes with light collectors. Costly astronomy is not only a question of telescopes and spacecraft.

There are other neutrino observatories, either already built or planned. In 2006, plans were announced for an American observatory to be buried in the Antarctic under several kilometers of ice—it is already known as the Ice Cube. Another, designed by a consortium of European scientists, a twin to the Ice Cube, will be situated on the seabed in the deepest part of the Mediterranean. Its name tells us its size: the Kilometer Cubed. The cost of about $265 million will be a minor problem, compared with that of allowing for disturbance by glowing crustaceans.

Plans notwithstanding, in 2001 and again in 2002, announcements were made of solar measurements made at the Sudbury Observatory. Non-electron-type neutrinos had been detected, demonstrating directly that neutrinos can indeed change flavor. The theory of matter used in the old calculations was evidently incomplete—a matter of high importance not only to astrophysicists but to physicists, too. To the great satisfaction of most astronomers, the "standard solar model" seemed to have survived unscathed, granted a little renaming of particles. When the number of muon and tau neutrinos was added to the count of electron neutrinos, the total was found to be more or less as the model predicted. In the opening years of the millennium, the stage was clear at last for a resumption of solar neutrino astronomy, to investigate what takes place deep inside the solar core.

EVOLVING STARS AND THE H-R DIAGRAM

The work of Weizsäcker and Bethe on CNO nuclear reactions, and that of Gamow, Hoyle, and others on nucleosynthesis—to be considered in the next chapter—had an obvious bearing on the wider problem of stellar evolution, but before any of this material was available Bengt Strömgren had made significant progress in interpreting the H-R diagram. We recall the common premise—assumed by Russell and opposed by Eddington—that giant stars were younger than dwarfs, a premise on the basis of which Trumpler at Lick had worked out the ages of star clusters. (If the stars in a cluster formed at the same time, then the cluster's age could be expected to increase with the proportion of dwarfs in it.) This common assumption was finally put aside as the result of studies by the young Strömgren in 1932 and 1933: as we saw earlier, he argued that hydrogen was predominant in stellar interiors, and not only in their surface layers. Making the very general but plausible assumption that the power supplies of stars consume hydrogen, he showed that the places of stars on the H-R diagram were determined by their masses and the proportions of hydrogen in their contents. He showed that as a star used up its hydrogen it would leave the main sequence, moving at right angles across lines of constant hydrogen abundance, in the general direction of the red giants on the now fairly standard diagram. He did not go into much detail about the circulation of materials within stars, and he assumed that there was insufficient time in the history of the universe for mass-loss to have been significant, so that evolutionary tracks on the diagram would be lines of constant mass. He decided that as the hydrogen content of a star decreases, the star expands.

Before long, Strömgren was recruited, along with Subrahmanyan Chandrasekhar, by Otto Struve, for the University of Chicago's Yerkes Observatory, of which Struve was the newly appointed director (the appointment was temporary, and he returned to Denmark in 1938). With an earlier recruit, Gerard Kuiper, Yerkes became overnight a leading center for this kind of study. Led by Kuiper, they reinterpreted Trumpler's work on clusters, and showed that while the main sequence stars in the clusters fell neatly together in the lower part of the diagram, stars of higher luminosities diverged greatly. The richer stars were in hydrogen, the longer they seemed to be able to cling to the main sequence. For the time being, not enough was known about the physics of stellar energy for Kuiper and his colleagues to draw any deep conclusion, but his descriptive analysis, published in 1937, was in itself exciting and challenging, and the old assumption that giants were younger than dwarfs began to be questioned more widely.

Before Strömgren returned to Denmark in 1938, he and Chandrasekhar attended a conference in Washington organized by George Gamow and Edward Teller on the subject of stellar energy. Weizsäcker's recent work on fusion mechanisms were known, and Hans Bethe was there to present his own new ideas on the CNO cycle. Gamow's enthusiasm was infectious and helped enormously to bring physicists and astronomers together to address these problems. He continued to throw his energies into devising new models, and it was at this period that he proposed the phenomenon of what he called "shell burning," hydrogen fusion in a shell just outside the enriched helium core of a star, after the hydrogen in its core is exhausted. He tried several other hypotheses with a bearing on whether or not white dwarfs were—as Chandrasekhar thought—at a final stage of stellar evolution, but he was faced with too many obstacles, not excluding Kuiper's insistence that white dwarfs still had hydrogen to burn. Ernst Öpik, however, was drawn to Gamow's shell burning idea, for which he found an application in explaining the nature of red giants. Making use of some of Robert Atkinson's ideas on energy generation, Öpik explained the differences between giants and dwarfs in terms of the way they became stratified after their core hydrogen was exhausted.

Following a period at Harvard, Öpik had returned to Tartu as director of the observatory there, and so unfortunately found it hard to communicate with others in this general area. The war years ahead only made matters worse for him in this respect, although they did not put a stop to his astronomical work. An Estonian, he had studied at the universities of Moscow and Tartu. As an opponent of the Bolshevik Revolution of 1917, he had volunteered for the White Russian army, so that when Soviet wartime occupation of Estonia seemed imminent he thought it expedient to move to Hamburg. In 1948 he moved to Armagh Observatory in Northern Ireland, where he remained until 1981. Öpik's astronomical range was as varied as his geographical location.

Chandrasekhar in the early 1940s was another who was impressed by Gamow's idea of shell burning, and he began to search for ways of introducing it into more complex stellar models—models with convection and lack of chemical homogeneity, for example. An important step had been taken when he and Mario Schönberg, a young Brazilian at Yerkes, worked out an upper limit to mass of a star's helium-filled core that can support the overlying layers of hydrogen against gravitational collapse. As the inert helium core contracts, regions just outside it heat up until they reach temperatures where they can fuse hydrogen to helium. This "shell burning" produces more energy than normal hydrogen burning in the core phase did originally, and luminosity rises, although some of the energy goes into expanding the star. Its radius rises, its surface temperature drops, but its core of helium keeps on contracting to keep up pressure and support the upper layers. So far, it is treated as though subject to the laws for a perfect gas. A point is eventually reached, however, at which the density becomes so high that the core no longer acts as a perfect gas. (It becomes "degenerate," in a certain quantum mechanical sense of the word, with an equation of state which is no longer temperature dependant.) It was evident that once the amount of mass in the helium core reaches a certain fraction of the star's total mass—the "Schönberg-Chandrasekhar limit," today variously quoted as between 8 and 15 percent—some new kind of change would have to occur. Chandrasekhar refused to speculate about any later configuration, since he could find none that seemed likely to be stable. Gamow, however, put his finger on the next evolutionary stage: the star could head for red giant status.

Gamow actually investigated many different models, varying the masses and the proportions of hydrogen to helium, and even their rotations, to see how they might evolve. He found various possibilities, including Wolf-Rayet stars, which he thought might be turned into white dwarfs through a planetary-nebula phase. Red giants and supergiants, however, on the right-hand side of the upper part of the main sequence on the H-R diagram, were also distinct possibilities. Taking matters a stage further, and invoking the "Urca process" that he and Schönberg had devised in 1941, by which the star's core might cool by emitting neutrinos, Gamow even showed that an unstable situation might follow with a nova-like outcome.

Schönberg is often described as Brazil's greatest nuclear physicist, and the Urca process is certainly an important one in its own right. It is a cycle of nuclear reactions in which a nucleus loses energy by first absorbing an electron and then reemitting a beta particle plus a neutrino-antineutrino pair. Gamow chose the name after Schönberg remarked that the energy disappears from the nucleus of a supernova as quickly as money from a roulette table—Mount Urca having given its name to a famous Rio de Janeiro casino.

For all that they learned from one another, Gamow and Chandrasekhar disagreed on very many points of detail. Gamow, for instance, thought that the shell-burning region might migrate through the interior until it reached the surface, while Chandrasekhar thought otherwise. Gamow spent much time trying to understand the evolution of red giants, and introduced a number of subdivisions into his classification. He often found the Schönberg-Chandrasekhar limit thwarting his aims, and while not dismissing it outright, often qualified it to suit his own purposes. Chandrasekhar had young researchers examine Gamow's procedures, and they were all critical of Gamow on this last question, even critical of some of his mathematical working. From 1944 onward, Gamow enlisted the help of observers from Mount Wilson and elsewhere, but they were often unable to provide what he needed. He seized on Walter Baade's discovery of stellar populations when he learned of it in 1944, and matched it to his own ideas on the relative ages of stars. The crucial fact remained, that the Chandrasekhar approach was hostile to his idea of red giants as stars that had evolved from the main sequence. Throughout the latter half of the 1940s and beyond, Chandra and the Chicago school voiced their criticism of Gamow loudly—and slanderously, as Gamow began to think. So bitter did their disagreement become that it frightened off several astronomers who might otherwise have studied the red giants problem themselves. It was almost as though the problem of stellar modeling was doomed to be divisive, for friction at Royal Astronomical Society discussions in Britain was becoming legendary, involving Eddington, Jeans, and Milne; and Chandrasekhar had his own unhappy memories of the 1935 meeting at which Eddington, his Cambridge mentor, had dismissed his ideas so abruptly.

Complicating the whole question at this time was an idea proposed in 1942 by Raymond Lyttleton and Fred Hoyle—an idea touched on obliquely above, in connection with their "sand bank model" and cometary theory (p. 534). They suggested that giant stars simply accrete interstellar material as they sweep through space. The advantage of this hypothesis—or disadvantage, if a neat and tidy universe is one's aim—is that it can explain the inhomogeneous chemical nature of giants, rich in hydrogen in their outer layers, without recourse to a variety of theoretical models of stellar structure. The accretion hypothesis continued to be studied by Hoyle, McCrea, and Bondi, but rather as a side issue to the Chicago-Washington debate over shell-burning.

Other evidence that might have been brought to bear on the evolution problem at this time was Baade's, on his two stellar populations. As we saw earlier, he was reluctant to enter the debate, although year by year it was growing clearer that his distinction was one between young and old stars. In the late 1940s and early 1950s, however, various astronomers began to bring dynamical arguments to bear on the evidence from the composition of stars of Baade's two populations and to suggest that Population I had formed relatively recently, from interstellar clouds rich in grains of heavy elements, whereas Population II stars were very much older, and had formed in some entirely different way. In fact Gamow was pleased to think that the red giant branch of the latter fitted well with his theory of massive shell-burning giants, for both on his theory and from observation it seemed that both were branching off the main sequence around absolute magnitude 2. The great age he wished to assign to them, around three billion years, was not well received, but by the 1950s his evolutionary theories were looking more plausible.

GIANTS AND THE BRANCHING OF THE H-R DIAGRAM

Just as the genial Baade at this period was at the hub of a network of observing astronomers working across a wide range of subjects, so in theoretical work Martin Schwarzschild had a rather similar role, combining, as he did, approachability with high competence. Born in 1912 in Potsdam, Germany, he was the son of the great astrophysicist and pioneer of general relativity

Karl Schwarzschild, who died in military service when the boy was only four. After taking his doctorate from Göttingen, and working for short periods in the Netherlands, Norway, and England, Martin Schwarzschild occupied a fellowship at Harvard for a period, before in 1940 taking up his first permanent appointment at Columbia University in New York. Returning there after overseas war service in the U.S. Army, he moved on in 1947 to a post at Princeton, simultaneously with his near-contemporary and friend Lyman Spitzer, Jr.—so filling a gap that had been left by the retirement of Henry Norris Russell. While in Oslo, before the war, Schwarzschild had worked under Svein Rosseland, who had introduced him to analog computing methods, which he later introduced to Columbia. With the German invasion of Norway, Rosseland actually followed Schwarzschild to the United States, returning home after the war. (He had worked at Mount Wilson in the 1920s.) Their computing techniques would today seem primitive, but they greatly speeded up computation. They did so at the cost of removing the need to seek for elegant mathematical solutions—for instance, in the search for solutions to differential and integral equations. (Huckleberry Finn might have compared their technique to Tom Sawyer's use of a pick in place of a case knife, while insisting on calling it a case knife.) Chandrasekhar—with whom Schwarzschild had much contact in his days at Harvard—greatly approved of his use of the new computing methods in stellar research, and of course, in the long run, they won the day.

Despite Schwarzschild's strong computational approach, he and Spitzer insisted on keeping up Russell's regular research visits to Mount Wilson, and this helped to make him the important catalyst he was, in combining observation and theory. From Baade, who was working hard at improving the redshift-distance parameters in the early 1950s, Schwarzschild learned much about the photometry needed for work on the H-R diagram. With his wife, Barbara Cherry Schwarzschild, he did much observing on his own account. Conversely, Princeton also acted as host to Baade, Greenstein, and Sandage, who were primarily observers, and on the theoretical side to Hoyle and Stromgren; and all of these collaborated with Schwarzschild in serious projects. With its snowballing complexity, astronomy was almost of necessity entering a new cooperative dimension, just as was happening elsewhere in the sciences of the postwar era.

Schwarzschild's first papers were on pulsating stars, and soon after the war he studied the structure of the Sun, using his knowledge of nuclear physics to determine its composition and to explain its differential rotation. His best known work, however, was on the structure and evolution of normal stars. What happens when the hydrogen in the central regions of a star becomes exhausted? This quite fundamental astrophysical question was still regarded by most astronomers as unanswered. Here he was fortunate to work with Allan Rex Sandage, a talented young American astronomer working at Palomar. Born in Iowa City in 1926, he took his doctorate from the California Institute of Technology in 1953. On

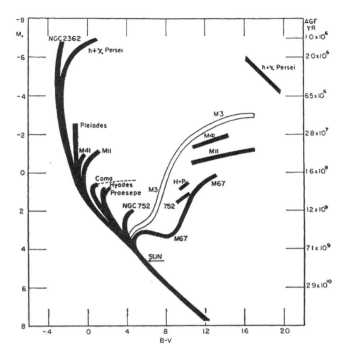

231. A. R. Sandage's "composite color-magnitude diagram of open clusters of all ages." Absolute
magnitude (and also estimated age) is on the vertical scale and color index (B–V, a measure corresponding roughly to spectral type) on the bottom scale. The common main sequence is the
stem of the tree. Each branch belongs to a different cluster. The styles of their names (NGC2362,
M3, M41, and so on) differ for historical reasons. Departure from the main sequence begs to
be interpreted in terms of evolution. Note that most of the branches have missing parts, the so-
called Hertzsprung gaps. Sandage noted how the area left by such gaps is wedge shaped.

the practical side, they collaborated with two other young astronomers,
Halton C. Arp and William Baum, all three making good use of Baade's
plates from the great 200-inch telescope. One serious problem with
interpreting the H-R diagram was the sheer scatter of points on the main
sequence. Sandage, by considering a series of diagrams, each for the stars
of a single cluster, found the scatter in each case to be much reduced. He
and others found that several such refined diagrams for different clusters
could be superimposed, but that when the lines picking out their main-
sequence stars were made to coincide, for different clusters stars started
to branch off the main line at quite different points. Assuming that all the
stars of a given cluster have much the same age and chemical composition, it seemed qualitatively obvious that the position of any one star on
the diagram must depend only on its mass and on the age of the entire
cluster. This suggested that the branching behavior of a cluster could be
seen as an indication of the cluster's age. The first detailed theoretical interpretation of this outline argument was contained in a 1952 paper by
Sandage and Schwarzschild, but a summary statement Sandage published

in 1956 includes an easily appreciated H-R diagram showing what he interpreted as the evolutionary behavior stars in several open clusters of different ages (fig. 231).

Received wisdom at this period, as we have already seen, was that after hydrogen exhaustion in the interior of a star, a helium core would be formed; and that, surrounded by Gamow's hydrogen-burning shell, the core would continue to grow until it reached the size set by the Schönberg-Chandrasekhar limit. Sandage and Schwarzschild considered that the next stage would be gravitational contraction of the core, and they constructed detailed models, showing how this would provide an additional energy source, a weak source, but enough to affect the star's structure enormously. As the core contracts, on this model, the radius of the whole star becomes much greater, and a red giant is formed. Schwarzschild's computations fitted well with observation, particularly for the earlier stages in the evolution of red giants in globular clusters. They had not only provided a good working model, but one that could be taken further to explain the formation of heavier elements by helium burning. As a bonus, assuming that the model proved to be acceptable, they had provided a basic method for determining the ages of star clusters by looking for the branch point on the main sequence. But more than that: the sheer refinement of Sandage's H-R diagrams for clusters was remarkable, for it showed continuous bridges between different areas of the diagram, with a subgiant branch connecting at the blue end with the main sequence, and at its red end with the giant branch.

After this, Schwarzschild continued to work on stellar evolution, generating whole families of stellar models and building into them the best of recent theories, such as Salpeter's triple-alpha process, about which more will be said in the following chapter. Schwarzschild persuaded himself that the mixing of the contents of stars by rotation was not as important as others had been suggesting, and he helped to turn Hoyle's thoughts away from the mechanism of accretion. Hoyle had a better grasp of nuclear processes than he, and was able to improve some of his earlier models. Working together in 1955, they showed how the observed increase in luminosity along the red-giant branch in globular clusters could be understood in terms of an outer convection zone and a new boundary condition at the photosphere. Devising model after model, they eventually accounted for most of the H-R diagram for a globular cluster of Population II stars. All this was done by old-fashioned hand calculating. When eventually high-speed electronic computing methods became available, their models, or variants of them, proved invaluable, and could be improved by others relatively painlessly.

Another important contribution was one Schwarzschild made jointly with Richard Härm—an Estonian-born colleague—in a series of papers beginning in 1961. Here they developed a theory according to which a "helium flash" can occur, a sudden and explosive beginning to helium burning. It occurs, they decided, at the tip of the globular cluster red-giant branch when additional energy due to helium burning is supplied to degenerate core material, which cannot take up the energy by expansion until degeneracy is removed, so that a rapid rise in temperature occurs. Energy output jumps by a factor of many billions, but only for a matter of seconds, before it settles down again. (There are further situations when a helium flash can occur, as other astrophysicists later showed.)

Schwarzschild was not yet at the halfway mark of his career, which yielded many other important results. Later, with Spitzer, he investigated the gravitational interaction of stars in the galactic disk with gas clouds, so helping to explain the evolution of the Galaxy in dynamical terms. And as a reminder that knowledge needs to be passed on, we have his outstanding textbook, *Structure and Evolution of the Stars* (1958), which is comparable in stature and influence to Eddington's classic of over thirty years before. Schwarzschild's professional life was a fine instance of what peaceable cooperation between observers and theoreticians can produce. In his awareness of those two worlds, he reminds us of Eddington, but with one great difference: Eddington was much more of an individualist.

OTHER REFINEMENTS TO THE H-R DIAGRAM

Although the main sequence on the H-R diagram accounts for a majority of stars, we have already come across other important types that do not lie on it, including the red giants and supergiants—cool and yet very luminous, on account of their great surface area—and white dwarfs, at first sight utterly distinct from the red giants, but many of them sharing a family history. The white dwarfs are degenerate stars of low luminosity, which have exhausted all possible fuel for thermonuclear fusion. We mentioned them earlier, in connection with the Chandrasekhar limit and their collapse into neutron stars or black holes. The first to be recognized, as early as 1910, was 40 Eridani B, a star with such a low luminosity that it was thought to be smaller than the Earth. Another dwarf that has achieved a certain fame is Sirius B—the companion star mentioned above, which had first been seen in 1862 without its character being properly recognized at the time. Relatively few white dwarfs—a matter of some hundreds—have been recorded, merely because they are of such low luminosity. It is thought that they might nevertheless number a tenth of the stellar population. Their spectra are complex, and of many subtypes, and a new classificatory scheme for them alone was proposed in 1983, by E. M. Sion and his collaborators. (The 1980s, as it happens, saw a number of new fashions in spectral classification. Even the concept of "main sequence" was redefined on several occasions, some astronomers restricting it to a range of luminosities, others preferring to define it in terms of theoretical structure or evolutionary stage.)

There are many dozens of stellar types, each with its own evolutionary history, some accounted for, and others still unexplained. Red giants are found to be rich in carbon or rich in oxygen. "Carbon stars," as the former are called, were originally given spectral types R and N in the Harvard classification; later they were shunted into classes K and M; and in the 1940s, Morgan and Keenan devised a whole new sequence of classes to describe them. Rare in our Galaxy, many thousands have been found in the Magellanic clouds.

A pair of star types that have been much studied, in the hope that they may shed light on the processes of stellar evolution, first came to general notice early in the twentieth century, at a time when the distribution of the bright stars was being investigated by Jacobus Kapteyn and Anton Pannekoek in the Netherlands, and by N. H. Rasmuson in Sweden. It became clear that O- and B-type stars are not distributed randomly in the sky, and are not gravitationally bound to one another, but tend to be associated, inasmuch as they follow the arms of our Galaxy. In 1949, assisted by Bart Jan Bok's earlier studies of their dynamics, Viktor Amazaspovich Ambartsumian showed them to be young, much younger than the rest of the Galaxy—a conclusion confirmed later, when they were fitted into color-magnitude diagrams. Interest in them grew. In the 1960s, Adriaan Blaauw investigated their links with interstellar matter, and others followed suit in the 1970s and 1980s, demonstrating that "OB associations"—Ambartsumian's expression—are usually located in or near regions where star formation is taking place. This fact of course makes them especially interesting. Their youth implies that star formation is still taking place in the Galaxy. (For a Hubble Space Telescope photograph related to this question, see plate 15.) The evidence that accumulated during the twentieth century for the existence of a rich interstellar medium, of gas and dust, helped to show how this was possible. At the same time, it was slowly realized that direct observation of the process will inevitably be thwarted optically, and indeed at most wavelengths, by the fact that most star-formation is taking place inside dark clouds. Hopes are presently pinned on instruments using infrared and millimeter wavelengths.

As we have seen so often, astronomical objects and properties have frequently been postulated on the basis of theory, before they were discovered empirically. In making a selection of stellar types, we must not overlook one highly significant class of stars that illustrates this. In the 1970s, it was being asked whether stars may exist with such small masses—the theoreticians settle on eighty times the mass of Jupiter—that they are unable to raise their central temperatures sufficiently to enable hydrogen fusion, and in this sense can never become true stars. Such objects are now known as "brown dwarfs," a term coined by Jill Tarter in 1975 (they were earlier called "black dwarfs"). None was known at the time of the opening discussion. It was admitted that they would be difficult to identify: the smallest true stars (red dwarfs) may, for instance, be so cool as to be indistinguishable from such objects. During the next two decades, several candidates were proposed, without being unequivocally accepted, but then, yet again, theory came to the aid of the searchers. It was shown that

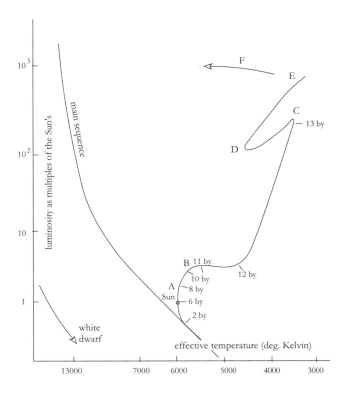

232. The evolutionary path of a star similar to the Sun, in a broadly "Hertzsprung-Russell" form, is not strictly historical but gives an overview of converging opinions. It shows the evolution of a star of one solar mass. The left-hand half of the scale of temperature (which can be related to spectral type) is distorted, since our main interest is in the right-hand side of the figure. The age of the star is given in billions of years (*by*). At *A*, there is still combustion of hydrogen in the core. Once the star exhausts its core supply of hydrogen, at *B*, after perhaps ten billion years of life, nuclear reactions decline rapidly. The helium-rich core begins to contract, since radiation pressure can no longer hold back gravitation, but the contraction releases large amounts of gravitational potential energy, which causes increased hydrogen-burning in a shell around the core. For a star of solar type, this leads to an expansion of its outer layers, and a rapid increase in its luminosity. Regardless of its original spectral type, its temperature drops, but its vastly expanded surface area—in size the giant will grow to be of the order of the Earth's orbit or more—will more than compensate. As the inner core gains mass (from the production of helium in the shell, at 11–12 billion years), it contracts, and the envelope expands (12–13 billion years), producing a red giant. At *C*, the *central* temperature exceeds 10^{8}°K. Reactions fusing helium start violently (the "helium flash"). This continues until fusion reactions involving helium restore a temporary equilibrium, with gravity once again balanced by internal pressure. The star leaves the red giant domain, although not permanently. From *D* to *E*, the core becomes increasingly inert, rich in carbon and oxygen, the products of helium fusion. Energy output diminishes, followed by contraction of the star as a whole, but by expansion of the inner layer where hydrogen is fused into helium. (The outer regions are rich in hydrogen.) The star resumes red giant status. From *E* to *F*, through the burning of helium and hydrogen in the outer layers, a very rapid change takes place. The outer atmosphere of the star is expelled, so producing a planetary nebula, and leaving behind a stellar remnant in the form of a white dwarf (lower left). The size of the star at the various stages of its evolution are not marked on this figure, but (taking the present solar radius as unit) its size at *B* is roughly 2 units; at *C* it approaches 30 units; and at *D* is about 15.

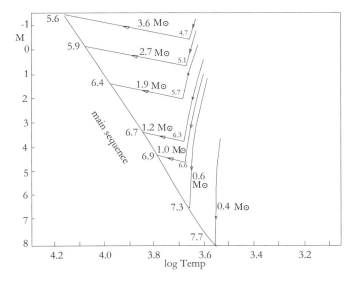

233. Evolutionary tracks, as calculated by Chushiro Hayashi in 1961. The masses of the stars are marked in multiple of the solar mass, the logarithms of their ages (in years) being marked twice, at the turning points and when they finally join the main sequence. The scales are of absolute magnitude (M) and the logarithm of the temperature.

the detection of lithium in the atmosphere of a candidate would ensure its brown dwarf status. In 1995, three incontrovertible substellar objects were identified by the presence of spectral lines of lithium. Before long, some hundreds were identified, and some astronomers came to believe that brown dwarfs may be the most numerous type of body in the Galaxy. It must have seemed that the nobility of the heavens was in danger of being reduced to dust and brown dwarfs, but as history would show, there were other pretenders to the title.

By the early 1960s, theories of stellar evolution had reached a certain plateau, and astrophysicists were beginning to work more amicably toward a consensus theory. It would be a mistake to suppose that the only relevant activity was in the Western world. There were significant developments being made at this period in Japan, especially by the astrophysicist Chushiro Hayashi and his students and colleagues. While much of the theoretical underpinning of their work will not be introduced here until late in the following chapter, it will be convenient to refer to a synthesis he presented in 1962 of work done on stellar evolution up to that time. He explained in some detail how a star could leave the main sequence, become a red giant, shrink again, then return to red giant status, before finally expelling its envelope and finishing life as a white dwarf. For a more detailed sketch of these processes, see figure 232. (More massive stars follow a more complex evolutionary track away from the main sequence. They fuse heavier elements in their cores, and may eventually end their lives in supernova explosions, violently ejecting much of their original mass, rich in heavy elements. Their remnants may include neutron stars or black holes.) While this survey rested largely on the work of others, Hayashi had added some highly significant contributions of his own.

Chushiro Hayashi was born in Kyoto, where in the 1950s and 1960s he did his most important work in devising models for stellar evolution. One of his early achievements was to show that at the high temperatures characteristic of the very early universe, electron-positron pair production had to be taken more seriously, and this led him to revise estimates of the early neutron-proton ratio, and to provide a better value for the abundance of helium in the universe. Later he worked on the formation of low-mass stars, and with Takenori Nakano made one of the earliest studies of what, as we have seen, are now called brown dwarfs. He is best known, however, for the account he offered in 1961 of what happens during the birth of stars. He described and explained a period of extreme activity when these proto-stars are many tens of times brighter than a main-sequence star. Demonstrating that they are fully convective, and not in a kind of hydrostatic equilibrium (radiative equilibrium), he explained how they move along what are now called "Hayashi tracks" on the H-R diagram, going downward (the stars contracting as they do so at near-constant surface temperature) until they suddenly turn up and head for the main sequence (fig. 233). The "Hayashi phase" of stellar birth filled a prominent gap in the theory of stellar evolution, and can be taken as rounding off the first great period of an astrophysical understanding of what lies behind the bare "Hertzsprung-Russell diagram." On this diagram, by the 1960, few traces of Hertzsprung or Russell remained.

The Renewal of Cosmology

THE ORIGINS OF RELATIVISTIC COSMOLOGY

The ancestry of almost all modern theories of the overall structure of the universe can be traced back in part to the ideas of Albert Einstein, which he developed between 1905 and 1915. A paper he published in that final year contained his general theory of relativity in a developed form. Einstein's two theories of relativity, the special theory and the general, are too complex to be briefly sketched, but central to both is the idea that the physical laws governing a system of bodies should be essentially independent of the way an observer studying those bodies is moving. In the special theory of 1905, Einstein considered only frames of reference moving at constant relative speed. Into this theory he introduced a very important principle: the measured velocity of light in a vacuum is constant and does not depend on the relative motion of the observer and the source of the light. He drew several important conclusions. One was that if different observers are in relative motion, they will form different conclusions about the relative timing and separation of the things they observe; and that instead of distinguishing sharply between space and time coordinates, we should consider all together as coordinates in a combined space-time. Another conclusion was that the mass of a body increases with its velocity, and that the speed of light is a mechanical upper limit that cannot be crossed. Perhaps the best-known of all his conclusions, summarized in the famous equation $E = mc^2$, was that mass and energy are equivalent and interchangeable.

All of these principles were anticipated to a greater or lesser degree by earlier physicists, but it was Einstein alone who bound them into a single and elegant physical system. The conversion of nuclear mass into nuclear energy is of course a fact of modern life, but understanding the conversion of mass to energy has also been of the greatest importance to an understanding of the production of energy in stars—a subject already touched upon in this book.

Einstein's general theory of relativity was very much more his own creation than the special theory had been. In it he considered how laws of physics are changed when referred to frames of reference *accelerating* relatively to one another. In his special theory, space-time was somewhat like the simple space of Euclidean geometry. It was "flat." You can work out separations in space-time by simply applying an extension of Pythagoras's theorem for spatial geometry. From the 1830s onward—and there are even earlier traces of the similar ideas—mathematicians had been developing theories of non-Euclidean geometries where space was said to be "curved," by analogy with the geometry on the surface of a sphere. On the surface of a sphere, of course, Pythagoras's theorem works only as an approximation for very small triangles.

Einstein required the space-time in his general theory of relativity to be curved. He laid down principles by which the curvature—which might vary from place to place, as it does on the surface of almost any physical object in the familiar world—is produced by matter. One of his most brilliant and important ideas was that which concerns the behavior of particles moving freely through this curved space-time. They move, he said, along geodesics. A geodesic in space-time is analogous to the shortest line between two points in Euclidean space. On a two-dimensional flat surface, this is a straight line. If we confine ourselves to the surface of a sphere, the shortest distance—and also the longest—will be along a "great circle" between the points. In Einstein's theory, the space-time geodesics are such that a particle that is moving freely under gravity follows them. There is no need for an extra law of force to account for gravity. Gravity is built into the geometry. There are also in general relativity certain special kinds of geodesics, of null length, and these are the tracks of light. Again, the geometry takes care of the physics, as it were, and this in a very elegant way.

We shall later have occasion to mention various models of the universe, in which assumptions made about the distribution of matter have consequences for "the geometry of space-time." The universe, the sum total of all matter, may of course change—time and change are aspects of space-time. One of the most conspicuous properties of Einstein's theory of gravitation is that small-scale gravitational problems cannot be solved in principle until the geometry of the space-time is known, and this requires a knowledge of the entire material system. The new theory of gravitation, therefore, quite inevitably became implicated in the development of a cosmological view. The first relativistic model for the universe was in fact announced by Einstein on 8 February 1917. It was called a "cylindrical" model, but the word here has an extended meaning, based on a mathematical analogy. (It treated space as a three-dimensional surface of a cylinder in four dimensions.)

We often make use of the phrase "the geometry of space-time," and it is occasionally useful to think of this as nothing more than an oblique reference to the set of rules for calculating intervals between "points," but points with space and time coordinates. Using familiar geometry, as mentioned earlier, to find the distance between points in space with known coordinates, we may make use of the theorem of Pythagoras. The non-Euclidean geometries—whether of space alone or of space-time—use more complicated principles in place of this, more complicated rules of calculation; and the masses and energies of the system of necessity enter into those rules.

Einstein was by no means the first to use non-Euclidean geometries in physics. Four earlier instances deserve brief mention. In the 1830s, Nicolai Lobachevskii, one of the three mathematicians chiefly responsible for the final development of non-Euclidean geometry—the others were János Bolyai and Carl Friedrich Gauss—had proposed an astronomical test for the curvature of space. This required a knowledge of the parallaxes of remote stars, and yet no reliable parallaxes were then available. Another attempt to link the new type of geometry with the physical world was that of the German mathematician Lejeune Dirichlet, who studied the law of gravitation in non-Euclidean space toward the end of 1850. This seems to have been generally considered as no more than an interesting mathematical excursion. The astronomer Karl Schwarzschild, toward the end of the century, was able to use similar arguments to Lobachevskii's, with later parallax measurements, to set upper limits to a parameter known as the space curvature, for two different sorts of geometry. Auguste Calinon had as early as 1889 gone so far as to suggest that the discrepancy between our space and Euclidean space might vary with time. There had been other sporadic attempts made to forge empirical links with some newer types of non-Euclidean geometry that had been developed in the course of the nineteenth century, but Einstein was the first to make use of the idea that gravitation is explicitly related to the geometrical structure of what is called "Riemannian" space-time. The name comes from the fact that he followed methods established by the German mathematician Georg Friedrich Bernhard Riemann in the 1850s for an analytical treatment of non-Euclidean geometry.

When Einstein's general theory was first published, the Western world was divided by war, as it would be once again within scarcely more than two decades. The interwar period was a golden age for the development of an extraordinary number of new and exciting cosmological ideas, but even before then, as we saw in the last chapter, a few of the best astronomers were coming round to the idea that the nebulae were "island universes," galactic systems comparable with our own. We have seen how Easton thought that our Galaxy is like the spirals, but that did not take matters far enough. Eddington, for example, who was as well-informed an authority as any, in his *Stellar Movements and the Structure of the Universe* (1914) had to confess that direct evidence for the nature of the spirals was entirely lacking. He could not say whether they were within or without our own stellar system, but he was of the opinion that the island universe theory was "a good working hypothesis." He still spoke of the structure of the sidereal universe: the significant unit member of the universe was for him the star, and not until he was within twenty pages of the end of his book did he touch briefly on the nature of the spirals. We have seen how this intuition was confirmed by careful observational work in the 1920s. What is so remarkable about the historical situation is that the new relativistic cosmology was already waiting in the wings, fully prepared for those new empirical discoveries.

OLD ASSUMPTIONS QUESTIONED

The general theory assumes importance chiefly on the large and very large scales. Of the numerous astronomical predictions made by Einstein and others, differences between them and the Newtonian theory—on which it was to a large extent modeled—are slight, but when very large stellar and galactic masses are concerned, and when distances are comparable with those between galaxies, then predictions are usually very different. An important stimulus to Einstein's researches was an awareness of problems even on the scale of the solar system. One puzzling fact was that, while Newton's theory of gravitation—when subtle perturbations of planetary motions by other planets were taken into account—seemed to be able to account for most planetary observations to a very high degree of accuracy over centuries, and even millennia, the perihelion of Mercury was found to be advancing at a rate that could not be explained. The smallness of the advance gives a measure of how extraordinarily refined astronomical techniques had become. Newcomb had derived a figure of 43 seconds of arc per century for the unexplained part of the advance of the perihelion—a quantity so small that it would take over eight millennia to move through just one degree. Einstein's theory yielded a figure only a second of arc smaller than this.

There were other problems, some of which have been touched on in earlier chapters. When applied to cosmological problems assuming an infinite universe, ordinary Newtonian theory, based on the familiar (Euclidean) geometry, seemed to lead to inconsistencies. Why does the matter in the universe not coalesce under gravity into a single mass? In fact, in the mid-1890s, Carl Neumann and Hugo von Seeliger, to name only two, tried to modify the Newtonian law of gravity to remove such difficulties. In doing so, strangely enough, they introduced what was effectively a cosmic repulsion—one that they supposed worked against the much more powerful gravitational attraction—that was to have its counterpart in later relativistic cosmology. Such theoretical difficulties might seem remote from astronomy, but they were at least as important to future cosmology as the realization that the spirals were comparable in status to the Galaxy.

Accounting for the behavior of matter of small and finite average density, uniformly spread over an infinite universe, led all too easily to paradox. There were paradoxes of infinite gravitational force (or gravitational potential), but there were others, not at first thought to connect with the gravitational properties of the world, such as the old paradox of the dark night sky. Some tried to avoid these problems by modifying physical laws slightly, while others—such as the Swedish astronomer Carl Charlier, in 1908 and 1922—wished to preserve the laws, and accordingly modified the assumptions generally made as to the distribution of matter in the cosmos. Charlier replaced the notion that matter is on average homogeneous throughout the universe with an assumption—reminiscent of Lambert's—that the universe is arranged hierarchically, as a series of systems within systems. Independent empirical evidence for such an idea was virtually nonexistent, although it remained an open question as long as the status of globular clusters and spirals was uncertain. Yet a third way of modifying the traditional treatment of large-scale gravitational problems was to dispense with the idea that the space we inhabit obeys the rules of ordinary Euclidean geometry. As we know, this was the course followed by Einstein.

In all these cases, the approach was truly cosmological, in the sense that it concerned the totality of matter in the cosmos. Moreover, what was gradually dawning on those concerned with these questions was something that had already come under discussion in the post-Copernican period, namely, that there is a strong element of convention in the path one chooses to follow when drafting theories. If one happens to feel strongly about Euclidean geometry, and if the empirical evidence seems to be placing it under threat, then—as Henri Poincaré realized—you may be able retain the geometry if you are prepared to modify, for example, the laws of optics. Those who feel strongly about the need to preserve the simplicity of the Newtonian law of gravitation may likewise be able to do so by modifying the geometry. Cosmology, true to its ancestry, showed the way to a liberal treatment of the whole notion of scientific truth, and the message has not been lost on the other physical sciences, where it took longer to penetrate.

Before the second quarter of the twentieth century, Newtonian and Euclidean principles were, for the great majority of practicing scientists, unquestionable truths, and resistance to changing either was very strong indeed. To take another example of an almost universally held belief from the first three decades of the century, and of course before then: almost everyone took the universe to be on average *static*. The generally unchanging pattern of stars and galaxies seemed to be guaranteed by centuries of observations. It seemed only marginally less obvious than that classic of the obvious, the darkness of the night sky. In the absence of large proper motions, who would have expected large radial velocities? The postwar period was a time when increasingly numbers of large radial velocities were being found among the nebulae, *as judged by Doppler shifts*, but so strong was the conviction that we inhabit a static universe that, long after this discovery, there was a veritable industry in finding alternatives to the usual interpretation of the spectral shifts as indicating real velocities. As will be seen, even Hubble, a central figure in this episode, had his doubts.

Explanations of the anomalous advance of Mercury's perihelion were, roughly speaking, of three sorts. Some, such as Le Verrier in 1859, thought to postulate invisible or barely visible matter, such as asteroids round the Sun. Others thought zodiacal light might suffice. Still others tried modifying Newton's law of gravitation. Asaph Hall was probably the first to do so in this context, in 1894. Ironically, in view of his cosmological objections to Newton's law, it was Seeliger who first proposed the zodiacal light hypothesis, in 1906. A third group attempted to introduce other physical forces than gravitation—electrical forces, for example. These various hypotheses were subject to intense discussion, especially between the years 1906 and 1920, and many new ideas were tried out. Not only was gravitational absorption, for instance, discussed by theoretical astronomers, but they attempted to detect it in elaborate experiments. It even provided a cosmological apprenticeship for Willem de Sitter, who in 1909 and 1913 made a critical study of the principle.

During these early years of the century there was far more professional contact between the applied mathematicians and the astronomers than is often supposed. Einstein was much helped, on astronomical questions, by Erwin Freundlich. De Sitter and Eddington in particular both helped to develop the general theory of relativity and to integrate it into astronomy. Both were eminently qualified for the task. De Sitter was yet another of Kapteyn's many influential pupils, perhaps the greatest of all. He worked under David Gill at the Royal Observatory in Cape Town between1897 and 1899, and then returned as assistant to Kapteyn in Groningen. He moved to Leiden in 1908. Eddington, as we have seen, had served at the Royal Observatory at Greenwich, where he remained from 1906 to 1913. He had led an eclipse expedition to Brazil in 1912, and it is not surprising that he was the man who led one of the two British expeditions in 1919 that offered the first empirical support for one of Einstein's predictions—that light grazing the Sun's surface would be deflected by a specified amount. (A dramatic demonstration of this effect is illustrated in a photograph taken from the Hubble Space Telescope in 1990. See fig. 245 on p. 736.) Both De Sitter and Eddington were experienced in the statistical analysis of proper motions and star counts in the Kapteyn tradition and in galactic modeling on that basis, and both were fully aware of the latest astronomy of large scale. It was, however, within the context of Einstein's general theory that each made his first important contribution to cosmology proper.

From about 1911, De Sitter occupied himself with the potential repercussions of the relativity theory on practical astronomy. He interested himself in a variety of fundamental problems, such as the interrelationship of Mach's principle and general covariance two ideas with a strong bearing on general relativity), the Ritz theory of light emission (a rival to Einstein's special theory), and the astronomical relevance of the relativity of time (which featured in both of Einstein's theories). He was regularly in correspondence with Eddington during these years, and met and discussed problems of common interest with Einstein and the physicists Paul Ehrenfest and Hendrik Lorentz in Leiden. Einstein valued this contact, because it allowed him—through two papers by De Sitter in 1916—to make his ideas known in Britain, while the war was still raging. Eddington pacified his colleagues with the message from De Sitter that Einstein was anti-Prussian.

MODELS OF THE UNIVERSE

Eddington appreciated the revolutionary nature of Einstein's new work as soon as he learned of it from De Sitter. He threw himself wholeheartedly into a study of the mathematical underpinning of the general theory, and wrote the masterly *Report on the Relativity Theory of Gravitation* on it in 1918, the proof sheets of which were read by De Sitter. We may see this as a test run for Eddington's *Mathematical Theory of Relativity* of 1923, a work described by Einstein in 1954 as the finest presentation of the theory in any language. toward the end of his report, in the context of De Sitter's model—which will be mentioned again shortly—Eddington was able to refer to "the very large observed velocities of spiral nebulae, which are believed to be sidereal systems," and to add that it "is not possible to say as yet whether the spiral nebulae show a systematic recession, but so far as determined up to the present, receding nebulae seem to preponderate." This reference to the fact that many nebulae had been found to have velocities away from the Sun, and that the recession might be "systematic," was a portent of things to come.

In 1916, Paul Ehrenfest suggested to De Sitter that some of the difficult problems associated with infinity in understanding our universe might be avoided if one took instead a *closed* model. This refers to a type of non-Euclidean space-time that can be thought of as analogous to a sphere in ordinary space, which has a closed and finite surface, but which allows one to move endlessly around a great circle on its surface. It is unbounded, and in that sense infinite, without end. In 1917, Einstein tried to find a static model with the same sort of spatially finite character, but could not do so without introducing a notorious "cosmical term." This gives rise, even in empty space, to a repulsive force which would allow a model universe to remain static if the cosmic repulsion exactly balanced the gravitational attraction of the matter in it. The term was considered to be a universal constant of unknown but necessarily very small value. He gave various interpretations of this constant, and for more than a decade there were many surprisingly strong views aired as to its virtues and its vices. De Sitter retained it, but referred to it as "a term which detracts from the symmetry and elegance of Einstein's original theory, one of whose chief attractions was that it explained so much without introducing any new hypothesis or empirical constant." What they did not realize was that the cosmological constant, as it is now always called, was a veritable Trojan horse, carrying within it a solution to a cosmological phenomenon as yet undiscovered. We shall meet the parameter again in a modern setting in chapter 20. It is

still known by the Greek symbol Λ (lambda), which Einstein chose for it
in his new field equations.

As explained, one of Einstein's new principles in the general theory
was that gravitational masses affect the entire system. Conversely, he believed that gravitational behavior required matter—for how can the geometry be modified without it? He thought that there should be no solution
of the field equations—the equations that forge a link between matter and
geometry—that describe a universe empty of matter. He was very soon
shown to be mistaken in supposing that his field equations would not allow this, for De Sitter presented a trio of solutions that seemed especially
strange. (He stipulated that his model should be isotropic—meaning that
appearances be the same in all directions—and static. He also demanded
that the spatial part of space-time satisfy certain conditions of constant
curvature.) One of these models was Einstein's own. It had matter in it of
finite density (related to the cosmical term) but zero pressure. Another
model had zero density, zero pressure, and zero cosmical term. The third
was a solution we now know by De Sitter's name. Density and pressure in
this were both zero, and the model had a strikingly interesting property:
it suggested that an observer should see a reddening of distant sources of
light within it—overlooking the fact that the model did not actually allow
for any objects with mass capable of sending out light in the first place! At
a time when redshifts in the spectra of the spirals were being found in increasing numbers, this seemed as though it might be highly relevant to the
real world, despite the fact that the model seemed inadequate to represent
mass in the universe.

Despite its potential relevance to the world of the spirals, De Sitter's
model alarmed some of his readers, including Einstein, for another reason. It revealed a "horizon" for every observer in it, a certain distance at
which any finite space-time interval between two events would correspond
to an infinite value of their time-coordinate interval. As then interpreted,
Nature would appear to be there at rest. In Eddington's words, "the region
beyond . . . is altogether shut off from us by this barrier of time." Horizons
of various similar kinds have played an important part in cosmological discussions ever since De Sitter's example. Several initial confusions as to the
nature of clock time and coordinate time needed to be sorted out, but here
again De Sitter did useful service, as did Eddington.

The reddening of distant sources, known as the "De Sitter effect," was
not strictly interpretable as a Doppler effect. Had it been so, it would have
seemed immediately to indicate a general recession of the spirals. However, both De Sitter and Eddington did argue, along a different route, for
recessional movements. They calculated that a number of particles initially
at rest in the De Sitter world would tend to scatter, up to a certain limit, at
which their velocities would be comparable with that of light. The conclusion was much criticized in the 1920s, and there was in the same decade
much more polemic over the interpretation of other points in the same

theory. Interest in De Sitter's solution soon waned, but only because it was soon supplemented by others that were no less exciting, and that seemed much more plausible.

It might be worth adding here that as late as the 1970s there were a few physicists—in Paris in particular—supporting the idea of a reddening of light in the course of its travel over long distances. This hypothesis of *photons fatigués* (tired photons) has now been generally abandoned. It usually went with a belief in a static universe. The reality of the recession of distant galaxies is now generally taken to be guaranteed by observations of a type of supernova (Type 1) that brightens and fades in a clock-like way. Such "clocks" are found to run slow, so to speak, when they recede from us, and their slowing is directly related to the redshift. For almost every modern astronomer, redshifts mean true recession; but reaching this conclusion certainly did not come quickly or easily.

THE EXPANDING UNIVERSE

Around the time that Einstein was putting together his general theory, there was a consensus among the best astronomers that Kapteyn's picture of the stellar universe, or Karl Schwarzschild's rewriting of it, was more or less acceptable. Through their statistical methods, especially as extended in scope by Oort, astronomers were becoming "universe-minded" just at the time that general relativity entered its cosmological phase. The evidence being gathered with the help of the new American telescopes suddenly turned the highly mathematical investigations into more than academic exercises. We have seen how in 1918 Shapley argued for a vast increase in the diameter to be assigned to the Galaxy, and in the distance of the Sun from its center. Distances for the spirals were beginning to come in, but what of their motions?

H. C. Vogel had observed Doppler shifts in starlight as early as 1888, and William Huggins had tried as early as 1874 to measure the radial motions of the "nebulae" spectroscopically, but his apparatus was not equal to the task. The first real successes were in 1890–1891, when James Edward Keeler found the radial velocities of ten planetary nebulae. Keeler, an American who had also studied in Germany before taking up various appointments in the United States, had the great advantage at Lick of highly sensitive photographic plates, as well as a fine concave diffraction grating by the finest maker of such things, Henry Rowland of Baltimore. Before his sudden death in 1900, Keeler had time to make several further measurements of a similar sort. After Keeler's death, his mantle was passed to William Wallace Campbell, not only as director of the observatory but as a leading spectroscopic observer. As early as 1893, a wealthy trustee of the observatory, Darius O. Mills, had given funds for a spectrograph, designed to Wallace Campbell's specifications and worthy of the great 36-inch Lick refractor. (Campbell had a way with millionaires: Phoebe Apperson Hearst, Randolph's widow, later gave him an automobile worthy of the observatory.) He made important radial velocity measurements not only there but in Chile, where he set up a southern station. After he turned sixty—Campbell was born in Ohio in 1862—his astronomical work went into partial eclipse, when he was made president, first of the University of California, and then of the National Academy of Sciences. His most notable work, however, was on stellar radial velocities. Beginning at Lick in 1896, when he inaugurated a program of cataloging such data as a means to establishing the Sun's path through the stars. This resulted in a collection of more than 25,000 spectrograms for more than 2,770 stars over 30 years. Time was nevertheless set aside for the measurement of the radial velocities of 101 "gaseous nebulae," which were used for a study of movements of what turned out to be "our" Galaxy.

It was Vesto Melvin Slipher of the Lowell Observatory in Flagstaff, Arizona, who, toward the end of 1912, was the first to measure the radial velocity of a spiral nebula. A graduate of Indiana University, he had been at the Lowell Observatory since 1903. His telescopes were small and his spectrographs slow, but he became highly practiced in spectrographic work, spending several years measuring the periods of planetary rotation by this means. (He was the first to detect bands in the spectra of Jupiter's satellites, a true test of his artistry.) He found that M31, the great "nebula" in Andromeda, was approaching the Sun at about 300 kilometers per second—the highest radial velocity then recorded. By 1914, Slipher had thirteen velocities—or spectral shifts, if we wish to reserve judgment, and consider the possibility that they were the De Sitter effect. By 1923 he had compiled a list of velocities for 41 nebulae (galaxies), 36 of which were receding from us (spectra shifted to the red). One of them was moving away from us at 1,800 kilometers per second, an astonishing speed; and later he added more. Such enormous velocities as he was finding made it quite clear to the few who pondered the mechanics of the case that the nebulae concerned must be well outside the gravitational influence of our own Milky Way system.

When Eddington had asked whether there was "systematic" recession of the nebulae, he seems to have had in mind a double effect: the De Sitter effect and the scattering of nebulae of which we have spoken in the same connection. Hubble later picked up this idea. There were astronomers who simply hoped to derive the *Sun's* velocity from Slipher's results. Carl Wirtz and Knut Lundmark were among these, but they soon realized that they were dealing with a more mysterious phenomenon. In 1925, Lundmark actually used data from forty-four nebulae to write down a law connecting velocity with distance. It was the results laboriously obtained by Edwin Hubble, however, and announced by him in 1929, that finally persuaded most of the astronomical community that universe is indeed systematically expanding. Oddly enough, even this classic paper by Hubble was largely concerned with the solar motion.

As we saw in chapter 16, Hubble made use of Henrietta Leavitt's discovery that the luminosities of bright Cepheid variable stars can be correlated with their periods of variation, the slowly changing ones being intrinsically the brightest. A star's periodic time gives us its intrinsic brightness, and this, taken together with its apparent brightness, allows us to calculate the star's distance (for the standard relationship, see p. 560). Another criterion for distance used by Hubble was an assumption that the brightest stars in galaxies were all of the same luminosity. On occasion, he assumed that galaxies themselves were of much the same luminosity. As we have already seen, using the 100-inch telescope at Mount Wilson he found Cepheids in M31 and other so-called nebulae, showing by 1924 that they are not part of our own Milky Way galaxy. This was surprising and important in itself, but his 1929 announcement—using all the criteria mentioned here—that *their distances were proportional to their recessional velocities* was even more surprising to those who had not been following the discussion in relativistic cosmology; and to those who had, it was a much-appreciated gift. Hubble gave 500 kilometers per second per megaparsec as an approximate figure for the constant of proportionality. He was sufficiently confident in his results to use it in an inverted manner, estimating the distances of (very distant) nebulae from their more easily measured radial velocities.

A word is in order here, to emphasize that the apparent tendency of the spirals to recede from the Sun does not mean that the Sun has been reinstated as the center of the universe. A lump of dough filled with currants, for example, expanding as it is baked in the oven, will be such that every currant will perceive the others to be receding. It is perhaps also worth commenting on posterity's treatment of Hubble's findings, although this will be discussed further at a later stage. Seeing objects at much greater distances than Hubble could, that is, farther back in time, it has become evident that an expansion that appears to be linear close at hand (with velocity directly proportional to distance) has been changing over time, and that in fact it has apparently been accelerating. It is also the case that Hubble's distances are now known to have been seriously in error, in part because he was not aware of the two types of Cepheid, and in part because neither he nor anyone else was then aware of how serious are the problems of light absorption when using his "brightest star" criterion. (We shall return to the calibration problem.) The simple proportionality question, however, is not drastically affected by such errors, since they affected his results in roughly the same way at all distances.

The law of universal expansion, with velocity proportional to distance, has become associated with Hubble's name, and his exceptionally thorough observational work was certainly important in making it possible. We should not overlook the fact, however, that it needed velocities as well as well as distances for its formulation, and that Hubble was able to make use of Slipher's velocities for his 1929 article. He was in the process of adding the measurement of velocities to his program, however, at this time; and with the help of Milton Humason at the 100-inch reflector he was soon to be adding galactic data on velocities, as well as distances, that could not have been found with the same precision with any other telescope. Humason's part in this exercise should not be underestimated. His career was a curiously shaped affair: two years younger than Hubble, Humason's career could hardly have been more different. Having dropped out of school at 14, in the metropolis of Dodge Center, Minnesota, Humason was promoted to assistant astronomer at Mount Wilson in 1919, but only after rising through the hierarchy, including the posts of mule driver—for the pack trains up to the observatory during its construction—and observatory janitor (shades of Jean Louis Pons in Marseilles). Helping Hubble with his program of collecting spectrographic data, he proved to have great practical talent, and it was he who made most of the exposures and plate measurements. Even at the end of the twentieth century, the list of velocities of 620 galaxies he published in 1956 still accounted for a substantial fraction of the known radial velocities of normal galaxies.

Large telescopes create large reputations, and Hubble's is the name now remembered, but he was in reality playing only one part in a complex intellectual operation. The very interpretation of a spectral shift depended heavily on an underlying theory—whether general relativity or some other. Hubble gave relatively little close attention to the latest theoretical developments, although a year earlier he had discussed De Sitter's model with its author. He now decided that the simple proportionality he had found might indicate the De Sitter effect, and that it might even be just a first approximation to the "apparent" acceleration of which De Sitter had been speaking. In correspondence with De Sitter afterwards, Hubble said that while in his paper he had spoken of "apparent" velocities, he would leave the question of interpretation "to you and the very few others who are competent to discuss the matter with authority." Later, when commenting on the vital need to obtain reliable distances, he wrote: "This fact, together with perhaps a natural inertia in the face of revolutionary ideas couched in the unfamiliar language of general relativity, discouraged immediate investigation." It so happens that in a paper published by Howard Robertson in 1928 we find a well-argued claim for there being a linear relationship between assigned velocities and distances of the extragalactic nebulae. Even earlier, Georges Lemaître had put forward a similar idea. One could say that "Hubble's law"—its now usual name—is an unusual law, being named after the person who confirmed it rather than the person who first proposed it.

THEORIES IN WAITING

To return to De Sitter's earlier trio of solutions to Einstein's equations: after they were presented, De Sitter, Eddington, Ludwik Silberstein, Hermann Weyl, Richard Tolman, and others began to investigate the physical aspects of the De Sitter model in particular. At this time there was a young applied mathematician in Russia who was making considerable progress with Einstein's equations. Aleksandr Aleksandrovich Friedmann was one of the founders of modern theoretical meteorology and aeronautics, and was not without practical experience, having served as an aviator-meteorologist at the northern front during the First World War. In 1920, returning to Petrograd as head of the mathematics department in the Academy of Sciences there—a post he vacated for that of director of the observatory, shortly before his early death five years later—he soon turned to the cosmological problem in general relativity. He mastered the few writings in general relativity that he could lay his hands on. They were very few, since the Soviet Revolution and the subsequent blockade of Soviet Russia had cut off the flow of scientific literature from outside the country. In 1922, Friedmann published an outstanding paper, the first study in the general theory of relativity to appear in Russia, in which he drew attention to the possibility of a cosmological model that was nonstatic. In this model, the curvature of space—the non-Euclidean analog in three dimensions of the curvature of a two-dimensional spherical surface—changed with time. Here was one of the first signs of a serious break with one of the great tacit presumptions of the past.

In 1924, the year before his premature death at the age of thirty-six, Friedmann investigated yet further possibilities, cases of stationary and nonstationary worlds having the geometrical property known as negative curvature. From these beginnings he derived a complete set of new models. He showed how it was possible to introduce matter into those models, and so cast off the embarrassment of De Sitter's empty world. His early death meant that he never knew of his future fame; but neither did he have to see the political purges in which over twenty of the most distinguished of Soviet astronomers perished, between March 1936 and July 1937, bringing many research programs to a halt. (Even in 1965, a Soviet reviewer of my own first book could write that its only fault was in failing to mention that great cosmologist of the century, Vladimir Ilich Lenin. It turns out that Lenin did indeed write a short, semimathematical, essay on the subject. Of course, he died in 1924, long before the purges.)

At first, Friedmann's work received surprisingly little attention from the scientific community. Einstein criticized it in a short note, later retracting his criticism, which had been based on an arithmetical error of his own. Tragically, Friedmann's fame was posthumous, and came in the wake of a renewal of interest in these matters, aroused by the work of Georges Lemaître and H. P. Robertson. As for the status of scientific cosmology in the Soviet Union, between the late 1920s and Stalin's death in 1953, it was the subject of various attacks, on such naive grounds as that the very idea of linking time and space was "anti-dialectical," and therefore ideologically unsound. Some of Friedmann's scientific supporters were among those who died in Stalin's purges of the mid-1930s. Others who escaped them were simply disparaged as "Lemaître's agents," an expression with an anti-Catholic barb.

Lemaître was in fact a Belgian Jesuit priest, educated in engineering, mathematics, and physics at Louvain University before and after the First World War. During the war he served in the Belgian army, and in fact won the Croix de Guerre. In 1923–1924 he studied with Eddington in Cambridge, and when he left there it was to stay for nine months at the Harvard College Observatory. From America he wrote his first paper on cosmology, raising objections to De Sitter's 1917 model. The model lacked matter, of course, but Lemaître disliked it for a different reason: since space had no curvature in the model, its extent was infinite. The education of cosmologists had come a long way when one of them could be so readily dismissive of that idea, which is a part of our everyday Euclidean geometry.

Lemaître objected further against De Sitter's model that it was presented in a way to suggest that the universe has a center. He went on to develop a model of his own, reminiscent of Friedmann's. Lemaître, however, just as Silberstein had done the year before in another connection, derived a formula making the redshift of spectra proportional to distance. A "Hubble's law" of sorts is seemingly to be found everywhere we look. If it did not attract much attention, perhaps this has something to do with the fact that those who were working in general relativity were constantly exploring various mathematical devices for transforming one model into another, with apparently different properties from the first. Thus, even Hubble and Milton Lassell Humason, of the Carnegie Institution, writing in 1931, emphasized that the strange behavior of the distant nebulae may be "only apparent," and that it may in some way be an illusion inherent in observation over vast distances. Humason was obtaining many new values for the redshifts, and—assuming that they meant real velocities—he obtained values up to a seventh of the velocity of light, an astonishing statistic in its own right.

If the universe really is expanding, does this not mean that there was a time in the past when it was a small compact mass? Friedmann's and Lemaître's models seemed to allow for this possibility. But was an expansion from such an "initial singularity" not simply an illusion, created by the mathematics? Lemaître had been ordained abbé in 1923. His science had strong theological relevance for him. An initial singularity was not something to be avoided, but a positive merit, a token of God's creation of the world. We find him in 1927 investigating more complex models, taking radiation pressure into account, for example. This is characteristic of his style. Where Friedmann and most of those who worked in general relativity during its first two decades were at heart mathematicians, Lemaître was a physicist, and he showed this clearly in his subsequent writings.

It is curious to learn that in 1927 Lemaître met Einstein, who told him that while his paper was mathematically sound, he did not believe in the expanding universe that it entailed. Lemaître said later that he had the feeling that Einstein was not aware of the latest astronomical facts, and indeed, Einstein does not seem to have accepted a nonstatic universe until a visit to California in 1930, when he discussed the matter with Hubble. It was from

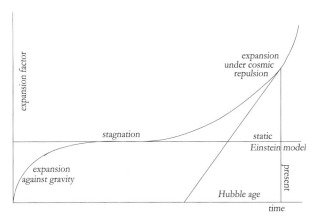

234. The Lemaître-Eddington model of the expanding Universe and its relationship with the static Einstein model.

Einstein, on the occasion of their meeting in 1927, that Lemaître learned for the first time of Friedmann's earlier work, and thereafter he was always very reticent about his own achievements. When he read a statement by Eddington, however, advocating more attention to nonstatic relativistic models, he was prompted to write to remind his former mentor of his 1927 paper. Eddington was at the time working with his research student George Cunliffe McVittie on the problem of the instability of the Einstein spherical world, and now he immediately appreciated the implications of Lemaître's work, which he had apparently quite forgotten. The Einstein model is intrinsically unstable. If our universe ever indeed resembled it, then we ought to be expecting the very expansion that was currently being suggested by the redshifts of the distant galaxies. Eddington convinced De Sitter of this, and through these two widely respected figures, the astronomical world learned of the important theoretical developments that had lain dormant for three years and more. Einstein gave the expanding universe his blessing, and many popular works brought it to the notice of the press, for whom "relativity" was at last becoming something of a cult.

Eddington developed Lemaître's model further, and a new "Lemaître-Eddington model" became a standard interpretation of the latest data from Mount Wilson. This pictured a universe evolving from a stagnating Einstein world of indeterminate age (fig. 234). The age of the universe was, and remained, a problem. "Hubble's law," naively stated, makes the velocities of the galaxies simply proportional to their distances. The multiplying factor of the distances has the dimensions of (1/time). We usually call this the Hubble factor, H. On a simple interpretation, $(1/H)$ was taken to be a measure of the age of the universe, the time that has elapsed since the galaxies were all together. Accepting Hubble's data for a simple expansion, Eddington supposed that it began about two billion years ago. The same rough answer was available to various different relativistic models. When, at a later stage, there were claims that—based on radioactivity in rocks—the Earth was four billion years old, roughly twice as old as the "Hubble time," the Lemaître-Eddington model seemed to have a distinct advantage over other models. It was very useful to conceive of the initial "Einstein state" as having existed for an arbitrarily long duration.

The situation changed around 1952, when Walter Baade, using the 200-inch Hale telescope at Palomar, and A. David Thackeray, using the 74-inch Radcliffe telescope at Pretoria—then the largest in the Southern Hemisphere—discovered that Hubble had made a mistake in regard to galactic distances. Astronomers had been worried for some time that Hubble's distances were implying that our own Galaxy was considerably larger than any other known galaxy; and the age of the universe implied by a simple interpretation of the Hubble expansion was turning out to be less than the age of radioactive isotopes. In 1952, Baade resolved these difficulties at a single stroke, with an announcement that he had recalibrated the period-luminosity relation for the Cepheids, that the old estimates of galactic distances were underestimated by a factor of two, that the radii of distant nebulae should be doubled, as should the time scale of the universe. This all came from Baade's studies of the two stellar populations, each with its own pulsating variables, to which different period-luminosity parameters must be applied. (On p. 574 we saw how Shapley's distances were in error for the same reason.) Over the next decade, cooperation between the Mount Wilson, Palomar, and Lick observatories—in a program led by Milton Humason, Nicholas Mayall, and Allan Sandage—changed the Hubble factor yet again, and many revisions have been made since then. Roughly speaking, if we quote the Hubble parameter as usual in kilometers per seconds per megaparsec, then it hovered between 450 and 550 in the prewar period, fell from about 300 to 100 or less between 1952 and the early 1960s, and has shuttled back and forth in the range 50 to 100 ever since. A widely accepted figure at the turn of the millennium was about 72, with an error of about 10 percent—but the error itself can be assessed in many ways.

After Baade's 1952 announcement, the Lemaître-Eddington world no longer had the edge over its rivals. (It experienced a revival in the 1960s, when it was used to explain the concentration of the redshifts of quasi-stellar objects in the neighborhood of 2. The *changes* in the wavelengths were the result of recession.) Even in the 1990s, however, there were still cosmologists worrying lest they find a universe that was younger than its contents. It is poetic justice that one of the most important functions of the Hubble Space Telescope has been to improve the accuracy of the parameters of the expanding universe. It has been deployed, for instance, to observe Cepheid variables in galaxies, even as far away as the Virgo cluster, and so to improve the accuracy of the cosmic distance scale.

THE PHYSICS OF THE UNIVERSE

Lemaître soon lost faith in his theoretical model and began to work on other things, but his paper turned the interests of several other theoreticians to examine new physical aspects of the expansion. Richard Chase Tolman is a good example of a new type of cosmologist. A graduate of the Massachusetts Institute of Technology who had also studied in Germany, he taught mathematical physics and physical chemistry at the California Institute of Technology. The author of the first American textbook on (special) relativity, Tolman took a strong interest in the observational work of Edwin Hubble—who in cosmological terms was one of his nearest neighbors—and he wrote a brilliant and influential study of ways in which thermodynamics could be introduced into relativistic cosmology. Others with similarly broad physical interests were Arthur Eddington, George McVittie, and William McCrea. Most mathematical cosmologists tended to concentrate on the geometrical aspects of their subject, taking Einstein's theory as a datum. But then there were the observing astronomers. Hubble was not exactly typical. While the relativists had been waiting for reliable distances, intense discussions of the significance of the redshifts had been going on, and it seems to have been Tolman who made Hubble nervous about treating the redshifts as indicative of velocities. Most working astronomers had no such scruples, and as a result, the "expanding universe" became for many people a byword for the most exciting astronomical discovery of the time. Some, presumably on the grounds that bigger is better, have insisted that it was the most important of all time.

At the other extreme of theoretical sophistication comes Eddington, who had a remarkable way of uniting seemingly all branches of mathematical physics—with the result that many of his colleagues were deeply suspicious of him. His greatest ambition was to unite general relativity with quantum theory. In 1931, Lemaître, on reading an address by Eddington on this theme, replied to it. Eddington—an active member of the Society of Friends (the Quakers)—had written that "the notion of a beginning of Nature" was repugnant to him. (We recall that the Lemaître-Eddington model had an infinite past. Looking backward, it tends asymptotically toward an Einstein state.) Lemaître explained why he believed that quantum theory suggests a beginning of the world very different from the present order of Nature. Thermodynamic principles, he said, require that (1) energy of constant total amount is distributed in distinct units (quanta); and that (2) the number of distinct quanta is ever increasing. If we go back in the course of time, he said, we must find fewer and fewer quanta, until eventually we find all the energy of the universe packed into a few, *or even into a unique quantum.*

So was born, in 1931, the idea of the Primeval Atom, a unique atom that might be seen as having an atomic weight equal to the entire mass of the universe. If one wanted to name the first modern cosmological model that incorporated a true "Big Bang," a definite beginning, mathematically speaking, then Lemaître's has as good a claim as any. In 1929, James Jeans had talked about a beginning "at a time not infinitely remote," but that was all very vague. Responding in May 1931 in the journal *Nature* to a paper by Eddington on the origin and end of the world that had appeared in the

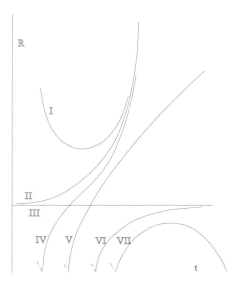

235. Seven distinct mathematical forms of relativistic models of the Universe. I contracts to a minimum and then expands to a so-called De Sitter model. II is the Lemaître-Eddington model, expanding from the Einstein state (III) to a De Sitter state (see the right-hand part of figure 234). IV expands steadily from a singular state. This was the case favored by Lemaître, who took the initial state to be his Primeval Atom. IV is similar, but has no "point of inflexion" in its arch. This is the Einstein-De Sitter model. VI is a limiting case of V. Finally, VII shows a typical oscillatory model, with the Universe expanding and contracting, perhaps repeatedly.

March 1931 issue of the journal, Lemaître explained how his concept of a Primeval Atom might be used to give more precise physical content to a relativistic model. Such a "unique quantum," he said, would be unstable, and "would divide in smaller and smaller atoms by a kind of super-radioactive process." He went on to say that "some remnant of this process might, according to Sir James Jeans's idea, foster the heat of the stars until our low atomic number atoms allow life to be possible." In November 1931, Lemaître returned to the theme, explaining in broad outline how cosmic radiation, with its enormous energy content, might be understood as "glimpses of the primeval fireworks of the formation of a star from an atom of atomic weight somewhat greater than that of the star itself." (For a way in which this provided a launching pad for one out of the general set of possible relativistic model universes, see figure 235.) Having developed his theory further until 1935, he put it aside until 1945, when he published a book on it, which, popular though it was, had serious undertones.

To support his ideas, Lemaître needed a theory of nuclear structure applicable to atoms of extreme weights. He also needed better information about cosmic radiation, of which very little was then known. He thought his hypothesis of super-radioactive cosmic radiation was supported by Arthur Holly Compton's claim that cosmic radiation consists of charged particles; and his faith in his idea was increased when it was gradually revealed that the energies involved in cosmic radiation were much greater than had at first been suspected. There is something almost prophetic about all this. He died in 1966, a year after the discovery of the cosmic microwave background radiation by Arno Penzias and Robert Wilson. His successor at Louvain, Odon Godart, was able to keep him informed about it up to the end. Cosmology had changed considerably in the meantime, but Lemaître clearly saw that there was much more to cosmology than gravitational geometry. Star formation, galaxy formation, and the relative abundance of the various chemical elements, all figured in his scheme. We shall return to the problem of abundances later in the chapter (beginning on p. 649 below), but only after taking note of several cosmological theories of this period that were more or less independent of the mainstream tradition of general relativity.

ALTERNATIVE COSMOLOGIES

The 1930s saw a movement in cosmology of great value to scientific practice generally. It was not for nothing that Eddington's general writings aroused great interest among the philosophers, especially about the nature of theoretical entities. Of course some of the problems raised came directly from fundamental physics and the theories of relativity, but the question of whether or not the spectral redshifts were true Doppler shifts, indicative of velocities, was seen to depend on what was meant by distance. As soon as this question was pondered, the entire network of interrelations between observational data and the concepts of cosmological theory was seen to be highly problematic. In no other branch of science was so much care given to the analysis of the concepts employed in it, and here the names of Eddington, E. T. Whittaker, R. C. Tolman, E. A. Milne, and G. C. McVittie are among those deserving mention. Milne is an especially interesting case, for with the help of W. H. McCrea, and later, of G. J. Whitrow, he showed how Einstein's ideas could be sidestepped and how a form of "Newtonian cosmology" might be revived.

Edward Arthur Milne first systematically investigated his "kinematic relativity" in 1932, three years after moving from Cambridge University to a chair in mathematics at Oxford. With its help, he constructed a world model in the familiar space of ordinary (Euclidean) geometry. It contains a system of fundamental particles moving in uniform motion relative to each other, all of them having been together at the zero of time. From the point of view of one of these particles—or rather, of an observer associated with the particle—the others seem to recede from it. It is easy to see how, at any given time, the whole system can be contained within a sphere, and that this sphere, whose size is determined by the fastest particles, will expand in time. Milne presented his model in two different ways. He found that he could change the scale of time in a particularly interesting way—making one sort of time vary logarithmically with the other—and that doing so turned it into a stationary system, each particle then being associated with a fixed point of space. On doing this, however, space ceases to be Euclidean—it is of the non-Euclidean "hyperbolic" type—and the time-scale is not what would be kept by an atomic clock.

Milne developed these ideas further, first with William McCrea and then with Gerald Whitrow, with consequences that closely resembled those of general relativity. This may be broadly explained by reference to an important step in Einsteinian relativistic cosmology that had been taken by H. P. Robertson in 1929, and independently by Arthur G. Walker some years later. The expression "space-time metric" refers to a formula by which intervals between two events (points in space-time) may be evaluated in one of the various Riemannian (non-Euclidean) geometries. It is an extension of the theorem of Pythagoras in ordinary geometry, as mentioned earlier. Robertson considered what would be the consequence if he were to place two restrictions on the new geometry. He required that space-time be (1) homogeneous, with all places alike, and that it be (2) isotropic, that is, the world should look the same in all directions from all points at a given (cosmic) time. The resulting space-time metric, which will not be given in detail here, was interesting because it included a certain time factor, a term by which the *spatial* interval between events had to be multiplied. In other words, by taking Einstein's theory, and adding the two conditions of cosmic symmetry, an "expanding universe" of sorts follows quite naturally. As we have seen, it had indeed been found by several theoreticians before the vital astronomical observations were made.

As a shorthand for the expansion factor referred to here, we shall use the usual expression for it, $R(t)$. This term, which depends on the time, is sometimes spoken of as an expanding radius of space; but here we shall not look deeper into its various interpretations. It is enough for the moment to make the general point that exactly the same metric could be obtained in Milne's theory as that found earlier, by Robertson and Walker, for Einstein's brand of relativity.

In both types of theory, the metric is really only the beginning of a cosmology. How the universe expands, that is, how the expansion factor $R(t)$ behaves in time, can only be judged by relating the matter in the observed world, with all its physical properties, to the model. In 1934, Milne and McCrea found that, with a simple mathematical trick for treating an unbounded system, Newtonian mechanics and gravitation led to precisely the Friedmann-Lemaître equations for a model without pressure (with or without cosmical repulsion of the sort Einstein had introduced with his cosmic term). On a local scale, the model should behave mechanically more or less exactly as in general relativity. On a larger scale, the fitting together of the local pieces of space was done differently in the two theories. One important lesson, though, was that there seems to be something very natural about the expanding universe, even when understood in a "classical" way.

As we saw when considering the early history of relativistic cosmology, the form of the expansion factor varies from model to model. It varies, for instance, between the Einstein model and De Sitter's, between Lemaî-tre's and Eddington's variant of it, and so forth. An interesting theoretical model was introduced in 1932 by Einstein and De Sitter jointly. In this, the expansion factor was proportional to the ⅔ power of the time. This has the effect of making the present age of the universe only two-thirds of the Hubble time. Another model, introduced by P. A. M. Dirac in 1938, was interesting because it did not require space to be curved, and it made the expansion factor proportional to the ⅓ power of the time, so that the age of the universe is just a third of the Hubble time. Without attempting to judge on the merits of the various alternatives, we can only repeat that it is obviously wrong to treat the Hubble factor as indicating the age of the universe, in the way that was usually done in the years following the discovery of the general recession of the galaxies. As for opinion in the 1930s on the question of the actual age of the universe, this is a separate issue, but we recall the difficulties into which Hubble's observations had led astronomers.

Other alternatives to Einstein's general theory of relativity were the theories of gravitation developed by G. D. Birkhoff, A. N. Whitehead, and J. L. Synge. All of them had cosmological implications. They were symptomatic of a period of great intellectual vitality. They were no doubt partly motivated by a desire to create something comparable with what Einstein had produced. Some ideas of a very different kind were then being put forth by Hermann Weyl, Eddington, and Dirac—the first two in 1930, and Dirac in 1937–1938. They seemed to many to be suggesting that cosmological observation was superfluous, and that all could be deduced from the constants of physics. Eddington, for instance, thought that all the dimensionless constants (pure numbers) obtained by suitably multiplying and dividing powers of the constants of physics—the mass of a proton, the charge on an electron, and so forth—turn out to be close to unity, or of the order of 10 raised to the power 79. This vast number he thought might characterize the number of particles in the universe. Eddington's monumental study setting out his ideas went under the title *Fundamental Theory*, although it was edited by Whittaker and published from Eddington's notes in 1946, two years after his death. It left some few cosmologists with a sneaking suspicion that there may be a numerological link between the physics of the very small scale and that of the cosmos, but most were content to ignore it. At an early stage, Eddington actually claimed to derive a value for "Hubble's constant" close to Hubble's. When the observed value had to be revised, seemingly refuting his methods, it was found that he had made a mistake, and that he should have obtained a constant more or less equal to the revised figure. With later revisions, however, no such rescue has proved possible.

In the short term, what mattered more to observational cosmology than Eddington's brilliant number magic was the ongoing drive, by Tolman, McCrea, McVittie, Otto Heckmann, and the rest, to derive theoretical relations between observable quantities. By means of these—between apparent magnitude and redshift, between apparent magnitude and counts of galaxies, and so forth—the various cosmological models could in principle be put to observational test. In practice, doing so was much delayed by the coming of the Second World War.

In 1918 and 1925, to remove the paradox of the dark night sky, W. D. MacMillan had put forward the idea that radiation, when traveling through empty space, may suddenly disappear, and that it may reappear in the form of hydrogen atoms. R. A. Millikan, a pioneer of cosmic radiation and a colleague of MacMillan at the University of Chicago, thought that this might be the explanation of the origin of cosmic rays. That idea was soon disproved by A. H. Compton. There was nothing particularly mysterious about the "creation" of atoms on this theory, for there was a named source of the energy responsible for them. Tolman and others likewise considered the conversion of radiation into matter, and what he wrote did not at first attract much attention on this account.

There were others, however, whose suggestions as to the creation of matter were far more radical. In 1928, James Jeans, searching for an explanation of the spiral character of the nebulae, suggested that the centers of the spirals might be places at which matter is "poured into our universe from some other, and entirely extraneous, spatial dimension." He added that to us, therefore, they appear to be "points at which matter is being continually created." Milne later noted that his kinematic relativity conformed with the speculations of Jeans.

When, in 1937, the Cambridge physicist P. A. M. Dirac picked up some of those ideas of Eddington's that we have just mentioned, it was to argue that large numbers of the sort that occurred in Eddington's theory, if they were proportional to the epoch, should increase in time. The "age of the universe" is an unusual sort of physical "constant," for of course it increases in time, but it is not a dimensionless number. It can be made into one by taking its ratio to some fundamental time of physics, such as the Planck time. (This quantity, 10^{-43} seconds, can be thought of as a kind of atom of time.) Dirac thought that the number of protons and neutrons in the universe, which occurred in Eddington's theory as seemingly related to the epoch, might therefore grow in time—implying continual creation—and that the gravitational constant might likewise change. In Japan, at much the same period, some mathematical physicists of the Hiroshima school, investigating the De Sitter model, claimed that the number of particles in that too increases with time. In Germany, Pascual Jordan took over some of Dirac's work, and claimed that stars, perhaps in the form of supernovae, and even entire galaxies, might be the created matter. He thought that Dirac might have been afraid of violating the principle of energy conservation, and Jordan was careful to provide a potential source for the new matter. It was, he said, provided by a loss of gravitational energy from the universe.

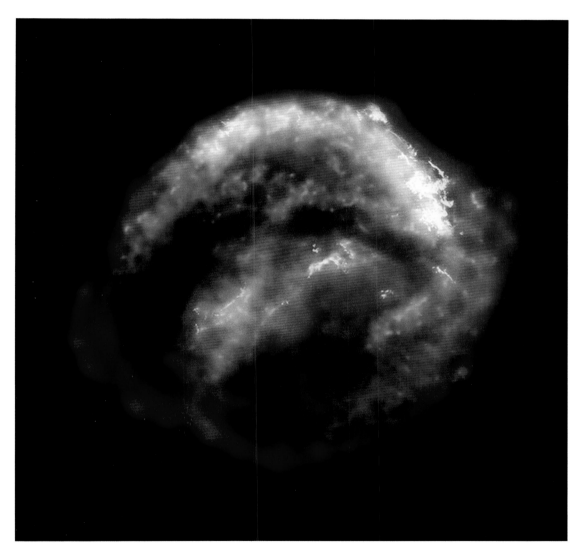

11. Kepler's supernova remnant, as seen from space by the Chandra X-Ray Observatory. Each color in this composite image represents a different region of the electromagnetic spectrum. X-rays and infrared light, invisible to the human eye, are color-coded so they can be seen in the final image. The bubble of gas that makes up the remnant appears different in various types of light. The hottest gas (blue and green for high and low energy, respectively) radiates X-rays. The Hubble Space Telescope shows the brightest, densest gas (yellow), apparent in visible light. NASA's Spitzer Space Telescope shows heated dust (here, red), radiating in the infrared.

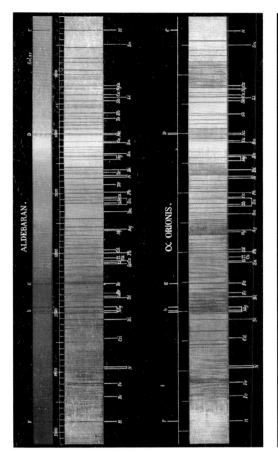 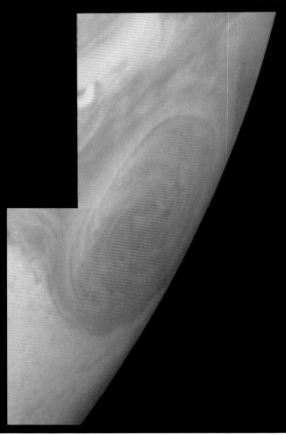

12. (*Facing*) The Whirlpool galaxy, M51 (or NGC 5194), is easily viewed from Earth with a good amateur telescope and has been much studied since Rosse first discovered its spiral form (see fig. 197). This image combines Hubble Space Telescope images with ground-based data from the Kitt Peak National Observatory. It shows visible starlight as well as light from hydrogen in the spiral arms, associated with the most luminous of young stars. The galaxy is close to another (NGC 5195), just off the upper edge of this image, and the gravitational effect of the companion galaxy is triggering star formation in M51—note the luminous clusters of young stars, highlighted in red by their hydrogen emission. This image allowed a research group, led by Nick Scoville of the California Institute of Technology, to define the structure of both the cold dust clouds and the hot hydrogen and to link individual clusters to their parent dust clouds, in which intricate structures can be traced. The new images reveal a dust disk in the nucleus, which may provide fuel for a nuclear black hole.

13. (*Above, left*) Spectra of the stars Aldebaran and Betelgeuse, recorded by William Huggins and William Allen in May 1864. Color-printing procedures of the time were surprisingly good, but not especially relevant to the scientific value of such records.

14. (*Above, right*) This view of Jupiter's Great Red Spot combines two images taken by the *Galileo* spacecraft on 26 June 1996, using two filters, violet and near-infrared, at each of two camera positions. The spot is a storm in Jupiter's atmosphere, moving counter-clockwise at about 400 kilometers per hour, and its many changes have been observed for over three centuries. In size, this feature exceeds the diameter of the Earth.

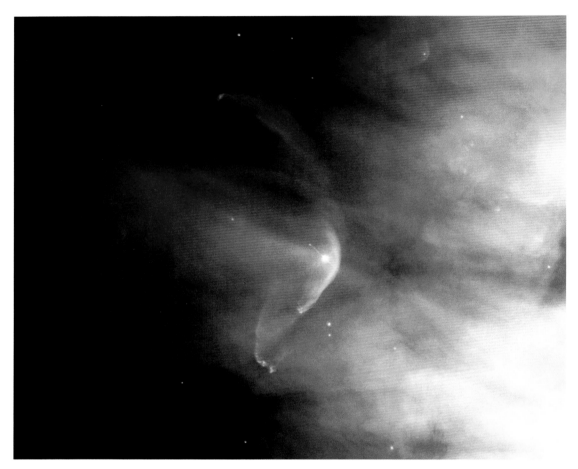

15. The rich star-forming region known as the Great Nebula in the Orion constellation reveals a very young star within a crescent-shaped wave, resembling that at the bow of a moving ship. This composite image of it was taken in 1995 by the Hubble Space Tele-scope, using filters that identify oxygen, nitrogen, and hydrogen emissions. Such a wave can be created when two streams of gas collide. In this case, one stream (far more energetic than our own solar wind) from the star LL Orionis collides with a stream of slow-moving gas emerging from the center of the Orion Nebula (at the lower right of the image). Numerous similar shock fronts have been found in this complex star-forming region.

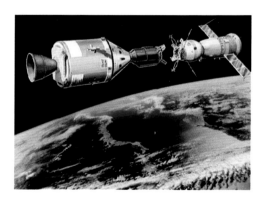

16. The final spaceflight of the Apollo program was the first in which spacecraft from different nations docked in space. In July 1975, a U.S. *Apollo* spacecraft carrying a crew of three docked with a Soviet *Soyuz* spacecraft with its crew of two. This artist's impression was produced two years earlier, after an outline plan had been agreed upon between the two nations.

17. Saturn and four of its moons: Tethys, Dione, and Rhea are visible against the dark background, and Mimas is visible against Saturn's cloud tops, just below the rings on the left. The shadows of Mimas and Tethys are also visible on the cloud tops, and that of Saturn on the rings. The rings are mostly ice particles, ranging in size from dust to boulders. The disk of rings is a mere 100 meters thick, but the outermost ring visible here is 272,400 kilometers in diameter. Note the Cassini division, which is 3,500 kilometers wide. This image was synthesized from images taken with Voyager's orange, blue, and ultraviolet filters, and its color is exaggerated to improve definition. Instruments on the *Cassini* spacecraft produce much higher resolution images, but the difference would not be obvious at this scale.

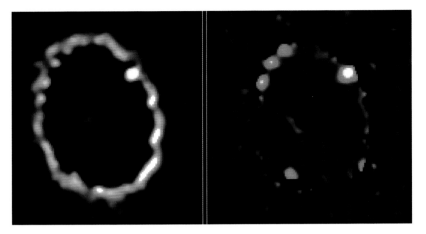

18. In November 2003, more than sixteen years after the detection of supernova 1987A, this image of it was taken by the Advanced Camera for Surveys aboard the Hubble Space Telescope. It shows many bright spots around a ring of gas, spots produced as shock waves unleashed during an explosion strike the ring at something approaching two million kilometers per hour. The collision is heating the gas ring, causing its innermost regions to glow. The first of these hot spots was noticed in 1996. It is expected that the entire ring will eventually be ablaze with them, yielding new information on how the star ejected material before it exploded. The elongated and expanding object in the center of the ring is debris form the supernova blast, heated by radioactive elements that were created in the explosion, principally, titanium 44.

19. (Right) Adjustments being made to the *Cassini* spacecraft, 1997. The silver-suited technicians are not dressed for space travel. (Above) Saturn's C and B rings, an image taken from the Cassini spacecraft's orbital insertion on 30 June 1997 shows definite compositional variation within the rings. Lettering the rings begins from the inside, with the D, C, B, and A rings followed by the F, G, and E rings. The general pattern is from "dirty" particles, indicated by red, to cleaner ice particles, shown in turquoise, in the outer parts of the rings. The image was taken with the ultraviolet imaging spectrograph instrument, built and organized by a team at the University of Colorado, Boulder. It resolved features less than 100 kilometers across.

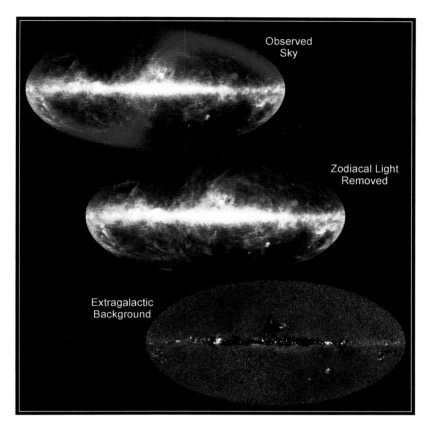

20. Maps of the full sky as seen in infrared light, illustrating the steps taken to find the cosmic infrared background. The data mapped here were compiled by NASA's Cosmic Background Explorer (COBE) between December 1989 and September 1990. The cosmic infrared background is a fossil of the early Universe in the form of cumulative starlight, which now appears in the infrared for various reasons, including the cosmic redshift and absorption and reemission by dust. The upper two are composite images, taken in wavelengths of 60 (here shown as blue), 100 (green), and 240 micrometers (red). The bottom image shows just the 240-micrometer brightness after foreground light from the solar system and our Galaxy has been removed. The bright yellow-orange line across the center of the topmost image is caused by interstellar dust in the plane of the Galaxy. Above and below this line are wispy clouds of interstellar dust (in red), while interplanetary dust in the solar system appear as S-shaped blue. The middle picture represents a view of the sky after the foreground glow of the solar system dust has been extracted. The remaining image is dominated by emission from interstellar dust in our Galaxy. (The two bright objects in the center of the lower right quadrant are nearby galaxies, the Large and Small Magellanic Clouds.) After removing the infrared light from these two sources, what remains is a uniform cosmic infrared background.

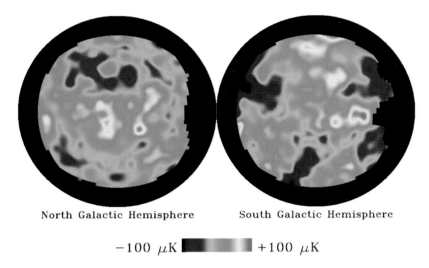

North Galactic Hemisphere   South Galactic Hemisphere

−100 μK ▮▮▮▮▮ +100 μK

21. These false-color images of the two hemispheres show tiny variations in the intensity of the cosmic microwave background, as measured over a period of four years by radiometers on NASA's Cosmic Background Explorer (COBE). The blue and red spots correspond to regions of matter of greater or lesser density in the early Universe, before that matter formed into stars and galaxies. The initial discovery of variations in the intensity of the cosmic microwave background were made by COBE in 1992. The scale of the map, which covers the visible Universe, is large: even the largest features seen by optical telescopes—such as the "Great Wall" of galaxies—would fit within the map's smallest details.

Despite the fact that their authors were of the highest credentials in fundamental physics, these speculations were little heeded at the time, perhaps because they were mostly put forward in a half-hearted way, and no doubt partly because they seemed quixotic. Dirac soon abandoned his idea, although Jordan stuck to his own resolutely into the 1950s. The concept of continual creation did not spring to the attention of a wide public until after the war, when Hermann Bondi, Thomas Gold, and Fred Hoyle incorporated it into a new cosmological theory, or rather, a class of theories. By this time, attitudes regarding the universe had changed very substantially from those of the 1920s and 1930s. The old assumption that it was an overall static entity had been almost entirely whittled away, and there was wide agreement that the redshifts do indicate a real expansion of the system of galaxies. Step by step, more and more physics was being woven into the story, making it harder for old diehards to oppose it.

As we shall see shortly, Lemaître, Gamow, and others had by now shown that cosmology might hold the key to the origin of the elements. When Eddington had been writing his work on the constitution of the stars, the importance of hydrogen and helium was scarcely perceived, but after Eddington, McCrea had developed the first acceptable model of a large hydrogen star. Hoyle introduced Bondi and Gold—who had both come to Britain from Nazi-occupied Austria—to astrophysical questions, first when all three were working at the Admiralty during the war, and later when they were all in Cambridge together, especially from 1948. They were dissatisfied by Lemaître's model, which seemed to overcome the embarrassingly short figure for the "age of the universe" then obtained from Hubble's work, but only by sleight of hand—that is, by the indeterminate "stagnation period" (see fig. 235 above). Gold had the idea that, if there were a process of continual creation of matter, it should be possible for the universe to continue in a steady state, despite its general expansion, so that the "age" problem would disappear. It is said that the three men came up with such an idea after seeing a film called *The Dead of Night*, which concludes with a return to the opening scene.

From these beginnings, Bondi and Gold moved in one direction and Hoyle in another, although they shared many ideas. Bondi and Gold presented their work as a consequence of a very general law that they called the "perfect cosmological principle," according to which the universe "presents on the large scale an unchanging aspect." Milne, whose work was of much influence here, had earlier worked with a "cosmological principle" that made the large-scale aspect of the universe independent of the *place* of observation; but now also the *time* of observation was said to immaterial. This meant that the average density of matter and radiation remain the same, and that the age distribution of nebulae remains constant. In steady-state cosmology, the average age of objects is only a third of the "Hubble age," so in this theory it seemed that our Galaxy must be much older than average. Since the data then available for distances led many to think that our Galaxy was unusually large by comparison with other galaxies, it was thought not improper that its age might indeed be greater than the average.

The earliest publications by the three all came in 1948. Hoyle's methods seemed less radical. He took Einstein's general theory of relativity as his starting point and was influenced by a principle that had first been enunciated by Hermann Weyl in the 1920s. According to this, there is a preferred velocity at each point in space-time, and only a set of fundamental observers sees the universe in the same way. Hoyle now simply added a term to Einstein's field equations that he interpreted as indicating the creation of matter. This, however, seemed to violate the conservation of energy, one of the most sacred of physical laws. In 1951, McCrea showed how Einstein's equations could be salvaged by reinterpreting Hoyle's new "creation field," and in the 1960s Hoyle revised his earlier treatment in collaboration with J. V. Narlikar. The heyday of steady-state cosmology, however, was the late 1940s and 1950s. Hoyle gave a celebrated series of talks on the new "Third Programme" of the BBC, and for a time his ideas became national talking-points in Britain, with many of the English clergy writing letters of protest to the press and preaching sermons on the audacity of those who seemed to wish to deny the Christian account of a single creative act. Matters were not helped by Hoyle's use of that pejorative phrase "Big Bang" for all rival models that began from a compact origin at a particular point in the past—although Gamow was one of his chief targets, for reasons that have to do with their rival explanations for the relative abundances of the chemical elements. The hopes of the steady-state cosmologists were raised for a time, as the result of the work by Hoyle, Fowler, and the Burbidges, on nucleosynthesis in stars, a subject that will be covered later in this chapter. Their work had been prompted in large part by Hoyle's hopes for steady-state cosmology, with which it fitted well, but in time it became clear that their new theories of nucleosynthesis could be perfectly well adapted to other cosmological models. Indeed, there they flourished.

The first observational evidence that worked against the steady-state theory came from counts of celestial radio sources, done primarily by Martin Ryle in Cambridge, with several colleagues in Britain, as well as by Bernard Yarnton Mills and John G. Bolton in Australia, and by Marcello Ceccarelli in Italy. It seemed that the number of faint radio sources was far too large, compared with the number of bright sources, for the steady-state theory's predictions to be acceptable. The explanation that came most easily to hand was one in terms of a model of Big Bang type: this was that the (intrinsically) bright sources are relatively near, since radiation from distant bright sources has not had time to reach us. In other words, on this evolutionary model, the further we go out into space the less evolved, and thus intrinsically fainter, will be the sources we observe.

Ryle's first evidence, with what to some was its clear message, was presented in 1955 at the same time as his latest survey of 1,936 radio sources. He had then been working seriously at the cosmological problem for about two years. The human effect of his 1955 claim, presented by Ryle and his student Peter Scheuer, was electric. Ryle had kept his cards close to his chest with good reason, but even he was unprepared for the onslaught of the steady-state theorists, and the popular press in Britain made much of the controversy. Joseph Pawsey and Bernard Mills in Sydney soon found radio results of their own that were hard to square with Ryle's, and he had to face up to many acrid exchanges with the advocates of steady-state on the one hand and radio astronomers on the other. For seven or eight years afterwards, there were independent judges of the situation who considered that Ryle's observations might be subject to errors. In fact, what revisions were made tended at first to make the case less decisive.

The situation in cosmology around the year 1960 was therefore by no means clear to those who had no commitment to a particular cause. There were several relativistic models over and above those mentioned here—for example the oscillatory models that expanded (for billions of years) before contracting, then perhaps entering again into an expanding phase. Such models had bizarre thermodynamic properties that made them generally unpopular with astronomers. There were attempts to develop cosmologies based on the different theories of gravitation founded, as already mentioned, in the 1920s by the mathematicians A. N. Whitehead (extended by J. L. Synge) and G. D. Birkhoff, but they were isolated, and attracted ever-diminishing attention. Einstein's general theory of relativity was still far from being widely accepted, and many who worked entirely within it admitted to having doubts. When the tide turned, it turned rapidly, and this largely as a result of a series of new discoveries made with the radio telescopes that were transforming the entire face of astronomy.

COSMOLOGICAL ABUNDANCE RATIOS

Lemaître, with his Primeval Atom, had been too far ahead of his time. No real progress on this front was possible before the advances in nuclear physics in the middle and late 1930s, and then came the war, during which most of the physicists with the relevant expertise were concerned with more mundane matters. As early as 1942, however, seven years after his Ohio paper on the potential role of the neutron in the generation of energy and the synthesis of elements in stars (see p. 607 of chapter 16), George Gamow returned to his earlier study of nucleosynthesis. There was a class of what became known as "equilibrium theories" of relative abundances, in which it was assumed that the presently observed data would follow from some state of thermodynamic equilibrium among nuclei at high density and temperature, but none of those who tried their hand at the calculations was successful. The first sentence of a study by Subrahmanyan Chandrasekhar and Louis R. Henrich in 1942 reads "It is now generally believed that the chemical elements cannot be synthesized under conditions now believed to exist in stellar interiors." Gamow found his way blocked by what he called "the heavy element catastrophe":

fitting temperature and density to what was known of elements of lower atomic mass was possible only at the expense of predicting too small a fraction of heavier elements. He therefore changed direction, with a theory vaguely similar to Lemaître's. His idea was that elements of higher and higher atomic mass would be built up by a process of neutron agglomeration until uranium was reached, after which fission would occur—as it was to occur before long at Alamogordo, Hiroshima, and Nagasaki—with the resulting fragments starting the process all over again. This process was considered in the setting of a universe expanding along the lines that had been described by the those working in general relativity, not to mention his one-time teacher in Leningrad, Aleksandr Friedman, and of course now Georges Lemaître, author of the Primeval Atom concept.

In 1946, addressing the problem of the background cosmology more carefully, Gamow found numerous difficulties. Above all, he found that current wisdom gave an initial expansion rate too great for an equilibrium state ever to be established. It was in the same year, however, that Ralph Asher Alpher began to develop these ideas as the subject of a doctoral dissertation for Gamow, at The George Washington University. An essential part of his study required a knowledge of thermonuclear reaction rates, but information about such things was then militarily highly sensitive, and so classified. By a stroke of good fortune, however, Donald James Hughes—a nuclear physicist who had worked on the Manhattan Project before moving to Brookhaven, New York where he investigated numerous elements for potential use in nuclear reactors—published data of which Alpher could make use. It appeared that the likelihood of neutron capture increased as abundance decreased. (There was an exponential decrease of abundance for atomic mass up to about 100, after which abundance was roughly constant.) On this basis, Alpher developed a "neutron-capture model" for nucleosynthesis in the early universe, presenting his results in his dissertation in 1948. In this he drew attention to several unresolved difficulties, chief of which was that while his model was designed to explain the density of matter in the expanding universe, the density of radiation would have been inordinately greater.

It was during this period that Alpher and Gamow joined forces with Robert Herman, who had accumulated expertise in physics across an extremely wide range, practical and theoretical. After the war he was at the Applied Physics Laboratory of Johns Hopkins University, working on condensed-matter physics, when Gamow, Alpher, and he combined forces to extend Alpher's model. As a program for future research, rather than a finished statement, they produced a paper of just a single page that was published in a leading journal, *Physical Review*, in 1948. This was signed by Alpher, Bethe, and Gamow, as an alphabetic pun on αβγ. (Bethe had little to do with it, and Herman was said to have been omitted by the puckish Gamow only because he refused to change his name to Delter.) The vital nuclear processes were here viewed as taking place in approximately the first hour of the expansion. After explaining how their process of nucleosynthesis would proceed, Gamow and Alpher calculated likely abundances, and predicted that the mass of helium in the universe would be about 25 percent. This was a remarkable achievement, as later observation would prove. They did not, however, work out the abundances of light elements in much detail, nor did they take into account the preponderance of radiation density in the early universe.

The αβγ paper spawned a whole series of similarly quantitative studies that linked the new particle physics with earlier geometrical model universes, some of these studies taking Lemaître's work as starting point. One of the most impressive of contributions, in retrospect, was that made by Alpher and Herman later in the same year. Gamow had explained in a previous paper that the early universe would pass through a "decoupling time" at which, from being radiation-dominated it would become matter-dominated, and transparent to radiation. After studying the implications of this decoupling, Alpher and Herman came up with a prediction that radiation surviving from the early history of the universe should still exist, with a present temperature of about 5°K. Such radiation was later detected by A. A. Penzias and R. W. Wilson in 1965. The fact that it is often described as one of the two most important cosmological discoveries of the century helps to put the earlier prediction in perspective.

Finally, two questions of terminology. As we have seen, Lemaître's notion of the Primeval Atom was not identical with that of the primordial state of matter envisaged by Gamow and his collaborators, and the latter in any case changed in character with successive accounts. Roughly speaking, Lemaître's was a nuclear fluid of moderate temperature, whereas the later versions were of a hot and dense gaseous state of the universe comprising neutrons, protons and electrons in a sea of radiation. For this, Alpher chose the name "Ylem," a word with an odd history. It is now occasionally spelled "Yelm," which in a way it deserves, since it derives from a sentence in which the fourteenth-century English poet John Gower was using Aristotle's Greek word *hyle* (matter, substance, as opposed to form) in the accusative case. ("That matere universall, Which hight Ylem in speciall.") Alpher, having apparently found the word in Webster's *New International Dictionary*, as meaning "the primordial substance from which the elements were formed," added that "it seems highly desirable that a word of so appropriate a meaning be resurrected." In fact this completely misses the point that Aristotle was making with his matter-form distinction—but then, that was a long time ago.

That the distinction between Primeval Atom and Ylem was later often lost has much to do with the phrase "Big Bang," which began to be applied fairly indiscriminately to any theory that made the universe begin suddenly from some primordial state. As we have seen, the last expression was coined by Fred Hoyle in 1949, in a broadcast talk advocating his own steady state theory. (One wonders whether he had been reading Eddington, who in

1928 wrote "I simply do not believe that the present order of things started off with a bang.") Like such pejorative terms as "Quaker," and "Methodist," the expression is now embraced by the believers themselves—and in this case by believers from many creeds. It is unfortunate in one respect, insofar as it persuades some people to picture relativistic theories of a Big Bang in terms of a Primeval Atom, ylem, fireball, or whatever, exploding outward into a preexisting empty space. As we have seen, in cosmologies based on Einstein's general theory of relativity it is space itself that is expanding, along with the matter in it—the dough and the currants together, to use the previous analogy.

THE BATTLE OF THE ABUNDANCES

The idea going back to Harkins, Payne, and others—that there was a more or less fixed and universal abundance ratio, with obvious local adjustments here and there—had encouraged Gamow and his colleagues in their ambition to show that all chemical elements were made in the hot early universe, when the temperatures and pressures were high enough to promote universal nucleosynthesis. By the 1940s there were many who took it to be virtually axiomatic that the chemical elements could not be synthesized in stellar interiors. Almost simultaneously with Gamow's earlier paper of 1935, when he was still hoping for a stellar solution, a small number of unusual stars were being discovered that would eventually help to shatter the general complacency. It was in 1935 that Walter Sydney Adams—the Syrian-born son of American missionary parents, who was then director of the Mount Wilson Observatory—together with colleagues found "intermediate white dwarfs" of puzzling hydrogen-line spectral types that seemed not to fit expectations. (The modern term for these stars, "subdwarf," was coined by Gerard Peter Kuiper in 1939.) In 1946 Luis Rivera Terrazas and Luis Munch in Mexico found other puzzling spectra in RR Lyrae stars. Yet other cases were added to the list of stars that seemed to violate the old assumption. No adequate explanation was found for them until 1951, when Joseph W. Chamberlain and Lawrence Hugh Aller showed how line weakening could be a consequence of low metal abundances relative to hydrogen. Those who had made these discoveries presented their evidence for differences in chemical abundances as between stars, but so entrenched was the idea of a cosmic constancy of abundance ratios that for three or four years it was generally disregarded. A referee's report on the paper by Adams and his colleagues even caused them to revise their conclusions in a later paper, and it is now agreed that they were wrong to do so (shades of Russell's treatment of Payne?). Eventually, the question of how, when, and where the chemical elements were formed began to be studied more widely and more intensively. It was found that the abundance ratio of heavy elements to hydrogen in stars might vary by a factor of the order of a thousand, depending on the ages of the stars and their evolutionary history. New ways of detecting and measuring the differences in heavy-element abundances were found and applied to unusual spectra that had often previously been set aside as unexplained oddities.

Subdwarfs are to be found in our Galaxy's halo, with very high velocities relative to the Sun. They emit a higher percentage of ultraviolet light than Population I stars of the same spectral type, and this "ultraviolet excess" is a consequence of their low metallicity, which allows more of their ultraviolet light to escape. Their outer layers, being less opaque, make for a lower radiation pressure, leading to a smaller, hotter star for a given mass. The ultraviolet excess was discovered, first by Nancy Grace Roman at Yerkes in 1954, then a year later also in stars in globular clusters, by Allan Sandage and Merle Walker at Mount Palomar Observatory, and subsequently by many others. It so happens that Baade had earlier conjectured that the types of stars seen as high-velocity subdwarfs in the field and as main-sequence stars in globular clusters are essentially the same, and now low metal criterion established by Chamberlain and Aller in 1951 supported that conjecture.

The mid-1950s saw much discussion of the topic of relative abundances, especially in the Sun and similar Population I stars. An important survey by Hans E. Suess and Harold Clayton Urey, for instance, appeared in 1956. New series of observational programs were set in motion, especially with the chemical evolution of the Galaxy as their focus. At Mount Wilson and Palomar, large numbers of high-dispersion plates were studied, and by the mid-1960s there was no longer any room for doubt: the constancy of abundance ratios had been an understandable but gross mistake. Before the 1950s, however, a new player had entered the arena, with a new theory of stellar evolution through nucleosynthesis that had a direct bearing on the problem of abundances.

NUCLEAR REACTIONS IN STARS

We have already seen how Bethe and Weizsäcker had shown that helium can be synthesised from hydrogen by thermonuclear reactions in stars. Several talented physicists looked into the question of mechanisms for the formation of higher elements, encountering one difficulty after another. Enrico Fermi and Anthony L. Turkevich, did so, for example, and decided that nothing beyond helium could have emerged from the primordial explosion of the universe. How do they come into existence? Why are atoms of gold and other precious metals rare, while carbon is common? Why are magnesium and silicon less abundant than oxygen? Why is iron relatively common, like its neighbours in the periodic table? At temperatures of 15 to 20 million degrees Kelvin, chain reactions such as the proton-proton reaction or the CNO cycle could synthesize helium from hydrogen, but what of the heavier nuclei? The heavier the nuclei, the more difficult it is to overcome the strong repulsion between them, for they carry bigger electrical charge—iron, with atomic number 26, has 26 protons, for instance, while hydrogen has only 1, and helium, 2. The random velocities of atoms in a gas increase with temperature, and with very much higher temperatures heavier nuclei might be expected to participate in reversible reactions in an equilibrium situation, but what are the necessary temperatures and densities, and could they be attained in stars? In the 1942 paper by Subrahmanyan Chandrasekhar and Louis R. Henrich, mentioned earlier, they gave reasons for thinking that to be possible, and soon their conclusion was to be reinforced by the work of Fred Hoyle in Cambridge.

It was in spare moments during his work in the Second World War on the development of radar that Hoyle puzzled over this general problem of the abundances of the elements. He calculated that to produce iron, a temperature a hundred times this would have been needed. This is not a temperature attained in the Sun or similarly modest stars, but it might, he thought, be attained in more massive stars. Hoyle published his elegant argument in 1946, after the war ended, and some would say that this was the first convincing application of the concept of nucleosynthesis in stars. He certainly set the scene for much important new work, by him and others, over the next decade. Especially after he had expanded his ideas and published his results in another important study in 1954, he convinced a growing number of astronomers that there is a recycling of the nucleosynthesis products that are produced in stellar interiors, these passing into the interstellar medium, where they provide for the formation of new stars. The mechanism whereby chemical elements heavier than hydrogen and helium are spread throughout the interstellar medium, by mass loss, is from stars on the so-called asymptotic giant branch on the H-R diagram, and from exploding supernovae. Since the recycling is taken to be a continuous process, it is expected that the metallicity of newly formed stars will exceed that of earlier stars, although there will be variation from place to place. A study of metallicity as a function of age, position, and velocity, in the Galaxy, could henceforth be expected to provide information about the Galaxy's history, and even of the evolution of galaxies in general.

Hoyle's idea of recycling of the heavy elements that are made in the stellar interiors was entirely new, and not only those like Gamow, who were investing their energies in theories of primordial abundances, were skeptical. There was some doubt as to whether the loss of mass from stars could occur at the rates Hoyle required, but in 1956 and 1960 A. J. Deutsch published important evidence in Hoyle's favor, for many different types of star, and further evidence has continued to accumulate ever since. It became accepted that mass loss is an extremely important process in evolving stars, of both high mass and luminosity, and especially in low-mass "asymptotic giant branch" stars, which make a high contribution to the interstellar material needed for the birth of new stars.

Not only did support for stellar nucleosynthesis began grow in the astrophysics community during the latter half of the twentieth century, but opposition from those supporting primordial nucleosynthesis began to decline. One of the chief reasons for this was the problem of the "mass

5 gap." There is no stable nucleus of atomic mass 5, which presents a difficulty in explaining the synthesis of the heavier elements from hydrogen and helium, and this difficulty seemed easier to overcome when working with nucleosynthesis in stars rather than in the early universe. The first important breakthrough in this case was by Ernst Julius Öpik, in a series of papers between 1938 and 1951. (As we mentioned in this last chapter, this was a period of great personal disturbance in Öpik's life, during which he moved from Tartu to Hamburg to Armagh.) He argued that after the temperature of the contracting core of a giant star reaches about 400 million degrees Kelvin, all its helium can be converted into carbon by certain triple collisions of helium nuclei, so crossing the mass 5 gap.

Aware of Hoyle's 1946 paper, but apparently unaware of Öpik's publications, in 1951 Edwin E. Salpeter effectively solved this problem. During a visit to work with William Fowler and Jesse Greenstein at the California Institute of Technology, he had been made aware of the potential relevance of the knowledge of quantum electrodynamics and nuclear physics he had acquired in England under Rudolph Peierls at Birmingham University. How could helium fusion be used to explain that giants were evolved stars, as his hosts believed possible? Salpeter made a concise but definitive calculation, showing that a triple alpha-particle reaction would work in a stellar interior. Two of the alpha particles (helium nuclei, of two protons and two neutrons) first combine to form beryllium (atomic mass 8), which is very unstable, but it has a good enough chance of combining with a third alpha particle for this final capture to be made in statistically significant numbers. The result is a carbon nucleus (6 protons and 6 neutrons; atomic mass 12). There are other candidates for the title of discoverer of the importance of this mechanism, which even Salpeter apparently considered to be common knowledge, but it was he who showed that it would work, and that beryllium 8 would survive long enough to capture the third helium particle and form stable carbon at temperatures around 200 million degrees Kelvin.

Hoyle was himself fortunate to enlist the help of Fowler, who was then performing a series of laboratory experiments at the California Institute of Technology with potential importance for a whole range of astrophysical problems. One of Fowler's first checks was on a speculation by Hoyle, who had pointed out that there is a so-called resonance in the carbon nucleus, with energy matching that of the beryllium and helium nuclei. For the triple alpha particle reaction to take place, the carbon must go through an excited state. In 1957 Fowler and three colleagues demonstrated in the laboratory that such a state exists, and an invaluable partnership was formed. Together with their colleagues Geoffrey and Margaret Burbidge, in 1957 they composed a book-length article on "cosmic nucleogenesis," which soon became a classic in its own right, important enough to have been accorded its own acronym, based on the initials of its four authors: B2FH. (It was originally published in *Reviews of Modern Physics*.) The human importance of carbon in the universe cannot be exaggerated, of course, since it is an essential ingredient of life-forms as we know them; but in this case it was a stepping stone to the synthesis of the heavier elements. What is more, the triple alpha particle reaction leading to it was simply not available to the theory being put forward by Gamow and his associates. It required temperatures of as low as a billion degrees Kelvin, for above this the rate of disintegration annuls all synthesis, and yet by the time this temperature is reached, the density of the "ylem"—if that is the right word for the thin soup of the time—was too low for the triple alpha-particle reaction.

In an autobiographical work *My World Line* that Gamow published in 1970, he made fun of Hoyle in a parody of the book of Genesis. In the beginning God created Radiation and Ylem. After going through the elements ("Let there be mass two," "Let there be mass three," and so forth up to the elements beyond uranium) God noticed that he had made a slip, and had omitted mass five. "And so God said, 'Let there be Hoyle.' And there was Hoyle. And God saw Hoyle, and told him to make heavy elements in any way he pleased." Hoyle made them in stars, of course, and spread them around with supernova explosions. And the story ends with neither Hoyle nor God, nor anyone else, being able to figure out how it was done, so complicated was Hoyle's method.

Hoyle's first position was as lecturer in mathematics at Cambridge. In 1958 he was made Plumian Professor of Astronomy there, although he continued to spend long periods in the United States. At Princeton University, for instance, he collaborated with Martin Schwarzschild on the evolution of low-mass stars through to the red-giant branch of the H-R diagram. He continued his collaboration with Fowler on nuclear processes in stars and supernovae; and when, in 1983, Fowler was awarded the Nobel Prize in Physics, many felt that it should have been shared with Hoyle. In the 1960s he published extensively on supermassive objects and high-energy astrophysics; many of his articles were coauthored with the Burbidges. Rather as Sommerfeld had been in nuclear physics, Hoyle was a born collaborator, but paradoxically, he had an uncommon knack for ruffling feathers—indeed, never more so than when he insisted, in a book published in 1986 with Chandra Wickramasinghe, that the feathers on the British Museum's famous archaeopteryx fossil were a modern fake. The two men were arguing for some strange hypothesis about mutating life-forms continually arriving from space, by courtesy of a super-intelligent civilization that wished to seed our planet.

In 1966 Hoyle helped to create an Institute of Theoretical Astronomy in Cambridge, modeled on the University of California's Institute of Geophysics and Planetary Physics in La Jolla. He organized an extensive visitors' program in his Institute, and there he fostered work by visiting American colleagues on the next important advance after B²FH, concerning elements higher up the periodic table than iron. In creating still heavier nuclei, such as lead and uranium, energy has to be supplied, rather than be released. For this, he and his collaborators proposed a process of "explosive nucleosynthesis," occurring during a supernova explosion, when material suddenly heated by a shock wave explodes and blasts a path outward through the star. This was another example of work that had been going on in two centers with little interaction between them, for similar ideas been put forward independently by the Canadian nuclear physicist Alastair G. W. Cameron. Cameron studied at the universities of Manitoba and Saskatchewan, with interludes working at Chalk River, Ontario, on an atomic energy project. He taught for a time at Iowa State College (now Iowa State University), returned to Chalk River, then moved to the United States in 1959. (He finished his academic career at Harvard.) In 1957 Cameron published studies that showed how nuclear fuels might burn explosively during supernova explosions. His low early profile no doubt explains why his ideas made little impact at first, but work in the 1970s confirmed several of the ideas he had put forward, and the role of supernovae in the recycling of elements throughout the universe became generally accepted.

THE HELIUM PROBLEM

Hoyle's early ideas on nucleosynthesis were inevitably entangled with his belief at that time in a steady-state world. This required him to believe that the formation of the various elements must be going on through-out history, and could not be assigned to a primeval explosive event, no matter how that was to be described in detail. Having sown the seeds of a new doctrine in B²FH, however, he and his colleagues found that their proposed stellar processes could not explain the abundances of all ele-ments as well as they had hoped. Accounting for one element in particular, namely helium, gave astrophysicists considerable difficulty. They saw that if a galaxy started out as hydrogen, only a small fraction would have been converted into the higher elements by the time the solar system—on the standard view of the matter—was born. As things stand, however, the old-est of celestial objects, in which heavy elements comprise only one or two percent of the total mass, actually contain nearly a quarter of their mass as helium, while the Sun and its like have rather more.

It was Hoyle and another colleague, Roger Taylor, who first appreciated the significance of these high helium abundances. Pondering the recently discovered quasars, in 1963 they asked whether an explosion far greater than that of an ordinary supernova might be the explanation. This led them to calculate how much energy might be generated in the explosion of stars millions of times more massive than the Sun—assuming that such objects exist. They found that if such a star began at a temperature greater than ten billion degrees Kelvin, about a quarter of its material would be con-verted into helium in the first hundred seconds. Cooling would take place in such a way that this process would end, and the problem of helium abun-dances would be very elegantly and precisely solved. On the other hand, this conjured up another difficulty. Where are the large numbers of super-stars needed to process the helium of the entire universe? In 1964, after the discovery of the cosmic microwave radiation by Penzias and Wilson, the astronomical community—which was not enamored of the steady-state model—turned toward notions reminiscent of Lemaître's Primeval Atom. Even Hoyle lent his weight to the shift, joining forces yet again with Fowler, and a new collaborator, Robert Wagoner. Together they computed the complex of nuclear reactions that might occur in a "hot Big Bang."

After this extremely important breakthrough on the nuclear physics side of cosmology, it was the turn of others to supply more detail, and to extrapolate back to the putative beginnings of the process. Atomic nuclei in a world a billion times hotter than now require that the universe's density was a billion cubed (10^{27}) times its present value, but even that is no greater than the density of air, and nuclear physics could cope thus far with the problem. It was the precise agreement of theoreticians' predictions with the findings of astronomical observers, however, that finally brought over a majority to the side of hot Big Bang nucleosynthesis. We met earlier with the figure of 25 per cent helium, obtained by Gamow and his colleagues in 1948. Hoyle and his true supporters could not at first, of course, openly go along with any such work, done in the context of a Big Bang theory. The prediction now was that stars and galaxies should never have less than 23 percent helium, and observations suggested that the figure for the oldest objects is 23 or 24 percent—a figure that has not needed revision. Deuterium and the higher elements similarly began to fall into line. Assumptions made by the theoreticians about the present density of matter were less critical, but again they chimed well with observation. The biggest objection to the idea of cosmologists being able to extrapolate back to a time when the universe had been expanding for only a second or so was a psychological one, and—beginning in the 1960s—that objection faded away with surprising rapidity.

We shall shortly see how steady-state theories lost ground, and—at a later stage—how the discovery of cosmic background radiation persuaded the great majority of interested parties that a hot Big Bang theory, within a framework of general relativity, was essentially correct, although in need of supplementation for the later evolutionary history of the universe's contents. Loosely called "the standard model," it was anything but standard when further developed, and especially when cosmologists began to consider earlier phases than the first second, the first millisecond, or even the first microsecond of the history of the universe.

Radio Astronomy

Just as William Herschel, with the help of a thermometer, discovered infrared radiations from the Sun that were invisible to the human eye, and just as photography allowed an extension of the spectrum beyond the violet, so too radio receivers have made possible the detection of electromagnetic radiation between wavelengths of roughly 1 millimeter and 30 meters. (The wavelength of yellow light, by comparison, is a little under $6/10,000$ millimeters.) The bridge between the theory of light and classical theories of electricity and magnetism was assembled only gradually during the course of the nineteenth century, the most significant single contribution being that made by James Clerk Maxwell. Maxwell united electricity and magnetism into a single theory, and his equations made it seem likely that electric waves might be propagated through space as an extension of light radiation. In 1882, the Dublin physicist George Francis Fitzgerald, basing his argument on Maxwell's theory, insisted that the energy of varying currents might be radiated into space, and a year later he described an apparatus—a magnetic oscillator—by which such an effect might be produced. This led to further theoretical work, but their practical repercussions were slight. There were attempts by Oliver J. Lodge to detect electromagnetic waves in wires, while Joseph Henry, Thomas Alva Edison, and others all showed that it was possible to send electromagnetic action over appreciable distances, but their claims were usually dismissed, the effects being assumed no different from local electromagnetic effects, "inductions." The Anglo-American David Edward Hughes, a prolific inventor with strong interests in telegraphy, showed conclusively that clicking signals from a sparking induction balance—the spark triggered repeatedly by clockwork—could be detected by a telephone receiver at distances of several hundred meters, and very clearly up to a hundred. In 1879 and 1880 he demonstrated his spark transmitter to representatives of the Royal Society in London and to officials from the post office there, but again, the effects were dismissed as having been caused by induction. Disappointed, he refrained from publishing on the subject.

The world of physics was finally convinced in the late 1880s, with some outstanding work by Heinrich Rudolf Hertz. Hertz had worked with Helmholtz but was now at Karlsruhe. He had spent much time in attempting to justify Maxwell's equations on theoretical grounds and in carrying out a number of related experiments. It is not clear whether he knew of the work of his predecessors when in 1888 he produced "electric waves." He did so with an open circuit connected to an induction coil, and the waves were detected with a simple loop of wire with a gap in it. This, in a strong sense, marks the birth of a new technology, which not only functioned, but was understood. For many years, however, advances were slow. Detection was difficult, and a great breakthrough here came in 1904 with John Ambrose Fleming's invention of the thermionic valve. Fleming's credentials were excellent: he had worked with Maxwell in Cambridge, he was for a time consultant to the London branch of Thomas Alva Edison's company and was inspired by some of Edison's work, he had experimented with radio transmission from the 1880s, and he had helped to design the transmitter used by Guglielmo Marconi to span the Atlantic in 1901.

Although radio, therefore, is usually regarded as a product of the twentieth century—since most of us are vaguely conscious of the beginning of regular sound broadcasting in the 1920s—it was moderately well established before the turn of the century. Since the connection between optical and electromagnetic radiations was by this time an accepted part of physics, it is not surprising that several astronomers considered the *Sun* as a potential source of radio waves. As early as 1890, Edison, with colleague A. E. Kennelly, discussed the design of a suitable antenna, settling for a cable on poles around a massive iron core. Oliver Lodge set up an ambitious receiver in Liverpool in the period 1897–1900, hoping to detect solar radio waves, but any signals there might have been were swamped by electrical interference from the city. Other unsuccessful attempts were made by J. Wilsing and J. Scheiner in Potsdam in 1896 and by C. Nordman in 1901—unfortunately, during a minimum of solar activity. Nordman set up good apparatus high in the French Alps, and used a long antenna. Had he persevered for longer than a day he would very probably have been successful; but his results, like the rest, came to nothing, and the idea that the Sun is a transmitter went into almost total eclipse.

The first unequivocal detection of radio waves from the Sun was made by James Stanley Hey, on 27 and 28 February 1942, when investigating what was thought to be the jamming of British radar by the Germans. That the Sun was the source was confirmed by comparing data from stations in various towns in Britain. Later it was found that the observatory at Meudon in France had (visually) recorded strong solar flares at the same time. While the emission was, as Hey said, a hundred thousand times more than had been expected, he noted that it would not have been registered had the Sun been as near as our nearest stellar neighbor. Only exceptional stars can reach out to us across galactic distances at radio wavelengths. Hey's discovery is often called fortuitous, but one should hesitate before calling it that, in view of Hey's record. He was the first to discover radar reflections from meteor trails—investigated for the same reason as what turned out to be solar radiation. He it was who first discovered a radio galaxy, Cygnus A, after the war. Only gifted observers make chance discoveries.

GALACTIC RADIO WAVES

Radio astronomy was well and truly incorporated into astronomical practice only in the late 1950s, but its successful practice goes back as far as 1932. In that year, Karl G. Jansky, an American radio engineer at the Bell Telephone Laboratory, was studying interference on the newly inaugurated trans-Atlantic radio-telephone service when he discovered that much of the interference came from extraterrestrial sources. His receiver, set up at Holmdel, New Jersey, was working at a frequency of 20.5 megahertz (that is, in the 15-meter band), and he had a steerable antenna with moderately strong directional properties. His records show that he distinguished between three main sources of noise: local thunderstorms, distant thunderstorms, and a background hiss that varied with the time of day. At first he thought he had detected a radio emission from the Sun, but as the Sun changed its celestial position and the radio source did not, he realized that he was concerned with quite another phenomenon. That year was one with a minimum of solar activity, which made it easier for him to study the hiss of background "static." He did so for a year and established that the signals were associated with the Milky Way, and were strongest in the direction of the center of the Galaxy. (He later made another conjecture, that the signals were from the solar apex, the point toward which the Sun is heading in space, but in this he was mistaken.)

Jansky published his findings over the next three years, and then moved on to other research. His work attracted little attention at the time. Grote Reber, however, a radio engineer from Wheaton, Illinois, built himself a paraboloidal antenna of a sort now very familiar, and in his spare time repeated much of Jansky's work. Receiving at higher frequencies and with a more strongly directional antenna, in 1939 he was able to detect strong radiation at 160 megahertz. Between 1940 and 1948, Reber produced contour maps of intensities from the sky, paralleled only by work of which he was then unaware, which was being undertaken by J. S. Hey and others at the Army Operational Research Group in the United Kingdom. (They were working at 64 megahertz, using newly developed radar equipment.) Reber's dedication to this infant subject is very impressive, and the same must be said of his antenna, bearing in mind his amateur status. It is often pointed out that radio astronomers can work by day as well as by night. Reber found that matters are not so simple: to avoid interference with spark emissions from automobiles, he needed to work in the dead of night. His original 9-meter parabolic dish antenna has been preserved, and now stands in the grounds of the National Radio Astronomy

Observatory at Green Bank, West Virginia, alongside a full-scale replica of Jansky's antenna.

Theories explaining the "galactic static" were numerous. Several astronomers proposed new classes of stars, rich in radiations at radio frequencies. Nuclear physics was giving increasing attention to the synchrotron as a means of accelerating charged particles, so it is not surprising to find a theory being proposed—first in 1950 by K. O. Kiepenheuer—that radio waves are transmitted by electrons with very high velocities, spiraling through the magnetic field of the Galaxy. As electrons meet with a magnetic field they spiral around it, not in flat spiral, but in the form a wire would take if wrapped round a pencil. The circling movement of electrons round the "pencil" (the field) implies a continuing acceleration toward the center of rotation, and this acceleration results in a continuing emission of (polarized) radiation. Electrons at velocities approaching that of light may give signals at radio frequencies that are powerful enough to travel astronomical distances. In the process of radiating, however, they lose energy, and if such synchrotron radiation is to be maintained, the supply of electrons must be kept up. They may in fact be supplied from a whole range of energetic sources, among which are supernova remnants, quasars, and various forms of active galactic nuclei.

A similar sort of theory was proposed in the same year by H. O. G. Alfvén and N. Herlofson, who thought that there might be celestial synchrotrons on a much smaller scale, perhaps even comparable in size with the solar system, in which case we might be receiving radio waves might from innumerable discrete stellar sources, rather than only from galactic centers. These ideas were infectious. Vitaly L. Ginzburg and Josef S. Shklovsky in the Soviet Union immediately began to develop the synchrotron theory, as applied to radio galaxies, and their detailed predictions proved highly successful, in particular those relating to the polarization of radiation.

This entire subject was nowhere more dramatically applied than in the case of the Crab Nebula, which provided astronomy with a testing ground for numerous theories of stellar evolution. In 1921, and again in 1939, photographic plates of the Crab Nebula—taken in 1909, 1921, and 1938— were carefully compared by John C. Duncan of the Mount Wilson Observatory, and he came to the conclusion that it was expanding at about a fifth of a second of arc per year. Its current size seemed then to imply that it was about 800 years old. In 1937, Nicholas Mayall at the Lick Observatory measured the velocities of expansion of parts of the cloud, using spectroscopy and the Doppler effect. The results he obtained were astonishing. The velocities were greater than any previously recorded, well over 1,000 kilometers per second. In 1942, Oort and Mayall published a joint paper in the United States, arguing for the identity of the Crab Nebula with the supernova of AD 1054, a date that matched the data rather well. Their paper was accompanied by a survey of relevant information from early Chinese chronicles, by the Dutch Orientalist J. J. L. Duyvendak, and others have since followed in his historical footsteps on this very subject. From the various measurements then made, it was possible to say that the nebula was about 5,000 light-years away, and this figure allowed a first estimate to be made of the energy presently being radiated. In 1949, John G. Bolton and Gordon Stanley in Australia found that the Crab Nebula is a powerful radio source. Five years later, in the Soviet Union, Viktor A. Dombrovsky and M. A. Vashakidze studied the polarization of light from the object— I. M. Gordon and V. L. Ginzburg had independently predicted that this should be observable. At last the link with Kiepenheuer's conjecture could be made. With supplementary information from Palomar provided by Walter Baade, as well as studies done in Leiden by Jan Oort and Theodor Walraven, it became clear that both the visible light from the Crab Nebula and its radio emissions were being generated by electrons or other charged particles emitting synchrotron radiation.

NEW RADIO TECHNIQUES

Reber's published results reached the Netherlands during the war, and gave Jan Oort the idea that a spectral line at radio wavelengths would be an important tool for mapping the rotation and structure of our Galaxy. This was something he had been studying, as we have seen, using ordinary optical information, or rather, such information as could penetrate the frustrating clouds of dust in the galactic plane. He realized that radio waves would penetrate that dust; and that Doppler shifts in the radio frequency of any spectral line emitted by galactic gases would make measurement of their velocities possible. Oort assigned to his student Hendrik C. van de Hulst the task of calculating what radio spectral lines there might be, and their likely frequencies. In view of the abundance of hydrogen, Van de Hulst started with this, and found that there should be radiation at a wavelength of about 21 centimeters, a frequency of 1420 megahertz. (The line arises from a so-called hyperfine transition in the ground state of neutral hydrogen.) The prediction was published in Dutch in 1945, but the economic state of the country was such that six years passed before it could be tested. Eventually, with help provided by the Dutch PTT (the post office and telephone service) and the firm of Philips, a 7.5-meter German radar reflector was installed at the telecommunication station at Kootwijk, with a suitable receiver. After modification of the receiver by C. A. Lex Muller, in 1951, he and Oort were able to observe the interstellar emission line they had been seeking—namely that of neutral hydrogen from the Milky Way. Unfortunately for them, they had been beaten to the post by E. M. Purcell and his doctoral student Harold I. Ewen, at Harvard, who used a crudely made but sensitive receiver. Almost simultaneously, the 21-centimeter line was detected in Sydney, Australia.

Radio astronomy received a great impetus from the ready availability, after the Second World War, of large quantities of surplus equipment and antennas, not to mention high expertise in radio communications and radar. Results came tumbling in, in no particular order, in a way reminiscent of what had happened when the telescope was first directed to the sky. An early result of great significance was obtained by Joseph Pawsey, who in 1946 found that the brightness temperature of the Sun at meter wavelengths is of the order of a million degrees Kelvin. This fitted well with optical observations of the corona, and the corona was thus seen to be all-important in this radio work—and to be opaque to radiations from below it at these wavelengths. Hey's discovery of a discrete radio source in Cygnus, while he was still working at his military establishment in 1945, just after the war, has already been mentioned. Many other radio stars were subsequently found. Ryle, in Cambridge, using his radio interferometer— of which more will be said in the following section—found an intense localized source in the constellation of Cassiopeia. In Sydney, in 1949, John G. Bolton and Gordon Stanley discovered three discrete radio sources that were also known from optical observations. One, which they called "Taurus A," was in the Crab Nebula, while the others "Virgo A" and "Centaurus A" were external galaxies.

At much the same time as all this, Bernard Lovell—who was working on cosmic radiation, radar reflections from meteors, aurorae, the Moon, and even the planets, as well as radio waves from space and the Sun— had begun to feel acutely the need for some sort of steerable antenna. In 1948 he launched plans for the first giant paraboloid, a type of antenna reflector with which most people are now familiar. His tenacity eventually yielded remarkable results. Finally completed in 1957 at Jodrell Bank not far from Manchester in northern England, the dish was 76 meters across, and weighed 1,500 tonnes (fig. 236). It was carried on turrets that had carried 15-inch guns on the British battleships *Royal Sovereign* and *Revenge*. This extraordinary engineering venture—which once nearly came to grief in a hurricane—seemed at times to be doomed to failure through lack of funds when it suddenly entered the public consciousness. In brief, the telescope proved to be useful for locating by radar the carrier rocket of the first Soviet *Sputniks*, the first and second of which were launched in 1957. Three years later, the American *Pioneer* 5 was released from its carrier rocket by a signal from Jodrell Bank. Such uses, for which of course the telescope was never intended, prompted the gift of money that was the telescope's final salvation. An earlier campaign in which the press had tried to shame the British government—the *Sunday Dispatch* stated that "Schoolboys send pocket money to save our face"—had unfortunately not been successful. The Mark I telescope was later modified, but the pattern was set. Another vast project, the Parkes telescope in Australia with a 64-meter dish, followed in 1961. One other splendid example from a long list of fully steerable radio telescopes was that completed in 1972 at Effelsberg in Germany. This had a 100-meter dish and a surface accurate enough for observations at wavelengths almost as short as 1 centimeter.

At the end of the twentieth century, the largest radio telescopes, in terms of their reflecting area, were those at Arecibo in Puerto Rico and

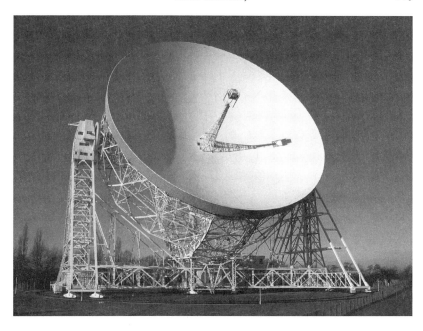

236. The 250-foot (76-meter) diameter dish of the radio telescope at Jodrell Bank, about 35 kilometers south of Manchester, England. This, the world's first giant dish radio telescope, was upgraded in 2003, with a new reflecting surface and a new high-precision drive system, which together allow it to be used over a range of frequencies four times greater than before.

Zelenchukskaya, in the Caucasus Mountains, although they were to very different designs. That at Arecibo is certainly the more dramatic. Its fixed bowl, with a diameter of 305 meters, is built into the ground of a mainly natural valley, and a cage with antenna and other materials is slung from three colossal pylons. The angle of view can be adjusted somewhat by shifting the position of the antenna in relation to the bowl, so that planets and asteroids can all be observed when the time is right. The Arecibo telescope, built between 1960 and 1963, owes its existence to a desire by the U.S. Department of Defense to track Soviet and other satellites, chiefly by plotting their ionospheric trails. The U.S. Department of Defense launched the project, funding it for seven years before passing it over to the National Science Foundation, as an arm of the National Astronomy and Ionosphere Center. Despite what Lincoln said, there are times when it is indeed permitted to swap horses while crossing the river.

Since its inception by James Hey and others during the war, studies of the solar system using large radio telescopes have depended heavily on radar techniques, and American astronomy was not the only benefactor of the Cold War, in this respect. Soviet astronomers soon made radar their specialty. Using an array of eight 16-meter dishes in the Crimea, they were the first to receive reflected signals from Mercury, Venus, and Mars in the early 1960s. This provided a context for the later Zelenchukskaya radio telescope, built by the Soviet Academy of Sciences, and given the name RATAN-600. This name was in an allusion to its diameter in meters and its being a "Radio Astronomical Telescope of the Academy of Sciences [Nauk]." Its 895 elements, each a parabolic plate, are arranged in four independent sectors of a circle 576 meters across, and each quadrant can be used as a self-contained unit. Its collecting power is comparable with that of the Socorro VLA in New Mexico, to be introduced shortly.

RATAN-600 belongs to the primary observing facility of the Russian state, and is a partner to the optical observatory at nearby Nizhnij Arkhyz. The latter is the location of what was, for a time, the world's largest single-mirror optical telescope, with its 6-meter objective. Perhaps not entirely unconnected with that world record was the fact that the observatory hosted five out of the first eleven International Astronomy Olympiads, a widely respected series of annual astronomy competitions for high school students, begun in 1996.

THE SEARCH FOR HIGH ANGULAR RESOLUTION

One of the most vital of early needs in radio astronomy was to improve resolution, that is, to narrow down the angle of view. The angular resolving power of the radio antennas available at that time was not sufficient to resolve the disk of the Sun, let alone locate the precise origins of the radio emissions detected by such as Jansky and Reber. At Cambridge University, however, Ryle and D. D. Vonberg, began to adapt war surplus radar equipment, and to develop new receiver techniques for meter wavelengths. Ryle was ideally suited to this kind of work, for he had played a distinguished part in British wartime operations involving all aspects of radar and radio. How could angular resolution to be improved? One early technique used with the Sun was to use a solar eclipse and to note the changes taking place as the Moon progressively blocked the "view" of the solar disk. This was the procedure adopted, for instance, by the Americans R. H. Dicke and R. Beringer in 1945, by A. E. Covington in Canada in 1946, by the Soviet astronomers S. E. Khaikin and B. M. Chikhachev, working from a naval vessel off the shore of Brazil in 1947, and by W. N. Christiansen, D. E. Yabsley and B. Y. Mills in Australia in 1948. Something better was clearly needed, however, for radio astronomy in general: a method that could be put to daily use.

Eventually, two other solutions were found, both depending on the phenomenon of interference, already much used in optics. We recall the early attempts by Stephan and Michelson to use an interferometer to measure the angular diameters of stars, and the first success at Mount Wilson in 1920, by Michelson, Pease, and Anderson. In the case of light there are many ways of obtaining interference effects besides the classic experiment of Thomas Young, with its two slits in a plate serving as the two interfering light sources. One variant of the arrangement, requiring only a single slit, was devised in 1834 by Humphrey Lloyd of Dublin. By viewing that slit very obliquely, in a mirror, he produced in effect a second slit source, the mirrored slit. Light from the mirror could interfere with light seen directly from the one real slit. The "Lloyd's mirror" technique was translated into radio terms and used for several years, especially in Australia by L. L. McCready, J. L. Pawsey, and R. Payne-Scott. In Cambridge, Ryle and Vonberg developed the radio analog of the Michelson interferometer, apparently without realizing at the time that this was what they had done. Their antennas were separated by distances of up to 140 wavelengths. A massive sunspot that occurred in July 1946 gave them an opportunity to put their apparatus to the test, and their observations showed conclusively that the radio emission was coming from a region on the surface of the Sun of much the same size as the sunspot. For those who had any doubts, Hey's discovery of the discrete source of radio emission in the constellation of Cygnus was tested by Ryle and Graham-Smith, who adapted their solar interferometer to observe the radio source in 1947. By 1950, Graham-Smith had measured the position of the source (Cygnus A) very precisely, showing it to be associated with a distant massive galaxy.

The resolution of a telescope increases with its aperture—in a radio telescope, for example, this will be the aperture of the large parabolic reflector, but it may also be the separation of elements of a complex of antennas. Resolution also depends on wavelength. A telescope with aperture of a can resolve a minimum angle equal to $1.22\lambda/a$. Clearly a radio telescope, working at much longer wavelengths than its optical counterpart, has to be proportionately larger, to match its results. The commonly used 21-centimeter wavelength is roughly 400,000 times as great as that of light in the middle of the optical spectrum. The disparity in the resolutions achievable is obviously enormous. There are obvious engineering limits to antennas that can be built in fully steerable form, but given the relative ease of building non-steerable antennas—by comparison, that is, with large mirrors—the advantage in the long term was with radio. Ryle showed how the advantage could be exploited.

His most important achievement was his development of the concept of "aperture synthesis." This is not something that lends itself to a simple explanation, but the technique makes possible the creation of images of radio sources by combining observations of interferometer type, using radio telescopes of modest dimensions at different interferometer spacings. It is necessary to measure both the relative amplitudes (strengths) and the phases (the stages reached in the wave cycle) of the incoming signals. These contain all the information needed to reconstruct the distribution of intensity of radio signals from across the accessible part of the sky, using widely separated telescopes. High resolution could be bought at a relatively low price. The signals from two or more widely separated telescopes were combined by certain mathematical procedures that could produce images with the same angular resolution as an instrument the size of the entire collection. In order to produce a high quality image, as many baselines as possible were required. One simple way of changing the baseline separating any two antennas was to let the Earth's rotation provide it, a method first considered in outline in 1950. In time, instruments used mechanical rotation of the interferometer array. Thus the VLA—an instrument to be introduced shortly—has 27 telescopes giving 351 independent baselines at once, providing images of correspondingly high quality.

So much for the theory. Putting it into practice in its early years was not easy. In 1956, the small Cambridge radio observatory moved to a disused wartime Air Ministry bomb store, and with a large grant from the firm of Mullard Ltd. it opened in 1957 as the Mullard Radio Astronomy Observatory. For more than a decade, the group concentrated on producing reliable catalogs of bright radio sources in the northern sky, and in doing so gradually forced a revision of the common view that most "radio stars" belong to our Galaxy. It became increasingly clear that most of them were associated with distant galaxies. Aperture synthesis had to wait until the end of the 1950s, when Ryle began to experiment with the rotation of the Earth as a means of moving the baseline around. Computing the resulting data now presented a problem but, by 1959, digital computers were fast enough to perform the necessary synthesis of data. Working with Ann Neville, Ryle created his first aperture synthesis map—by Earth rotation—of a region of sky around the north celestial pole. The angular resolution of the survey was 4.5 minutes of arc. He had not only shown the feasibility of the method, but had also pointed the way to aperture synthesis arrays using fully steerable antennas. The Cambridge One-Mile Telescope was completed in 1965, and the Five-Kilometer Telescope in 1972, both of them without any rival, for the time being. In 1974, Martin Ryle and his colleague Antony Hewish—another person whose wartime experience of telecommunications research had guided him into radio astronomy—jointly received a Nobel Prize for this and other contributions to the development of radio interferometry. Hewish had made an important contribution to the development of aperture synthesis. We shall shortly see how his studies of the flickering, or scintillation, of radio sources, caused by irregularities in the flow of the solar wind, led him and his graduate student Jocelyn Bell to the discovery of pulsars.

After the detection of the 21-centimeter hydrogen line in 1951, the Dutch had drawn up plans for a 25-meter telescope near the village of Dwingeloo, and in 1956 this was used for more comprehensive surveys of the sky at a wavelength of 21 centimeters. Early successes led to an attempt to set up an overambitious project, in which the Dutch and Belgian governments agreed to build what was to be called the "Benelux Cross." It proved to be too hard a cross for the Belgian government to bear, and Belgium pulled out of the scheme, leaving her neighbors to bear the cost. Three or four different schemes on a more modest scale were then drawn up. In a sense, the delay was fortunate, since in the meantime Martin Ryle's group in Cambridge had been developing a new technique that could eventually be used in what was to be known as the "Westerbork Synthesis Radio Telescope." The council of the village where it was situated insisted on the name, rather than that of the University of Groningen, in the hope of altering the public's perception of the place. Westerbork had been used as a holding point in the Second World War for many thousands of Dutch Jews and gypsies before their transportation to concentration camps in Germany and Poland.

After many years of negotiating funding for the abortive "Benelux Cross" project and debating its design, the Dutch astronomers were well prepared to make use of Ryle's new techniques. The Westerbork telescope— or rather, the array of telescopes—quickly took shape. Its original design was for twelve 25-meter steerable telescope dishes on an east-west baseline, extending over 1.5 kilometers. Construction began in 1966, and the twelfth telescope was in place before the end of 1968. In normal operation, when first used, two movable dishes on a 300-meter track at the east of the array were correlated with each of the ten fixed dishes, producing 20 interferometers. Two new movable dishes were added later, extending the array to 2.7 kilometers.

Using such relatively simple and—by comparison with what was to come—inexpensive arrays, here and elsewhere, more catalogs of radio sources were produced, supplementing those from Cambridge. Although arrays with ever greater separations of receivers gave problems with linkage by cable, ways were developed for recording observations, and only combining them at a later stage. This avoided the need for connecting cables. The Very Large Array (VLA), built on the plains of San Agustin, eighty kilometers west of Socorro, New Mexico, eventually took the lead in this field, achieving some remarkable results in the 1980s and after. Formally inaugurated in 1980, eight years after its funding was approved by Congress, the VLA is composed of 27 radio dishes in a Y-shaped configuration, each dish being 25 meters in diameter. These antennas are steerable in the usual sense—that is, that they can be pointed to any part of the sky—but they are also movable on rails, and their configuration is modified at regular intervals, according to need. The best potential resolution is that of an antenna 36 kilometers across, while the resulting sensitivity is that of a dish 130 meters (422 feet) in diameter. Actual resolution depends on the frequency of the received signal, but at the highest usable frequency of 43 gigahertz, the resolution is 0.04 arc seconds. This, as they boast, is equal to the angle subtended by a golf ball at a distance of 150 kilometers.

The VLA was the product of a collaboration calling itself the National Radio Astronomy Observatory (NRAO), a collaboration between nine American universities and the National Science Foundation. By the mid-1980s, its annual budget was around $15 million, which covered the running of the $80 million VLA. The first telescope funded by that group had been put into use in 1959, a 25.9-meter dish telescope that could be linked with two similar movable telescopes, forming an interferometric array of baseline 2.7 kilometers. Other fully steerable dishes followed.

When NRAO built a dish accurate enough to be used at millimeter wavelengths, it was situated at Kitt Peak, with a view to avoiding atmospheric absorption of the signals.

High as is the resolution of the VLA, it cannot compare with radio interferometers using VLBI—very long baseline interferometry. Here, the component antennas are separated by great distances, even of thousands of kilometers, and again, they are not linked electrically, or even by microwave radio, but the signals received are recorded on tape or disk, together with extremely accurate time signals. Analyzed in one laboratory, the resulting resolutions are of the order of thousandths of an arc second, although they cannot map directly with that accuracy. VLBI provides astronomers with their most precise view of the most energetic phenomena in the universe—such as supernovae, starburst galaxies, active galactic nuclei, pulsars, star-forming regions in molecular clouds, and gravitational lenses. This is a field where those who have no head for acronyms should not enter. VLBA (Very Long Baseline Array, unconsciously copying the Victoria Ladies' Bowling Association) is a North American network, set up with ten antennas, stretching from Hawaii to St. Croix in the north-east of Canada. It achieves resolutions of a fraction of a millisecond of arc. MERLIN, operated by Jodrell Bank Observatory, is—never mind the acronym's translation—an array of radio telescopes distributed around the United Kingdom, with separations of up to 217 kilometers. It achieves resolutions of better than 0.05 arc seconds—somewhat greater than the optical resolution of the Hubble Space Telescope. A European VLBI network was formed in 1980 by five founding institutes for radio astronomy, in Bonn, Bologna, Dwingeloo, Onsala (Sweden) and Jodrell Bank.

The first truly striking discovery made using the VLBI technique was that of so-called superluminal radio sources that appeared to be expanding at several times the speed of light. The first observations of this phenomenon were made in 1967 by David S. Robertson in Australia and Allen T. Moffet at Owens Valley in California. Three years later, Irwin Shapiro and colleagues, using facilities in Southern California and in Massachusetts, discovered a more spectacular example, 3C 279 (that is, item 279 in Ryle's third Cambridge catalog of sources), which was apparently expanding at ten times the speed of light. Since VLBI can pinpoint positions extremely accurately, it can also, given time, determine proper motions, motions transverse to the line of sight. Such motions, when the distance of the object and its radial velocity are known, can provide a figure for the transverse velocity, and so for the actual velocity. (Values well in excess of the speed of light were sometimes found, seeming to contradict Einsteinian relativity theory, but there are ways out of this dilemma.) Superluminal behavior has since been found in some radio galaxies and quasars. In 1994, I. F. Mirabel and L. F. Rodriguez announced the discovery of a superluminal source in our own galaxy, the cosmic X-ray source GRS1915+105. Within weeks, they saw blobs expanding in pairs by typically 0.5 seconds of arc.

All of these superluminal sources are now thought to contain a black hole, which is ejecting mass at high velocities.

In the 1980s, the VLBI idea was taken into space. Trials were successfully performed between 1986 and 1988, combining data from ground radio telescopes at Usuda (Japan) and Tidbinbilla (Australia) with data from a satellite telescope. In 1997, the VLBI Space Observatory Programme, led by the Institute of Space and Astronautical Science, in collaboration with the National Astronomical Observatory of Japan, launched a satellite from the Kagoshima Space Center. The satellite, carrying an 8-meter diameter radio telescope in a stable elliptical orbit round the Earth, makes baselines possible that are as great as three times the length of those available on Earth. By the end of the millennium, this kind of activity had reached such proportions that a new organization was instituted to coordinate organizations operating, or supporting, VLBI. Astronomers are more numerous than ever, but they have not lost any of their old conviviality.

THE DISCOVERY OF QUASARS

An early advance in radio technique in the postwar period was that by J. P. Wild and L. L. McCready, around 1950. They produced radio spectrographs, analogs of spectrographs at optical wavelengths, by scanning a wide range of frequencies very rapidly—from around 70 to 130 megahertz, scanned in less than a second. In the first instance, this allowed them to analyze bursts of emissions from the Sun in ways that greatly increased the understanding of its physical constitution. The ability to shorten the response time proved to be of the greatest importance to astronomy more generally, and made possible, as we shall see, the detection of two hitherto unknown types of object—quasars and pulsars. Leaving the discovery of pulsars for a later section, here we shall consider what, in its day, was one of the most exciting of all discoveries made with radio telescopes, that of "quasi stellar radio sources"—which soon became known simply as "quasars."

The first radio galaxies had been discovered in the early 1950s, but it was more than ten years before positions had been found with an accuracy that allowed them to be matched with visual objects. The astronomers at Jodrell Bank were beginning, nevertheless, to pay much attention to a number of radio sources with very small angular diameters. In 1960, Allan Sandage—Hubble's successor at the Mount Wilson and Palomar Observatories—took photographs of three regions containing such sources, and Thomas A. Matthews and John G. Bolton at the Owens Valley Radio Observatory found in each case that the only visual object within the error rectangle was a star. At the end of that year, Sandage announced—in an unscheduled paper at a meeting of the American Astronomical Society—that the photographic plates showed what seemed to be a bright star at the precise position assigned to a strong radio source, 3C 48. It was accompanied by a faint luminous wisp. If this was indeed a star, then it would be the first distant radio star to be discovered. When its spectrum was analyzed, however, it turned out to have numerous emission lines, and to be quite unlike that of any star then known. Sandage's talk was given a summary paragraph in the popular magazine *Sky and Telescope* in March 1961, but its significance was not properly appreciated at first, even by those responsible for the work.

Early in 1963, an even brighter star was identified with another radio source, 3C 273. This particular source had been very accurately located by Cyril Hazard, then at the University of Sydney, and his colleagues M. B. Mackey and A. J. Shimmins. Working at the large Parkes radio telescope, they used an occultation by the Moon to pinpoint the star's position and form, to a precision of about a second of arc—in fact the object turned out to be double. (A group of observers headed by H. P. Palmer at Jodrell Bank had determined as early as 1960 that its angular size was less than 4 seconds of arc.) Maarten Schmidt, a Dutch astronomer then working at Mount Wilson and Palomar, obtained a spectrum for it, and found hydrogen lines in it that were shifted toward the red by an amount that was so large— 16 percent—that it seemed more appropriate to a distant galaxy. It appeared point-like, however, and was certainly smaller than a galaxy, and yet it was much brighter than an entire galaxy should have been, at the great distance implied by its colossal redshift. The spectrum allowed Schmidt to spot oxygen and magnesium lines similarly displaced. His colleagues Jesse L. Greenstein and Thomas A. Matthews, armed with this information, then re-examined the spectra from 3C 48, and found that it had a redshift of 37 percent, suggesting an even more remarkable velocity of recession.

In 1964, Margaret Burbidge and T. D. Kinman began recording spectra with the 120-inch reflector at the Lick Observatory, and C. R. Lynds and his colleagues did likewise, using the 84-inch reflector at the Kitt Peak National Observatory. Ten such objects were known by the end of 1965, and thereafter the number increased rapidly. As it did so, even more extraordinary redshifts were measured: by the end of the following year, three were known with redshifts of greater than 200 percent.

These discoveries set astronomers the difficult problem of deciding whether the redshifts were "cosmological"—that is, indicative of objects moving with the general expansion of the world of galaxies—or due to some new intrinsic properties, or perhaps simply due to colossal local velocities, such as might follow from some sort of galactic explosion. A few influential physicists and astronomers, including James Terrell, Geoffrey and Margaret Burbidge, and Fred Hoyle, argued that they were within, or relatively near to, our Galaxy. Martin Rees and Dennis Sciama argued against this conclusion—although Sciama had at first believed the opposite, until he was persuaded by the findings of the optical astronomers. (Sciama had been a tenacious believer in steady-state theory, but abandoned that, too, at about the same time.) If the redshifts were a local velocity effect, then should we not also see approaching objects, with spectra shifted toward the blue? An analysis was made of the numbers, and it seemed that there were simply too many faint sources with large redshifts for them to be compatible with the steady-state cosmology that Hoyle was still defending in a rearguard action. A considerable body of evidence began to accumulate that the quasar redshifts were indeed cosmological. The nearby quasar 3C 206, for example, is seen to be within a *cluster* of galaxies, which all have identical redshifts to that of the emission lines of the quasar itself.

On a question of terminology: "quasar" was very quickly seen to be an ambiguous term, once quasi-stellar galaxies and "interlopers" were identified, and when optical studies gradually revealed the fact that only about one quasar in ten emits massive energy at radio wavelengths. Sandage began finding "radio quiet quasars" in 1965. "Quasi-stellar object" (QSO) is an uninformative designation but carries no implication that it covers only one sort of object. The term "quasar" is now usually reserved for an enormously energetic *star*-like source with such a large redshift that it is very remote—assuming, as the jargon goes, that it is genuinely "cosmological." The great majority of the redshifts of *galaxies* measured with the help of the Schmidt telescope at Palomar were of less than 20 percent. If the redshifts of the quasars were truly indicative of enormous distances, then each pours out around a hundred times more energy than an entire typical galaxy, although the most powerful radio galaxies, such as 3C 295, were found to have comparable luminosities. Allowing for a slowing down of the expansion of the universe, it can be said that when we observe quasars we may be looking back to a time when the universe was as little as a tenth of its present age. Whether or not the universe had an amorphous beginning, this suggested that structure appeared in it at that early stage in its history. There was a division of labor here. Explaining why that structure was there at all was a problem for cosmologists. A more obvious problem henceforth confronting astronomers was that of explaining the source of energy in quasars, the most luminous known objects in the universe.

Radio studies in the 1970s and 1980s showed that many of them have, at radio wavelengths, a double structure, typical of many radio galaxies. The optical "star" was then usually found to coincide with a powerful and compact radio component. Radiation from quasars was shown to be often partly polarized, and to include X-ray emissions. It was found that many are of variable output—in radio and optical terms—and that the time scale of the variation is of the order of a year. Measurements of their angular sizes, made with the help of interferometers with an intercontinental baseline, gave results of around a thousandth of a second of arc. This showed that, despite the great luminosity of quasars, the diameter of a typical galaxy is around ten thousand times as great as theirs. It was found that many quasars are embedded in galaxies, and twenty years after the first was discovered, it was shown that 3C 273 is embedded in a galaxy with a nebulosity similar to that of many giant elliptical galaxies.

The theory was thus developed that quasars are galactic nuclei. In due course this line of research linked up with another, stemming from a discovery made by Carl K. Seyfert of the Mount Wilson Observatory in 1943. Seyfert had found galaxies with bright compact nuclei and unusual spectra. (We mentioned these in passing, in chapter 16.) He noted that the spectra of the nuclei of his galaxies seemed to suggest emission lines of hot ionized gas streaming out at velocities of thousands of kilometers per second. Twenty years later, optical detectors were developed that made a closer study of Seyfert galaxies possible. They too were found to vary in brightness with time. Many of their properties were later found in quasars. There are galaxies with similar properties to Seyferts, but with less active nuclei, and that can be resolved. These N galaxies, so-called, include the BL Lacertae objects, a subgroup lacking strong emission lines in their spectra. These take their name from the object BL Lacertae, which was once considered a variable star, but was later found to be a radio source with an elliptical nebulosity around it. There is a whole class of objects with these properties, assumed to be peculiar types of active galactic nucleus. The study of galaxies with active nuclei became an important new development in the astronomy of the 1970s and after.

It was soon realized that the enormous luminosity of quasars gave them a very special place in cosmology. When we observe a distant galaxy we are observing it as it was in the past. There are few observable galaxies known at distances greater than 1,200 megaparsecs, but those are being observed as they were when their light left them around 3.6 billion years ago. There are quasars, however, that seem to be at more than ten times this distance. Here one cannot simply divide the distance by the velocity of light and produce a time of light-travel, for the cosmological model accepted is of crucial importance. In broad terms, however, we can say that as we look further and further into space, the redshifted objects seen do not simply go on increasing. In fact few are known with redshifts above 350 percent, which sets an upper limit to the age of the objects, say about 18 billion years. This is a substantial fraction of the age assigned to the universe based on most expanding models. In the 1980s, this age was usually set at about 20 billion years. Again, however, as already explained, this figure depends on the model accepted. Which model to accept was a critical and hotly debated question in the early 1960s. By 1965 there was much evidence inclining the astronomical community to the view that the universe was evolving, and not locked in a steady state, but there was no direct evidence that it had ever passed through the hot dense phase of an earlier existence, as had been claimed by many of the "Big Bang" adepts for so long. This evidence was provided in 1965, from an entirely unexpected quarter.

Before considering this evidence, it is worth noticing that the discovery of quasars had an interesting effect on the direction taken by fashions in astronomy. With some notable exceptions—Eddington was perhaps the best example—theoreticians who occupied themselves primarily with relativistic cosmology had always been regarded as something of a separate species, remote from the more conventional branches of astronomy. In some quarters they had never been taken very seriously, while in some places the bulk of the profession saw even them as a threat, especially when it came to making university appointments. The regard in which they were held improved significantly in the 1960s, and this seems to have been largely as a consequence of the problems set in the course of investigating quasars. Theorizing about the universe had been an admissible foible, but now cosmologists were seen to be sorely needed, if quasars were to be understood. Their services were called upon increasingly in the decades that followed, as yet other branches of astronomy became ever more closely integrated into a study of the universe as a whole.

THE COSMIC MICROWAVE BACKGROUND

The story of the new evidence with a bearing on the course taken by the expanding universe begins around 1930, with the work of Tolman on thermodynamics and radiation in an expanding world. In 1938, Weizsäcker tried to explain the production of the heavy elements from hydrogen, in an early "superstar" stage of the universe—that is, before its expansion. In 1948, Gamow pointed out that—according to general relativity—the universe could never have existed in a static, high-temperature state. He proposed instead that the elements were formed, and that radiation was emitted, during an early and very rapid expansion. A theory of the formation of the galaxies followed. He and his collaborators calculated that the density of radiation in that early universe was much greater than the density of matter, but Gamow did not consider the possibility that remnants of that phase, in the form of persisting radiation, might survive to the present day.

In 1949, as mentioned in our last chapter, Alpher and Herman—following the likely change in temperature over the history of the universe up to the present—predicted a universal background radiation temperature of 5°K. (See page 651 above for the context of this prediction.) They noted that there were no available observational data on the present density of radiation overall. Four years later, in a classic study with J. W. Follin, they extended their work on physical conditions at the time of the initial stages of the expansion, but they did not revise their earlier calculation, although the Soviet astronomers A. G. Doroshkevitch and I. V. Novikov later did so, deciding that the present temperature of background radiation throughout the universe is close to zero.

Gamow's ideas had a certain following, although a small one, but it would be a mistake to pretend that he had presented the astronomical community with a clear goal, or that everyone working in the same area came to the same conclusions. In 1950 C. Hayashi criticized his ideas, and calculated that in the first two seconds of the expansion of the universe the temperature would have been above the threshold for the creation of electron-positron pairs. Other calculations showed that, while helium would have been produced in the first phase, it was impossible to account for the heavier elements in the way Gamow had suggested. Added to this, new theories of element-creation within stars were soon being very successfully developed, and that very fact led to a general neglect of his theories. His work was waiting to be rediscovered, as surely as the radiation that occurs as one of its essential characteristics.

What about the empirical side of all this? In the 1950s, several radio astronomers in France and Russia had reported a background hiss that could not be traced to instrumental effects. At the end of that decade, plans were being made at the Bell Laboratories in the United States, especially at Holmdel, New Jersey, to work on communications satellites. Initial tests were to be made with the inevitably weak radio echoes from a balloon, and this required a receiving system with very low noise. Use was made of a so-called traveling wave maser working at very low temperatures (those of liquid helium) and a 6-meter horn reflector, vaguely resembling a square-shaped ship's horn. (More will be said about the maser in the following section.) Although not large by the standards of radio astronomy at that time, its properties were accurately measurable, and when no longer needed for echo work it was transferred to radio astronomical projects. In charge of these were Arno Penzias and Robert W. Wilson, who hoped to be able to calibrate radio sources more accurately than had been done previously. Even when used in 1961 with the Echo communications satellite, by a colleague, Edward Ohm, the temperatures obtained for the system—partly due to internal "noise" and partly coming from outside—had been consistently 3.3°K more than expected. Now Penzias and Wilson also found an excess over expectation. They thought at first that it was an antenna problem, but almost a year passed by, and even clearing out a pair of roosting pigeons made no difference. No matter which part of the sky the antenna pointed to, the radiation was there.

Whereas most experimenters would have ascribed the unexplained readings to unknown instrumental errors, they persevered tenaciously. They discussed the matter with R. H. Dicke, at Princeton, who was then considering an oscillatory model of the universe, and who was expecting something of the sort they had found. Dicke had been asking himself how, in such a model, he could escape the steady accumulation of nuclear waste at the "bounce" between cycles. Why was our universe not left with nothing but iron? His hypothesis was that, in the heat of each bounce, iron nuclei would be broken down into a brave new world of hydrogen and helium, and that the most recent of these cosmic fireballs might have left us with relics, for which he ought to search. Neither his nor the Bell group knew then of the highly relevant researches of Alpher, Hermann, and Gamow. Dicke, however, had himself already published a prediction of a background radiation at 10°K at a wavelength of 3 centimeters. His group—P. J. E. Peebles, R. G. Roll, D. T. Wilkinson—had the misfortune to be embarking on a program to search for the radiation with some highly sophisticated radio apparatus they had developed, when they heard of the work being done at the Bell Laboratories. It was with an accompanying letter from them that Penzias and Wilson published their own findings at last in 1965, in the *Astrophysical Journal*, announcing "excess antenna temperature at 4080 Mc/s." In 1978 the two shared a Nobel Prize in Physics for their work.

What Penzias and Wilson had found, working at a wavelength of 7.3 centimeters—two hundred times shorter than that used in the pioneering work of Karl Jansky at the very same laboratories—was that even a seemingly empty part of the sky gives off radio waves. It seemed that the fact could be explained in terms of a Big Bang, say ten or twenty billion years ago. The idea was that the energy of the primeval explosion became diluted as a result of the general expansion of the universe, so that it now corresponds to radiation from what physics terms a "black body" at a temperature of about 3°K. (In standard heat theory, a black body is one that absorbs all radiation falling on it, while black-body radiation is what such a body—often called a full or perfect radiator—radiates. Corresponding to a particular temperature, a theoretical spectrum can be drawn for the thermal radiation from a black body.) They quoted $3.5°K \pm 1.0°K$. There is a well-defined sense in which intergalactic space contains heat: every cubic meter contains about 400 million photons, quanta of radiation. (The number of atoms is probably on average only a billionth of this.) Strangely enough, Igor Novikov and Andrei Doroshkevich in Moscow had not only suggested in 1962 that relics of the radiation from the Big Bang might be detected, but they had even noted that the Holmdel antenna—later used by Penzias and Wilson—would have been well-suited to searching for it. Like Dicke, they came close to a discovery that has been generally ranked with that of the expanding universe, the difference being that Dicke was virtually ready for observational tests.

Penzias and Wilson were fortunate in their wavelengths. There is a "window" of wavelengths, roughly between 1 and 20 centimeters, through which radiation from the "primeval fireball"—to take one of several names for it—can be observed at the Earth's surface. At longer wavelengths, extragalactic signals are submerged by those from our own Galaxy, while at shorter wavelengths the Earth's atmosphere radiates too strongly. And theirs was not the only possible route to the discovery. In retrospect, it is possible to say that others—for example Haruo Tanaka in Japan (1951) and Arthur E. Covington and W. J. Medd in Canada (1952)—had anticipated them in different ways, but the accuracy of the earlier data, judged by later work, was poor. In retrospect, it was possible to make sense of an explanation offered in 1940 by Andrew McKellar of the Dominion Astrophysical Observatory in Victoria, Canada, for some puzzling absorption lines found at Mount Wilson: he had thought that they might be due to absorption by cyanogen molecules in space, at a temperature of 2.7°K. He even predicted the existence of another absorption line, and found it, but his ideas were little known, explicitly rejected by some, and like the others just mentioned had no wider theoretical repercussions.

In 1965, cosmologists reacted with great excitement, for now the case was different in one vital respect. Thanks to Dicke, the evidence was seen to have a bearing on a crucial question, one that was being asked by more and more astronomers. Here at last was evidence that could greatly narrow the choice of cosmological theories, out of the many then on offer. The steady-state theories could still perhaps be defended. Some of their supporters considered the possibility that *new* radiation comes into existence with newly created matter throughout space, but this was not how the discovery of the 3°K background radiation was generally perceived. From this time onward, most cosmologists turned their attention to the investigation of an evolving universe with a hot Big Bang initial phase whose evolution is determined by the laws of the physics of fundamental particles.

This new style in cosmology, what one might call the Lemaître-Tolman-Gamow style, was vigorously pursued in the 1960s and 1970s by such theoreticians as Fowler, Wagoner, Thorne, Sachs, Wolfe, Sacharov, Weinberg, Schramm, and Steigmann. Many new kinds of data were appearing with an important bearing on this theoretical stream, not only from the radio astronomers, but from such new styles as X-ray and gamma-ray astronomy. Parallel to this new theoretical field of interest, another was developing fast, with a more mathematical focus. Studies were being made of so-called horizon effects in cosmological models, physical effects with strange topological properties, effects that were leading some theoreticians to countenance the idea that the universe might not be homogeneous on a large scale. We shall return to these questions when we have examined the new evidence acquired at other wavelengths. As already intimated, not all of the steady-state cosmologists gave up their allegiance readily. In 1975 we find Fred Hoyle and Jayant Narlikar defending an expanding steady-state model, with a "scalar-tensor" version of the general theory of relativity, somewhat resembling one developed by Pascual Jordan in 1939. Particles were no longer said to be created, but existing particles were said to change in mass. The 3°K cosmic radiation was presented as starlight from an earlier phase turned into heat, having been scattered from atoms of very great size at the epoch when most particles are near to zero mass. These ideas have so far not acquired a large following.

COSMIC MASERS AND THE DISCOVERY OF PULSARS

Two other fundamental astrophysical discoveries were made around the time of the verification of the microwave background radiation, both having interesting aspects in common with it. In the first case, that of cosmic masers, we see the principle that discovery requires recognition, and recognition requires some sort of theoretical understanding. In the second case, the discovery of pulsars, we see yet again the principle that one does not brush aside the unexplained, however insignificant it may seem.

In 1964, a group at the University of California at Berkeley, led by Harold Weaver, while studying the Galaxy at microwave frequencies, found an extremely puzzling set of spectral lines in the great nebula in Orion, and these at first they simply ascribed to "mysterium." It was soon identified with radiation resulting from a change occurring in the hydroxyl radical—a pairing of an oxygen atom with a hydrogen atom (OH). From similarities with what can be produced in the laboratory, the object responsible was accepted as a cosmic equivalent of the maser. The *maser*, a device for amplifying microwaves of very narrowly defined frequency, had been invented in 1954 by Charles Townes and his associates at Columbia University in New York. Its name is an acronym for "microwave amplification through stimulated emission of radiation" (the underlying principle was later used at optical wavelengths for the better-known *laser*, "light amplification through stimulated emission of radiation"). The assumption is that there is strong amplification by some sort of natural processes of the microwave emission from the OH molecules in the source. Many other cases have since been recognized, some found at the Massachusetts Institute of Technology and at Jodrell Bank even earlier, before they were understood, and some involving other groups of atoms—such as water, silicon monoxide, formaldehyde, and methyl alcohol. Masers have been found in cool clouds in the vicinity of hot ionized gas, and have also been found associated with certain types of stars that are strong in infrared radiation.

Pulsars have a more specific point of discovery, beginning with an observation by a graduate student at Cambridge University, S. Jocelyn Bell. She noticed "a bit of scruff" on a 120-meter paper roll, recording one complete sweep of the sky over the (fixed) antenna. She was attempting to detect the way in which gas emitted by the Sun affects signals from radio sources, and here man-made radio noise was a great problem. The recording had been made on 6 August 1967, and it was in October that she noticed its unusual character, and that the signal, looking like a series of rapid pulses, remained fixed at a particular point of the sky. With Anthony Hewish, her thesis adviser, and three other colleagues, she carefully observed the signal further, and the pulses were found to hold their spacing very accurately indeed (at about 1.3 seconds). Before Christmas, Jocelyn Bell had found a second pulsating source, with a period only marginally shorter than the first. When the discovery was announced in the journal *Nature*, there was much annoyance in Cambridge that Hewish's close colleagues had kept their discovery to themselves. In fact, they seem to have been worried at the thought that the pulsations might have a simple explanation close at hand, a fault in the equipment, perhaps. At the Cambridge seminar at which the first pulsars were announced, it was mooted that they might carry signals from distant civilizations, and their provisional names, LGM 1–4, alluded to "little green men." Were they white dwarfs or neutron stars? Were they pulsating or spinning? The Cambridge group first opted for pulsating white dwarfs, and were mistaken on both scores. Whatever the answer, one of the most surprising things about them turned out to be the sheer precision of their timing. The periodicity of the first to be discovered is now confidently quoted to eight places of decimals!

It was Thomas Gold, then at Cornell University, who seems to have been the first to recognize that pulsars were the long sought-for neutron stars, which had been discussed on theoretical grounds since the 1930s. Other extensive theoretical studies of neutron stars were made after the discovery of the first pulsar (named CP 1919, that is, Cambridge Pulsar at right ascension 19^h19^m) and Gold's identification. Theory predicts that the magnetic fields of neutron stars are a thousand billion (10^{12}) times as powerful as that of the Earth, and that these play an essential part in the production of radiation from the stars. On the other hand, while most of the thousand or so neutron stars known in our Galaxy were first discovered as radio pulsars, almost all of the energy they emit is known to be in the form of very high energy photons, that is, X-rays and gamma-rays. It was an extraordinary stroke of good fortune that they were first recognized on the slender basis of their minuscule radio emissions. In radio pulsars it was decided that there is a rapidly rotating neutron star with a synchrotron mechanism—something we encountered earlier in this chapter (see p. 662). This produces radio waves from relatively near to the star's surface. As we shall see, pulsars emitting X-rays were later found in 1971 by satellite observation, and theory suggested that the radiation in pulsars of this type comes largely from the poles of the star. There are pulsars known at other wavelengths. In all cases, the pulsation we observe is simply a lighthouse effect, a pulse occurring every time the peak of the emission flashes across our field of view.

The incidental value of historical records in astronomy has already been illustrated in connection with the Crab Nebula of 1054. Interest in this object increased yet again when it found to contain a pulsar at its center. Rotating about thirty-three times a second, out of more than a thousand pulsars discovered before the end of the millennium, this remains one of the most rapid. Its rapidity is a consequence of its relative youth— theory predicts a slowing down with time, as energy is dissipated from the star, making the historical record an important datum for understanding the mechanism. The rates of change are extremely small but measurable— witness the accuracy with which that of CP 1919 is quoted. Not all changes are smooth, however. Discontinuities in timing are often observed, and have been found various explanations—for example, in terms of fractures of the star's rigid crust. Astrogeology suddenly extended its reach beyond the solar system, encompassing the study of starquakes. While there is at least a core of generally accepted theory, which has explained a striking number of phenomena, there is still much disagreement over the nature of neutron star mechanisms. This is hardly surprising, since the state of matter at the center of neutron stars presents some of the most difficult questions in fundamental physics, as may well be imagined when bearing in mind the enormous densities there.

Pulsars have offered new routes to that fundamental task of putting to the test Einstein's general theory of relativity—the relativity theory that covers gravitational effects. In the previous chapter we saw how Einstein solved—at least to most of his supporters' satisfaction—the very slight advance of the perihelion of Mercury. In 1974, Richard Hulse and Joseph Taylor discovered a stellar system in which the corresponding precessional advance would be ten thousand times as fast. This was a binary pulsar, a pulsar orbiting round another neutron star every eight hours. Taylor tracked the movements for twenty years, and in the course of doing so found that the orbit was gradually contracting. In this way he confirmed Einstein's theory twice over, on the assumption that the shrinking was due to the loss of energy from the system by gravitational radiation, another consequence of general relativity. (A moving system of gravitational mass generates vibrations in space itself, so losing energy.)

As more and more pulsars have been recorded, the record for rotational velocity has been often revised. In January 2006, a binary pulsar was recorded to be spinning at 716 revolutions per second, replacing the previous record of 642 revolutions per second, which had stood for 24 years. The new champion is one of 33 pulsars in the globular cluster Terzan 5. It follows a circular orbit, with a period 1.0944303 days. Those seven decimal places speak volumes for modern chronometry, but the real star of the show is a mass that is spinning faster than a bee can beat its wings.

ADVANCES IN OPTICAL ASTRONOMY

To appreciate the great changes that were taking place in astronomy, the world over, in the postwar period, it is as well to keep the optical tradition in perspective, at the risk of giving a bare catalog of events. Optical astronomy did not suddenly fade away when confronted with radio astronomy. The flagship of the subject at this time was the great 200-inch Hale reflector at Palomar, brought into service in 1948. Extremely important work was done with it from the very beginning—Baade's work on star populations has already been mentioned, and some of Sandage's too. Allan R. Sandage was an assistant to Hubble as a graduate student from 1950 and joined the staff of the Mount Wilson and Las Campanas observatories in 1952. At Hubble's death he inherited the mission to map the distances and expansion rate of the galaxies. He produced data, including a figure for the deceleration of the expansion, that greatly influenced cosmological thinking during the 1950s and 1960s (the "deceleration parameter" will be discussed again in chapter 20). Later from Palomar came Jesse Greenstein's studies of white dwarf stars, work by Eric E. Becklin and G. Neugebauer in the infrared, and observations of remnants of supernovae by Baade, Zwicky, and Minkowski.

The Soviet Union put into commission an altazimuth optical telescope of larger aperture—6 meters, or 236 inches—in 1975, at a time when size meant everything. Ground and polished in Leningrad, beginning in 1968, its mirror alone weighs 70 tonnes. The telescope operates under unfavorable conditions, however, at a site in the Caucasus mountains in southern Russia, and is not markedly superior to some smaller instruments. It is interesting to find Isaac Newton, in his *Opticks*, recommending mountain tops as places where tremors in the atmosphere may be avoided. There are better reasons, however. At very high altitudes, water vapor in the atmosphere is much diminished, making observations possible at infrared and submillimeter wavelengths. That advantage was not fully appreciated when the world's first high mountaintop observatory was established, at Pic-du-Midi in the French Pyrenees. That observatory, which was originally used chiefly for meteorological purposes, is at an altitude of 2,877 meters. It has been fitfully operative since 1882, but French state support for it has dwindled, and its chief claim to fame now depends on its tourist potential as the world's most craggily spectacular observatory. In some quarters, it is customary to speak of "the oldest *permanent* high-altitude observatory," so ensuring that the palm is given to that at Mount Hamilton, California (which has been in operation since 1888). In 1889, Harvard opened a more modest station in Arequipa, Peru. It remained there until 1927, when it was transferred to Bloemfontein in the Orange Free State, South Africa, finally closing down in 1966. Of several observatories built under the guidance of the astrophysicist Donald Menzel, his first Western example was the High Altitude Observatory at Climax and Boulder, Colorado. Begun in the 1930s, it was not the only one of his that earned a high reputation, but it is interesting as an example of material influence. Menzel's specialty was imaging the solar corona in the absence of an eclipse, and he had learned much of practical value at first hand from Bernard Lyot, whom he visited at the Pic-du-Midi.

Telescope technology is a subject in itself, but it is hard to pass by Newton's remarks about tremors in the atmosphere without mentioning the way in which the European Southern Observatory began to counter it in 2006, using "adaptive optics" to compensate for atmospheric turbulence. Adaptive optics is a way of altering the image by physically deforming the mirror in a carefully controlled way. A powerful laser beam is projected into the sky, in the direction of interest. The beam hits sodium atoms at a height of about 90 kilometers, creating a faint artificial star there, so allowing for the measurement and automatic correction of the atmospheric effects in that direction.

In the second half of the twentieth century, the importance of matching site to wavelength led to the establishment of yet other mountain observatories. All new large telescopes were by this time being erected at altitudes of around 2,000 meters or more—much more, in the case of those on Hawaii. There were mountain observatories built in Arizona (Kitt Peak), Hawaii (with the Canada-France-Hawaii Telescope, the NASA and United Kingdom infrared telescopes, and the Keck telescopes), Chile (one of three major institutions there is the European Southern Observatory at La Silla),

Australia (including the Anglo-Australian Telescope and the United Kingdom Schmidt Telescope), and Spain (including a German-Spanish Center
at Calar Alto, and the Herschel and Newton reflectors on the Atlantic island
of La Palma, one of the Canaries). The 2.5-meter Isaac Newton telescope
has a curious history, for it was originally erected at the Royal Observatory
in Herstmonceux, in Sussex, England, in 1967, but could never be properly exploited under English weather conditions. Duly upgraded in many
different ways, it was moved to La Palma. As for the 4.2-meter mirror of
the William Herschel telescope at La Palma, this has what may still count
as the most accurately figured mirror in world astronomy. Like most of
the very large telescopes of this period, the Herschel reflector uses the
mechanical advantages of an altazimuth arrangement, with Nasmyth foci.
(For an illustration of the modern arrangement, see p. 489 above.) The
genesis of the European Southern Observatory is interesting: it was set in
motion as a result of Jan Oort's efforts, but the idea was given to him by
the German-born astronomer Walter Baade, of the Mount Wilson and
Palomar observatories, while Baade was spending two months at Leiden
University in 1953.

   That there is still a place for non-governmental funding of such colossal
projects as these could be illustrated many times over, but nowhere better
than by the W. M. Keck Observatory, on the summit of Hawaii's dormant
Mauna Kea volcano, at an altitude of 4,205 meters. The site is remarkable
for several reasons—the lack of city lights, the dry and clear atmosphere,
and the extensive, thermally stable surrounding ocean. The instruments
there—one operating since 1993 and the other since 1996—are currently
the world's largest optical and infrared telescopes. They stand eight stories
high, each weighing 300 tonnes, and yet they operate with extreme precision. Each has a primary mirror 10 meters in diameter, comprising 36
hexagonal segments, but working together as a single mirror. Operated by
a Californian association, in partnership with NASA since 1996, this extraordinary enterprise was made possible through grants totaling more
than $140 million from the W. M. Keck Foundation.

   So much for a brief survey of the optical scene. By the early 1990s, there
were still only about a dozen optical telescopes with mirrors more than
3 meters in diameter. By the end of the century there were at least twenty-
five, even counting instruments working in harness as one telescope at a
single site. By 2006 there were six at Mauna Kea, six at sites in Chile, and
six in the continental United States. Under construction, there was a segmented mirror instrument, inspired by the Keck in Hawaii, of 10.4 meters
aperture, and a binocular telescope at Mount Graham, Arizona, with two
8.4-meter mirrors. In the first decade of the new millennium, however,
ambitions were beginning to grow faster than mirrors. A group of United
States institutes began to plan for their 25-meter Giant Magellan Telescope,
and a U.S.-Canada venture was aiming at 30 meters. Lund Observatory in
Sweden, in collaboration with Spain, Finland, and Ireland, was planning for

a 50-meter instrument; and the European Southern Observatory—a joint venture supported by eleven governments and operating in Chile—was talking of 100 meters. This last mirror was planned to be segmented in some way. (It is technically out of the question to build a single useful mirror much more than about 8 meters across.) Leapfrogging ESO's existing flagship instrument, namely the Very Large Telescope (VLT, a mere 4×8.2 meters), the new one was provisionally called OWL, for overwhelmingly large telescope. Which, one asked, would run out first—new money or new names? In any event, the review board set up to evaluate the idea agreed that it was feasible, but expressed concern over the projected cost, estimated at around 1.5 billion euros ($1.9 billion). The project was scaled down to one with a main mirror of 42 meters across, more modestly named Extremely Large Telescope (ELT).

At radio wavelengths, where much valuable work is possible even in climates where water vapor blocks optical wavebands, large radio telescopes multiplied more quickly in most of the world's astronomical centers, during the first two decades following the Second World War. Surveying the history of postwar funding, we may ask why it was that, despite several important American pioneering efforts, radio astronomy in the United States was slow to catch up with what was being done elsewhere—in the United Kingdom, the Netherlands, and Australia in particular. There is no doubt that the lavish funding of optical astronomy in the United States goes far to explaining why this was so; and the interplay of military interests and the building of observatories for use in space, doubtless played an even greater part. Once the dividends to be expected from work at radio wavelengths began to be obvious, the situation changed quickly—and the dividends flowed in.

Observatories in Space

Heavenly journeys to survey worlds beyond our own have a long history. We have already met with those by Cicero and Macrobius, Dante and Chaucer, and even that by Kepler, all of them leaving the principles of flight wrapped in mystery. The cleric Francis Godwin, son of a bishop, who himself became bishop of Llandaff, Wales, and then of Hereford, on the English side of the border, seems an unlikely follower in the tradition. He raised the level of plausibility, however, when in 1638 he published his Copernicanized fantasy *The Man in the Moone: or A Discourse of a Voyage Thither by Domingo Gonsales, the Speedy Messenger*. His hero Gonsales is carried to the Moon in a harness lifted by migrating geese, and there finds a utopia in which lunar women are of such beauty that men have no wish to commit adultery. Pure fiction. The book went into numerous editions and even provided the inspiration for a London theatrical production in 1706, *Wonders of the Sun*, by Thomas D'Urfrey, in which models of flying geese played a part. It is to balloons, however, not to geese, that we are to look for the beginnings of high-altitude astronomy. Balloons, filled with hot air, with hydrogen, and later with helium, have been used in the interests of various branches of science ever since the first flights by the brothers Montgolfier. The two brothers, Michel-Joseph and Étienne-Jacques, favored hydrogen in their first experiments of 1782 but achieved real success only with heated air. The first human flight was made on 20 November 1783.

We have already met with two astronomers who traveled by balloon. In the case of Janssen, he used one to make his successful escape from Paris in 1871, but he had no intention of observing his eclipse from the balloon itself. Dorothea Klumpke's journey of 1889, on the other hand, was made to observe the Leonid meteor shower of that year. The first data of astrophysical value obtained from a balloon flight were found almost by chance. The first studies of radioactivity, shortly before and after the beginning of the nineteenth century, revealed its presence in water and in the atmosphere. Shielded but charged electroscopes placed out of doors, for instance, had been found to lose their charge gradually, as though they were exposed to radiation. The general consensus was that the cause was radiation from radioactive elements in the ground, or possibly the radioactive gases they produce. A few measurements of ionization rates at various heights showed a decrease, and this was explained in what seemed a straightforward way, as due to absorption of the ionizing radiation by the intervening air. A more thorough investigation by balloon was called for.

In 1911 and 1912, the physicist Victor Franz Hess of the Vienna Radium Institute made a series of ten balloon ascents, performing experiments that destroyed this general assumption. On 7 August 1912, with the balloon commander and a meteorologist, he made his most valuable flight. Traveling from Aussig on the Elbe to Pieskow, their flight lasted about six hours and took them to an altitude of more than five kilometers. Hess took readings from three electroscopes during the journey to measure the intensity of the radiation that caused ionization of the atmosphere. One widespread belief was that its source was rocks in the Earth's crust. He found that, while it fell for the first 150 meters, it rose steadily thereafter, as the balloon gained in altitude, so that at 5,000 meters it had reached double its value at the Earth's surface. He had earlier found that at a given altitude it remained the same, by night and day, and so was not caused by the direct rays of the Sun. He eventually ruled out the Sun as source by making a flight by balloon during a solar eclipse. (His conclusion was later in need of some modification.) Hess published his findings later in the year, concluding that there was a highly penetrating radiation entering the atmosphere, but from outside the solar system—and not from the Earth below us, as so many had thought.

Through these experiments, Hess was effectively the founder of cosmic ray astronomy. His results were repeatedly verified by Werner Kohlhörster, who as early as 1913 had ascended by balloon to continue measurements to an altitude of over nine kilometers. By the mid-1920s, the phenomenon, although not its name, had gained wide acceptance. (The phrase "cosmic ray" was coined by R. A. Millikan in 1925.) In 1936, Hess shared the Nobel Prize in Physics. The distinction was not enough to protect him, after the *Anschluss*, the annexation of Austria by Nazi Germany in 1938. Hess's strict Catholicism led then to his being dismissed from his professorship at Graz University. He moved to the United States and continued his work there, making good use of the high altitude of the tower of the Empire State Building for some of his experiments.

The interwar period saw much important work done on cosmic radiation, and this proved to be an enormously important tool for an understanding of the physics of fundamental particles. Robert Millikan, for instance, who received a Nobel Prize in 1923 for his measurement of the charge on the electron, used U.S. Army balloons to make measurements of the Hess type. Millikan duly produced a theory explaining how Hess's rays might be photons of very high energy, generated during the creation of light elements out of hydrogen. Werner Kohlhörster and Walther Bothe, in 1929, performed experiments with Geiger counters, showing that much cosmic radiation was from charged particles of very high energy—often more than a thousand times that of radioactive material in the laboratory. Other highlights in cosmic ray research included Bruno Rossi's demonstration that primary particles produce showers of secondaries; and Carl David Anderson's momentous discovery of the positron (the positive electron) among the secondaries, in his cloud chambers. ("Primary" cosmic rays are those from beyond our atmosphere. On encountering it, their collisions with atomic nuclei produce "air showers" of secondary rays.)

Soon Millikan's theory of primary high-energy photons (gamma-rays) was disproved by experiments done under the direction of A. H. Compton with an army of assistants. They found that the intensity of cosmic rays at the Earth's poles was greater than at the equator, showing that the primaries must be charged particles, guided by the Earth's magnetic field. This raised many problems: if they were electrons or protons, they should be absorbed in the atmosphere more quickly than seemed to be the case. Anderson came to the rescue, again using cosmic ray experiments in the interests of fundamental physics. In 1937, his cloud chamber experiments brought to light a new secondary particle, the μ meson (or muon), which could slow down the expected absorption. The relation of this particle to the earlier meson theory developed by the Japanese theoretical physicist Hideki Yukawa is too complex for discussion here; suffice it to say that what he proposed as the carrier of nuclear force, now known as the π meson (or pion), was at first identified with Anderson's particle. That this was a mistake was shown in 1949, when experiments by Cecil F. Powell and his colleagues at Bristol University found a case where a pion turned into an electron and a muon. (The muon then disintegrated into another electron and an invisible neutrino. This was all observed using a photographic emulsion as the detecting device. Neutrinos are fundamental particles with spin, but with little or no mass and no charge. They are the subject of increasingly intense study in twenty-first-century astronomy.)

To return to the beginnings of this rich new subject, and methods of exploring the upper atmosphere: a notable early use of unmanned balloons to carry instruments to investigate radiation from space was made by the German physicist Erich R. A. Regener, an authority on the physics of the atmosphere and stratosphere. (In 1909 Regener had obtained a rather accurate value for the charge on the electron, and it is for this that he is now usually remembered. Millikan's measurement was slightly earlier in the same year, but the two methods were different.) To measure cosmic radiation in the early 1930s, Regener sent up unmanned balloons of rubber, and later of cellophane, some of them reaching as high as 30 kilometers. In 1933 he made a personal ascent, and on this occasion he found a connection between an eruption of the Sun and an unusually high degree of ionization of the atmosphere. This was an important discovery, for it showed that the stars—insofar as they are like our Sun—must be a source of cosmic radiation. Especially after the Second World War, unmanned balloons developed for meteorological purposes were used as a standard means of transporting cosmic ray instrumentation.

Conventional aircraft were later used with much success, when there was a need to avoid the blocking effect of the Earth's atmosphere at short wavelengths. Serious work began in 1966, when Frank Low and Carl Gillispie made measurements of the brightness temperature of the Sun—for instance at 1-millimeter wavelength—from a Douglas A3-B bomber, flown for fourteen missions. A year later, NASA funded studies of the planets from a Convair 990 aircraft, named *Galileo*. This crashed with great loss of life in 1973. Its successor, *Galileo II*, likewise met with misfortune, for it was lost in a fire on the runway in 1985, but this time without human injury. NASA funded the Lear Jet Observatory from 1968, which took a 30-centimeter reflector to over 15 kilometers altitude. This was the prototype for perhaps the most successful facility of this type, namely NASA's Kuiper Airborne Observatory (KAO), which was carried in a modified C-141 four-engine military jet transport. Operating from 1974 onward, the KAO typically carried a crew of three, up to seven experimenters, the telescope operator, the tracking telescope operator, and a computer operator. A flight might be six or seven hours in duration. It usually left from Moffett Field in California, but in some cases from Hawaii, Australia, and Japan. Great technical problems were overcome in this project, especially in achieving stability—through gyroscopes and the adoption of special flying techniques—and in avoiding air turbulence across the field of view of the instruments.

Airborne observatories were flown by several nations in the 1970s, including the United Kingdom, West Germany, India, and Japan. Cosmic radiation studies were much pursued, as were planetary studies. From the KAO came a notable series of spectral studies of planetary atmospheres. The rings of Uranus were discovered from it in 1977—simultaneously with a ground-based discovery at the time of a stellar occultation—and like other airborne observatories it yielded much information about the heat emitted from planets. It was found, for instance, that Jupiter, Saturn, and Neptune all emit more heat than can be accounted for in terms of the reflection of solar heat, showing that they possess active internal sources of heat. Uranus was found to have none of any significance. This type of high altitude astronomy was soon to be put into eclipse, however, by rocket-launched probes. As an example of progress over the next three decades: a joint effort using the Hubble Space Telescope and the Keck telescope in Hawaii, at the end of 2005, revealed twelfth and thirteenth rings around Uranus (the outermost was bright blue, like the outermost ring of Saturn—in both cases a consequence of the small size of the particles in the rings).

ROCKET-BORNE OBSERVATORIES AND
THE SOLAR SYSTEM

One of the secondary effects of the many successes in astrophysics and cosmology in the period before the Second World War was that academic work in celestial mechanics went out of fashion. When a resurgence of interest in the subject came in the postwar period, it was for reasons having at first little to do with astronomy, and it took place far from center stage. It came with the "Sputnik era," when, for the first time, observatories were carried by rockets beyond the limits of the Earth's atmosphere. To give the era such a title is of course to oversimplify events, but the "space race," for the world at large, had more to do with the transporting of warheads than of telescopes.

Military rockets have a long history, with Chinese origins. The first rockets to be used with success in modern warfare were those designed by the English engineer William Congreve. They were used in many campaigns in the Napoleonic wars, and were afterwards copied by most European armies, helped by the writings of Jacques-Philippe Mérignon de Montgéry—who in 1825 produced a well-documented history and theory of the rocket as a weapon of war. Rocketry remained an appendage of gunnery, albeit with a theory of its own that might have been developed more enthusiastically by artillery experts had propellants been more reliable. Konstantin Eduardovich Tsiolkovsky, a Russian schoolteacher, made several important contributions to the theory; Robert Hutchings Goddard of Worcester, Massachusetts, did likewise, and launched his first successful rocket powered by liquid fuel in 1926. Hermann Oberth in Germany was an admirer of Tsiolkovsky's work, and organized enthusiasts into a society for space flight, to which the young Wernher von Braun belonged. The Treaty of Versailles, at the end of the First World War, limited Germany to the deployment of artillery of small caliber, with the result that the country put much research effort into rocketry, calling on the services of several members of the society. It was Braun who led the remarkable German military program that culminated in the attacks on southern England by supersonic V-2 rockets in 1944 and 1945. A captured stock of these was shipped to the United States, and 25 of them were earmarked for scientific purposes.

At first the rockets were used for research in the upper atmosphere. These objects, 14 meters long and 14 tonnes in weight, could reach to an altitude of 120 kilometers and so were well suited to the task. Radio telemetry techniques were developed to send back data during flight, one of the early difficulties being that of reducing the speed of the rocket on landing, so as to salvage the instruments on board (hardly a serious problem for the German rocketeers). As early as October 1946, the first spectra of the Sun were obtained from one of these modified V-2s, spectra that were especially valuable, since they could be obtained from above the ozone layer that cuts out much of the ultraviolet. The group responsible, at the U.S. Naval Research Laboratory, was headed by Richard Tousey. The rocket

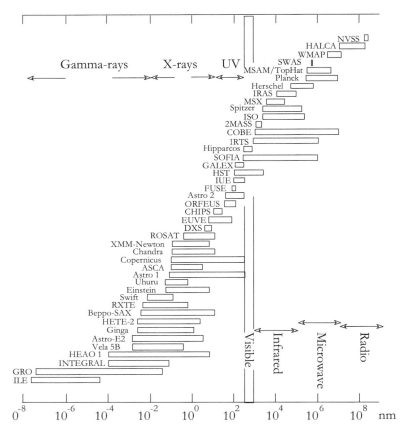

237. Intervals of wavelength covered by the instruments of several important flight missions. The listing is not comprehensive, and ignores most missions in solar physics and planetary astronomy. Relatively few missions are mentioned in our text. The wavelength scale is in nanometers (1 nanometer is 10^{-9} meters, a billionth of a meter—or 10 Ångstrom units, to use another common measure). Note that the scale is not linear but logarithmic. The main purpose of the figure is to show the relative coverage and spread of data-gathering devices. The central band covers the visible spectrum, with violet to the left and red to the right.

ascended to 80 kilometers, and it became immediately clear why so many attempts to obtain ultraviolet spectra of the Sun from balloons in the 1920s and 1930s had failed: the height of the ozone layer had been greatly underestimated.

It was to be several years before controls were developed that would allow pointing of the instruments accurately enough for most serious astronomical purposes, but when this was done, it became possible to obtain a solar spectrum down to X-ray wavelengths. Where the first half of the twentieth century had seen phenomenal astronomical gains won by building ever larger reflectors, in the second half of the century, in view of developments then taking place in radio astronomy, still larger gains were made by virtue of the vast extension of the range of wavelengths of radiation received. At a later stage, knowledge of the solar system snowballed by the simple device of visiting the objects of interest.

Both the Soviet Union and the United States employed large numbers of space technologists, and before long the whole enterprise became a focus of national pride, with scientific research as its showpiece. As early as 1954, a committee of the International Geophysical Year took advantage of the situation, and recommended that governments launch Earth satellites for scientific purposes. The "year," which was planned to run from 1 July 1957 to 31 December 1958, had the support of sixty-six nations. An important aspect of its plan was that of research into phenomena in the upper atmosphere. The Soviet Union put their contribution under the umbrella of military rocketry. In the United States, the Eisenhower administration, faced with a choice between derivatives of the Army's Jupiter missile and the Navy's Vanguard, chose the latter. On 4 October 1957, using a version of their Semyorka missile as launch vehicle, the Soviet Union launched *Sputnik*, the first artificial satellite to orbit the Earth. The preparation of Vanguard in the United States was well advanced, and those concerned were conscious of Soviet ambitions, but they were taken unawares by the successful launch. Only a month later, *Sputnik 2* carried the dog Laika into orbit, and the world's newspapers became even more excited. Technology could not return the dog alive, but the fortieth anniversary of the October Revolution had been well and truly celebrated. The mass of this *Sputnik* had exceeded half a tonne, and countries in the Western Bloc began to fear that Soviet engineers had developed some extraordinary new fuel. (They had not. They were simply working on a massive scale.) On 6 December of the same year, Vanguard, carrying a satellite of 1.5 kilograms, rose only a meter or so above the launch pad before falling back and exploding. What is truly astonishing is that American national pride, so badly dented, could be restored within a couple of months. On 31 January 1958, Wernher von Braun's Jupiter-C rocket was able to put a satellite—*Explorer 1*—into orbit, with some refined instruments that quickly soothed some of the heartbreak of previous months.

Later in 1958, new educational schemes were set up by the U.S. National Science Foundation and the newly created NASA. In the name of "national defense and space exploration," many hundreds of government employees, industrial research workers, college teachers, and students were introduced to the skills needed for orbit and trajectory calculation—celestial mechanics in a new guise. In all of this, the United States set a template for most of the world to copy, when setting up scientific, civilian, or military space programs. These are too numerous for us to do justice to them here, but it is worth distinguishing between totally independent national activities—as when France launched its *Astérix* satellite with a French rocket in December 1965—and collaborative ventures—as when the United Kingdom (1962), Italy (1964), and the European Space Research Organization (1968), launched satellites using American rockets. Some Eastern European countries collaborated with the Soviet Union. Japan and China launched satellites in 1970, and India, in 1975. A stage was soon reached at which the number of scientific satellites in space far exceeded the number of large telescopes on Earth. A sign that relations between the two sides of the Iron Curtain were relaxing came in 1973, with a United States-Soviet Union agreement to work toward a docking arrangement between *Apollo* and *Soyuz* (plate 16). The most ambitious collaborative project of all, however, was undoubtedly the International Space Station. (The physicists might challenge this, citing their colossal particle accelerators.) It has a strong claim to being the largest and most complex scientific project in human history, drawing on the scientific, technological, and manufacturing resources of sixteen nations. The station will have dimensions comparable with those of a football field, and almost an acre of solar panels to provide electrical power to its six laboratories.

The situation at the end of the twentieth century was plainly very different from that when NASA was created. At that time, astronomers had been profiting for more than a decade from budgets that had originally been justified in terms of national security. While they were trying to determine the nature of the cosmos, and in some cases to decide whether there is life elsewhere in the cosmos, they were often obliged to work with others whose purpose was to threaten life on Earth. Even ground-based astronomical observation was under threat, for a time. There were military plans to fill the upper atmosphere with vast quantities of copper needles for purposes of radar screening, needles that would have resulted in a permanent "fog" above all radio telescopes. Astronomers on both sides of the Iron Curtain were often used cynically to provide an intellectual cover for the development of new weapons. It was fortunate that in so many cases the astronomical mask became the reality, that astronomical aims were acknowledged as worthy of vast financial support, and that in the space of twenty or thirty years the face of astronomy became transformed out of all recognition.

One of the most important of early discoveries made possible by satellite was that there is a dense distribution of very energetic charged particles surrounding the Earth. The so-called Van Allen belts are named after their discoverer, James Alfred van Allen, one of a team working at the Johns Hopkins Applied Physics Laboratory in Baltimore. He was among those who had proposed the "International Geophysical Year" program of 1957–1958. As part of the U.S. contribution, he and his colleagues constructed instruments for the early Explorer satellites. By a stroke of good fortune, he detected the presence of the trapped particles in January 1958, with a radiation detector installed on *Explorer* 1—the first U.S. artificial Earth satellite—which had become overloaded by the unexpectedly high flux of ions of high energy trapped in the Earth's magnetic field. Van Allen redesigned the equipment, and was later able to chart the doughnut-shaped distribution of charged particles, helped by further data from *Explorer* 4 and *Pioneer* 3. It was eventually found that the inner belt extends from roughly 1,000 to 5,000 kilometers above the terrestrial surface, and the outer belt from some 15,000 to 25,000 kilometers. Van Allen's may be reasonably described as the first important discoveries of the new space age.

These discoveries confirmed to some extent ideas about the movement of charged particles in the Earth's magnetic field that had been developed in the course of Birkeland's 1896 studies of the polar aurorae—the spectacular lights in the night sky seen in latitudes between about 60° and 75°. Birkeland's suggestion that the aurorae might be caused by electrically charged corpuscular rays shot out by the Sun and drawn in by the Earth's magnetic field near the poles was examined in the 1930s by another Norwegian, the mathematician F. C. M. Störmer. He tried to calculate the paths the particles should follow, but his theory was only superficially successful, and for many decades the theory of what became known as the solar wind, responsible for the outer portions of the Van Allen belts, failed to explain the aurorae well. The first real progress came at much the same period, with Sidney Chapman's account of how the magnetic storms that disturb radio and telephone communications on the Earth may be due to clouds of ions ejected from the Sun. There followed the convergence of many different research programs that resulted in a much clearer idea of the solar wind and its effects; and ultimately there was confirmation by means of the satellites—*Lunik* 1, *Lunik* 2, *Mariner* 2, and *Explorer* 10, for instance. It was shown that the magnetosphere had the expected spiral pattern, apart from some irregularities; and that the energy involved in expanding the solar wind is not great. An insignificant million tonnes of hydrogen per second are thrown out, this act requiring only a millionth of the Sun's regular rate of energy output.

The Sun was one of the chief objects of modest rocket-aided study in the pre-Sputnik era. Between 1949 and 1957, solar spectra of high resolution were obtained at all optical wavelengths, and some others, using rocket-borne instruments. One surprising discovery was that the far ultraviolet and X-ray radiation from the Sun was extremely variable. In 1956, work was done by the staff of the Naval Research Laboratory in Washington, D.C., on early type stars in the Galaxy, and there was a suspicion that X-rays were being detected from outside the solar system. (We shall have more to say about this in the following section.) In 1962 there came the launching of the first Orbital Solar Observatory (OSO-1), one of a series of eight, covering seventeen years in all—three-quarters of a full double-cycle of solar activity, observed at various wavelengths simultaneously, almost without break. Using a coronagraph of the type developed by Bernard Lyot in 1930, an image of the Sun's corona, free from scattering by the Earth's atmosphere, was observed continuously for several months, out to a distance of ten solar radii from the limb—far better than during the most favorable eclipse as seen from the Earth.

The largest of the solar observatories was a manned example, and was given the name Skylab. Perhaps it should be described rather as a science and engineering laboratory. Launched into orbit around the Earth by a

Saturn v rocket in May 1973, it carried eight large telescopes, one with a coronagraph, and its crew brought back many thousands of photographs of the Sun's atmosphere from their three missions—between May 1973 and February 1974. During it lifetime, many hundreds of experiments were conducted aboard the second Skylab. It eventually fell from orbit on 11 July 1979, its parts falling mostly into the Pacific, and a few on thinly populated Australian territory. Three crews, each of three men, visited the station, their missions lasting between four and twelve weeks. A later satellite, SMM (Solar Maximum Mission), was sent up in 1980 to examine the Sun at the maximum of its cycle of activity. It acquired newspaper status when it was found necessary to carry out a repair, and when this was done by astronauts James Nelson and James van Hoften using a remotely controlled manipulation system from the space shuttle *Challenger*, on 11 April 1984. The problems Dondi had experienced in repairing his astrarium, or Herschel, his mirror, were by comparison mundane.

In 1985, astronomers were reminded of their debts to their military patrons. An American satellite named *Solwind* had been launched in 1981 to supplement the findings of SMM, and to monitor the Sun through a complete sunspot cycle. Among other things it revealed the presence of five Sun-grazing comets, all previously unobserved. In September 1985, however, its life was brought to an abrupt end when it was used as a target for an American antisatellite weapon (ASAT). It was important that astronomers should remember who was the paymaster.

Perhaps it should be said, in passing, that military need is not the only mantra by which funds can be justified to the public. Ecological arguments were already coming into fashion in the old millennium, but began to gather speed in the new. The European Space Agency's (ESA's) 2006 Venus Express mission—using design and manufacturing techniques shared with the earlier Mars Express mission, to save time and money—has regularly been justified on the grounds that, since Venus shares in so many of the Earth's properties, and yet suffers badly from runaway greenhouse warming, the mission may throw light on the dangers facing our own atmosphere. The argument would have been almost inconceivable two generations earlier, but even harder to imagine then would have been such an enterprise depending on a Soyuz rocket, launched from Kazakhstan. The talk then was of the dangers of nuclear fallout. In 2005, priorities had changed, as was evident from the Russian government's new ten-year space budget—approximately ten percent of the corresponding U.S. budget.

MEN ON THE MOON AND INTELLIGENCE BEYOND

Great advances were made in knowledge of the Moon, beginning in the late 1950s, as a result of the competition between superpowers, and their desire to demonstrate military superiority. Soviet probe *Luna 2* crashed on the far side of the Moon in September 1959, and in the following month *Luna 3* sent back pictures of that hidden face. One of the most surprising of Soviet discoveries was that the far side of the Moon is lacking in large maria—the vast "seas," plains of basalt that create the "Man in the Moon" image that is visible to us on the near side. Not until 1962 did the United States land a rocket—*Ranger IV*—on the far side of the Moon, and even then all internal power on board the spacecraft failed, two hours after launch, and the spacecraft failed to send back televised pictures. That it landed on the Moon was known only because it could be tracked using a tiny radio transmitter in the lunar capsule. The capsule also carried a seismometer, designed to be released from the rocket before impact, and to land on the Moon in a good enough state to measure the frequency of natural lunar earthquakes. In fact, the first six *Rangers* failed, some of them missing the Moon entirely. Finally, on 31 July 1964, *Ranger VII* sent back more than 4,000 excellent images—some showing objects only a few feet across—before crashing into the Moon's Mare Cognitum. *Rangers VIII* and *IX* repeated the feat, and the program was brought to an end in March 1965. The first soft landing was by *Luna* 9, in 1966, while five Lunar Orbiter missions followed, preparing the way for the most spectacular landing of all, that of the American astronauts.

The Apollo missions that first placed men on the Moon's surface were undoubtedly the best known of all the lunar expeditions. Cameras and seismometers may be more useful to science than humanoids, but the public's perception of the case was very different. The first human being to orbit the Earth had been the Soviet cosmonaut Yuri Gagarin, whose spacecraft was launched on 12 April 1961, and who parachuted safely to Earth after a single orbit. (He was to die in a training flight on a humble jet aircraft in 1968.) There was a strong popular feeling for human involvement in space travel, although with the passing of time the American episode has perhaps been reduced in most human memory to the walk on the lunar surface on 21 July 1969, by Neil A. Armstrong, and to his words: "That's one small step for a man, one giant leap for mankind." (The first reported version, where "a man" became "man," made less good sense.) Armstrong was accompanied on the Moon by Edwin E. "Buzz" Aldrin, Jr. The spacecraft was *Apollo* 11, and this remained in orbit round the Moon while they were landed by the lunar module Eagle. What has been largely forgotten is the sheer scale of manned operations, in the period from 1968 to 1972. No fewer than nine manned American spacecraft orbited the Moon then, all named *Apollo*, numbered 8 and 10–17. Six of them landed astronauts (not 8, 10, or 13), and between them they sent back over 380 kilograms of lunar samples. A scientifically invaluable visit was by *Apollo 12*'s lunar module, which on 20 November 1969 landed astronauts, allowing them to examine *Surveyor* 3, a craft that had soft-landed in April of the same year. An analysis of flakes of paint from the probe later yielded useful information about the solar wind, a most unexpected bonus.

Three unmanned Soviet vehicles, *Lunakhods*, were landed on the Moon, but they sent back only 0.3 kilograms of samples. In the Luna missions, a series of detectors was set in orbit round the Moon to pick up signals from a sensor on the surface. The same kind of geological sensing, with an automated station on the planet's surface, has since been used for the geological analysis of Mars, as part of the Viking project.

The samples of lunar rocks have been analyzed by numerous laboratories in many countries, and their comparison with meteorites has provided far more information about the origins of the Earth-Moon system than would ever have been possible without direct access to the terrain. Three theories in particular were previously current: that the Moon and Earth were a naturally formed planetary pair, that the Moon was captured before it crystallized, or that the Moon was a detached part of the Earth's mantle. In any event, the evidence allowed none of these theories to survive unscathed. Rather than enter into the minutiae of the many discussions that took place in the pre-Apollo period, we may consider just one, a vigorous debate that took place in the 1950s between the chemist Harold Urey and the astronomer Gerard Peter Kuiper, of "Kuiper Belt" fame, who were then colleagues at the University of Chicago. Urey had a high reputation. He had played an important part in the Manhattan Project, and after the war had become deeply interested in the Moon, the formation of the planets in general, meteorites, the abundances of the elements, and the origin of life. As a Nobel prizewinner, his influence was considerable, even on astronomy, especially through the support he gave to the manned mission to the Moon. In brief, his view was that the Moon, a leftover from the formation of the solar system, has large maria lacking in distinctive craters because large bodies bombarded the Moon *after* the smaller bodies that yield small craters. He thought the maria to be flows of molten lava, produced by the heat generated at impact. There was much skepticism at the time, largely based on ballistic evidence, which was that liquefaction would not have taken place at all. An extremely important study of lunar craters had been produced by R. B. Baldwin. Baldwin plotted the depth of craters against their diameters, taking into account craters produced by explosive shells, bombs, meteorites on Earth, and Moon craters; and he found that they all fitted a single curve. The formation of gas, not of liquid, was the key to understanding these phenomena. Kuiper had a much wider knowledge of astronomy than Urey. He based himself on observations with the Mac-Donald Observatory 82-inch reflector, and accepted Baldwin's arguments for impacts as the cause of lunar craters.

So the question was, hot Moon or cold Moon? Some very lengthy arguments were produced, based especially on studies of cratering and colors, but suddenly the old arguments came to a standstill. Spacecraft began to carry out photographic surveys at close quarters; then there was remote sensing from orbiting craft that allowed rough and ready mineralogical surveys; finally there came samples of rock, core tube samples, seismological data, heat and other physical measurements, all made possible by American and Soviet missions. Lunar chemistry, tectonics, and a dozen branches of lunar physics were all quickly established. Most of Kuiper's ideas survived, while those of Urey did not. What to the new lunar astronomers was an exciting climax, must have been exactly the opposite to those who had spent so much time in spinning theories out of very thin evidence.

After the Moon, all signposts were pointing to the remaining parts of the solar system. Venus was an obvious target, and one that presented many unsolved problems. It had long been reluctantly accepted that Venus was covered with a layer of cloud, impenetrable to telescopic view, and concealing even such a fundamental parameter as the rotation rate of the planet. Spectroscopic attempts to analyze the cloud had proved difficult to interpret. In 1929, Bernard Lyot had made studies of the polarization of light from the planet, and had decided that the clouds were chiefly of water vapor. In 1932, however, Walter S. Adams and Theodore Dunham, working at Mount Wilson, interpreted the puzzling Venusian spectral lines as coming from carbon dioxide at very high pressure. By the 1950s, there were numerous hypotheses being put forward. Fred Whipple proposed oceans of seltzer water covered with carbon dioxide clouds. Fred Hoyle even put forward an argument for oceans of oil, "beyond the dreams of the richest Texan oil-king." Radio observations were ambiguous. Did Venus have an extremely hot surface, or was its radiation coming from an ionosphere at much lower temperature?

The answer was not easily found. As early as 1962, the United States launched *Mariner 2* with 18 kilograms of scientific instruments, which it carried to within 3,500 kilometers of Venus, scanning the planet's surface with infrared and microwave radiometers. In this way, data was secured that showed the surface temperature to be about 425°C—a figure that has since been revised to about 470° C. On 12 June 1967, the Soviet Union launched *Venera* 4, the first probe to be placed directly into the atmosphere of Venus and to return atmospheric temperatures and pressures. Before it reached the surface of the planet it was crushed by the pressure of the atmosphere, some ninety times as great as that on Earth. Two days later, the United States launched *Mariner* 5 to Venus, where it arrived on 19 October 1967, just 36 hours after *Venera* 4. After studying the Venusian magnetic field, as well as the planet's atmosphere—but more cautiously than it Soviet predecessor—*Mariner* 5 entered a solar orbit (its radio survey suggested a surface temperature as high as 527° C and a pressure of 100 atmospheres). Not until 1970 was the first successful landing made on Venus, this by the Soviet *Venera* 7. The United States seemed to have lost interest in Venus even before *Mariner* 5, having begun to focus more attention on Mars. Perhaps this was in view of the fact that Venus seemed so hostile as an environment for life, past or future. Soviet scientists, however, kept up their interest in the hot and highly pressurized planet, with the help of instruments parachuted from *Venera* and *Vega* probes—some in the course of a mission to Halley's comet. By and large, these instruments transmitted their information for only an hour or so, before falling silent. *Venera* 9 and 10, late in 1975, sent back the first pictures of the rock-strewn landscape of Venus, before the television camera perished.

Planetary surfaces, which even the best land-based optical telescopes cannot resolve well, have been closely studied using space probes. The first close-up pictures of Mercury's surface, for example, were obtained by *Mariner* 10 in March 1974, showing heavily cratered regions. The Mariner series produced the first scientifically valuable missions to Mars. Indeed, as already pointed out, for most astronomers the Mariner and Viking probes marked the end of an era of fantasizing over the possibility that a civilization had been responsible for producing "canals" on Mars. With one exception, the lines Percival Lowell had drawn on his maps were shown to be the products of optical illusions. The exception was a giant canyon that he had called *Agathodaemon*, but that his compatriots preferred to rename *Valles Marineris*—presumably on the grounds that it had not been discovered until it had been seen at close quarters. By that argument, Columbus did not discover America.

PLANET SHALL SPEAK PEACE UNTO PLANET

The 1970s did not put an end to the desire for intelligent company in the universe—on the contrary. In the intervening century there had been many variants of the old argument put forward by Huggins and Miller in 1864, that the elements needed for life were demonstrably abundant throughout the universe. We might not be in a position to visit them in the near future, but it was suggested that if they have long been in a position to visit us, and are too shy to do so—and many thought that they had arrived already—then we should take a longer view of history. The *Pioneer* spacecrafts 10 and 11, launched in 1972 and 1973, respectively, carried gold-anodized aluminum plaques with a message that it was hoped might one day reach the gaze of intelligent beings far beyond the solar system. These beings were assumed to have sight and curiosity. The plaques, six by nine inches in size, were designed to be the longest lived works of mankind—say some hundreds of millions of years—and to locate our solar system in space with respect to fourteen pulsars. The plaques carried images of a couple resembling Tarzan and Jane, with what were supposed to be "panracial characteristics," as well as much implicit information that is bound to intrigue the finder—such as that Earth folk have not yet invented clothing, children, or the metric system (fig. 238). Another human project in much the same spirit was realized in 1974, when a complex message in binary code was transmitted toward the Great Cluster in Hercules from the vast antenna at Arecibo. In the midst of other information, this carried the chemical formulas for components of the DNA molecule. Not a century had passed since Camille Flammarion published his monograph on Mars (1892), in which he waxed eloquent about the discovery of new worlds, their inhabitants surrounded by the "work and noise of peace," with whom

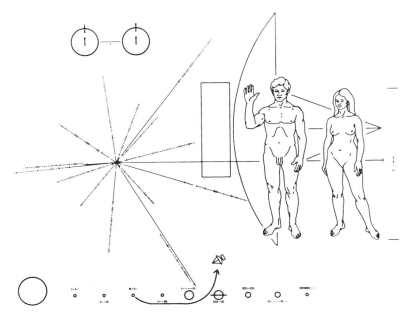

238. The plaque carried by *Pioneer 10*, with a series of symbolic messages intended for any intelligent being intercepting it in the remote future. Its design was engraved on a gold-anodized aluminum plate of 152 × 229 millimeters (6 × 9 inches), attached to the spacecraft's antenna support struts, to help shield it from erosion by interstellar dust. The spacecraft, launched in 1972, sent its last message to Earth in 2003.

we might some day unite. He had spoken of the Earth as a mere province of the universe, and of "unknown brothers" living in its infinite depths. He, at least, would have approved of the fact that by the end of the twentieth century there were groups of astronomers devoting a substantial part of their energies, and considerable sums of money, to the search for extraterrestrial intelligence.

The habit of sending messages out into the unknown is catching. In 2005, the Japanese space probe *Hayabusa* (Falcon) landed a target marker on the asteroid Itokawa, containing two aluminum sheets with the names of 880,000 people from 149 countries, including those of science fiction writer Arthur C. Clarke, film director Steven Spielberg, and *manga* artist Reiji Matsumoto. This of course had little to do with discovering extraterrestrial intelligence—it was more akin to throwing messages in bottles into the sea.

There is general assumption that truly intelligent beings will communicate on radio frequencies if they are as anxious to meet others as we are, since radio uses so much less power than light or X-radiation. There has been much discussion of tests that might prove an emission to be artificial—most favor number sequences, and hope that their remote counterparts will have thought likewise, albeit long, long, ago. In 1960, the distinguished Dutch logician Hans Freudenthal went so far as to write a whole book, the title of which speaks for itself—*Lincos: Design of a Language for Cosmic*

Intercourse. This he no doubt saw more as an intellectual exercise than as a plausible program, but others entertained greater ambitions, and in 1984 the SETI Institute was founded at Mountain View, California, dedicated to the search for extraterrestrial intelligence. This private, nonprofit organization eventually employed more than a hundred serious-minded scientists and promoters, and it has a far more willing potential audience than almost any conventional branch of astronomy. A leading figure in the SETI enterprise has been Frank Drake, author of a much-quoted formula for the number of extraterrestrial civilizations in our Galaxy with which we might expect to be able to communicate at any given time. This formula, a version of which he devised while working at the National Radio Astronomy Observatory in the 1950s, was expressed as a simple product of seven numbers, some astronomical (such as the rate of star formation in our galaxy), some biological (such as the average number of planets that can potentially support life for every star that has planets), and some social (such the expected lifetime of a civilization able and willing to communicate). Wildly different estimates of the values of the seven parameters have led to highly divergent estimates of the relevant number of civilizations, ranging from zero to many thousands. Human hope being what it is, the higher the number, the greater the funding it will yield for the great search.

Intelligence without life is scarcely conceivable, although perhaps not absolutely inconceivable, if we are to take *The Black Cloud* (1957) seriously. In this, the most successful of several novels written by Fred Hoyle in his spare moments, an interstellar cloud with a brain composed of complex networks of molecules visits the Earth's neighborhood. Finding our planet inhabited by an intelligent race, it proceeds to reveal a number of facts about intelligence elsewhere in the universe. Unfortunately, clouds do not make good television, and Hoyle's idea has not caught on. Intelligent life without a supporting planet has usually been assumed to be impossible, and the probabilities of the existence of the first are usually coupled with a discussion of the evidence for planetary systems around distant stars. There are protostars in plenty to be seen condensing in clouds of gas and dust, such as the Great Nebula in Orion, and some of these are surrounded by spinning disks of particles from which planets will be born. Planetary systems are too faint to be detected directly, but they are in principle detectable from a very slight oscillatory change in the position of the central star. (We recall that it is the center of mass of the system that is fixed, not the central star.) Many claims to have detected such an oscillation have been made in the past. In 1992, such a motion was found in a spinning neutron star. Alex Wolzczyan had been using the giant radio telescope dish at Arecibo, Puerto Rico, for several years, to monitor one particular pulsar, and he found that the times of arrival of its pulses were irregular. Having in mind an earlier—but mistaken—claim by Andrew Lyne, of the Jodrell Bank Radio Observatory, to have found a planet orbiting a pulsar, Wolzczyan was no doubt made more conscious of this possibility than he might otherwise have been. The inference he drew from his data, however, was remarkable: the irregularities in the pulsation were effects of two, or even three, planets orbiting the star.

Such planets, bombarded by fast particles and strong radio waves, would presumably be entirely unsuitable for life. In 1995, however, Michel Mayor and David Queloz, both of the Geneva Observatory, found evidence in a more conventional form. They found a fluctuating Doppler shift in the light from the star 51 Pegasus—a star similar to our Sun—indicating that it has a planet of roughly the size of our Jupiter circling round it. Within a few months, Geoffrey Marcy and Paul Butler in California discovered planets around other stars, some with temperatures at which water would not freeze or boil, making life as we know it at least conceivable. When honored by the American National Science Foundation in 2001, Marcy and Butler, with the help of colleagues, had discovered 38 of the 53 known extrasolar planet-encircled stars—in one case, a star with multiple planets. Some of their work has been done at the Lick Observatory, some at the Keck telescopes in Hawaii, some using the Anglo-Australian observatories, and some using the Magellan telescopes in Chile. By June 2005, there were 150 candidates for the status of exoplanet—to use the new jargon—when a serious claim was made to have obtained the first *image* of a planet orbiting another star. The image, in the near-infrared, was obtained using an adaptive optics instrument attached to one of the units making up the VLT at the top of Cerro Paranal (Chile). The star was GQ Lupi, a mere 400 light-years from Earth. Whether the reddish object going round the star is indeed a planet, or another star, such as a brown dwarf, has not yet been agreed upon.

Planets are seemingly as abundant as had long been suspected on theoretical grounds, but what of intelligent life? Is one of Hoyle's intelligent black clouds reading the Pioneer plaques at this very moment, without any need of a planet on which to rest its elbows? No, planets seem to be a universal *sine qua non*. One approach to the general question has been through biological theorizing, first about life, considering a planet's likely temperature, chemistry, and gravity. Some have tried to argue that there the reasoning must meet a stone wall, since (they say) intelligence on Earth has emerged after a series of extraordinarily rare accidents, involving a time-scale of the same order as the lifetime of the Sun. Even on the galactic scale, it is claimed, our own combination of circumstances is unlikely to have been repeated. For a variety of reasons, optimism seems to have slowly waned in the last quarter of the twentieth century. At the beginning, it was common to hear it said that our Galaxy alone must be home to millions of civilizations—the word being interpreted loosely, of course. By the end of the century, not only had the number fallen, but many of the old optimists were beginning to speak of life on other worlds as something likely to be infinitely more primitive than the Hollywood image of ET; indeed, if it were to exist at all, as more likely to be at the modest level of microscopic impurities in the family swimming pool, wholly lacking in an ambition to advertise its presence to the universe at large.

Against this, there is a new but small breed of optimists, many of them hailing from Russia, who continue to talk of rockets and boosters from other civilizations and all the resulting detritus that space-age pollution generates. Some of the microscopic bits and pieces, they suggest, having in the remote past been shed into intergalactic space, may have inadvertently found their way on to our Moon, so we should search for them—discounting, presumably, all those with stars and stripes on them. Yet others live in the hope of detecting laser pulses from stars, pulses with some kind of artificially superimposed pattern, rather like our Morse code. Paul Allen, one of the founders of Microsoft, has backed the construction of an array of some hundreds of radio telescopes in California, with a view to receiving the hoped-for signals. There is some virtue in relieving government agencies of the need to decide on such thorny questions, although there is a counterargument in favor of leaving funding to governments. We should recall the answer given by the English king George III, when he was asked by a French visitor why he rewarded William Herschel so handsomely for his astronomical discoveries. He said that he judged it preferable to spending money on killing soldiers. Unfortunately, as matters turned out, like many modern heads of state, the king had enough money to do both.

SPACECRAFT AND PLANETS AT CLOSE QUARTERS

The first planetary survey with great visual appeal was the near approach to Jupiter by the message-bearing *Pioneer* 10 in 1974. This crossed the asteroid belt, and transmitted data successfully—data that was reconstituted into images—even after experiencing high-velocity impacts with small particles in Jupiter's neighborhood. A year later, *Pioneer* 11 sent back fine images of Jupiter before moving on to Saturn, which it had reached by August 1979. Passing through the rings safely, it sent back images of Saturn. The *Voyager* probes, launched in August and September 1977, both passed near Jupiter and Saturn, and returned extremely fine and beautiful images of Saturn's satellites and rings, of their structure, of Cassini's division, and of marks crossing the rings like bent spokes. The trajectories of the *Voyager* probes would have made Isaac Newton proud. They had been so calculated that the near-encounter with the first planet pulled the probes round in the direction of the next. In the case of *Voyager* 2, this slingshot technique was used beyond Saturn: after launch (5 September 1977) and visiting Jupiter (July 1979), Saturn (August 1981), Uranus (January 1986), and Neptune (August 1989), the probe left the solar system for good (for an image of Saturn and four moons, synthesized from *Voyager* images with different color filters, see plate 17).

After a time, such launches tended to be viewed as commonplace, and what had filled newspaper headlines in large type became footnotes on the last page, before eventually ceasing to be mentioned altogether. The spacecraft *Pioneer* 10 was early enough to be capable of striking a romantic chord in many ordinary people, by virtue of the great distance from which it was sending information. Having sent back the first excellent close-up images of Jupiter, on 3 December 1973, it passed beyond the orbit of Neptune on 13 June 1983, and in doing so, becoming the most remote of all man-made objects. No longer in the limelight, *Pioneer* 10 continued to send valuable data from the outer regions of our solar system, until its mission was officially ended in March 1997. The last, very weak, signal from it was received in January 2003. By this time it was heading in the direction of the star Aldebaran in Taurus, with which it is due to rendezvous in over two million years time. Assuming that its plaque is there intercepted and decoded, and that a return message is transmitted, we can only hope that there will still be life on Earth when the message arrives, and that astronomy will still be a part of the curriculum.

As a measure of what had become possible by the 1980s, using interplanetary probes, the case of *Voyager* 2 is particularly instructive. When it flew past Uranus on 24 January 1986, for example, in one brief encounter that lasted only a matter of hours, the planet that for two centuries had been no more than a tiny disk of light was revealed as being at the center of a complex system of rings (two new ones were added to nine found in 1977) and satellites (in 1985–1986, ten previously unseen were added to the five already known; see fig. 239). Its rotational period was accurately found for the first time, likewise its axis of spin, tilted at a surprising 98°, and its strange magnetic field, tilted at a considerable angle (60°) to the rotational axis. This field was thought to be generated by a super-pressurized ocean of water and ammonia between the molten core and the atmosphere of the planet. The atmosphere, mainly composed of hydrogen and helium at a low temperature (–219°C), experiences winds at well over 500 kilometers per hour. Its moon Miranda is no less startling to the imagination: 500 kilometers across, it has canyons 20 kilometers deep, terraced layers, and cliffs of sheer ice as high as 16 kilometers. By comparison, the worlds depicted in *Star Trek* were positively suburban.

Before 1979, our knowledge of Saturn, based on telescopic observation, was very meager. The Voyager missions revolutionized the situation: the resolution of the images they obtained, of the planet and its satellites, were comparable with the resolution of images of the Moon obtained with Earth-based telescopes. The number of known satellites was found to be in excess of twenty. (There is some uncertainty about the growing number. Some of the smaller satellites are not confirmed, but as they are confirmed, others are likely to be added to their number.) A whole crop of new satellites was discovered in 1980. The largest satellite, Titan, previously well known to terrestrial telescopes, and bigger than our own Moon and Mercury, was then found to have its own atmosphere. It is now commonly spoken of as the most Earth-like place in the solar system, although that is not saying much, since its temperature is about –178°C, and since methane

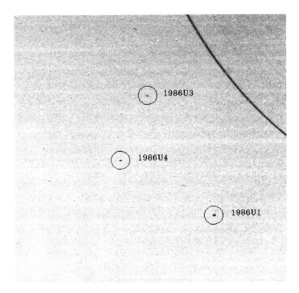

239. Three of the ten satellites of Uranus, discovered in 1985–1986 by *Voyager 2*. Five were known previously. This image is an inverted (negative) version of one taken in 1986, when the spacecraft was 7.7 million kilometers from the planet. All three satellites are outside Uranus's nine known rings, the outermost of which is seen at the upper right. The elongated form of the images resulted from the relatively long exposure.

on Titan appears to play the part that water plays on Earth. Enceladus, a mere 500 kilometers across—that is, under a tenth of the size of Titan—was found to show signs of geological activity.

This last discovery was supplemented by another, announced early in 2006, that there are liquid water reservoirs on Saturn's moon Enceladus. It was discovered during the Cassini-Huygens mission—*Cassini* referring to the spacecraft with its instrumentation, and *Huygens* referring to a probe with its own extensive collection of instruments. This combination was the most advanced ever made, a truly international project, to which seventeen nations contributed. *Cassini* was built and managed for NASA by the Jet Propulsion Laboratory, while the ESA contributed the *Huygens* probe. Many of *Cassini*'s instruments were contributed by the Italian Space Agency. The data received were analyzed by some three hundred scientists. *Cassini-Huygens* was launched by a Titan IV-B/Centaur launch vehicle on 15 October 1997. Its great mass—about six tonnes—meant that it could not be launched directly at Saturn, but that a "gravity-assisted"— or "gravitational slingshot"—trajectory had to be followed (fig. 240). It reached Saturn in July 2004. As it headed for the planet, *Cassini* fired its main rocket engine to brake its fall and put it into orbit round Saturn. At a later stage, the *Huygens* probe separated from the *Cassini* orbiter and descended into Titan's atmosphere, leaving the orbiter to study the Saturnian system more generally. There were good reasons for studying the planet in detail. Saturn—like Jupiter, Uranus, and Neptune—is a gas giant, composed of mostly hydrogen and helium. Posing difficult questions are

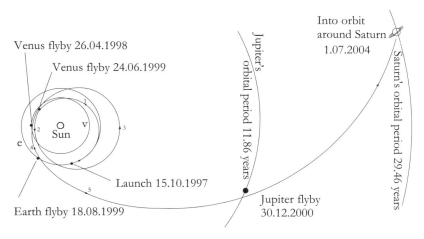

Venus flyby 26.04.1998

Venus flyby 24.06.1999

Jupiter's orbital period 11.86 years

Into orbit
around Saturn
1.07.2004

Saturn's orbital period 29.46 years

O
Sun

v

e

Launch 15.10.1997

Earth flyby 18.08.1999

Jupiter flyby
30.12.2000

240. The gravity-assisted course of the spacecraft *Cassini-Huygens*. The orbits of the Earth and Venus are marked as *e* and *v*, respectively. The small numerals (12345) are added to make following the trajectory easier. It looped round the Sun twice, passing close on both occasions to Venus, and on the second circuit passing close to Earth. On all three occasions it gained enough momentum from those planets to allow it to reach the outer solar system. Another gravity-assist from Jupiter, in December 2000, gave it the last burst of energy needed to project it to Saturn.

its huge magnetosphere and stormy atmosphere, with winds approaching speeds of 2,000 kilometers per hour near the equator. As for the effects of those extraordinary winds, it has been concluded that they combine with the heat rising from the planet's interior to create the yellow and gold bands that are visible in Saturn's atmosphere.

To return to Saturn's moon Enceladus, and the geological activity found on it: high-resolution images from *Cassini* show jets and plumes of water particles in large quantities and moving at high speed. Other moons in the solar system have liquid-water oceans covered by kilometers of icy crust, but not in pockets below the surface, as is implied by the astonishing Enceladus observations. The only three places in the solar system where anything similar might be occurring are the Earth, of course, Jupiter's moon Io, and possibly Neptune's moon Triton. The rare occurrence of liquid water so near the surface of Enceladus raised many questions. How could this be, when the planet's temperature is so low? Parallels were drawn with the Old Faithful geyser in Yellowstone, although the underground temperature on Enceladus must be much lower. Are conditions on Enceladus suitable for living organisms? One earlier question at least has been answered. As *Cassini* approached Saturn, it was found that the Saturnian system was filled with oxygen atoms. It is now clear that some of them are likely to be from the water thrown out by Enceladus, breaking down into oxygen and hydrogen.

Much of what the Cassini-Huygens mission revealed was by way of supplementing what had been learned about Saturn from earlier missions. By 1990, much had been learned about the nature of the planet's rings, which are composed mostly of ice particles, some no more than dust, others the size of boulders. These were all found to orbit the planet in a disk of the order of 100 meters thick—an extraordinary statistic, bearing in mind that the outermost easily visible ring has a diameter of more than 270,000 kilometers. The Cassini division—which, as its discoverer found, has the appearance of a dark central band in the rings—has relatively few ring particles within it, as half-expected. The rings in general revealed several unexpected properties, one of the most enigmatic being a set of what can best be described as spokes, the nature of which is still not well understood. No less surprising was the evidence produced by the Cassini mission, early in 2006, that there is an entirely new class of small moonlets residing within Saturn's rings, possibly numbering as many as ten million in a single ring. It is hoped that their existence will eventually help to settle an old question: were Saturn's rings formed through the disintegration of a larger body, or are they the remnants of the disk of material out of which Saturn and its moons were formed in the first place?

One of the many surprises yielded up by an even earlier survey of Saturn, that of 1980, was in the discovery that the planet has two smaller satellites in the same orbit as Dione. (Dione has twice the diameter of Enceladus.) The same was later found to be true of Saturn's satellite Tithys. The triple-orbits are stable, and the placement of the extra satellites fits perfectly with a theory laid down by Joseph-Louis Lagrange, more than two centuries earlier. They are at what are now called the "Lagrange points," which are important not only from the point of view of theoretical astronomy but also for very practical reasons in satellite technology—sufficiently important for us to say more about them here.

THE LAGRANGE POINTS

The Lagrange points—also known as "libration points"—are places where a body can remain motionless with respect to two more massive bodies that are orbiting each other under gravity, bodies such as the Sun and Earth. Taking this particular case, which is easy to visualize, there are five Lagrange points where a satellite may sit motionless relative to us. In figure 241, three of them are shown as L_1, L_2, and L_3, all suitably placed on the Earth-Sun line. The other two Lagrange points, L_4 and L_5, are in the same orbit as the Earth, but one is 60° ahead, the other 60° behind. Lagrange showed that bodies at the last two points may be in stable equilibrium, as long as the central body—the Sun, in the case discussed here—is more than 24.96 times as massive as its "planet"—Earth, in this case. The first three Lagrange points do not have this property of stability: bodies there may be in equilibrium, but when disturbed slightly, they will not move back to their starting points, as in the other cases. In the cases just mentioned, where Dione and Tithys have two partners each, those are at the stable points.

Lagrange's lengthy mathematical explanation of these far from obvious possibilities was given in 1772, in a prizewinning essay on the theory of

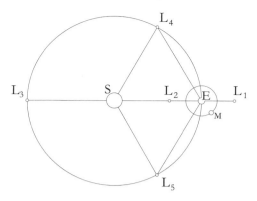

241. The five Lagrange points in relation to a planetary orbit, namely points at which an object may remain in stable (L_4, L_5) or unstable (L_1, L_2, L_3) equilibrium. S is here the central body and E a satellite to it—a planet, if S is the Sun. Since the Sun-Earth system is especially interesting, M is added to the figure to mark our Moon. Ignoring the Moon, S could stand for a planet (Saturn is of interest here), in which case E will be one of its satellites—Dione, for example, which is known to have partners at L_4 and L_5. Simple circular orbits are drawn here, not to scale.

the lunar orbit and incorporated into a renowned essay on the three-body problem. He shared the prize with Leonhard Euler, who had earlier discovered L_1, L_2, and L_3, but Lagrange's analysis was more complete and led him to the existence of points L_4 and L_5.

The stable Lagrange points are important, since matter will tend to accumulate at them. The most impressive case of this is at Jupiter's stable Lagrange points, each of which has hundreds of asteroids circulating in its neighborhood. The first of these asteroids was discovered by Max Wolf in 1906, by comparing photographs taken on successive nights, and comparing them to detect stellar (or similar) movement. They began being named for heroes in the Trojan War—Greeks at L_4 and Trojans at L_5, with an occasional mistake—but they are now so numerous that they are known simply as "the Trojan asteroids." Helene and Polydeuces, the moons of Saturn that lie at Lagrange points in relation to Saturn's moon Dione, are sometimes referred to as "Trojan moons"—a sad fate for a Greek brother and sister.

A question that has often been asked is whether our Moon has anything at its stable Lagrange points. In 1846, Frederic Petit, director of the Toulouse Observatory, claimed that he and several named colleagues had observed a second Earth's moon. The idea was little heeded, but after Jules Verne had taken up this idea in his novel *From the Earth to the Moon*, astronomers began to claim from time to time that they had detected one or more elusive object. Before long, the astrologers had added moons to their tool kits. One Moon hunter extraordinaire was undoubtedly Georg Waltemath of Hamburg, who in 1898 claimed to have discovered not one but an entire system of moonlets. While this claim was often scorned, the fundamental idea was not one that could be dismissed out of hand, and several attempts were made by main-line astronomers, in the course of the twentieth century, to detect the undetected. Eventually, important evidence was reported by the Polish astronomer Kazimierz Kordylewski of Kraków Observatory. He had started his search in 1951, naturally enough with a telescope. He found nothing until, in 1956, a colleague suggested that if there were numerous small bodies rather than a solid moon, naked-eye observations on a dark night, with the Moon below the horizon, might serve better than a telescope. Acting on this suggestion, Kordylewski found the evidence for which he had been seeking in October 1956. He and a graduate student saw a faint patch, about four times as large as the Moon would have been, in one of the Lagrange points. Two years later he detected its partner, another mere cloud. Not until 1961 did he successfully photograph the two clouds. Even then, for several years there were doubts about his findings. NASA's flying observatory failed to confirm them in 1967; but then at last, in 1975, the Orbiting Solar Observatory no. 6 (OS06), and Skylab's Zodiacal Light Analyzer, found both the L_4 and L_5 clouds in more or less the expected positions. It was later found that the clusters regularly wandered away from their Lagrange points, but their stable equilibrium meant that they returned home again. There have been few searches in the history of astronomy conducted with such tenacity for such ephemeral objects.

As for the Earth's own Lagrange points, they have come into their own with the space age. The L5 Society, which advocated the colonization of space at the appropriate point, no doubt has a noble history, but the Lagrange points as parking lots for artificial satellites have a more economically viable pedigree. In 1995, L_1 was made the home of the Solar and Heliospheric Observatory (SOHO), while in 2001, L_2 was used for NASA's Wilkinson Microwave Anisotropy Probe (WMAP). The latter was successor to the groundbreaking COBE satellite, both of them dedicated to the study of the cosmic microwave background radiation, mapping dark matter and dark energy in the universe, as we shall discuss in more detail in the following chapter.

The unstable Lagrange points may seem an odd choice for spacecraft use, but they have the advantage that they are easily entered and left. The Genesis mission, which—after collecting samples of solar wind—unfortunately crashed at high speed into the Utah desert on its return, used L_1 on its outward journey and L_2 on its return. There are many advantages in the use of these points for making fuel-saving journeys beyond them, and proposals have been made for a manned station at the Moon's L_1.

In September 2006, NASA announced that it had plans to go one better and establish a station on the Moon itself by 2020, and then to "extend human presence across the Solar System and beyond."

THE EXPLOSION IN SATELLITE RESEARCH

By the end of the twentieth century, the number of launches of space probes by all nations was approximately 160, a figure which included about 30 failures of one sort or another. In terms of numbers of launches and expenditure, the United States had a clear lead over the Soviet Union—later Russia—in the ratio of three to two. Before the millennium ended, Japan had launched four missions alone and one jointly; and Europe two alone and five jointly—like Japan, always partnered with the United States. In the new millennium, budgetary considerations forced one or two scientific programs to be canceled, their resources being used instead to fund vote-catching human space flight. Funding agencies have favored international cooperative ventures, but improvements in the efficiency with which data are collected is beginning to compensate. While the most spectacular of findings have not all been made by missions in what one might call the "grand series," a rough idea of the sheer scale of investment in those series can be had from their numbers alone: the Soviet Venera series reached a total of sixteen missions to Venus, while the United States launched nine Pioneer solar probes, seventeen Apollo missions to the Moon, and ten Mariner missions to Venus and Mars. These numerous examples, with their familiar names, account for less than a third of all missions. The number of artificial satellites and probes involved in astronomical enterprises is now so great that what began as a nine-day wonder has become an event that fails to make even the inside pages of most of the world's press, except in the case of disaster, which can come in many forms. NASA's *Galileo* planetary explorer, for example, launched in October 1989, had by the time of its first circuit of the Earth (December 1992) cost something in the region of $1.5 billion, and yet its main radio reflector was jammed by a piece of epoxy resin only a millimeter or two in thickness.

We have seen how competition between superpowers in the early years was so often to the benefit of astronomy—the *Sputniks*, the Moon landing, the crushed *Venera* 4, the cautious *Mariner* 5, the more successful *Venera* 7. Another instance of Soviet priority was in 1971, when *Mars* 3 (fig. 242) released the first successful lander to Mars. This relayed 20 seconds of video data to its orbiter, before the latter returned with the information to Earth. *Mars* 2 had earlier crashed into the planet. Both landers had rovers aboard, to no avail. The United States Venus orbiter *Magellan*, which operated between 1989 and 1994, mapped almost all of the Venusian surface by radar before obeying an order to enter the planet's atmosphere, and so end its life valiantly in the cause of astronomy. Entering a solar orbit is another fate, but whether it is a fate worse than death has not been very loudly debated. It was not peculiar to *Mariner* 5: *Mariners* 2, 4, 7, and 10 and *Vega* 2 are yet other illustrations of a not uncommon fate of spacecraft, probes, and debris in general, that have managed to avoid an uncontrolled planetary impact. As more distant targets are approached, human artifacts will be increasingly prone to leave the solar system entirely. *Pioneer 10*, for example, launched by the United States in 1972, and later passing Jupiter (1 December 1973) before crossing the orbit of Pluto (13 June 1983), was destined to leave the solar system.

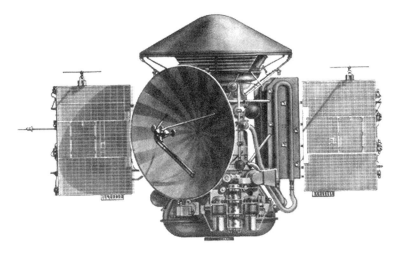

242. The Soviet *Mars 3* orbiter, with lander attached. The Mars 2 mission used an identical pairing of orbiter and lander. Launched by Tyazheliy Sputniks, the landers were the first human artifacts to reach the planet, although that from *Mars 2* crashed (27 November 1971) and that from *Mars 3* broadcast for only 20 seconds (2 December 1971). Both carried tethered rovers. The plan was to have them image the Martian surface and clouds, determine the temperature, topography, physical properties of the surface and atmosphere, and monitor the solar wind and the interplanetary and Martian magnetic fields.

In terms of targets, it is hardly surprising that our nearest sizable neighbor, the Moon, began as a clear favorite, with over sixty missions. Mars came in second place, but with only half as many, and Venus next with over twenty. Mariner 10 (1974–1975) was actually the first dual-planet mission, visiting Mercury and Venus, from where it sent back 10,000 images. Jupiter and Saturn were not ignored, but like the more remote parts of the solar system, they only came truly into their own with the new millennium. The space enterprise has served astronomy well. *Giotto*, *Ulysses*, *Galileo*, *Phobos*, *Vesta*, and the rest serve to remind us of the limitations of the telescope for planetary work. Before the advent of rocket-borne probes, most professional astronomers were content to leave the physical appearances of the planets to their amateur colleagues. When eventually satellite photographs of planetary surfaces were produced, some of them in such fine detail that they might easily be confused with photographs of the Earth's surface, what had been an unfashionable branch of astronomy suddenly came back into the limelight. As physical samples were obtained from increasingly remote parts of the solar system, even from asteroids, geophysics and traditional astronomy became more closely affiliated, to their mutual advantage. New disciplines were called into existence and new theories followed in profusion—theories concerning the mantles and cores of planets, their atmospheric, magnetic, geological, seismic, and other properties, and their likely evolution. In this way, planetary astronomy has been provided with many of its missing links.

The funding of such an expensive science, with few obvious material returns, requires a measure of public support, and above all else, good images—images that do not require reconstitution from data obtained by scanning, and so are instantaneously available, have a much stronger popular appeal than composite images. *Viking* 1 (1975-) *and Viking* 2 (1980-) were designed after the *Mariner* spacecraft, each of them composed of an orbiter and lander. Both provided detailed color panoramic views of the Martian terrain—with 52,000 images to their credit. Both also monitored the Martian weather. By the end of the millennium, images were being returned not only from a static lander but from "rovers." The U.S. Mars Pathfinder, having arrived at Mars on 4 July 1997, covered about a kilometer in sixteen bounces, before coming to rest. It eventually returned 16,000 images on its own account, while a six-wheel rover it dispatched, named Sojourner, added another 550. This was the second mission in what NASA termed its Discovery "low-cost" series—a political, rather than a scientific, turn of phrase.

Optical wavelengths do not satisfy astronomers, of course, as is shown by the merest glance at figure 237 on p. 690. The sky as a whole has been surveyed at infrared wavelengths by various means. In the mid-1970s, for instance, several American rockets were launched for this purpose, making flights of only short duration, and not as a means of placing instruments in orbit. The first almost complete survey of the sky was achieved by the infrared astronomical satellite (IRAS), launched in 1983 as a joint project of the United States, the United Kingdom, and the Netherlands. The satellite carried a 57-centimeter telescope cooled by liquid helium to less than 3°K, so that the telescope's own heat emission could be cut down to negligible proportions. IRAS made a number of important discoveries, including clouds in the Galaxy that emit infrared, a cloud around the star β Pictoris that has the appearance of a planetary system in formation, and a series of six hitherto unknown comets. (That the birth of a planetary system had indeed been observed around β Pictoris was challenged two years after it was announced, and the discussion ended on an inconclusive note.)

Observation beyond the other end of the visible spectrum, that is, at ultraviolet wavelengths, is needed for very hot sources—say, hotter than the Sun's photosphere. This includes the Sun's chromosphere, and interstellar gas heated by nearby hot stars. It also includes very massive stars, some of them even more than a hundred solar masses. Only in 1981 did this class of object enter astronomy. Satellites dedicated to observing the sky in the ultraviolet had by then included the second Orbiting Astronomical Observatory (OAO-2), which was launched in 1968 with no fewer than eleven telescopes—seven built at the University of Wisconsin and four at the Smithsonian Astrophysical Observatory. Its successor was OAO-3, launched in 1972 and named after Copernicus, who had been born in 1473, half a millennium earlier.

By the late 1970s, astronomy was becoming increasingly international in scope. We are reminded that not until the end of 1991 could it be said that the Soviet Union was irrevocably breaking up. Had astronomers ruled the world, perhaps the Iron Curtain would never have been there in the first place. Informal East-West collaboration was taking place long before the curtain suddenly wafted open. A textbook example for would-be peacemakers was the plan to study Halley's comet from space in 1985, a plan that involved the Giotto mission of the ESA, the Sakigake and Suisei missions dispatched by the Institute of Space and Aeronautical Science of Japan (ISAS), and Vega 1 and Vega 2 missions of the Soviet Union, not to mention contributions from Austria, France, Bulgaria, Czechoslovakia, both East and West Germany, Hungary, Italy, and the United States. This was by no means a purely nominal collaboration. Observations from the *Vega* spacecraft, for instance, were used to direct *Giotto* toward a close encounter with the cometary nucleus; and in return, ground-based observations of comet Halley, coordinated by the International Halley Watch (IHW), were used to direct the *Vega* spacecraft.

An outstanding earlier instance of collaboration on the western side of the curtain was the *International Ultraviolet Explorer* (IUE) launched in 1978. This highly successful satellite, with a 45-centimeter reflecting telescope, could record stars down to magnitude 16. Always visible from the Goddard Space Center near Washington, D.C., or the ESA's station near Madrid, it was controlled alternately from one or the other. Earlier ultraviolet telescopes had revealed an extraordinarily high rate of loss of mass by very massive stars—say a loss of one solar mass in a million years, for a star of thirty solar masses. IUE made possible a closer study of this problem of mass-loss, which radically changed the theory of stellar evolution then generally accepted, as regards stars in the upper part of the H-R diagram. Stars with phenomenally large luminosities, even a million times that of the Sun, could of course not previously be accounted for in terms of a model that entailed instability in stars above 60 solar masses. Mass-loss was in the 1980s seen as the answer to instability. It has been seen, too, as possibly offering assistance in understanding the Wolf-Rayet stars, that have been so difficult to fit into the H-R diagram. (As mentioned previously, these stars were described for the first time in 1867. They have peculiar spectra with strong broad-band emission lines, are exceedingly hot and luminous, and eject shells of gas at high velocities.)

SATELLITES AND X-RAY AND GAMMA-RAY ASTRONOMY

In the decades beginning in 1970, no branch of astronomy benefited more from rocket-borne satellites than high-energy astrophysics. This study makes use of radiations of very short wavelength—X-rays and gamma-rays, with wavelengths broadly indicated on figure 237 on p. 690 above. The name "high-energy astrophysics" comes from the fact that photons, packets of radiation, at these wavelengths have considerably more energy than photons of visible light.

The Sun was known to emit X-rays. In 1962, the most powerful source of X-rays in our sky, Scorpio X-1, was found in the constellation of Scorpio. The discovery was yet another on the fringe of professional astronomy. Bruno Rossi, a professor of physics at the Massachusetts Institute of Technology, was also chairman of a company—the American Science and Engineering Corporation—founded by a former student, Martin Annis. The company had taken on an Italian cosmic-ray physicist to start a program in space-science, and together with another MIT physicist, George Clark, they designed instruments for X-ray observations of the Sun, Moon, and certain stars—supernovae, for example. NASA turned down one of their proposals, but the Air Force supported an attempt to study X-ray fluorescence radiated by the Moon, and in the course of this it was discovered that the background X-ray radiation from the sky completely obscured any lunar fluorescence there might have been. X-ray sources were found, but it was their intensity rather than their existence that was surprising. In fact the team knew of an earlier unpublished report of a suspicion that an X-ray source had been found. This, by Herbert Friedman and James Kupperian, Jr., in 1957, was not published since it could not be confirmed by a later flight.

It had never been expected that individual stars in the Galaxy, which is what these X-ray sources are, could release such abundant energy as they proved to do. Some of them pour out say as much as 100,000 times the energy of our Sun. On 12 December 1970, another American satellite, now specifically intended for the study of X-ray stellar sources, was launched from Kenya. (Commemorating the country's independence, it was called *Uhuru*, the Swahili word for "freedom.") The event opened an era in which the positions of the most powerful X-ray sources were charted. *Uhuru*'s original purpose had been to discover X-ray sources beyond our own Galaxy, but fewer clusters of galaxies and quasars were found than had been expected, and its greatest successes were within the Galaxy. Among 339 sources identified with the help of the *Uhuru* satellite, one of the brightest in the entire Galaxy was discovered, namely, Cygnus X-1. This source is associated with an optically visible blue giant star of twenty solar masses and an invisible companion that was later estimated at 8.5 solar masses—which some have argued must be a black hole, since its mass exceeds the theoretical limit for a neutron star.

Another satellite with the same aim, HEAO-2 (successor to the similar HEAO-1), was launched in 1978 and eventually renamed *Einstein*, to celebrate the centenary of Albert Einstein's birth in 1879. The American and European satellites of the preceding period could not produce direct images, and were accurate to no more than about a degree. HEAO-2 carried a wide range of instruments, was capable of direct imaging, and could place X-ray sources to an accuracy of about 2 seconds of arc, usually allowing them to be identified with optical sources. Yet another satellite observatory for high-energy astronomy, launched at much the same time as HEAO-2, was the Japanese *Hakucho* (Cygnus) X-ray satellite, launched in 1979, and joined four years later by satellite *Tenma* (Pegasus).

Having begun so vigorously, the satellite age continued, although moderated by an economic recession. We have reached a stage, however, at which merely to catalog the host satellites used by astronomers would be as pointless as the cataloguing of telescopes in, shall we say, Herschel's day. Each satellite played its part in creating a substantially new vision of stellar structures and their place in the cosmic pattern. It was soon widely accepted that the X-ray sources correspond to systems that include highly compact stars of one sort or another—white dwarfs, neutron stars, or black holes—all with very high gravitational energy that can be converted to high-energy radiation, largely in the form of X-rays and gamma-rays. As in the case of the telescope, the technological challenges satellite instrumentation present are so great that it is easy to forget that it is not an end in itself, and that an interpretation of the data has to follow. Much of the ensuing debate over the existence of black holes centered on data relating to the powerful source, Cygnus X-1. In this particular case—and many other too—a binary system was found to be involved, the compact star component seemingly drawing in material from its partner, a normal star. This was found to take place in various ways, according to the size and proximity of the components.

X-ray sources were found to vary in output, often periodically, but occasionally with violent bursts lasting only hours or days. In some cases, periodic bursts were detected that lasted only for a matter of seconds. Even in some of these cases, known as "bursters," it was found that a binary system was involved, but the systematic eclipsing that had formerly been offered to explain the fluctuations in light intensity was no longer enough to explain the more complex patterns of X-ray pulsation. Relatively simple cases were found, with explanations that included a single star's rapid rotation, perhaps with an unsymmetrical disk around it, or a stream of ejected matter; but in other cases it was found that pulsation is rapid. A high concentration of X-ray bursters—stars showing themselves in sudden outbursts lasting perhaps a few days, but not periodic—has been found in the galactic plane. At an early stage of satellite exploration, globular clusters were also found to be powerful sources of X-rays.

Gamma-radiation, like X-radiation, is most easily observed from space, since most of it is intercepted by our atmosphere. Very high energies (above a few tens of billions of electron volts) can be detected from the ground, however, as the British physicist of Patrick M. S. Blackett realized at least as early as 1948. He knew that on entering the atmosphere, such radiation would create a cascade of particles, which would in turn produce a flash of blue light, namely Cherenkov radiation. This blue light—today perhaps most familiar from images of the water-cooled fuel rods of nuclear reactors—takes its name from the Soviet physicist Pavel Cherenkov, who investigated it in the 1930s. It is emitted when a charged particle, or high-energy radiation, travels through a transparent medium faster than the phase velocity of light in the medium, and has been likened to the sonic boom produced by an aircraft traveling faster than the speed of sound. Its direction depends only on the energy of the incident particle (or radiation) and the refractive index of the medium. The emerging light thus provides an indication of the speed and direction of the high-energy particles or radiation causing it. When stimulated by gamma-rays, it is very faint and very brief, lasting only a few nanoseconds. Cosmic radiation has a similar effect, and in far greater abundance, but the emerging Cherenkov flashes of light differ in the two cases, that from a gamma-ray meeting the ground in a neat ellipse, about the area of a football field. In the twenty-first century this is measured using arrays of telescope mirrors and photomultipliers. Studies of gamma-ray sources such as supernova remnants, for example, were begun using the HESS array in Namibia in 2003—an array named in honor of Victor Franz Hess, who was awarded a Nobel Prize for his discovery of cosmic radiation, discussed earlier. This array (the name of which has a second meaning: high energy stereoscopic system) was built with the support of eight European and African countries, especially Germany and France.

After Blackett had postulated its existence, the first successful Earth-based detection of Cherenkov radiation from particle showers was achieved in 1953 by two physicists at the British Atomic Energy Research Establishment at Harwell. The two men, W. Galbraith and J. V. Jelley, used a detector built out of a yet another army surplus searchlight mirror, a galvanized-iron domestic dustbin, and a photomultiplier tube. Better endowed pioneering work followed in the 1970s: in the United States, the Canary Islands, and Australia, under the guidance of Ted Turver, of the University of Durham; and at the Fred Lawrence Whipple Observatory, Arizona, led by Trevor Weekes. It was a member of the Weekes group—Michael Hillas, from Leeds University in England—who first suggested the all-important way of discriminating between the neat pattern of Cherenkov radiation from gamma-rays and the rag-bag of radiation and particles from cosmic rays.

The first detection of high-energy gamma rays from space was achieved in 1961 by the *Explorer XI* satellite, and eventually gamma-ray bursters were detected, although not until 1967. They were found not to be distributed in the same way as X-ray bursters. Those recorded seemed at first to be relatively near and faint, perhaps isolated neutron stars. The first of them were found by American military (*Vela*) satellites, which had been deployed to detect Soviet nuclear explosions—explosions that would have been puny by comparison. Some gamma-ray bursters were found to be associated with optically observable objects. One remarkable event, observed from no fewer than nine satellites on 5 March 1979, involved a very short pulse followed by a series of about two dozen pulses at 8-second intervals. There has been much discussion of a possible mechanism, perhaps involving a neutron star, but what made the incident especially interesting was that the source seemed to coincide with a supernova remnant in the Large Magellanic Cloud, that is, beyond our Galaxy.

Built by five European research institutes—the European Space Agency, France, The Netherlands, Italy and Germany—was a satellite with a gamma-ray detector that operated for what was then an unusually long period, from 1975 to 1982. With this, named Cos-B, a comprehensive map of sources was produced, many of them surprisingly powerful. Of the two dozen most powerful, almost all were found to be in or very near to the plane of the Galaxy. This colored the views of astronomers for several years, until in 1991 NASA launched an enormous orbiting probe, weighing 17 tonnes, sixty times as massive as Cos-B. Known as the Gamma Ray Observatory (GRO) and later renamed after the pioneering physicist Arthur Holly Compton, the ambition was to bring gamma-ray astronomy to the state that X-ray astronomy had attained two decades earlier. Supernovae in galaxies, pulsars, and active galactic nuclei were on the agenda, but also gamma-ray bursters. It was expected that these would be found concentrated in the plane of the Galaxy, where neutron stars are to be found; but this was not so. They appeared over the whole sky. Gamma-ray sources had seriously unsettled astronomy once again.

The massive GRO—its instruments alone weighing six tonnes—was the second of NASA's Great Observatories series, being preceded by the Hubble Space Telescope, and followed by the Chandra X-ray Observatory and the Spitzer Space Telescope. (Lyman Spitzer was instrumental in the design and development of the Hubble Space Telescope, lobbying for it, both with Congress and the scientific community as a whole.) The GRO produced the first all-sky map of gamma-ray sources in 1993, and revealed a surprisingly large number of active galactic nuclei (AGNs) among the sources recorded. (AGN is an inclusive designation for a very small region at the center of a galaxy with colossal energy output in the form of radiation or jets of high-speed particles. This energy may exceed that from the rest of the galaxy. It is now widely accepted that an AGN has a super-massive black hole at its center, and that the observed energy is generated as matter is accreted by it, and heated in very high-speed collisions outside it.)

"Pulsar" is a term that has been used for pulsating stars of various sorts, as we have already seen, with neutron stars in rapid rotation offering the best explanation of their behavior. The complexity of the physical problems they present became even more apparent when a pulsar of period less than a tenth of a second was found in the Large Magellanic Cloud. This had to be counted as the most extraordinary of all X-ray sources to have been discovered by satellite recording, for it was found to radiate more X-rays than all the known sources in our Galaxy combined. Its name? While many a modest lump of rock in the solar system bears a human name, the object AO 538–66 seems to have confounded the collective imagination of astronomers. It appears to include an ordinary star of about a dozen solar masses, perhaps combined with a black hole (needed to explain its energy) or neutron star (needed to explain the pulsation). The decision between the two alternatives, black hole and neutron star, will presumably be based ultimately on the mass deduced for it. If greater than roughly three solar masses, theory favors a black hole. In the first two decades of X-ray astronomy, only a handful of candidates for this distinction were found, and each case was the subject of much disagreement. The theory of black holes, however, may be found an application on a much larger scale, that is, it may be applied to the nuclei of galaxies, where the mass is of the order of a billion stars, and where the black hole consumes matter (gas) equivalent to several solar masses annually.

NOVAE AND SUPERNOVAE

X-ray astronomy soon proved to offer a way to a more complete understanding of the evolution of stars, and this through the study of supernovae. We have already seen that the phrase *stella nova* was at first reserved for any star showing a sudden increase in brightness, and that the new star of 1054, recorded in China, and those of 1572 and 1604, made famous by Tycho Brahe and Johannes Kepler, would now be classified as supernovae. Two completely different mechanisms were found to be responsible. In the case of a nova, only the outer layers of the star seem to be involved in the sudden flaring up, a relatively small fraction of the mass of the star is lost, and indeed, some of the mass involved comes in any case from an adjacent star. Accepting an idea mooted by Robert Kraft in 1964, novae, without exception, are now thought to be members of close binary systems—for example, a white dwarf with a cool companion. The change in absolute magnitude—and so also in apparent magnitude—is of the order of ten magnitudes or less.

By contrast, a supernova is an explosion on a much greater scale, involving most of the material of the star. It was impossible to appreciate the difference between the two classes of phenomena without first knowing the initial and final intrinsic brightness, and this required a knowledge of distances. The turning point in an understanding of the situation came with observations of novae (some later reclassified) in the "Great Nebula" in Andromeda, M31, and in particular, of a star later designated S Andromedae. First seen by C. E. A. Hartwig of the Dorpat Observatory on 20 August 1885—unless we count L. Gully of Rouen, who saw it three days earlier but thought it was to be explained by a blemished telescope—it brightened from the ninth to the seventh (apparent) magnitude before fading rapidly. It disappeared from view after 7 February, but not before its spectrum had been studied by at least five astronomers. Among these, Huggins recorded bright emission lines and Copeland bright bands. These were the first steps toward understanding the extraordinary events that had taken place in 1054, 1572, and 1604.

In 1895, a similar "nova" was found by Williamina Fleming at the Harvard College Observatory, in an unresolved nebula (NGC 5253) in Centaurus. It was later named Z Centauri. She found it from its peculiar spectrum, and this was classified by Annie Cannon as of spectral class R. She thought its spectrum resembled that of S Andromedae—a decision that was revised long afterwards by Cecilia Payne-Gaposchkin, who in 1936 realized that the spectral lines were unusually bright and wide.

Novae were classified loosely at that time in a scheme devised for variable stars generally by Pickering (1880, revised 1911). On this scheme, there were simply "normal novae" and "novae in nebulae." In addition, Pickering's classification of variables—which was introduced in chapter 16—included stars of the "U Geminorum type," which are now generally called "dwarf novae." Their outbursts are typically at intervals of a few months. The prototype star was discovered by J. R. Hind in 1855–1856. Not for forty years was another found—this time by Miss L. D. Wells of the Harvard College Observatory (SS Cygni). By 1922, the similarities of the spectra of this still small group to the spectra of novae was recognized by Adams and A. H. Joy.

Other novae in spiral nebulae were found in 1909 (by Max Wolf), and 1917 (by G. W. Ritchey), the latter prompting astronomers to study much more carefully photographs that had been taken with the Mount Wilson telescopes. As a result, many more were found in this same category. The last step on the road to an appreciation of their great brightness came with the realization that the spirals were indeed very distant "island universes." Only then, in the mid-1920s, was the highly energetic character of the novae that could be seen in them perceived. Data remained in short supply (by 1937, only five had been studied spectroscopically). The typical star in this class, when it is identifiable on earlier plates, is found to increase in brightness by at least fifteen magnitudes. In the sudden explosion, more energy is released than our Sun has radiated in its entire lifetime of four or five billion years. In 1925, Lundmark distinguished between "upperclass" and "lower-class" novae, and Baade and Zwicky in 1934 substituted the name "supernova" for the former, the extremely bright novae in distant galaxies. Their astonishing character followed from Hubble's finding that distances to these galaxies, "nebulae," were to be measured in millions of light-years, distances a thousand times larger than those of novae in the Milky Way, implying that they are a million times more energetic. It was left to Baade, in 1938, to insist that here we are not only dealing with widely differing luminosities but with completely different classes of object.

The study of novae and supernovae could hardly have been advanced at all without photography, since the stars concerned are almost always inconspicuous before the outburst. Afterwards, with the help of old plates, the pattern of their changing brightness can often be pieced together reasonably completely. The typical light curve shows a rapid rise, followed by a slow decline, and there is an element of luck in catching the star during its rise. The first photograph of a nova's spectrum during the rise to maximum was that for the nova in Perseus in 1901. That it was largely an absorption spectrum almost persuaded the Harvard astronomers that they had photographed the wrong star. Not until 1918 was a spectrum found for a nova before its outburst (nova Aquilae). The plates were from 1899.

Numerous attempts had been made to explain novae. Isaac Newton, for instance, had a collision theory, Laplace believed in some sort of surface conflagration, W. Klinkerfues argued for tidal eruptions through the near approach of another star, Seeliger thought a collision between a dust cloud and a star was involved, and Lockyer held that collision between meteorites in two intersecting streams were responsible. At least this last theory could be ruled out, once photographs had shown stars in place before the outbursts. What nova Aquilae provided in 1918 was a great deal of new information. From the spectral information, W. S. Adams and J. Evershed drew the conclusion that a shell of gas was being ejected from the star at high velocity, and that some of the spectral peculiarities were due to the fact that we see superimposed spectra from the near and far parts of the shell, spectra that undergo Doppler shifts in opposite senses.

Broadly speaking, this fits into the general picture of a nova pieced together later in the century, of a complex interaction between a relatively cool star and an adjacent dwarf star. A disk of accreted material forms round the cool star. In the case of an ordinary nova this is drawn to the dwarf and triggers certain highly energetic nuclear reactions there that explode the outer layers of the star. In the case of a dwarf nova, it seems that matter is drawn at supersonic speeds from the dwarf star, not completely into the other, but into the surrounding accretion disc. The disk is then heated at the place where the incoming material hits it, and can do so repetitively, with the release of much less energy than in the "ordinary" case. In both cases, the accretion disk plays a crucial role. Observation of this, using the ultraviolet and X-ray regions of the spectrum, would have been quite impossible before the advent of satellite observatories.

As explained, only after the great distances of the galaxies was established was it fully realized that there are two different classes of supernovae. A supernova in 1937, analyzed by Zwicky and Baade, had an intrinsic brightness about a hundred times greater than the entire galaxy in which it was found (IC 4152). Rudolf Minkowski studied its spectrum, and by 1941 could announce that fourteen supernovae he had studied were distinguished clearly by their spectra into two classes—they were distinguished primarily by the presence or absence of certain hydrogen lines. Those of one class (Type II) are reminiscent of ordinary novae, but of course are very much brighter, and those of the other (Type I) are considerably brighter, and have very unusual emission bands. Studies of the Crab Nebula provided much valuable data here, and we have already seen that radio studies of the same object added to them. In addition, in 1964 the Crab Nebula was found to be a powerful emitter of X-rays. In a pioneering rocket observation conducted by Herbert Friedman and his associates, from the U.S. Naval Research Laboratory, the gradual occulting effect of the Moon—gradual because the nebula is not a point-source—was used to prove that this highly photogenic nebula really was the X-ray source. Only later was the additional complication of the pulsar at its center appreciated.

In the 1950s, William A. Fowler and Fred Hoyle had suggested a mechanism to explain the source of its energy. Their complex picture was of a star built up of a series of shells, like the layers of an onion, each shell the product of nuclear reactions at a particular stage in the star's long history. They suggested that, as heavier elements are created, temperatures increase and alternate with gravitational contractions until a stage is reached when the star is in equilibrium, with a mixture of iron and nickel and other moderately heavy elements at its center. Gamma-ray photons then step in and enter into certain nuclear reactions with the iron and nickel, in processes that require heat. The star cannot remain in equilibrium, and so collapses. The temperature rises, and after further stages have been gone through, nothing but protons, neutrons, and electrons remain. The protons absorb the electrons, and the core contracts further very rapidly, until a point is reached where a certain nuclear force that causes neutrons to repel one another brings the contraction to a halt. The star is then very compact, with the neutrons of the order of 10^{-13} centimeters apart. This is the theoretical state of the neutron stars that we have referred to so often. The core collapse takes only a matter of minutes. As it takes place, the outer layers fall inward, are accordingly compressed, and heated. Reactions are accelerated, and the layers explode. And this is the barest of summaries of the theory of supernova explosions according to Fowler and Hoyle.

Other theories were offered later, although in broad outline the pioneering study of Fowler and Hoyle held the allegiance of most astronomers for many years. The variant theories had much in common with theirs. They explained the formation of very heavy elements such as uranium by the nuclei in the outer shells capturing neutrons during the final neutron phase. In this phase too, in the very high temperatures involved—say ten billion degrees—extremely large numbers of neutrinos are generated.

Supernova explosions are much more than astonishing firework displays. They spread heat energy and the products of nucleosynthesis around the universe, and so in particular influence the evolution of the galaxies in which they are situated. Especially important for this evolution are the heavy elements, and in the 1980s much attention was given to the theory of their fundamental role in the formation of new stars. A particularly important supernova was found in 1986 in the radio galaxy Centaurus A, important to astronomers since it is the nearest radio galaxy to Earth. It was found by the Rev. Robert Evans, an amateur astronomer in New South Wales, who had already discovered a dozen or so supernovae in galaxies with his 40-centimeter reflector, noticed it first as a bright star in Centaurus A. After half an hour it had not moved, so was not an asteroid in the line of sight. Evans rang the Siding Spring Observatory, and within three hours it was confirmed as a supernova. (It later proved to be of Type I.) For its X-ray, gamma-ray, and radio properties to be better understood, its distance was needed. Fortunately Evans had discovered it before its maximum brilliance was reached, so allowing it to be monitored through the critical stages. The evidence was enough to place it among the Local Group of galaxies.

On 23 February 1987, another supernova was observed under extremely difficult conditions, and with poor equipment, by Ian Shelton of the University of Toronto, who was working at the Carnegie Institution of Washington's Las Campañas Observatory in Chile. This supernova in the Tarantula Nebula, in the outer part of the Large Magellanic Cloud—the small neighboring galaxy of ours—provided one of the most spectacular stellar discoveries of modern times. Its nearness made it doubly important, and it has been much studied ever since it was discovered. Its brightness in visible light peaked in May 1987 at a magnitude of 2.8, characteristic of a dimmer, Type II supernova, rather than the Type I originally predicted. Not only was it the brightest supernova since 1595, and the first easily visible naked-eye specimen since Kepler's of 1604, it was the only one that had ever been identified with a star known before its outburst—a blue giant component in the triple star system Sanduleak $-69°$ 202, a star with a mass estimated at about twenty times the mass of our Sun. It was some time before astronomers at Siding Spring, Australia, and Harvard, Massachusetts, pieced together the evidence for precisely which star was the true progenitor of the supernova. That it was a blue supergiant was in itself a great surprise, for according to earlier theories, blue giants are too young and too dense to produce Type II supernovae, which must come from old red supergiants.

Early on the day of its discovery, the supernova gave rise to the first bursts of neutrinos ever to be detected from outside the solar system, bursts reported by Carlo Castagnoli of the Istituto di Cosmogeofisica in Turin, Italy, from the neutrino observatory under Mont Blanc. On the basis of the neutrinos received, it has been claimed that, for a second or so, the luminosity of the star in neutrinos alone equaled the luminosity of the rest of the universe combined.

Subsequent optical, ultraviolet, and X-ray observations of the supernova allowed astronomers to piece together the following account of its progress. About ten million years ago the star formed out of a dark, dense cloud of dust and gas. Roughly a million years ago, it lost most of its outer layers, in the form of a slowly moving stellar wind that formed a vast cloud of gas around it. Before the star exploded 160,000 years ago, a high-speed wind blew off its hot surface and carved out a cavity in the cool gas cloud. (Such cavities have often been recorded elsewhere.) It is assumed that the intense flash of ultraviolet light from the supernova illuminated the edge of this cavity, so producing the bright ring seen by the

Hubble Space Telescope (plate 11). The optical hotspots and X-ray producing gases, on the other hand, are thought to be the result of collisions
with dense fingers of cooler gas streaking inward from the inner ring. The
supernova explosion sent a shock wave rumbling through the cavity, as recorded from the Chandra Observatory in 1999; and within three or four
years, evidence was mounting that this event—of extreme importance for
an understanding of stellar evolution—had begun.

One more event from the astrophysical diary that seems set to provide
much new information about the eruption of a certain type of neutron
star took place on 27 December 2004. It is thought to have been the largest explosion observed by astronomers—but not from Earth—since the
supernova observed by Kepler in 1604. The flash was seen from space
to have lit up the Earth's atmosphere, the Moon having helped by acting as a reflector. It lasted for more than a tenth of a second, and was detected by NASA's newly launched orbiting X-ray observatory SWIFT, by
European satellites, and by several radio telescopes. Radiation was brightest at gamma-ray wavelengths. The star, SGR 1806–20, is identified as a
magnetar, more precisely a "soft gamma repeater" (hence the "SGR"), with
a massive magnetic field, and rotating once every 7.5 seconds. On the far
side of our Galaxy in the constellation of Sagittarius, its distance was estimated at about 50,000 light-years and its absolute magnitude at about −29.
The magnetar released more energy in its tenth-of-a-second moment of
glory (1.3×10^{39} joules) than our Sun has released in 100,000 years.

The precise mechanism responsible for such a starquake was assumed
to involve a surface crust of iron nuclei, suddenly deforming violently
under the action of magnetic forces. The event stimulated discussions of
many kinds. Inevitably, members of the press concentrated on the likely
consequences of such a thing happening close at hand (such as the headline
in the London *Times* that exclaimed, "Earth 49,990 Light Years From Disaster"). Others pondered the question of whether a close burst of gamma-
radiation could have been responsible for one of the several extinctions
of life on Earth. By the twenty-first century, at any event, some questions
had more or less left the realm of controversy. It was by this time accepted
that, by means of supernovae, such elements as carbon, nitrogen, calcium
and iron are thrown into interstellar space, where they enrich clouds of
hydrogen and helium that are in the process of giving birth to new stars.
They also assist in the creation of the heavier elements—gold, silver, lead,
uranium, among others. It was also agreed that supernovae, by generating the cosmic rays which contribute to the mutation of living cells, and so
their evolution, are one of the most important of the many signposts to the
path taken by the evolving universe—and, at the same time, to life itself.

COMETS AND SPACE PROBES

Before the astronomical application of spectroscopic analysis, it seemed
that a knowledge of the chemical nature of extraterrestrial matter would

be forever unobtainable. After a century of spectroscopic and orbital study, theories of the physical and chemical nature of comets had reached a high level of sophistication. As we have seen, the dirty snowball theory had repercussions on both the chemical and the physical side of cometary study. It helped, for instance, to underpin Bessel's "fountain theory," put forward after he studied Halley's comet on its 1835 reappearance, and saw that material thrown out by the nucleus toward the Sun was being pushed back so as to present a fountain-like head (see p. 513 above). Fred Whipple proved the merits of the snowball concept many times over. With it he could explain how sunlight, falling on a rotating comet, could reduce or increase its orbital speed by a process of vaporization. He found that it could explain how meteor streams may be generated by sunlight blowing off solid pieces. It could explain the tidal effects that are in some cases capable of splitting up comets. By the late 1950s, the behavior of light on fine dust—even in supersonic situations—was well understood, enabling two aerodynamics experts, Michael Finson and Robert Probstein of the Massachusetts Institute of Technology, to perfect the fountain model.

Working along similar lines in 1968, Ludwig Biermann—of solar wind fame—predicted that sunlight breaking down the water molecules thrown out by comets should emit a certain spectral line from the hydrogen that might be expected to survive in an atomic and neutral state at about the Earth's distance from the Sun. The line, the fluorescent Lyman Alpha line, cannot pass through our atmosphere. To detect it, a spacecraft would be needed. The time was fast approaching when this would be available. As we saw earlier in this chapter, the first orbital flights with a view to astronomical observation were given the name Orbital Astronomical Observatory (OAO), and carried telescopes designed and built at the University of Wisconsin under the guidance of Arthur D. Code. Unfortunately, the high-voltage devices for star tracking OAO-1 began a series of troubles that ended when the system became so hot that it exploded. OAO-2, later in the same year, was highly successful, operating for more than four years and returning data on planets, comets, stars, and galaxies, all data being collected in the ultraviolet part of the spectrum. It is at these wavelengths that the atoms and molecules of comets radiate their most valuable information. In January 1970, Code was the first to detect a huge bright halo of hydrogen atoms around comet Tago-Sato-Kosaka. Later in the same year, the fifth Orbiting Geophysical Observatory (OGO)—another United States initiative—showed that the hydrogen cloud had a diameter at least three times that of the Sun, and that it was rapidly expanding. The first of the OGOs had been launched on an Atlas rocket in September 1964, and the fifth was sent up on a later version of the Atlas, in March 1968. These had orbits with low perigees (low minimum altitudes) and high apogees (high maximum altitudes), making them especially useful for terrestrial and nearby solar system study.

As we know, comets are usually visible only when near the Sun. The comet Kohoutek of 1973–1974 attracted much attention because it was visible at an unusually great distance from the Sun. Great things were therefore expected of it, as far as public spectacle was concerned. They failed to materialize, but its spectrum was unusual—there were carbon and oxygen lines never before detected in a comet—and it had a massive hydrogen cloud around it. Again, the best views were from space, and here Skylab made a useful contribution. Valuable sketches of comet Kohoutek were also drawn by one of the Skylab astronauts, E. G. Gibson.

At much the same time as this, radio telescopes were being used for cometary research, proving themselves useful for tracking down complex molecules. Radio was also used in a radar capacity. In 1980, P. G. Kamoun and his colleagues finally gave the proof that the dirty snowball supporters had wanted, proof that comets had hard cores as their nuclei. The nucleus of comet P/Encke was measured by radar techniques as lying between 0.4 and 4 kilometers—a wide range, but this was an astonishing achievement, made with the 1,000-foot diameter dish at Arecibo in Puerto Rico. It was astonishing in several respects, not in the least because the radar reflections were from an object 50 million kilometers away. In 1983, comet IRAS-Araki-Alcock approached the Earth to a tenth of this distance. Radar observations were not enough to provide an image of the surface, but they did prove it to be irregular, and they even proved the presence of small objects—a few meters in size at most—moving independently of the nucleus. The flying rubble was a concession to the sandbank theorists, but the sandbanks were not the nucleus. The snowballers had long been confident that they were right, and there was no longer any room for doubt.

NASA was not the only player in the satellite game at this period. The *International Ultraviolet Explorer* (IUE) was a satellite launched in 1978 by a consortium of the ESA, the Scientific and Engineering Research Council of the United Kingdom, and the NASA Goddard Space Flight Center. IUE was, before the launch of the Hubble Space Telescope in April 1990, one of only two astronomical telescopes working in orbit. Of the many discoveries made with its help, the detection of sulfur in the nucleus of a comet was important. Others of still more telling value were the result of the detection of a massive halo of hot gas surrounding our Galaxy and the continuous monitoring of the supremely important supernova 1987A. Another telescope in orbit was that in the Soviet spacecraft *Soyuz*13, launched in 1983 with a two-man crew. As in all such devices, reference stars were used to stabilize the telescope.

The year 1983 was an especially important year in cometary studies, for it also saw NASA's launch of a satellite with infrared sensors. This *Infrared Astronomical Satellite* (IRAS) was sent out to search for asteroids, or indeed any other emitters of infrared (heat) radiation. It astonished even its designers, by the fact that it turned up six or more comets in its short life of eight months. Even then, by the rules of the game, five independent amateur discoverers of these IRAS comets had their names used in the final designation. Thus IRAS-Araki-Alcock was one of the new bunch.

It happened to be very fast-moving, and came closer to Earth than any comet since 1770 (Lexell's comet). Supplementing the IRAS findings, astronomers using powerful telescopes to survey interstellar clouds or other phenomena, began to turn up photographic plates with characteristically blurred images that once again turned out to be comets, but in these cases extremely remote.

Comet Halley has been dear to astronomers, ever since the periodicity Halley predicted for it was confirmed. A rendezvous with that comet, around the time of its nearest approach to the Earth in 1985–1986, was the aim of no fewer than five satellite launches from Europe, the Soviet Union, and Japan. After the tragic explosion of the NASA shuttle *Challenger* on 28 January 1986 and the death of its seven crew members, came the cancellation of the NASA Halley mission, planned for March in the same year. The ESA's probe *Giotto*, however, which had been launched in July 1985, passed within 600 kilometers of the comet on 14 March 1986 and sent back images of the comet's nucleus, showing it to be an irregular solid object with two dust jets issuing from its surface. *Giotto* was damaged in its encounter but passed on to meet up with a second comet. (The name *Giotto* was chosen in view of the fact that the great thirteenth-century Italian artist Giotto di Bondone painted a comet, in lieu of the "star of Bethlehem," in his fresco *The Adoration of the Magi*, at Padua.) The first images of the nucleus had been taken a few days earlier, however, the first from the Soviet probe *Vega* 1 (6 March) and the second from *Vega* 2 (9 March). The latter gave a very clear image, superior to *Giotto*'s but matching it closely. One very surprising discovery by *Vega* 2 was that the comet's nucleus had an extremely low albedo (reflectance): it reflected even less light than a lump of charcoal would have done. The Moon's albedo is more than three times as great, and the only similar items of any size in the solar system then known were certain types of asteroid and Phobos, a satellite of Mars. Both *Giotto* and *Vega* were able to analyze the chemical composition of cometary dust, taking advantage of the high speed (70 kilometers per second) at which particles were encountered. Impacting on a metal target of silver or platinum, the dust was completely vaporized. The molecules having decomposed, its atomic constituents were then analyzed, using a mass-spectrometer. (There are various forms of this instrument, but all work by separating ions on the basis of their charge-to-mass ratio, so allowing their masses and chemical properties to be analyzed.) One of the more surprising discoveries made as a result of these surveys was that light elements, such as hydrogen, carbon, nitrogen, and oxygen, were predominant. This might not seem surprising, in view of previous talk of the presence of water (H_2O), carbon dioxide (CO_2), and hydrogen cyanide (NCN), but it was generally agreed that these compounds should have evaporated by the time they reached the probe. The only viable alternative was that they came from organic grains of low volatility, and that they had been *interstellar grains*, gathered into the cometary nucleus in some way.

243. The view from the *Deep Impact* spacecraft as it turned back to look at comet Tempel 1, fifty minutes after the spacecraft's probe had been run over by the comet. That collision kicked up the plumes of ejected material seen here.

A more recent rendezvous with a comet was of a more violent kind, although it revealed unprecedented visual and chemical detail. Comet 9P/Tempel 1 had been discovered on 3 April 1867 by Ernst Wilhelm Leberecht Tempel of Marseille. It was faint—then of the ninth magnitude—and in the course of its later history it presented those who calculate cometary orbits with many problems, for a time even seeming to be lost to those working from the data provided by the calculators. When its path was once again well charted, it was recognized that, as a near visitor, it would repay a visit; and in January 2005, for the NASA Discovery mission a craft named *Deep Impact* was launched with a view to reversing the traditional role of snowballs and actually bombarding it. Before passing within 500 kilometers of the comet in the following July, the spacecraft released a probe ("impactor") which eventually crashed into it, throwing up a plume of icy debris. (For a remote view of the impact as seen from the mothership, see figure 243.) In the moments before impact, the probe itself relayed some remarkable images of the 14-kilometer-wide mass of ice, dust, and rock that made up the comet's nucleus. Large circular craters could be seen on the comet's surface, removing all possible doubt that the dirty snowball was a reality. The last image was sent just three seconds before impact.

After millennia of marveling at comets, astronomers were becoming increasingly aware of their importance. Deep-frozen from times before the formation of the solar system, they are time-capsules, that allow us— almost literally—to look at that earlier situation. A century ago, no one could have imagined how many other uses would be found for them. They help with the study of the solar wind and the Earth's magnetosphere; they help with an understanding of interstellar matter, plasma physics, quasars, black holes, radio stars; and indirectly, therefore, they assist cosmology in its very widest sense. They are certainly not "airy nothings," to use a phrase not uncommon a century ago.

COMETS AND LIFE ON EARTH

The 1986 studies of comet Halley stimulated much interest in cometary studies, and made further exploration seem highly desirable. Interest in comets around this time was enhanced by the renewal of a debate on the possibility that, being rich in carbon compounds, they were responsible for bringing these down to the Earth, and so initiating the earliest forms of life. During most of human history, life has been generally—but not universally—regarded as created by a deity. Aristotle stood in a different tradition, arguing for spontaneous generation, but in 1864, experiments by Louis Pasteur put an end, for a time, to all serious pursuits along such lines. With the advent of Darwinism in particular—*The Origin of Species* was published in 1859—scientists began to search further afield, although seldom into deep space. Two physicists of note who considered space a possible source of life forms were William Thomson (Lord Kelvin) and Hermann von Helmholtz. Following in their footsteps in 1908, the Swedish physical chemist and Nobel prize-winner Svante Arrhenius published a book, *Worlds in the Making*, in which he postulated the existence of living spores in space, spores which were spread around by light pressure, and which had initiated life on Earth. Many wild theories grew out of this and other vaguely similar suggestions, branching off into directions that need not be pursued here.

The idea that microorganisms or simpler precursors of life are present in space, able to initiate life on Earth or other favorably situated planets, gradually disappeared from serious scientific view, but the old idea of spontaneous generation made a surprising comeback in the 1920s, albeit in new versions. The Russian biochemist Alexander Oparin and English geneticist J. B. S. Haldane, writing independently, decided that there is a spontaneous generation of life, but that it does not necessarily occur on Earth. This general idea received little support until 1953, when the American chemists Harold Urey and Stanley Miller showed experimentally that simple molecules such as hydrogen, methane, water, and ammonia, could combine into amino acids under certain conditions—such as obtain during electrical discharge, a common enough phenomenon in space.

The Oparin-Haldane theory, with many modifications, has survived well, but there was a serious rival contribution to the whole subject in the late 1970s which cannot be ignored, in view of the high reputations of its two proposers, Fred Hoyle and Nalin Chandra Wickramasinghe. Both were then at Cambridge, but Wickramasinghe, originally from Sri Lanka, later moved to Cardiff University, Wales, where in due course he established a center for astrobiology. The two argued that there was evidence for traces of life in interplanetary dust, and they proposed that life on Earth had its origins in material carried by comets. Accepting the dirty snowball theory, and the proven diversity of materials in comets and their tails, they considered comets to be an excellent nursery for life forms, possibly generated in pools of water kept at suitable temperatures by radioactivity. (Whipple had proposed the existence of such pools, more than a decade before.) Comets, in short, were supposed to carry bacterial life across galaxies, and to—no less important—to protect it from radiation damage along the way.

These new ideas were not well received, in particular when the two men went on to suggest that the process has not ended, and that viruses, even that for influenza, are visited on us by comets. (Does influenza not require a host animal?) The general idea, however, was not without supporters. Amino acids were a recognized route to life forms, but few astronomers acknowledged that there could be suitable conditions for their formation *inside* cometary nuclei. The consensus among most astronomers was that the newly formed Earth was a far more likely place for their formation. Hoyle died in 2001, but Wickramasinghe has nevertheless continued to fight for his and Hoyle's ideas with much skill. He has been encouraged by numerous pieces of evidence—the proof that bacteria can survive unchanged for many millions of years, even under bombardment with cosmic radiation much stronger than ours on Earth. (In 1984, Mayo Greenberg of Leiden demonstrated, to the satisfaction of many, that they would not survive ultraviolet radiation or other hazards in space for more than a thousand years or so.) Various claims were made in the late 1990s, and subsequently, to have found fossilized evidence of ancient life in meteorites. Very large organic molecules were being detected in space. In 2004, a NASA source issued photographs of fossilized cyanobacteria in a meteorite. Cyanobacteria live in water and are one of the largest groups of bacteria here on Earth. They usually have a single cell, although they may form colonies large enough to be visible to the naked eye. Their fossils are the oldest known, exceeding 3.5 billion years, a substantial fraction of the age of the Earth.

Without the outreach of space probes, most of these findings would not have been possible. Dust, with the advent of the new millennium, had gradually become big business. That business reached an important juncture with the project named Stardust. This was the first U.S. space mission dedicated solely to the exploration of a comet, and the first robotic mission designed to return extraterrestrial material from anywhere outside the orbit of the Moon. Before saying more about this project, it must be made clear that Earth satellites and radio observation did not exhaust the techniques introduced into astronomy during the rich period of cometary research with which he twentieth century ended. High-flying aircraft and stratospheric balloons were also used to collect dust. Among those who made intensive studies of the material collected was Donald E. Brownlee of the University of Washington, in Seattle. In a program guided by him, particles small enough to decelerate without melting through atmospheric friction were collected from the stratosphere and studied in the laboratory. These "Brownlee particles" are usually fluffy aggregates of minerals and organic matter—and it is now generally accepted that they are the residuals of cometary dust, void of volatile matter. They were, for a couple of decades, useful evidence for a variety of competing theories of the chemical composition of comets. It was generally agreed that they were probably frozen at low temperatures from a mixture of elements comparable with those present in the Sun. A great deal of effort was devoted to this problem in laboratories. Mayo Greenberg devised experiments to show how interstellar dust grains could grow, and he used his findings to interpret many other phenomena than cometary dust—the gas in interstellar dust clouds, for example, novae and supernovae, and the irradiation of dust grains in space generally. Remote stars seen through dust clouds appear redder than they would otherwise. Increasingly, pressure was brought to bear on those controlling funding for space missions to launch either a mission to return cometary samples, or at least one that would allow for an analysis of cometary samples in the spacecraft itself. With the Stardust mission, the idea became a reality.

The *Stardust* spacecraft was launched on 7 February 1999, from Cape Canaveral, Florida, aboard a Delta II rocket. It was the fourth mission in the NASA Discovery program that had been approved by Congress as early as 1992, and it owed much to the planning of Peter Tsou, of the Jet Propulsion Laboratory of the California Institute of Technology. Other missions in the Discovery program included Mars Pathfinder, the Near Earth Asteroid Rendezvous (NEAR) mission, and the Lunar Prospector mission. Like them, it entailed close collaboration with American universities and industry, costs being kept to a minimum. Contributions were also made by Germany's Max Planck Institute and Britain's Open University.

The primary goal of Stardust was to collect dust and carbon-based samples during a close encounter with comet Wild 2 (235 kilometers ahead of the nucleus), as well as interstellar dust. The aim of the latter task, while perhaps less spectacular than the former, was to collect materials surviving from before the formation of the solar system, as well as dust streaming into it from the direction of Sagittarius. While in orbit, counts of comet particles were made, as were real-time analyses of their composition. Samples were captured without damaging them, the catcher resembling a tennis racket, covered with a silica-based substance called aerogel. This remarkable material, developed by the Jet Propulsion Laboratory, combined properties of airy lightness with surprising strength, and an ability to arrest the cometary particles—moving at a relative speed six times that of a rifle bullet—without letting them pass through it completely. On 15 January 2006, *Stardust* returned its cargo by parachuting a reentry capsule into the Utah desert. Under the direction of NASA's principal investigator for this project, Donald Brownlee, an initial analysis of some of the hundreds of grains captured suggested a high concentration of complex molecules of the kind thought necessary for the evolution of life. There was talk at one stage of allowing anyone who found a grain to give it a name. Whether this will require the International Astronomical Union to establish a dust-naming commission remains to be seen. What also remains to be seen, or rather, debated, is how the molecules actually form inside comets, and whether they indeed delivered organic material to Earth, before life here began.

TILTING AT WINDMILLS IN SPACE

There are two headline-grabbing subjects that stem from cometary analysis. In the more sedate branches of journalism, "Aliens Have Landed" might refer to nothing more than a few cometary particles, or even to particles that are said to have had a cometary origin, such as those found after blood-red rain fell over the Indian district of Kerala in July and August 2001. (Godfrey Louis and A. Santhosh Kumar, of Mahatma Gandhi University in Kottayam, India, put forward the cometary hypothesis, but an Indian government commission ascribed the color to spores from aerial algae.) The other eye-catching cometary theme goes under such titles as "Doom of the Dinosaurs," and while it is occasionally been mixed up with the sort of unsupportable notions put forward by Immanuel Velikovsky in 1950 in his *Worlds in Collision*, and in later popular books, it has much well-respected science behind it.

In 1980, Luis and Walter Alvarez (father and son), Frank Asaro, and Helen Michel, of the University of California at Berkeley, analyzed strata of deep-sea limestones in Denmark, Italy, and New Zealand. To their surprise, they found a stratum with relatively high levels of iridium, rare on Earth, but relatively less so in meteorites. They put forward the idea that at the end of the Cretaceous period, 65 million years ago, an asteroid or comet crashed into the Earth, throwing out so much dust that sunlight around the world was blocked for months, or even years. Temperatures would have plummeted, causing much vegetation and many life forms to perish, including the dinosaurs. The headline was made, but it was not to be the last, for evidence was found in 2006 for a far more massive crater under the Antarctic ice. Using gravity measurements made from satellites, and radar images from aircraft, a group from the United States located a crater 480 kilometers wide, at a depth of two kilometers under the East Antarctic icesheet in Wilkes Land. This was said to have been responsible for the biggest mass extinction in world history, 250 million years ago, the event that ended the Permian age and heralded the Jurassic era, the age of the dinosaur.

The 1980 discovery produced deep divisions in several scientific communities, but opposition dwindled when the iridium anomaly was found at site after site, around the world. In 1985, a group of chemists at the University of Chicago found evidence for a global firestorm that raged at about the time of the dinosaurs' disappearance. Deposits of carbon from this period were 10,000 times as great as in the layers above and below, and some rough calculations hinted at an impact of the order of 100 million megatons, in what became the Bering Sea, sending out a fireball that incinerated all plant and animal life, far into North America and Asia. Details apart, the basic idea soon became so well accepted that paleontologists began to assume that the same mechanism—cometary or asteroid impact—was responsible for *all* of the major extinctions they had previously identified. (Some, it is true, are more enamored of the idea of gamma-ray bursts from outer space, depleting the ozone layer for a long period, leaving only deep-sea creatures relatively immune.)

That of 1980 was not the first astronomical explanation to be offered. In 1970, Craig Hatfield and Mark Camp of the University of Toledo, Ohio, had suggested that the great extinctions were periodic, and were connected with the way in which the Sun's path (the ecliptic) oscillates, being sometimes above and sometimes below the plane of the Milky Way. The period of that oscillation is thought to be about 33.5 million years, and the extinctions are estimated by different authorities to occur at intervals of between 30 and 36 million years. The original idea was that molecular clouds are more concentrated near the plane of the Milky Way, that is, where the danger from them is greatest. Soon afterwards, many different hypotheses were put forward as to mechanisms whereby the interstellar clouds could cause extinctions. Victor Clube and Bill Napier, both then of the Royal Observatory in Edinburgh, devoted much time to this problem, and in 1979 and later publications worked cometary theory into the story, and in the process launched an important attack on conventional wisdom concerning the status of the Öpik-Oort cloud of comets. They considered it necessary to revise the idea that the cloud is primordial and permanent, as usually accepted, arguing that the number of "near" approaches by massive nebulae in its long history of billions of years would have depleted it long ago. They objected to the idea that the present comets had to grow along with planets in the dense regions round the early Sun, and proposed instead that a star and its comets are formed at the same time, each star carrying off an attendant cloud of comets, but that these are thrown into interstellar space by planets or passing nebulae, until an equilibrium is reached when sporadic captures and losses are equal. This, they thought, is our present situation. One consequence of their theory was that interstellar comets are commonplace, and that comets originally coming from other stars, or from their birthplaces in cold, dense, interstellar clouds, may be captured by our solar system. Capture, they said, is especially likely when the solar system passes through a spiral arm of our Galaxy. An interesting subsidiary conjecture, made in the context of this wide-ranging study, was that the surprisingly large number of short-period comets in the solar system—objects not only with short periods but short lifetimes of a few thousand years at most—may be due to a single large comet fragmenting during capture by Jupiter.

Having presented a plausible general cometary mechanism, they put it to work, showing first that asteroid and comet bombardment of Earth is much more frequent than was generally recognized. To such bombardment they ascribed catastrophic mass extinctions, but also such geological phenomena as sea-level changes, the ice ages, and mountain building in the course of "plate tectonic episodes." In support, they brought forward evidence in the form of the dates of large impact craters from the last 350 million years. In view of the relatively small number involved, such a rough statistical approach is clearly never likely to be much improved upon, except in the finer details of dating. It leaves open the question of special events, and the evidence for them. One that deserves mention was proposed by Marc Davis and Richard Muller of the University of California and Piet Hut of Princeton. They supposed that a companion star to our Sun—they called the star Nemesis—with a period of 26 million years, disturbed the comets in the Öpik-Oort cloud, and sent a billion or so of them into the solar system at intervals of about a million years. Of these, very many would have struck the Earth, causing the extinctions and general mayhem.

The number of variants of such hypotheses as these grows at a steady pace, and controversy here is as lively here as in any other branch of astronomy. It is helped along not only by the paleontologists, but by some types of religious creationism. (Should we discount the role of Scientology, and its account of the warlord Xenu, who was said to have moved 13.5 trillion beings to Earth from 76 overpopulated planets? Duly vaporized by Xenu, their radioactive souls were said to have been infected by organized religion and rational science before attaching themselves to receptive human beings.) It is hardly surprising that the last part of the twentieth century was marked by the emergence of a large number of serious academic ventures into astrobiology. By the turn of the century, at a rough count, the number of projects from around the world—if we may lump together university departments, government-funded projects, societies, and journals—was of the order of a hundred. In the wilder territory of film, television, and popular science writing, the theme touches a sensitive nerve. The cosmos is all very well, but it is all the more interesting if it can be brought to life. For years, French "comet wines" of 1811 were sold on the strength of a belief that the comet of that year was responsible for an exceptional vintage. Now we can choose between the idea that molecular life once landed, and the alternative, that of molecular disease. After the Plague that devastated London in 1665, the comets of 1664 and 1665 were blamed; and throughout history there have been claims that pandemics have a cometary source. But this is small fry, compared with the possibility of an extinction of life by some future impact.

There is much capital to be had from the perceived danger that a comet may collide with the Earth at any time. Of the many instances of panic induced by the public's misunderstanding of an astronomer's utterances, there is none more absurd than that which took hold of Paris around 20 May 1773, after a lecture on the subject a month earlier by Lalande. The occasion was satirized by Voltaire, in his "Letter on the alleged comet." In the last few decades, much attention has been paid to the statistical likelihood of such an event. In 1984, Donald Yeomans and Zdenek Sekanina analyzed the orbits of comets that can be shown to have passed within 2,500 Earth radii of us, and decided that "a fair-sized, active comet" could be expected to strike the Earth once every 33 to 64 million years. Such statistics are cold comfort, even if the chance of a collision is so very small, and they do not prove that a comet will not land here tomorrow. In 1989, the Apollo asteroid 4581 Asclepius passed the Earth at a distance less than that of the Moon, and was only discovered by the Spacewatch survey *after* it had made its closest approach to Earth. In 1994, a comet striking Jupiter created a hole in that planet the size of the Earth. What if we find such a thing heading our way? Should we simply nuke any approaching object, and cross our fingers in the hope that the fragments will not come our way? Or should we try to divert it, with the help of rocketry? And who are the "we" who decide, and the "we" who implement the plan? In 2006, a team of scientists at what had been a British government defense research agency began work on a "space ram," a high-velocity rocket that it was hoped would be capable of deflecting any asteroid or comet that might seem to threaten us. (The project came under the aegis of the ESA.) The mission was named Don Quijote and the craft was named *Hidalgo*— after Cervantes' impoverished Hidalgo, famous for tilting at windmills. A test launch has been planned for the year 2011. In the interval, matters are being left in the hands of a higher power.

THE HUBBLE TELESCOPE

The numerous early satellite observatories were short-lived—with lives of months more often than years. The first plans for a versatile optical observatory that could be kept working in space for many years were being laid at the time of building the OAO spacecraft in the early 1970s. The key figure in its conception was Riccardo Giacconi, director of the research team working toward it, initially within the private company American Science and Engineering. Giacconi was made director of an independent Space Telescope Science Institute, which was given responsibility for coordinating its various tasks, and limited proprietary rights to the data obtained. Many other organizations eventually became involved, and the boundaries of responsibility shifted as the project continued. The institute was operated by the Association of Universities for Research in Astronomy, Inc., on behalf of NASA. The venture is somewhat reminiscent of Tycho Brahe's advancement of learning through the astute combination of technology and capitalism, with the support of an enlightened state. Occasional government irritation is a part of the formula.

The space telescope, eventually named after Hubble, was to have been put into orbit in 1985 by NASA's Space Transportation System—the "space shuttle." The first serious delay in the launch followed the temporary suspension of the whole space shuttle program, in the wake of the *Challenger* catastrophe of 28 January 1986. The telescope project was plagued with misfortune, even before the instrument left the ground. One of its most important parts was a charge-coupled device (CCD) that converts light to electrical signals that are then reconstituted as a picture. Such CCDs are a now familiar part of digital cameras, but were then in their infancy. By chance, it was discovered that the CCD chip's sensitivity to blue light depended on the area on which the light fell, and the extent to which the chip had been exposed previously. At least this problem was discovered in time, but unfortunately it was only one of several.

The advantage of the shuttle system is that a payload can be put into orbit and can be repaired and updated to a certain extent for many years. An early plan was to visit the Hubble telescope at three-year intervals over fifteen years. The instruments are carried within a tube roughly 4 meters in diameter and 13 meters long, and the basic telescope is a Cassegrain of modified design, of 2.4 meters aperture (fig. 244). This would have been a sizable instrument even for a ground-based observatory, but above the atmosphere it is, in principle, far superior even to a ground-based 4-meter telescope. When calls for proposed scientific programs to be carried out on the satellite were made, the project was oversubscribed six times. A selection was made, and the Hubble Space Telescope was provided with its secondary instrumentation. Many of the component parts were so arranged that they could be moved into the focal plane of the main instrument when needed, there detecting wavelengths from the ultraviolet to the infrared. The original configuration included three instruments used for cometary observations in the ultraviolet. To give a rough idea of the complexity of operating the whole system, we note that one of these three instruments was replaced on a service mission in December 1993, while the other two were removed on the second service mission, in February 1997. These facts hint at the high usage of the Hubble platform, requiring a system of advance booking—a problem for cometary observation, in view of the fact that comets have a way of arriving unannounced.

The 2.4-meter primary mirror was to be the most precise large optical surface ever made, so it was all the more tragic that, only after launch, the mirror was found to suffer from certain distortions that might have been discovered earlier had the testing apparatus itself not been faulty. Other faults developed, in the gyroscopes and in the solar panels, which were not rigid enough. Despite these painful problems, the telescope from the outset began to return much useful information that was not available from any other source. Communication to and from the telescope was through a network of communications satellites, stationary above the Earth. Data transmitted through them was passed to the Goddard Space Flight Center, and so to the Space Telescope Science Institute, from which commands could be returned via the same links. The data were analyzed by resident astronomers at the latter Institute, and distributed if necessary to others elsewhere. The European Coordinating Facility assisted in various ways, providing instruments and astronomers.

Progress was much slower than anticipated, and when the telescope launch finally took place in April 1990, the object launched—even disregarding the cost of launching it—had the dubious distinction of being the most expensive scientific instrument in human history. Many lessons had been learned, not all of them about astronomy. The slow progress, and

244. An artist's impression of the Hubble Space Telescope in orbit, but with the interior exposed to show some of its contents.

overruns in the cost of the project, had led to delays in other space-science programs, such as the Advanced X-ray Astronomy Facility (AXAF), and some harsh words were spoken. On the other hand, the eventual repair of the Hubble telescope by a team transported to it in the space shuttle *Endeavor* offered such dramatic publicity that if anything it did more than redeem the original enterprise. (That mission lasted from 2 to 13 December 1993.) Giacconi is reported as having commented that seven space telescopes could have been launched for the total cost of the Hubble and its repair. When at last the telescope began to function as originally planned, however, it opened a completely new era in astronomical observation. The quality of the images it produced was unparalleled, and the data it yielded began immediately to affect astrophysics in very many fundamental respects.

By the time of the Hubble launch, the HIPPARCOS satellite of the ESA was beginning a successful astrometric program, the first satellite to be aimed at measuring star positions very precisely. Here too there were problems: two of its gyroscopes failed, and the apogee booster motor failed to put it into the desired "geostationary" orbit, so that it crossed the Van Allen belts four times a day, making it unusable for a third of the time. While its name pays tribute to an activity pioneered by Hipparchus more than two millennia previous, the acronym had another interpretation: <u>hi</u>gh <u>p</u>recision <u>pa</u>rallax <u>co</u>llecting <u>s</u>atellite. Whereas air turbulence means that observations with the very best ground-based telescopes can be accurate to no more than a tenth of a second of arc, the precision of HIPPARCOS was one or two milliseconds of arc—say the angle subtended by a centimeter at a distance of one or two thousand kilometers. Parallaxes and proper motions, and all that depend on them—such as the scale of the universe, no less—have been obtained with unprecedented accuracy as a result. In addition, the satellite has measured star brightnesses to a very high accuracy, allowing for improvements in the light-curves of many known varieties of variable star, not to mention the discovery of new varieties.

Such achievements as these come at a price. The total cost of the Hubble Space Telescope, before running costs were taken into account, approached one percent of the U.S. annual budget for national defense. In

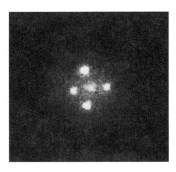

245. Photograph of a very distant quasar, which is more or less in line with a relatively nearby galaxy. The latter, by acting as a gravitational lens (refractor of light rays) has produced four images of the quasar. The upper and lower images are 1.6 arc seconds apart. The image was taken from the Hubble Space Telescope by the European Space Agency's extremely sensitive "Faint Object Camera." Among its successes were the first ever photograph of a star's surface (Betelgeuse), as well as images of stars in globular clusters and the Pluto-Charon system.

cosmic terms, that is a mere nothing; and it is a moot point as to which was the better investment. By the year 2006, however, there was a widespread feeling that Hubble was dying, so that out of NASA's total budget of $16.5 billion, only $93 million was allocated to it. Should it be left to die in orbit, or be brought down to Earth safely? Many argued that it should be serviced in space, but a mission to service Hubble using human, rather than robotic, methods, ran foul of the recommendations of the *Columbia* accident investigation board. The only certainty was that the final decision would not please everyone.

PROPULSION METHODS AND HUMAN EXPLORATION

Astronomers—at least since the days of Copernicus—have not been great economists, although they have often displayed a taste for scientific restraint and simplicity. It was only after much delay, however, that they made use of the ion drive, a particularly elegant piece of technology that can power spacecraft at very low cost. Ion drives use electrical power, first to create charged ions, and then a magnetic field by which they can be accelerated and expelled from the rear of the spacecraft, so propelling it like any other rocket. The rate at which mass is expelled is small, and therefore so is the reactive force that moves the craft. (It is roughly equivalent to the weight of one or two small coins in one's hand.) In the virtually frictionless environment of space, however, high velocities can be built up slowly over time and then easily sustained. The ion drive appears to have first been described in 1947, in a short story entitled "The Equalizer," written by Jack Williamson for the magazine *Astounding Science Fiction*. George Lucas picked up the idea for his *Star Wars* of 1977: the small maneuverable spacecraft of the evil Imperial Starfleet were known as TIE Fighters, the acronym indicating that they were powered by Twin Ion Engines. A similar idea was used by American astronaut Edward Gibson, in his novel *Reach*, in 1989. Of these fictional examples, only Williamson's anticipated reality, for the basic principles of the ion drive were first tested in 1959, at NASA's Glen Research Center, and a working drive was built there in 1970.

This beautifully simple idea was occasionally used on a small scale from the early 1980s for gently correcting the orbits of communications satellites, but the first spacecraft with an ion drive was *Deep Space* 1, an experimental vehicle launched by NASA in 1998, half a century after Williamson's story. It made flybys of asteroid Braille and Comet Borrelly in the following year. This was followed in 2003 by *SMART*-1, an ESA craft that was sent into orbit around the Moon. The ionized gas was in this case xenon, and there was enough on board for 7,000 hours of use. Solar panels produced more than a kilowatt of power to ionize and propel it. Launched on an Ariane-5 rocket from French Guiana, *SMART*-1 was first put into orbit round the Earth, spiraling slowly outward under its own minuscule power, until eventually breaking free from the Earth's gravity. After thirteen months, and spiraling through a journey of 80 million kilometers, it reached the Moon, there to enter a lunar orbit over the poles, and survey the lunar surface. It finally impacted on the surface in a controlled way in September 2006. *Apollo* 11, on its lunar landing mission, had taken only four days to reach the Moon, following a fairly direct track—but a track that was only one two-hundredth of the distance. The most telling statistic, however, was one of cost: putting what strongly resembled a large washing machine on the Moon cost just $85 million, the price of a couple of fighter planes.

There is an element of theater in astronomy, for which astronomers should count themselves fortunate, since it so has often loosened the purse strings of state. Landing men on the Moon was an instance of this. When NASA, in 2004 and after, affirmed that it was guided by President George W. Bush's vision for space exploration, American astronomers once again began to cut their cloth accordingly. Soon there was a detectable shift in emphasis toward human exploration, with the possibility of returning astronauts to the Moon, and perhaps then moving on to Mars. The United States set up a program with the title Vision for Space Exploration, looking into the possibilities for robotic and human exploration of the solar system. The hope was that this would include long-term stays on the lunar surface, robotic missions to search for life on Jupiter's moons, exploration of the outer reaches of the solar system, and human expeditions to Mars.

The ion drive is not suitable for spacecraft with human passengers, where time is of the essence. One other disadvantage is that it depends on reasonable proximity to the Sun. The future is presumably with nuclear electric power, heat from a reactor being used to produce electricity by direct thermoelectric or thermionic conversion, using solid-state devices. NASA has entered into partnership with the Department of Energy's Office of Naval Reactors, with a view to developing and supporting a suitable "space reactor." The creation of a Prometheus Nuclear Systems & Technology program in this connection attracted much interest among astronomers, and a multi-billion dollar project going under the title Jupiter Icy Moons Orbiter (JIMO) mission is already sheltering under its wing. The original idea was to go into orbit around the giant planet and its moons, perhaps putting landers on their surfaces in the way that the *Cassini* spacecraft did with

Huygens on Titan. The intention was to launch this mission in or around the year 2015, if only as a demonstration of what was possible. But those who pay the piper still call the tune, and even NASA's purse is not bottomless. In 2006, the JIMO mission was declared to be too ambitious an undertaking for the purposes of an initial demonstration. Only history will reveal the outcome.

✳ 20 ✳

Macrocosm and Microcosm

Throughout the long history of theorizing about the universe, there have always been important principles at stake that were far removed from the province of the observatory. There have always been considerations of simplicity, harmony, and aesthetics, often masquerading under the name of philosophy, and often dictated by strongly held religious beliefs. The steady-state theories, for example, were often charged with having taken away all sense of purpose from the universe and having put it into a monotonous, endless, and pointless state. While such notions are not usually visible in the mathematical equations of rival theories, hidden preferences that helped to shape them are often revealed when their authors enter an informal discussion. When someone insists that a universe beginning in a primordial fireball and dwindling into nothingness is intrinsically even less attractive than the steady-state idea, it becomes plain that we cannot discount the place of the human psyche in modern cosmology.

It is also easy to forget how deep are the potential connections between cosmology and the other sciences. Take theories of stellar evolution: no one will deny that the component parts of the universe are in a process of change. Fred Hoyle, in a retrospective view offered in 1988, went so far as to say that to deny evolution, in the sense given to the term by the Big Bang cosmologists, "is no loss at all, for the issue of whether galaxies have grown a little brighter or a little fainter, or a little larger or a little smaller, over the past [ten billion] years is singularly uninteresting." "The only evolutionary processes of real subtlety in astronomy," he went on, "are those which relate to stars, and strikingly enough stellar evolution proceeds essentially without reference to cosmology." To think about evolution cosmologically, he thought, one should look at problems of a "superastronomical order of complexity," such as the problem of the origin of the biological order; and to have any hope of solving such a problem as this, a particular type of cosmology—the steady-state theory, as he believed—may be required. In short, no cosmologist can afford to be a cosmologist and nothing else.

The fact that all cosmology, all astronomy, should ideally be made to connect with other acceptable sciences—and that their histories must therefore also be linked—is not a new idea, but in an increasingly complex intellectual world the principle of the division of labor tends always to force them apart. Those in the past who have tried to draw them together have often been undisciplined fantasists, although a few have had the highest formal credentials: Aristotle, Kepler, Newton, Einstein, and Eddington together make up a peerless quintet. Generality has always been acknowledged as one of the highest of scientific virtues. One of the most remarkable aspects of astronomy in the twentieth century was the high degree to which it was able to ally itself with an ever-increasing number of branches of science. Biology apart, we have seen how theories of gravitation, optics, electromagnetism, and spectroscopy became allied, and how the last-named introduced considerations of physics at the subatomic scale. From the 1930s, thermodynamics played an increasingly important part in cosmology, and the new theories of stellar structure and the transformation of chemical elements would have been impossible without the new quantum physics. We have mentioned too some attempts to link the physics of the very large and the very small through the fundamental physical constants. From the 1960s onward, there have been a number of exciting ideas that bind together quantum physics and relativistic (gravitational) cosmology and thermodynamics in entirely new ways. At the heart of this new movement was the notion of a black hole, which bound these subjects together, simply by being in need of them all.

SPHERICAL MASSES AND BLACK HOLES

We have already tentatively introduced the notion of a black hole, and have given some examples of the ways in which it insinuated itself into the affections of astronomers (recall that the phrase itself was coined only in 1968). It should hardly be necessary to point out that, while one may use ordinary words to describe the strange properties of black holes, the essential geometrical ideas can only easily be conveyed by using analogies and that these are potentially misleading. It is necessary to have a rough idea of what is meant by a *singularity*, in the sense in which the word is used in geometry. Suppose one quantity depends mathematically on another—for instance, y may be equal to x^2, or z may be equal to $1/x$. Producing a simple graph of the relationship in the first case will give rise to no great problems, but in the second we shall find that when x is zero the function z behaves badly. In this case we may loosely say that it "becomes infinite" when x is zero. In other cases of "bad behavior," we may find that a relationship we wish to represent in a graph is discontinuous, or that we cannot specify its direction unambiguously. The values of x where the trouble occurs are the "singularities" for the function in question.

We came across one sort of singularity in certain relativistic models of the expanding universe, models that—when naively interpreted—seemed to point back to a universe of zero dimensions. Another case was in Karl Schwarzschild's treatment of the gravitational field surrounding a spherical mass. We discussed this briefly in connection with the Sun, in chapter 16. (The Schwarzschild singularity is at a distance from the center equal to twice the solar mass, if we choose suitable relativistic units.) Schwarzschild's calculations were made half a century before the discovery of neutron stars, the first objects which allowed these characteristically Einsteinian ideas to be tested.

It was in the context of Karl Schwarzschild's account that Eddington, in 1924, found a way of avoiding that singularity problem—although he did not place any emphasis on what he had done. The device was simply to choose a different coordinate system—say, to redefine x, in the simple example given above. For example, if x is redefined as $x+1$, the singularity will be moved elsewhere, where it might be physically less troublesome; but if x is redefined as $1/x$, the singularity will disappear. The second case may seem like cheating, and it certainly looks trivial, but in a physical theory coordinates are not always directly interpreted. For instance, we might wish to interpret x as a length, but it might make sense in a particular theory to treat it as something that indicates a length very indirectly. Coordinates are not sacrosanct. This was well appreciated by those who manipulated the coordinate systems used in De Sitter's cosmological model, and Lemaître in 1933 first stressed this point in regard to the Schwarzschild singularity. He and Eddington had found coordinate systems that took them through the troublesome region. Later, in 1950, J. L. Synge provided an improved coordinate system, and at least four other mathematicians did the same independently over the next ten years or so.

As a result of their work some very strange results were obtained concerning "Schwarzschild geometry." One of the most surprising concerned the case where there was no central "star." Instead of the geometry being that of a point-mass at the center, it turned out to be more like a wormhole joining two universes, a wormhole capable of expanding and recontracting. It is not possible here to go into more detail, except to make the point that work done on the Schwarzschild geometry was making it abundantly clear that space-time geometry has many strange and unexpected properties when gravity becomes very strong, and that it must be taken into account in the physics of stars that collapse under gravity and in the physics of black holes.

Once the Schwarzschild solution was more closely scrutinized with a view to applying it to real objects of interest, it became clear that it was in need of amendment. Most stars are spinning, and so—like the spinning Earth—are not perfectly spherical, as was assumed of the central mass in the Schwarzschild analysis. In 1963, the New Zealand mathematician Roy Kerr generalized Schwarzschild geometry to include rotating stars and rotating black holes. Since most stars are rotating, it is natural to suppose that newly formed black holes will be behaving likewise. Kerr discovered a more general solution of Einstein's equations, representing collapsed rotating objects. He found that—as in the Schwarzschild case—a horizon shields the interior from our view, but that in the new case space has a vortex-like quality. Anything having the misfortune to fall into the hole would be swirled round, as in a whirlpool. Kerr's worlds were found to have still stranger singularities than Schwarzschild's. In the next decade, Kerr and others found that his was not an untypical, idealized, solution, but that it applied to all black holes, however they came about. They found that there are no means of distinguishing them observationally in any other respect than their masses and their spins. They are cut off from our world, leaving only a gravitational footprint in the space they have left behind. It is through this, or rather through its effects on nearby stars and gas, that hopes of detecting them were pinned.

Turning to astronomy now, to remind ourselves of the objects that were considered to be candidates for a relativistic treatment of the sort here sketched. Had we asked an astronomer around the year 1970 for a historical summary of the situation, we should have had information given to us about at least five categories of object, four of which had not been observed in any direct sense at all, but had been put forward on theoretical grounds. Here is what we should have been told:

(1) *White dwarf* stars have a radius of about 5,000 kilometers, and they are of about one solar mass—we recall that Chandrasekhar before 1930 had shown that their masses must be below a certain limiting value not much greater than the Sun's mass. They are of density about one tonne per cubic centimeter. While they are collapsing under gravity, they are supported by an outward pressure of degenerate electrons, following Pauli's Exclusion Principle. (R. H. Fowler was perhaps the first to appreciate this possibility, in 1926.) They have stopped burning their nuclear fuel, and are gradually cooling as they radiate heat. In 1949, S. A. Kaplan had shown that relativity predicts that they are unstable if they are under a certain radius—he said 1,100 kilometers.

(2) *Neutron stars* are of much the same mass but with much smaller radii, say, about 10 kilometers, making their densities enormous—around 100 million tonnes per cubic centimeter, comparable with the density of an atomic nucleus. They are supported against gravitational collapse by two forces, the pressure of neutrons and certain nuclear repulsive forces ("strong-interaction forces"). The energy they radiate is partly heat and partly obtained at the cost of their energy of rotation. They were implicit in the theory of L. D. Landau (1930), as mentioned earlier, and were called upon by Baade and Zwicky (1933–1934) to explain supernovae. The outburst of a supernova was said to be the consequence of a normal star collapsing to a neutron star. In 1939, J. R. Oppenheimer and G. Volkoff used relativity to work out the process in

detail, and in doing so set the stage for later relativistic theories of star structure. A certain star was singled out by Baade and Minkowski in 1942 as the remnant of the Crab supernova. After Gold suggested in 1968 that pulsars are rotating neutron stars, it was shown in the following year by W. J. Cocke, H. J. Disney and D. J. Taylor that the same star is in fact a pulsar, flashing on and off thirty times a second. Nothing but a neutron star, tightly bound by its gravitation, could spin so fast without disintegrating. This seemed to tie up the presumed connection conclusively.

(3) Although the idea of a *black hole*, a region in which there is so much mass that light cannot escape, is one with a long history, it would have been pointed out to us that within general relativity the foundations of the theory of black holes were laid only in 1939. It was then that Oppenheimer and Hartland Snyder showed how, when all thermonuclear sources of energy are exhausted, a sufficiently massive star will collapse. It would have been noted that the collapsing sphere will cut itself off from the rest of the universe, and that there is a "horizon," a "surface of the black hole," that will not reflect light from outside it and through which light from the "collapsed star" cannot pass to the outside. No interaction with the outside world was deemed to be possible for anything inside it. (Notice that this account, which considered the viewpoints of both internal and external observers, differed radically from any account that would have been given in a pre-relativistic theory of the nonescape of light from a massive object.)

(4) *Supermassive stars*, as conceived by Fred Hoyle and William Fowler in 1963, would have been clear candidates for the new relativistic treatment. S. Chandrasekhar and R. P. Feynman had very soon afterwards developed a theory of the pulsations and instability of supernovae. The original idea was that they might be associated with the nuclei of galaxies, and be the energy sources for quasi-stellar objects, quasars, then newly discovered. They would be of hot plasma, and less dense than ordinary stars, supported by the pressure of light (photons) mostly trapped within. Their putative masses were estimated at between a thousand and a billion solar masses. ("Plasma" is a word we have met often before, for a fluid containing a large number of free negative and positive electrical charges, such as our ionosphere or in this case the gases where a fusion reaction takes place.)

(5) *Relativistic star clusters*, clusters so densely packed that Newtonian physics cannot be used to explain their behavior, would also have been candidates. They were analyzed in 1965 by Yakov Boris Zel'dovich and M. A. Podurets. In 1968, J. R. Ipser showed how, when the cluster becomes great enough, it can begin to collapse to form a black hole. It should be noted that the greater the mass packed into a black hole, the less tightly it needs to be packed, that is, the less its minimum density.

So much for the situation around 1970. Now, although our astronomer of that time knew that any star containing more than about three solar masses should theoretically collapse to form a black hole, the number of references to the idea in astronomical literature was relatively small. There are many stars whose masses are known to be in excess of three solar masses. Gravity is in them balanced by the outward pressure from nuclear reactions, and it was widely taken for granted that when the nuclear process stopped there would be an explosion that would reduce the central mass to less than the critical mass—perhaps producing a white dwarf. In short, it was supposed by many that black holes need not be taken too seriously.

Not all were of the same mind. In 1964, Zel'dovich and O. Kh. Guseynov began a search for black holes, looking through catalogs for stars known from spectroscopic evidence to be binaries, but where only one component can be seen. The masses could be estimated, and the two men sought examples where the mass of the invisible companion is more than three times the mass of the Sun. They knew that a black hole in orbit round a star would accrete gas from it, which would swirl inward at an ever-increasing velocity, turbulence in the heated gas causing flicker in the X-rays that would be generated. The X-ray source Cygnus X-1 seemed to be the most likely candidate, for it orbited round another star, as did several neutron stars that were known emitters of X-rays, but it differed from them in vital respects. Its X-radiation was erratic, and its mass, estimated at over six solar masses, was too great for it to be a neutron star of a white dwarf. For several years the evidence remained extremely tentative, and there was much controversy over its interpretation, but slowly attitudes were changing, and the very dispute helped to create an atmosphere favorable to the idea that black holes might exist. Other reasons were about to intrude, and while they came from a highly theoretical direction, they had entirely unexpected consequences as to other ways in which black holes might be detected.

GALACTIC BLACK HOLES

In the 1980s, in parallel with searches for black holes representing the tombstones of stars—these being objects with radii well below a hundred kilometers—an even more exciting possibility began to be investigated. It was realized that the whirlpools of gas and stars at the centers of many galaxies may surround black holes with masses amounting, not to half a dozen, but to many millions of solar masses. It was realized that these may be seen as quasars, and that they might be detectible in at least two ways: they would distort light in their neighborhood, in accordance with general relativity, and they would lead to rapid motions of nearby stars and gas. Various candidates were proposed, for example M 87, in the Virgo Cluster of galaxies, a giant galaxy which was known to have a dark central mass; and even our massive and familiar neighbor M31 in Andromeda was often recommended for study. Radio astronomers using the interferometric arrays introduced in chapter 18 were beginning to map the centers of galaxies more accurately than could be done optically, even by the Hubble Space Telescope, and looking at the nearby galaxy NGC 4258, a disk of gas orbiting its center suggested the possible existence of a black hole of 36 million solar masses. In view of this evidence, which some wished to explain away as caused by quasars, it was asked whether we should not turn to our own Galaxy, where there was certainly no quasar.

In 1988, the physicist Charles Townes of the University of Berkeley, California, and his colleagues, including his postdoctoral student Reinhard Genzel, from Munich, expressed their belief that there is a massive black hole at the center of our Galaxy. Townes and Genzel estimated the mass of the supposed black hole at our galactic core, based on infrared tracking of the motion of ionized hydrogen gas clouds around it, as about 4 million solar masses, later revising the figure to 3 million. Objections were raised in various quarters to the whole idea, on the grounds that the favored object, radio source Sagittarius A*, emitted less than a ten-thousandth of what would be expected from a gas flowing into such a vast black hole. Eventually, early estimates of the expected emissions were revised by Eliot Quataert—also of Berkeley—who showed that gas flowing toward a black hole does not all stream through its horizon and augment the hole, as had usually been assumed, but that much of it would be swirled around and thrown outward in the form of jets.

For more than a decade, the whole idea remained unconvincing to the astronomical profession as a whole, but at last the necessary evidence was forthcoming. In 2002, Genzel and his colleagues at the Max Planck Institute for Extraterrestrial Physics in Munich, reported virtually conclusive new evidence that they had been right. Using the European Southern Observatory's Very Large Telescope (VLT) in Chile, Genzel had spent more than ten years tracking a star (named S2) that orbits our galactic center in a highly eccentric path. Only with the advent of the system of adaptive optics (for which see p. 682 above) had it become possible to single out individual stars with short enough periods to make such a survey possible in a reasonably short time. The star's period was only 15.2 years. The mass deduced for the object around which it is orbiting was found to be the equivalent of some 3 million solar masses, much the same as had been expected. It was from the rapid velocity of S2 as it came close to the neighborhood of the radio source Sagittarius A*, however, that the team were finally able to identify this previously favored candidate as the black hole at the galactic center. The Schwarzschild radius of a black hole of 3 million solar masses is about a tenth of the Earth-Sun distance, while the approach of the star to the object had been as close as roughly three times the distance of Pluto from our Sun.

Since there were known to be some twenty or thirty stars orbiting even closer to the galactic core than S2, the new discovery at once called forth a new research program, using stellar infrared interferometry, not available at the time, to follow their paths more minutely and put general relativity through new tests. There is a substantial difference between black holes of stellar mass and those in galactic nuclei. The former, following from the collapse of stars to enormous densities, make great demands on nuclear physics which are hard to satisfy at those concentrations. Galactic black-hole cores, on the other hand, need be no denser than air. Their effects on whatever falls into them would nevertheless be devastating. Assuming a rotating hole, tidal forces would "spaghettify" the unfortunate object. Several experts have speculated on far more drastic effects, suggesting that the incoming object would erupt into a new space, even a new universe. Black holes lead not only into new worlds, but into new cosmologies.

While it is to be expected that black holes are present at galactic centers in general, although the chances of detecting most are likely to remain slender, that at the center of M31 in Andromeda was at last captured unambiguously on images obtained by NASA's Spitzer Space Telescope in 2005. Independent ground-based studies had claimed this as early as 1988, but the central dark object—seen then far less distinctly than in infrared light from a space platform—was open to several rival explanations, which could at last be eliminated. The 2005 data allowed an estimate of the mass of the black hole, which was said to be 140 million solar masses. This was larger by a factor of three than the most ambitious of previous estimates.

DARK MATTER, MACHOS, AND WIMPS

In chapter 16 we saw how Jan Oort concluded, on the basis of dynamical calculation, that there was much invisible material in the Galaxy and its surroundings. A decade earlier, Jeans had briefly expressed a belief that there may be three dark stars in the Galaxy for every bright star; and Kapteyn had also spoken of dark matter, although he did not think it considerable. We have already seen how, in the mid-1930s, Zwicky and Sinclair Smith had been led to believe that there was a dominant dark component in galaxies. In 1939, the American astronomer Horace W. Babcock completed a doctoral dissertation at the Lick Observatory, in which he correlated rotation speeds with positions in the galaxy M31. He found such a vast difference between his figures for M31's mass and mass-to-luminosity ratio and the figures accepted at the time for our own Galaxy that he was discouraged from publishing his findings, but the drift of his work was in the same direction as Sinclair Smith's: the masses of galaxies were evidently very much greater than the masses deduced from their luminosities. The critical reception Babcock was given at a meeting of the American Astronomical Society was such that he spent the rest of his career in solar astronomy.

During and immediately after the Second World War, little was added these several intimations of "missing mass," but the estimated mass of the Galaxy crept upward, and by the 1960s it was becoming increasingly

obvious to many astronomers—although there were dissentient voices—
that rich clusters of galaxies were bound by vast quantities of dark matter,
far in excess of what was visible, and that the same might be true of superclusters of galaxies. Gradually, other observational material was introduced
into what was beginning to crystallize into a research subject in its own
right. The tendency of most of this observational material was to rule out
candidates for the missing mass, rather than to positively locate any such
thing. There were still sizable groups of skeptics, some doubting the accuracy of observations, others prepared to criticize the assumption that all
regions of the universe are expanding from its initial state in similar ways,
and yet others proposing ingenious modifications of Einsteinian gravitation theory. In 1983, an Israeli physicist, Mordehai Milgrom, proposed a
modification of Newton's inverse square law of gravitation as a possible
alternative to the hidden-mass hypothesis. His MOND theory (modified
Newtonian dynamics) found a measure of support, chiefly among those
who refused to acknowledge the supremacy of Einstein's replacement for
Newton's theory across a much wider canvas. Few astronomers have been
prepared to turn the clock back so far.

A very different line of argument started from the need to account for the
very fact that galaxies and clusters of galaxies exist at all. Why does a cluster
of many hundreds of galaxies exist as such, rather than as a vast assemblage
of stars in a single unit? (Shades of Bentley's question to Newton.) Are we
to seek the seeds of the present situation in fluctuations that were already
present in the universe in its earliest moments? While explaining the origin
of existing structures was found to be extremely difficult, it was generally
agreed that finding plausible theoretical models was much harder without
dark matter than with it; and by incorporating selected hypotheses about
the nature of the mysterious material, and by then checking the resulting
models against observation, light could in principle be thrown back on
those hypotheses. With the advent of more powerful computers, this approach became ever more attractive, but for the majority of astronomers,
seeing was believing. As for the theoreticians, much attention was paid to
the role of black holes. To suppose that dark matter was what remained
from heavy stars that had died long ago presented a difficulty: they should
have left behind far more carbon, oxygen, and iron than was to found. A
way around this dilemma was offered: ultra-massive stars might implode
into the residual black holes. That there are no such stars to be seen forming now, stars with masses hundreds of times as great as the Sun's, required
an extra hypothesis as to their existence in the remote past. This did not
discourage supporters of the idea, and in the 1980s such remnants joined
with brown dwarfs as the leading candidates for dark matter.

By the 1990s, several groups had formed with the problem of dark matter as their focus. Since then, broadly speaking, there have been two main
lines of attack. In the first, a search has been made for a group of objects
known as MACHOs—a generic term covering black holes, neutron stars,

brown dwarfs, or even white dwarfs, and very faint red dwarfs—indeed, almost anything of ordinary (baryonic) matter that emits no radiation and that might contribute to a solution of the dark matter problem. The name is an acronym for <u>m</u>assive <u>c</u>ompact <u>h</u>alo <u>o</u>bject, and was first used of dark matter in galaxy haloes.

In the 1990s, searches for MACHOs began at several centers using microlensing—gravitational lensing caused by small objects, rather than by large aggregations of mass in galaxies or clusters of galaxies. In Chile, a French group searched for the them in the halo, while a Polish group in another Chilean observatory searched in the general direction of the galactic center for lensing effects. The most ambitious group was one led by Charles R. Alcock, a British astronomer working at the Lawrence Livermore National Laboratory in California. Alcock knew about certain advanced technologies that could track numerous fast-moving objects simultaneously, thanks to the fact that Livermore had been involved in President Reagan's Star Wars initiative. The team linked powerful computers to what were at the time the largest digital cameras ever built for astronomical use. The imaging system was mounted on the Great Melbourne Telescope in Canberra, Australia, a refurbished hundred-year-old instrument. All told, the MACHO Collaboration was typical of a brave new style of pooling international resources, with hope of nothing more than intellectual returns. No star wars now; the collaborative was made up of Australian computing experts and astronomers, as well as physicists from Australia's Mount Stromlo Observatory, the European Southern Observatory, universities from the United States, Canada, and Britain, and of course Livermore.

With their highly innovative arrangement, they were able to image and measure the brightnesses of as many as 600,000 stars at a time, a figure that eventually rose to two million. They collected microlensing data for eight years, beginning in 1993. Among their first successes were images of microlensing caused by objects in the halo of our own Galaxy, these bending the light from stars in the Large Magellanic Cloud, as the former passed in front of the latter. Depending on the mass of the MACHO and its distance from us, there is a period of brightening of the background star which can last days, weeks or months. In fact, before the year was out, the group had made their first sighting, which is usually reckoned to have been the first indisputable evidence for dark matter. In eight years, they registered more than 400 lensing events, but as they proceeded it was becoming increasingly clear that MACHOs do not make up most of the dark matter in the Galaxy.

One of their claims was to have found microlensing at a level high enough to ascribe perhaps 20 percent of the dark matter in the galaxy to MACHOs with masses around 0.5 solar masses. That figure for the masses suggested that the objects in question might be white or red dwarfs, but since those emit some light, they should have been detected by the Hubble

Space Telescope, and that was not the case. Another group, the EROS2 collaboration, which had more sensitive equipment, failed to confirm the claimed levels of microlensing. The Hubble telescope showed that about six percent of the galactic halo is composed of brown dwarfs. The problem of missing mass was clearly not to be entirely solved by any of these types of MACHO. This was in itself an important finding, but the team could also be credited with two other important results. The distribution of the MACHOs they recorded toward the center of the Galaxy provided new, quantitative data that are contributing to a revised mapping of the Galaxy's spiral structure. Perhaps the team's most impressive achievement, however, was to have provided vast quantities of data on variable stars, in fact more than had been previously accumulated by all other astronomers throughout history.

During this same period, cosmological arguments were being added to observational, casting doubt on the likelihood that a search for normal, baryonic matter would ever provide a full answer to the dark matter problem. (Baryons are a class of subatomic particles that include the proton and the neutron and certain unstable, heavier, particles called hyperons. The word comes from the Greek for "heavy"—which the particles are, relative to particles in the other chief groups in particle physics. Baryons experience the strong nuclear force, and obey the Pauli exclusion principle.) It was assumed that all baryons present in the first few minutes of the universe played a part in the nuclear reactions responsible for the hydrogen, helium, and lithium seen today in the oldest stars. The calculated numbers of baryons involved—as deduced from observations of present abundances—were not enough, however, to account for the observed dynamical masses of galaxies and clusters. Partly for this reason, several cosmologists turned their attention away from these and other candidates for MACHO-type dark matter, such as black holes, neutron stars, and brown dwarfs, and began to enlarge upon earlier speculation about the existence of various exotic particles.

The neutrino captured the stage for a time, after 1979, when Velentin Lyubimov, in Moscow, claimed to have measured the mass of the neutrino. It had previously been thought to have zero rest mass, and so was judged to be of little gravitational importance. Small though Lyubimov's figure was, the fact that there are hundreds of millions of neutrinos for every atom in the universe meant that they had to be taken seriously—and they continued to have a following after Lyubimov's findings were generally rejected, or were at least revised. The attractions of the neutrino hypothesis were that neutrinos move so fast that they can explain the smoothing out of any variations that might have been present in the distribution of other matter in the infant universe. Moving at high random speeds, they offered a "hot dark matter" (HDM) solution to galactic formation: the neutrinos start in the denser regions, but move so as to smooth out fluctuations smaller than superclusters. The latter then fragment into galaxies. This is the reverse of the "cold dark matter" (CDM) option, where everything evolves upward hierarchically. Observation was not kind to HDM. It showed that clusters are rarer at high redshifts, that is, in the more distant past. As a consequence, most attention was paid to CDM theories after the late 1980s.

The neutrino did not hold the stage alone. More than twenty other particles were proposed at various times, many of them purely hypothetical, and most with odd names and odd properties which it would be impossible to describe briefly. There were binos and dinos, inflatons and preons, photinos and gluinos, hedgehogs and cosmions, as well as dozens of particles with equally imaginative names, some of them synonyms, just to confuse matters. Many of these particles were accommodated under the generic name of WIMPs. (The complementary nature of the acronyms MACHO and WIMP should not escape notice. WIMP was the earlier term.)

To qualify as a WIMP, certain basic properties are called for. WIMPs, weakly interacting massive particles, are—if they exist at all—particles that interact through the weak nuclear force and gravity, and possibly through other weak interactions, but not with electromagnetic forces. The last requirement guarantees that they cannot be "seen" directly, that is, through normal electromagnetic channels—visual, radio, and so forth. Since we should recognize them more or less directly if they were to react strongly with atomic nuclei, it is required of them that they do not interact with the strong nuclear force. Compared with normal particles, they must have large masses or be extraordinarily numerous—their enormous contribution to the mass of the universe being paramount. Dark and invisible, and not interacting with normal matter, their large mass will make them relatively slow moving, and so of low temperature. (Such CDM offered a more promising solution to the problem than HDM, which was ruled out as being incompatible with the structure of galaxies on a large scale.)

Because WIMPs can only interact with other matter through gravitational and weak forces, they will inevitably be very difficult to detect. How the question of detection was approached depended on the fine detail of the underlying theoretical model. Of each model, there were usually many variants, and it is out of the question that we even begin to list the theories, let alone sketch their character. (This topic is one of the most actively pursued in contemporary astronomy, with several hundred research papers on it being published each year.) Many proposals for astrophysical and laboratory tests have been made, and some work has begun. WIMP capture or annihilation, for instance by the Sun or in the Earth, is expected to form other particles, including neutrinos, which may in turn be detectible by such instruments as the Super-Kamiokande detector in Japan (for which see p. 612 above). Other techniques that are being tried include the use of a scintillating material that will generate light pulses from any atoms which WIMPs have affected. A detector of this sort was built in Italy, and there have been tentative claims that the expected kinds of signals have been detected, but these have not been confirmed by other groups.

An entirely different approach has been that mentioned above, in which an acceptable theory of galaxy formation is posited, and the type— or combination of types—of dark matter that are compatible with it are then investigated. Early in 2006, a team from the Cambridge Institute of Astronomy, led by Gerry Gilmore, announced some new findings that will require many old assumptions to be revised, even those about galaxy formation. Using data from several of the largest telescopes, including the VLT in Chile, the group made detailed maps in three dimensions of twelve dwarf spheroidal galaxies around the edge of our galaxy, using the movements of their component stars to assess the distribution and mass of the dark matter around and among them. Some entirely unexpected results emerged.

The galaxies were found to contain dark matter to the tune of four hundred times that of normal matter. The velocity of the dark matter particles was even more surprising, at about 9 kilometers per second, equivalent to a temperature of the order of 10,000°K. If confirmed, the existence of this hot dark matter will have implications for the mechanisms by which galaxies and clusters form, favoring the formation of larger as against smaller galaxies. The most common of earlier assumptions on this score were challenged in other ways. Near the center of one dwarf spheroidal, in the constellation Ursa Minor, the team found a clump of slow-moving stars, which they interpreted as the remains of a globular cluster. According to the most widely supported "lambda cold dark matter" model, dark matter should rapidly increase in density toward a galaxy's center, which in this case would imply a dispersal of the star cluster.

Until the outer limits of these dwarf galaxies are determined, their total mass will be uncertain, but first indications—based on the slow movement of the galaxies' outermost stars—were that dark matter does not extend as far beyond them as was previously thought. Another conclusion drawn from this study was that our own galaxy is more massive than previously believed, and that it is now to be ranked as the largest galaxy in the Local Group, larger than even M31 in Andromeda. We are more important than we thought.

ZEL'DOVICH, PENROSE, AND HAWKING

Black holes were a subject of intensive mathematical study before astrophysicists fully appreciated how important they might be. Not to be outdone, Chandrasekhar, at the age of 72, published a 667-page book on the subject: *The Mathematical Theory of Black Holes* (1983). The number of people involved in this activity was not small, but in our all too short account, we must confine our attention to three outstanding figures: Yakov Zel'dovich, Roger Penrose, and Stephen William Hawking. Zel'dovich was the head of an Institute for Physical Research at the Soviet Academy of Sciences in Moscow, a man with enormous energy who led a team of very talented younger physicists working on the problem of black holes, and especially their interaction with light. In their work in the early 1960s, they viewed black holes in a way hinted at by the name they then gave to them, "frozen stars," stars that had contracted until brought to a halt at the Schwarzschild radius.

Penrose was an applied mathematician, and was at the time at Birkbeck College, London University. (He had studied at University College there, and had later worked in Cambridge and in the United States.) Around 1965 he showed how it was possible to introduce new kinds of coordinates ("Eddington-Finkelstein coordinates") in which the star's collapse does not slow down, but continues all the way to a singularity—a region of zero volume with matter of infinite density—leaving behind it a "horizon" at the Schwarzschild radius. In fact, it was in response to this new kind of language that John Wheeler chose the words "black hole" to describe the curved and empty space-time left behind with the horizon.

At much the same time, Denis Sciama was leading a group doing research on relativity and cosmology in the Department of Applied Mathematics and Theoretical Physics at Cambridge. It was he who had led Penrose to look into cosmological matters. Among a succession of Sciama's doctoral students there were George Ellis, from South Africa, Stephen Hawking, Brandon Carter, and Martin Rees. Also at Cambridge at this time were Hoyle and Narlikar, and there were active contacts between Sciama's group and Bondi, Penrose, and others in London. It was when returning from a meeting in London that Hawking conceived an idea that gave his doctoral thesis its importance. In it, he applied Penrose's so-called singularity theory, not to a star but to the universe as a whole. The central idea was that of reversing the time and treating Penrose's point singularity as the beginning of the universe, rather than as the end of a star's collapse.

In 1970, Hawking and Penrose wrote a joint paper claiming that a universe exhibiting the expansion, as well as containing matter that ours is observed to have, must have begun in a singularity. This was not readily accepted at first, but most members of the community of cosmologists were gradually converted. Hawking, however, later changed his position on the question of the initial singularity, when he found ways of working quantum mechanics, the theory of the very small, into his account.

Hawking studied physics at Oxford before his move to Cambridge in 1962. His doctoral study had hardly begun when it became clear that a motor-neuron disease he had developed—amyotrophic lateral sclerosis, or, ALS—was serious and was leading to a situation where he would soon be largely incapable of speech and movement. His subsequent battle with fate would have been noteworthy enough on its own account, but he was at the same time making some of the most remarkable scientific advances of the century, and they were to bring him wide acclaim. He was to succeed to the Lucasian Chair in Mathematics at Cambridge in 1979, a latter-day successor to Newton, and soon became one of the best-known scientific figures of his time.

In the 1960s, black holes had a theory of dynamics developed for them, an important pioneer in this respect being the mathematical physicist Werner Israel. Israel—born in Berlin, brought up in South Africa, a graduate student in Ireland, but by this time working in Canada—had unraveled some of their physical properties in 1967, but was dealing then only with *static* black holes. He believed that they have to be spherical, and that only spherical objects can collapse to form them. Penrose and Wheeler argued that the demand for perfect sphericity was not particularly stringent, since in contracting, an irregular star would give off gravitational waves and become more perfectly spherical, until in the end it would be a truly perfect sphere.

Israel had shown that the collapsing stars may have external fields (gravitational and electrical) that are determined by their masses and charges. At Cambridge, Carter (in 1970) and Hawking (in 1971–1972) modified and extended the principle, adding angular momentum as a third property, and showing that the shapes need not in any case be spherical. Between them, the three had shown that black holes could have no other distinguishing characteristics than those mentioned. In a phrase coined at the time, they "had no hair" to distinguish them. The chemical nature of the matter falling into them, for example, ceases to be relevant to what we can learn about them from the outside.

The chances of detecting black holes seemed slim, and yet it was realized that they might, even so, conceivably make up a significant fraction of the universe. In 1966, Zel'dovich and Novikov wrote about what would now be termed black holes that might have formed as perturbations in the matter of the universe as it began to expand. A black hole can have any mass: the smaller the mass, the greater the pressure needed on it, but great pressures were no problem in the early universe, according to this theory. Some of the old miniature black holes might by now have sucked in matter and radiation to such an extent that they have become as massive as a million galaxies, they thought. Hawking later argued in 1971 that some of them might have remained unchanged, and still be only a few millionths of a gram in mass.

In 1969 Penrose showed that a black hole could lose energy and slow down, and so might energize electromagnetic radiation—light, radio waves, and so forth—nearby. But what of its size? It was in the course of pondering this question at the end of 1970 that Hawking made one of his most fruitful discoveries. If nothing can get out of a black hole, then the surface area of its "horizon" could not decrease; and if anything—matter or radiation—were to fall into it, or it were to combine with another black hole, then the area would actually increase. This in itself may seem an innocuous enough point, but its implications, in Hawking's thoughts, were dramatic.

The nondecreasing behavior of the black hole's area was reminiscent of the behavior of the physical quantity entropy, which was discussed in chapter 16 (see p. 541). This is a concept used together with others—such as temperature, pressure, heat energy, and so on—in specifying the thermal state of a system. It may be measured in terms of the heat that we need to add in order to transform a system from a given state to the state considered. Conversely, it can be seen as a measure of the "quality" of heat energy, that is, as the amount of energy in a system that is available for doing useful work. Another way of considering entropy is as a measure of disorder, for instance, among the atoms making up a system. (This interpretation is easily abused. "Entropy" is a well-defined scientific concept, but terms such as "order," "disorder," "organized," and "disorganized" have a wide range of meanings, as every student of T-shirt slogans will be aware.) The second law of thermodynamics states that the entropy of a closed system, one that does not interact with its surroundings, never decreases, or—in some accounts—that its decrease is extremely improbable. It is a generalization of the principle that heat cannot, in and of itself, pass from a colder to a warmer body. Without outside interference, a glass of water does not suddenly start to boil in one region while forming ice at another, to provide the heat for the boiling. Some might wish to say that there is an extremely remote possibility that this will happen, but all will agree that the probability is vanishingly small. Open a bottle of perfume in one corner of the room, and an hour later molecules of scent will be detectable in all parts of the room. The probability that they will all later simultaneously be back in their original highly ordered state, in the bottle, is so small that it can be ignored. The entropy of the system has increased.

How probability comes into the picture is not usually considered an astronomical theme, but it is of profound importance in astronomy, on both the large and the small scale. Probabilistic arguments were introduced into thermodynamics by James Clerk Maxwell in 1871, but in this brief account we begin with Ludwig Boltzmann, in the mid-1870s, and his response to a difficulty that had been pointed out earlier by William Thomson and Joseph Loschmidt. The two men noted that the existing laws of mechanics were time-reversible, that is, they recognized no asymmetry in regard to time, nothing like the thermodynamic arrow of time. But are the laws of mechanics not at the root of the laws of thermodynamics? How can that be so, if the latter subject has an asymmetry in its time variable? Boltzmann's explanation was that large-scale systems are aggregates of microstates, which are to be considered equally probable. He then identified the entropy of a system with a certain (logarithmic) function of the probability of its macroscopic state—this in turn depending on the number of microstates. Once this definition is accepted, the second law can be rewritten: thermodynamic systems have a *tendency* to evolve toward more probable states. Decreasing entropy is no longer considered impossible, but merely very improbable. Every day, for example, the Sun radiates 3.5×10^{31} joules of energy into space, powered by the nuclear reactions in its core, and in due course it will die, having in the process increased the entropy in the universe enormously. The probability that the genie will go back in the bottle and that the energy will be reordered in the shape of the Sun is countenanced by nobody. The behavior of countless dying suns in the universe contributes greatly, needless to say, to the heat death predicted by Helmholtz.

HAWKING AND BLACK HOLE EMISSIONS

So much for suns, but do black holes violate the second law of thermodynamics? What if matter with a certain entropy were to fall into a black hole? The entropy outside the hole would decrease. And inside? We cannot look inside to see, but might it not be that there is some indirect way of judging? A Princeton research student, Jacob Bekenstein, had suggested that the area of the event horizon of a black hole could be a measure of its entropy. Since this would increase as matter fell into the hole, we are encouraged to think that the second law of thermodynamics is preserved for the total system.

Hawking here raised the objection that if a black hole has entropy, it should have temperature, and in that case it must emit radiation. But black holes, by the standard definition, emit nothing. In 1972, with Carter and an American colleague, Jim Bardeen, Hawking dismissed Bekenstein's idea, but later saw how he could make use of it. It was not until Hawking was visiting Moscow in 1973 that he was made aware of a proof by Zel'dovich and Aleksandr A. Starobinsky, published in 1971, that a *rotating* black hole might create and emit particles. When Hawking later tried to improve the mathematics of the proof he found to his "surprise and annoyance" that even non-rotating black holes should create and emit particles at a constant rate. He thought there was an error in his working, until he realized that the emission was of precisely the amount required by thermodynamics, that is, such as would prevent violations of the second law. The black hole behaves as though it has a temperature: the higher the mass, the lower the temperature.

Hawking's explanation of this escape from the supposedly absolute security of the black hole was that the particles are coming from just outside its event horizon. There are electrical and magnetic fields there, which we should be inclined to set at zero were it not that quantum mechanics forbids us to do so. Heisenberg's "uncertainty principle" (1927) in quantum mechanics denies us the possibility of measuring both the position and momentum of a particle simultaneously, with complete precision. The more certainty we have in the one, the less we shall have in the other. Niels Bohr in the same year—in his "complementarity principle"—extended this to experimental knowledge of other aspects of physical situations. In the case of black holes there will be a certain minimum of uncertainty in the value of the field. Hawking proposed that we think of the quantum fluctuations in the value of the field as pairs of particles of light (or gravity) that appear together at some time, move apart, and then—in reuniting—annihilate one another. In some cases, however, one of the two "virtual particles," a particle or its "antiparticle," might be captured by the black hole, and if the other is not captured too, and has positive energy, it might escape. In this case it will seem to have come from the black hole, and since the particle of negative energy that enters the black hole will reduce the mass of the latter, this completes the illusion—if that is the best way of describing an aspect of a theoretical discussion—that *the black hole is emitting particles.*

As a black hole loses mass, its temperature rises and its emissions increase. This process is thought likely to accelerate until the thing finally explodes with extraordinary violence. It is possible to calculate what might be called the "evaporation time" of such objects, and in the case of a black hole of mass equal to that of the Sun, this will be far in excess of the supposed age of the universe. The primordial black holes proposed by Zel'dovich and Novikov, on the other hand, could have evaporated by now, had their initial masses been less than about a billion tonnes. Those of somewhat greater mass might still be radiating X-rays and gamma-rays profusely— black holes, but white-hot, as Hawking has remarked. His calculations based on the observed background radiation of gamma-rays set an upper limit to this phenomenon, if it is a real one. It emerged that primordial black holes cannot make up more than a millionth of the matter in the universe, although they would be expected to congregate near large masses, such as the centers of galaxies. In this context, the extent and nature of the background gamma-ray radiation became a question of great concern. If the early universe had been significantly irregular, it would have produced many more primordial black holes than seem to be set by the limit stated here. A universe that was smooth, uniform, and at high pressure would be expected to produce fewer than a universe irregular or at low pressure.

These ideas were received with general incredulity when Hawking first tried them out on his colleagues, but after an initial period of skepticism they were widely accepted. Their implications are of enormous importance, for previously it had been thought that a black hole was a one-way sink to another world, as it were. After Hawking, it became clear that there is a process of cosmic recycling going on through their agency. From a wider theoretical viewpoint, we can say that with Hawking's work something still more fundamental was happening: general relativity itself was being modified, or extended, in the sense that quantum physics was now shown capable of eliminating some of the singularities that relativity predicted. But would singularities reappear in a completely combined quantum theory of gravitation?

From 1975, Hawking turned to problems of quantum gravity, using the so-called sum-over-histories approach devised by the American mathematical physicist Richard Phillips Feynman. Feynman said that, instead of thinking of a particle as we ordinarily do, as something having a single history that takes it along a particular path in time, we should consider it to have every possible history, taking it over every possible path in space-time. Not all paths will be equally probable, however, and ways of computing the probabilities were needed, using the rules of quantum mechanics. With each "history" there are two numbers associated, one representing the intensity of a wave, the other the phase of the wave, and at each point in space-time the probability of finding the particle there is found by summing all the waves for all possible histories. In most places, the probabilities—more strictly, the probability amplitudes—more or less cancel out, but there will be places where they build up significantly. As an example: there are very high probabilities that an electron going round the nucleus of an atom will be in one of a limited number of orbits, the orbits physicists already assumed them to be in.

We recall that in Einstein's theory of gravitation, a free particle follows a "geodesic" line in curved space-time, the equivalent of a straight line in Euclidean space, the shortest distance between two points. Applying Feynman's sum-over-histories approach here, what corresponds to a clear-cut history of a single particle is something involving the whole of space-time, the whole of the history of the universe. As we have seen, in the classical theory of general relativity, various models had been developed that described the history of the universe, following on from its initial state. In Hawking's new theory of quantum gravity, we cannot be specific about the way the universe began, although the probabilities of some of the results ensuing will be greater than that of others.

If the universe resembles a black hole, then everything may be pictured as expanding outward from the initial singularity—commonly referred to as "the beginning of time"—and after a certain stage it may all collapse back on itself, in what is now graphically referred to as the Big Crunch, ending at "the end of time," and mirroring the Big Bang. There were no boundaries as far as the spatial arrangement of the accepted models was concerned: these were finite in space, without edges, in the way the two-dimensional space that surfaces a ball is finite and unbounded. Hawking disliked the fact that there seemed to be time boundaries, however, for what seem at first to have been aesthetic reasons. Whatever his reasons, testable predictions ensued.

How the boundaries in time were dissolved is something that requires a careful mathematical description, but Hawking has provided a graphic analogy with circles of latitude on the Earth's surface. These are to be envisaged as representing the spatial size of the universe, and distance from the North Pole as representing time. The universe begins as a point (the North Pole) and grows as a circle of latitude, until it reaches the equator, after which it shrinks until reaching the South Pole, where it becomes a point again. The universe has zero size at the poles, but these are not singularities any more than the Earth's poles are singular. This is to picture the universe as completely self-contained, and having no edges or boundaries, as it were. There is no place in the model for negative time, no time before the Big Bang, none after the Big Crunch. But is the model an acceptable one?

Under his "no boundaries" assumption, Hawking found—approximately in 1981, and more precisely with the help of Jim Hartle of the University of California in 1982—that, out of the histories that the universe may have followed, one particular group is very much more probable than the others. It was calculated that there is a very high probability that the present rate of expansion of the universe is almost the same in all directions—a result that seemed to be borne out by observations of the microwave background radiation. This led on to an investigation of the sizes and consequences of departures from uniform density in the early stages of the universe. The uncertainty principle suggested the absence of complete uniformity at the beginning, ten or twenty billion years ago, and that during expansion the initial irregularities would have been amplified, that is, the initial fluctuations would have led to the existence of clouds, stars, and galaxies—and, as a sideshow, to human beings. How could astronomers detect that primordial granulation of matter, which made the universe such an interesting place?

COSMIC HIERARCHIES:
FROM STARS TO SUPERCLUSTERS

The key to detecting it was likely to be in the microwave background radiation, which it was realized might somehow show ripples as residuals of the Big Bang. Considerable effort was spent on the search for such irregularities, beginning in the 1970s, and numerous claims were made that were short-lived, judging by the astronomical consensus. That there is massive inhomogeneity in the distribution of *galaxies* cannot be doubted, and an aside on this narrower problem is in order here, for clearly the two must be somehow connected.

Our Galaxy contains stars and gases, mostly concentrated in a spinning disk, with the highest concentration in its central bulge, which is thought to harbor a vast black hole. This is typical of other galaxies, one of which will be found, on average, in a cube of side about 10 million light-years. (Our nearest large galactic neighbor, M31 in Andromeda, is about 2 million light-years away. The small spiral M33 is slightly nearer.) Their distribution is not random, however, and in our neighborhood there is the cluster of about twenty galaxies, the Local Group, located at intervals of a few million light-years. Gravitation is not to be neglected over these distances: M31, for instance, is drawing toward us under gravity at about 100 kilometers per second. Some clusters are much more numerous. Our Local Group is near the much larger Virgo Cluster, which contains some two thousand members, and represents the physical center of our Local Supercluster. (The latter is sometimes called the Virgo Supercluster.) Its enormous mass influences all the galaxies and galactic groups, but the strong gravity of the Virgo Cluster is enough to accelerate its own members to velocities of more than 1,500 kilometers per second with respect to its center of mass. Much attention was paid in the last quarter of the twentieth century to the

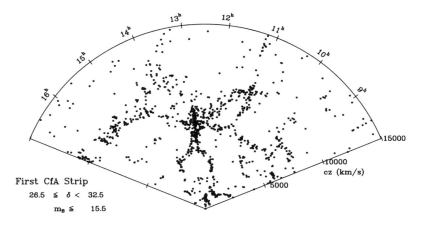

246. In 1988–1989, Margaret Geller and John Huchra were investigating data from a survey of 1,060 galaxies when they discovered what they called a "Great Wall." This sheet of galaxies 250 million light years away measures about 200 million × 600 million light years in area, but has a thickness of only 20 million light years. (Figures quoted vary widely.) It stretches more than a third of the way round the sky (from 8^h to 17^h in right ascension, and from + 26° to + 32° in declination). At 10^{16} solar masses, the Coma and Hercules superclusters make up the bulk this wall.

complex dynamic structure of all of these vast but irregular aggregates of galaxies.

In 1975, a study by G. Chincarini and H. J. Rood claimed to have detected a lumpiness in matter over a unit of about 20 million light-years. At an IAU symposium held in 1977, several astronomers reported that they had found spaces more or less empty of galaxies, voids that were several hundred million light-years across. A year later, S. A. Gregory and L. A. Thompson gave evidence for a large assemblage of galaxies—more than three thousand in all—known as the Coma Supercluster, surrounded by relatively empty space. (The names are confusing: note that this supercluster is largely made up of two smaller clusters, the Coma and Leo clusters. They all take their name from the constellation in which they are found.) Gradually a picture was built up of a world of galaxies that is very lumpy, indeed, and by the late 1980s it began to seem that the distribution followed something like that of a foam of soap bubbles, some large, some small, each with galaxies around its surface and a space relatively empty of galaxies within. The first clear details of this three-dimensional picture were largely worked out by Margaret Geller, John Huchra, and Valerie de Lapparent of the Harvard-Smithsonian Center for Astrophysics. One curious structure they later found was a "Great Wall" of galaxies, a sheet-like array presenting a vast area to us, but strangely thin (see fig. 246). When they first found it, there was much doubt expressed as to whether gravity would have permitted such a large formation, and there were similar doubts about an even larger structure found by the Sloan Digital Sky Survey, completed in 2005, which was mapping a million galaxies over a quarter of the sky visible from the Apache Point Observatory in New Mexico. In both cases, those doubts were eventually laid to rest by computer simulations. Needless to say, the Sloan Great Wall attracted worldwide astronomical and mathematical interest. Size evidently matters.

On the subsidiary question of clustering, it is worth asking how high the hierarchy goes. There are many objects in the natural world—feathers, ferns, our bronchial tubes, and so forth—which present a pattern that is repeated on a smaller scale when examined under a microscope, and this in a hierarchical fashion, often several times over. Starting from much earlier studies, the mathematician Benoit Mandelbrot investigated such patterns at a very general level, describing them in 1980 by the term "fractal." The ease of generating images of "Mandelbrot sets" on a personal computer brought the idea to popular notice, and eventually led to a flourishing internet cult. Can it be that the universe is fractal, in the sense that there are clusters, clusters of clusters, and so forth, perhaps even endlessly so? (Charlier's ideas on this question, a century ago, were introduced on p. 629 above.) As far as is known, there are no structures of greater clustering complexity than those introduced above, a finding of much relevance to modern studies of the history of the universe. By the end of the twentieth century it was widely agreed that, if samples were to be taken on a reasonably large scale, say counting galaxies within a cube of size half a billion light-years, their numbers, and the numbers of their structures, would be the same everywhere.

If clusters and superclusters of galaxies are spread uniformly—on the grandest of scales—then their gravitational effects should cancel out, but the Virgo and Hydra-Centaurus superclusters, our relatively near neighbors on the cosmic scale, are not balanced by anything comparable in the opposite directions, and so should influence our Galaxy gravitationally. We have seen how, since the 1920s, knowledge of the Sun's movement around our own Galaxy has been ascertained, but what of the Galaxy's own movement? In the 1970s, George Smoot and his colleagues looked for signs of it in the microwave background radiation. The NASA *Cosmic Background Explorer* satellite (COBE) was still in the future, and they flew their detectors at high altitude in a U2 "spy" aircraft. They discovered that our Galaxy has a motion of about 600 kilometers per second in the direction of the Virgo Cluster. They were retreading Herschel's path to the solar apex, as it were, on a much vaster scale.

Can we go no further? The Virgo Cluster must clearly be pulled toward more distant clusters and superclusters, but our Galaxy is no longer directly involved in the motion. In 1994, Alan Dressler, of the Carnegie Institution in Pasadena, California, published *Voyage to the Great Attractor: Exploring Intergalactic Space*, a work in which this problem is addressed. He describes a five-year project in which he and six colleagues—calling themselves the "Seven Samurai"—discovered that several hundred galaxies, including ours, are being drawn in a coordinated way toward some enormous concentration of mass. The name they chose for this, the "Great Attractor," has joined the ranks of the genus "Great Wall" in the public's imagination. The coordination of the motions over very large scales, rather as houses on a floating island, shows that the unseen mass is very large indeed. The unanswered question is then whether there are other comparably great attractors distributed throughout the universe, and what part dark matter has to play in the phenomenon.

A more amenable problem is that of the smoothness of the observed cosmic background microwave radiation. In 1992, some exciting new evidence on inhomogeneity was produced by the COBE. John Mather had been a young graduate student at Columbia University in 1974 when he sent a modest design for such a probe to NASA, who decided to back the idea. The plan was to search for possible variations, with direction, in the strength of signals of the sort first found by Penzias and Wilson in 1964. One of the biggest problems facing Mather and his colleagues at the Goddard Space Center—George Smoot was again leader of the team— was to prevent detectors in their satellite from picking up microwave radiation from within our own Galaxy. Three different detectors were used to monitor radiation levels at slightly different wavelengths, and the results were then compared by computer (plates 20 and 21). In December 1991 they had data showing, it seemed, clear signs of fluctuations, of patchiness in their radiation map, as they looked out into the past, so to speak. Within four months, they had run a number of convincing checks and counterchecks. The difficulties were immense, for the ripples detected were only denser than the background radiation by one part in a hundred thousand. Even greater, therefore, would have been the difficulties of detecting them from below the Earth's atmosphere—as various groups of astronomers, notably the Jodrell Bank group working in Tenerife, had planned to do.

COBE's success in detecting these faintest of signs of irregularities in the gravitational field of the early universe was a technical accomplishment of the first order. When the results were announced, at an American Physical Society meeting in April 1992, those concerned were suitably ecstatic. Smoot was reported as having remarked at their press conference that "if you're religious, it's like seeing God," qualifying his words defensively in the following month with a theological question: "It really is like finding the driving mechanism for the universe, and isn't that what God is?" *Newsweek* carried the headline "The Handwriting of God," as other astronomers fueled the excitement: "They have found the Holy Grail of cosmology" (Michael Turner); "It's like *Genesis*" (Stephen Maran); and "the greatest discovery of the century, if not of all time" (Stephen Hawking). The fluctuations were not unexpected, however, and in view of irregularities in the later history of the universe it would have been surprising had something of the sort not turned up in the survey. While the whole episode might remind us of the enthusiasms of an earlier age, it also reminds us that there are societies in which dubious theology can make good economic sense.

Astronomers had few genuine causes for complaint about funding, during the following decade. The Holy Grail?

Studies of cosmic radiation independent of—but complementary to—COBE's were made soon afterwards at the South Pole. Physicists Mark Dragovan and Jeffrey Peterson, both from Princeton University, made crucial observations using facilities at the University of Chicago's Center for Astrophysical Research in Antarctica (CARA). They announced their results in 1993 at an American Astronomical Society meeting, but in a lower key than that of their predecessors. Using two specially designed radio telescopes during the southern summer of 1992–1993, on one of them they detected temperature fluctuations as small as thirty parts in a million—considerably smaller than COBE's. There were various advantages of working in the Antarctic, compared with more comfortable environments, although in principle a satellite should always be able to reveal finer detail. Their radio detectors had to be cooled as near as possible to absolute zero, something that was marginally easier than it would have been in their home laboratories. No break was necessary in observing a chosen patch of sky from the latitude of the Pole, and this made for a more efficient use of telescopes there. (Of course, only half the sky is visible from the Pole, and in this respect a satellite has the advantage.) Most important of all, the extremely dry Antarctic air is almost completely free of water vapor, which ordinarily obscures earthbound observations of the microwave radiation. The chief advantage was in the matter of cost.

After the COBE satellite experiment ended in 1996, other ground-based and balloon-based experiments followed, measuring irregularities with increasingly fine resolution. These measurements were able to rule out cosmic strings as an adequate theory of cosmic structure, and strengthened belief in cosmic inflation. (These two theories will be mentioned again shortly.) Over the next decade, similar programs multiplied rapidly, to a total of at least seventeen, based at places between Saskatchewan and the South Pole, all of them sharing the aim of improving knowledge of irregularity (anisotropy) in the power spectrum of the cosmic microwave background radiation. They differed in various respects, chiefly in the scale of the fluctuations measured, the sizes of the regions of the sky surveyed, and the angular resolution achieved. Some measured polarization, which will help in deciding when the first stars formed, and perhaps provide clues about events that transpired in the earliest phases of the universe's career. One of the most ambitious programs was NASA's Wilkinson Microwave Anisotropy Probe (WMAP), a satellite project which mapped the relative cosmic background temperature over the full sky, with an angular resolution of at least 0.3°, and a high sensitivity. To do this, an apparatus was used that could measure temperature differences between two points in the sky very accurately. (As noted on p. 706 above, WMAP was placed in orbit at the L_2 Sun-Earth Lagrange point.)

Data from WMAP observations over three years were released in March 2006, and these data were interpreted as providing stronger evidence for the standard lambda-CDM model, which was by this time attracting growing support, though only weeks earlier it had been questioned by Cambridge astronomers (see p. 751). On the question of the "fingerprint" of the early universe, responsible for the present organization of matter on the largest scale, and revealed by an analysis of temperature fluctuations in the cosmic microwave background, two independent groups had in January 2005 almost simultaneously announced broad agreement with earlier interpretations of COBE and WMAP data. These were the Two Degree Field Galaxy Redshift Survey (2dFGRS), using a robotic telescope in New South Wales, and the Sloan Digital Sky Survey (SDSS), with an observatory in New Mexico. The former team set the figure for baryonic matter in the universe at 18 percent, with dark matter at 82 percent. On the question of CDM, the jury is out.

Q, Ω, AND Λ

Turning the clock back to 1990, preliminary COBE results announced in that year provided an improved figure for the temperature of the cosmic background radiation, 2.735°K as against the 3.5 ± 1.0°K found by Penzias and Wilson seven years earlier. (There had been several other measurements made in the intervening period. For the COBE graph of intensity against wavelength, see figure 247.) As a measure of the fluctuations in temperature, ripples in a generally smooth world, a parameter Q was used. Q gives the energy in the fluctuations as a fraction of the total energy in the universe, and it could be expected to help determine the time at which such structures as clusters and superclusters would have condensed out.

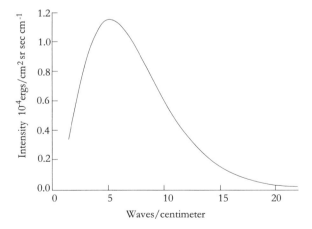

247. The cosmic microwave background is black-body radiation that fills the Universe, with a 2.725°K thermal spectrum (see p. 677). This graph shows the very precisely measured black body spectrum obtained by the FIRAS (far-infrared absolute spectrophotometer) instrument on the COBE satellite. It peaks at a wavelength of 1.9 millimeters (a frequency of 160.4 gigahertz).

It should be obvious that the wider the angle over which samples were taken, the smoother the radiation would appear to be. COBE surveyed irregularities over an angle of seven degrees at best, and found a value for Q of about 10^{-5}, a number so small that it speaks not only for the technical achievement of COBE but for the very great uniformity in the radiation being measured. Much the same figure had already been arrived at on the basis of calculations of the masses of those vast material structures that provide the peaks in the broadly homogeneous universe—Great Walls, superclusters, and so forth. Dark matter was an unknown quantity, and not until the later measurements were made was it known how a change of scale would affect Q.

Findings from COBE and other surveys were more surprising in another respect. It appeared that ripples were formed in the first 300,000 years of the universe's history, which in turn suggested that the universe is much denser than was generally believed. (The greater the mass, the more quickly would fluctuations have formed. Abundances of extraterrestrial deuterium, detected in 1973 in Jupiter, the Orion Nebula, interstellar clouds, and elsewhere had been thought to indicate a low-density universe.) This was precisely the sort of evidence needed by those who had argued for a "Big Crunch" version of the universe's future, as opposed to the "heat death" models that go on expanding and thinning out for ever, producing a world filled with an ever-diminishing haze, formed from the ashes of burnt-out stars. It now appeared that the expanding universe is being gravitationally braked by its total mass. This being so, another important parameter is its density. It was calculated that the expansion would eventually grind to a halt, and be reversed, if the cosmic density was about five atoms per cubic meter, an extremely low figure, but not vastly different from what was generally estimated. The density parameter generally used, and invariably labeled omega (Ω), is the actual density as a fraction of the critical density, the latter being assumed to have been correctly calculated. Estimates for Ω varied, and continued to be debated, but in the early 1990s the theoreticians had a strong liking for the value 1.0, despite the fact that they could only scrape together evidence for a contribution of about 0.1 from ordinary atoms and possibly 0.2 from dark matter. (The former figure can be related to the line of argument mentioned at the end of chapter 17, concerning the conversion of primordial matter to helium and deuterium in the first minutes of the universe.)

The parameter Ω is clearly connected with deceleration in the expansion, and we have already mentioned efforts by Sandage and others to assess the value of the "deceleration parameter." Graphically, it will manifest itself in the curvature of the velocity-distance graph for the galaxies, as found, for instance, from correlations of the (intrinsic) magnitudes of galaxies with their redshifts. Rival values moved to and fro for over twenty years and more, until eventually Beatrice M. Tinsley—a New Zealander who in 1967 completed a doctoral thesis at the University of Texas—persuaded the profession that the *evolution of galactic sources* made a direct measurement of the deceleration virtually impossible, at least for the time being. She pointed out that we cannot consider galaxies to be standard light sources, since they change as they age. She therefore analyzed the spectra of previously studied nearby galaxies, deducing the mixture of stars in them, and applied existing knowledge of stellar evolution to work out what they would have been like when they were younger—remote galaxies, by the time they are seen, being younger than those seen near at hand. This was an important beginning, although many imponderables remained, and still remain, in knowledge of galactic evolution, without which the deceleration parameter cannot be properly estimated in the way Sandage and others had tried to do. Other ways of finding it were conceivable, such as the use of quasars, but it was soon realized that these were even more problematic than galaxies.

A higher Ω than had generally been supposed, that is, a more massive universe, fitted well with a set of cosmological models that had gained favor in the 1980s, and that stemmed from the so-called inflationary model proposed by Alan Guth of the Massachusetts Institute of Technology in 1980. As we shall see, supporters of such models were hoping for a value of this parameter at, or close to, unity. They were thus opening themselves up to what in principle was as severe a test as scientific theories ever have to face. Unfortunately, the uncertainties of dark matter stood in the way.

UNIFIED PHYSICAL THEORIES

Around 1978, long before the COBE measurements, cosmology was invaded by a number of leading experts in particle physics, motivated by a desire to find a way of applying a new class of theories known as "grand unified theories." These theories were meant to bring under a single theory the forces of gravitation, electromagnetism, the strong nuclear force and the weak force. (The strong force binds protons and neutrons in atomic nuclei and has a range of only about the size of the nucleus. Strong interactions account for the energy released in the explosion of a hydrogen bomb. The weak force is seen in the radioactive decay of nuclei and is also responsible for scattering neutrinos. Its range is about a hundredth that of the strong force.) Those who promoted them were concerned with particle energies entirely outside the range of any conceivable present or future Earth-based laboratory, energies that might be reproduced if only we could build a particle accelerator from here to a place well on the way to the nearest star. The early universe was the theoreticians' only available laboratory, a place where densities and temperatures were phenomenally high. They soon became involved, therefore, in the task of extrapolating the history of the universe back to long before the time when the laws of physics as we know them were judged to be reliable, say a microsecond, 10^{-6} seconds, after the Big Bang; and they hoped, in fact, to take matters back to as early as 10^{-35} seconds. (The Large Hadron Collider, built by an international group at CERN in Geneva, can simulate energies typical of the situation calculated for the universe at 10^{-14} seconds.)

To appreciate this fresh new approach, we need to look back to the general drive toward unification that has been present in the sciences throughout history. James Clerk Maxwell brought electricity and magnetism under one banner in the nineteenth century. Quantum electrodynamics (QED) provided a complete quantum description of the electromagnetic force in the 1940s. Ideas taken from this theory were later adopted for other forces. In 1967, the Pakistani scientist Abdus Salam, and two well-known American physicists, Sheldon Glashow and Steven Weinberg, independently proposed similar theories, unifying electromagnetism and the weak nuclear force. Their theory was given experimental support by the discovery at CERN, in 1983, of particles that they had postulated. (The particles were the Z and W bosons. The experiments were conducted by Carlo Rubbia's team.)

Before this confirmation, the next step on the road toward unification had been taken. This was to include the strong interaction with the electroweak force. At this stage, theory was called the "grand unified theory" (GUT). A quantum theory of the strong force ("quantum chromodynamics," QCD) had been developed in the 1970s, a theory based on the hypothesis that all strongly interacting particles are composed of quarks. Using this intricate—but later widely acknowledged—theory, Sheldon Lee Glashow and Howard Georgi proposed in 1974 the first of the grand unified theories, applicable to energies above 10^3 giga-electronvolt. (One giga-electronvolt, 10^9 electron volts, is roughly equal to the rest energy of a proton. It would keep a 100-watt bulb alight for about a minute, which may not impress, but the mass of a proton is only 1.67×10^{-27} kilograms!) Since then, there have been several proposals for GUTs, although none is in a finished state. Other theories to improve on Glashow's and Georgi's were proposed in due course, some aiming to handle energies up to 10^{14} giga-electronvolts or more. To test them, it would be necessary to do so indirectly, through their consequences in our world of low-energy particles. One of several successes to their credit is that they can explain why the charges on the electron and proton are equal. If experiment should show that they differ by as little as one part in 10^{-24}, the theories would need to be abandoned.

When a satisfactory result is obtained at this level, the final stage in unification will be to include gravitation in an acceptable theory of the other three forces. (Gravity is not important when the number of elementary particles is small.) Since almost everyone concerned in the mammoth search accepts Einstein's general relativity as the requisite theory of gravitation on the large scale, that means reconciling it with quantum mechanics at scales where both are called for. The quantization of gravity remains a prime target for many theoretical physicists.

In the search for a quantum theory of gravitation, string theory soon became a leading candidate. This is no doubt familiar—if only from the titles of books—as leading to a "theory of everything." Developed in the late 1960s, it aims to account for the properties of space-time over very small distances, distances of the order of the Planck length, 10^{-35} meters. At those very short distances, quantum fluctuations in energy have local effects on the curvature of space-time that disrupt it in increasingly violent ways, as the scale is reduced. John Wheeler described this effect as producing a "space-time foam." String theory replaces the point-like quanta of earlier theories—quarks, gluons, leptons, vector bosons, and so on—by string-like objects of the order of the Planck length. One of the consequences of making quantum string theory consistent is that its strings can vibrate, and that the modes of vibration can be used to identify them with different species of particle. Among these particles there is one with the properties of the quantum of the gravitational field, the graviton. It is this that makes the theory especially interesting to theoretical physicists, and in particular to those with cosmological interests.

Another property of the theory is that, if it is to be kept consistent and finite, it requires a space-time of more than four dimensions. While, for the most part, its further development has had few repercussions on cosmology, some of the directions in which it has gone may be worth mentioning briefly. It is found that in order to provide descriptions of bosons (such as photons and *W* and *Z* particles) and fermions (such as quarks and leptons), a kind of symmetry beyond that of the special theory of relativity is needed. This "supersymmetry" in turn requires space-time to have six dimensions, in addition to the ordinary four. It is the behavior of these six additional dimensions at every (four-dimensional) point that characterizes individual versions of string theory—of which there are several. In the analogies used to describe the highly complex possibilities, these six extra dimensions are said to be "curled up structures" named after their chief investigators, "Calabi-Yau spaces." It was shown at an early stage that while there are only five consistent theories, there are some thousands of ways of arranging the six extra dimensions. Important progress was made in 1995 when ways were found of going beyond earlier approximative methods. To the surprise of all concerned, the new techniques gave rise to the recognition of new types of objects in string theory, in particular the so-called *p*-branes, a term coined out of the word "membrane." They also yielded symmetries showing how the five superstring theories could be related, and relating all five to a sixth theory, known as supergravity, requiring eleven dimensions. It seems that the other superstring theories may be approximations to this more comprehensive theory, which has been given the name of M-theory.

INFLATIONARY THEORIES

Cosmologists who extrapolated Big Bang theories backward in time in the 1980s and 1990s were generally in agreement that in the first 10^{-36} seconds, and not before, the particles in the universe would be interacting at energies of 10^{15} giga-electronvolts, the figure for the critical energy at which unification occurs, according to those who were trying to take the penultimate step in the sequence sketched in the last section. Ambitions did not stop there, however. Some discussed what might happen if the entire world visible to us were to be compressed into a space as small as a single atom. It was calculated that this would have happened in the first 10^{-43} seconds (the "Planck time"). At this stage, both quantum and gravitational effects would be significant. The trouble with pronouncing on the situation in this ultra-early phase is that the next phase after it (up to 10^{-36} seconds) might have wiped out all vestiges of it.

It is interesting to see how different scientific groups were approaching different periods in the history of the universe in the last thirty years of the twentieth century. The following is a very rough summary of the situation.

Working back from the present to about a million years ago, when the first stars, galaxies, and clusters condensed out of what went before, we are in the realm of mainstream astronomy and mainstream physical laws.

From about a millisecond to that million-year-old universe, the cosmic expansion—accelerated, decelerated, or neither—was deemed to be normal, and the physics needed to describe such phenomena as cosmic abundances, especially critical figures for helium and deuterium, was well-supported in the laboratory. This was the province of such astrophysicists as Hoyle and Fowler—who were not of course confined to it.

The first millisecond, from the Planck time onward, was the hunting ground of a select group of mathematical physicists, cosmologists in the broadest sense of the term, some of whom hoped to use the remote past as a test-bed for their physical ideas. For reasons to be explained, perhaps this period should be divided into two, at 10^{-36} seconds, after which time supporters of inflationary theory argued that ripples of the type found by COBE had been generated. (It was for this reason that Hawking was so rhapsodic about COBE.)

Of questions to which the grand unification theories outlined in the last section could provide answers, or at least a framework for future answers, one in particular was of immediate cosmological interest. It was first suggested by Andrei Sakharov in 1967 that those theories may provide a value for the ratio between the number of photons and the number of baryons in the universe. This ratio is about 10^{10}. Strictly speaking, the "baryon number" as technically defined is not a simple number totaling protons and neutrons, but it deducts the numbers of antiprotons and antineutrons, and adds contributions from a few short-lived particles. The advantage of talking about baryon numbers, and not of the individual particles by name, is that in interconversion between them—according to what was generally believed before the arrival of the grand unified theories—the baryon number is left unaltered (it is currently estimated to be about 10^{78}). Calculations using the new theories, however, suggested that at temperatures of 10^{27}°K and higher, changes in baryon number would be common. This, it was decided, would only be possible if a certain distinction could be made between matter and antimatter. By a piece of historical good fortune, the necessary small difference between the two types of matter had been discovered experimentally in 1964, by Val L. Fitch of Princeton University and James W. Cronin of the University of Chicago. While calculation of the precise baryon number to be expected in the universe required a knowledge of the values of certain parameters as yet unavailable, this was the first cosmological success of the new ideas.

After 1981 it was followed by others, based on the class of models of the universe known as "inflationary." The first step was taken by Alan H. Guth, in a paper entitled "Inflationary Universe: A Possible Solution to the Horizon and Flatness Problems" (1981). Guth tried to explain why our universe, after a period of the order of 10^{10} years, is still expanding with a value of Ω not far from unity. Why did it not collapse earlier, before stars had time to form and evolve? Alternatively, why did it not expand faster, with an energy outstripping the gravitational braking effect, a situation that would not have allowed time for galaxies to condense out? Why did the initial impetus more or less match the decelerating effect of gravity?

Guth's "horizon problem" concerns the universe's homogeneity and isotropy, that is, why its average qualities are much the same everywhere, and looking in all directions from everywhere. Why, when causal influences could be exchanged less easily in the past than now, do remote regions appear so similar, on average? Why were the temperatures measured by COBE and its successors the same, over the entire sky? Guth saw that this problem could be solved if it could be shown that the early universe went through an accelerating phase. Causal interaction would have been easier, allowing remote parts to homogenize before they accelerated apart.

Guth's original paper contained a technical flaw, which he pointed out but did not correct there. The flaw was avoided in an independent inflationary model worked out in 1982 by Andrei D. Linde of Moscow, and by Andreas Albrecht and Paul J. Steinhardt of the University of Pennsylvania. The key idea is that there is a switch, a phase transition, in the very early history of the universe, when the temperature was about 10^{27}°K—the temperature mentioned above. (The chief innovation in the new inflationary model was that it involved a slow dissolution of the phase transition, as opposed to Guth's original idea that this happened very rapidly.) At temperatures above such a value, their theories suggested, there would be one unified type of interaction between particles, one unified force. Below the critical temperature, however, the spell would be broken, and the weak, the electromagnetic, and the strong forces would all acquire their separate modes of action, as we know them today.

This was an important idea behind the inflationary theory, but there were several possibilities as to the manner of the crucial transition. Did it occur suddenly, or after a delay in which a situation occurred that was analogous to the situation in which water can be supercooled well below freezing point without solidifying? Or was there some intermediate solution? Calculation and reference to some of the unacceptable outcomes of the alternatives led to the conclusion that the universe underwent a period of extreme supercooling. Heat would have been released during the transition, in the way water releases latent heat when it freezes, and this was said to be the heat that survived—although it has steadily cooled as it has become diluted in the universe's expansion—and now survives as the cosmic microwave background radiation.

The basic theory—of which there were many variants—ran along the following lines. Immediately before the phase transition began, that is, the beginning of the inflationary period, the material that is in the *presently observed* universe—a region with a radius reckoned in round numbers at about 10 billion light-years—was a billion times smaller than a proton. (The theory made no pronouncement about the nonobserved regions, which were said to be potentially infinite.) This "universe" was about 10^{-35} seconds old, but it would have been prolonged through a period, the "inflationary era," during which an expansion by an extraordinary factor, quoted as 10^{50} or even much greater, stretched the region to the size of the observed universe. The era of inflation was said to have lasted until the universe was 10^{-30} seconds in age. One merit claimed for the theory was that the sources of the microwave background radiation presently observed in all regions of the sky were once in close contact, with time to reach a common temperature before the inflationary era began.

The acceleration of the expansion was ascribed to a cosmical repulsion that called for some new ideas. It was said that the gas filling the universe supercooled during the inflationary era to temperatures far below that of the phase transition. In doing so it was supposed to approach a state of matter that was called a "false vacuum," a state never observed and never likely to be observed, but a state of the lowest possible energy density appropriate to the prevailing physical laws. (The energy density was attributed not to particles but to "Higgs fields," but it is not necessary to enter into such detail here.) Combining its properties with general relativity, the false vacuum was found to promote a gravitational *repulsion*, and so to accelerate the expansion for as long as it lasted. As the universe inflated, the energy in the false vacuum became greater and greater. The false vacuum was seen to have similar effects to those of Einstein's cosmological constant Λ.

Eventually, the phase transition would have been completed, the temperature during the inflationary era having dropped and then risen again to the critical value of 10^{27}°K. The false vacuum would have been switched off after the phase transition had taken place—that was the nature of such a transition—but by then it would have kick-started the expansion of the universe as we know it. Supporters of inflationary theory took pride in the thought that they were the first to explain the initial momentum of the expanding universe, which for adherents of the older theories had merely been an unexplained initial condition. They thought that after the inflationary stage (the acceleration stage) a deceleration would set in, with gravitation its cause. From this point on, the old cosmologies that had ruled the roost during the previous six decades could be allowed to take over.

After Guth's paper began to attract attention, cosmologists became aware of the fact that others had put forward vaguely similar ideas beforehand—a common enough phenomenon in science. (As the father of the pioneer in non-Euclidean geometry Johann Bolyai wrote to his son, urging him to publish his ideas, things have a way of turning up in several places at the same time, "just as violets appear everywhere in spring.") Alexei Starobinsky in the Soviet Union, Richard Gott in the United States, and Katsuoko Sato in Japan had all touched on the possibility of early acceleration, and they had expressed other opinions somewhat resembling those of Guth, Linde, Albrecht, and Steinhardt, making them natural allies. Others were quick to join forces, and a select band of them came together at a Cambridge workshop lasting three weeks in 1982, where they developed many new ideas collectively. One problem that had worried them was that inflation theory would require an entirely smooth universe. One of the participants was Stephen Hawking, who pointed out that the situation could be saved by applying quantum theory, with its probabilistic predictions: quantum effects occurring at a very small scale in the early stages might be supposed stretched to the sizes of galaxies and clusters in the course of inflation. (We shall return to such ideas in the next section.) They found that the spectrum of irregularities in density would be much the same on all scales of astrophysical significance—something that had been proposed in the early 1970s by Edward Harrison and Yakov Zel'dovich, independently of one another, for use in modeling galaxy formation. This is where the quantity Q, introduced earlier in this chapter, enters the picture. The (extremely small) empirical value for Q found from surveys of the microwave background, as well as from surveys of galaxies, clusters, and other matter, confronted inflationary theory with the difficult task of accounting for it.

As a result of this early activity, there was a broad consensus among adherents of inflationary models on several key issues. The matter and energy in the universe was seen as something produced during the inflationary process. The density of the universe was taken to be close to the critical density of unity, and thus the geometry of the universe was taken to be flat. Fluctuations in the primordial density in the early universe were accepted as having the same amplitude on all physical scales; and it was believed that there should be, on average, equal numbers of hot and cold spots in the fluctuations of the cosmic microwave background temperature. Here were assertions enough to occupy observational astronomy for years to come, and indeed one of the leading projects with them in mind was NASA's Wilkinson Microwave Anisotropy Probe, as mentioned earlier in this chapter.

From such ideas came a stream of new cosmological ideas. "Hawking Radiation," originally related to the horizon around a black hole, was now found a new role. With Gary Gibbons, Hawking showed that the same sort of radiation might originate with other kinds of horizon—for instance, the sort that prevents regions of the universe communicating with light signals when they are moving apart (with the expansion of space) faster than the speed of light. They showed that this might have been significant in the early history of the universe. Again, these ideas were developed and led to a multiplicity of new alternatives. The observational evidence was hardly equal to the task of deciding between them, but the COBE results at least marked a new phase in this process, which still continues, although not without casting a few seeds of doubt on the inflationary picture.

From the outset, even among adherents, there was room for difference of opinion. Linde asked whether the phase-transition might have come in different places at different times—an analogy with champagne bubbles was useful. His variant scheme of 1983, which acquired a following, was known as "the chaotic inflationary model." Models of this type differed notably in the temperature variations they required the microwave background radiation to have. The chaotic versions picture the universe as possibly infinite and eternal, but having some regions ("universes," in some writers' terminology) that are expanding, others that are contracting, some at high and some at low temperature. It was suggested that we might inhabit, as it were, a bubble that has not merged with another bubble, as it might in principle have done; or that there might be laws forbidding the merging of such disparate regions. In any events, it is usually supposed that our bubbles have a common ancestry, even though our Big Bang is our very own. Some would say that different universes may even be ruled by different physical laws; and others have suggested that collapsing black holes may be suitable seeds for new expanding universes. Those who regret that the steady-state model is dead and gone might take heart from those versions of inflationary theory that envisage our world as a single island among many in a steady-state background.

One of the more paradoxical of consequences of the inflationary models is that they see our universe, ten billion light-years and more across, and perhaps extending even further beyond our present horizon, as having had its origins in a minuscule space less than the size of a proton. Such humble origins become possible, if the idea is to be accepted at all, because *the total energy is almost zero*. This seems to flout the principle that nothing can be created out of nothing, *ex nihilo nihil fit*. Such a situation was held to be possible, however, once the gravitational energy of the universe's contents are understood as negative energy, counterbalancing their rest energy (the mc^2 in Einstein's famous and familiar equation $E = mc^2$).

MONOPOLES AND COSMIC STRINGS

One of the successes of inflationary theory was an explanation it made possible of why we do not observe monopoles (single magnetic poles) in the universe. In 1931, Paul Dirac asked himself why electrical charges, positive and negative, as on protons and electrons respectively, can exist in isolation, while magnetic poles, north and south, cannot. Divide a magnet into two, and two smaller magnets result, dipoles still. He finally produced an explanation of why electrical charges were quantized in this way, but in doing so had to postulate the existence of monopoles with a certain mass—an unknown quantity, but assumed to be small—and magnetic charge. He saw that they cannot be numerous, or they would neutralise the magnetic fields we find at many levels, from the Earth to the Galaxy. Dirac's monopoles nevertheless secured a place in later quantum theory.

When, in the early 1970s, attempts were being made to produce a grand unified theory, several proposed solutions were found to indicate the presence of a magnetic monopole particle, or rather a range of particles known as dyons, of which a monopole is the most basic state. Their magnetic charge was predicted. Eventually, monopoles were introduced into cosmology, where they were presented as "knots" in the vacuum of the early universe, at the stage when the vacuum had energy. Andrei Polyakov and Gerard 't Hooft, working independently, found that massive monopoles, 10^{15} times as large as ordinary baryons, should be expected to have been created and to have survived. This was doubly embarrassing, for not only would they destroy, for instance, the magnetic field of our Galaxy, they would have provided more mass than that in the rest of the universe, dark matter estimates included. One of Guth's early successes was that he explained away this "monopole problem," claiming that according to inflationary theory the monopoles would be diluted during the inflationary period to such an extent that the chances of meeting even one, in our whole Galaxy, would be virtually zero. True believers in inflationary theory who were by profession astronomical observers could hardly be expected to search for signs of something so rare. However rare, they should be expected to seek out magnetic fields, and one choice target would be neutron stars, where fields are strong and where matter was packed densely enough for encounters with particles to be probable. It was calculated that much heat would be generated in such encounters, and that this should manifest itself as X-rays. No such evidence was found, and monopoles were relegated to the sidelines of astronomy.

Mathematically, monopoles are represented as point-like, zero-dimensional, knots or flaws in space. Two-dimensional defects in the vacuum were also considered, but these "domain walls" were found to be even less likely to exist. The creation and survival of one-dimensional defects in the vacuum, "cosmic strings," were taken very seriously. The idea was pioneered by the physicist Thomas W. B. Kibble, of the Imperial College of Science and Technology, a part of the University of London. In a paper published in 1976, he addressed the question of what should happen in the brief interval of time over which the separate physical forces were replacing the overweening superforce. (This was of course before the inflationary theory had come on the scene, but his ideas were easily transferred to that.) His model suggested that the rapid cooling after the explosion of the universe caused flaws that were string-like—comparable to the cracks often seen on a glazed dish, but in three-dimensional space. Kibble described them as slender strands, tubes thinner than a proton, of highly concentrated mass-energy. They might be expected to stretch across the universe, carrying as much as 10^{17} tonnes per meter. The strings could have no ends, but might be small if in the form of closed loops.

Kibble's ideas were not widely known until, in the early 1980s, Yakov Zel'dovich and Alexander Vilenkin, at Tufts University in Massachusetts, realized that in them might lie the secret of the coherent clumping of matter in the early universe, evident today in stars and galaxies. This was the question that cost early supporters of inflationary theory so much trouble, and the string idea had compelling attraction to those who puzzled over the existence of sheet-like groups of galaxies and the Great Attractor, for it was argued that ours and other galaxies are possibly being drawn by gravitational or magnetic fields caused by cosmic strings. Kibble, Vilenkin, and Neil Turok, considered ways in which loops of cosmic string might act as places around which galaxies would have formed, although this idea presented several difficulties. It was then generally abandoned, after Paul Steinhardt and Neil Turok developed mathematical and computational techniques which decisively disproved cosmic strings as a mechanism for galaxy formation.

In the late 1980s, numerous theoretical studies were published, adding to the theory for which no direct empirical evidence had ever been found, but which had great mathematical attraction. It was shown how cosmic strings could evolve by vibrations causing part of a string to detach itself, making strings of all sizes possible. It was suggested that they should oscillate near the speed of light and give off gravitational waves, ripples in space-time. By vibrating and giving off energy, they would ultimately disappear. If they are never found, then—like monopoles—they will no doubt continue to be supported by those who can show why they have all disappeared, or that they are indeed present, but are much more slender than was expected. How might they be found? As propagators of gravitational waves, they might be found indirectly, in the course of more general surveys with such waves in mind. The most promising way of detecting the strings was thought to be by gravitational lensing, for a galaxy behind a long cosmic string would be expected to appear as two identical images, one on each side of the (invisible) string. If such were found, then the mass per unit length of the string would be calculable. The first serious claim to have found just such a lensing effect—which would be different in character from the lensing caused by a compact gravitational intermediary—was presented in 2005 in a paper by Mikhail Sazhin of the Capodimonte Astronomical Observatory, Naples, and the Sternberg Astronomical Institute, Moscow. He and his colleagues found this evidence in a double image of the object CSL-1, found using the European Southern Observatory's VLT at Paranal, Chile. The two images seemed to them to match so closely that—if this was indeed a case of lensing—they could not have been caused by a single compact concentration of matter intervening between us and the source.

The inflationary theories held center stage into the next millennium, despite murmurings in plenty from a few working astronomers, deeply suspicious of an invasion by people hailing from institutes of theoretical physics. Before long, however, most astronomers working in areas touching on cosmology began to accept the concept of inflation and the predictions it spawned. It is not easy to define a term like "cosmological theory" narrowly, but by most definitions it would be said that the number of papers in the subject published annually in the early 1960s was fewer than a hundred, while by 1980 it was around six hundred, most of them by physicists and mathematicians, few of whom were in close contact with the observational procedures needed to confirm their ideas. There were, of course, those in the subject who still objected even to theories of Big Bang type altogether. Halton Arp, for example, was an eloquent rearguard skeptic with much observing experience at a high level who doubted that redshifts on a cosmological scale are primarily indicators of recession velocities. He was disturbed especially by the case of quasars, and some of his challenging photographs were far too glibly dismissed by his colleagues. For most astronomers, however, this part of the debate belonged to ancient history. When the cosmological boat was severely rocked by astronomical observations made toward the end of the millennium, the evidence came from a very different direction.

For decades, astronomers had been measuring the expansion rate, the Hubble "constant," and even into the 1990s some of their results were differing by a factor of two. (They varied between roughly 50 and 100 kilometers per second per megaparsec. NASA's space telescope at one stage announced a figure of 70, with an uncertainty of ten percent. Using a very different approach, based on fluctuations in the structure of the cosmic microwave background, WMAP narrowed the odds to $71 + 0.04 / - 0.03$.) The density parameter Ω was likewise regularly estimated, with a view to deciding on the probable rate of the deceleration that almost everyone had grown used to accepting. And then, in 1998, two teams of astronomers independently announced new results for those parameters, doubly uncomfortable results, since they were in broad agreement.

As distance indicators, both teams used Type 1 a supernovae, extremely bright exploding stars that, as we have seen, were thought to be reliable guides, since they have nearly identical intrinsic peak luminosity. One group, the Supernova Cosmology Project, headed by Saul Perlmutter of the Lawrence Berkeley National Laboratory in California, reported measurements of the apparent brightness and red shift of each of forty-two such supernovae. The High-Z Supernova Search Team, headed by Brian Schmidt of the Mount Stromlo and Siding Spring Observatories in Australia, reported on sixteen such stars. Both teams came up with the astonishing result that, not only is the rate of expansion not decelerating, it also appears to be *accelerating* slightly.

These announcements produced shock waves in the profession, and quickly led to various revisions of existing theories. (The Chandra X-ray Observatory later confirmed the original findings, through observations of gas in galactic clusters at multimillion degrees Kelvin. In 2006 it was announced, more precisely, that the acceleration began six billion years ago.) In 1998, when the two bombshells fell, there was one concept already waiting in the wings, namely the self-repulsive gravitation, of which use had been made in inflationary theory. As we have seen, this concept, associated with the quantum activity of the vacuum, had only been used for understanding the inflationary era, not for a picture of the later universe. The discovery of acceleration in the expanding universe was nevertheless now commonly interpreted to mean that the bulk of the energy in it is neither matter nor radiation, but a gravitationally self-repulsive component. Following a coining by Michael Turner, it became known as "dark energy." Newtonian gravitation did not recognize such a thing, but—as we saw earlier—Einstein's theory allowed it in the guise of the cosmological constant. The fact that this became the favored solution shows how ingrained the acceptance of inflationary theory had become by the end of the century.

It was not the only solution offered, however. That old stand-by, a modification of the laws of gravitation at large distances, was another. Some even went so far as to ask whether dark energy is an illusion, created by the fact that we do not understand gravity. Another possibility considered was that the dark energy is connected with the extra dimensions predicted by superstring theory, and that it might help in putting that to the test. Other authors adjusted inflationary theory itself, or rather, that part of it that dealt with those perturbations in the primordial density that later showed up in the cosmic microwave background. Out of the several explanations offered, we shall here single out just one, to illustrate how precariously the superstructure of theoretical cosmology was balanced on observation. It was a theory put forward by Paul Steinhardt of Princeton University, in 2002. This is doubly interesting, since he had been one of the prime movers of inflationary theory, twenty years before.

Steinhardt's proposed new model was one undergoing a cyclic evolution, with a potentially infinite past and future. As we know, there had been cyclic models before, in the earliest years of relativistic cosmology, but there were serious problems with them, such as infinite temperatures and densities at successive "crunches," and also problems with entropy. Other difficulties had been found to remain, when attempts were made to blend cyclic models with inflationary theory.

Steinhardt took it as read that the past was a period of deceleration, and that matter and radiation had only been losing their significance in a relatively recent slice of the universe's history. He assumed that whereas, formerly, structures were forming on ever-larger scales, accelerated expansion was bringing this process to an end. Describing a typical cycle, starting from "the bang," he put the temperature at $20^{10°}$ K, a finite maximum. While ideas were borrowed from older inflationary models, there was no inflation in his new model, which passed directly to a radiation-dominated universe, in which are formed the usual nuclear abundances; then to a matter-dominated universe, in which the atoms and galaxies and larger scale structures form; and then to a phase of the universe dominated by dark energy. It is this dark energy which ultimately drives the universe, and in fact drives it into cyclic evolution. The acceleration it produces is a hundred orders of magnitude smaller than the acceleration required in the inflationary models. Given enough time, however, it can achieve the same results. Over time, acceleration thins out the distribution of matter and radiation, making this more and more homogeneous and isotropic, even perfectly so, driving it into a vacuum state. At the same time as the universe is made homogeneous and isotropic, it is also being made geometrically flat. Euclidean geometry is back with us.

Rather as in inflationary theory, the dark energy only survives for a finite period, but Steinhardt explains how, in his new model, it triggers a series of events that result in the start of a new period of expansion of the universe. The fine detail of all this is worked out using the concepts of superstring theory. His virtuoso description of the cavortings of "branes" cannot be described here, but what emerges at the end of it all is that the expected spectrum of fluctuations in the distribution of mass, energy, and temperature is essentially the same in both this and the old inflationary model. And this despite the fact that the assumed physical processes are very different, and the time scales likewise: the work of 10^{-30} seconds now takes place over billions of years.

The historical situation in the new millennium was thus strongly reminiscent of that in the 1930s, when it was shown that neo-Newtonian cosmology could predict redshifts similar to those predicted by general relativity. In both cases, however, the physics underlying competing theories was different enough for it to be possible to find tests discriminating between them. In inflation, there are fluctuations created in energy and temperature, but also in space-time itself, namely, gravitational waves. This is not the case in Steinhardt's newer model, with its sedate progress through time. It remains to be seen whether this decisive difference will ever be put to the test. It is not the only point on which Steinhardt might be challenged, for he made assumptions about superstring theory that might be questioned. And the more points at which cosmological theory connects with the world, the happier astronomical observers will be.

MANKIND AND THE UNIVERSE

Just as in the late 1940s the steady-state theories aroused much theological hostility, so it was in the 1980s, when various asides by cosmologists—excited by some of the new ways of describing the universe—led to a series of exchanges redolent of those in which Hoyle had earlier played a large part. In much the way it had been for at least three thousand years, the situation was a three-cornered one, involving God, mankind, and the universe. For part of that long period, relations involving God had been treated in three different ways. Some had maintained that one could prove God's existence on the basis of what was known of the universe. (The most famous attempts were Thomas Aquinas's "five ways," his "cosmological argument" being that the universe is an effect which must have a first cause, which we call God.) Some had held that God's existence could be positively disproved, using similar evidence. And some had taken an intermediate position. The middle group was not a single cohort. Some were simple agnostics, but some were firm believers. It is interesting to observe how the Abbé Georges Lemaître avoided drawing parallels between his Primeval Atom and the biblical story of Genesis, and refused to enter into traditional philosophical discussions of a creation out of nothing, to support his faith. Far from being attracted to his theory because it somehow bolstered the biblical story, Lemaître actually took strong exception to the way Pope Pius XII gave it such a role, in 1951. It was enough for Lemaître that his theories were compatible with his belief. God was not to be regarded as a cosmic impresario who had pressed the button that started the fireworks.

The middle ground was occupied by a large number of people, working in and around scientific cosmology, perhaps even a majority, who gave very little thought to these questions. The marginalizing of mankind in spatial terms by Copernicanism and its successors had helped to weaken theology's traditional links with the subject, and theologians who wished to keep their heads below the parapet were content to insist that, while the scientist explains *how* things happen, *how* the universe evolves, they, the theologians, had the answer to *why* things happened. For more than a century, they had become used to attacks from atheistic and agnostic evolutionists, whose arguments might have run along the following lines:

THEOLOGIAN: By God's good grace, animals act considerately toward their offspring.

RESPONDENT: No. Only those species of animals with this property survive to produce later generations.

Or, more pointedly:

THEOLOGIAN: Only by God's good grace do we inhabit a universe perfectly suited to our needs, that is, satisfying the conditions necessary for our existence.

RESPONDENT: God may or may not be responsible, but we should not be surprised that we encounter conditions suited to our existence. If they did not exist, we should not exist. Our existence proves nothing.

There were to be echoes of this last exchange in late twentieth-century cosmology, when several writers tried to explain some fundamental aspects of the universe on the basis of human existence.

Throughout history there have been those who have argued that the form of the world as described by us is dependent on categories and principles that we ourselves create. The Mind is in some respects the Maker of Nature—as the philosopher Immanuel Kant, among others, had insisted. What we see is likewise governed to some extent by what we are, as biological entities. Astronomers have long been conscious of the limitations of the human senses, and of the fact that there is a world of radiations—Herschel's infrared, Ritter's ultraviolet, and so forth—of which we can only be aware with the help of instruments. Many of our relationships with the universe depend on the character of lifeforms—of language, of our physiological nature, and so forth—but there are others that have to do with our very material existence. The elements needed for the existence of the living organisms known to us include nitrogen, oxygen, phosphorus, and above all, carbon. At the primordial explosion of the universe, it is now said, hydrogen and helium were synthesized, and only after a very long process were they converted into the heavier elements, chiefly in the interiors of stars. The death of stars is followed by the dispersion of those elements, and by their conversion into planets and lifeforms. All told, a time of the order of at least ten billion years is needed before this can happen, and it can therefore be seen as not in the least surprising that we—a carbon-based form of life—find ourselves in a universe more than ten billion years old. To this extent there is be nothing inherently occult or mystical in saying that there is a connection between mankind and the universe. A sufficient cause, however, is not a necessary one, and there were those who wanted to make it necessary, and say that the presence of human life constrains the physical conditions that guarantee the universe's habitability.

Most attempts to assess the probability that life exists elsewhere in the universe have rested heavily, not only on conditions elsewhere, astronomically judged, but also on what we know of the biological properties of Earthbound life. This does not address the (highly ambiguous) question "why are we here?" but it does provide a context for studying the conditions under which life could have developed, and, for example, how different physical circumstances—even changes in the expansion rate or the timescale of the universe—would have affected the result. This is not new. Thomas Chamberlin, for instance, believed in some sort of atomic power supply for the Sun, simply because the known alternatives would have made the Sun's life too short for other known processes, evolutionary and geological, to have been completed. The steady-state models were similarly designed to save the embarrassment of a short time scale, in this case one of a whole universe younger than its contents. What was new in the last quarter of the twentieth century, however, was an awareness of how extremely fine-tuned the universe is as a repository of life.

The roots of the new discussion go back to the 1930s, when Eddington, Dirac, and others were trying to find relationships between the fundamental constants of physics, and when some of Dirac's ideas suggested that there might be a slow change with time in some of the "constants" of Nature. Robert Dicke spent several years examining astronomical and geological evidence for such a change. In 1957, he argued that the number of particles in the observable part of the universe, and Dirac's numerical coincidences, were not random, but were conditioned by biological factors. If the universe were appreciably older than it is, he said, all of the stars would be cold, and under those conditions mankind would simply not be around to survey the scene. We live at a very special time; but if it were not special, we should not be alive to know it.

In 1961 and later, Dicke presented his ideas with more detailed calculations, proposing the principle that the conditions necessary to produce life involve certain relationships between the fundamental constants of physics. (Whether this should be called the "weak anthropic principle" is a moot point. That expression, as we shall see, came later.) For example, a slight change in the charge on the electron, with other physical constants held unchanged, would have made the mechanism of nuclear fusion in stars unworkable, and so would have made our form of life impossible. Martin Rees, a Cambridge astronomer of wide experience who was eventually appointed Astronomer Royal, spent much time from the 1970s onward investigating the detailed and quite dramatic consequences for the evolution of the universe if no more than the gravitational constant were to be slightly changed. Life might have appeared, but not at all as we know it. Parallels might be drawn with the evolution of the universe as described by later Big Bang theories: they require a carefully chosen blend of initial conditions in order to make possible a continuation of expansion, the creation of stars, galaxies, and so forth. Or do they? Varying initial conditions and laws in certain ways might have stood in the way of our present situation coming into being in the way we think it did; but might our present situation not have come about in another way? The game is to rule out all possibilities but one. We are almost back in the sixteenth-century debate about alternative explanations for planetary motions.

In 1974, Brandon Carter coined the expression "anthropic principle" to describe what amounts to a simple logical requirement: intelligent beings cannot find themselves in regions that are uninhabitable by intelligent beings. No one will doubt the logic. One could say that behind the principle there is a recommendation that astrophysicists and others should beware of ignoring it when speaking of the fine-tuning that makes our world fit for life. Are they likely to do so? One could say that Carter was removing the causes of our astonishment at our own good fortune. His was a principle that is easily extended to cover anything whatsoever that is a necessary physical condition of life or intelligent life, and if it has stirred little debate, that is surely because it has been seen as self-evident.

There is an argument with a different starting point, concluding that the physical nature of the universe is such that it must have within it, at some stage, living beings, and mankind in particular. Carter gave the name "strong anthropic principle" to the principle that the universe must be knowable, and must at some stage "admit the creation of observers within it." Its name is unfortunate, for it is not a stronger form of the simple logical principle at all. It comes in various forms. In 1986, John Barrow and Frank Tipler made a resolute defense of a strong teleological principle, arguing that the universe is habitable simply in order that intelligent life can evolve in it. Habitability is a goal, the end of the universe, so the latter must be habitable, where "must" has a strength that Aristotle would have recognized, but that is not the "must" of simple logic. Books, rather than a few sentences, are needed to discuss the legitimacy of teleological explanation, which is easiest to accept when a divine guiding force is pushing events along. Immanent tendencies are no longer in fashion. Barrow and Tipler went even further, in hinting at a "final anthropic principle," that life must never disappear. (The word "anthropic" is used very loosely in all of these debates. It should be used to refer only to human life, and there was no suggestion that human life will survive to the end of time.)

Some of those who have favored a form of the strong principle have proceeded by trying to describe, in very broad but acceptable physical terms, all possible worlds, some inevitably incapable of supporting life, and some capable. The next step then is to try to pinpoint structural qualities that are necessary, if observers are to be generated. However remote the chances of this program's success might seem, much of the work done in cosmology and astrophysics in the second half of the twentieth century, while done without any thought for this general problem, can be fitted into the scheme of the argument. Fred Hoyle, for example, was led to consider the anthropic principles quite naturally, as a consequence of his discovery of a series of coincidences in the relations between certain properties ("nuclear resonance levels") of biologically important chemical elements. His best-known example concerns the properties of the nucleus of the carbon atom, fine-tuned in a way that allows carbon to be manufactured from helium inside stars. If the theoretical properties of carbon had differed by even a very small amount, that process could not have taken place, there would be no carbon in the universe, and no element with atoms heavier than those of carbon. Living creatures, of course, are largely built out of molecules containing carbon atoms. No fine-tuning, no life.

This type of evidence can be fitted into all of the arguments considered here, even into religious arguments that Hoyle had no wish to support. Of the principles considered here, it is hard to see how those that are not simply logically true may function scientifically. They may alert us— or rather force us to alert ourselves—to certain characteristics of the world we live in, and may even help in the design of hypotheses, by ruling out certain options. Do they lend themselves to confirmation or refutation? Ways have been suggested, starting from the idea that ours is a privileged universe in a multiple universe. There are several ways of introducing the last concept, some people speaking as though many universes actually exist, perhaps seeded from black holes, while others speak of them as merely existing potentially. The choice mechanism for singling out the best of the potential bunch presents obvious problems, which we shall ignore, but an idea put forward by the American cosmologist Lee Smolin is worth mentioning. It brings us back full circle, in a sense, to the old question of animal and plant evolution. He suggested that, in the course of a black hole collapsing, another might sprout from its interior, with physical laws differing from those in the parent, but differing only to a very slight degree. After many generations, the game would go to the universes that were most efficient at generating offspring. The test would be to decide whether we live in a universe that is in some sense or other maximally efficient at spawning new universes, through Smolin's black hole mechanism or something similar to it. What future historians will make of such twentieth-century speculation is anyone's guess. (Is the mother of our universe perhaps also capable of sustaining life?) The theory does, at least, offer hostages to observational fortune.

Judging by the literature they have generated, much of which is directed to the general public, "anthropic" principles are capable of attracting an enormous amount of interest. This is no doubt because they are thought to touch on deep, even religious, truths. Those who claim to answer the question "Why are we here?" do seem to be either supplementing or challenging religion, but most would not admit to this. There have been some, however, who have done one or other of those things—in some cases seriously, in others puckishly. The most widely known and persistent antagonist in this respect has been Stephen Hawking. His record-breaking publication of 1988, *A Brief History of Time*, provided an opportunity for him to explain his skepticism at the same time as explaining away a conceptual difficulty people have with a first moment in time.

One of the classic objections to the Big Bang models was that in the conditions of high density obtaining in the neighborhood "time zero," we simply do not know what physical theories can be applied. We recall the model within quantum cosmology that was worked out in the 1980s by Stephen Hawking and James Hartle, in which they found a clever way of avoiding talk of a beginning to the universe, and explained it using an analogy with the situation at the north pole. There is no latitude above 90°, but one is not at an extreme of the Earth's surface. You can even walk through the pole without being subjected to any special physical force. By analogy, as he wrote, "The quantity that we measure as time had a beginning, but that does not mean spacetime has an edge." By removing the edge from time, his cosmological theory avoided the deadly singularity that others feared meant a breakdown in the laws of physics, but what conclusions did he then draw? He made plain his belief that, since his universe has no beginning or end, no boundary or edge, therefore there is no place in it for a Creator.

Hawking announced some of his technical findings on this subject to a meeting held at the Vatican in 1981. Despite his expressed skepticism about the existence of God, he was not afraid later to liken his ideas to those of St. Augustine, who said that God made time with the universe. What did he mean by drawing parallels between himself and an influential Christian writer who believed that God was the Creator of the universe? Augustine is not a physical or philosophical construct. He lived, and believed, and was a highly coherent thinker who explained why God is needed, not just at the beginning of the world, but at every subsequent point in time, to keep it in existence. To return, however, to Hawking: one of the most often quoted of his statements is one that he made at the very end of that same book. If we do ever discover a complete theory, he suggested, then we shall all be able to take part in the discussion of why it is that we and the universe exist. If we find the answer to that, it would be the ultimate triumph of human reason—for then we would know the mind of God.

Nowhere does he explain what he means by the question of why we exist. To the typical theist, human life has no meaning in a universe without God, so that the question of why we exist contains a covert assumption of God's existence. It seems that when Hawking—like many other cosmologists—refers to a God in which he professes not to believe, he does so only for rhetorical purposes, and falls back on the language with which most of his readers were brought up. We are all of us, by our upbringing, bound up with history, apparently even to the point of giving answers to questions that we believe to be strictly meaningless. When cosmologists make casual reference to God the Creator, for or against, they seem usually to have in mind a theology of creation that demands a sudden beginning in time. This is the common religious view, based on the biblical story of Genesis, but it is not a necessary part of the theology of creation. As Aquinas pointed out long ago, even a world without a beginning needs a reason why it should exist at all. At all events, in an Oxford lecture given eighteen years after Hawking's book appeared—a lecture being heard also by an audience in Trieste, in Italy, through a video link—Hawking was still to be heard using theological language. After again explaining his neat polar analogy, he went on to say that "We are the product of quantum fluctuations in the very early universe." And then, echoing Einstein's dictum, "God really does play dice." God?

One should not exaggerate the numbers, but cosmologists do appear to be more prone to make excursions into theology than other scientists. They are drawn as a moth to a candle. And to explain the moth's movements, a knowledge of the moth is perhaps in the long run more useful than a knowledge of the flame. Personal biography cannot be entirely ignored as a determinant even of modern science. James Jeans, Cambridge-educated mathematician, seeing the universe as something built on a complex mathematical pattern, insisted that it must be the product of the mind of a divine mathematician. This is almost pure Plato. Jeans did not exactly claim to know God's mind, but he used the existence of highly successful mathematized science as evidence for the existence of a Creator, and so as a prop to his religious belief. One could produce a hundred instances of writers who had come to a similar conclusion before him. For the Quaker Herbert Dingle, the steady-state theory violated energy conservation, therefore it was miraculous, therefore it was wrong. For the theist Sir Bernard Lovell, it was miraculous, therefore if it was right it required God. The Benedictine Stanley Jaki diagnosed the three authors as having a deep-seated pagan longing for an eternal universe. Words like "creation" can still have an explosive effect, when they land in the right place.

The great majority of astronomers today are of a more sober cast of mind, content to see themselves as offering no more than a scientific account of the world we inhabit, an account essentially no different from those offered by the other physical sciences. Astronomy has been many things in its time, not least of which is a prototype for the more fundamental sciences. It has provided them with their methods, with laws of motion, and with empirical data, later borrowing back from the other sciences with interest. Astronomy remains for some people what it has been throughout its long history, a playground for unrestrained metaphysical and theological speculation, but this is no longer at its core. When modern astronomers in one breath profess themselves to be agnostic, and in the next describe the ripples revealed by COBE as "traces of the mind of God," this tells us only that theological sophistry is not what it was. Astronomy's roots have long been entangled with those of religion, and are still not completely free. It is one of the ironies of history that the study of such a vast and impersonal subject matter should, from beginning to end, have been so intimately bound up with principles of human nature.

Bibliographical Survey

The following bibliography is highly selective—it would not be difficult to amplify it a hundredfold. It is confined mainly to books rather than articles, and to books with a broad coverage of the subject matter, wherever possible, rather than to narrow monographs. While it includes a few references to scientific texts, its chief aim is historical. It is aimed at an English-speaking audience, although to avoid undue bias, many works without equivalents in English have been included.

This book was meant to be more or less self-sufficient for those most likely to read it, but for those who are relative strangers to astronomy and who wish to look deeper, two suggestions are offered. We may divide astronomy texts very roughly into two classes: mathematical "spherical astronomy," where at an introductory level a standard text of a century ago differs little from its recent equivalents; and physical astronomy, where current wisdom changes almost daily. The first is mathematically well covered by W. M. Smart's *Text Book of Spherical Astronomy*, 6th ed. (Cambridge: Cambridge University Press, 1985; orig. publ. 1931). In the latest edition, edited by Robin M. Green, the language has been updated where necessary to conform with current astronomical practice. For a single source of information on mainline physical astronomy that is beautifully illustrated and readable with a minimum of prior scientific knowledge, one could seek out the latest edition of *The Cambridge Atlas of Astronomy*, 3rd ed. (ed. Jean Audoze and Guy Israel, Cambridge University Press, 1994). This originated in *Le Grand atlas de l'astronomie*, edited by Jean Claude Falque under the general editorship of Jean Audouze and Guy Israel (Paris: encyclopedia Universalis, 1983), and the second English-language edition appeared in 1988. For a more historical approach, but with useful surveys of the scientific content of cosmology up to the time of its publication, see Norriss S. Hetherington (ed.), *Encyclopedia of Cosmology: Historical, Philosophical, and Scientific Foundations of Modern Cosmology* (New York: Garland, 1993); and for astronomy, more narrowly defined, see John Lankford (ed.), *History of Astronomy: An Encyclopedia* (New York: Garland, 1997).

If a star atlas is wanted, the choice is great when locating brighter objects in the sky, but the 18th edition of *Norton's* might be found useful, with its maps that are now brought up to date through the new millennium: Ian Redpath (ed.), *Norton's 2000.0*

Star Atlas and Reference Handbook (London: Longman; and New York: Wiley, 1989; 20th ed., 2003). Few readers who are not already familiar with professional catalogs of stars and other celestial objects are likely to want references to them. These number in the hundreds and would require a bibliography of their own. A recent tendency is to publish them in digital form only. A warning should be given about the tangled history of star names. R. H. Allen's *Star Names, their Lore and Meaning* (1899; repr. New York: Dover, 1963) has done valiant service, but is not altogether reliable. It should be supplemented where possible by the excellent but very brief work by P. Kunitzsch and T. Smart, *Short Guide to Modern Star Names and their Derivations* (Wiesbaden: Harrassowitz, 1986), where further bibliographical sources will be found. See also the essays by Paul Kunitsch in his *The Arabs and the Stars* (Aldershot: Variorum, 1989).

Since astronomy makes extensive use of data from the past, historical studies of a basic kind were often composed by astronomers themselves in the past, even in the ancient world. One of the genre that dedicated historical scholars will wish to consult was P. J. B. Riccioli's truly monumental *Almagestum novum* (Bologna, 1653), but this is in Latin, and it will not be found in many of even the best libraries today. Other classical works of history are those by J. B. J. Delambre, beginning with his *Histoire de l'astronomie ancienne* in 2 volumes (Paris, 1817) and including further volumes on medieval (in one volume, 1819), "modern" (two volumes, 1821), and eighteenth-century astronomy (one volume, 1821). These are fundamental sources and have been reprinted (New York and London: Johnson Reprint Corporation, 1965–9, in the original French). Delambre's work was motivated by the need to use history as an astronomical tool, but ended by setting high standards in intellectual history as such. Not in the same class, but certainly encyclopedic, is Pierre Duhem's weighty *Le Système du monde. Histoire des doctrines cosmologiques de Platon à Copernic*, 10 vols. (Paris, 1913–1959). Excerpts from this work have been issued in an English translation by Roger Ariew under the title *Medieval Cosmology* (Chicago: University of Chicago Press, 1985). In many ways more reliable is J. L. E. Dreyer's much shorter general work of the same period, *A History of Astronomy from Thales to Kepler* (London, 1912, reprinted New York: Dover, 1953).

As modern examples of the successful use of history for modern astronomical purposes, see F. Richard Stephenson and David A. Green, *Historical Supernovae and their Remnants* (Oxford: Oxford University Press, 2002); and F. Richard Stephenson, *Historical Eclipses and Earth's Rotation* (Cambridge: Cambridge University Press, 1996). Robert R. Newton published much along similar lines, as for example *Ancient Planetary Observations and the Validity of Ephemeris Time* (Baltimore: The Johns Hopkins University Press, 1976). He had an astringent way with past astronomers, and their successors may well shudder at the thought that they will be similarly handled in two thousand years' time. There is a growing body of computer software available that is capable of handling historical data with the help of modern parameters. One excellent example is Rainer Lange's "Alcyone Ephemeris" software, produced with the collaboration of Noel Swerdlow (visit http://www.alcyone-ephemeris.info/ for comprehensive documentation).

Far more extensive in scope than its title suggests, and containing the answers to most questions the reader is likely to have on the relations of astronomy and the

Christian ecclesiastical calendar, is G. V. Coyne, M. A. Hoskin and O. Pedersen (eds.), *Gregorian Reform of the Calendar: Proceedings of the Vatican Conference to Commemorate Its 400th Anniversary, 1582–1982* (Vatican City: Pontifical Academy of Sciences and Specola Vaticana, 1983). *Calendrical Calculations* by Nachum Dershowitz and Edward M. Reingold (Cambridge: Cambridge University Press, 1997) provides a description of 14 calendars of historical importance from several cultures, with algorithms to allow them to be programmed for computer use.

Articles on the history of astronomy, as a part of the history of science, will be found in the many journals—several hundreds, in all—devoted to that wider subject. Most of the journals supported by national organizations are now effectively international in character—for example, *Isis,* the journal of the American History of Science Society, and the *British Journal for the History of Science.* Both of these are written almost entirely in English. *Centaurus* is another established journal with much astronomical content from many periods. *Suhayl,* published by the Faculty of Philology of the University of Barcelona, Spain, includes many articles (largely in English) on the history of Islamic and related Western astronomy. Articles in five or six world languages will be found in the journal of the International Academy of the History of Science, *Archives internationales d'histoire des sciences* (published by Brepols of Turnhout, Belgium, formerly published by the Institute of the Italian Encyclopedia, Rome). A specialist publication of great value is the *Journal for the History of Astronomy* (Chalfont St. Giles, England: Science History Publications), which from 1979 to 2002 was published in tandem with a supplement titled *Archaeoastronomy*. Such material is now incorporated in the main journal, and the title *Archaeoastronomy* is now in use by another journal.

Among international bodies responsible for organizing conferences of relevance to the subject, the IAU (International Astronomical Union, with closed membership) publishes extensively, for instance with Kluwer (Dordrecht) and Springer-Verlag (Berlin). For its history, see Adriaan Blaauw, *History of the IAU: The Birth and First Half-Century of the International Astronomical Union* (Dordrecht: Kluwer, 1994). The IAU has a historical commission. The historical division of the IUHPS (International Union for the History and Philosophy of Science, having countries rather than individuals as members) has plenary meetings every four years; and since it often publishes papers given at them, may provide an idea of work in progress. One of the most useful of all bibliographical sources is the ongoing *Isis Critical Bibliography*. First published as occasional supplements to the journal *Isis,* a series of large bound volumes edited by Magda Whitrow was begun by the History of Science Society (of the United States), combining in the first instance bibliographies for the years 1913–1965 in three volumes (London: Mansell, 1971–1976). These and continuing volumes list a very large proportion of books and articles in the history of science from many countries. (For the latest information on the society and that important bibliography, visit http://www.hssonline.org/.)

There are various surveys that simplify the task of navigating through the wider subject. The advent of the Internet, however, has changed the situation dramatically, and the use of such search engines as Google is often a reasonable substitute for reference books—although some skill is needed, to avoid secondhand—indeed, even tenth-hand—nonsense. The online catalogs of such national libraries as the U.S. Library of Congress, the British Library, and the French Bibliothèque Nationale, not to mention leading university libraries, are invaluable bibliographical aids. Biographical information in relation to specific persons can often be found more quickly in various national biographies (such as the *Oxford Dictionary of National Biography,* now available online to subscribers at http://www.oxforddnb.org/). For short biographies, Thomas Hockey, ed., *The Biographical Encyclopedia of Astronomers* (New York: Springer, 2007), with its 1550 entries on astronomers from antiquity to the twentieth century, will be found chiefly in well-endowed libraries.

Biographies of scientists from a much wider field, together with partial bibliographies, have long been available in German in an important compendium usually known simply as "Poggendorff." Johann Christian Poggendorff's first two volumes of *Biographisch-Literarisches Handwörterbuch* (Leipzig: Sächsische Akademie der Wissenschaften) appeared in 1863. After his death, a third volume appeared in 1898, covering the period 1858–1883, and a fourth was published in 1904. There have been many reprints and other versions since. Another German venture worth remembering is the *Encyklopädie der mathematischen Wissenschaften* (Leipzig: Teubner, 1898–1935), of which section VI includes two volumes of astronomical material, edited by Karl Schwarzschild, Samuel Oppenheim, and Walther von Dyck. George Sarton, in his *Introduction to the History of Science,* 3 vols. in 5 parts (Washington: Carnegie Institution, 1927–1948), made an ambitious attempt to give a bio-bibliographical outline of the subject for the whole of history, but only reached to the second half of the fourteenth century. These weighty volumes are now dated, in tone and in fact, but are still useful.

By far the most important single English-language biographical source of scientists is now Charles C. Gillispie (ed.), *Dictionary of Scientific Biography,* 15 vols. (New York: Charles Scribner's Sons, 1970–1978). There are plans to revise and update this seminal work. An abbreviated version appeared in one volume in 1981, but errors crept in during the abbreviation, and it is best avoided. The same publisher issued a series of illustrative volumes to supplement the *Dictionary,* under the general title *Album of Science,* and all are of some relevance to our theme: John E. Murdoch (ed.), *Antiquity and the Middle Ages*; I. Bernard Cohen (ed.), *From Leonardo to Lavoisier, 1450–1800;* L. Pearce Williams (ed.), *The Nineteenth Century*; and Owen Gingerich (ed.), *The Physical Sciences in the Twentieth Century.* For the barest outlines of the biographies of some 30,000 scientists, half of whom were living in 1968 at the time of publication, see Allen G. Debus (ed.), *World Who's Who in Science* (Chicago: Marquis Who's Who, 1968).

There are several modern histories of astronomy covering long historical periods. For the ancient world, the most complete survey of astronomy with a mathematical content is Otto Neugebauer's *A History of Ancient Mathematical Astronomy,* 3 vols. (New York: Springer-Verlag, 1975), referred to below as *HAMA.* Willy Hartner's *Oriens-Occidens,* 2 vols. (Hildesheim: Olms, 1968, 1984) covers such a great range of history that it may be placed in this category. *Astronomy Before the Telescope* (ed. Christopher Walker, London: British Museum, 1996) is attractively illustrated in color, and its title explains its wide coverage. Other general works are mentioned below, including odd volumes of *The General History of Astronomy,* a project now abandoned. Published under the auspices of the IAU and the IUHPS, this was originally meant to comprise four volumes, each in more than one part. Some were abandoned before publication, but relevant here are R. Taton and C. Wilson (eds.), *Planetary Astronomy from the Renaissance to the Mid-Nineteenth Century: Part A* (Cambridge: Cambridge University Press, 1989) and O. Gingerich (ed.), *Astrophysics and Twentieth-Century Astronomy to 1950: Part A* (Cambridge: Cambridge University Press, 1984). A far more ambitious publication, authoritative and beautifully illustrated in color throughout, is the *Storia della Scienza* in ten large volumes, edited by Sandro Pertruccioli and numerous collaborators (Rome: Istituto della Enciclopedia Italiana, 2001–2004). This was planned by the International Academy of the History of Science and the Istituto jointly. There are hopes that the Italian edition will eventually be issued in some form in English, perhaps online. A more modest volume, in scope as well as subject matter, but one that contains much history of astronomy off the beaten track, is *Encyclopaedia of the History of Science, Technology, and Medicine in Non-Western Cultures* (Helaine Selin, ed., Dordrecht: Kluwer Academic Publishers, 1997).

CHAPTER 1

A historical awareness of the astronomical concerns of prehistoric peoples is itself ancient, but little work of lasting value was done before the end of the nineteenth century. Norman Lockyer's works are still of some value, and their shortcomings can often be put down to excessive enthusiasm. See *Dawn of Astronomy* (London: MacMillan, 1894; repr. Cambridge, MA: MIT Press, 1964) and *Stonehenge and Other British Stone Monuments Astronomically Considered* (London: MacMillan, 1909; repr. Cambridge, MA: MIT Press, 1965). Works on prehistoric astronomy should if possible be read together with works giving the archaeological context—such as R. J. C. Atkinson, *Stonehenge* (London: Penguin Books and Hamish Hamilton, 1956, 1979), for that monument. Jean-Pierre Mohen's *The World of Megaliths* (London: Cassell, 1989) is not concerned with astronomy, but provides a well-illustrated survey of megalithic monuments across the world, from Colombia to Japan.

   Archaeoastronomy first reached a high standard with the work of Alexander Thom, whose books include *Megalithic Sites in Britain* (Oxford: Oxford University Press, 1967) and *Megalithic Lunar Observatories* (Oxford: Oxford University Press, 1971). Several articles by him and his son A. S. Thom appear in the *Journal for the History of Astronomy* and its original supplement *Archaeoastronomy* (see above). For the astronomy of Stonehenge, see John North, *Stonehenge: Neolithic Man and the Cosmos* (London: Harper Collins, 1996). See also F. R. Hodson (ed.), *The Place of Astronomy in the Ancient World* (Oxford: Oxford University Press, for the British Academy, 1974). For Stonehenge models, visit http://www.stonehenge3d.co.uk/. *In Search of Ancient Astronomies* (E. C. Krupp, ed., London: Chatto and Windus, 1979), gives a good overall survey at an introductory level of European, American, Egyptian, and other early astronomy of largely nonliterate peoples. The book includes a useful bibliography, although many of the items in it are from the lunatic fringe. See also E. C. Krupp, *Skywatchers, Shamans and Kings: Astronomy and the Archaeology of Power* (New York: Wiley, 1997). *Records in Stone: Papers in Memory of Alexander*

Thom (C. L. N. Ruggles, ed., Cambridge: Cambridge University Press, 1988), includes a representative selection of mostly material from the Northern Hemisphere. Details of the new discoveries in Brazil and the Andes, mentioned in this text, have not yet been published. The current journal *Archaeoastronomy* is issued by two bodies, the Center for Archaeoastronomy and the ISAAC (the International Society for Archaeoastronomy and Astronomy in Culture).

For the nuraghes, see Mauro Zedda and Juan Antonio Belmonte, "On the Orientations of Sardinian Nuraghes," *Journal for the History of Astronomy* 35 (2004): 85–107; and for the Menorcan situation, M. Hoskin, *Temples, Tombs, and Their Orientations: A New Perspective on Mediterranean Prehistory* (Bognor Regis: Ocarina Books, 2001).

CHAPTER 2

Pyramidology has long had its astronomer-adepts, most of them best forgotten. Of interest for its own sake, or rather because it was written by no less a man than the Astronomer Royal for Scotland, is Piazzi Smyth's *The Great Pyramid. Its Secrets and Mysteries Revealed* (London: Bell, 4th ed., 1880; repr. New York: Outlet Book Co., 1990). Note also Norman Lockyer's *Dawn of Astronomy* (see chapter 1 bibliography) as an influential source from the same period. Egyptian technical competence in astronomy has long been exaggerated, but for a more balanced view see Otto Neugebauer's *HAMA* (see the introductory bibliography), his *The Exact Sciences in Antiquity*, 2nd ed. (Providence, RI: Brown University Press, 1957; reissued New York: Dover, 1969), and his *Astronomy and History. Selected Essays* (New York: Springer-Verlag, 1983). With R. A. Parker he produced a magnificently printed study that will be available in some larger libraries: *Egyptian Astronomical Texts,* 3 vols. (Providence, RI: Brown University Press, 1960–1969). More readily available is Parker's topical essay in volume 15 of the *Dictionary of Scientific Biography* (see the introductory bibliography), which includes further bibliographical information. The first port of call now, however, should be Marshall Clagett, *Ancient Egyptian Science,* 3 vols. (Philadelphia: American Philosophical Society, 1992–1995), although he is silent on the question of astronomical alignments.

CHAPTER 3

A compact and authoritative survey is Asger Aaboe, "Babylonian Mathematics, Astrology, and Astronomy," in John Boardman and others (eds.), *The Cambridge Ancient History,* 2nd ed., vol. III, part 2 (Cambridge: Cambridge University Press, 1991: 276–92). Longer works with valuable sections on Babylonian astronomy include B. L. van der Waerden, with contributions by Peter Huber, *Science Awakening,* vol. 2: *The Birth of Astronomy* (Groningen: Wolters Noordhoff, 1950; and Leyden: Oxford University Press, 1974; see especially chapters 2–8); A. Pannekoek, *A History of Astronomy* (London: Dover, 1961, trans. from the 1951 Dutch edition and repr. 1989; see especially chapters 3–6); O. Neugebauer, *The Exact Sciences in Antiquity* and his collected essays (see chapter 2 bibliography); H. Hunger and D. Pingree, *MUL.APIN: An Astronomical Compendium in Cuneiform* (Horn, Austria: F. Berger, 1989). These works, as well as O. Neugebauer, *HAMA*, contain numerous references to fundamental studies by such scholars as T. G. Pinches, J. N. Strassmaier, J. Epping,

F. X. Kugler, A. J. Sachs, and A. Aaboe. H. Hunger has continued the task of putting cuneiform records into print, in a series published by the Austrian Academy of Sciences. On Babylonian eclipse prediction, see J. M. Steele, *Observations and Predictions of Eclipse Times by Early Astronomers* (Dordrecht: Kluwer, 2000).

Bartel van der Waerden contributed a lengthy entry titled "Mathematics and Astronomy in Mesopotamia" to volume 15 of the *Dictionary of Scientific Biography* (667–80; see the introductory bibliography). Note that he supports Willy Hartner's controversial views on the earliest constellations, as does E. Krupp, in "Night Gallery: The Function, Origin, and Evolution of Constellations," *Archaeoastronomy* 15 (2000): 43–63. An important paper on the zodiac is Lis Brack-Bernsen and Hermann Hunger, "The Babylonian Zodiac: Speculations on its Invention and Significance," *Centaurus* 41 (1999): 280–92.

The more traditional approach to Babylonian astronomy tended to concentrate on those methods that were recognizable as sources of later formal astronomy. For a wider historical approach, see Erica Reiner, *Astral Magic in Babylonia* (Philadelphia: American Philosophical Society, 1995); Noel Swerdlow, *The Babylonian Theory of the Planets* (Princeton, NJ: Princeton University Press, 1998); Francesca Rochberg, *Babylonian Horoscopes* (Philadelphia: American Philosophical Society, 1998); and H. Hunger and D. Pingree, *Astral Sciences in Mesopotamia* (Leiden: Brill, 1999). Among the many good introductory histories of the region is Joan Oates, *Babylon* (1979; repr. London: Thames and Hudson, 1986).

CHAPTER 4

To get an idea of how essentially primitive Greek cosmology was before Eudoxus, even in the brilliant *Timaeus* of Plato, see the translation of that, with running commentary by F. M. Cornford, in his *Plato's Cosmology* (London: Routledge, 1937). D. R. Dicks, *Early Greek Astronomy to Aristotle* (London: Thames and Hudson, 1970), reexamines some of the traditional views of its subject. Pioneering work by Delambre has been mentioned earlier, as has Neugebauer's *HAMA*. Among the various classic works by Thomas L. Heath, an important instance, far more general than its title suggests, is *Aristarchus of Samos. A History of Greek Astronomy to Aristarchus, together with his Treatise on the Sizes and Distances of the Sun and Moon* (1913; repr. Oxford: Clarendon Press, 1959). Dreyer's *A History of Astronomy from Thales to Kepler* is still valuable over a wide historical range. (We have mentioned Dreyer's own astronomical work at several points in our text.)

The technicalities of Aristotle's homocentric astronomy are well treated in the works here mentioned. His complete writings, as befits their enormous influence over two millennia, are readily available in numerous editions and translations. For a general commentary on his system of natural philosophy, see G. E. R. Lloyd, *Aristotle: The Growth and Structure of his Thought* (Cambridge: Cambridge University Press, 1978).

The first important study of the Antikythera device (and still essential reading) is Derek de Solla Price, *Gears From the Greeks: The Antikythera Mechanism—A Calendar Computer From ca 80 BC* (Philadelphia: American Philosophical Society, 1974). A brief but far-ranging analysis, with a good bibliography that includes references to the many works of M. T. Wright on the subject, is by T. Freeth (and sixteen other authors), "Decoding the Ancient Greek Astronomical Calculator Known as the Antikythera Mechanism," *Nature* 444 (November 2006): 587–91.

To properly appreciate the great stature of Ptolemy it is almost essential to see his own writings, in particular his *Almagest,* which is now available in a fine English translation by G. J. Toomer (London: Duckworth, 1984). This could be supplemented for Ptolemy's physical views by Bernard R. Goldstein, *The Arabic Version of Ptolemy's Planetary Hypotheses* (Philadelphia: American Philosophical Society, 1967). See also Goldstein's collected papers, *Theory and Observation in Ancient and Medieval Astronomy* (Aldershot: Variorum, 1985). An excellent synoptic guide to the intricacies of Ptolemy's *Almagest* is Olaf Pedersen's *A Survey of the Almagest* (Odense, Denmark: Odense University Press, 1974). At a simpler level, the work by Pedersen and M. Pihl, *Early Physics and Astronomy* (London: MacDonald, and New York: American Elsevier, 1974; repub. Cambridge: Cambridge University Press, 1993), takes the story to the Middle Ages in an eminently readable way.

James Evans, *The History and Practice of Ancient Astronomy* (New York and Oxford: Oxford University Press, 1998) is an excellent textbook, encompassing history and astronomical instruction in early techniques. For an important and readable study by him of ancient practice, see his article "The Astrologer's Apparatus: A Picture of Professional Practice in Greco-Roman Egypt," *Journal of the History of Astronomy* 35 (2004)1–44. *Measuring the Universe: Cosmic Dimensions from Aristarchus to Halley* by Albert van Helden (Chicago: University of Chicago Press, 1985) *On the Distances Between Sun, Moon and Earth According to Ptolemy, Copernicus and Reinhold* by Janice Adrienne Henderson (Leiden: Brill, 1991) handle the distance question.

Auguste Bouché-Leclercq, *L'Astrologie grecque* (Paris: Leroux, 1899; repr. Brussels: Culture et Civilisation, 1963) has been superseded in a number of minor respects, but it remains the best single source on Greek astrology, with no equivalent in English, although Otto Neugebauer and H. B. van Hoesen, *Greek Horoscopes* (Philadelphia: American Philosophical Society, 1959) includes much of general interest, and an invaluable glossary. In German, Wilhelm Gundel and Hans Georg Gundel, *Astrologumena: Die Astrologische Literatur in der Antike und Ihre Geschichte* (repr. Wiesbaden: Steiner, 1966), and F. Boll, C. Bezold and W. Gundel, *Sternglaube und Sterndeutung: Die Geschichte und das Wesen der Astrologie* (repr. Darmstadt: Wissenschaftliche Buchgesellschaft, 1977) continue ancient history into later periods. In English, Jim Tester, *A History of Western Astrology* (Bury St. Edmunds, Suffolk: The Boydell Press, 1987), is useful and eminently readable. Of classical astrological texts, the most important is Ptolemy's *Tetrabiblos,* and this is available in parallel translation in a standard Loeb edition (London: Heinemann, and Cambridge, MA: Harvard University Press, 1940).

Astronomy continued to be cultivated in Byzantium, and a series entitled *Corpus des Astronomes Byzantins* is dedicated to publishing relevant texts. See, for example, Alexander Jones (ed.), *An Eleventh-Century Manual of Arabo-Byzantine Astronomy* (Amsterdam: J. C. Gieben, 1987), which is number 3 in the series.

CHAPTER 5

The history of Chinese science generally was placed on completely new footing with the lifework of Joseph Needham, whose multivolume *Science and Civilization in*

China (Cambridge: Cambridge University Press, 1954–), continued by others before and after his death, will be found in many large libraries. Colin A. Ronan, *The Shorter Science and Civilization in China,* 3 vols. (Cambridge: Cambridge University Press, 1978–1986), is an abridgement of Needham's original volumes. Needham was not interested in mathematical astronomy, nor impressed by its importance, and for this see Nathan Sivin, *Cosmos and Computation in Early Chinese Mathematical Astronomy* (Leyden: Brill, 1969); and Christopher Cullen, *Astronomy and Mathematics in Ancient China: The Zhou bi suan jing* (Cambridge: Cambridge University Press, 1996), and especially his "The First Complete Chinese Theory of the Moon: The Innovations of Liu Hong *c.* AD 200," *Journal for the History of Astronomy,* 33 (2002): 21–39. Works peripheral to Needham's are numerous, but note Ho Peng Yoke, *Li, Qi and Shu: An Introduction to Science and Civilization in China* (Hong Kong: Hong Kong University Press, 1985) and Wang Ling, Joseph Needham, and Derek J. De Solla Price, *Heavenly Clockwork* (Cambridge: Cambridge University Press, 1960; 2nd ed. with supplement by J. H. Combridge, 1986). Ho Peng Yoke's works written in Chinese may be traced through library catalogs if needed.

Outside the Needham tradition, see articles in Hartner's *Oriens-Occidens* (cited in the introductory bibliography), and others by Yasukatsu Maeyama (largely in English) in his collected essays, *Astronomy in Orient and Occident: Selected Papers on its Cultural and Scientific History* (Hildesheim: Olms, 2003). Richard J. Smith, *Fortune-Tellers and Philosophy* (Boulder: Westview Press, 1991) touches on astrology. There is a four-volume work titled "History of Chinese Astronomy" in Chinese by Z. Chen: *Zhongguo Tianwenxue* (Shanghai: Shanghai Renmin Chubanshe, 1980–1989). Those who understand Japanese may consult Michio Yano's "Astrology of Esoteric Buddhism": *Mikkyōno senseiitsu* (Tokyo: Tokyo Bijitsu, 1986).

There is little in English on Tibetan astronomy, but see Winfried Petri, "Tibetan Astronomy," in *Vistas in Astronomy* 9 (1968):159–64; and especially the entry by Yukio Ōhashi in the encyclopedia edited by H. Selin (see the introductory bibliography), pages 136–39. Most of the works in Ōhashi's bibliography are in Tibetan or Japanese, however.

While earlier histories regarding astronomy in early Japan are generally inaccessible to most of us for reasons of language, it should be noted that Japan publishes an excellent journal in the history of science written largely in English, *Historia Scientiarum,* alias *Japanese Studies in the History of Science.* Articles in this by K. Yabuuchi, especially on Chinese and Japanese calendars, are of high importance. (Yabuuchi's main book-length study of 1969 is available only in Japanese.) An excellent survey of Japanese astronomy, paying much attention to the historical background, is Shigeru Nakayama, *A History of Japanese Astronomy. Chinese Background and Western Impact* (Cambridge, MA: Harvard University Press, 1969), and this includes useful bibliography to that time.

Most general accounts of the Jesuit period are written without much astronomical awareness. An exception is Noël Golvers and Ulrich Libbrecht, *Ferdinand Verbiest, S. J. (1623–1688) and the Chinese Heaven: The Composition of the Astronomical Corpus, its Diffusion and Reception in the European Republic of Letters en zijn Europese Sterrenkunde* (Leuven: Leuven University Press, 2003). This was first published in Dutch in 1990. Even without knowledge of that language, one cannot fail to enjoy the magnificently illustrated catalog prepared for a Brussels exhibition held under the title *China, Hemel en Aarde. 5000 Jaar Uitvindingen en Ontdekkingen* [China, Heaven and Earth: 5,000 years of Inventions and Discoveries] (Brussels: Vlaamse Gemeenschap, Trierstraat 1000, 1988). Nathan Sivin shows that the failure of the Jesuits to promote Copernicus was a consequence of their having—under the shadow of the Church—garbled his work badly. See his "Copernicus in China" in *Studia Copernicana*, 6 (1973): 63–122. (The first volume of this series, which is still in progress, was published in Warsaw [Polish Academy of Sciences].)

CHAPTER 6

Two classic studies on Mayan writing are J. E. S. Thompson, *Maya Hieroglyphic Writing* (Norman: University of Oklahoma Press, 1960) and his *A Commentary on the Dresden Codex* (Philadelphia: American Philosophical Society, 1972). On Mayan ideas of time, see Barbara Tedlock, *Time and the Highland Maya* (Albuquerque: University of New Mexico Press, 1982). An excellent introduction to the Mayan astral sciences is Susan Milbrath, *Star Gods of the Maya: Astronomy in Art, Folklore, and Calendars* (Austin: University of Texas Press, 1999). See also earlier studies by Anthony Aveni, *Archaeoastronomy in Pre-Columbian America* (Austin: University of Texas Press, 1975); *Native American Astronomy* (Austin: University of Texas Press, 1977); and *Archaeoastronomy in the New World* (Cambridge: Cambridge University Press, 1982). For a history of the first Spanish contacts with the New World see T. Todorov, *The Conquest of the Americas* (New York: Harper and Row, 1982). For the Mayan civilization more generally, see M. Coe, *The Maya* (New York: Praeger, 1973).

General studies of the Aztecs are A. Demarest, *Religion and Empire* (Cambridge: Cambridge University Press, 1984) and M. Léon-Portilla, *Aztec Thought and Culture* (Norman: University of Oklahoma Press, 1963). N. Davies, *The Ancient Kingdoms of Mexico* (Harmondsworth: Penguin Books, 1982), is readable and accessible.

For the fall of the Incas, see J. Hemming, *The Conquest of the Incas* (New York: Harcourt Brace, 1970). A general study of Inca culture is A. Kendall, *Everyday Life of the Incas* (London: Batsford, 1973). Concepts of cosmic religion are touched upon in a chapter by L. Sullivan in R. Lovin and F. Reynolds, *Cosmogony and Ethical Order* (Chicago: University of Chicago Press, 1985).

Accounts at a general and introductory level are included in E. C. Krupp (ed.), *In Search of Ancient Astronomies* (London: Chatto and Windus, 1979), as well as in his *Beyond the Blue Horizon* (New York: Oxford University Press, 1991). For the cultural background to all these groups, see C. A. Burland, *Peoples of the Sun. The Civilizations of Pre-Columbian America* (London: Weidenfeld and Nicolson, 1976).

CHAPTER 7

The gargantuan task of listing the scattered historical sources has been undertaken in different ways. David Pingree spent much of his life compiling a *Census of the Exact Sciences in Sanscrit* (Philadelphia: American Philosophical Society, 1970–). See also S. N. Sen, *Bibliography of Sanskrit Works on Astronomy and Mathematics* (New Delhi: Indian National Science Academy, 1966). For a comprehensive history, see

S. N. Sen and K. S. Shukla, *History of Astronomy in India* (New Delhi: Institute of History of Medicine and Medical Research, 1985); and, for source material, B. V. Subbarayappa and K. V. Sarma, *Indian Astronomy: A Source-Book* (Bombay: Nehru Centre, Worli, 1985). There are a number of points on which historians of Indian astronomy are divided, as will be noticed when consulting the next three important histories, of which the first is the most readily accessible: D. Pingree, "History of Mathematical Astronomy in India," in the *Dictionary of Scientific Biography,* vol. 15 (1978: 533–633); R. Billard, *L'Astronomie indienne* (Paris: Ecole française d'extrème Orient, 1971); and S. N. Sen, "Astronomy," in *A Concise History of Science in India* (Calcutta: Indian National Science Academy, 1971: 58–135). For the division of opinion to which I referred on p. 179 however, see D. Duke, "The Equant in India: The Mathematical Basis of Ancient Indian Planetary Models," *Archive for History of Exact Sciences,* 59 (2005):563–76. Duke contrasts the views of Neugebauer with those expressed briefly by Van der Waerden, and improves on the latter.

The classic source of Indian cultural (and astronomical) history, as seen by an outsider, is al-Bīrūnī's. For an English translation of the Arabic text, see Edward C. Sachau, *Alberuni's India* (Delhi: S. Chand, 1964). A similarly rich early source is *The Astrological History of Māshā'allāh,* trans. and ed. E. S. Kennedy and D. Pingree (Cambridge, MA: Harvard University Press, 1971). Anyone with an interest in the growth of a modern Western awareness of Indian and Chinese astronomy will consult J. B. Biot, *Etudes sur l'astronomie indienne et sur l'astronomie chinoise* (1862; repr. Paris: Blanchard, 1969).

On astronomical activities by Europeans in India, a subject relatively little investigated, see S. M. Razaullah Ansari, *Introduction of Modern Western Astronomy in India during the 18th–19th Centuries* (New Delhi: Institute of History of Medicine and Medical Research, 1985). On Persian astronomy, see articles in W. Hartner's *Oriens-Occidens* (see the introductory bibliography) and his chapter "Old Iranian Calendars," in *The Cambridge History of Iran,* vol. 2: *The Median and Achaemenian Periods,* ed. Ilya Gershevitch (Cambridge: Cambridge University Press, 1985: 714–92). For further studies of relevance to the present chapter, see the bibliography for chapter 8.

CHAPTER 8

Bernard Lewis (ed.), *The World of Islam* (London: Thames and Hudson, 1976), is a beautifully illustrated collection of authoritative chapters on the main aspects of Islamic faith, people and culture, Eastern and Western. Anna Caiozzo, *Images du ciel d'Orient au Moyen Âge* (Paris: Presses de l'Université de Paris-Sorbonne, 2003) is a systematic history of Islamic manuscript illustrations of the zodiac, but it deals with far more, and is both well illustrated and authoritative. For an excellent short survey of the Greek inheritance of Islam, see Abdelhamid I. Sabra, "The Appropriation and Subsequent Naturalization of Greek Science in Medieval Islam," *History of Science* 25 (1987): 223–43. Medieval Islamic astronomy has been studied extensively, and two important collections of papers, the first more easily found than the second, are: David A. King and George Saliba (eds.), *From Deferent to Equant: A Volume of Studies in the History of Science in the Ancient and Medieval Near East in Honor of E. S. Kennedy* (New York: New York Academy of Sciences, 1987); and E. S. Kennedy (with colleagues and former students), *Studies in the Islamic Exact Sciences* (Beirut: American University of Beirut, 1983). George Saliba, *A History of Arabic Astronomy: Planetary Theories During the Golden Age of Islam* (New York: New York University Press, 1994) is a collection of essays centered on the Marāgha astronomers, and has much information on the problem of the unacceptable equant. *Oriens-Occidens* (see the introductory bibliography) contains important studies spanning three continents. A significant gap in the literature on Muslim cometary records is largely filled by David Cook, "Muslim Material on Comets and Meteors," *Journal of the History of Astronomy,* 30 (1999): 131–60.

On observatories, see Aydin Sayili, *The Observatory in Islam* (Ankara: Turkish Historical Society, 1960); to which subject Paul Kunitsch, *The Arabs and the Stars* (Aldershot: Variorum, 1989) is also relevant. Studies of astronomical artifacts are numerous, but Francis Maddison and Emilie Savage-Smith, *Science, Tools and Magic,* 2 vols. (London: The Nour Foundation and Oxford University Press, 1997), is a work well worth consulting and is richly illustrated. On mathematical and instrumental techniques, see David A. King, *Islamic Mathematical Astronomy* (Aldershot: Variorum, 1986), *Islamic Astronomical Instruments* (Aldershot: Variorum, 1987), and especially the two-volume collection of his writings, weighing in at 7.5 kilograms, in: *In Synchrony with the Heavens: Timekeeping and Instrumentation in Medieval Islamic Civilization,* vol. 1: *The Call of the Muezzin;* and vol. 2: *Instruments of Mass Calculation* (Leiden: Brill, 2005). These are well illustrated and contain considerable bibliographical information. See also Richard P. Lorch, *Arabic Mathematical Sciences: Instruments, Texts, Transmission* (Aldershot: Variorum, 1995). A standard work on zījes is E. S. Kennedy, *A Survey of Islamic Astronomical Tables,* 2nd ed. (Philadelphia: American Philosophical Society, 1989). A collection of papers by Raymond Mercier in his *Studies on the Transmission of Medieval Mathematical Astronomy* (Aldershot: Variorum, 2004) is largely devoted to the spread of zījes across half the world, from south and central Asia to Latin Europe.

Representative texts in astrology—and there is much to be extracted from their editors' commentaries—include David Pingree (ed.), *The Thousands of Abu Ma'shar,* (London: Warburg Institute, 1968); Edward. S. Kennedy and David Pingree (eds.) *The Astrological History of Māshāʾallāh* (see the chapter 7 bibliography); Charles Burnett, *Abu Maʿshar: The Abbreviation of the Introduction to Astrology, together with The Medieval Latin Translation of Adelard of Bath* (Leiden: Brill, 1994).

Various journals have specialized in Arabic science—as of 2006, there are at least 18 doing so. For the one most relevant to astronomy, *Suhayl,* see the introductory bibliography. An active IUHPS commission on Islamic science issues a regular and valuable newsletter, and can be found at http://islamsci.mcgill.ca/. Two substantial journals that are now well established are *Arabic Sciences and Philosophy* (Cambridge: Cambridge University Press) and *Zeitschrift für Geschichte der arabisch-islamischen Wissenschaften* (Frankfurt: published annually from 1984; ed. F. Sezgin). Most of the articles in the first publication and many in the second are in English, and astronomy is a recurrent theme in both. Sezgin has also published an important series of bio-bibliographical aids: Fuat Sezgin, *Geschichte des arabischen Schrifttums,* Bd VI: *Astronomie bis ca. 430 H* (Leiden: Brill, 1978), and Bd VII: *Astrologie-Meteorologie und Verwandtes bis ca. 430 H* (Leiden: Brill, 1979). The same publisher issues the magnificent *Encyclopaedia of Islam*, a long-established international work of scholarship, with many authoritative entries on astronomical and astrological subjects. New CD-ROM and online versions based on (and including revisions to) a new printed edition published from 1999 onward, edited by P. Bearman and others, are available to subscribers. Many large libraries have a version on CD-ROM for consultation. Brill also publishes a *Concise Encyclopaedia of Islam,* edited by H. A. R. Gibb and J. H. Kramers. For the latest information about these important series, visit http://www.brill.nl.

CHAPTER 9

Western Islam has been fortunate in its historians, largely centered in Barcelona. The best early literature is largely in Spanish, but recently much has also appeared in English. Classics in the field include two works by J. M. Millás Vallicrosa: *Estudios sobre Historia de la Ciencia Española* (Barcelona: CSIC, 1949) and *Nuevos Estudios sobre Historia de la Ciencia Española* (Barcelona: CSIC, 1960); both have been reissued together by Julio Samsó (ed.), with introduction by Juan Vernet (Barcelona, 1987). Of the many valuable works by Vernet, a general history of the European debt to Spanish Arab culture has been translated into French: *Ce que la culture doit aux Arabes d'Espagne* (Paris: Sindbad, 1978). Recent general work of the Barcelona school, with a strong bearing on astronomy, includes Julio Samsó, *Las ciencias de los antiguos en al-Andalus* (Madrid: Mapfre, 1992); and Julio Samsó and Josep Casulleras, *From Baghdad to Barcelona: Studies in Islamic Exact Sciences in Honour of Prof. J. Vernet* (Barcelona: Instituto M. Vallicrosa de Historia de la Ciencia, 1996). Julio Samsó, *Islamic Astronomy and Medieval Spain* (Aldershot: Variorum, 1994) offers bibliographical keys to the many valuable editions and studies, written largely in Spanish, by members of the Barcelona institute.

On specific technical matters, much can be learned from O. Neugebauer, *The Astronomical Tables of Al-Khwārizmī, Translated with Commentaries on the Latin Version edited by H. Suter* (Copenhagen: Historisk-filosofiske Skrifter, Det Kongelige Danske Videnskebernes Selskab, 1962). This could be supplemented by the following edition, with translation and commentary: Bernard R. Goldstein, *Ibn al-Muthanna's Commentary on the Astronomical Tables of al-Khwārizmī. Two Hebrew Versions* (New Haven: Yale University Press, 1967). The two most influential astronomical documents from medieval Spain were the Toledan and Alfonsine tables. The former are now available in a monumental survey: Fritz S. Pedersen, *The Toledan Tables: A Review of the MSS and the Textual Versions, with an Edition,* 5 vols. (Copenhagen: Royal Danish Academy of Science and Letters, 2002). An earlier brief overview was G. J. Toomer, "A Survey of the Toledan Tables," *Osiris,* 15 (1968): 1–174.

On the Alfonsine Tables, the standard work is now José Chabas and Bernard R. Goldstein, *The Alfonsine Tables of Toledo* (Dordrecht: Kluwer, 2003; references to their work on the Zacut material will also be found there). A useful version of the tables, close to the early printings of John of Saxony's version, is E. Poulle (ed.), *Les Tables Alfonsines, avec les canons de Jean de Saxe* (Paris: Editions du CNRS, 1984). A survey of the history of the tables generally will be found in J. D. North, "The Alfonsine Tables in England," reprinted as chapter 21 in his *Stars, Minds, and Fate* (London: Hambledon, 1989). Articles in various languages on the Alfonsine theme are in M. Comes, R. Puig, and J. Samsó (eds.), *De Astronomia Alphonsi Regis. Proceedings of the Symposium on Alphonsine Astronomy held at Berkeley, August* 1985 (Barcelona: Instituto "Millás Vallicrosa," 1987). A finely illustrated work on science in Al-Andalus (Muslim Spain), with authoritative but readable topical essays, was produced under the direction of J. Vernet and J. Samsó in connection with an exhibition held in Madrid in 1992: *El legado científico Andalusí* (Madrid: Ministerio de Cultura, 1992).

On the Jewish contribution to astronomy in medieval Spain, see Bernard R. Goldstein's *Ibn al-Muthanna's Commentary,* listed above, and his *The Astronomy of Levi ben Gerson* (New York: Springer, 1985). José Luis Mancha, *Studies in Medieval Astronomy and Optics* (Aldershot: Variorum, 2006) has much on Levi ben Gerson, as well as Western versions of Arabic material; and note his studies of pinhole images. There is much relevant material in Y. Tzvi Langermann, *The Jews and the Sciences in the Middle Ages* (Aldershot: Variorum, 1999), and for the closely related Jewish community across the Pyrenees, see Gad Freudenthal, "Science in the Medieval Jewish Culture of Southern France," *History of Science* 33 (1995): 23–58.

CHAPTER 10

For a scholarly overview of medieval science, including much astrology and some astronomy, from the perspective of surviving manuscript evidence, see Lynn Thorndike, *A History of Magic and Experimental Science,* 8 vols. (New York: Columbia University Press, 1923–1958). Extracts from numerous texts will be found in Edward Grant (ed.), *A Source Book in Medieval Science* (Cambridge, MA: Harvard University Press, 1974). Note especially the material from Sacrobosco and the common *Theorica planetarum* treatise. The medieval handling of the physics of the Aristotelian cosmos is a recurrent theme in Grant's collected essays, *Studies in Medieval Science and Natural Philosophy* (London: Variorum Reprints, 1981), but see especially his *Planets, Stars and Orbs: The Medieval Cosmos 1200–1687* (Cambridge: Cambridge University Press, 1994). For early influences, see Bruce Eastwood and Gerd Grasshoff, *Planetary Diagrams for Roman Astronomy in Medieval Europe, ca. 800–1500* (Philadelphia: American Philosophical Society, 2004), and Bruce S. Easwood, *The Revival of Planetary Astronomy in Carolingian and Post-Carolingian Europe* (Aldershot: Ashgate, 2002). The classic early European calendar treatise is *Bede: The Reckoning of Time,* translated, with commentary, by Faith Wallis (Liverpool: Liverpool University Press, 1999). Stephen C. McClusky, *Astronomies and Cultures in Early Modern Europe* (Cambridge: Cambridge University Press, 1998) is a concise survey with wide range. The collection in North, *Stars, Minds, and Fate* (see the bibliography for chapter 9) covers many medieval and Renaissance astronomical themes, and includes an elementary introduction to the astrolabe.

Illustrations of astrolabes from East and West may be found in many museum catalogs. Important collections are held by the Museum of the History of Science (Oxford), National Maritime Museum (Greenwich), British Museum (London), Smithsonian Institution National Air and Space Museum (Washington), Adler Planetarium (Chicago), and the Museo di Storia della Scienza (Florence). These all have a wide variety of medieval astronomical instruments. An outstanding and copiously illustrated catalog of an exhibition that drew on over 70 collections (notably that of the German National Museum in Nuremberg) is Gerhard Bott (ed.), *Focus Behaim Globus,* 2 vols. (Nürnberg: Germanisches Nationalmuseum, 1992). The first volume includes topical essays (all in German), some on astronomical themes, some on the terrestrial globe of Martin Behaim, the focus of the exhibition. An important work on astronomical instruments, covering the eleventh to the eighteenth centuries, is Ernst Zinner, *Deutsche und niederländische Astronomische Instrumente des 11.-18. Jahrhunderts* (Munich: Beck, 1956; rev. ed. 1967). Among well-illustrated surveys of a general sort are Henri Michel, *Scientific Instruments in Art and History* (London: Barrie and Rockliff, 1967); Harriet Wynter and Anthony Turner, *Scientific Instruments* (London: Studio Vista, 1975); Anthony Turner, *Early Scientific Instruments: Europe 1400–1800* (London: Sotheby's Publications, 1987); Gerard L'E. Turner, *Antique Scientific Instruments* (Poole, Dorset: Blandford, 1980) and *Renaissance Astrolabes and Their Makers* (Aldershot: Ashgate Variorum, 2003); and Roderick Webster and Marjorie Webster, *Western Astrolabes* (Chicago: Adler Planetarium and Astronomy Museum, 1998).

The processes of higher education are the subject of Hilde De Ridder-Symoens, *A History of the University in Europe*, vol. 1: *Universities in the Middle Ages* (Cambridge: Cambridge University Press, 1991). For more detail on Oxford, see J. I. Catto and T. I. R. Evans (eds.), *The History of the University of Oxford,* vol. 2 (Oxford: Oxford University Press, 1992). On the general position of astrology in medieval and Renaissance culture, the following are now somewhat dated but are still of value: Theodore Otto Wedel, *The Mediaeval Attitude Toward Astrology, Particularly in England* (New Haven: Yale University Press, 1920; repr. North Haven: Archon Books, 1968) and Don Cameron Allen, *The Star-Crossed Renaissance: The Quarrel about Astrology and its Influence in England* (Durham, NC: Duke University Press, 1941). For a different perspective, see Eugenio Garin, *Astrology in the Renaissance: The Zodiac of Life,* translated from the Italian by C. Jackson and J. Allen (London: Routledge and Kegan Paul, 1983). Astrology at the English court and university in the later Middle Ages is the subject of Hilary M. Carey, *Courting Disaster* (London: MacMillan, 1992). Several medieval and Renaissance themes are discussed in Patrick Curry (ed.), *Astrology, Science and Society: Historical Essays* (Woodbridge, Suffolk: Boydell Press, 1987). The example of the astronomer Kratzer's influence on the artist Holbein is explained in John North, *The Ambassadors' Secret: Holbein and the World of the Renaissance* (London: Hambledon and London, 2002; repr. London: Phoenix, 2004).

Perhaps the most widely disseminated astronomy textbook of all time was Sacrobosco's. For the Latin text, with translation into English, see Lynn Thorndike, *The Sphere of Sacrobosco and its Commentators* (Chicago: University of Chicago Press, 1949). For an introduction to the main doctrines of medieval Western astrology, as they were inherited from Islamic writers, and for Chaucer's literary use of them, see J. D. North, *Chaucer's Universe* (Oxford: Clarendon Press, 2nd ed., 1990). This work also explains the principles of the astrolabe and Chaucer's equatorium. For an extensive history of the medieval equatorium see E. Poulle, *Equatoires et Horlogerie planétaire du XIIIe au XVIe siècle,* 2 vols. (Geneva: Droz, 1980). For a shorter account in English, see the description of the albion and its historical context, in J. D. North,

Richard of Wallingford, 3 vols. (Oxford: Oxford University Press, 1976). These volumes contain the earliest surviving description of any mechanical (and astronomical) clock, and also (in appendix 31) a fuller outline of medieval planetary theory than could be given in the present work. For a life of this astronomer as well as an account of the medieval university curriculum in the sciences, see part IV of John North, *God's Clockmaker: Richard of Wallingford and the Invention of Time* (London: Hambledon and London, 2005). For the Dondi astronomical clock, see Emmanuel Poulle (ed.), *Johannis de Dondis Paduani Civis Astrarium: Édition Critique de la Version A* (Padua: 1+1, and Paris: Les Belles Lettres, 1988).

CHAPTER 11

Copernicus' own writings are available in many editions, but the most complete is that by the Polish Academy of Sciences, of which the first volume was a reproduction of Copernicus' manuscript *De revolutionibus* (ed. Pawel Czartoryski, trans. Zygmunt Niesda and Erna Hilfstein, Warsaw, 1973). The series includes Latin texts and English translations. Another translation is A. M. Duncan, *On the Revolutions of the Heavenly Spheres* (Newton Abbot, Devon: David & Charles, 1976). The *Commentariolus* is well translated by Noel Swerdlow (*Proceedings of the American Philosophical Society,* 117 [1973]: 423–512). It is available with translations of two other minor Copernican works (the *Letter against Werner* and the Rheticus's *Narratio Prima*) in Edward Rosen, *Three Copernican Treatises,* 3rd ed. (New York: Dover, 1971). The latter contains a Copernicus bibliography with over a thousand items and short critical—often hypercritical—descriptions of each. A Polish bibliography produced in 1958 by H. Baranowski contained nearly 4,000 items, and that number was greatly swelled by the five hundredth anniversary celebrations in 1973. The fortunes of the various early editions and copies of the *De revolutionibus* are the subject of the article that provides a title for Owen Gingerich's *The Great Copernicus Chase and other Adventures in Astronomical History* (Cambridge, MA: Sky Publishing, 1992). The results of the chase are given in his *An Annotated Census of Copernicus' De Revolutionibus* (Leiden: Brill, 2002), and in narrative form in his *The Book Nobody Read: Chasing the Revolutions of Nicolaus Copernicus* (New York: Walker, 2004).

The most complete analysis (in a single work) of the mathematical aspects of Copernicus' work is Noel M. Swerdlow and Otto Neugebauer, *Mathematical Astronomy in Copernicus' De Revolutionibus,* 2 vols. (New York: Springer, 1984). There are many articles speculating on the link between Copernicus and Muslim astronomy. A short article that should not be overlooked is Graziella Federici Vescovini, "The Place of the Sun in Medieval Arabo-Latin Astronomy: The *Lucidator dubitabilium astronomiae* (1303–10) of Peter de Padua," *Journal for the History of Astronomy,* 29 (1998): 151–5 (see there also references to the author's critical edition of Peter of Abano).

Earlier works should not be ignored. For an example of a writer representing Copernicus' work as a relatively superficial geometrical transformation of Ptolemy, a thesis no longer as fashionable as it was, see Derek J. de S. Price, "Contra Copernicus," in M. Clagett (ed.), *Critical Problems in the History of Science* (Madison: Wisconsin University Press, 1959: 197–218). A good general Copernicus biography in English, up to 1973, is M. Biskup and J. Dobrzycki, *Copernicus, Scholar and Citizen* (Warsaw:

Interpress, 1973). A widely known but somewhat fanciful account of his life is in Arthur Koestler, *The Sleepwalkers. A History of Man's Changing Vision of the Universe* (New York: Grosset and Dunlap, 1959). This book does more justice to Kepler than to Copernicus and Galileo. Among the better collections of studies dating from the 1973 celebrations is Jerzy Dobrzycki (ed.), *The Reception of Copernicus' Heliocentric theory. Proceedings of a Symposium held by the IUHPS in Torun, 1973* (Dordrecht: Reidel, 1973). For an indication of the progress made over the following thirty years on this theme, see the well-informed account of the reception of Copernicanism in one influential culture in Rienk Vermij, *The Calvinist Copernicans: The Reception of the New Astronomy in the Dutch Republic, 1575–1750* (Amsterdam: KNAW, 2002).

A valuable series of monographs going far beyond Copernican studies proper is produced by the Polish Academy of Sciences under the title *Studia Copernicana*. A later parallel series has been published in Leiden with the qualification "Brill Series," which includes Janice Henderson's work on distances (see the bibliography to chapter 4).

CHAPTER 12

The works of Tycho Brahe were edited in 15 volumes by J. L. E. Dreyer (*Tychonis Brahe Dani Opera omnia*, Copenhagen: Danske Sprog-og Litteraturselskab, 1913–1929). Dreyer wrote a standard biography which is still valuable: *Tycho Brahe. A Picture of Life and Work in the Sixteenth Century* (1890; repr. New York: Dover, 1963). Dreyer's general history (see bibliography to chapter 4) is still useful and is strong on this period. He issued a facsimile of Tycho's work on instruments, which was later translated into English and edited by H. Raeder, E. Strömgren and B. Strömgren as *Tycho Brahe's Description of his Instruments and Scientific Work as given in Astronomiae Instauratae Mechanicae* (Copenhagen: Ejnar Munksgaard, 1946). The best biographies of Tycho are now Victor E. Thoren, *The Lord of Uraniborg: A Biography of Tycho Brahe* (Cambridge: Cambridge University Press, 1990) and John Robert Christianson, *On Tycho's Island: Tycho Brahe and his Assistants, 1570–1601* (Cambridge: Cambridge University Press, 2000). Also relevant, but more specific, is Owen Gingerich and Robert S. Westman, *The Wittich Connection: Conflict and Priority in Late Sixteenth-century Cosmology* (Philadelphia: American Philosophical Society, 1988). On the comet question, see the rich collection of materials in Tabitta Van Nouhuys, *The Age of Two-Faced Janus: The Comets of 1577 and 1618, and the Decline of the Aristotelian World View in the Netherlands* (Leiden: Brill, 1998).

Much of the important work of Alexandre Koyré centered on this period. While now largely superseded, the following were very influential: *Etudes Galiléennes* (Paris: Hermann, 1966); *Galileo Studies,* trans. John Mepham (Hassocks, Sussex: Harvester Press, 1978; not identical to the French work); *From the Closed World to the Infinite Universe* (Baltimore: The Johns Hopkins University Press, 1970); *The Astronomical Revolution: Copernicus-Kepler-Borelli,* trans. R. E. W. Maddison (London: Methuen, 1973). The last is heavily weighted toward Kepler.

Fundamental texts on the discovery of the telescope were long available only in Dutch editions, but see now Albert van Helden, *The Invention of the Telescope* (Philadelphia: American Philosophical Society, 1977). For a general history of the instrument, see Henry C. King, *The History of the Telescope* (New York: Dover, 1979). Like King's book, Richard Learner, *Astronomy Through the Telescope* (New York: Van Nostrand Reinhold, 1981), is now dated, but it is well illustrated and gives an insight into the optics and engineering of large telescopes.

On Galileo's science in general, see the relevant sections of A. C. Crombie, *Styles of Scientific Thinking in the European tradition,* 3 vols. (London: Duckworth, 1994). For a well-rounded and accessible study of Galileo and his achievements see Stillman Drake, *Galileo at Work: His Scientific Biography* (Chicago: University of Chicago Press, 1978; repr. New York: Dover, 1995). The works of Galileo were published in an Italian edition by A. Favaro (Florence: Barbèra, 1890–1909) and have been reprinted with additions (Florence: Barbèra, 1929–1939, 1968). English translations are numerous, beginning with Thomas Salusbury's of 1661. Relevant to astronomy are, for example: Stillman Drake and C. D. O'Malley (trans.), *The Controversy of the Comets of 1618.* [texts by Galileo Galilei; Horatio Grassi; Mario Guiducci; Johann Kepler] (Philadelphia: University of Pennsylvania Press, 1960); S. Drake, *Dialogue Concerning the Two Chief World Systems* (Berkeley: University of California Press, 1953; rev. ed., 1967); S. Drake, *Discoveries and Opinions of Galileo* (New York: Doubleday Anchor, 1957). The last includes a work on sunspots and parts of the *Sidereal Messenger,* among other items. A complete translation of the latter, with comment, is by Albert van Helden: *Sidereus nuncius, or the Sidereal Messenger* (Chicago: University of Chicago Press, 1989). Two important works on the Galileo trial are Maurice A. Finocchiaro, *Retrying Galileo,* 1633–1992 (Berkeley: University of California Press, 2005) and Ernan McMullin (ed.), *The Church and Galileo* (Notre Dame, IN: University of Notre Dame Press, 2005). A more wholesome involvement of the Roman Catholic Church in astronomy is explored in J. L. Heilbron, *The Sun in the Church: Cathedrals as Solar Observatories* (Cambridge, MA: Harvard University Press, 1999).

On Harriot, see John W. Shirley (ed.), *Thomas Harriot, Renaissance Scientist* (Oxford: Oxford University Press, 1974), as well as articles in J. D. North and J. J. Roche (eds.), *The Light of Nature: Essays in the History and Philosophy of Science presented to A. C. Crombie* (Dordrecht: Nijhoff, 1985). Shirley is the author of the standard work *Thomas Harriot: A Biography* (Oxford: Oxford University Press, 1983). Recent studies appear in Robert Fox (ed.), *Thomas Harriot: An Elizabethan Man of Science* (Aldershot: Ashgate, 2000). Since Harriot was interested in astronomical techniques of navigation, this is a suitable point at which to mention David Waters, *The Art of Navigation in England in Elizabethan And Early Stuart Times,* 3 vols. (Greenwich: National Maritime Museum, 1978). A natural supplement to this is E. G. R. Taylor, *Mathematical Practitioners of Tudor and Stuart England* (Cambridge: Cambridge University Press, 1967).

A new edition of the complete works of Kepler has long been in preparation (Munich, 1937–), and translations into English were for many years only piecemeal. Kepler writes at a fairly difficult mathematical level, but among the less difficult studies of him are: Johannes Kepler, *Kepler's Somnium: The Dream, or Posthumous Work on Lunar Astronomy,* translated, with a commentary by Edward Rosen (Madison: University of Wisconsin Press, 1967); and *Mysterium cosmographicum: The Secret of the Universe,* trans. A. M. Duncan, with an introduction and Commentary by E. J. Aiton

(New York: Abaris Books, 1981). For the text of the *Harmonices mundi,* see Johannes Kepler, *The Harmony of the World,* translated with an introduction and notes by E. J. Aiton, A. M. Duncan, and J. V. Field (Philadelphia: American Philosophical Society, 1997). Lengthy translations from the *Astronomia nova* are available in Koyré's *Astronomical Revolution* (trans. R. E. W. Maddison; see above) and a full version appears in William H. Donahue, *Johannes Kepler, New Astronomy* (Cambridge: Cambridge University Press, 1992). Analyzing the work in *The Composition of Kepler's Astronomia nova* (Princeton: Princeton University Press, 2001), James Voelkel draws extensively on Kepler's correspondence and manuscripts. Earlier excellent studies are Bruce Stephenson, *Kepler's Physical Astronomy* (New York: Springer, 1987; repr. Princeton: Princeton University Press, 1994) and his *The Music of the Heavens: Kepler's Harmonic Astronomy* (Princeton: Princeton University Press, 1994); and Rhonda Martens, *Kepler's Philosophy and the New Astronomy* (Princeton: Princeton University Press, 2001). Judith V. Field, *Kepler's Geometrical Cosmology* (London: Athlone Press, 1988) takes a well-balanced view of the mathematical mysticism in Kepler.

An accessible biography of Kepler is James Voelkel, *Johannes Kepler and the New Astronomy* (Oxford: Oxford University Press, 2001). A more comprehensive standard biography, first published in German in 1938, is Max Caspar, *Kepler* (trans. C. D. Hellman; London: Abelard-Schumann, 1959; repr. New York: Dover, 1993). The reprint has a new introduction and extensive references by Owen Gingerich, and invaluable bibliographical citations by him and Alain Segonds. Caspar prepared a basic bibliography (1936) which was revised by Martha List (Munich, 1968); but more easily accessible recent bibliographies in English include that in the Caspar reprint, that in Gingerich's Kepler biography in the *Dictionary of Scientific Biography,* vol. 7 (1973: 308–12), and that in Field, *Kepler's Geometrical Cosmology* (see above). A number of useful articles appeared in Arthur and Peter Beer (eds.), *Kepler: Four Hundred Years. Proceedings of Conferences Held in honor of Johannes Kepler* (Oxford: Pergamon, 1975). Related subjects are dealt with in Curtis Wilson, *Astronomy from Kepler to Newton: Historical Studies* (London: Variorum, 1989) and Albert van Helden, *Measuring the Universe: Cosmic Dimensions From Aristarchus to Halley* (Chicago: Chicago University Press, 1985).

There is surprisingly little by way of recent biography of the Cassini family, but for an overview of its four astronomically engaged members, see the *Dictionary of Scientific Biography*, vol. 3 (1971: 100–9). See also works on the Paris Observatory in the chapter 14 bibliography, below. On Hevelius, see *Dictionary of Scientific Biography*, vol. 6 (1972: 360–4). For seventeenth-century solar theories, see Yasukatsu Maeyama, "The Historical Development of Solar Theories in the Late Sixteenth and Seventeenth Centuries," *Vistas in Astronomy* 16 (1974): 35–60. A comprehensive survey of star maps from the advent of printing to 1800 is found with a number of reproductions of maps, rare and not so rare, in Deborah J. Warner, *The Sky Explored. Celestial Cartography, 1500–1800* (Amsterdam: Theatrum Orbis Terrarum, 1979). See also the rich illustrations in Peter Whitfield, *The Mapping of the Heavens* (London: The British Library, 1995). For an authoritative work on lunar mapping, see Ewen A. Whitaker, *Mapping and Naming the Moon: A History of Lunar Cartography and Nomenclature* (Cambridge: Cambridge University Press, 1999).

Many of the literary consequences of advances in cosmology are more widely covered in this book than the title suggests: Francis Johnson, *Astronomical Thought in Renaissance London* (Baltimore, MD: The Johns Hopkins University Press, 1937). As a measure of the progress of popular astrology at this time and later, see Bernard Capp, *Astrology and the Popular Press: English Almanacs 1500–1800* (London: Faber, 1979). That astrology was still very much alive, and even developing new mathematical techniques, is evident from J. D. North, *Horoscopes and History* (London: The Warburg Institute, 1986).

The perennial question of truth and hypothesis, as seen through astronomers' eyes, is the subject of Nicholas Jardine, *The Birth of History and Philosophy of Science: Kepler's "A defense of Tycho Against Ursus" and Essays on its Provenance and Significance* (Cambridge: Cambridge University Press, 1984). For Tycho's influence on the style of the first European observatory in the New World, see J. D. North, "Georg Markgraf: An Astronomer in the New World," in E. van den Boogart (ed.), *Johan Maurits van Nassau-Siegen, 1604–1679: A Humanist Prince in Europe and Brazil* (The Hague: Johan Maurits van Nassau Stichting, 1979: 394–423).

CHAPTER 13

The cosmological views of Descartes are well surveyed in Eric J. Aiton, *The Vortex theory of Planetary Motions* (London: Macdonald, and New York: American Elsevier, 1972). Those who wish to study Descartes closely should not overlook Gregor Sebba's excellent annotated bibliography covering the period 1800 to 1960: *Bibliographia Cartesiana* (The Hague: Nijhoff, 1964), supplemented by the *Isis Critical Bibliographies* (see introductory bibliography). Newtonian bibliography to a slightly later period is covered by Peter and Ruth Wallis, *Newton and Newtoniana, 1672–1975* (London: Dawson, 1977), which may be supplemented by the *Isis* lists and the excellent biography by R. S. Westfall, *Never at Rest. A Biography of Isaac Newton* (Cambridge: Cambridge University Press, 1980). Newton's *Principia* is readily available in English translation in numerous editions and printings, many based ultimately on Andrew Motte's translation of 1729 (for example, Florian Cajori's 1934 revision is frequently reprinted by the University of California Press). The new standard is set in Isaac Newton, *Principia: Mathematical Principles of Natural Philosophy,* trans. I. Bernard Cohen and Anne Whitman, with the assistance of Julia Budenz, and preceded by "A Guide to Newton's Principia" by I. Bernard Cohen (Berkeley: University of California Press, 1999). Newton's mathematical papers, where much theoretical astronomy resides, have been edited, translated and commented upon in a masterly way in D. T. Whiteside, *The Mathematical Papers of Isaac Newton,* 8 vols. (Cambridge: Cambridge University Press, 1967–80).

As an indication of contemporaneous practical astronomy and a useful counterweight to Newtonian abstractions, see John Flamsteed, *The Gresham Lectures of John Flamsteed,* edited and introduced by Eric Forbes (London: Mansell, 1975). Before his early death, Forbes also edited two volumes of Flamsteed's correspondence. Flamsteed's instrumentation is among the topics dealt with in Allan Chapman, *Astronomical Instruments and Their Uses* (Aldershot: Variorum, 1996). For the longitude problem, see William J. H. Andrewes (ed.), *The Quest for Longitude* (Cambridge, MA:

Harvard University Press, 1996). A working tool for future historians of practical astronomy, which takes its title from the place where it was prepared, is Derek Howse, *The Greenwich List of Observatories. A World List of Astronomical Observatories, Instruments and Clocks, 1670–1850* (Chalfont St. Giles: Science History Publications; being *Journal of the History of Astronomy* 17 [1986]).

CHAPTER 14

The reprinted papers in Michael A. Hoskin, *Stellar Astronomy. Historical Studies* (Chalfont St. Giles: Science History Publications, 1982) shed light on aspects of the astronomy of this period. On instrumentation, the literature is very extensive, but—in addition to items listed for chapter 13—see J. W. Bennett, *The Divided Circle: A History of Instruments for Astronomy, Navigation and Surveying* (London: Phaidon-Christie's, 1987) and King, *The History of the Telescope* (see chapter 12 bibliography). Maurice Daumas, *Scientific Instruments in the Seventeenth and Eighteenth Centuries* (trans. M. Holbrook; London: Batsford, 1972) covers a wider field than astronomy, but reflects well on French practice. The notes in the English version are chaotic, and so the original French edition is to be preferred. For the work and influence of Short, see David J. Bryden, *James Short and his Telescopes* (Edinburgh: Royal Scottish Museum, 1968). For a conspectus of the all-important English trade in instrument-making at this time, see E. G. R. Taylor, *Mathematical Practitioners of Hanoverian England* (Cambridge: Cambridge University Press, 1966). How a new observatory was assembled in Palermo, Sicily, at the end of the eighteenth century, is well brought out in Giorgia Foderà Serio and Ileana Chinnici, *L'Osservatorio astronomico di Palermo: La storia e gli istrumenti* (Palermo: S. F. Flaccovio, 1997). Note that the observatory is hung with portraits of Lalande, Herschel, and—between them—Ramsden. No rude mechanical he. Bologna has a more ancient history, but note instruments by Sissons, Graham, Dollond, and Ramsden in the important catalog: Enrica Baiada, Fabrizio Bónoli, and Alessandro Braccesi (eds.), *Museo della Specola: Catalogo Italiano-Inglese* (Bologna: Bologna University Press, 1995).

On Halley, see C. A. Ronan, *Edmond Halley—Genius in Eclipse* (New York: Doubleday, 1969; London: Macdonald, 1970), and for his comet see the bibliography in Bruce Morton, *Halley's Comet, 1755–1984: A Bibliography* (Westport, CT: Greenwood Press, 1985). For the context of much French astronomy in the period, see Roger Hahn, *The Anatomy of a Scientific Institution: The Paris Academy of Sciences, 1666–1803* (Berkeley: University of California Press, 1971). See also his "Les observatoires en France au XVIIIe siècle," in *La Curiosité scientifique au XVIIIe siècle: cabinets et observatoires* (Paris: Hesmann, 1986: 653–59). The works of Guillaume Bigourdan are still useful in this connection, for instance his *Histoire de l'astronomie d'observation et des observatoires en France,* 2 vols. (Paris: Gauthier-Villars, 1918–1930). For astronomical architecture, see P. Müller, *Sternwarten in Bildern: Architektur und Geschichte der Sternwarten von den Anfangen bis ca. 1950* (Berlin, 1992). A special issue of the *Journal for the History of Astronomy,* 22, part 1, (1991) treated of national observatories in several countries. A historical introduction to astronomy in the southern hemisphere, although not strong on Australasia, is David S. Evans, *Under Capricorn: A History of Southern Hemisphere Astronomy* (Bristol: Hilger, 1988). It should be

supplemented by Raymond Haynes, Roslynn Haynes, David Malin, and Richard Mc-
Gee, *Explorers of the Southern Sky: A History of Australasian Astronomy* (Cambridge:
Cambridge University Press, 1996).

A key text in the new cosmology is Thomas Wright, *An Original Theory or New
Conception of the Universe,* edited and introduced by Michael Hoskin (London:
Macdonald, 1971). This edition includes *A Theory of the Universe* (1734). Immanu-
uel Kant's cosmological views are often submerged beneath a sea of hardly relevant
philosophy. For an English text, see his *Universal Natural History and Theory of the
Heavens,* translated, with introduction and notes, by Stanley L. Jaki (Edinburgh: Scottish Academic Press, 1981). Lambert's contribution to the unfolding picture of island
universes can be seen at first hand in Johann Heinrich Lambert, *Cosmological Letters
on the Arrangement of the World-Edifice,* translated, with introduction and notes, by
Stanley L. Jaki (New York: Science History Publications, 1976). William Herschel's
collected papers were edited by J. L. E. Dreyer and printed with other biographical
material in *The Scientific Papers of Sir William Herschel,* 2 vols. (London: Royal Astronomical Society, 1912). Long excerpts from the papers are included, with historical commentary by M. A. Hoskin and astrophysical notes by K. Dewhirst, in *William
Herschel and the Construction of the Heavens* (London: Oldbourne, 1963). See also
Hoskin's *The Herschel Partnership as Viewed by Caroline* (Cambridge: Science History Publications, 2003) and relevant studies in his *Stellar Astronomy* (listed above).
Full-length biographies are available by J. B. Sidgwick (*William Herschel, Explorer
of the Heavens,* London: Faber and Faber, 1953), Angus Armitage (*William Herchel,*
London: Nelson, 1962), and Günther Buttmann (*Wilhelm Herschel, Leben und Werk,*
Stuttgart: WV, 1961 [in German]). Buttmann's biography of John Herschel is available
in English as *The Shadow of the Telescope* (New York: Charles Scribner's Sons, 1970;
London: Lutterworth, 1974).

On the possibility of a plurality of worlds see Steven J. Dick, *Plurality of Worlds: The
Origins of the Extraterrestrial Life Debate from Democritus to Kant* (Cambridge: Cambridge University Press, 1984) and Michael J. Crowe, *The Extraterrestrial Life Debate
1750–1900: The Idea of a Plurality of Worlds from Kant to Lowell* (Cambridge: Cambridge University Press, 1986); and for reprints of two important texts in the debate,
see John Ray, *Wisdom of God Manifested in the Works of Creation* (New York: Garland,
1979) and Bernard Le Bovier, sieur de Fontenelle, *Conversations On the Plurality of
Worlds* (trans. H. A. Hargreaves; Berkeley: University of California Press, 1990).

The extraordinary additions to Newtonian planetary theory in the eighteenth
and early nineteenth centuries, especially by French astronomer-mathematicians,
are central to the classic study by Robert Grant, *History of Physical Astronomy from
the Earliest Ages to the Middle of the Nineteenth Century* (London: Robert Baldwin,
1852). Grant is occasionally too generous in the credit he gives to Laplace, and his
book is in many respects outdated, but it is still worth consulting for the history of
celestial mechanics—a subject to which historians have rarely paid much attention.
For the same theme, in its classic period, see the last of Delambre's histories (see the
introduction to this bibliography). Perhaps the most useful single historical survey
of Laplace is the entry in the first *Supplement to the Dictionary of Scientific Biography,* vol. 15 (1978: 273–403). (That the entry is seven and a half times as long as the

one on Einstein, twenty-seven times that on Eddington, and eighty-seven times as long as that on Richard of Wallingford, illustrates a law that would have defeated Laplace's analytical powers.) More general but more reliable replacements for Grant are to be found in the later chapters of C. M. Linton, *From Eudoxus to Einstein: A History of Mathematical Astronomy* (Cambridge: Cambridge University Press, 2004), which sports a massive bibliography; and Stephen G. Brush, *A History of Modern Planetary Physics,* 3 vols. (Cambridge: Cambridge University Press, 1996). Harry Woolf, *The Transits of Venus. A Study of Eighteenth-Century Science* (Princeton: Princeton University Press, 1959), provides some of the background to the scaling of the solar system. For a good general survey of Newtonian mechanics and theories that preceded and followed it, René Dugas, *A History of Mechanics* (trans. J. R. Maddox; London: Routledge, 1955) is still worth reading, though it was written at a moderately technical level. A companion volume by the same author covers mechanics in the seventeenth century (*Mechanics in the Seventeenth Century,* trans. F. Jacquot; Neuchatel: Griffon, 1958). A journal worth combing systematically for articles on this general theme is *Archive for History of Exact Sciences.* The subtitle "A new perspective on eighteenth-century advances in the lunar theory" well describes Eric G. Forbes, *The Euler-Mayer Correspondence (1751–1755)* (London: Macmillan, 1971), which might lead interested readers into the deeper and richer waters of Euler's collected works. J. D. North, *The Universal Frame: Historical Essays in Astronomy, Natural Philosophy and Scientific Method* (London: Hambledon Press, 1989) contains chapters on calendar reform, Tycho, Harriot, the satellites of Jupiter, Markgraf, and time in special relativity.

CHAPTER 15

A bibliography of over 1,400 entries is to be found in David H. DeVorkin, *The History of Modern Astronomy and Astrophysics* (New York: Garland, 1982). A convenient and brief source of biographical information on the important astronomer Bessel is Jürgen Hamel, *Friedrich Wilhelm Bessel* (Leipzig: Teubner, 1984). See also P. Brosche et al., *The Message of the Angles: Astrometry from 1798 to 1998* (Thun: Harri Deutsch, 1998) and P. C. van der Kruit and K. van Berkel, *The Legacy of J. C. Kapteyn: Studies on Kapteyn and the Development of Modern Astronomy* (Dordrecht: Kluwer, 2000).

A well-balanced and classic historical work, where "popular" in the title does not mean "superficial," is Agnes M. Clerke, *A Popular History of Astronomy During the Nineteenth Century* (London: Black, 1893; 4th ed., 1902). Her historical interests give added value to her general astronomical surveys in her *The System of the Stars* (London: Black, 1890) and *Problems in Astrophysics* (London: Black, 1903). At a more elementary level, Camille Flammarion, *Popular Astronomy* (London: Chalto & Windus, 1880; trans. J. E. Gore, London, 1891) is instructive. Since so many instrumental advances were made in Germany at this time it is useful to consult in the same vein such a work as Rudolf Wolf, *Geschichte der Astronomie* (Munich: R. Oldenbourg, 1877), or for a modern bird's-eye view, Dieter B. Herrmann, *Geschichte der Astronomie von Herschel bis Hertzsprung* (Berlin: VEB Deutscher Verlag der Wissenschaften, 1975), translated and revised by K. Krisciunas as *The History of Astronomy from Herschel to Hertzsprung* (Cambridge: Cambridge University Press, 1984).

This last popular work has a useful bibliographical guide to further sources. (The numerous references to Friedrich Engels need not be taken too seriously.) It is impossible in a short space to pay much attention to biographies, of which there are now some hundreds of relevance, but library catalogs easily available on the Internet will provide quick answers to searches by name. One that might be missed is *The Reminiscences of an Astronomer* (Boston: Harper, 1903), in which Simon Newcomb brings to life a vanished era. Orthodox astronomers are often the least interesting subjects for biography, but Lockyer was far from orthodox. See A. J. Meadows, *Science and Controversy: A Biography of Sir Norman Lockyer* (London: Macmillan, 1972).

On instrumentation, as well as the birth of astrophysics, see Owen Gingerich (ed.), *Astrophysics and Twentieth-century Astronomy to 1950: Part A.* (Cambridge: Cambridge University Press, 1984). This is vol. 4A of the *General History of Astronomy* mentioned in the introductory bibliography. The work includes a checklist of refractors and reflectors, 1850–1950. Compare D. Howse's "Greenwich List" of observatories and instruments under the bibliography to chapter 13. Histories of observatories are legion and are easily tracked down through library catalogs on the Internet, but are mostly acts of piety. For Palermo and Bologna, see the bibliography to chapter 14. Pulkovo is well described in A. N. Dadaev, *Pulkovo Observatory. An Essay on its History and Scientific Activity* (trans. Kevin Krisciunas; Springfield, VA: NASA, 1978). Some interesting social determinants in the creation of an observatory will be found in J. A. Bennett, *Church, State, and Astronomy in Ireland: 200 Years of Armagh Observatory* (Armagh: Armagh Observatory & Institute of Irish Studies, 1990). Krisciunas later wrote a survey of observatories, especially useful for the later period (it has a good chapter on Pulkovo, and others on Harvard, Lick, Yerkes, Mt. Wilson, and Palomar, for instance) titled *Astronomical Centers of the World* (Cambridge: Cambridge University Press, 1988). A comprehensive list of histories of individual observatories is out of the question here, but for its spread, it easy narrative style, and the importance of its subject, see the history of Lick (founded 1888) by Donald E. Osterbrock, John R. Gustafson, and W. J. Shiloh Unruh, *Eye on the Sky: Lick Observatory's First Century* (Berkeley: University of California Press, 1988).

For a good general account of statistical cosmology, with Kapteyn at center stage, see Erich Robert Paul, *The Milky Way Galaxy and Statistical Cosmology, 1890–1924* (Cambridge, Cambridge University Press, 1994). A comprehensive study of spectroscopy that well summarizes the essentials of the original scientific sources is J. B. Hearnshaw, *The Analysis of Starlight. One Hundred and Fifty Years of Astronomical Spectroscopy* (Cambridge: Cambridge University Press, 1986). See also R. L. Waterfield, *A Hundred Years of Astronomy* (London: Macmillan, 1938). For solar astrophysics, see A. J. Meadows, *Early Solar Physics* (Oxford: Pergamon, 1970). For photography, see Gérard de Vaucouleurs, *Astronomical Photography, from the Daguerrotype to the Electron Camera* (London: Faber & Faber, 1961) and Dorrit Hoffleit, *Some Firsts in Astronomical Photography* (Cambridge: Cambridge University Press, 1950). A fine collection of photographs taken by the Schmidt telescopes at Tautenberg and the European Southern Observatory, with a biography of Bernhard Schmidt, is S. Marx and W. Pfau, *Astrophotography with the Schmidt Telescope* (trans. P. Lamle; Cambridge: Cambridge University Press, 1992). Hale's remarkable achievements—his solar work as well as his campaigning for astronomy—are reported with a wealth of illustration in Helen Wright, Joan N. Warnow and Charles Weiner, *The Legacy of George Ellery Hale. Evolution of Astronomy and Scientific Institutions in Pictures and Documents* (Cambridge, MA: MIT Press, 1972).

On real and imagined discoveries, see Morton Grosser, *The Discovery of Neptune,* (1965; repr. New York: Dover, 1979) and William Graves *Hoyt, Lowell and Mars* (Tucson: University of Arizona Press, 1976).

CHAPTER 16

In addition to the bibliography of the previous chapter (including Clerke, Gingerich [ed.], Hearnshaw, and Meadows), the following original texts are useful historical sources for stellar physics: J. Scheiner, *A Treatise on Astronomical Spectroscopy* (1890; trans. E. B. Frost, Boston: Ginn, 1894); Arthur Stanley Eddington, *The Internal Constitution of the Stars* (Cambridge: Cambridge University Press, 1930); Ejnar Hertzsprung, *Zur Strahlung der Sterne* (ed. D. B. Herrmann; Leipzig: Ostwalds Klassiker, 1976); Martin Schwarzschild, *Structure and Evolution of the Stars* (Princeton: Princeton University Press, 1958; repr. New York: Dover, 1965); and Subrahmanyan Chandrasekhar, *An Introduction to the Study of Stellar Structure* (New York: Dover, 1967).

The source books mentioned in the introductory bibliography are useful. On solar theories, see Karl Hufbauer, *Exploring the Sun: Solar Science Since Galileo* (Baltimore: The Johns Hopkins University Press, 1991), and compare Ronald E. Doel, *Solar System Astronomy in America: Communities, Patronage & Interdisciplinary Research* 1920–1960 (Cambridge: Cambridge University Press, 1996). On the famous chart showing stellar evolution, see B. W. Sitterly, "Changing Interpretations of the Hertzsprung-Russell Diagram, 1910–1940," in *Vistas in Astronomy,* 12 (1970): 357–66. For historical background to the question of geochronology, see Francis Haber, *The Age of the World, Moses to Darwin* (Baltimore: The Johns Hopkins University Press, 1959).

For the discovery of spirals, see Charles Parsons (ed.), *The Scientific Papers of William Parsons, Third Earl of Rosse, 1800–1867* (London: Charles Parsons, 1926). The background to the discovery is well depicted in Patrick Moore, *The Astronomy of Birr Castle* (London: Mitchell Beazley, 1971). Not easily found, but useful, is W. Strohmeier "Variable Stars, their Discoverers and First Compilers, 1006 to 1975," *Veröffentlichungen der Remeis-Sternwarte Bamberg,* no. 129 (1977). The role of the Harvard College Observatory in this story is so important here that Solon I. Bailey, *The History and Work of Harvard Observatory, 1839–1927* (New York: McGraw-Hill, 1931) is worth seeking out.

CHAPTER 17

For a comprehensive history of cosmology in the first half of the twentieth century, and of the mathematical background to it, see J. D. North, *The Measure of the Universe. A History of Modern Cosmology* (Oxford: Oxford University Press, 1965, 1967; New York: Dover, 1990). Helge Kragh, *Cosmology and Controversy: The Historical Development of Two Theories of the Universe* (Princeton: Princeton University Press, 1996) and Pierre Kerszberg, *The Invented Universe: The Einstein-De Sitter Controversy (1916–17) and the Rise of Relativistic Cosmology* (Oxford: Oxford University Press,

1989), cover much common ground in their detailed historical studies. The best separate history of the paradox of the dark night sky is Edward Harrison, *Darkness at Night: A Riddle of the Universe* (Cambridge, MA: Harvard University Press, 1987). The subtitle of Robert Smith, *The Expanding Universe: Astronomy's "Great Debate" 1900–1931* (Cambridge: Cambridge University Press, 1982) is self-explanatory. Edward R. Harrison, *Cosmology: the Science of the Universe* (Cambridge: Cambridge University Press, 1981) is only semi-historical but is simple and instructive. Norriss S. Hetherington, who also discusses previous centuries in his *Science and Objectivity: Episodes in the History of Astronomy* (Ames: Iowa State University Press, 1988), is rather harsh in his onslaughts on astronomers' professionalism and integrity, but raises some interesting questions.

Not new, but valuable still, is Harlow Shapley and Helen E. Howarth, *A Source Book in Astronomy, 1900–1950* (Cambridge, MA: Harvard University Press, 1960). It should be supplemented by Kenneth R. Lang and Owen Gingerich, *A Source Book in Astronomy and Astrophysics, 1900–1975* (Cambridge, MA: Harvard University Press, 1979), which is of wide scope and has excellent commentaries.

A readable survey of changing conceptions of space will be found in Edmund Whittaker, *From Euclid to Eddington* (Cambridge: Cambridge University Press, 1949). Of the many biographies of Einstein, Abraham Pais, *Subtle is the Lord. The Science and Life of Albert Einstein* (Oxford: Oxford University Press, 1982) is one of the most comprehensive. Jeremy Bernstein, *Einstein* (New York: Viking Press, 1973) concentrates on Einstein's physics for the general reader. For a selection of classical papers by H. A. Lorentz, A. Einstein, H. Minkowski and H. Weyl, see *The Principle of Relativity. A Collection of Original Papers on the Special and General theory of Relativity,* (trans. W. Perrett and G. B. Jeffery; New York: Dover, n.d.). Texts of a cosmological character that should help to give an idea of the events of the first half of the century are these: Arthur Stanley Eddington, *Stellar Movements and the Structure of the Universe* (London: Macmillan, 1914), *Space, Time and Gravitation* (Cambridge: Cambridge University Press, 1920), *The Internal Constitution of the Stars* (Cambridge: Cambridge University Press, 1926), *The Expanding Universe (Cambridge: Cambridge University Press, 1933)* and *Stars and Atoms* (Oxford: Oxford University Press, 1927), the last being his only popular account of his astrophysical work; James Jeans, *The Mysterious Universe* (Cambridge: Cambridge University Press, 1930); Edwin Hubble, *The Realm of the Nebulae* (Oxford: Oxford University Press, 1936) and *The Observational Approach to Cosmology* (Oxford: Oxford University Press, 1937); and Georges Lemaître, *The Primeval Atom: A Hypothesis of the Origin of the Universe,* (trans. B. H. Korff and S. A. Korff; Toronto: Van Nostrand, 1950). More difficult, but no less important to the subject, are R. C. Tolman, *Relativity, Thermodynamics and Cosmology* (Oxford: Oxford University Press, 1934), Otto Heckmann, *Theorien der Kosmologie* (Berlin: Springer, 1968), and Herman Bondi, *Cosmology* (Cambridge: Cambridge University Press, 1960).

For a biography of Friedmann, see E. A. Tropp, V. Ya. Frenkel and A. D. Chernin, *Alexander A. Friedmann: The Man Who Made the Universe Expand*, translated from the Russian edition (Moscow: Nauka, 1988; trans. A. Dron and M. Burov, Cambridge: Cambridge University Press, 1993). For a deep insight into Lemaître's work, see O. Godart and M. Heller, *Cosmology of Lemaître* (Tucson, AZ: Pachart Publishing House, 1985). (The Belgian astronomer Godart had been his assistant.) The dispute between Eddington and Chandrasekhar is the subject of Arthur I. Miller, *Empire of the Stars: Friendship, Obsession and Betrayal in the Quest for Black Holes* (New York: Houghton Mifflin, 2005). For a good account of Hoyle's complex mindset, see Jane Gregory, *Fred Hoyle's Universe* (Oxford: Oxford University Press, 2005).

John Lankford, *American Astronomy: Community, Careers, and Power, 1859–1940* (Chicago: University of Chicago Press, 1997) is a penetrating study of how American observational astronomy took the lead in the first half of the twentieth century, and gives an insight into key personalities. For a survey of astronomy in the Soviet Union, from the Revolution to Stalin's purges, see E. Nicolaïdis, *Le Développement de l'astronomie en l'U. R. S. S., 1917–1935* (Paris: Observatoire de Paris, 1984); and also a special issue, "Astronomy under the Soviets," of *Journal for the History of Astronomy,* 26 (1995): 279–368. H. Van Woerden, R. J. Allen and W. B. Burton (eds.), *The Milky Way Galaxy: Proceedings of the 106th Symposium of the IAU held in Groningen, 30 May–3 June, 1983* (Dordrecht: Reidel, 1985), has a useful historical component. B. Bertotti and others (ed.), *Modern Cosmology in Retrospect* (Cambridge: Cambridge University Press, 1990) is an invaluable document, for it includes many chapters of a scientific but autobiographical nature by those who contributed to cosmology in the earlier part of the century (including R. A. Alpher and R. Herman, H. Bondi, W. McCrea, F. Hoyle, R. M. Wilson, and M. Schmidt). Wolfgang Yourgrau and Allen D. Breck (eds.), *Cosmology, History and Theology. Based on the Third International Colloquium Held at Denver, 1974* (New York: Plenum, 1977), has a similar value, including as it does papers by H. O. Alfvén, P. G. Bergmann, W. H. McCrea, C. W. Misner, A. Penzias, Kenji Tomita, and others.

CHAPTER 18

The most thorough history of the first decades of radio astronomy is W. T. Sullivan III, *The Early Years of Radio Astronomy* (Cambridge: Cambridge University Press, 1984), which could be supplemented by the source material in his *Classics in Radio Astronomy* (Dordrecht: Reidel, 1982). David O. Edge and Michael J. Mulkay, *Astronomy Transformed: The Emergence of Radio Astronomy in Britain* (New York: Wiley Interscience, 1976) and Gerrit L. Verschuur, *The Invisible Universe Revealed: The Story of Radio Astronomy* (New York: Springer-Verlag, 1987) both deal with the various drastic changes in astronomical technique. Readable autobiographical material by pioneers in radio astronomy are J. S. Hey, *The Radio Universe* (Oxford: Pergamon Press, 1983) and Bernard Lovell, *The Voice of the Universe. Building the Jodrell Bank Telescope,* revised ed. (London: Praeger Press, 1987).

An institutional view of the new science—and much else—from a European perspective is given in Adriaan Blaauw, *Early History: The European Southern Observatory, from Concept to Reality* (Munich: ESO, 1991). Peter Robertson, *Beyond Southern Skies. Radio Astronomy and the Parkes Telescope* (Cambridge: Cambridge University Press, 1992), tells of the building and operation of the important Parkes radio telescope in New South Wales (completed 1961). W. Patrick McCray, *Giant Telescopes: Astronomical Ambition and the Promise of Technology* (Cambridge, MA:

Harvard University Press, 2004) is an institutional history of its subject, starting from the 1930s. The fine biography by Donald E. Osterbrock, *Walter Baade: A Life in Astrophysics* (Princeton: Princeton University Press, 2001), is more than a life. It is the story of the Mt. Wilson and Palomar Observatories at their zenith.

CHAPTER 19

There has been a steady flow of published material on space astronomy over the last half century, helped greatly by its photogenic appeal, but relatively little of its has been historical. Two works by David H. DeVorkin on pioneering efforts in the United States, with much on previous efforts elsewhere, are: *Race to the Stratosphere: Manned Scientific Ballooning in America* (New York: Springer-Verlag, 1989) and *Science with a Vengeance: How the Military Created the U.S. Space Sciences after World War II* (New York: Springer-Verlag, 1992). K. Krisciunas, *Astronomical Centers of the World* (Cambridge: Cambridge University Press, 1988), mentioned earlier in connection with telescope locations, included much on space astronomy up to 1988. Richard Hirsch, *Glimpsing an Invisible Universe: The Emergence of X-Ray Astronomy* (Cambridge: Cambridge University Press, 1983), is still valuable, although thin on work outside the United States, which was by no means insignificant. Robert W. Smith and others, *The Space Telescope. A Study of NASA, Science, Technology and Politics* (Cambridge: Cambridge University Press, 1989) was published before the launch of the Hubble telescope, but was an important study of the background to that and many other enterprises in this new category. David DeVorkin and Robert W. Smith, *The Hubble Space Telescope: Imaging the Universe* (Washington, DC: National Geographic Society, 2004) tells of later history. With the rise of the Internet and the need for NASA and ESA to keep the public's goodwill, printed and online materials have been issued gratis, some of the downloadable literature being of book-length. Steven J. Dick is currently NASA's chief historian, and Steve Garber its history Web curator. Visit http://history.nasa.gov/index.html for access to bibliographies and online histories. ESA is less history-conscious, but much can be pieced together, starting at http://www.esa.int/esaCP/index.html. For a comprehensive history of European space flight (but only likely to be found in large libraries) see Kevin Madders, *A New Force at a New Frontier: Europe's Development in the Space Field* (New York: Cambridge University Press, 1997).

On comets, a standard of reference, especially important for conventions of naming, is Brian G. Marsden and Gareth V. Williams, *Catalog of Cometary Orbits,* 15th ed. (Cambridge, MA: Central Bureau for Astronomical Telegrams and Minor Planet Center, Smithsonian Astrophysical Observatory, 2003). Two valuable works by experienced astronomers are Fred L. Whipple, *The Mystery of Comets* (Cambridge: Cambridge University Press, 1985) and Donald K. Yeomans, *Comets: A Chronological History of Observation, Science, Myth and Folklore* (New York: Wiley, 1991). For a history of an understanding of meteorites, which made rapid progress in the nineteenth century, see John G. Burke, *Cosmic Debris: Meteorites in History* (Berkeley: University of California Press, 1987). The title of the next work says it all: Gerrit L. Verschuur, *Impact! The Threat of Comets and Asteroids* (New York: Oxford University Press, 1996).

CHAPTER 20

Relatively few historical studies of the modern period are as yet available, and the line between biography and hagiography is as thin as ever. Twenty-seven lengthy interviews with cosmologists are to be found in Alan Lightman and Roberta Brawer, *Origins: The Lives and Worlds of Modern Cosmologists* (Cambridge, MA: Harvard University Press, 1990), which includes a good select bibliography of cosmology up to that time. John D. Barrow and Frank J. Tipler, *The Anthropic Cosmological Principle* (Oxford: Clarendon Press, 1986), is not history, but it has a valuable historical element. Stephen W. Hawking and Werner Israel (eds.), *Three Hundred Years of Gravitation* (Cambridge: Cambridge University Press, 1987), does not pretend to be historical, despite its title, although Israel writes on the evolution of the idea of dark stars, and C. M. Will surveys experimental gravitation from Newton to the twentieth century. The rest of the volume will in the fullness of time be of extreme historical value. It has chapters by R. Penrose on quantum theory and reality, A. H. Cook on gravitational experiment, R. D. Blandford on astrophysical black holes, K. S. Thorne on gravitational radiation, and M. J. Rees on galaxy formation and dark matter. S. K. Blau, A. H. Guth, and A. Linde write on inflationary cosmology, while S. W. Hawking writes on quantum cosmology. On strings (not cosmic strings) see B. Greene, *The Elegant Universe: Superstrings, Hidden Dimensions and the Quest for the Ultimate Theory* (London: Cape and New York: Norton, 1999). Stephen Hawking, *A Brief History of Time from the Big Bang to Black Holes* (London: Bantam Press, 1988) hardly needs an introduction, since it has broken numerous best-selling records. Steven Weinberg, *The First Three Minutes* (New York: Basic Books, 1976) had much popular success a decade earlier, with the physics of the Big Bang. His *Gravitational Cosmology* (New York: Wiley, 1972), was a successful textbook that a reader might use to gain an entry to the historical field, although D. W. Sciama, *Modern Cosmology* (Cambridge: Cambridge University Press, 1971) and Michael Rowan-Robinson, *Cosmology* (Oxford: Oxford University Press, 1977) both make fewer mathematical demands. An earlier classic work that long held the field as a textbook is G. C. McVittie, *General Relativity and Cosmology*, 2nd ed. (London: Chapman and Hall, 1965). Ya. B. Zel'dovich and I. D. Novikov, *Relativistic Astrophysics*, vol. 1, *Stars and Relativity*; vol. 2, *The Universe and Relativity* (Chicago: University of Chicago Press, 1971–1974) throws much light on important Soviet work.

While not overtly historical, the following excellent collection of essays has a vital historical core: Robert John Russell, Nancey Murphy, and C. J. Isham (eds.), *Quantum Cosmology and the Laws of Nature: Scientific Perspectives on Divine Action* (Vatican City: Vatican Observatory, 1993). On theological attitudes with respect to cosmology, one might consult essays by Charles Misner, Philip Hefner, David Peat, Arthur Peacocke, or Stanley Jaki in Yourgrau and Breck (eds.), *Cosmology, History and Theology* (see the bibliography to chapter 18). See also the earlier work by Stanley Jaki, *Science and Creation: From Eternal Cycles to an Oscillating Universe* (Edinburgh: Scottish Academic Press, 1974), which marries theology to the history of certain cosmological views. P. Davies, *God and the New Physics* (London: Dent, 1983), is highly relevant, and has been influential. See also R. Jastrow, *God and the Astronomers* (New York:

W. W. Norton, 1978); and for a short, accessible, and part-historical argument for a super-intelligent Creator, by an astronomer who finds genetics more challenging than astronomy, see Owen Gingerich, *God's Universe* (Cambridge, MA: The Belknap Press, 2006). For 18 historical views that between them cover a broader spectrum—that is, the natural sciences and Christian theology over two millennia—see David C. Lindberg and Ronald L. Numbers, *God and Nature* (Berkeley: University of California Press, 1986). It becomes clear on reading this that cosmology has had more than its due share of influence on the shaping of theology, and that for some strange reason it continues to have it.

Index

Sosigenes, 74

Soter (Ptolemy I), 105, 106

soul: Plato on, 53, 74; in Zoroastrianism, 53

South America. *See* Central and South America

Southern Cross, 16

Soviet and Russian space program, 516, 691–92, 694, 695–96, 697–98, 709, 712, 725, plate 16

Soyuz and *Apollo* docking, 692, plate 16

space missions, 708–11; to comets, 697, 712, 725–26, *726*, 728–29; ion drives for, 736–37; to Moon, 694–98; to planets, 697–98, 702–6; to Pluto, 484; solar wind and, 516. *See also* NASA (National Aeronautics and Space Administration); satellites, artificial

space race, 689, 691–92, 694, 709

space shuttle, 694, 725, 733, 734, 735

space stations, 692, 693–94

space-time foam, 767

space-time metric, 644

Spain: Jesuit astronomy in, 155

Spain, Muslim, 189, 215–22; astrolabe and, 132, 133, 216, 217, 223–26; water clock and, 226; zīj of al-Khwārizmī in, 197. *See also* Alfonsine Tables; Toledan Tables

special theory of relativity, 544, 625, 626, 630, 641

speckle interferometry, 592

spectrobolometer, *505*

spectrography: computer processing in, 524; at radio wavelengths, 671; stellar velocities and, 582

spectroheliograph, 548, *549*

spectrohelioscope, 548

spectroscopic binaries, 557, 566

spectroscopic parallax, 559–60, 589

spectroscopy, 474–76; of comets, 526, 531–32, *532*, 533, 534; Huggins's researches in (*see* Huggins, William); of novae and supernovae, 717–19; Saha equation and, 604; solar, 501–2, 503–4, *505*, 506, 512–13, 517–18, 550, 555, 557, 605; stellar (*see* stellar spectroscopy); of Uranus, 512; of Venus, 697

speculum metal, 490–91, 493, 494

spherical astronomy: Aristotle and, 82–84; of Autolycus, 72, 84; of Callippus, 81, 83; Copernicus on, 311; of Eudoxus, 73–74, 75–81, 82; Greek treatises on, 72; John of Sacrobosco on, 251–52, 270; Menelaus of Alexandria and, 108

spherical trigonometry: al-Battānī on, 198; Greek methods in Indian texts on, 173; Ibn Yūnus's use of, 201; Regiomontanus on, 274–75; Richard of Wallingford and, 258–59; translations into Chinese on, 153

Sphujidhvaja, 172–73

Spielberg, Steven, 699

Spinola, Ambrogio, 361, 362

spiral galaxies, *575*, *576*; Baade's Type II stars in, 574; distances of, 577–78, 587, 588–89; Easton's Milky Way model and, 574–75, *576*, 581, 627; evolution of, 586–89, *588*; form and dynamics of, 578–86; Hubble's classification and, 587, *588*; Hubble Space Telescope image of, plate 12; island universe theory and, 627; motions of, 633–36; redshifts in spectra of, 632. *See also* Milky Way

spiral nebulae, 574–78, *575*; Eddington on velocities of, 631; Jeans on creation of matter in, 646; M51, 491, plate 12; novae in, 718; radial velocities of, 634; Rosse's observations of, 491, *492*; velocities of matter along arms of, 576–77. *See also* Andromeda Nebula, M31; nebulae; spiral galaxies

Spitzer, Lyman, Jr., 616, 617, 619, 716

Spitzer Space Telescope, 716, 746

spontaneous generation, 727

Sputniks, 664, 689, 691, 709

Srīpati, 180

Ssuma Chhien, 134, 139

Stadius, Georg, 321, 338

St. Albans clock, 107

Stalin, 637

standard model, cosmological, 658

Stanley, Gordon, 663, 664

star atlases. *See* star maps and atlases

star calendars, Greek, 71

star catalogs: in Alfonsine Tables, 230;

Walker, S. C., 483
Wallenstein, Albrecht von, 345
Wallingford. *See* Richard of Wallingford
Wallis, John, 392, 409, 410
Walraven, Theodor, 663
Waltemath, Georg, 707
Walther, Bernhard, 255, 274, *275*, 309, 320
Wan-li, 153
Ward, Seth, 401, 402
Warren, John, 174
Wartmann, Louis François, 478–79
Wassenius, Birger, 503
water: in comets, 535, 536, 723, 725; in electrical discharge experiments, 727; microwave radiation and, 762; in solar system, 705
water clocks: British monasticism and, 237; Chinese, 138, *146*, 146–47, 148, 149; Egyptian, 30, 31; Greek influence on Arabs and, 226; Indian, 172; in Islamic Spain, 218, 226–27, 228; Moroccan, 226; in Tower of the Winds, 88, 124, *125*; unequal seasonal hours and, 45; Vitruvius on, 124
Waterston, John James, 539
weak anthropic principle, 780
weak nuclear force, 611, 750, 765, 766, 769
weather. *See* meteorological phenomena
Weaver, Harold, 678
Wedgwood, Thomas, 497
week: Christian introduction of, 232; Egyptian, 27; Jewish, 185; Muslim, 185
Weekes, Trevor, 715
Wei dynasty, 142
Weierstrass, Karl, 74
Weinberg, Steven, 678, 766
Weizsäcker, Carl Friedrich von, 545, 563, 596, 608–10, 612, 613, 653, 675
Wells, H. G., 531
Wells, L. D., 718
Welser, Mark, 367, 371, 373–74
Wenlock, H. E., *29*
Werner, Johann, *283*
Werner, Johannes, 308, 309
Wesley, W. H., 503

Westerbork Synthesis Radio Telescope, 668–69
Weston, Thomas, 466
Westphal, James, 534
wet collodion process, 499, 500
Weyl, Hermann, 636, 645, 648
Wheeler, John, 596, 752, 753, 767
Whewell, William, 509, 510
Whipple, Francis, 523
Whipple, Fred, *515*, 535, 536, 697, 723
Whiston, William, 381, 402, 408, 416
white dwarfs, 620; Chandrasekhar on, 596–97; as dark matter candidate, 748–49; Fowler on, 594; Greenstein's studies of, 681; on Hertzsprung-Russell diagram, 560, *561*, 563; neutron stars and, 597–98; novae and, 368, 717, 719; relativistic treatment of, 742; Sirius B as, 555; stellar evolution and, 614, 615, *622*, 623, 744; subdwarf, 652, 653; X-rays from, 714. *See also* dwarf stars
Whitehead, A. N., 645, 649
Whitrow, G. J., 643, 644
Whittaker, E. T., 643, 645
Wickramasinghe, Nalin Chandra, 656, 727–28
Widmanstadt, Johann Albrecht, 308
Wien's law, 552
Wiesel, Johann, 387
Wild, J. P., 671
Wilkins, John, 409
Wilkinson, D. T., 676
Wilkinson, Lancelot, 182
Wilkinson Microwave Anisotropy Probe (WMAP), 708, 762–63, 772, 775
William IV, Landgrave of Hesse-Kassell, 324, 325, 329, 335
William of Normandy, 422, *422*
William of Orange, 385
William of Saint-Cloud, 254, 255–57
Williamson, Jack, 736, 737
William the Englishman, 220–21, 290
Willibrord, 294
Wilmington Man, 14, plate 3
Wilsing, Johannes, 592, 660
Wilson, Alexander, 269, 502
Wilson, Robert W., 643, 651, 658, 676, 677, 761, 763